Editing and Analysing Numerical Tables

Ptolemaeus Arabus et Latinus

Studies

Volume 2

Ptolemaeus Arabus et Latinus

Ptolemaeus Arabus et Latinus (PAL) is a project of the Bavarian Academy of Sciences and Humanities and the University of Würzburg. As part of the Academies' Programme, PAL is jointly funded by the Federal Republic of Germany and the Free State of Bavaria.

PAL is sponsored by the Union Académique Internationale (UAI).

Editing and Analysing Numerical Tables

Towards a Digital Information System for the History of Astral Sciences

Edited by

Matthieu Husson
Clemency Montelle
Benno van Dalen

Cover design by Hilde Verhelst, T'Hi, Lier, Belgium.

Cover illustration:
Woodcut (detail) from *La geografia di Ptolemeo Alessandrino*, Venice, 1548.
Copy from Munich, Bayerische Staatsbibliothek.

D/2021/0095/218

ISBN 978-2-503-59606-8

E-ISBN 978-2-503-59607-5

DOI 10.1484/M.PALS.5.125692

Printed in the EU on acid-free paper.

Contents

List of Colour Plates

Acknowledgements

This volume was made possible by the generous support of several institutions: the Paris Sciences et Lettres – Observatoire de Paris, the project Ptolemaeus Arabus et Latinus at the Bavarian Academy of Sciences and Humanities (Munich), and the University of Canterbury (Christchurch, New Zealand). We are also grateful to the many individuals who contributed to the realisation of this book: to our academic colleagues who provided helpful and expert advice via workshops, conferences, and seminars, and to our technical and administrative support teams who contributed to the design and development of the TAMAS project more broadly. Many thanks to Galla Topalian, Antonin Penon, Ségolène Albouy, Leni Gauffier, Svyat Kovanov, and Stefan Müller — the DISHAS database would not have been possible without you.

In the book's production, we are indebted to Claudia Dorl-Klingenschmid (Ptolemaeus Arabus et Latinus) for her patience, persistence, and attention to detail in editing the individual articles and preparing the indexes. We also acknowledge the fine job carried out by Megan Foster and Michael Maudsley in crafting contributions into clear and crisp English. Finally, we are enormously grateful to Bart Janssens, who worked diligently with the teams at Brepols and Cultura to produce the end product.

Editing and Analysing Numerical Tables: Towards a Digital Information System for the History of Astral Sciences

Introduction

Matthieu Husson, Clemency Montelle, and Benno van Dalen

Dip into any corpus of scientific texts and you are likely to come across numerical tables. The practice of storing numerical data in compressed form was embraced by many cultures of inquiry from the very earliest literate societies, and our appetite for tabulating data has not waned even in the digital age. Yet, despite the prevalence of numerical tables as a genre in many historical contexts, detailed and thorough scrutiny of them has been lacking. Until recent decades, historians of science have been largely drawn to proposition-proof, analytical forms of technical expression, usually cast in prose, which echo the modern disciplinary ideals of mathematics, considering other forms of mathematical activity, such as worked computations, tables, diagrams, instrumentation and the like to be less worthy of consideration. However, as our purview of mathematical practice broadens, so too has our understanding of the importance and richness of the historical insight that the analysis of numerical data and computations can bring.

Indeed, beyond this issue of status, it is true that numerical tables pose particular challenges to historians. They can be large and unwieldy. Some sources contain thousands of data points, each of which, in principle, needs to be checked. They can appear monotonous and dull, and can present themselves as impenetrable and intractable to even the most persistent analysis. Accompanying text, if any, is usually not helpful in revealing the details of their construction; the underlying algorithms and parameters often lie buried in the data. The conventions and simplifying assumptions made by the human computer are almost never made explicit; all decisions of execution and display are bound up in the numerical data. And as a textual genre, tabular data are prone to particular types of copying errors or even, in some specific contexts, alterations at the hands of computer-scribes.

But leaving aside the potential historic minefield, numerical tables are testament to a rich treasury of practices which are directly relevant to the history of science. These data points are the closest the historian can come to the actual hand computational practices and priorities of the compilers of historical tables and of their users. More broadly, they account for a line of numerate culture

Editing and Analysing Numerical Tables: Towards a Digital Information System for the History of Astral Sciences, ed. by Matthieu Husson, Clemency Montelle and Benno van Dalen, PALS 2 (Turnhout, 2021), pp. 1–16
 10.1484/M.PALS-EB.5.127693

which is often overlooked by investigations: the development and emergence of sophisticated data management practices in computational contexts. As such they are of undeniable value and provide substantial illumination into the practices of historical computers.

Over the last decade, a series of publications have brought the study of numerical tables into sharper focus. Recent surveys of historical tabular cultures were carried out in collective volumes edited by Martin Campbell-Kelly and colleagues and by Matthieu Husson and Clemency Montelle.[1] These initiatives have collected specialised contributions on numerical tables from various historical contexts ranging from Near Eastern and other pre-modern cultures of inquiry, through a variety of Eurasian and north African milieus, right up to the present day.

Research dedicated to tables in specific cultures of inquiry is more advanced in some areas than in others. Tabular culture in the ancient Near East has been well documented thanks to the pioneering efforts of both Assyriologists and historians of science. The basic characteristics of Babylonian astronomical tables were explained in the foundational work by Otto Neugebauer.[2] Many further types of tables were analysed and edited by John Britton and Lis Brack-Bernsen[3] and more recently by John Steele and Mathieu Ossendrijver.[4] In Greek astronomy, Ptolemy's most important theoretical work, the *Almagest*, was edited around 1900 by Johan Ludvig Heiberg and was later authoritatively translated into English by Gerald Toomer.[5] A full analysis of the tables in the *Almagest* was provided by Glen Van Brummelen in his doctoral dissertation.[6] The more practical *Handy Tables* was studied in the doctoral dissertation of William Duane Stahlman and is currently being edited, translated and commented upon by Anne Tihon and Raymond Mercier.[7] Two Byzantine almanacs were edited and analysed by Alexander Jones and Raymond Mercier.[8]

Important groundwork on Islamic astronomical tables, the corpus of some 200 known *zīj*es, was carried out in the 1956 survey by Edward S. Kennedy,

[1] Campbell-Kelly et al., *The History of Mathematical Tables*, and Husson and Montelle, *The Transmission of Arabic Astronomical Tables*.

[2] Neugebauer, *Astronomical Cuneiform Texts*; for an overview, see 'Book II. Babylonian Astronomy' in Neugebauer, *HAMA*, vol. I, pp. 345–555.

[3] See, for example, Britton, 'Studies in Babylonian Lunar Theory, Parts I–III' and Brack-Bernsen, *Zur Entstehung der Babylonischen Mondtheorie*.

[4] See Steele, 'Newly Identified Lunar and Planetary Tables' and Ossendrijver, *Babylonian Mathematical Astronomy: Tabular Texts*.

[5] Halma's translation from the 1810s was superseded by Heiberg, *Syntaxis mathematica*. Presently, the most commonly used English translation is Toomer, *Ptolemy's Almagest*.

[6] Van Brummelen, *Mathematical Tables in Ptolemy's Almagest*.

[7] Stahlman, *The Astronomical Tables of Codex Vaticanus Graecus 1291*; Tihon, *Les Tables Faciles 1a*, and Mercier, *Ptolemy's Handy Tables 1b*.

[8] Jones, *An Eleventh-Century Manual* and Mercier, *An Almanac for Trebizond*.

which was consolidated by David A. King and Julio Samsó, and is currently being expanded by Benno van Dalen.[9] Full editions and studies of some important *zīj*es were published early on by Carlo Alfonso Nallino and Heinrich Suter; recent studies of tables in *zīj*es were carried out by Carlos Dorce and Benno van Dalen.[10] Sanskrit astronomical tables have been gaining prominence amongst Indological scholarship thanks to David Pingree's ground-breaking inventories and more recently through a detailed study by Clemency Montelle and Kim Plofker.[11] At the same time, several sets of tables were edited and commented upon by Anuj Misra and others.[12] The Latin and early-modern European tabular culture, including tables in Hebrew, has recently been reinvigorated by the monumental edition of the *Toledan Tables* published by the late Fritz S. Pedersen[13] and an uninterrupted series of descriptions of works with tables by José Chabás and Bernard R. Goldstein, culminating in surveys of the contents of these works and of the different sets of tables.[14] Work on the *Alfonsine Tables* that will go far beyond the edition of the Parisian tables by Emanuel Poulle[15] is currently being carried out under the auspices of the project ALFA (see below).

Scholarship in western languages on Chinese numerical tables has been advanced by Christopher Cullen with his publication of several astronomical systems from the Han dynasty and by Nathan Sivin's full study of the thirteenth-century *Season-granting System*.[16] A particularly interesting case of

[9] Kennedy, 'A Survey of Islamic Astronomical Tables' and King and Samsó, 'Astronomical Handbooks and Tables'. Van Dalen's new survey will appear in the book series *Ptolemaeus Arabus et Latinus – Studies* in the coming years.

[10] See Nallino, *al-Battānī sive Albatenii opus astronomicum*; Suter, *Die astronomischen Tafeln des Muḥammed ibn Mūsā al-Khwārizmī* (with supplementary material and English translations in Neugebauer, *The Astronomical Tables of al-Khwārizmī*); Dorce, *El Taŷ al-azyāŷ de Muḥyī al-Dīn al-Magribī*, and van Dalen, *Ptolemaic Tradition and Islamic Innovation*.

[11] Pingree, *Sanskrit Astronomical Tables in the United States*; Pingree, *Sanskrit Astronomical Tables in England*, and Montelle and Plofker, *Sanskrit Astronomical Tables*.

[12] See, for example, Misra et al., 'Eclipse Computation Tables'; Misra et al., *The Sanskrit Astronomical Table Text*, and the articles in Keller and Montelle, *Special Issue on Numerical Tables*.

[13] Pedersen, *The Toledan Tables*.

[14] Chabás and Goldstein, *A Survey of European Astronomical Tables*; Chabás, *Computational Astronomy in the Middle Ages*. For a selection of reprinted articles, see Chabás and Goldstein, *Essays on Medieval Computational Astronomy*. For descriptions and analyses of two specific sets of tables, see Chabás and Goldstein, *The Alfonsine Tables of Toledo* and Id., *The Astronomical Tables of Giovanni Bianchini*.

[15] Poulle, *Les Tables Alphonsines*.

[16] See Cullen, *The Foundations of Celestial Reckoning*, and Sivin, *Granting the Seasons*. An English overview of Chinese astronomical tables from the Han to the Qing dynasties can be found in Yabuuti, 'Astronomical Tables in China, Han to T'ang', and Yabuuti, 'Astro-

transmission of astronomical tables is the Persian *zīj* from the Yuan period that was translated into Chinese as the 'Islamic Astronomical System' (*Huihui lifa*). Most recently, Shi Yunli, Benno van Dalen and Li Liang have studied the methods and tables in this work.[17] Broadly speaking, then, numerical tables are receiving ever more scholarly attention, and researchers are becoming increasingly focused on efficient and novel techniques to explore them.

As part of these efforts, various collective research initiatives have also emerged. One example is the long-term collaborative project *History of Numerical Tables* led by Dominique Tournès, which organised a series of workshops in Paris. HAMSI (History of Astronomical and Mathematical Sciences in India)[18] is an ongoing scholarly project that explores, prioritises and studies Sanskrit sources and related vernacular traditions in the exact sciences in second millennium India. One of the key goals of this project is to document, digitally encode, edit and study the overwhelming corpus of numerical tables from this period. Various digital tools, such as CATE (Computer Assisted Tables Editor (see pp. 166–67), which automates the bulk of the critical editing process, have been developed to assist in this aim, but their scope and application need not be limited to Sanskrit traditions.

ALFA (Alfonsine Astronomy: Shaping a European Scientific Scene)[19] is devoted to the history of Alfonsine astronomy, a tradition of mathematical astronomy which became dominant in Latin sources between the fourteenth and the early sixteenth centuries. The central goal of ALFA is to build a detailed and deep understanding of this tradition of mathematical astronomy, in which the works of the astronomical 'scientific revolution' are rooted. This is done by embracing approaches to the sources from the history of manuscript cultures, the history of mathematics and the history of astronomy. The *Alfonsine Tables*, in the multiple forms they took during these two and a half centuries, are central to the Alfonsine tradition, and their edition and the analysis of the astronomical practices they fostered are essential to the ALFA project.

---

nomical Tables in China, Wutai to Ch'ing'. Martzloff, *Astronomy and Calendars* presents an overview of the technical contents of Chinese astronomical systems. Besides publications in western languages, there is a huge amount of important literature on Chinese astronomical systems in publications in Chinese or Japanese.

[17] See, for example, Yano and van Dalen, 'Tables of Planetary Latitude I and II'; Shi Yunli, 'The Korean Adaptation', and Li Liang, 'Arabic Astronomical Tables in China'.

[18] See http://www.hamsi.org.nz. P.I.s Clemency Montelle and Kim Plofker. HAMSI is hosted at the University of Canterbury, Christchurch, New Zealand, and is supported by a five-year grant from the Royal Society of New Zealand.

[19] See http://alfa.hypotheses.org. P.I. Matthieu Husson. ALFA is a project of the European Research Council (ERC CoG 723085) and is hosted by the Paris Observatory.

Another long-term endeavour is the ground-breaking scholarly enterprise Ptolemaeus Arabus et Latinus (PAL) in Munich.[20] PAL's main objective is to produce catalogues of Arabic and Latin Ptolemaic manuscripts as well as critical editions of all Ptolemaic works on astronomy and astrology and a selection of commentaries in Arabic and Latin. This includes full editions of the tables in the *Almagest* in both languages. But another important goal of the project is to survey tables and horoscopes in the medieval Ptolemaic tradition and to provide electronic tools for their digitisation, critical edition and mathematical analysis. For this purpose, PAL has been actively involved in the table-related projects carried out under the auspices of ALFA and has undertaken to host the present volume in its *Studies* series.

These very different projects share a common interest in the analysis and edition of a large corpus of astronomical tables with digital tools. Indeed, much of the tedium and complexity surrounding table analysis can be managed and alleviated by appropriate computer-assisted technology, and the momentum in this area is growing. Given these new attitudes towards historical numerical data, and the growing pool of expertise and digital tools amongst a diversity of projects dealing with different cultural areas, the editors of this volume realised that it was high time to bring together these efforts and to share and build on them collectively.

This cooperation took place under the aegis of the TAMAS project (Tables Analysis Method for the history of Astral Sciences).[21] Over the course of several years, no less than seven small intensive workshops were held, in which the same core group of international specialists presented case studies from their own domain of inquiry, and shared their views on the practices, features, challenges, and aspirations of their numerical tables. From this basis, collective discussions were initiated to address both the inner workings and the broader aims of database design. Among other relevant topics, issues such as developing a common language and encoding practices, metadata standards, and shared digital tool development were discussed in detail.

The database DISHAS (Digital Information System for the History of Astral Sciences, https://dishas.obspm.fr/) arising from this work is now available online and its initial datasets were under construction at the time this

[20] See http://ptolemaeus.badw.de. Project leader Dag Nikolaus Hasse, research leaders David Juste and Benno van Dalen. PAL is a project of the Bavarian Academy of Sciences and Humanities and the University of Würzburg and is funded for a period of 25 years jointly by the Federal Republic of Germany and the Free State of Bavaria.

[21] See http://tamas.hypotheses.org. P.I. Matthieu Husson. TAMAS was funded by the 'Jeunes Chercheurs/nouvelles équipes' grant from the Université PSL (Paris Sciences et Lettres) from 2017 to 2019. LOCOMAT (https://locomat.loria.fr), an earlier and somewhat different project for creating a repository of historical numerical tables, was inspirational for this joint initiative.

introduction was written. DISHAS is designed to accommodate the diverse tabular layouts that the historical corpus attests to, as well as to handle different numerical systems, errors and variants, and to provide a variety of digital outputs for the circulation and dissemination of scholarly studies. Various digital tools, such as CATE (Computer Assisted Tables Editor), DTI (DISHAS Tables Interface) or DIPS (DISHAS Interactive Parameter Squeezer) were developed in the context of the various projects participating in TAMAS and integrated in the common information system. New tools are currently under development especially in the direction of opening up DISHAS datasets to Machine Learning and Artificial Intelligence applications. DISHAS has the potential to transform the way scholars study numerical tables; in fact, the intense collective work that was undertaken in shaping DISHAS has already given rise to new insights, questions and approaches to astronomical tables. From this, the desire naturally arose to bring together these insights and the newly gained expertise in a collective volume.

Our aspirations with this volume, then, are to build on this momentum, to harness the power of digital tools to edit, process, and analyse the corpus of numerical data, focusing on astronomical tables. As the potential of data science and automated analysis grows, careful consideration of the development of digital tools and their benefits and pitfalls needs to be made, so that we can remain sensitive to the nuances and subtleties of data analysis. Digital tools offer the potential to customise the editing process to the individual user, but must be applied with caution.

The first steps towards this aim involve collecting and considering the issues related to numerical data both in the editing process and in the process of technical analysis. This preliminary work is crucial for developing robust digital systems which can shift the focus from painstaking and time-consuming manual analysis to automated ones, so that historians can focus on higher level questions. In this way editing and analysing may even become dynamic processes rather than static ones. Clearly articulating the problems and challenges associated with these processes, and offering strategies and solutions to resolve them, are important aims of this volume.

The resulting collection of papers testifies to the range and scope of the scholars involved and provides a strong coverage of the Eurasian continent and North Africa. They include Sanskrit, Chinese, Latin, Hebrew, and Islamicate cultures of inquiry in a number of different domains, such as scientific and monastic contexts, and astronomical, mathematical, and geographical branches. These different sources and contexts, however, are connected in multiple ways. First and foremost, they are technically related to the style of mathematical astronomy developed in the Hellenistic world, of which Ptolemy is the most famous and successful example. Arabic sources and the Latin ones that directly depend on them are deeply connected to the *Almagest* and the *Handy Tables*.

In successive episodes from the thirteenth to the sixteenth centuries, the Arabic/Persian and Latin astronomical traditions were transmitted to India and China and began a shared history with already well-developed traditional astronomical disciplines in Sanskrit and Chinese. However, long before these contacts with Arabic/Persian and Latin sources (no later than the beginning of the first millennium of the common era) pre-Ptolemaic Hellenistic astronomical concepts had reached India and from there, in the second half of the first millennium, circulated to China through the channel of Buddhist transmissions. These connections between the various cultural milieus, beyond the methodological interest of joining forces in the study of astronomical tables through digital tools, also justify the alliance of our different projects in shaping DISHAS and the collection of studies of tables from each of the areas in a single volume. Each of the contributions presented here contains key insights that will be useful in determining future directions in the field.

An overview of the contributions

The contributions are organised in four sections. The first group presents state-of-the-art approaches to table cracking that are used to derive different types of historical conclusions. The second group of contributions focuses on the relations between the critical edition of tables and their analysis. The third group shares a common concern with computational practices in relation to astronomical tables. The fourth and last group explores new paths and approaches to table analysis.

Classical approaches to table cracking

The first chapter in this group is a survey by **Glen Van Brummelen, Matthieu Husson and Clemency Montelle** of different techniques that historians have used to 'crack' numerical tables, and of the historical implications of these techniques. Relying on specific cases taken from a range of cultural contexts, the authors demonstrate the variety of situations in which table cracking has been used successfully, including the recovery of an author's sources and methods, the reconstruction of missing entries, the 'squeezing' of numerical parameters underlying computed tabular values, the determination of dependencies between tables, the reconstruction of the computation scenarios of a data set, the inference of a table's purpose or its intended users, and the like. This chapter also explores larger methodological issues arising from table cracking. It argues for a reflective approach when applying any of the many analytic tools available to historians in order to produce balanced and historically sensitive analyses of astral science data.

In the second chapter of this group **José Chabás and Bernard Goldstein** reveal how a close description of explicit features of tabular data along with some elementary methods of recomputation can provide insight into the origin of sets of tables. In this spirit, the authors survey the *Almanac* of Jacob ben Makhir (*c.* 1236–*c.* 1305), a Jewish astronomer from southern France. Their analysis relies on a detailed description of the table layout, the selection by Jacob ben Makhir of astronomical quantities used as arguments and entries, and the identification of a few key values in each table. They then associate these features with other known sources and underlying theoretical models. In this way, the authors are able to establish that the *Almanac* is computed from the *Toledan Tables*, and that it improves the accuracy of earlier almanacs by computing every value to the precision of arcminutes. Their analysis also allows them to reflect on how Muslim, Jewish, and Christian scholars contributed to the transmission and circulation of this work.

In a further article in this group, **Seb Falk** highlights how standard modern mathematical techniques can be adapted in order to reveal underlying data structures and dependencies on other historical sources. Falk analyses a small set of tables compiled in the late fourteenth century by John Westwyk as he headed from Oxford to a northern monastery of his order in Tynemouth. Westwyk relied predominantly on the works of Richard of Wallingford, especially his *Albion*. Because of the change in local circumstances, Westwyk had to adapt some of the tables, modifying them for the much larger terrestrial latitude. To illuminate this process, Falk focuses on Westwyk's table of oblique ascensions. Using standard statistical techniques, he is able to reconstruct the circuitous route of computation scenarios that Westwyk followed to recast this table from its original latitude of 51;50° (the latitude of Oxford) to 55° (Tynemouth). In this way, he demonstrates how modern mathematical algorithms can be used to reveal computational choices that are left unstated by the historical table compilers.

The last paper in this group uses the facility of a modern spreadsheet to reveal the underlying sets of tables that historical compilers used to construct new ones. **Kailyn Pritchard** investigates a relatively understudied aspect of the history of trigonometry in Europe: the computation of tangent tables from sine tables. It is almost never made explicit which sine tables these table compilers used, but the numerical signature of a specific sine table can be detected in certain tangent values. Pritchard's methodology relies on an analysis of the (for this purpose) most interesting part of the tangent table, namely, values for arguments approaching 90°, which, in modern terms, increase without bound. The accuracy of these values offers a window into the original sine function. Using this observation, Pritchard is able to identify the underlying sine and cosine tables for the three earliest important tangent tables in Latin sources: those of Giovanni Bianchini, Regiomontanus, and Rheticus. More broadly, her study offers important insights into the choice of tables used by historical computers, and into the (lack of) consistency in these choices.

Editing and analysing analysing astronomical tables

A second group of case studies are based on the mutual consideration of the processes of table cracking and table editing. In particular, these studies focus on the ways in which table reconstruction techniques can help achieve a sensitively constructed critical edition. These studies thus advance the fundamental issues arising from table editing by combining them with the process of table cracking.

In the first article in this group **Clemency Montelle** approaches an Indian set of tables via three levels of analysis, culminating in the critical edition of the tables: 1) a presentation of the manuscripts, 2) a technical description of the algorithms underlying the tables, and 3) the critical edition itself, along with a critical apparatus. Montelle uses this approach to produce a critical edition of a sixteenth-century Sanskrit table text, the *Candrārkī* of Dinakara, a source used by Indian practitioners to produce the annual calendar (*pañcāṅga*). Invoking salient features of a set of manuscripts of this work, she develops a comprehensive account of the ways in which Sanskrit numerical table texts can vary in layout, style, and content, and offers a variety of editorial resolutions to address each of them in a systematic way.

In the second chapter in this set, **Anuj Misra** explores table dependency and proposes some new and innovative ways to detect it, as well as to address discrepancies in precision. Misra tackles table dependency in the context of a seventeenth-century Sanskrit table text, Nityānanda's *Amṛtalaharī*, known through a single manuscript witness. His analysis focuses on six elementary interrelated trigonometric tables. He advances a mathematical analysis that not only aims to recompute the tables in full, but also considers each individual difference between the attested and recomputed values with a view to assessing whether the source of the discrepancy can be identified as computational or scribal. Misra thus connects the mathematical analysis of a table to the generation of a critical edition in a mutually affirming way.

Computational practices and table cracking

The third group of papers also approaches the processes of table cracking and editing as complementary, but adds additional criteria, most notably the prioritisation of the procedures of historical computers over modern analytical tools.

The first article in this group reveals how historically sensitive analyses can reveal contrasting underlying computational models used to construct tables for the same phenomenon. **Li Liang** considers the case of tables for sunrise and sunset in the Chinese astronomical systems of the Yuan and Ming periods (1271–1644). He discovers that two different types of tables were used to compute these times, which ultimately relied on different computational models. Using the accuracy of the numerical data, he shows how a more sophisticated and accurate approach was used for the sets of tables covering sunrise and sunset times for the capital cities of the two dynasties, and that a cruder method was used for other, less prominent locations. This same methodology also allowed him to successfully analyse how these sunset and sunrise tables were adapted in the Korean kingdom of Joseon (1392–1910).

A similar kind of approach to the editing and analysis of astronomical tables is developed by **Glen Van Brummelen** in the second article in this group. Focusing on a highly original set of double-argument tables for planetary latitude from Jamshīd al-Kāshī's fifteenth-century *Khāqānī Zīj*, Van Brummelen uncovers the underlying planetary latitude models on which these tables are based and accounts for the way in which they were transformed by Islamic astronomers. This approach allows him to explore the variety of computational scenarios that reveal the sensitivity arising from the computational choices the historical computers had to make. He analyses these sensitivities, focusing on the most computationally and historically significant choices. In this way he is able to identify six original features of this set of tables and in turn curates a critical edition of the tables.

In the third paper in this group, **Sho Hirose** investigates how an in-depth study of textual tabulated data and attested historical algorithms and the process of editing this data can mutually benefit from each other. He relies on a close analysis of the planetary equation tables of Parameśvara (*c.* 1360–1460) contained in the work *Dṛggaṇita* and the relevant algorithms presented there, allowing him to reconstruct possible computation scenarios for the tabular values. What is particularly interesting about these tabular data is that they are expressed in verse employing an alphanumeric form of encoding called *kaṭapayādi*. The analysis of the ways this particular numerical notation system interacts with the tabular values and their variants produces particularly valuable results both in table editing and in reconstruction.

The last article in this group addresses this same issue, but with respect to Latin astronomical tables. **Richard Kremer** considers the tables of true syzygies that were compiled around 1340 by the two Parisian astronomers John of Murs and Firmin of Beauval, and were included in the work known as *Tabulae permanentes*. In the process of providing an analysis and critical edition of this historically important double-argument table, Kremer carefully outlines the different steps and choices he makes, and progressively refines his computation scenarios through exploratory data analysis. This allows him to reconstruct plausible algorithms with confidence, and also to anticipate places where the underlying tables behave unexpectedly.

Pushing approaches to table analysis further

The last group of papers pushes further the ambitions of the modern table analyser and editor. It explores new approaches and poses new questions that can be addressed thanks to emerging digital tools, and even broadens the purview of the types of sources that are relevant for table analysis.

In the first paper in this group, **Matthieu Husson** surveys a range of manuscript witnesses for a single table and probes how the variation between sources might affect their performance as computational tools. Husson's contribution focuses on a simple set of solar and lunar velocity and equation tables belonging to the *Tabule magne* (*c.* 1325) of the Parisian astronomer John of Lignères. These tables formed the basis for the computation of true syzygies. Relying on a description for this computation provided by John of Lignères in the canons to the same work and on a mathematical analysis and critical edition of the tables, the study then focuses on the variants and their significance. In the process of analysing the variants as clues for understanding scribal practices, the computational consequences of these variants are simultaneously addressed both quantitatively and qualitatively.

A second study in this group embraces the full power of modern mathematical analysis to assess historical numerical data. **Johannes Thomann** analyses a fragment of an Arabic ephemeris using the standard modern statistical tool for parameter analysis, least squares estimation. This method is usually applied to complete tables depending on a limited number of parameters after they have been conscientiously purged of outliers (scribal or computational) and interpolated values. However, Thomann develops a methodology allowing him to successfully apply least squares estimation in a case where the number of parameters is higher than usual and the data are partial and messy. His approach is first tested on synthetic data, and then relies on an iterative use of least squares on carefully selected subsets of the data in the ephemeris in order to establish the source on which it was based.

The last article in this group explores methods of table cracking appropriate for non-computational tables. **Benno van Dalen** considers the geographical table from a thirteenth-century Arabic astronomical handbook, the *Shāmil Zīj*. In a context where the tabular values were not produced by computation, van Dalen's analysis provides novel insights into purely scribal phenomena in the transmission of large numerical texts, and allows corrections to be guided in part by a quantitative lens. For this he makes use of the huge database of Islamic geographical coordinates published by Edward S. Kennedy in 1987, and for each locality establishes the most likely original coordinates in several traditions. In this way, van Dalen proposes editorial resolutions for norms and metrics to measure scribal variants. This holistic approach enables him to resolve many ambiguities in the text and understand how they occurred by bringing to light their dependence on the larger stemma of manuscript relations.

Overall, the aims of a collection of studies of this nature are to bring together experts in numerical tables from various cultures of inquiry, to present a thorough scrutiny of the aspects and features of table editing and analysis related to their particular cultural domain, and to document them with recourse to specific examples. In this way, the arcane and specialised skills of the human editor can be paired with the power and efficiency of the computer to deal with the overwhelming corpus of tables. Such shared scrutiny and collaboration will also add new dimensions to our understanding of the practices and priorities of historical scientific cultures in unprecedented and exciting ways.

It is clear that the numerical data in historical tables lend themselves readily to digitisation in ways that no other historical sources do. With due caution, substantial progress can be made by embracing the synergies between data scientists and historians. Harnessing modern mathematical modelling and data science techniques can enable historians to establish dependencies and underlying parameters in a way that would be almost impossible to achieve by manual methods. These technologies can help highlight connections and trace cases of transmission which may remain largely invisible when the sources are examined on a small scale. They can uncover patterns of circulation of knowledge and reveal the ways in which individual historical compilers experimented and modified inherited technical knowledge, and they can provide plausible explanations for these modifications, be they due to the introduction of empirical data or to new or simplifying computational techniques. They can also reveal the techniques that historical compilers used to mass compute data along with interpolation preferences, and highlight issues surrounding granularity, numerical precision and accuracy, bypasses and hacks for improving computation speed, and the like. Digital tools offer the potential to add new dimensions to our understanding of historical computational culture and substantially widen the range of questions we can pose as we investigate the brilliance and virtuosity of these historical scientific milieus.

Bibliography

Brack-Bernsen, Lis, *Zur Entstehung der Babylonischen Mondtheorie. Beobachtung und theoretische Berechnung von Mondphasen*, Stuttgart: Franz Steiner, 1997.

Britton, John P., 'Studies in Babylonian Lunar Theory: Part I. Empirical Elements for Modeling Lunar and Solar Anomalies', *Archive for History of Exact Sciences* 61 (2007), pp. 83–145.

Britton, John P., 'Studies in Babylonian Lunar Theory: Part II. Treatments of Lunar Anomaly', *Archive for History of Exact Sciences* 63 (2009), pp. 357–431.

Britton, John P., 'Studies in Babylonian Lunar Theory: Part III. The Introduction of the Uniform Zodiac', *Archive for History of Exact Sciences* 64 (2010), pp. 617–63.

Campbell-Kelly, Martin, Mary Croarken, Raymond Flood and Eleanor Robson, *The History of Mathematical Tables. From Sumer to Spreadsheets*, Oxford: Oxford University Press, 2003.

Chabás, José, *Computational Astronomy in the Middle Ages. Sets of Astronomical Tables in Latin*, Madrid: Consejo Superior de Investigaciones Científicas, 2019.

Chabás, José, and Bernard R. Goldstein, *The Alfonsine Tables of Toledo*, Dordrecht: Kluwer, 2003.

Chabás, José, and Bernard R. Goldstein, *The Astronomical Tables of Giovanni Bianchini*, Leiden: Brill, 2009.

Chabás, José, and Bernard R. Goldstein, *A Survey of European Astronomical Tables in the Late Middle Ages*, Leiden/Boston: Brill, 2012.

Chabás, José, and Bernard R. Goldstein, *Essays on Medieval Computational Astronomy*, Leiden: Brill, 2015.

Cullen, Christopher, *The Foundations of Celestial Reckoning. Three Ancient Chinese Astronomical Systems*, London/New York: Routledge, 2017.

Dalen, Benno van, *Ptolemaic Tradition and Islamic Innovation: The Astronomical Tables of Kūshyār ibn Labbān*, Turnhout: Brepols, 2021.

Dorce, Carlos, *El Taŷ al-azyāŷ de Muḥyī al-Dīn al-Magribī*, Barcelona: Instituto 'Millás Vallicrosa' de Historia de la Ciencia Árabe, 2002–2003.

Heiberg, Johan Ludvig, *Syntaxis mathematica*, 2 vols, Leipzig: Teubner, 1898–1903. [*Claudii Ptolemaei opera quae exstant omnia*, Volumen I.]

Husson, Matthieu, and Clemency Montelle (eds), *The Transmission of Arabic Astronomical Tables in Sanskrit, Latin, and Chinese*, Barcelona: Universitat de Barcelona, 2014 [special issue of *Suhayl*, vol. 13].

Jones, Alexander, *An Eleventh-Century Manual of Arabo-Byzantine Astronomy*, Amsterdam: Gieben, 1987.

Keller, Agathe, and Clemency Montelle (eds), *Special Issue on Numerical Tables in Sanskrit Sources*, Delhi: Indian National Science Academy, 2014 [*Indian Journal for History of Science*, vol. 49/2].

Kennedy, Edward S., 'A Survey of Islamic Astronomical Tables', *Transactions of the American Philosophical Society* New Series 46 (1956), pp. 123–77. [Reprinted: 1989, with pagination 1–55.]

King, David A., and Julio Samsó, 'Astronomical Handbooks and Tables from the Islamic World (750–1900): An Interim Report', *Suhayl* 2 (2001), pp. 9–105.

Li Liang, 'Arabic Astronomical Tables in China: Tabular Layout and Its Implications for the Transmission and Use of the *Huihui lifa*', *East Asian Science, Technology, and Medicine* 44 (2016), pp. 21–68.

Martzloff, Jean-Claude, *Astronomy and Calendars – The Other Chinese Mathematics. 104 BC–AD 1644*, Berlin/Heidelberg: Springer, 2016.

Mercier, Raymond P., *An Almanac for Trebizond for the Year 1336*, Louvain-la-Neuve: Academia, 1994.

Mercier, Raymond P., *Πτολεμαίου Πρόχειροι κανόνες. Ptolemy's Handy Tables. Volume 1b. Tables A1-A2. Transcription and Commentary*, Louvain-la-Neuve: Université Catholique de Louvain, Institut Orientaliste, 2011.

Misra, Anuj, Clemency Montelle and Kim Plofker, 'Eclipse Computation Tables in Sanskrit Astronomy: A Critical Edition of the Tables of the *Karaṇakesarī* of Bhāskara (*fl. c.* 1681)', *History of Science in South Asia* 4 (2016), pp. 1–79.

Misra, Anuj, Clemency Montelle and Kim Plofker, *The Sanskrit Astronomical Table Text* Brahmatulyasāraṇī. *Numerical Tables in Textual Scholarship*, Leiden/Boston: Brill, 2020.

Montelle, Clemency, and Kim Plofker, *Sanskrit Astronomical Tables*: Cham: Springer, 2018.

Nallino, Carlo Alfonso, *al-Battānī sive Albatenii opus astronomicum. Ad fidem codicis Escurialensis arabice editum. Latine versum, adnotationibus instructum*, 3 vols, Milano: Ulrico Hoepli, 1899–1907.

Neugebauer, Otto, *Astronomical Cuneiform Texts. Babylonian Ephemerides of the Seleucid Period for the Motion of the Sun, the Moon, and the Planets*, 3 vols, London: Lund Humphries, 1955. [Reprinted New York: Springer, 1983.]

Neugebauer, Otto, *The Astronomical Tables of al-Khwārizmī. Translation with Commentaries of the Latin Version edited by H. Suter supplemented by Corpus Christi College MS 283*, København: Det Kongelige Danske Videnskabernes Selskab/Munksgaard, 1962.

Neugebauer, Otto, *HAMA = A History of Ancient Mathematical Astronomy*, 3 vols, New York/Heidelberg/Berlin: Springer, 1975.

Ossendrijver, Mathieu, *Babylonian Mathematical Astronomy: Tabular Texts*, New York: Springer, forthcoming.

Pedersen, Fritz S., *The Toledan Tables. A Review of the Manuscripts and the Textual Versions with an Edition*, 4 vols, Copenhagen: Reitzels, for Det Kongelige Danske Videnskabernes Selskab, 2002.

Pingree, David, *Sanskrit Astronomical Tables in the United States*, Philadelphia: American Philosophical Society, 1968. [*Transactions of the American Philosophical Society*, New Series 58.]

Pingree, David, *Sanskrit Astronomical Tables in England*, Madras: The Kuppuswami Sastri Research Institute, 1973.

Poulle, Emmanuel, *Les Tables Alphonsines avec les canons de Jean de Saxe. Édition, traduction et commentaire*, Paris: Éditions du Centre National de la Recherche Scientifique, 1984.

Shi, Yunli, 'The Korean Adaptation of the Chinese-Islamic Astronomical Tables', *Archive for History of Exact Sciences* 57 (2003), pp. 25–60.

Sivin, Nathan, *Granting the Seasons. The Chinese Astronomical Reform of 1280, With a Study of Its Many Dimensions and an Annotated Translation of Its Records*, New York: Springer, 2009.

Stahlman, William Duane, *The Astronomical Tables of Codex Vaticanus Graecus 1291*, PhD dissertation, Brown University, 1959 [accepted 1960].

Steele, John, 'Newly Identified Lunar and Planetary Tables from Babylon in the British Museum', *SCIAMVS* 11 (2010), pp. 211–39.

Suter, Heinrich, *Die astronomischen Tafeln des Muḥammed ibn Mūsā al-Khwārizmī in der Bearbeitung des Maslama ibn Aḥmed al-Madjrīṭī und der latein. Uebersetzung des Athelhard von Bath*, København: Det Kongelige Danske Videnskabernes Selskab/Høst, 1914.

Tihon, Anne, *Πτολεμαίου Πρόχειροι κανόνες. Les Tables Faciles de Ptolémée. Volume 1a. Tables A1-A2. Introduction, Édition critique*, Louvain-la-Neuve: Université Catholique de Louvain, Institut Orientaliste, 2011.

Toomer, Gerald J., *Ptolemy's Almagest*, London: Duckworth/New York: Springer, 1984. [2nd ed. Princeton: Princeton University Press, 1998 (with a foreword by Owen Gingerich).]

Van Brummelen, Glen, *Mathematical Tables in Ptolemy's Almagest*, PhD dissertation, Simon Fraser University, 1993.

Yabuuti, Kiyosi, 'Astronomical Tables in China, from the Han to the T'ang Dynasties', in Kiyosi Yabuuti (ed.), *Chūgoku chūsei kagaku-gijutsu-shi no kenkyū (Researches on the History of Science and Technology in Medieval China)*, Tokyo: Kadogawa, 1963, pp. 445–92.

Yabuuti, Kiyosi, 'Astronomical Tables in China. From the Wutai to the Ch'ing Dynasties', *Japanese Studies in the History of Science* 2 (1963), pp. 94–100.

Yano, Michio, and Benno van Dalen, 'Tables of Planetary Latitude in the *Huihui li* (I) and (II)', in Yung-Sik Kim and Francesca Bray (eds), *Current Perspectives in the History of Science in East Asia*, Seoul: Seoul National University Press, 1999, pp. 307–15 and 316–29.

Part 1

Classical Approaches to Table Cracking

Tools of the Table Crackers: Using Quantitative Methods to Analyze Historical Numerical Tables

Glen VAN BRUMMELEN, Matthieu HUSSON
and Clemency MONTELLE

1. Introduction: Defining the Issues

The content of numerical tables presents technical and historical challenges for those who seek to analyse them. However, such endeavors can be rewarding to historians. Table values provide a unique insight into the practices of historical table makers, beyond the usual textual content. As well as illuminating the practices of historical table-makers, tabular entries can also reveal something of the users of the tables and their priorities.

Tabular manuscripts can reveal new historical information that may not be explicit in other historical sources. One might be able to retrieve information about the original numerical methods, algorithms, and hand computations used to construct the entries. Other historical information might also be restored, including the author's sources, the abilities of those who compiled them and their priorities, the intended audience (who used it), the table's application (how it was used), the underlying theory (astronomical, physical or mathematical), transmission both to and from other sources, and dependence on other tables and traditions. In some cases, analyses of tables can even prompt scholars to propose new chronologies or dating.

Typically, table analysis efforts are directed primarily toward ancient and early sources, rather than modern. Generally the older the table, the less information we have surrounding its numerical data. With the passing of time, historical documents may be destroyed, damaged, separated, or corrupted. Documentation that surrounds the tables might not survive; tables might be incomplete. Furthermore, ancient cultures of inquiry may have had different priorities and incentives from their modern counterparts in presenting and promulgating scientific results. Often historical table makers were little motivated to reveal their techniques and methods of construction. All of this makes recourse to the results of table cracking even more valuable.

To reflect the complexity of the process of seeking the underlying structure of a table, as well as the epiphany once the content has been unlocked, table analysis has often been dubbed 'table cracking'. Table cracking involves

Editing and Analysing Numerical Tables: Towards a Digital Information System for the History of Astral Sciences, ed. by Matthieu Husson, Clemency Montelle and Benno van Dalen, PALS 2 (Turnhout, 2021), pp. 19–52

10.1484/M.PALS-EB.5.127694

the manipulation of a table's entries to retrieve information about the table. Typically this includes the use of modern quantitative methods, arithmetical or statistical, to glean underlying features of the numerical data. This line of inquiry is often far removed from the original historical context in which the table was constructed. Nonetheless these features can be relevant to a proper historical understanding of the table and its context.

Despite the new information table cracking offers, many historians remain skeptical. Some feel uneasy or unqualified to assess the appropriateness of the sophisticated statistical procedures that are often employed. Others object to the number or scope of the assumptions that must be made, the apparent *ad hoc* nature of the process, or of the occasional imprecision or excessive precision of the results. Some are wary of the historical interpretations that have been proposed as a result of table cracking analysis.

We seek here to consider and classify the various processes that have fallen under the rubric of 'table cracking'. We will analyze through selected examples the procedures involved and the results they have generated. We will outline some of the advantages and pitfalls of this approach, advance some preliminary general standards for those who want to appreciate table crackers' results, present guidelines to evaluate them critically, and provide references to help readers begin to develop their own skills.

1.1. Accounting for Tabular Errors

The manual copying of literary texts invariably introduces errors or discrepancies; textual critics use this principle when they produce a critical edition of a text and determine the stemmatic relation between manuscripts. Errors also can be illuminating to table crackers.[1] Numerical tables are subject to the same discrepancies when they are copied. These types of error can help determine stemmatic relations between various copies of the same table and other related tables. Moreover, these discrepancies can take on additional significance. Indeed, errors in tabular values can be caused by significant factors other than unintentional copying mistakes.

The most obvious errors in tabulated data are scribal errors. These are accidental mistakes and alterations that are introduced when a table is copied. Usually the result of a moment of inattention or carelessness by the scribe, they can reveal something of the process of copying and other aspects of the original sources, such as common confusions between various symbolic notations or poor layouts. Another class of error is computational errors. These discrepancies can indicate idiosyncratic decisions made at intermediary steps

[1] See van Dalen, *Ancient and Mediaeval Astronomical Tables*, pp. 12-18; Neugebauer, *Astronomical Cuneiform Texts*, p. 27.

of computation, simple arithmetic mistakes, rounding conventions, the precision desired, or even differences in the nature of the function supposed to have been tabulated. These errors can be evidence of the difference between the algorithm expressed in the text and its tabular implementation.

Common ways of detecting errors include: comparing a recomputed tabular value with the original, examining first (or sometimes second) differences, and plotting the function values on a graph and looking for aberrations or irregularities. Sometimes one can trace the cause of an error back to some intermediary computational step, although given the variety of decisions that are made when implementing an algorithm, this is often very difficult in practice. Correcting errors can be more contentious; in some situations (especially where there may be several alternative computational models), it can be unclear precisely what the correct value corresponding to a given entry should be. In fact, computational discrepancies can challenge the notion that there is an original correct table underlying the existing one. One can then imagine the challenges that might arise in producing a critical edition of such a table.

2. Table Cracking I: Restoring the author's sources and methods

The term 'table cracking' has been interpreted in a variety of ways; almost the only constant between them is the use of the table's entries to gain historical information, usually in a quantitative manner. These meanings divide roughly into two categories. Firstly, one may learn about the means by which the table was constructed. This includes ascertaining the theoretical models underlying the table, determining the use of certain historically attested numerical parameters or other underlying tables, and reconstructing the process of computation. These goals reflect the table's origins, the scientific activity and authorial process that resulted in the table's production. Secondly, table crackers attempt to restore information about how the table must have been used or evaluated *after* it was constructed. Activities related to authorial process, in particular, tend to share certain methodological features that deserve close attention. Thus we survey efforts to uncover a table's mathematical origins in Section 2.2 and make evaluative methodological observations in Section 2.3, before proceeding to restoring the table's applications in Section 2.4.

2.1. Understanding the theoretical models underlying a table

Reconstructing the theoretical model according to which a table was computed is often challenging. The text that accompanies the table, known as paratext, can offer some evidence about the underlying structure. But the task can be more difficult in the typical case when the historical sources

Tabular data				First differences	
Date	*sign*	*deg*	*min*	*deg*	*min*
1 Jan	9	20	22		
2 Jan	9	21	24	*1*	*2*
3 Jan	9	22	25	*1*	*1*
4 Jan	9	24	27	*1*	*2*
⋮	⋮	⋮	⋮	⋮	⋮
1 March	11	20	55		
2 March	11	21	55	*1*	*0*
3 March	11	22	54	*0*	*59*
4 March	11	23	54	*1*	*1*
⋮	⋮	⋮	⋮	⋮	⋮
1 July	3	18	26		
2 July	3	19	23	*0*	*57*
3 July	3	20	20	*0*	*57*
4 July	3	21	17	*0*	*57*

Table 1: An excerpt from Cortés' *Table of the true place*, with first differences

describing the tables give contradictory, vague, or misleading reports. The numerical entries themselves are usually the starting point; they often provide table crackers with sufficient information to recover the table's mathematical framework. We consider in detail an example of a reconstruction of the model underlying a pair of tables in which the layout of the tables proves to be misleading.

The Spanish cosmographer Martin Cortés de Albacar's (1510–1582) *Arte de Navigar* (1551) became one of the most popular navigation manuals by the beginning of the seventeenth century. Many aspects of astronomical navigation are detailed in this treatise. We concentrate here on two tables designed for the computation of the sun's true longitude at noon, a necessary step in the determination of one's terrestrial latitude.

In this example, as with most others in this chapter, the author was working in a geocentric astronomical system inspired by Ptolemy's *Almagest*. In this tradition, the standard procedure to determine the true solar longitude involves two steps. The first employs a set of tables that give the sun's mean longitude. Since it is assumed that the sun travels along the ecliptic at a constant velocity, the table computes a linear function with respect to time. The second step uses a table that computes a correction factor, called the solar equation, that adjusts the longitude from mean to true based on the sun's position in its orbit (so that both the argument and the function are measured in degrees). A geometric model (see Figure 2 on p. 31) allows the solar equation to be calculated trigonometrically for any argument. Did Cortés follow this model of computation in the *Arte de Navigar*?

Tabular data			First differences	
years	*deg*	*min*	*deg*	*min*
1545	1	0		
1546	0	45	0	−15
1547	0	30	0	−15
1548	0	15	0	−15
1549	1	2	0	+47
1550	0	47	0	−15
1551	0	32	0	−15
1552	0	18	0	−14
1553	1	4	0	+46
1554	0	49	0	−15
1555	0	34	0	−15
1556	0	19	0	−15
1557	1	5	0	+46

Tabular data			First differences	
years	*deg*	*min*	*deg*	*min*
1558	0	50	0	−15
1559	0	35	0	−15
1560	0	21	0	−14
1561	1	7	0	+46
1562	0	52	0	−15
1563	0	37	0	−15
1564	0	23	0	−14
1565	1	9	0	+46
1566	0	54	0	−15
1567	0	39	0	−15
1568	0	25	0	−14
1569	1	11	0	+46

Table 2: An excerpt from Cortés' 'solar equation' table, with first differences

The caption of the first table in the Spanish edition is *Tabla del verdadero lugar del sol*, which was incompletely translated in the first English edition (1561) as *The table of the true place*. Just as in mean motion tables, this table's argument is in time units and its entries are in degrees of arc, giving values for each day of an unspecified year. However, a quick examination of the first differences of the entries verifies that the tabulated function is not linear (see Table 1).

The caption of the second table is *Tabla de las equaciones del sol*, correctly translated in the first English edition as *Table of the equations of the sunne*. But this table is nothing like a solar equation. Its argument is in calendar years (running from 1545 to 1688), rather than degrees; and the table's entries form a roughly linear trend rather than trigonometric (see Table 2). Thus, despite the titles, Cortés does not follow Ptolemy.

Cortés instructs his readers how to use the tables using a worked example. To determine the sun's true longitude on 22 February 1568, select the entry corresponding to 22 February in the first table; then select the entry corresponding to 1568 in the second table. Add these two entries together, and the result is the sun's true longitude. This method of computation confirms what we just saw: Cortés is not using Ptolemy's tabular approach. However, this does not imply that the underlying geometrical model is fundamentally different. Further exploration is needed. When Cortés' tables were compiled, the Parisian Alfonsine tables were dominant in Europe and would have been readily available to Cortés or his sources. The models underlying Alfonsine astronomy are well known.[2] So we begin by hypothesizing that the tables can

[2] See for example Chabás and Goldstein, *The Alfonsine Tables of Toledo*.

Tabular data				Alfonsine position (1483)			Difference
Date	*sign*	*deg*	*min*	*sign*	*deg*	*min*	*min*
1 Jan	9	20	22	9	20	23	1
2 Jan	9	21	24	9	21	24	0
3 Jan	9	22	25	9	22	25	0
4 Jan	9	23	26	9	23	27	1
⋮	⋮	⋮	⋮	⋮	⋮	⋮	⋮
1 March	11	20	55	11	20	57	2
2 March	11	21	55	11	21	57	2
3 March	11	22	54	11	22	56	2
4 March	11	23	54	11	23	56	2
⋮	⋮	⋮	⋮	⋮	⋮	⋮	⋮
1 July	3	18	26	3	18	27	1
2 July	3	19	23	3	19	24	1
3 July	3	20	20	3	20	21	1
4 July	3	21	17	3	21	18	1

Table 3: Sample of Cortés' *Table of the true place* compared with the true position of the sun computed according to the *editio princeps* of the Parisian Alfonsine Tables

be reconstructed from Alfonsine material. If this hypothesis proves correct, then both the model and parameters originate in Alfonsine astronomy.

Let us begin with the *Table of the true place*, which gives the position of the sun for each day of a full year. The first differences of the entries in this table vary, indicating that the table does not provide mean positions. But it may provide true solar positions. In fact, it appears that it does. We recomputed the true solar position for each day of the year 1545, the first argument of the equation table, according to the Parisian Alfonsine Tables as they are presented in the 1483 Ratdolt edition (*princeps*). The difference between these recomputations and the values given by Cortés are within 2 arcminutes (see Table 3).[3] More advanced table-cracking techniques may help us analyse whether the discrepancies between Cortés' values and the recomputed ones show any significant pattern,[4] but this fit is good enough to demonstrate that the table computes Alfonsine true solar positions for 1545.

Turning to the entries of the equation table, we find a clear pattern: each value is about 15 minutes smaller than the preceding one for a sequence of three entries; then, at the fourth entry (a leap year) the value increases by 46 or 47 minutes; and the cycle repeats. So this table does not represent

[3] In this table we have expressed the Alfonsine positions using 30° increments in order to follow Cortés' practice and to avoid confusion. The Parisian Alfonsine Tables, however, use 60° increments.

[4] For instance, one may notice that all differences are positive. This may be caused by a different epoch value.

a solar equation, but rather a quantity that makes an adjustment every leap year. It seems likely that the table gives the total displacement of the mean (or true) sun on the ecliptic for the year: that is, the difference between the mean (or true) sun's longitude at the beginnings of years n and $n+1$. The numerical effect of the distinction between mean and true displacement of the sun is only around 2 arcminutes in longitude. So, to confirm the link between Cortés' tables and those of the Parisian Alfonsine tradition, we may chose either hypothesis. Tentatively, we guess mean displacement.[5]

A variety of methods have been developed to derive mean motion parameters from mean motion tables.[6] Here we adopt a simple calculation, dubbed 'squeezing' by Neugebauer and Kennedy,[7] and refer the reader to the next section for descriptions of more sophisticated parameter derivation techniques. Computing the difference between mean solar positions over the largest possible range consisting of a multiple of four years in Cortés' equation table (1545 to 1685), we can derive a daily mean motion parameter by simply adding this difference to 50,400 (360×104, i.e., the 140 complete revolutions of the sun in 140 years) and dividing the result by 51,135 ($365;25 \times 140$, i.e., the number of days in 140 Julian years). This gives us a parameter value of 0;59,8,19,38 °/day. If we assume that the table values were rounded to the last place, we obtain lower and upper bounds for the solar mean motion parameter of 0;59,8,19,36 °/day and 0;59,8,19,40 °/day.[8] The parameter used in the Alfonsine tradition of 0;59,8,19,37,19,13,56 °/day lies within this interval.[9] On the other hand, the second plausible historical parameter, the sidereal motion of 0;59,8,11,28,27 °/day used in the Toledan tradition, does not lie within the interval. The same holds for Ptolemy's 0;59,8,17,13,12,31 °/day and al-Battānī's 0;59,8,20,47 °/day (corresponding to a solar year of 365;14,26 days). Therefore this rough estimate is enough to confirm our hypothesis that the equation table depends on Alfonsine mean motions.

Historians often encounter considerably more complex cases. When studying a set of fourteenth-century Latin tables by John Vimond, José Chabás and Bernard R. Goldstein were faced with peculiar layouts, headings, and

[5] The hypothesis that the equation table gives the true position (rather than the mean) is less likely because it would make little sense to use a position specific for the first day of the year as an adjustment for every position of that same year.

[6] See, for instance, Mielgo, 'A Method of Analysis', and van Dalen, 'Origin of the Mean Motion Tables'.

[7] See, for example, Kennedy, 'A Survey', p. 20.

[8] These bounds do not constitute a confidence interval as they are not derived from statistical analysis. However they can be heuristically used in a similar fashion.

[9] This number may seem unreasonably precise to us but it actually reflects the actors' practices.

entries, and no accompanying instructions.[10] These tables includes elements to compute latitudes and longitudes of the planets and luminaries, syzygies, and material on the fixed stars. Using mathematical tools similar to those described below (reconstruction of computations, parameter estimation, comparison of results with historically attested sources), Chabás and Goldstein were able to uncover the underlying theoretical model and parameters. This analysis revealed that Vimond's set is the earliest known Parisian tabular material related to the Alfonsine tradition.

In other cases, issues can arise from the number and complexity of the possible underlying models. For instance, in sixteenth-century Vienna, John Angelus computed ephemerides on the basis of Peurbach's planetary tables. Richard Kremer and Jerzy Dobrzycki analyzed the discrepancies between the ephemerides and the planetary positions derived from the standard Alfonsine tradition.[11] The differences revealed that Peurbach must have known of some of the geometrical models developed in Maragha, at least in the form of a diagram.

Restoring missing entries in tables

Especially in ancient studies, the historian may be confronted with a numerical table in a fragmentary state. Cuneiform tables, for instance, are frequently broken, manuscripts can be torn or crumbling, or the surface of a document may be damaged by wear, water, or mould, rendering part of the text illegible. Despite this, table crackers often are able to reconstruct the missing entries. This is because the table's numerical content, unlike (say) the text of prose or poetical works, generally has an underlying theoretical model and computational algorithm that determines the entries. Thus, in principle, it can take as little as one entry and its argument (and sometimes partial at that) to establish the underlying pattern, and consequently restore the missing entries.

The mathematical corpus of Mesopotamia has benefited especially from such reconstruction efforts. Cuneiform texts from this region were inscribed on clay tablets, by now several millennia old. When excavated, more often than not they are broken (sometimes into many pieces), the clay surface is worn, and the imprinted cuneiform signs are illegible. In the case of numerical tables, reconstruction of the missing content is sometimes a trivial task (e.g., fragmentary multiplication tables or reciprocal tables); however, in cases where the underlying mathematical relation is unexpressed or unclear, it can be difficult or impossible. If the table is the only one of its kind, scholars usually are able to restore it only after prolonged and painstaking analysis, if at all.

[10] See Chabás and Goldstein, 'Early Alfonsine Astronomy in Paris'.

[11] Dobrzycki and Kremer, 'Peurbach and Maragha Astronomy?'.

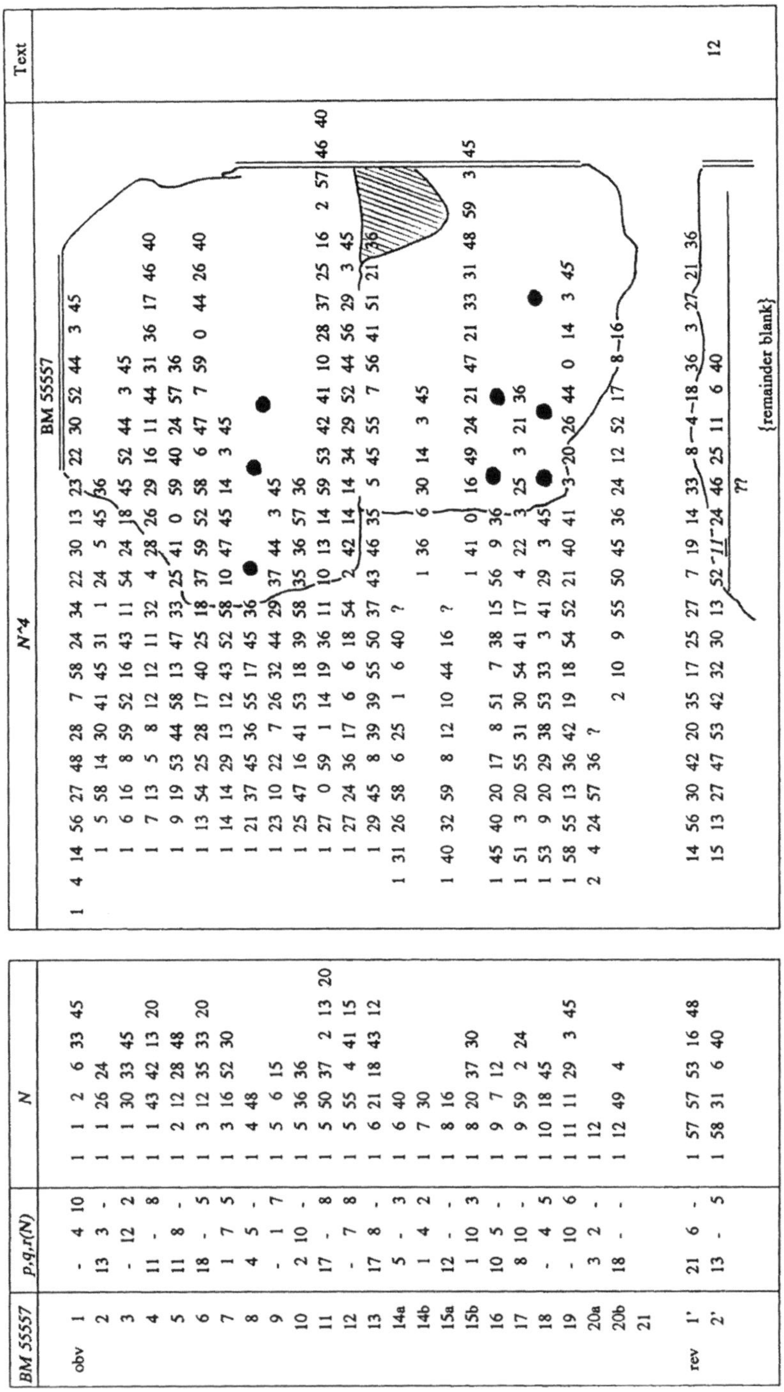

Figure 1: John Britton reconstructs a table of fourth powers based on a very small fragment. Reproduced from the *Journal of Cuneiform Studies* 43–45 (1991–93), p. 75.

				2]3	22	30	52	44	3	45				
				]36										
			]18	45	52	44	3	45						
			]26	29	16	11	44	31	36	17	46	40		
		]41	0	59	40	24	57	36						
1]8	37	59	52	58	6	47	7	59	0	44	26	40		
5]8	10	47	45	14	3	45								
	]37	44	3	45										
	]35	36	57	36										
	1]0	13	14	59	53	42	41	10	28	37	25	16	2	57
	]2	42	14	14	34	29	52	44	56	29	3	45		
				]5	45	55	7	56	41	51	21			
				]30	14	3	45							
				]16	49	24	21	47	21	33	31	48	59	3
				]25	3	21	36							
					]20	26	44	0	14	3	45			

Table 4: A transcription of the extant numbers on BM 55557

One impressive reconstruction was a small fragment of a cuneiform tablet (BM 55557) dating from some time in the first millennium BCE. The significance of the numerical entries in this fragment eluded scholars until recently (see Figure 1 and Table 4). Smoothness on two contiguous edges of the fragment revealed that the extant piece came from the top right hand corner of the original tablet. Strings of numbers remained on both sides of the tablet, evidently the tail ends of long entries in sexagesimal numeration. After some careful number crunching, data matching, failed attempts, and a few key hypotheses, assyriologist John Britton discovered a surprising relation between these strings.[12]

The key to Britton's reconstruction was a pattern in the tail ends of the sexagesimal strings. He noticed that almost all the terminal values were 36, 45, or 40; a significant number of those ending in ...45 ended in ...44,3,45. These and other similar observations led him to conclude that the numbers are 'regular'; that is, they have only 2, 3, or 5 as prime factors. Regular numbers played an important role in Mesopotamian mathematics partly because their reciprocals have a finite representation in the sexagesimal number system. There is much evidence of their properties being explored by ancient

[12] Britton, 'A Table of Fourth Powers'.

practitioners; for instance, there exist in the cuneiform corpus tables of reciprocals of regular numbers, as well as tables of the squares of regular numbers.

Armed with this insight, Britton consulted a massive tabulation of 11-digit regular sexagesimal numbers $N = 2^p 3^q 5^r$ and their reciprocals compiled by Gingerich, over 36 pages in length, listed according to various values of the powers p, q, r.[13] For those N that ended in the string ...44,3,45, Britton noticed similarities in the corresponding values of p, q, r. In particular, the value $N(0, 16, 40) = \ldots 23{,}22{,}30{,}52{,}44{,}3{,}45$ matched line 1 of his text. In the same way, Britton identified the numbers in lines 3 and 9, and from this followed yet more lines. Once a handful of numbers had been identified, Britton noticed that the values of p, q, r were all divisible by 4. This leads naturally to the conclusion that the numbers are the tail ends of fourth powers of regular sexagesimal numbers. At this point it was easy to reconstruct the leading sexagesimal places (around eight) of the broken entries, and complete entries that had been broken off of the table altogether. Furthermore, based on the tablet's physical dimensions he also proposed that there were three additional columns that preceded the fourth powers.[14] These columns, he argued, included a row count, the regular numbers themselves, and their factorisation details.

Because of the damage to the tablet (only 2 partial edges remained: top right and far right edge), Britton could only hypothesize where the fourth powers began and where they ended. Now that the nature of the table was confirmed, he could compare with tables of similar functions (such as reciprocal tables and the so-called 'double six place' tables). This led to the surmise that the original tablet contained fourth powers of regular numbers running from 1,1,2,6,33,45 to 1,58,31,6,40. Britton noted that while contextual considerations such as these can be helpful, they can also raise more questions than they solve. BM 55557 is the only fourth power tablet of its kind in the extant record, and there are no mathematical problems in the entire corpus that call for the computation of a fourth power or its root. Thus, despite the reliable reconstruction of the tablet's numerical content, issues concerning its purpose still remain outstanding.

Mathieu Ossendrijver recently faced a similar challenge.[15] The fragments he was studying appeared to be without parallel in the primary literature and exhibited remarkable mathematical virtuosity, seemingly unconnected to any practical application. Some tablets contained numbers with up to 30 sexagesimal places, making them the longest numbers appearing in the cuneiform

[13] Gingerich, *Eleven-Digit Regular Sexagesimals*.

[14] This was supported by the curvature of the existing fragment, which suggested the overall size of the original table. See Britton, 'A Table of Fourth Powers', pp. 71–72.

[15] Ossendrijver, 'Powers of 9'.

corpus (and possibly antiquity!). Ossendrijver noticed that the final sexagesimal places in one of these lists of numbers, so-called text A, alternated between 9 and 21, suggesting a relation to powers of 9 expressed in sexagesimal form. From this clue, he identified text A as a special sort of factorization table for 9^{46}, namely sexagesimal representations of $9^n, 0 \leq n \leq 46$.[16] In a like manner, the final places of a similar text, the so-called text B, alternated between 12, 36, 48, and 24, which suggested factors of 5. (The factors of 5 had emerged because 12 is the reciprocal of 5 in base 60.) Text B turned out to be a factorization table for $9^{11} \cdot 12^n$ $(n \geq 39)$. These tablets reveal a new sort of mathematical activity in the ancient Near East, hitherto unknown to modern scholars.

A contrasting technique for reconstructing the contents and circumstances of cuneiform tabular texts was developed by Neugebauer when he was working on what would eventually be called the Astronomical Cuneiform Texts (ACT) in the 1930s. Neugebauer was faced with many fragments of tabular astronomical ephemerides, some dated, some not. In cases where he had both a dated fragment and a fragment of the same type but undated, he developed a method of dating the undated fragment called the 'Linear Diophant.'[17] In a nutshell, Neugebauer computed the function tabulated on the dated text forward and backward in time until he reached a value (or modulo thereof, for periodic functions) that corresponded to a value in the undated text. From this he could calculate the time interval from the dated tablet to the undated one, and thus establish the date of the undated fragment. Of course, many of the fragments concerned functions that were periodic. These were typically computed via linear zigzag functions or step functions, so that the Linear Diophant method would furnish infinitely many possible dates for the undated fragment. However, almost always only one solution was historically plausible. Concerning his procedure and its connection to the tasks of the historian of astronomy, he was quite emphatic:

> The method ... has nothing to do with astronomy, nor with history. It only fulfills a task for a certain group of astronomical cuneiform texts which would otherwise fall on the custodian of a museum [namely, joining tablet fragments] ... It is essential to emphasize that the solution of this task becomes possible here without any hypothesis about the content of the texts, since nothing else is used but the generative laws of the series of numbers that are empirically derived from the fragments.[18]

[16] Here, Ossendrijver postulates that a table such as this one could have been a way to double check 9^{46} is correctly computed by repeatedly multiplying it by 6;40, i.e. the reciprocal of 9, until 1 is reached; this is the sense in which 'factorisation' is invoked.

[17] For details, see Neugebauer, *Astronomical Cuneiform Texts*, pp. 35–37.

[18] Ossendrijver, 'Translating Babylonian Mathematical Astronomy', p. 335.

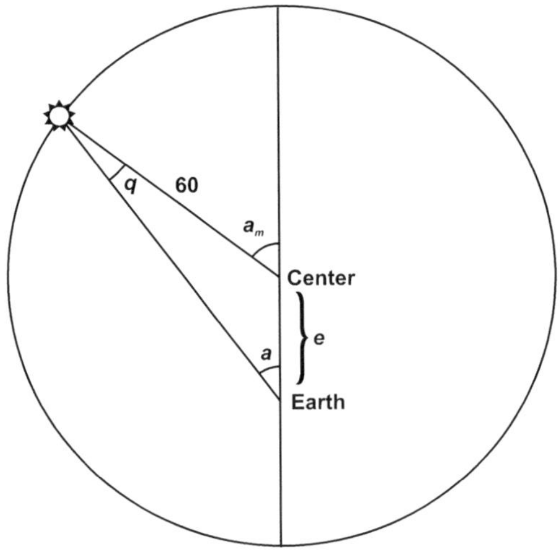

Figure 2: The solar equation $q(a_m)$

2.2. Reconstructing a table's numerical parameters

Most mathematically computed astronomical tables rely on functions with built-in numerical parameters. For instance, the solar equation, the difference between mean solar motion and true solar motion, is found in Ptolemaic astronomy according to

$$q(a_m) = \arctan \frac{e \sin a_m}{60 + e \cos a_m}, \quad (1)$$

where the radius of the deferent (the circle on which the sun orbits the earth) is set equal to 60, the mean anomaly a_m increases at a constant rate, and e is the distance from the earth to the center of the deferent (see Figure 2). Different astronomers used different values for e; Ptolemy used 2;29,30, while al-Battānī used 2;4,45. A common parameter in Islamic tables, related to the *Zīj al-ʿAlāʾī*, was to choose e so that the maximum solar equation (which occurs at $a_m \approx 92°$) has exactly the value 1;59; this occurs when $e \approx$ 2;4,35,29,51. However, medieval authors did not often report their parameter values; they simply presented the completed table. Determining the parameter used by a certain astronomer from his table of the solar equation would help to place that astronomer into a tradition of astronomical inquiry.

Van Dalen has developed a pair of statistical tools that allow the scholar to input a historical table and receive back an estimate of a numerical parameter embedded within it.[19] Briefly, the first method works as follows: take the 180 entries in the solar equation table, substitute them one at a time with

[19] van Dalen, 'A Statistical Method'.

the appropriate value of a_m into (1), and solve for e. The resulting values of e will differ slightly from each other, due to errors caused by rounding and approximation at various stages of the calculation. Some values will be more reliable than others; for instance, $q(1°) = 0;2,0$ is a small number and rounding will have a larger relative effect on it than on other entries. Van Dalen thus computes a weighted average of the estimates, with the weights chosen corresponding to a measure of the reliability of the estimate. He then uses a technique related to least squares[20] to generate an interval around the estimator that is 95 % likely to contain the true value of e (provided certain statistical assumptions are satisfied[21]).

Van Dalen's second approach is a maximum likelihood estimator. Suppose that the table has $q(1°) = 0;2,0$. Presumably this is the rounded result of a calculation that produced a number somewhere between 0;1,59,30 and 0;2,0,30. Back calculating from these two values produces an interval of possible values for e. In a perfect world, intersecting the intervals produced in this way from every entry in the table provides a very small interval of values of e that could have led to this table. Unfortunately, due to errors in computation and rounding, this seldom occurs. Instead one takes an estimate derived from the values of e that correspond to the largest number of these intervals. Van Dalen points out that this criterion is most effective for tables with few errors, and especially for mean motion tables (for which the underlying function is linear).

Van Dalen has applied these methods to a number of situations with success. We report here one case in his original paper, the solar equation table in the popular 13th-century *Shāmil Zīj*. For the solar eccentricity his weighted estimator yields $e = 2;4,35,29,29$, with the 95 % confidence interval (2;4,35,26 , 2;4,35,35). The maximum likelihood estimator yields similar results, with an interval of (2;4,35,29,29 , 2;4,35,32,56). This clearly rejects both e values given by Ptolemy and al-Battānī, but confirms strongly the parameter generated by a maximum solar equation of 1;59. Van Dalen has published

[20] Other table cracking researchers have used least squares techniques; see for instance Kremer, 'Marcus Schinnagel's Winged Polyptych' (cf. p. 45). The basic idea is to fit a mathematical model to a set of data, and choose the parameters of the model in order to minimize the sum of the squares of the deviations of the data from the model's prediction.

[21] The most important of these assumptions is statistical independence of the individual estimates for e, which can lead to problems. For instance, if a stretch of ten entries was computed by interpolating linearly between the two 'node' entries preceding and following them, then the estimates of e generated by these ten entries will be related to each other for two reasons: all ten entries are affected by the errors in the entries at the two nodes, and the quantity being tabulated is a linear function between the nodes rather than the function itself. In this situation these estimates would fail the assumption of statistical independence. Van Dalen is careful to check for the use of interpolation before applying his methods.

results based on these and similar methods elsewhere,[22] and has also reconstructed several parameters at once from al-Khwārizmī's table for the equation of time.[23]

2.3. Determining dependences between tables

The nature and process of the transmission of knowledge is fundamental to the study of the history of science. However, transmission of numerical tables often occurred through means that left no documentary evidence behind: for instance, an author finds a table of interest to his current project in his personal library, and uses it to compute a new table for some other purpose. Seldom is the debt to the original table acknowledged in the manuscripts. Such a transmission might happen within or across cultures, between two different stages of a single author's career, or even within a collection of tables. The latter is the case for the sine and tangent tables in the 13th-century *Baghdādī Zīj*, which van Dalen ascribes to Abū l-Wafā'.[24] As the argument of the tangent function approaches 90°, its values grow without bound. If the tangent is calculated the conventional way, according to $\tan\vartheta = \sin\vartheta / \sin(90^\circ - \vartheta)$, the values for the sine in the denominator become very small, and rounding produces a large relative error. Through recomputation, van Dalen shows that the large errors in the tangent table (up to more than eleven units in the second-last sexagesimal place) are almost completely accounted for by computation from the values in the sine table. This verifies that the tangent table derived from the sine table.

Usually, however, the errors in table entries are much smaller, and the dependence is not so obvious. One such situation is found in the works of 14th-century Syrian astronomical timekeeper Shams al-Dīn al-Khalīlī, whose occupation was to use mathematical astronomy to guide the Muslim faithful to pray at the appointed times of day, in the direction of Mecca (the *qibla*). Al-Khalīlī composed one of the most accurate and thorough tables for the *qibla* of the medieval period.[25] Separately, he constructed a set of *auxiliary tables*, which in various combinations allow the reader to solve a variety of problems in spherical astronomy, including the *qibla*.[26] Al-Khalīlī's auxiliary functions were

$$f(\varphi,\vartheta) = \tfrac{\sin\vartheta}{\cos\varphi}, \quad g(\varphi,\vartheta) = \sin\vartheta\tan\varphi, \quad \text{and } G(x,y) = \arccos\tfrac{x}{\cos y}. \tag{2}$$

[22] For examples, see van Dalen, *Islamic Astronomical Tables*.

[23] van Dalen, 'Al-Khwārizmī's Astronomical Tables'.

[24] van Dalen, 'Islamic and Chinese Astronomy', pp. 349–51.

[25] King, 'Al-Khalīlī's Qibla Table'.

[26] King, 'Al-Khalīlī's Auxiliary Tables'.

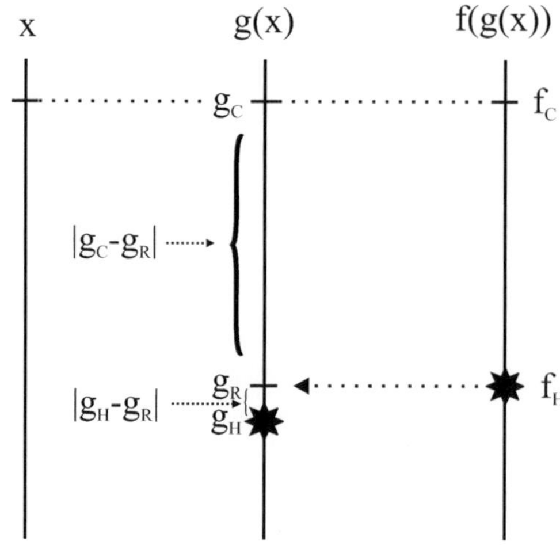

Figure 3: The dependence of the values in one table on the values of another. Consider the three vertical number lines aligned so that moving horizontally between them corresponds to precise computation. The ✷ symbols represent the historical table values; g_R is back computed from f_H. In this instance g_R is much closer to g_H than to g_C, providing evidence for dependence.

Van Brummelen[27] hypothesized that al-Khalīlī's table for f was computed from the entries in the table for g, according to the relation

$$f(\varphi, \bar{\varphi} \pm n) = \cos n \pm g(\varphi, n).$$

If this were to be verified, it would effectively demonstrate that Muslim astronomers were sufficiently aware of issues in numerical computation to devote intellectual resources to seeking out mathematical techniques for improving computational speed: the same sort of thinking that in Europe produced prosthaphairesis and, eventually, logarithms. Since the relation is mathematically correct, recomputation of the entries proves nothing — any mathematically valid method should produce correct function values. Rather, to demonstrate the hypothesis one must locate traces of the errors in the entries of the table for g in the entries of the table for f.

A general procedure to test hypotheses such as this was developed by Van Brummelen.[28] Applied to this example, it works as follows. Let $g(x)$ be the function of the purported underlying table, and let $f(g(x))$ be the function of the hypothesized dependent table with argument $g(x)$. For each entry of g and the conjectured dependent entry in f, define the following quantities:

- f_C and g_C, the correct values of the functions;

[27] Van Brummelen, 'The Numerical Structure'.

[28] Van Brummelen and Butler, 'Determining the Interdependence'.

- f_H and g_H, the values of the entries in the historical tables;
- $g_R = f^{-1}(f_H)$, a reconstructed value of the underlying g derived from f_H.

The question hinges on whether g_R lies closer to g_H or to g_C (see Figure 3).

- If g_R is closer to g_H than to g_C, a trace of the error in g is present in f, and we have evidence for dependence.
- If g_R is closer to g_C than to g_H, the dependent entry is more accurate than one would expect from the use of the underlying entry, and we have evidence that the dependent table is too accurate to derive from the underlying table.

For each pair of underlying and dependent entries the quantity $|g_H - g_R| - |g_C - g_R|$ is computed, and the resulting data are tested for a mean different from zero. For this purpose Van Brummelen chooses the non-parametric Wilcoxon signed-rank test; while it is less powerful than the traditional statistical procedures described elsewhere in this article, it is robust against the possibility of scribal errors and other disturbances in the data.

Each page of the auxiliary tables gives entries for f and g for a fixed value of φ and $\vartheta = 1°, 2°, \ldots, 90°$, so a separate test was performed for the columns with $\varphi = 5°, 10°, \ldots, 55°$ (the highest value of φ in the table). The test for $\varphi = 5°$ gave a p-value of 0.2 % in favor of dependence,[29] and for $\varphi = 45°$ the p-value was 1.4 %; all other p-values were less than 0.05. Thus Van Brummelen concluded strongly in favor of dependence.

This episode has a surprising epilogue. In 2000, David King discovered another set of tables authored by al-Khalīlī. In a rare instance of statistical methods verified by later historical sources coming to light, the new manuscript confirmed that al-Khalīlī had computed his tables according to the method asserted by the statistical procedure.

2.4. Reconstructing the process of computation

For some tables, it is possible to reconstruct the computational process that generated the table in more detail than identifying parameters or underlying tables. The possibilities here are endless and depend on the context. They might include deciding between several possible mathematical paths to the solution, detecting the use of interpolation, or identifying the use of an approximate method at a certain moment in the process of computation. Since most tables represent functions defined mathematically, the errors in the entries are usually the only basis on which to decide these questions, along with

[29] A p-value measures the probability of obtaining a result at least as far from the expected value as the observed result, under the null hypothesis that no effect exists.

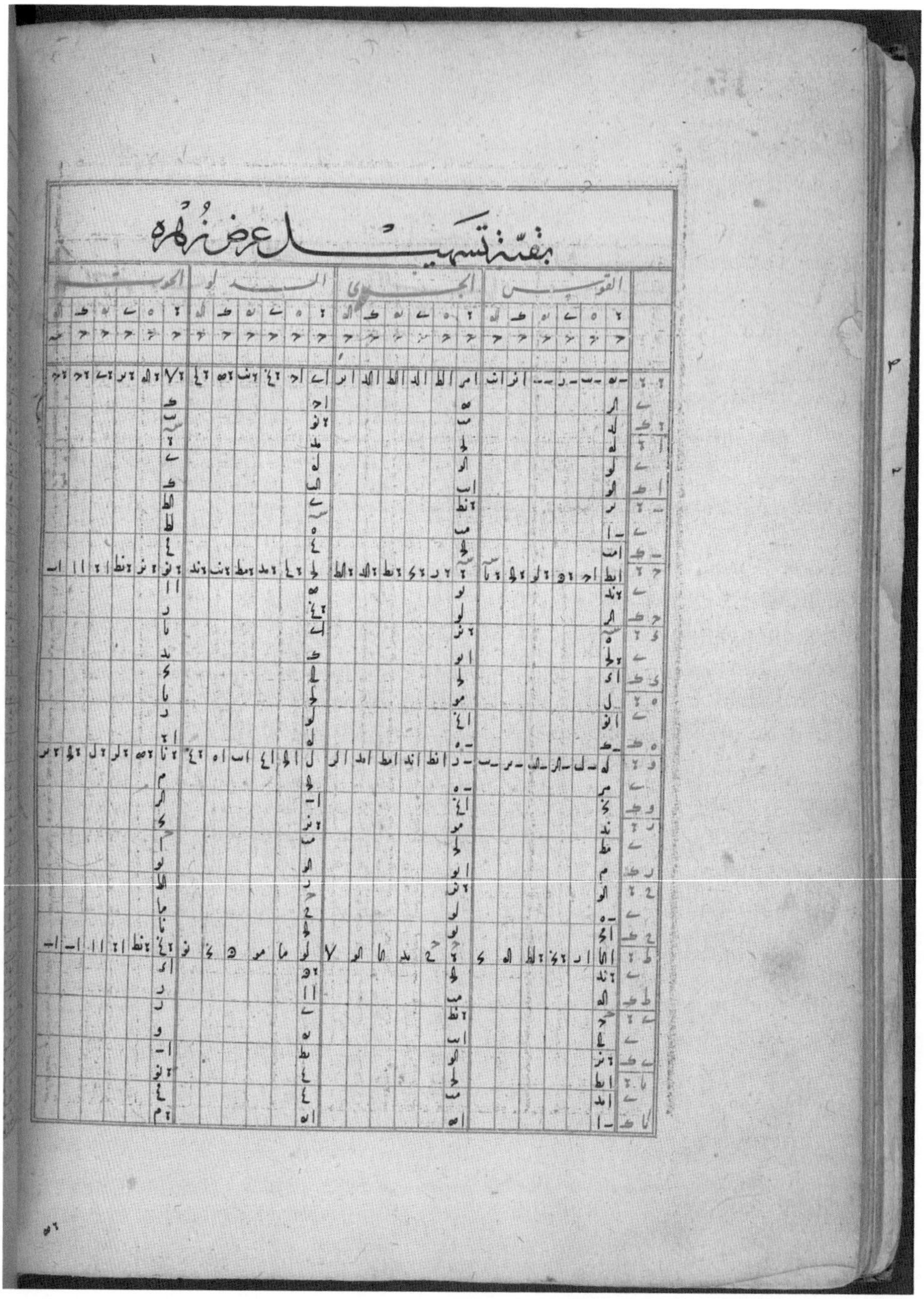

Figure 4: Third and last page of al-Kāshī's incomplete double-argument table of the latitudes of Venus. For the first page, see Plate 11; for the second page, see p. 307. © The British Library Board, MS India Office 430, fol. 154v.

the collection of historically plausible techniques that bear on the problem. We seldom read of table construction in the primary literature or see a table partly completed. One nice exception to this (see Figure 4), part of a double-argument table of latitudes for Venus in Jamshīd al-Kāshī's early 15th-century *Khāqānī Zīj*, illustrates that we may not assume that a computer started with the first entry and simply worked his way through to the end.[30]

Since the possibilities for how a table may have been constructed are so varied, general techniques are not usually available. Usually researchers work by comparing the results of the various plausible methods of calculation with the pattern of errors in the table. It is both easy and difficult to measure the success of a certain method: easy, since one may declare victory by choosing the method that best fits the entries among the historically plausible alternatives; difficult, because one's confidence in that assertion cannot be measured easily in a quantitative way. Fortunately, in many cases the result of the re-computation is so clear that an unequivocal decision is easy to make.

One simple example of this is the sine table in a manuscript of a Latin translation of al-Khwārizmī's *zīj*. This table uses a base circle radius of $R = 150$, of Indian origin. But Benno van Dalen has pointed out that almost all the entries in the table end in 0, 2, 5, or 7.[31] This suggests that the table was generated by multiplying a sine table with the Ptolemaic parameter $R = 60$ by 2½.[32] In this case the generated table is in fact precisely equal to the original table scaled by a constant. Other instances have been discovered where a new table was generated from an existing table by means of a scaling factor, but the results are only approximately correct.[33] In these cases, presumably the table's author either wanted to save time and effort, or was not capable of computing the new table directly.

Another example of a clear reconstruction is the sine table in al-Samaw'al's 12th-century *Exposure of the Errors of the Astronomers*. Al-Samaw'al criticizes traditional sine tables for using approximative methods, forced on table makers by the use of a circle divided into 360 parts, i.e., degrees. Al-Samaw'al's table uses 480°, seemingly bypassing the problem. However, a distinctive error pattern (every fourth entry correct, and errors bulging in the negative direction between them) is almost precisely matched by computation of the table by interpolating between entries in a traditional sine table.[34]

[30] See Van Brummelen, 'The Tables of Planetary Latitudes'.

[31] van Dalen, 'Al-Khwārizmī's Astronomical Tables', p. 206.

[32] Hogendijk, 'Al-Khwārizmī's Table of the "Sine of the Hours"', p. 11, reconstructs the sine table used by al-Khwārizmī to generate his table of the 'sine of the hours'; this underlying sine table actually does use $R = 150$.

[33] See for instance Van Brummelen, *Mathematical Tables*, pp. 176-79, and Dorce, 'The *Tāj al-azyāj*'.

[34] Van Brummelen, Mimura and Kerai, 'Al-Samaw'al's Curious Approach'.

c	$g(c)$	Error
12	0,24	[−7]
24	2,16	[+12]
36	4,32	[−6]
48	8,18	[+7]
60	12,26	[−11]
72	17,44	[−7]
84	23,24	[−19]
96	29,49	[−10]
108	36,14	[−9]
120	42,38	[+1]
132	48,18	[−1]
144	53,12	[+2]
156	56,36	[−17]
168	59, 4	[−8]
180	60, 0	0

Table 5: Excerpt from the *Almagest* lunar interpolation table. (The full table is given for arguments in multiples of 6° up to $c = 90°$, and in multiples of 3° thereafter.) Errors, given in square brackets, are in terms of the last place. Thus, for instance, the correct value of $g(c)$ is 0,31.

A typical example of a slightly more sophisticated reconstruction comes from Ptolemy's *Almagest*. Most of the *Almagest*'s tables are difficult to study for their computational secrets, because the tables' entries tend to be accurate to within one or two units in the last place. This is not true of one group of tables. To determine the longitudes of the moon or one of the planets, the *equation of anomaly p* must be tabulated. However, it is a function of two arguments. To avoid tabulating a gigantic rectangular grid of entries, Ptolemy devises an approximative process that requires the tabulation of four single-argument functions. One set of these tables, the *interpolation coefficient* $g(c)$, contains unusually large errors in the last place (see Figure 5).

Now, the function $g(c)$ is computed fairly directly from $p_{\max}(c)$, the *maximum equation of anomaly*. For the lunar table, Van Brummelen[35] back computed a set of values of $p_{\max}(c)$ from the values of $g(c)$, and discovered a clear pattern: the third sexagesimal places of these values cluster strongly around 0, 15, 30, and 45 (see Figure 5). Thus Ptolemy used values of $p_{\max}(c)$ rounded to units of 0;0,15. (Further examination revealed that this step size was caused by an application of linear interpolation; Ptolemy had directly computed values of $p_{\max}(c)$ only for multiples of 12°.) Thus Van Brummelen was able to reconstruct a small table of $p_{\max}(c)$ for the moon. Similar results were obtained when the same method was applied to the planetary tables.

[35] Van Brummelen, 'Lunar and Planetary Interpolation Tables'.

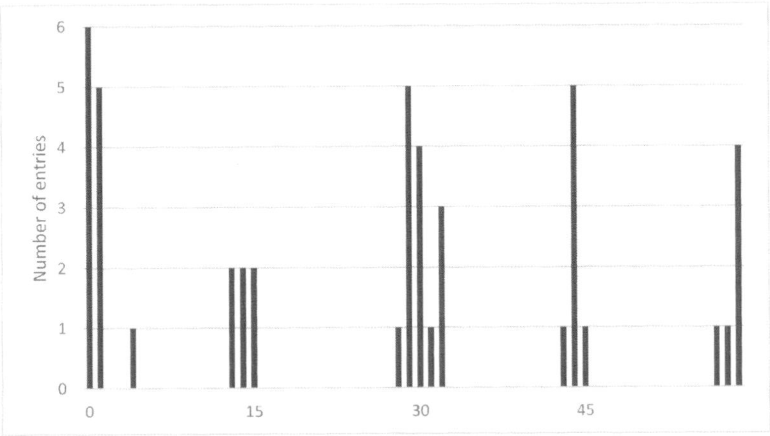

Figure 5: Clustering of the third sexagesimal place of back-computed values of $p_{\max}(c)$.

3. Methodological Considerations

3.1. The moment of decision

Mathematical methods applied to answer questions in the history of science have been at a minimum controversial, and at worst have been ignored by the historical community. This is due partly to the technical 'smoke screen' that often makes it very difficult to understand the procedures without months of training. Skepticism has only increased when some of the conclusions reached through quantitative methods have been wildly at odds with results achieved by conventional historical inquiry. It is irrational to reject these methods outright simply because one does not understand them; nevertheless, it is incumbent on quantitative researchers to make their results, and the strengths and weaknesses of their conclusions, available to the community. This has seldom been done, so we provide a road map here.

The general practice of a table cracker usually follows the same path. The researcher hopes to draw a historical conclusion from a set of historical quantitative data (say, the reconstruction of an underlying table of p_{max} in the *Almagest* planetary interpolation tables in the preceding section). This data is manipulated in some way, often to undo mathematical processing of the data that had been performed by the historical author (e.g., back computing from the interpolation table's values to a set of values for p_{max}). The transformed data reveals, or does not reveal, a pattern (e.g., clustering of the last sexagesimal place of the reconstructed p_{max} values around multiples of 15). A moment of decision is reached. The researcher concludes that the pattern is not a coincidence, and must therefore be explained (if it is decided the pattern isn't there, the result usually is not published). Finally, the researcher asserts a cause for the pattern (e.g., the p_{max} values originally had multiples of 15 in the last place), and makes a historical interpretation (e.g., Ptolemy used the underlying table of p_{max}).

The moment of decision must be parsed carefully to evaluate the reliability of the researcher's conclusion. Two distinct stages must be considered. The first is an evaluation of the assumptions made when concluding that the pattern is a genuine artifact; the second is a consideration of the integration of the pattern with the available historical context. The quality of a result seldom relies on the details of the researcher's computations: the devil is either in the assumptions made prior to the calculations, or in the historical interpretations when they are done.

Every identification of a pattern involves certain assumptions, implied or not. Although many of these assumptions are benign, they are not always obvious, and it is difficult to bring them to the surface. In the case of Ptolemy's interpolation tables, we assumed that the back calculation of p_{max} should produce a uniform distribution of values in the third sexagesimal place. More subtly, we assumed that each back calculation is an independent witness to the underlying event. If for some reason the fact that the last sexagesimal place of the first p_{max} value is close to a multiple of 15 would imply that the next one will also be, then we might look at Figure 5 with more skeptical eyes.

To agree upon a conclusion, the pattern must be sufficiently clear that the researcher and the reader agree that it must be there. In the case of general statistical procedures such as van Dalen's parameter estimation and Van Brummelen's table dependence test, the assumptions embedded in the moment of decision are more explicit — part of the discipline of statistics — and thus are associated with standard practices of evaluation. Usually they are the following:

- The data points are statistically independent; that is, the value of any one data point does not influence the value of another.
- The data points are identically distributed.
- The data set is normally distributed.

The first of these assumptions is the most difficult in practice. For instance, if a table was computed with the aid of linear interpolation, then most entries might be in error in the same direction, and van Dalen's parameter estimation may produce a confidence interval for the parameter that is systematically too low or too high. Thus, before employing his method, van Dalen checks for the use of interpolation procedures. The second assumption is often safer in table cracking than it is in, say, matches of historical data to physical phenomena, but it must still be considered. For instance, in Van

Brummelen's table dependence test, the back calculation from f_H to the underlying g_R, rather than the more intuitive forward calculation of g_H to an f_R value, is done to avoid the potential skewing of distributions that might result from the computation from g to f. Finally, the normal distribution assumption is not difficult to satisfy due to the Central Limit Theorem; if a data set contains 40 or more data points, this assumption is almost always benign. However, if the data set is small, or if scribal errors possibly affect a number of tabular values, the assumption can be bypassed using non-parametric tests (such as that used by Van Brummelen with respect to table dependence).

Statistical conclusions come in two varieties, both often misinterpreted. A 95 % confidence interval (such as that produced by van Dalen's parameter estimation) gives an interval of values that contains the correct parameter 95 % of the time the procedure is performed, *as long as* the assumptions are met. A test (such as Van Brummelen's table dependence procedure) concludes with a p-value. This number reflects the probability that a result as unusual as the observed one would arise by random chance. It is *not* the probability that the result is false. In typical scientific practice, a p-value of $< 5\,\%$ is considered strong enough evidence to reject the random chance hypothesis.

No historical investigation, quantitative or not, is free of assumptions or doubt. The advantage of quantitative methods is that the reliability of the assumptions may often be evaluated directly. It is the responsibility of the researcher to make these assumptions as explicit and verifiable as possible. Studies of tabular data tend to be the most straightforward in this respect, since the data have a strict mathematical structure and are not usually subject to the vagaries of physical phenomena.

The second stage, the historical interpretation of the pattern, depends on the specific situation being studied, and usually cannot be framed in a quantitative analysis. Often the interpretation is obvious. In the case of the p_{max} calculations, one may question whether or not Ptolemy actually compiled the reconstructed p_{max} values into tabular form, or whether it was Ptolemy himself who performed these computations, but there is no further controversy. Generally, one hopes that the quantitative result is coherent with historical evidence from other sources. In some cases the result may lean against the weight of previously-established historical analysis. In these situations, *both* the quantitative and the historical assumptions must be examined for a resolution. In some cases the result might be explained by more than one historical interpretation.[36] The more dramatically the quantitative result varies from

[36] See, for example, a debate most recently between Chabás/Goldstein, Samsó/Castelló, and Poulle on the origin of a value for precession of 17;8° in the star catalogue in the *Libro de las estrellas de la ochuaua espera*, a 13th-century text. Poulle, 'The Alfonsine Tables', assigns the parameter to a pre-Alfonsine theory of precession; Samsó and Castelló, 'An Hypothesis

established wisdom, the clearer that result must be; as the aphorism says, extraordinary claims require extraordinary evidence.

3.2. General versus specific techniques

By now the reader may have recognized an important methodological distinction. Some procedures (Sections 2.2 and 2.3) apply generally to a class of tables, while others (especially from Sections 2.1 and 2.4) apply to specific situations. General techniques can be applied uniformly to multiple contexts, can sometimes be automated, and allow easy comparison between different sources. Specific techniques have complementary advantages: they are adaptable to particular sources and particular historical scenarios.

Each type of procedure has its cautions. While general methods tend to rely on clearly stated assumptions, a single unstated and flawed assumption might undermine many conclusions at once. Also, while interpreting results applied to a spectrum of tables, one runs the risk of assuming (dangerously) that all the tables were constructed using similar computational norms. Finally, general procedures have tended to be more cautious and conservative, less likely to make historically dramatic assertions.

On the other hand, the reliability of *ad hoc* methods designed to apply to specific situations can be inherently difficult to evaluate. Assumptions are not often stated clearly, and the moment of decision must occur through the instinct of the table cracker rather than the result of a statistical test. Nevertheless, it should be noted that disagreements have seldom arisen in practice due to this shortcoming; when a specific method has identified a pattern, respondents usually have agreed that the pattern is in fact there. When controversies arise they are almost always at the stage of historical analysis, in cases where more than one explanation may be proposed to reconcile the pattern with the historical narrative.

The choice between the two types of method will always depend on local conditions (the nature of the table, the question that is being asked). Arguments will be more convincing if they rely on well tested general methods whenever possible, and use specific ones only when no other option is available. Regardless of the procedure used, the researcher must make explicit the underlying assumptions and demonstrate their validity (or at least their insignificance).

on the Epoch', p. 118, shows that it might arise from an erroneous dating of Ptolemy's star catalogue by the Alfonsine astronomers; while Chabás and Goldstein, *The Alfonsine Tables of Toledo*, pp. 234-35, derive it from a Castilian source.

4. Table Cracking II: Restoring Tabular Context

Coming to terms with a historical numerical table requires the determination of the processes that led to its creation, but it is also crucial to understand the role that the completed table played for its author and its users. This information can be elusive; sometimes all that remains in the manuscript record is the table itself (or just part of it), with little or no trace of the paratext. Even when the paratext exists it must be treated with caution. (One example is al-Kāshī's double-argument planetary latitude table, part of which is shown in Figure 4; the table's instructions have the arguments reversed.) Finally, even an intact and correct paratext may not be sufficient to answer contextual questions: the author may not have felt the need to describe the users and purpose of the table, or the users may themselves have found their own new applications.

Secondly, many tables were designed to model or represent some sort of physical phenomenon, often (but not always) astronomical. The extent of a table's predictive success often can provide useful information concerning the table's context. What specific phenomenon was the table meant to reproduce? How concerned would the authors and users have been about successes and failures in the table's predictive power? What concerns, other than accuracy, might have motivated judgments about a table's quality? This section considers the activities of table crackers in answering these contextual questions.

4.1. Reconstructing a table's purpose

Many numerical tables from early sources have come down to us as they are, with no accompanying information. In these cases, establishing a plausible interpretation of the tabular data is not always straightforward: sometimes, numerical patterns may correspond to multiple interpretations. As the following example illustrates, table crackers have had difficulties for centuries. In the eighteenth century, the Sanskrit astronomer Kevalarāma was commissioned by Jayasiṃha, regent of Jaipur (1699-1743), to translate Philippe de La Hire's *Tabulae Astronomicae*, which had come to the court of Jayasiṃha in the possession of Portuguese astronomer Pedro da Silva, into Sanskrit.[37] The part of Kevalarāma's translation (*Dṛkpakṣasāriṇī*, 1725) on how to compute true planetary positions with de La Hire's tables was a disaster. Kevalarāma failed to understand the proper use of de La Hire's tables, especially their use of logarithms. Kevalarāma, unfamiliar with logarithms (and probably poorly advised by the visiting European scholars), simply ignored the steps that invoked them. As a result, his reasoning and procedures became completely

[37] See Pingree, 'Philippe de La Hire's Planetary Theories', for an account.

senseless. Since de La Hire's tables of logarithms came without any explanation of the rules of how to manipulate them, Kevalarāma's confusion wasn't entirely his own fault.[38]

More recently, perhaps the most famous mathematical document subjected to this sort of table cracking is the Old Babylonian tablet Plimpton 322. The surviving fragment contains four columns of numbers. Neugebauer[39] instigated interest in the tablet by noticing that three of the columns (once the first has been heavily reconstructed) follow the pattern

$$\frac{d^2}{l^2}, b, d,$$

where b and d are Pythagorean numbers (integer solutions to $d^2 = b^2 + l^2$).

This was pursued further by Aaboe, Neugebauer, and Sachs, arguing for an underlying theory of generating functions where the columns were sums and differences of certain squares (p and q), their selection subject to mathematical criteria.[40] Other scholars, perhaps prompted by a short remark by Neugebauer and Sachs, argued that Plimpton 322 was in fact a trigonometric table, since it appears to measure the lengths of sides of right triangles.[41]

Eleanor Robson has taken these accounts to task for various technical and historical shortcomings.[42] However, before delving into the arguments, she enumerates six conditions that must be satisfied in any hypothesis that purports to explain any tabular text: historical sensitivity, cultural consistency, calculational plausibility, physical reality (respect for the physical dimensions of the original archaeological artifact), textual completeness (the explanation should account for the paratext as well as the entries), and tabular order (sensitivity to the logical order of the columns).[43] None of these criteria should be emphasized at the expense of another. These categories reflect a more general movement in the history of mathematics in recent decades toward a greater sensitivity to contexts outside of the purely scientific content of the text.

Robson's interpretation of Plimpton 322 (taking a cue from work by Bruins and Høyrup) considers the context of administrative tabular documents written in the early second millennium BCE. Readings of the column titles

[38] Not long afterward, de La Hire's work was translated again twice. One of these translations, *Phiraṅgicandracchedyopayogika* (1734 or 1735), correctly reported the procedures, possibly due to the influence of Father Boudier, who was visiting Jayasiṃha's court at the time.

[39] Neugebauer, *The Exact Sciences in Antiquity*, pp. 36 ff.

[40] Aaboe, *Episodes*, pp. 30–31; Neugebauer and Sachs, *Mathematical Cuneiform Texts*, pp. 38–41.

[41] See Joyce, 'Plimpton 322', and Calinger, *A Contextual History*.

[42] Robson, 'Neither Sherlock Holmes nor Babylon'; Robson, 'Words and Pictures'.

[43] Robson, 'Neither Sherlock Holmes nor Babylon', p. 176.

turned out to provide significant inspiration in her analysis. She argues that the numbers in the table relate to reciprocal pairs (numbers whose product is 60), shows that they apply in procedures of the so-called concrete geometry,[44] and concludes that they provide numerical examples to help teachers generate problems for their students. Robson's work has been followed by analyses also inspired by contextual considerations, such as Friberg and Britton, Proust, and Shnider.[45] In the latter, the authors argue that Plimpton 322 is a problem text that includes a complete worked solution. The problem concerns 'normalized sexagesimal rectangles' — that is, this tablet was produced from a desire to generate a series of finite sexagesimal rectangles from a basic algorithm. Their interpretation is that Plimpton 322 was generated via the following problem:[46] 'Make a list of all the rectangles with length equal to 1 and width and diagonal equal to finite sexagesimal numbers, and represent the dimensions in reduced form, without a common sexagesimal factor.' The mathematical theory underlying the numbers uses the diagonal rule for rectangles (how to find the diagonal of a rectangle in terms of the sides) and the process of completing the square. Their account is supported by a comparison with related non-tabular mathematical documents such as MS 3971, which describes related procedures based on reciprocal pairs and produces a sexagesimal rectangle.

In the case of Plimpton 322, since so little information accompanies the table, questions regarding its purpose arise mostly by examining its numerical content. In other cases, an abundance of accompanying information can (paradoxically) lead to even more issues for the table cracker. Such is the case with the 1489 polyptych[47] by Marcus Schinnagel, recently studied by Richard Kremer.[48] Fully opened, the polyptych is more than 3 meters wide; its central panel measures 140 by 130 cm (see Plate 1). This polyptych is filled not with the usual Christian iconography, but rather with calendrical, astronomical and astrological tables and related material. Such a unique document gives rise to several questions. What sources participated in its realization? What could have been the author's and supporters' purpose? Was it intended as a practical astronomical tool?

Kremer begins his study of the polyptych not with table cracking, but with an analysis of the prose found within it. Other than Schinnagel's signature in

[44] Cut-and-paste geometrical procedures that manipulate geometrical shapes in a concrete manner.

[45] Friberg, *A Remarkable Collection*, and Britton et al., 'Plimpton 322: A Review'.

[46] Britton et al., 'Plimpton 322: A Review', pp. 558–559.

[47] A polyptych is a large artistic display, usually a collection of paintings, divided into panels. Common in early modern central Europe, polyptychs are often found as altarpieces in cathedrals.

[48] Kremer, 'Marcus Schinnagel's Winged Polyptych'.

Latin at the top of the central panel, the entire text is in German; it constitutes a primer for the horoscope-maker or the phlebotomist (a medical specialist on blood-letting). Next, turning to the tables, Kremer identifies their layout and a sequence of oddities and inconsistencies, mostly consequences of graphical constraints in a panel's layout or of mistakes by the artist. Once the tables have been identified, Kremer sets about determining the sorts of computations that can be done with them (along with a geometric instrument also found on the polyptych). He concludes that the polyptych may be used to determine planetary longitudes (to degrees) from 1489 to 1526, times of true syzygy (to minutes) from 1475 to 1512, and times and magnitudes of eclipses from 1489 to 1551. However, certain parts of the polyptych are inconsistent with practical use. For instance, without a solar mean motion table, the solar equation table that appears on the polyptych is useless. Next, using techniques similar to those outlined in Section 2.2, Kremer identifies the sources and numerical parameters underlying the tables. He concludes that the various sources were compiled with little care for mathematical and astronomical coherence. When predictions can be achieved, the polyptych produces results comparable to the Parisian Alfonsine tables with an accuracy of degrees rather than minutes. In this respect it matches the accuracies of almanachs and ephemerides made in Europe at this time for horoscope making and medicine.

Next, Kremer turns to the history of art to gauge the cultural meaning of the polyptych, concluding that the work was commissioned by the von Reischach family to convey a message about the harmony of the cosmos. As for Schinnagel, his mission was likely one of self-promotion, seeking patronage as he attempted to establish himself in Swabia.

The two examples of this subsection illustrate that in very different settings, table cracking may be used to understand the role that the table played among its practitioners, the table's link to other tables of the period, and the mathematical practices that would have been needed in its implementation. Along with insights gained by more familiar historical techniques, this information helps the researcher to frame a proper historical interpretation.

4.2. The table's fit with its associated physical phenomenon

A number of modern studies of early astronomical tables compare the results produced by the table with actual, physical celestial positions. This is done for several reasons. The most obvious is to gauge a measure of the table's predictive success. For instance, one might compare a table of computed eclipse possibilities with actual eclipses, or compare planetary visibility tables with the actual planets' visibilities. Some celestial phenomena are more amenable to such comparisons (for instance, did an eclipse occur or did it not?), while some are more delicate (such as acronychal risings, planetary positions, and stationary points; these are impossible to observe directly with any precision). Researchers who seek such comparisons usually recompute estimates of these quantities using modern scientific theory (taking into account effects such as gravitational influences, refraction, and so on) and match these, sometimes value for value, with those given in a historical table. The results allow investigators to make conclusions regarding the effectiveness of the historical astronomical models and procedures.

However, caution must be applied when pursuing this sort of inquiry. Conformity with a physical phenomenon was not always the primary goal. In some cases the historical scientist might have been attempting to conform instead with a dominant scientific theory, or even with observational data generated by themselves or illustrious predecessors. In these cases it is all too easy to leap to a conclusion of scientific fraud. As in modern times, historical authors lived and worked in cultures where the interactions between theory, observation, and authority were more nuanced than the textbook account of the scientific method would have one believe.

One such case is the planetary theory and related tables of Ibn al-Shāṭir, a 14th-century Syrian astronomer. Ibn al-Shāṭir's work is part of a lengthy scientific tradition within medieval Islam questioning Ptolemy's planetary models. Their critique was not the fit of these models to the planetary data; rather, astronomers aimed their criticisms at Ptolemy's violations of Aristotelian physics. In particular, to reproduce certain planetary phenomena Ptolemy had been forced to introduce a new point in his models, called the *equant*. The center of the epicycle was asserted to move around its orbit circle (the *deferent*) uniformly, not around its own center (as required by Aristotle), but around the equant point. Ibn al-Shāṭir's tables instead built upon a geometric model of his own invention that avoids any such violations. Ibn al-Shāṭir's tables were considered an improvement on Ptolemy not because they fit the observations any better, but because they conformed more closely to the general principles of natural philosophy.

In another case, the priority of computability outranked that of fit with the phenomenon. In the first half of the second century BCE, Hypsicles of Alexandria composed the *Anaphorikos*, within which he set out a scheme to compute a table of all oblique ascensions for a given local latitude (see Figure 6).[49] Hypsicles' scheme was based on the assumption that rising times can be computed arithmetically — that is, by a linear sequence of values. The results produced by such a scheme could not hope to do more than model the true oblique ascensions qualitatively, but the values would be very easy to

[49] For a translation and technical commentary of this work, see Montelle, 'The Anaphoricus of Hypsicles'.

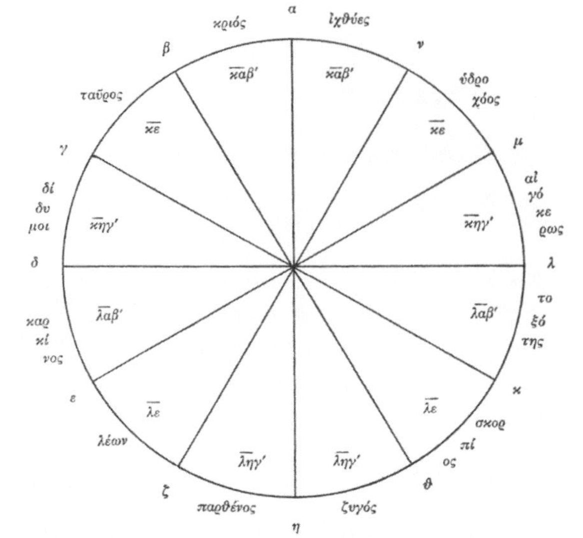

Figure 6: A transcription of the circular table displaying oblique ascensions for Alexandria found in MS Vat. gr. 204, f. 135v, from Hypsicles' *Anaphorikos*. Reproduced from V. de Falco and Max Krause, 'Hypsikles. Die Aufgangszeiten der Gestirne', *Abhandlungen der Akademie der Wissenschaften in Göttingen, philologisch-historische Klasse*, Dritte Folge, Nr. 62 (1966), p. 37.

compute. Using the local latitude of Alexandria, Hypsicles presents a worked example; for successive zodiacal signs (and applying symmetry for signs 7 to 12), they are:

sign	rising time	sign
1	21;40	12
2	25	11
3	28;20	10
4	31;40	9
5	35	8
6	38;20	7

Note that the difference between oblique ascensions for successive signs is the constant value 3;20.

Clearly, this scheme was very rough indeed.[50] However, to dismiss it for this reason is to neglect the elegance of the approach and its primary purpose

[50] Several of the factors that contribute to this 'roughness' include: i) it is a linear arithmetical scheme; ii) the value for the obliquity of the ecliptic ε is not made explicit by Hypsicles; and iii) the ratio of 5 : 7 for the shortest to longest day (equivalent to a latitude of $\varphi \approx 35;32^\circ$) is too high for Alexandria, which is closer to $\varphi = 30^\circ$. For the purpose of comparison, we have recomputed the oblique ascensions using spherical trigonometry, assuming ε to be Ptolemy's 23;51°. For $\varphi = 30^\circ$, there results: 20;59, 24;17, 29;57, 34;35, 35;31, 34;41, 34;41. When $\varphi = 35;32^\circ$, there results: 19;21, 22;54, 29;21, 35;11, 36;54, 36;19, 36;19.

of computability. Hypsicles and many after him modelled rising times linearly, and continued to do so even when more accurate spherical approaches became available. Astrologers used it to compute oblique ascensions for their local circumstances using only a single empirically derivable fact (the ratio of longest to shortest day). Therefore, rather than rejecting a table or computational scheme that poorly reproduces a physical phenomenon as scientifically inferior, the table cracker should evaluate the historical context to consider the table maker's goals in the light of the computing tools and empirical data available, as well as the quantitative abilities of the table's audience.

As delicate as the matter is, comparing tabular data with physical phenomena has produced important results, especially when comparing two different historical techniques against each other. For instance, scholars such as Steele have investigated Babylonian tables used for the prediction of eclipses, to determine whether the later methods improved in their ability to predict the timing and circumstances of eclipses over the last millennium that they were astronomically active. Using modern retrodictions, comparisons, and statistical trends, Steele concluded, surprisingly, that there is no evidence for an improvement in accuracy as time progressed.[51] Building on Steele's work, Montelle considers the reckoning of eclipse possibilities in the so-called 'ACT' tabular cuneiform sources. Her comparison with actual eclipse possibilities also suggests that these later ACT methods (which could contain up to 18 columns of intermediary tabulated data) were not able to predict correctly the timing and circumstances of eclipses.[52] In fact these tabulated predictions were no more accurate than the earlier non-tabular sources, despite being more deliberate, taking into account more factors, working to greater precision, being technically more elaborate, and showing more mathematical reasoning. This reveals the prodigious difficulty early investigators were facing in producing schemes that fit actual observed eclipses.

5. Concluding Remarks: Approaching a Table with Due Caution

When used appropriately, table cracking can reliably enhance existing historical methods and studies. It can help inform historians in their key lines of inquiry: how tables were computed, how they were read and understood, and how they were used. In some instances, table cracking is the only recourse we have when generating historical information about the table's creation, purpose, and effectiveness. However, table cracking techniques are most powerful when they are used in tandem with other historical information. Forwarding a claim based on analytical means alone can lead to unwarranted conclusions. Table crackers may initially produce results in apparent conflict with

[51] Steele, *Observations and Predictions*.

[52] Montelle, *Chasing Shadows*, pp. 94–97 and Appendix B.

other more traditional inquiries. Through a process of balancing, evaluating, and re-interpreting these sources of information within a wider context, the table cracker can form a clearer impression of the contents and use of the table.

Clearly, reliability of quantitative results will aid the researcher in convincing her colleagues of the validity of her conclusions. General procedures can accomplish this by using tried and tested statistical techniques. However, every table is unique. When general methods fail to adapt to the local conditions of the table or to the question asked by the researcher, techniques can be adjusted to fit the situation.

Beyond the arrays of numbers lies the diligence of a compiler, the assiduousness of a scribe, the expectations of a patron, the industry of a user. The table embodies a combination of these influences in both direct and nuanced ways. Traces of these are detectable to table crackers; however, no amount of quantitative analysis will be able on its own to reconstruct fully the historical circumstances of a table. Thus table cracking efforts form but a part of the tools historians can draw upon when investigating a numerical table.

Bibliography

Aaboe, Asger, *Episodes from the Early History of Mathematics*, Washington, DC: Mathematical Association of America, 1964.

Britton, John P., 'A Table of Fourth Powers and Related Texts from Seleucid Babylon', *Journal of Cuneiform Studies* 43–45 (1991–93), pp. 71–87.

Britton, John P., Christine Proust and Steve Shnider, 'Plimpton 322: A Review and a Different Perspective', *Archive for History of Exact Sciences* 65 (2011), pp. 519–566.

Calinger, Ronald, *A Contextual History of Mathematics: To Euler*, Upper Saddle River, NJ: Prentice Hall, 1999.

Chabás, José, and Bernard R. Goldstein, *The Alfonsine Tables of Toledo*, Dordrecht: Kluwer, 2003.

Chabás, José, and Bernard R. Goldstein, 'Early Alfonsine Astronomy in Paris: the Tables of John Vimond (1320)', *Suhayl* 4 (2004), pp. 207–294.

Dalen, Benno van, *Ancient and Mediaeval Astronomical Tables. Mathematical Structure and Parameter Values*, doctoral dissertation, Utrecht University, 1993.

Dalen, Benno van, 'A Statistical Method for Recovering Unknown Parameters from Medieval Astronomical Tables, *Centaurus* 32 (1989), pp. 85-145.

Dalen, Benno van, 'Al-Khwārizmī's Astronomical Tables Revisited: Analysis of the Equation of Time', in Josep Casulleras and Julio Samsó (eds), *From Baghdad to Barcelona. Studies in the Islamic Exact Sciences in Honour of Prof. Juan Vernet*, Barcelona: Instituto 'Millás Vallicrosa', 1996, pp. 195-252.

Dalen, Benno van, 'Origin of the Mean Motion Tables of Jai Singh', *Indian Journal of History of Science* 35 (2000), pp. 41-66.

Dalen, Benno van, 'Islamic and Chinese Astronomy under the Mongols: A Little-Known Case of Transmission', in Yvonne Dold-Samplonius, Joseph W. Dauben, Menso Folkerts and Benno van Dalen (eds), *From China to Paris: 2000 Years Transmission of Mathematical Ideas*, Stuttgart: Steiner, pp. 327–56.

Dalen, Benno van, *Islamic Astronomical Tables. Mathematical Analysis and Historical Investigation*, Farnham UK: Ashgate, 2013.

Dobrzycki, Jerzy, and Richard L. Kremer, 'Peurbach and Maragha Astronomy? The Ephemerides of Johannes Angelus and their Implications', *Journal for the History of Astronomy* 27 (1996), pp. 73–123.

Dorce, Carlos, 'The *Tāj al-azyāj* of Muḥyī al-Dīn al-Maghribī (d. 1283): Methods of Computation', *Suhayl* 3 (2002), pp. 193-212.

Friberg, Jöran, *A Remarkable Collection of Babylonian Mathematical Texts. Manuscripts in the Schøyen Collection. Cuneiform Texts I*, New York: Springer, 2007.

Gingerich, Owen, *Eleven-Digit Regular Sexagesimals and Their Reciprocals*, Philadelphia: American Philosophical Society, 1965. [*Transactions*, New Series, vol. 55/8.]

Hogendijk, Jan, 'Al-Khwārizmī's Table of the 'Sine of the Hours' and the Underlying Sine Table', *Historia Scientiarum* 42 (1991), pp. 1-12.

Joyce, David E., 'Plimpton 322', available at https://mathcs.clarku.edu/~djoyce/mathhist/plimpnote.html, Department of Mathematics and Computer Science, Clark University, 1995.

Kennedy, Edward S., 'A Survey of Islamic Astronomical Tables', *Transactions of the American Philosophical Society. New Series* 46 (1956), pp. 123-177.

King, David A., 'Al-Khalīlī's Auxiliary Tables for Solving Problems of Spherical Astronomy', *Journal for the History of Astronomy* 4 (1973), pp. 99-110.

King, David A, 'Al-Khalīlī's Qibla Table', *Journal of Near Eastern Studies* 34 (1975), pp. 81-122.

Kremer, Richard L., 'Marcus Schinnagel's Winged Polyptych of 1489: Astronomical Computation in a Liturgical Format', *Journal for the History of Astronomy* 43 (2012), pp. 321–346.

Mielgo, Honorino, 'A Method of Analysis for Mean Motion Astronomical Tables', in Josep Casulleras and Julio Samsó (eds), *From Baghdad to Barcelona: Studies in the Islamic Exact Sciences in Honour of Prof. Juan Vernet*, Barcelona: Instituto 'Millás Vallicrosa', 1996, pp. 159–179.

Montelle, Clemency, 'The Anaphoricus of Hypsicles of Alexandria', in John M. Steele (ed.), *The Circulation of Astronomical Knowledge in the Ancient World*, Leiden: Brill, 2016, pp. 287–315.

Montelle, Clemency, *Chasing Shadows: Mathematics, Astronomy, and the Early History of Eclipse Reckoning*, Baltimore: Johns Hopkins University Press, 2011.

Neugebauer, Otto, *Astronomical Cuneiform Texts*, 3 vols, London: Lund Humphries, 1955; 2nd edition, New York: Springer, 1983.

Neugebauer, Otto, *The Exact Sciences in Antiquity*, 2nd edition, New York: Dover, 1969.

Neugebauer, Otto, and Abraham J. Sachs, *Mathematical Cuneiform Texts*, New Haven: American Oriental Society / American Schools of Oriental Research, 1945.

Ossendrijver, Mathieu, 'Translating Babylonian Mathematical Astronomy: Neugebauer and Beyond', in Alexander Jones, Christine Proust and John M. Steele (eds), *A Mathematician's Journeys. Neugebauer and Modern Transformations of Ancient Science*, Cham: Springer, 2016, pp. 333–42.

Ossendrijver, Mathieu, 'Powers of 9 and Related Mathematical Tables from Babylon', *Journal of Cuneiform Studies* 66 (2014), pp. 149–165.

Pingree, David, 'Philippe de La Hire's Planetary Theories in Sanskrit', in: Yvonne Dold-Samplonius, Joseph W. Dauben, Menso Folkerts and Benno van Dalen (eds), *From China to Paris: 2000 Years Transmission of Mathematical Ideas*, Stuttgart: Steiner, 2002, pp. 429–53.

Poulle, Emmanuel, 'The Alfonsine Tables and Alfonso X of Castille', *Journal for the History of Astronomy* 19 (1988), pp. 97–113.

Robson, Eleanor, 'Neither Sherlock Holmes nor Babylon: A Reassessment of Plimpton 322', *Historia Mathematica* 28 (2001), pp. 167–206.

Robson, Eleanor, 'Words and Pictures: New Light on Plimpton 322', *American Mathematical Monthly* 109 (2002), pp. 105-20.

Samsó, Julio and Francisco Castelló, 'An Hypothesis on the Epoch of Ptolemy's Star Catalogue According to the Authors of the Alfonsine Tables', *Journal for the History of Astronomy* 19 (1899), pp. 115–20.

Steele, John M., *Observations and Predictions of Eclipse Times by Early Astronomers*, Dordrecht: Kluwer, 2000.

Van Brummelen, Glen, 'The Numerical Structure of al-Khalīlī's Auxiliary Tables, *Physis* 28 (1991), pp. 667–97.

Van Brummelen, Glen, *Mathematical Tables in Ptolemy's Almagest*, Ph.D. dissertation, Burnaby, Canada: Simon Fraser University, 1993.

Van Brummelen, Glen, 'Lunar and Planetary Interpolation Tables in Ptolemy's *Almagest*', *Journal for the History of Astronomy* 25 (1994), pp. 297–311.

Van Brummelen, Glen, and Kenneth Butler, 'Determining the Interdependence of Historical Astronomical Tables', *Journal of the American Statistical Association* 92 (1997), pp. 41–48.

Van Brummelen, Glen, 'The Tables of Planetary Latitudes in Jamshīd al-Kāshī's *Khāqānī Zīj*', in this volume.

Van Brummelen, Glen, Taro Mimura and Yousuf Kerai, 'Al-Samaw'al's Curious Approach to Trigonometry', *Suhayl* 11 (2013), pp. 9–31.

The *Almanac* of Jacob ben Makhir

José Chabás and Bernard R. Goldstein

Jacob ben Makhir Ibn Tibbon (*c.* 1236 – *c.* 1305) was a member of the prominent Ibn Tibbon family established in Languedoc, in southern France, well known for their many contributions to the transmission of Arabic culture in the Iberian Peninsula to a Jewish audience. Its members, most of them physicians, translated into Hebrew scientific texts in Arabic, some of which had previously been translated from Greek into Arabic. Jacob was also known by his vernacular name, Profeit Tibbon, later rendered in Latin as Profatius. Although probably born in Marseille, he was active in Montpellier both as an author and as a translator. Among his translations, almost exclusively on mathematical and astronomical matters, are the *Elements* by Euclid (Alexandria, third century BC), *On the Configuration of the World* by Ibn al Haytham (Egypt, eleventh century), and the *Iṣlāḥ al-Majisṭi* (*Correction of the Almagest*) by Jābir ibn Aflaḥ (al-Andalus, twelfth century). Jacob is also the author of two original works, a text on a new version of an astronomical instrument, called the quadrant, and a set of tables entitled *Almanac*, which is the subject of this paper.

The text on the quadrant, *Explanation of the Instrument Called the Quadrant of Israel*, was completed in 1288 and then revised in 1301.[1] It is also extant in Latin, translated by Armengaud Blaise, otherwise called Ermengol Blasi (Montpellier, *c.* 1264–1312), under the supervision of Jacob ben Makhir himself. In Latin Jacob's quadrant was usually referred to as *quadrans novus*, in contrast to the *quadrans vetus*, which was already in use at the time and associated with Robertus Anglicus who taught at the University of Montpellier (*c.* 1270).[2]

The main part of Jacob's *Almanac* is devoted to tables for the true positions of the celestial bodies at intervals of a few days, from which their true positions at any time in between can easily be determined, following the pattern set in other almanacs such as that ascribed to Azarquiel (al-Andalus, eleventh century).[3] It also contains a series of tables for computing the circumstances of solar and lunar eclipses, and among them are two extensive and unprecedented tables that are closely related. The purpose of the first is to provide over a long period of time the true anomaly of the Moon and the minutes of proportion

[1] For the Hebrew versions, see Steinschneider, *Die hebraeischen Übersetzungen*, p. 608.

[2] Millás Vallicrosa, *Estudios sobre la historia*, pp. 65–110.

[3] Millás Vallicrosa, *Estudios sobre Azarquiel*, pp. 72–237. See also Boutelle, 'The Almanac'.

Editing and Analysing Numerical Tables: Towards a Digital Information System for the History of Astral Sciences, ed. by Matthieu Husson, Clemency Montelle and Benno van Dalen, PALS 2 (Turnhout, 2021), pp. 53–78
© BREPOLS PUBLISHERS 10.1484/M.PALS-EB.5.127695

ultimately as a function of the double elongation of the Moon from the Sun, whereas the second is a double argument table for the complete equation of the Moon (that is, based on Ptolemy's second lunar model), which in itself is an innovative approach in the West, with the characteristic that the entries have been displaced vertically to avoid subtractions. These features may help to explain the success of these tables.

The associated text, originally written in Hebrew, was mainly diffused in Latin. As already noticed by Steinschneider, it was translated into Latin in two versions.[4] Both have a prologue (a general non-technical introduction) and canons, that is, a set of rules explaining the use of the tables. In the old literal version the prologue begins *Quamquam multi homines velint astrologie scientiam et eam habere desiderant* ... (prologue A), and in the paraphrastic version it begins *Quia omnes homines naturaliter scire desiderant* ... (prologue B). In some manuscripts we find variants of these incipits. The Hebrew prologue is significantly different from both versions of the prologue in Latin, as already noted by Steinschneider.

The Latin texts say nothing about the translator, but we suggest that the translator was also Armengaud, again with the help of Jacob himself, for in the Latin version changes were introduced that are unlikely to have been made by anyone other than Jacob. These changes include, for example, references to the Toledan Tables in prologue B, and entries for year 1300 for some planets. Still another change is the inclusion in the incipit of prologue B of an implicit reference to the opening sentence of Aristotle's *Metaphysics* I.1 ('All men by nature desire to know'), which is not in the Hebrew text. A similar case, where a Latin translator of a Hebrew astronomical text benefited from information supplied by the author that was not in the original text, involves Peter of Alexandria who translated the *Astronomy* by Levi ben Gerson (Orange, France, d. 1344). As Mancha persuasively argued, 'Levi's participation in the actual translation process is most probable, although it is not possible to be completely certain...'.[5] Still another example is Juan de Salaya's translation into Castilian in 1481 of the astronomical treatise, *ha-Ḥibbur ha-gadol* (*The Great Composition*), by Abraham Zacut (Salamanca, 1452–1514),[6] where the manuscript containing the translation explicitly states that it was made from the Hebrew original with the help of Zacut himself.[7]

Our analysis is based on the following manuscripts (we have added 'h' or 'l' to the sigla of the manuscripts in Hebrew and in Latin, respectively):

[4] Steinschneider, 'Prophatii Judaei'.

[5] Mancha, 'The Latin Translation', p. 34.

[6] Chabás and Goldstein, *Astronomy in the Iberian Peninsula*.

[7] Cantera, 'El judio samantino', p. 236.

Mh: Munich, Bayerische Staatsbibliothek, MS Heb. 343, 202b (prologue), 203a–269a (tables);

Ph: Paris, Bibliothèque nationale de France, MS Heb. 1046, 1b–2a (prologue), 2b–80a (tables);

Vh: Vatican, Biblioteca Apostolica, MS Heb. 393, 9b–99a (tables), 103a–b (prologue);

Kl: Bernkastel-Kues, Cusanusstiftsbibliothek, MS 215, 32r–84v and 92v–94v (tables), 85r–87r (prologue A and canons), 88r–92r (prologue B and canons);

Ml: Madrid, Biblioteca Nacional, MS 9288, 15r–88v (tables), 89r–92v (prologue B and canons);

Pl: Paris, Bibliothèque nationale de France, MS lat. 7408A, 2r–v and 74r–77v (prologue A and canons), 3r–73r (tables).

Additional Hebrew manuscripts:

Cambridge, University Library, MS Add. 1741, 2/7, 99b–107b;[8]

London, British Library, MS Or. 10725, 7 (Merhav: online catalogue of the National Library of Israel);

Parma, Biblioteca Palatina, MS Heb. 2112, 2, formerly de Rossi MS 1181, brief extracts only;[9]

Parma, Biblioteca Palatina, MS Heb. 2113, 1, formerly de Rossi MS 1374;

Parma, Biblioteca Palatina, MS Heb. 2770, formerly de Rossi MS 749;

Oxford, Bodleian Library, MS Marshall Or. 95.[10]

Among the Latin manuscripts are:

Bergamo, Biblioteca Civica Angelo Mai, MS 388, 113r–120v (planets only);

Brussels, Bibliotèque royale de Belgique, MS 281–83, 77r–96r;

Cambridge, Gonville and Caius College, MS 141/191, 387–533;

Cracow, Biblioteka Jagiellońska, MS 613, 36v–37r, 99v–100v, 138r–154r, 158r–159r;

Erfurt, Bibliotheca Amploniana, MS 4° 379, 63r–99v (tables), 100r–1011v (canons, prologue A);

[8] Reif, *Hebrew Manuscripts*, pp. 336–37.

[9] For the three Hebrew manuscripts in Parma, see Richler and Beit-Arié, *Hebrew Manuscripts*, pp. 433, 432, and 429, respectively.

[10] Neubauer, *Catalogue of the Hebrew Manuscripts*, col. 700 (No. 2041).

Florence, Biblioteca Medicea Laurenziana, MS Plut. 18.1, 2r–61r (prologue B, canons, and tables);

Florence, Biblioteca Medicea Laurenziana, MS Plut. 18.2, 115r–119r (prologue B: edited by Boffito and Melzi d'Eril in 1908);

London, British Library, MS Harley 267, 179r–210v;

Munich, Bayerische Staatsbibliothek, MS Clm 83, 22r–45v;

Naples, Biblioteca Nazionale, MS VIII.C. 19, 356r–381r (prologue A);

Oxford, Bodleian Library, MS Bodley 464, 1r–4r (prologue B), 4v–57v (tables);

Oxford, Bodleian Library, MS Digby 114/191, 38r–52v (tables);

Oxford, Bodleian Library, MS Laud. misc. 594, 6r–13v (tables, incomplete);

Oxford, University College, MS 41, 47r–51v (prologue B), 52r–73v (tables, incomplete);

Paris, Bibliothèque nationale de France, MS 7272, 68r–84v (only prologue A);

Paris, Bibliothèque nationale de France, MS 7286B (prologue B, no tables);

Paris, Bibliothèque nationale de France, MS 7300, 1r–55v (tables);

Paris, Bibliothèque nationale de France, MS 10263, 92r–94v (prologue A);

Rennes, Bibliothèque municipale, MS 593, 9r–41r (tables), 41v–42v (canons, summary in French);

Vatican, Biblioteca Apostolica, MS Pal. lat. 1387, 8v–40v (tables);

Vatican, Biblioteca Apostolica, MS Pal. lat. 1436, 23v–24r (canons), 24v–36v (tables, planets only).

Texts

Steinschneider transcribed prologue A and the first half of prologue B, as well as the Hebrew prologue, based on four Hebrew manuscripts: Oxford, Munich, Parma (MS 2113), and Paris.[11] He also included his own Latin translation of the Hebrew version of the prologue, headed *versio mea*. Renan and Neubauer treated the works of Jacob extensively, including excerpts of the prologue to the *Almanac* in both Latin versions as well as the prologue in Hebrew in the Oxford manuscript.[12] In 1908 Boffito and Melzi d'Eril transcribed prologue

[11] Steinschneider, 'Prophatii Judaei', pp. 607–14.

[12] Renan and Neubauer, *Les rabbins français*, pp. 616–20. The text was composed by Neubauer and then edited by Renan.

B in its entirety and the canons, beginning *Quando per istud almanach scire uolueris loca...*[13]

In prologue B, but not in the corresponding passage in prologue A, there are two references to the 'tables of Toledo', after mentioning Azarquiel.[14] However, it is not clear whether the author of prologue B meant that the Almanac of Azarquiel or what nowadays is known as Toledan Tables served as Jacob's model in compiling his almanac. Prologue B also provides information on the coordinates of Montpellier: the longitude is given as 148° from the East and 32° from the West, and the latitude as 43°.[15] These data are not included in the Hebrew manuscripts we have consulted or in the printed versions of the Hebrew prologue. Next there is a striking remark indicating that the almanac is based on one compiled by Ptolemy for his daughter Cleopatra! This remark also does not appear in the Hebrew text. This strange statement is probably to be linked to the reference earlier in the prologue to a certain Armenius or Ammonius, a disciple of 'King' Ptolemy, who compiled a similar almanac about 600 years before Azarquiel (Ph 1b). This Ammonius is probably the same as Humeniz, sometimes called the son of Ptolemy.[16] It was common in the Middle Ages to confuse Ptolemy the astronomer with members of the Egyptian Ptolemaic dynasty. We are then told that the revolutions of the planets begin in year 1300 on March 1, contradicting what is stated at the very beginning of the canons in Latin, where we are instructed that, to enter the tables for the true positions, one has to subtract 1300 from the year for which the positions are sought, clearly indicating that the first tabulated year is 1301 for all planets. The prologue to the Hebrew only says 'the beginning of the tables is after year 1300 of the Incarnation' (Ph 1b). However, in the canons for the mean motion tables for the outer planets, the Hebrew has 'Cast off 1300 years of the years of the Christians, and enter with the remainder...' (Ph 21a). Note that the columns in the Almanac are headed with the number of years in the cycle beginning with 1 (not a year number such as 1301).

On the basis of our analysis of the tables, we can confirm the claim[17] that Jacob depended on the Toledan Tables for computing the entries in his *Almanac*.[18] It would be surprising that Jacob depended on a source in Latin, since he only refers to texts in Hebrew and Arabic. However, the Toledan Tables were translated from an Arabic version which is not extant; hence, it is pos-

[13] Boffito and Melzi d'Eril, *Almanach Dantis Aligherii*, pp. 1–8.

[14] Boffito and Melzi d'Eril, *Almanach Dantis Aligherii*, p. 2, lines 18 and 22.

[15] See, e.g., Ml 89v, and Boffito and Melzi d'Eril, *Almanach Dantis Aligherii*, p. 2, lines 31–32, where we find erroneously 22° instead of 32°.

[16] Millás, *Estudios sobre Azarquiel*, p. 379.

[17] Toomer, 'Prophatius Judaeus'.

[18] For the Toledan Tables and their Arabic origin, see Pedersen, *The Toledan Tables*, pp. 11–20. See also Toomer, 'A Survey of the Toledan Tables'.

sible that the Arabic version was still available at the time of Jacob. A related possibility is that there was a copy of the Arabic version in Hebrew characters. The existence of a unique copy, dated 1327, of the Tables of Novara (an adaptation of the Toledan Tables, compiled in the mid-thirteenth century) in Arabic written in Hebrew characters supports the possibility that Jacob may have had access to a version of the Toledan Tables.[19] On the other hand, there is no indication that the Toledan Tables were ever translated into Hebrew. The astronomical tables in the zij of al-Battānī (Raqqa, Syria, d. 929) were the main representative of the Ptolemaic tradition in al-Andalus and then in other parts of Europe, beginning in the twelfth century.[20] One relevant text that depended on al-Battānī's zij was the astronomical tables of Abraham Bar Ḥiyya (Barcelona, d. 1136); it was the first set of such tables written in Hebrew and widely copied in the Middle Ages.[21] We refer to one copy of Bar Ḥiyya's unpublished tables: Paris, Bibliothèque nationale de France, MS Heb. 1046.

Tables

The *Almanac* uses signs of 30° and tropical coordinates, whereas the Toledan Tables and its derivative, the Tables of Toulouse, use signs of 30° and sidereal coordinates. We agree with previous scholars, in particular Toomer (see note 17, above), who identified the source for the computed true planetary longitudes as the Toledan Tables, despite the difference in the coordinate system, noted above, and we provide additional evidence to support this claim. Jacob's computation of true tropical longitudes of the Sun and the five planets for any given time began with the computation of their true sidereal longitudes according to the Toledan Tables, to which must be added a value for precession, the difference between a sidereal longitude and a tropical longitude at the given time. To compute the appropriate value for precession, he used the tables ascribed to Thābit (see note 22, below), where the correction table has entries at 5°-intervals that require interpolation. Small errors (in the seconds) can easily be the result of approximation in these interpolations. Note that in these tables the year begins on March 1 rather than on January 1; hence, dates in January and February in the tables belong to the following year in the usual reckoning.

1. True positions of Saturn (Mh 203a–207b; Ph 2b–7a; Vh 9b–14a; Kl 32r–34r; Ml 15r–19v; Pl. 3r–7v).

For days 10, 20, and last of each month in a cycle of 59 years, the *Almanac* displays entries given in signs, degrees, and minutes. The first entry in the

[19] Paris, Bibliothèque nationale de France, MS Heb. 1102. See also Goldstein, 'The Survival', pp. 34–35.

[20] Nallino, *Al-Battānī sive Albatenii.*

[21] For a partial description, see Mercier, 'Astronomical Tables', pp. 165–81. See also Millás, *La obra Séfer Ḥeshbón*, pp. 109–32.

Hebrew manuscripts, Leo 21;10° (Leo 20;10° in Vh), corresponds to March 10, 1301, whereas in the Latin manuscripts we have examined it is Leo 7;7°, corresponding to March 10, 1300. Indeed, in the Latin copies of this table there is an extra column for 1300, and 60 columns altogether. When comparing the last column (for 1359) with the column for 1300, it turns out that the positions of Saturn after a cycle of 59 years result from adding 1;31° to those for 1300, despite the fact that the canons give the value 1;30°: see Table A, below.

2. True positions of Jupiter (Mh 208a–214b; Ph 7b–14a; Vh 14b–21a; Kl 34v–37v; Ml 20r–26v; Pl. 8r–14v).

For Jupiter, the cycle lasts 83 years, and the frequency of the entries is the same as for Saturn, every ten days. Also, as was the case for Saturn, the Latin manuscripts provide data for 1300, so that the first entry in the Latin manuscripts we consulted is Ari 23;35° corresponding to March 10, 1300, whereas in the Hebrew manuscripts we examined it is Tau 23;0° corresponding to March 10, 1301. The difference between corresponding entries for 1383 and 1300 is −0;30°, in agreement with the canons.

3. True positions of Mars (Mh 215a–221a; Ph 14b–21a; Vh 21b–28a; Kl 38r–41r; Ml 27r–33v; Pl. 15r–21v).

The cycle of Mars is 79 years, and the entries are given with the same frequency as for Saturn and Jupiter. In Latin the first entry is Psc 11;2° (March 10, 1300), whereas in Hebrew it is Leo 8;47° (March 10, 1301). Again, comparison between entries for 1300 and 1379 show that 1;40° has to be added to the true positions after a cycle of 79 years, in agreement with the canons.

4. True positions of Venus (Mh 222a–b; Ph 21b–22b; Vh 28b–29b; Kl 32r–34r; Ml 34r–35r; Pl. 22r–v).

For Venus the table displays entries for days 5, 10, 15, 20, 25, and last day of each month in a period of 8 years. The first entry, Aqr 4;40°, is the same in both the Hebrew and Latin manuscripts and corresponds to March 5, 1301, indicating that, in contrast to the rest of planets, the Latin manuscripts do not begin in 1300, but in 1301. In this case, it is not possible to compare entries 8 years apart, but the canons indicate that in order to determine the true position of the planet after one cycle one has to add 1;30° to the initial position.

5. True positions of Mercury (Mh 223b–231a; Ph 23b–31a; Vh 30b–38a; Kl 43r–46r; Ml 35v–43r; Pl. 23r–30v).

The cycle of Mercury lasts 46 years, and the entries in the table are displayed for days 5, 10, 15, 20, 25, and last day of each month, as was the case for Venus. In agreement with the tables for the other planets, except Venus, the

Planet	Cycle	Correction before 1600	Correction after 1600
Saturn	59 years	+1;30°	+0;30°
Jupiter	83 years	−0;30°	−1;30°
Mars	79 years	+1;40°	+0;40°
Venus	8 years	+1;30°	–
Mercury	46 years	−2;45°	–

Table A: Cycles and corrections to be applied to the true positions after a complete cycle (Mh 221b, 223a; Ph 21a, 23a, 30b; Kl 89r–v; Ml 90r–v).

Latin version has a column for year 1300, not found in the Hebrew version. The first entry in Latin manuscripts is Ari 2;33° (March 5, 1300) and that in the Hebrew manuscripts, Ari 11;30° (with the Latin manuscripts and Mh; Ph: Ari 12;30°) for March 5, year 1 (= 1301). The amount to be added to the initial true positions after a cycle of 46 years is −2;45°, as stated in the canons, although this does not hold for the differences between entries in the columns for 1300 and 1346.

In the *Almanac* there are more than 12,000 entries for the true positions of the five planets, given in signs, degrees, and minutes. Recomputation of a random set of these positions with the Toledan Tables produces close agreement globally, and the differences between text and computation are generally a few minutes. This indicates that Jacob ben Makhir computed the entries with this set of tables. Sometimes, however, the differences between text and computation exceed a degree, which suggests that interpolation was used extensively to generate the entries, as is the case for the solar positions (see section 6, below).

The canons also explain that the corrections to be added or subtracted to the true positions of the planets are only valid for increasing values of the equation of the eighth sphere, and that after AD 1600 (some Latin manuscripts read 1690), when the equation begins to decrease, other corrections apply (see Table A). To be sure, the maximum value for the equation of the eighth sphere in Pseudo-Thābit's table for trepidation, 10;45°, occurs at about 1600, when the equation changes its algebraic sign.[22] Although the relevant Latin text is ascribed to Thābit Ibn Qurra (d. 901), his authorship is considered doubtful.[23]

[22] See Toomer, 'A Survey of the Toledan Tables', p. 118. On Pseudo-Thābit's theory of trepidation, see Neugebauer, 'Thâbit ben Qurra'; Goldstein, 'On the Theory of Trepidation'. On the history of trepidation, see Comes, 'Accession and Recession'; Goldstein, 'Historical Perspectives'; Mercier, 'Accession and Recession', and Ragep, 'Al-Battānī, Cosmology'.

[23] See Neugebauer, 'Thâbit ben Qurra', p. 299; and Ragep, 'Al-Battānī, Cosmology', pp. 267–68.

It may be worth comparing various key elements in Jacob's *Almanac* with those in other almanacs, such as the Almanac of Azarquiel and the Almanac of 1307.[24] The entries are given to degrees in these two almanacs, whereas in Jacob's the precision is to minutes. This in itself is a major improvement. As for the frequency of the entries, in both the Almanac of Azarquiel and the Almanac of 1307 the entries are computed for exactly one day after those in Jacob's *Almanac*. In all three almanacs the length of the cycles is the same, but the amounts to be added after a cycle differ, except for Venus and Mercury. All this suggests that Jacob ben Makhir could have used the Almanac of Azarquiel as his model, but did not compute his entries from it.

It is not possible to account for the values for Saturn and Jupiter listed in Table A from historical values of their mean motions.[25] On the other hand, the increment for Mars (+1;40°) is compatible with al-Battānī's value for its mean motion. For the inferior planets, the corresponding increments (+1;30° and –2;45°) are compatible with the values of their mean motions in both the zij of al-Battānī and the Toledan Tables.

6. True positions of the Sun (Mh 231b–235a; Ph 31b–35a; Vh 38b–44a; Kl 46v–50r; Ml 43v–47r; Pl. 31v–35r).

There are four subtables with the daily positions of the Sun given to seconds, covering a period of 4 years (1301–1304), where the leap year is 1303, which means that Feb. 29 was the last day of year 3. The first entry (March 1, 1301) is the same in all manuscripts consulted: Psc 17;51,58°. The last entry (February 28, 1305, corresponding to Feb. 28, 1304 in the year that begins in March) is Psc 16;51,15°. The entry of the Sun into Aries occurs on March 13, indicating that tropical coordinates were used for the entries (Table 6). However, when subtracting the position corresponding to the last entry in the table from that of the first, which are separated by 1460 days, the mean motion we obtain is 0;59,8,11,23°/d, corresponding to a sidereal mean motion, close to that used in the Toledan Tables (0;59,8,11,28,27°/d), and very different from al-Battānī's tropical value (0;59,8,20,46,56,14°/d). This indicates that, in order to compute his solar table, Jacob used a parameter underlying the Toledan Tables to which he applied a correction for precession. Indeed, our recomputations of Jacob's solar positions demonstrate that he used the tables for precession ascribed to Thābit that were associated with the Toledan Tables.

[24] On the Almanac of Azarquiel, see Millás, *Estudios sobre Azarquiel*, pp. 149–237. On the Almanac of 1307, see Chabás, 'El almanaque perpetuo'.

[25] Chabás and Goldstein, *A Survey of European Astronomical Tables*, p. 59.

Day		Solar position (°)				
1	Psc	17;51,58	Ph 17;51,57			
2		18;51,39				
3		19;51,20	Boff: 19;50,20	Pl: 19;50,32		
4		20;51, 1				
5		21;50,42	Boff: 21;51,42			
6		22;49,46				
7		23;48,49	Boff: 23;47,49			
8		24;47,53	Boff: 24;46,53			
9		25;46,55	Boff: 25;45,55			
10		26;45,58	Boff: 26;44,57	Kl: 26;45,57		
11		27;44,50	Boff: 27;43,54	Kl: 27;44,54		
12		28;43,41	Boff: 28;42,51	Kl: 28;43,51	Pl: 28;43,5	Vh: 27;43,41
13		29;42,32	Boff: 29;41,32			
14	Ari	0;41,24	Boff: 0;40,24	Kl: 0;42,24		
15		1;40,15	Boff: 1;39,15	Kl: 1;41,15		
16		2;39,24	Boff: 2;38,29	Kl: 2;39,29	Mh 2;39,22	
17		3;38,32	Boff: 3;37,32			
18		4;37,40	Boff: 4;36,40			
19		5;36,48	Boff: 5;35,58			
20		6;35,57	Boff: 6;34,47	Kl, Pl, Mh: 6;35,47		
⋮						

Table 6: True solar positions beginning in March 1301 (excerpt). The manuscripts consulted contain many copyists' errors; the most reliable witness seems to be Ml, which is the closest to the Hebrew manuscripts and therefore it has been taken as the base manuscript.

Inspection of successive entries in Table 6 shows that Jacob computed only one out of five entries, and found the intermediate values by interpolation. As an example, consider the entries for June 1301 listed in Table 6 and December 1301 (see Table B). Instead of 1301, the Hebrew reads year 1 (of a four-year solar cycle): Ph 32a.

Between two computed entries, marked here in bold type, the author divided the difference into five equal parts to determine the intermediate entries. The different strategies used when the difference was not a multiple of 5 seconds are shown in Table B. The trouble with this system for computing entries 5 days apart is that a mistake in one computed entry affects nine successive entries in the table (four previous and four subsequent entries). Hasty computation entails high risks.

We have recomputed selected entries in this table for the meridian of Toledo, at noon and at two moments of time before noon to account for the difference

Day	Solar position March (°)	Successive differences (°)	...	Solar position December (°)	Successive differences (°)
1	Psc 17;51,58			Sgr 16;34,52	
2	18;51,39	0;59,41		17;36,24	1;1,32
3	19;51,20	0;59,41		18;37,56	1;1,32
4	20;51, 1	0;59,41		19;39,28	1;1,32
5	**21;50,42**	0;59,41		**20;40,59**	1;1,31
6	22;49,46	0;59, 4		21;42, 1	1,1, 2
7	23;48,49	0;59, 3		22;43, 4	1;1, 3
8	24;47,53	0;59, 4		23;44, 7	1;1, 3
9	25;46,55	0;59, 2		24;45,10	1;1, 3
10	**26;45,58**	0;59, 3		**25;46,13**	1;1, 3
11	27;44,50	0;58,52		26;47,28	1;1,15
12	28;43,41	0;58,51		27;48,43	1;1,15
13	29;42,32	0;58,51		28;49,58	1;1,15
14	Ari 0;41,24	0;58,52		29;51,13	1;1,15
15	**1;40,15**	0;58,51		Cap **0;52,29**	1;1,16
16	2;39,24	0;59, 9		1;53,44	1;1,15
17	3;38,32	0;59, 8		2;54,59	1;1,15
18	4;37,40	0;59, 8		3;56,14	1;1,15
19	5;36,48	0;59, 8		4;57,29	1;1,15
20	**6;35,57**	0;59, 9		**5;58,43**	1;1,14
⋮					

Table B: Interpolations for the solar longitude in March and December, 1301.

in longitude between Toledo and Montpellier. Table C shows the recomputation for various dates evenly spaced in the period 1301–1304. Note the last entry in Table C, for February 1, 1305, corresponds to 1304 in the *Almanac*. The value of precession for March 1, 1301 is 9;27,5°, and for Feb. 1, 1305 it is 9;28,54°.

The agreement between the entries in the text and recomputation is best when the distance in longitude between Toledo and Montpellier is taken to be 1;15h (= 18;45°), for the magnitude of most of the differences is in seconds, except for some entries where the difference reaches a little over a minute. As far as we can determine, the value of 1;15h for the distance between Toledo and Montpellier is not attested in any medieval text. According to modern data, this distance amounts to about 8°, or about 0;32h, corresponding to about 0;1,20° in solar longitude. In many medieval sources the longitude of Toledo is 28;0° from the western limit, and in prologue B the longitude of Montpellier is 32;0° from the western limit, for a difference of 4;0°, or 0;16h, corresponding to about 0;0,40° in solar longitude. However, in the *editio princeps* of the

Date	Text	Computation		T – C
March 1, 1301	Psc 17;51,58°	Noon:	347;55,10°	–0; 3,12°
		–1h:	347;52,42°	–0; 0,44°
		–1;15h	347;52, 5°	–0; 0, 7°
May 1, 1301	Tau 17;18,27°	Noon:	47;21,30°	–0; 3,13°
		–1h:	47;19, 6°	–0; 0,39°
		–1;15h:	47;18,30°	–0; 0, 3°
June 1, 1302	Gem 16;40,51°	Noon:	76;45,23°	–0; 4,34°
		–1h:	76;43, 0°	–0; 2, 9°
		–1;15h:	76;42,24°	–0; 1,33°
August 1, 1302	Leo 14;59,47°	Noon:	135; 2,50°	–0; 3, 3°
		–1h:	135; 0,25°	–0; 0,38°
		–1;15h:	135;59,49°	–0; 0, 2°
September 1, 1303	Vir 14;49,43°	Noon:	164;53,11°	–0; 3,28°
		–1h:	164;50,45°	–0; 1, 2°
		–1;15h:	164;50, 8°	–0; 0,25°
November 1, 1303	Sco 15;34,46°	Noon:	225;37,51°	–0; 3, 5°
		–1h:	225;35,19°	–0; 0,33°
		–1;15h:	225;34,41°	+0; 0, 5°
December 1, 1304	Sgr 16;50, 7°	Noon:	256;54, 2°	–0; 3,55°
		–1h:	256;51,29°	–0; 1,22°
		–1;15h:	256;50,51°	–0; 0,44°
February 1, 1305	Aqr 19;53,11°	Noon:	319;56,45°	–0; 3,34°
		–1h:	319;54,13°	–0; 1, 2°
		–1;15h:	319;53,36°	–0; 0,25°

Table C: Recomputation of the solar longitude.

Parisian Alfonsine Tables, the difference in longitude between Montpellier and Toledo is 15° (26;0° – 11;0°), which is equivalent to 1h.[26]

In a complete set of astronomical tables one expects to find a table for the equation of time. However, it is not clear if Jacob included it, for such a table is only found in two of the six manuscripts we examined (both in Hebrew): Ph 82b–83a which is not in the part of this manuscript that contains Jacob's tables, and Vh 96b–97a. The same table for the equation of time occurs in the earlier tables of Abraham Bar Ḥiyya and later in Bonfils's Tables for 1340.[27]

[26] Ratdolt, *Tabule astronomice*, m5r; see also Kremer and Dobrzycki, 'Alfonsine Meridians'. The longitude of Montpellier in the Tables of Barcelona is 32;10°. On the Tables of Barcelona (second half of the fourteenth century), see Chabás, 'Astronomía andalusí'; see also Millás, *Las tablas astronómicas*, p. 238. In the list of coordinates of cities in the Toledan Tables, the longitude of Toledo is 11° but Montpellier is not mentioned: see Pedersen, *The Toledan Tables*, pp. 1512–13.

[27] On Bonfils's Tables for 1340, see Goldstein and Chabás, 'Analysis of the Astronomical Tables', pp. 76–78. On the equation of time, see Chabás and Goldstein, *A Survey of European Astronomical Tables*, pp. 37–41.

7. Mean motion in lunar longitude in days in a Julian year (Mh 236a–b; Ph 35b–36b; Vh 46a–47a; Kl 51r–v; Ml 48v–50r; Pl. 36r–v).

This table displays the mean motion in longitude traveled by the Moon in a year, where the argument is the month and day. The first entry is for March 1: 0s 13;10,35°. The last entry is for February 29: 4s 22;32,48°. The underlying daily mean motion in lunar longitude can thus be calculated as (13 · 360 + 4s 22;32,48)/366 = 13;10,34,53,6,53°/d, a value which is close to that underlying the Toledan Tables (13;10,34,52,48,47°/d) and, again, far from al-Battānī's value (13;10,35,2,7,17,10°/d). Indeed, the first 30 entries in this table agree, but for scribal errors, with those in the table for the mean motion of the Moon in days in the Toledan Tables.[28]

8. Mean motion in lunar longitude in collected Julian years (Mh 235b; Ph 37a; Vh 45b; Kl 50v; Ml 47v–48r; Pl. 35v).

The entries in this table represent the mean motion in longitude for 76 consecutive years. Before the entry for year 1 (= 1301) of 11s 11;36,11° (with Ph; Kl reads 11s 11;36,44°), there are two other entries (2s 22;51,41° and 7s 2;14,21°), corresponding to 1299 and 1300, respectively. While the entry for 1299 is the same in all manuscripts, that for 1300 has variants: 7s 2;14,52° in Mh, and 7s 2;13,23° in all three Latin manuscripts, whereas it should be 7s 2;13,56°, as it is in Vh. These two entries are not found in Ph. Comparison of the following entries in the different manuscripts reveals many scribal errors.

The canons indicate that, after one cycle of 76 years, the mean positions of the Moon have to be increased by 3°.[29] To be sure, this value results from subtracting the radix for 1300 (7s 2;14,21°) from the entry for year 76 (7s 5;14,12°). The result is 2;59,51° ≈ 3°. The text adds that this increment is only valid when the equation of the eighth sphere is increasing, and that 1° is to be applied for decreasing values of the equation (Ph 37a, Mh 235b, and Vh 45b agree on the 3° but all three manuscripts have 1;55°, instead of 1°, to be applied for decreasing values of the equation).

9. True lunar anomaly (Mh 247b–255a; Ph 37b–66b; Vh 47b–84a; Kl 52r–75v; Ml 50v–74v; Pl. 37r–61r; see Plate 2).

This table is unprecedented and it is the largest table in this set, for it contains more than 8500 entries for the true lunar anomaly on a daily basis for a period of about 24 years, from March 1300 to December 1323 (see Table 9). At the left of the table in Latin there are successive integers for days, from 1 to 30 or 29. There are altogether 294 columns, 156 of them with 30 entries and other 138 columns with 29 entries, totaling 8682 days. Each column represents a lunar month (which is not named), and for each day we are given two entries:

[28] Pedersen, *The Toledan Tables*, p. 1155.

[29] Boffito and Melzi d'Eril, *Almanach Dantis Aligherii*, p. 4, lines -9 and -8.

	3		5		6		...
	22		21		20		...
	March	min	April	min	May	min	...
1	5s 4;28	3	6s 8; 5	5	6s 25; 6	2	
2	5s 21; 1	9	6s 24;30	17	7s 10;14	6	
3	6s 10;48	20	7s 10;20	25	7s 27;49	19	
⋮							
15	11s 4;39	0	0s 8;17	2	0s 25;16	0	
⋮							
28	4s 18;15	4	5s 21;47	2	6s 8;48	4	
29	5s 4;46	0	6s 8;27	0	6s 25;21	0	
30	5s 21;33	1			7s 12; 4	0	

Table 9: True lunar anomaly in degrees and minutes of proportion, beginning on March 22, 1300 (excerpt).

one for true lunar anomaly, in signs, degrees, and minutes, and another for the minutes of proportion. In Ph the columns are numbered consecutively from 1 to 294. At the head of the column for each such month we are given information on the weekday of the first day of the lunar month, the day of the Julian month when the lunar month begins, and the name of the month in the Julian calendar. For example, for the first column, we find 3 (weekday), 22 (day in the Julian month for the beginning of the lunar month), and March (name of the Julian month). This is to be understood as the lunar month beginning on Tuesday, March 22 (Nisan 1, 5060 AM in the Hebrew calendar, and Rajab 1, 699 AH in the Hijra calendar). For that specific day, the entry for true anomaly is 5s 4;28° and that for the minutes of proportion, 3 min, in all manuscripts.

In the table for the lunar equations in the Toledan Tables there is also a column for the minutes of proportion, which serves for interpolation purposes between extremal values. The same column is already found in Ptolemy's *Almagest*. The entries for the minutes of proportion given by Jacob correspond in each case to the equation of center of the Moon associated with the tabulated values of true anomaly.

To recompute the values of the true anomaly, α, first one has to determine the mean longitudes of the Sun and the Moon and the mean lunar anomaly, $\bar{\alpha}$, and then to compute the equation of center of the Moon, $c_3(2\eta)$, from the double elongation of the mean Moon from the mean Sun: $\alpha = \bar{\alpha} + c_3(2\eta)$. See Table D for computations of selected entries corresponding to the first half of the first column (lunar month) in the table for the true lunar anomaly. The first entry corresponds to March 22, 1300, that is, Rajab 1, 699 AH in the civil calendar but, for astronomical purposes, the day begins at noon, rather than at sunset as in the civil Hijra calendar.[30] Hence noon on March 22, 1300 cor-

[30] Neugebauer, *The Astronomical Tables of al-Khwārizmī*, pp. 10–11.

1300	March 22	March 25	March 28	March 31	April 3
Mean anomaly Noon:	151; 5,37°	190;17,19°	229;28,58°	268;40,42°	307;52,24°
True anomaly Noon: –1h: –1;6h	 155;13,38 154;31,50 154;27,49	 203; 3,25 202;27,42 202;24, 8	 231;29,14 231;17,54 231;16,46	 255;33,22 254;59,41 254;56,1 9	 302; 9,37 301;27,49 301;23,38
Text	5s 4;28°	6s 22;24°	7s 21;17°	8s 14;57°	10s 1;23°
Diff. T – C Noon: –1h: –1;6h	 –0;46° –0; 4° 0; 0°	 –0;36° –0; 4° 0; 0°	 –0;12° –0; 1° 0; 0°	 –0;36° –0; 3° +0; 1 °	 –0;47° –0; 5° –0; 1°

Table D: Recomputation of the true lunar anomaly.

responds to the beginning of Rajab 2, 699 AH according to the astronomical convention. All values for the mean motions and the lunar equation of center were taken from the Toledan Tables.[31]

The agreement in Table D is excellent when the distance between Montpellier and Toledo is taken to be 1;6h (= 16;30°). It is indeed an excellent result because the Moon, being the swiftest object, is likely to produce the worst results. Note, however, that Table 6 for solar longitude seems to have been computed for a different, and greater, distance between the two cities. A few explanations come to mind, but none is supported in the text.

10. Complete lunar equation (Mh 256a–261b; Ph 67b–73a; Vh 85a–90b; Kl 76r–81v; Ml 75r–80r; Pl. 61v–67r; see Plate 3).

This is a double argument table, also unprecedented in the West, and directly linked to the previous table for the true anomaly of the Moon (Table 9). The minutes of proportion at the head of the table are given at intervals of 5 minutes from 0 to 60 minutes, and they represent one variable. The other variable, at the left of the table in Latin, is the true anomaly, from 0s 1° to 11s 30°, at intervals of 1°, a quantity obtained from Table 9. The first entry (for 0 minutes of proportion and a true anomaly of 1°) is 7;35°, as shown in Table 10.

In Table 10 only selected entries, based on the Latin manuscripts, have been displayed (the variants in the Hebrew manuscripts are minor), but they are sufficient to demonstrate the role of a key entry, 7;40°, corresponding to the rows for values of the true anomaly of 5s 30° (= 180°) and 11s 30° (= 360°). It also shows that the entries in rows equally distant from the central one (for true anomaly 180°) add up to 15;20°, which is twice 7;40°. To understand the meaning of this number, we need to recall how the complete lunar equation,

[31] Pedersen, *The Toledan Tables*, pp. 1147–48, 1154–55, 1159–60, and 1253–58.

	0 (°)	...	15 (°)	...	30 (°)	...	45 (°)	...	60 (°)
0s 1°	7;35		7;34		7;33		7;33		7;32
⋮									
2s 30°	2;40		2; 1		1;22		0;43		0; 4
3s 1°	2;40		2; 1		1;21		0;42		0; 3
⋮									
3s 5°	2;39		2; 0		1;20		0;41		0; 1
⋮									
3s 8°	2;39		2; 0		1;20		0;40		0; 0
⋮									
5s 30°	7;40		7;40		7;40		7;40		7;40
⋮									
8s 22°	12;41		13;20		13;59		14;40		15;20
⋮									
8s 25°	12;41		13;20		14; 0		14;39		15;19
⋮									
8s 30°	12;40		13;19		13;59		14;37		15;17
⋮									
11s 29°	7;45		7;46		7;46		7;47		7;48
11s 30°	7;40		7;40		7;40		7;40		7;40

Table 10: Complete lunar equation (excerpt).

c, to be applied to the mean lunar longitude to obtain its true longitude, is computed in Ptolemy's second lunar model:[32]

$$c = c_4(\alpha) + c_5(\alpha) \cdot c_6(2\eta),$$

where

$$\alpha = \bar{\alpha} + c_3(2\eta).$$

The subscripts refer to the number of the column in Ptolemy's table for the lunar equations in *Almagest* V.8, and the two variables, α and 2η, are the true lunar anomaly and the double elongation, respectively. In the expressions above, $c_3(2\eta)$ is the equation of center, $c_4(\alpha)$ is the equation of anomaly, $c_5(\alpha)$ is called the increment and represents the effect of bringing the epicycle closer to the observer in the second lunar model than in the epicyclic version of the first lunar model, and $c_6(2\eta)$ corresponds to the minutes of proportion for purposes of interpolation. In *Almagest* V.8 the maximum values for the equation of center, equation of anomaly, and increment are 13;9°, 5;1°, and 2;39°, respectively. In the table in the *Almagest* the argument is given at intervals of 3° and 6°, but in his *Handy Tables* Ptolemy expanded it to intervals of 1°, where the maximum value of the increment was taken as 2;40°. The latter was the basis for the corresponding tables in the zij of al-Battānī and the Toledan Tables.

[32] Chabás and Goldstein, *A Survey of European Astronomical Tables*, pp. 68–70.

The entries in Jacob's double argument table represent the complete lunar equation for each pair of values of α and 2η, increased by 7;40°, which is the maximum resulting from adding corresponding values of the two quantities depending on the true anomaly, $c_4(\alpha)$ and $c_5(\alpha)$. They can be computed by means of the expression:

$$c + 7;40° = c_4(\alpha) + c_5(\alpha) \cdot c_6(2\eta) + 7;40°.$$

The inclusion of a vertical shift of +7;40° in all entries ensures that no negative numbers are involved or, to say it in a non-anachronistic manner, to avoid the cumbersome rules used by medieval astronomers to deal with all possible values and signs of the equations. This makes Jacob's table for the lunar equation exceptional, for it is the first known example of a double argument table in the West with a vertical shift. On displaced tables with vertical and horizontal shifts in Arabic, Hebrew, and Latin astronomical tables, see Chabás and Goldstein, 'Displaced Tables in Latin', and the references there.

To recompute the entries, let us first consider the case where $c_6(2\eta) = 0$. The expression for computing the entries then reduces to $c_4(\alpha)$ + 7;40°. For α = 1°, the value obtained in the table for the lunar equations in the Toledan Tables is –0;4,50° + 7;40° ≈ 7;35°, in agreement with the entry in the table. For α = 180° and 360°, the result is 0° + 7;40° = 7;40°. For intermediate values of the true anomaly such as 95° and 265°, we find –5;1° + 7;40° = 2;39° and +5;1° + 7;40° = 12;41°, respectively, both in agreement with the entries in the table.

Consider now the case where $c_6(2\eta) = 60$. The expression for computing the entries then reduces to $c_4(\alpha) + c_5(\alpha)$ + 7;40°. For α = 1°, the value we obtain from the table for the lunar equations in the Toledan Tables is –0;4,50° – 0;3° + 7;40° ≈ 7;32°, in agreement with the entry in the table. For α = 180° and 360°, the result is 0° + 0° + 7;40° = 7;40°. For intermediate values of the true anomaly such as 95° and 265°, we find –5;1° – 2;38° + 7;40° = 0;1°, and + 5;1° + 2;38° + 7;40° = 15;19°, respectively, both also in agreement with the entries in the table.

For intermediate values of the minutes of proportion, consider the case where $c_6(2\eta) = 30$. The expression above becomes $c_4(\alpha) + \frac{1}{2} \cdot c_5(\alpha)$ + 7;40°. With the Toledan Tables, for α = 1° we obtain –0;4,50° – 0;1,30° + 7;40° = 7;33,40°, rounded to 7;33° in the table. For α = 180° and 360°, the result is again 7;40°. For intermediate values of the true anomaly such as 95° and 265°, we find –5;1° – 1;19° + 7;40° = 1;20° and + 5;1° + 1;19° + 7;40° = 14;0°, respectively, both in agreement with the entries in the table.

It follows that Jacob divided the procedure for computing the key quantities associated with the Moon into two phases: the first deals with double elongation to compute the true lunar anomaly and is presented in the form of a table for lunar months (Table 9), whereas the second uses the true lunar anomaly to compute the lunar equation and it is presented in the form of an unprecedented double argument table with a vertical shift (Table 10). The geometrical model and the parameters underlying these two tables are Ptolemy's, probably transmitted via the Toledan Tables or the zij of al-Battānī. In short, this is a new and clear example of a typical approach for constructing astronomical tables in the Middle Ages: no changes were made in the model or the underlying parameters; rather, enhancing 'user-friendliness' is attained by means of innovation in the presentation. By 'user-friendliness', we mean that the compiler of the table does more calculations so that the user has to do fewer and simpler computations, thus saving time and avoiding possible computational mistakes. In this case, the presentation of the table is very ingenious.

This kind of double argument table for the lunar equation is also found in the zij of Joseph Ibn Waqār, composed in Arabic in Toledo in 1359/1360 and then translated into Hebrew by the author himself. Moreover, Moses Farissol Botarel (Avignon, late fifteenth century), whose astronomical tables are in Hebrew, adapted this presentation to the Alfonsine Tables, with an equation of lunar anomaly reaching a maximum of 4;56°, rather than 5;1° as is the case in Jacob's *Almanac*.[33]

11. Mean and 'true' motion of the lunar node in collected Julian years (Mh 262b; Ph 73b; Vh 91a–b; Ml 81r; Kl 83r; Pl. 67v).

This table displays the mean motion of the lunar ascending node for 93 consecutive years. In each case we are given the mean motion and the 'true' motion, which is taken as the complement in 360° of the mean motion. The entries for year 1 in MS Kl are 1s 29;37,45° and 10s 0;22,15°, respectively, which add up to 12, as expected. Although it is not stated in the table or the canons, year 1 corresponds to 1301. In the Latin version there are also values for years 1299 (0s 20;57,35° and 11s 9;2,25°) and 1300 (1s 10;17,40° and 10s 19;42,20°).

12. Daily mean motion of the lunar node in a Julian year (Mh 263a–b; Ph 74b–75a; Vh 92a–b; Kl 83v–84r; Ml 81v–82r; Pl. 68r–v).

The entries in this table represent the daily mean motion of the lunar ascending node, where the argument is the month and the day in a year. The first entry is for March 1: 0;3,11°. The last entry is for February 29: 19;23,45°. The resulting daily mean motion is 19;23,45/366 = 0;3,10,46,43°/d, which is close to that in the Toledan Tables (0;3,10,46,42,33°/d) and, again, far from al-Battānī's value (0;3,10,37,19°/d). Indeed, the first 30 entries in this table agree, but for scribal errors, with those in the table for the mean motion of the lunar node in days in the Toledan Tables.[34]

13. Lunar latitude (Mh 264a; Ph 75b; Vh 93a; Kl 84r; Ml 87r; Pl. 69v).

The entries in this table are given to minutes for each integer degree of the argument, and reach a maximum of 5;0° at 90°. The table in the *Almanac* is

[33] On the tables of Ibn Waqār, see Chabás and Goldstein, 'Ibn al-Kammād's *Muqtabis* zij', p. 607. On the tables of Moses Farissol Botarel, see Goldstein and Chabás, 'The Astronomical Tables of Moses'.

[34] Pedersen, *The Toledan Tables*, p. 1165.

Argument of lunar latitude 0s / 6s (°)	5s / 11s (°)	Digits of eclipse (d)	Minutes of immersion (′)	Half-totality (′)
11;42	18;18	0	0; 0	0; 0
10;58	19; 2	1	18;22	0; 0
10;26	19;34	2	25; 1	0; 0
9;54	20; 6	3	30;33	0; 0
⋮				
5;38	24;22	11	52;19	0; 0
5; 6	24;54	12	53;40	0; 0
4;34	25;26	13	43;15	12; 1
4; 2	25;58	14	39;20	16;18
⋮				
1;22	28;38	19	33;43	25;41
0;50	29;10	20	33;27	26;18
0;18	29;42	21	33;26	26;36
0; 0	29;42	21	33;30	26;40

Table 14: Lunar eclipses in MS Kl, 84v (excerpt).

based on the corresponding table in *Almagest* V.8, where the entries are also given to minutes, but only at intervals of 3°. Later tables for the same purpose, such as those in the zij of al-Battānī and the Toledan Tables used the same value for the maximum lunar latitude, but displayed entries to seconds and at intervals of 1°.

14. Lunar eclipses (Mh 264a; Ph 75b; Kl 84v; Ml 87r; Pl. 69v).

In contrast to most of the sets of tables dealing with eclipses which present two separate tables for lunar eclipses, one for greatest and another for least distance of the Moon, Jacob's *Almanac* has a single table for any lunar distance (see Table 14). It consists of five columns. The first two columns are for the argument of lunar latitude in degrees and minutes, to be added to 0s and 6s (first column) and to 5s and 11s (second column). For reasons of symmetry, the corresponding entries in these two columns should add up to 30°. The third column, which displays the fraction of the diameter of the eclipsed disk, i.e., the digits of eclipse, is the real argument in this table. They are given in integer digits from 0 to 21. The fourth and fifth columns display the minutes of immersion and the minutes of half-totality of the lunar eclipse. The last two columns in this table could have been derived from the table in Ptolemy's *Almagest*[35] by computing the arithmetical mean of the entries in the columns for greatest and least distances, which differ by only a few minutes.

[35] Toomer, *Ptolemy's Almagest*, p. 307.

15. Hourly velocities of the Moon and the Sun and the Moon, and length of half-day light (Mh 268b; Ph 76a; Vh 93b–94a; Kl 84v; Ml 87v; Pl. 69r).

In several manuscripts, both in Hebrew and in Latin, between the columns for the lunar velocity and the solar velocity, we find other columns for interpolation. In particular, one ranges from 0 to 60 min and another from 0 to 12, and correspond to columns already found in Ptolemy's tables, as well as in other sets of medieval tables, such as the zij of al-Battānī and the Toledan Tables, under the title *tabula attacium* (see section 19, below).

16. Hourly velocity of the Moon relative to the Sun (Mh 264b–265a; Ph 76b–77a; Vh 94b–95a; Kl 82r–v; Ml 82v–83r; Pl. 70r–v).

This table consists of 13, 14, or 15 columns, depending on the manuscript, labeled 1 to 13, 14, or 15. The first column displays the difference between the hourly velocities of the Moon and the Sun, from 0;27,50°/h to 0;33,20°/h (or 0;33,30°/h) at intervals of 0;0,10°/h, and the rest of the columns display multiples of the entry in the first column up to 13, 14, or 15 hours. The aim is to facilitate calculations when computing the time of true syzygy.

17. Parallax (Mh 265b–267a; Ph 77b–78a; Vh 98a–99a; Kl 92v–93r; Ml 83v–84v; Pl. 71r–72v).

There is only one table for parallax and it is for geographical latitude 43°, corresponding to Montpellier. The title of this table in Ph 77b is 'Lunar parallax for latitude 43° which is the latitude of Montpellier (*ha-har*), and its hours are 15;32h as found in the table of ascensions by Ibn Ezra', but in Ph 78a the title is 'Lunar parallax for 43° derived from the two tables found in the book by ha-Nasi' [i.e., Abraham Bar Ḥiyya] for the fifth and sixth [climates]'. In Mh 265v it is simply 'Lunar parallax for latitude 43° and the longest daylight is 15;32h'. In Vh the title is 'Table for lunar parallax for the fifth climate, latitude 43° and [longest daylight] 15;18h'. All three Latin manuscripts mention the name of the city, Montpellier, and its latitude, 43°. This table is intended to be used in computing the circumstances of a solar eclipse; it corrects the true position of the Moon to its apparent position for an observer at a given geographical latitude. In fact, computations using this type of table were made very rarely. Levi ben Gerson was one of a few exceptions to the general pattern, for he observed four solar eclipses from 1321 to 1337 and computed their circumstances.[36]

Jacob's table has the same structure as the corresponding tables in Ptolemy's *Handy Tables*, the zij of al-Battānī and the tables of Abraham Bar Ḥiyya, etc.

[36] Goldstein, 'Medieval Observations', pp. 123–30. For discussion of similar tables, see Chabás and Goldstein, *A Survey of European Astronomical Tables*, pp. 127–38.

It consists of 12 subtables, one for each of the zodiacal signs, displaying the components of parallax in longitude and latitude, as a function of time. But, unlike those tables which give the components of parallax to minutes, Jacob gives them to minutes and seconds. This is quite unusual, but this precision also occurs in the tables of Levi ben Gerson.[37]

Each of the six manuscripts examined has a single table for parallax. However, in the three Latin manuscripts and Vh the maximum time of half-daylight is 7;39h, implying a longest daylight of 15;18h, whereas in two Hebrew manuscripts, Mh and Ph, the corresponding time is 7;46h, which agrees with the value of the longest daylight of 15;32h in the title. Because of this discrepancy, we examined another copy of this table in Hebrew (Parma, MS Heb. 2113, 78b–79b), and we found that it agrees with the three Latin manuscripts and Vh. Thus, we encounter two different tables with entirely different entries. The formula for computing longest half-daylight is

$$\sin \gamma = \tan \varepsilon \cdot \tan \phi,$$

where γ is the increment in longest half-daylight (in degrees) over 90°, ε is the obliquity of the ecliptic, and ϕ is the geographical latitude.[38] With ε = 23;51° and ϕ = 43°, we find γ = 24;21° = 1;37,24h. Longest half-daylight is then 6h + 1;37,24h = 7;37,24h, and longest daylight is 15;15h, which is close to the value of 15;18h in one version of this table. Lower values for the obliquity of the ecliptic produce worse agreement. The other version, however, with longest daylight of 15;32h, is not consistent with a geographical latitude of 43°, which is that of Montpellier. Moreover, according to the title in Ph, the table was derived from two subtables for the fifth and sixth climates due to Abraham Bar Ḥiyya, which are the same as those in al-Battānī's zij, but for minor variants.[39] However, in al-Battānī's zij the length of daylight for the sixth climate is 15;30h, corresponding to a geographical latitude of 45;22° (Bar Ḥiyya, Paris 1046, 32a: length of daylight 15;30h, latitude 45;0°) which implies that the longest daylight in Ph of 15;32h would have to correspond to a latitude even greater than 45;22°, a latitude that is much too high for Montpellier. It follows that the parallax table in Mh and Ph does not correspond to Montpellier, whereas that in Vh, Parma 2113, and the Latin manuscripts does. The title for this table in Ph 77b (translated above) refers to a table for (oblique) ascensions by Abraham Ibn Ezra, but we are not aware of any such table: it is likely that this is a mistake introduced by a copyist. Moreover, the reference in the title of the parallax table in Ph 78a to the tables for the fifth and sixth climates does

[37] Goldstein, *The Astronomical Tables of Levi*, pp. 184–207.

[38] Chabás and Goldstein, *A Survey of European Astronomical Tables*, pp. 31–32.

[39] See Paris, MS Heb. 1046, 31b–32a; Nallino, *Al-Battānī sive Albatenii*, vol. II, pp. 99–100.

Greatest distance			Least distance		
Latitude (′)	Digits of eclipse (d)	Min. of immersion (′)	Latitude (′)	Digits of eclipse (d)	Min. of immersion (′)
32; 0	0	0; 0	34;30	0	0; 0
29;20	1	12,32	31;28	1	12;56
27;40	2	17;19	28;48	2	17;54
⋮					
5;20	10	30;54	7;28	10	32;33
2;40	11	31;13	4;48	11	33;11
0; 0	12	31;20	2;16	12	33;16
			0; 0	12	33;20

Table 18: Solar eclipses in MS Kl 87r (excerpt).

not seem compatible with the entries in Jacob's table, and it too may have been added by a copyist. We conclude that the table with longest daylight 15;18h was probably in Jacob's original Hebrew text, and that an early copyist in the Hebrew tradition replaced it with a table for longest daylight of 15;32h, preserved in some (but not all) copies in Hebrew. The Latin text was probably translated from Hebrew soon after the time of composition and was unaffected by the replacement of the parallax table.

18. Solar eclipses (Mh 267b; Ph 78b; Vh 97b; Kl 94v; Ml 87r; Pl. 73r).

There are two subtables, one for greatest distance of the Moon and another for least distance. The argument is the lunar latitude in minutes and seconds of arc, following the pattern of Ptolemy's *Handy Tables*, in contrast to the table for lunar eclipses (see section 14) where it is the argument of lunar latitude, as is the case in Ptolemy's *Almagest*.[40] The other two columns are for the digits of eclipse and the minutes of immersion (see Table 18). The same table (with variants) appears in Bar Ḥiyya's tables: Paris, MS Heb. 1046, 26b.

19. Table for division (Ph 80a; Vh 44b; Kl 94v; Ml 88r; Pl. 31r).

The three subtables in the Latin manuscripts have the same format and their purpose is to divide integer numbers from 1 to 60 by 10, 5, and 6. Only two subtables for division, those by 10 and 5, are found in the Hebrew manuscripts.

[40] Chabás and Goldstein, *A Survey of European Astronomical Tables*, p. 167.

Conclusion

Jacob ben Makhir computed his *Almanac* with the Toledan Tables, and not with any other set of tables available at the time, such as the Almanac of Azarchiel or the tables of al-Battānī. In his *Almanac*, Jacob used tropical coordinates, and thus had to compute precession to adjust the sidereal coordinates found in the Toledan Tables or a version of them. Jacob computed the positions of the planets to minutes, thus increasing the precision found in previous almanacs and enhancing the user-friendliness of the table. Moreover, Jacob adjusted the true positions of the celestial objects to Montpellier, by applying a distance from that city to Toledo that we have not found in the previous literature. In the *Almanac*, two unprecedented features are introduced: a table for true anomaly, the largest in this set, using both the Julian and the Hebrew calendars, and a double argument table for the complete lunar equation, with a vertical shift. We also learn about contacts among astronomers in various religious communities: the Toledan Tables were originally composed in Arabic by Muslim scholars; translated into Latin by a Christian scholar; used to compose an almanac in Hebrew by a Jewish scholar; and this almanac was then translated into Latin by a Christian scholar.

Acknowledgments

We are grateful to Niran Garshtein (Bar Ilan University, Israel) for sending us an electronic copy of the parallax table in Parma, MS Heb. 2113, and to Matthieu Husson (ALFA, Observatoire de Paris) and Benno van Dalen (Ptolemaeus Arabus et Latinus, Munich) for their valuable comments on a draft of this paper. We also thank Richard L. Kremer (Dartmouth University) for his advice on medieval lists of geographical coordinates.

Bibliography

Boffito, Giuseppe, and Camillo Melzi d'Eril, *Almanach Dantis Aligherii sive Profhacii Judaei Montispessulani.* Almanach Perpetuum *ad annum 1300 inchoatum*, Florence: Olschki, 1908.

Boutelle, Marion, 'The Almanac of Azarquiel', *Centaurus* 12 (1967), pp. 12–20.

Chabás, José, 'El almanaque perpetuo de Ferrand Martines (1391)', *Archives Internationales d'Histoire des Sciences* 46 (1996), pp. 261–308.

Chabás, José, 'Astronomía andalusí en Cataluña: Las Tablas de Barcelona', in Josep Casulleras and Julio Samsó (eds), *From Baghdad to Barcelona. Studies in the Islamic Exact Sciences in Honour of Prof. Juan Vernet*, Barcelona: Instituto Millás Vallicrosa de Historia de la Ciencia Árabe, 1996, Vol. I, pp. 477–525.

Chabás, José, and Bernard R. Goldstein, *Astronomy in the Iberian Peninsula. Abraham Zacut and the Transition from Manuscript to Print*, Philadelphia: American Philosophical Society, 2000.

Chabás, José, and Bernard R. Goldstein, *A Survey of European Astronomical Tables in the Late Middle Ages*, Leiden: Brill, 2012.

Chabás, José, and Bernard R. Goldstein, 'Displaced Tables in Latin. The Tables for the Seven Planets for 1340', *Archive for History of Exact Sciences* 67 (2013), pp. 1–42.

Chabás, José, and Bernard R. Goldstein, 'Ibn al-Kammād's *Muqtabis* zij and the Astronomical Tradition of Indian Origin in the Iberian Peninsula', *Archive for History of Exact Sciences* 69 (2015), pp. 577–650.

Cantera Burgos, Francisco, 'El judio salmantino Abraham Zacut', *Revista de la Academia de Ciencias Exactas, Físico-Químicas y Naturales de Madrid* 27 (1931), pp. 63–398.

Comes, Mercè, 'Accession and Recession Theory in al-Andalus and North Africa', in Josep Casulleras and Julio Samsó (eds), *From Baghdad to Barcelona. Studies in the Islamic Exact Sciences in Honour of Prof. Juan Vernet*, Barcelona: Instituto Millás Vallicrosa de Historia de la Ciencia Árabe, 1996, Vol. I, pp. 349–64.

Goldstein, Bernard R., 'On the Theory of Trepidation According to Thābit b. Qurra and al-Zarqāllu and its Implications for Homocentric Planetary Theory', *Centaurus* 10 (1964), pp. 232–47.

Goldstein, Bernard R., *The Astronomical Tables of Levi ben Gerson*, New Haven: Archon, 1974.

Goldstein, Bernard R., 'Medieval Observations of Solar and Lunar Eclipses', *Archives Internationales d'Histoire des Sciences* 29 (1979), pp. 101–56. Reprinted in Bernard R. Goldstein, *Theory and Observation in Ancient and Medieval Astronomy*, London: Variorum Reprints, 1985, ch. XVII.

Goldstein, Bernard R., 'The Survival of Arabic Astronomy in Hebrew', *Journal for the History of Arabic Science* 3 (1979), pp. 31–39. Reprinted in Bernard R. Goldstein, *Theory and Observation in Ancient and Medieval Astronomy*, London: Variorum Reprints, 1985, ch. XXI.

Goldstein, Bernard R., 'Historical Perspectives on Copernicus's Account of Precession', *Journal for the History of Astronomy* 25 (1994), pp. 189–97.

Goldstein, Bernard R., and José Chabás, 'Analysis of the Astronomical Tables for 1340 Compiled by Immanuel ben Jacob Bonfils', *Archive for History of Exact Sciences* 71 (2016), pp. 71–108.

Goldstein, Bernard R., and José Chabás, 'The Astronomical Tables of Moses Farissol Botarel', *Suhayl* 15 (2016–2017), pp. 29–65.

Kremer, Richard L., and Jerzy Dobrzycki, 'Alfonsine Meridians. Tradition Versus Experience in Astronomical Practice *c.* 1500', *Journal for the History of Astronomy* 29 (1998), pp. 187–99.

Mancha, José Luis, 'The Latin Translation of Levi ben Gerson's Astronomy', in Gad Freudenthal (ed.), *Studies on Gersonides. A Fourteenth-century Philosopher-scientist*, Leiden: Brill, 1992, pp. 21–46.

Mercier, Raymond, 'Accession and Recession. Reconstruction of the Parameters', in Josep Casulleras and Julio Samsó (eds), *From Baghdad to Barcelona. Studies in the Islamic Exact Sciences in Honour of Prof. Juan Vernet*, Barcelona: Instituto Millás Vallicrosa de Historia de la Ciencia Árabe, 1996, Vol. I, pp. 299–347.

Mercier, Raymond, 'Astronomical Tables of Abraham Bar Ḥiyya', in Sacha Stern and Charles Burnett (eds), *Time, Astronomy, and Calendars in the Jewish Tradition*, Leiden/Boston: Brill, 2014, pp. 155–207.

Millás Vallicrosa, José María, *Estudios sobre Azarquiel*, Madrid/Granada: Consejo Superior de Investigaciones Científicas, Instituto Miguel Asin, Escuelas de Estudios árabes de Madrid y Granada, 1943–1950.

Millás Vallicrosa, José María, *Estudios sobre la historia de la ciencia española*. Barcelona: Consejo Superior de Investigaciones Científicas, 1949.

Millás Vallicrosa, José María, *La obra Séfer Ḥešbón mahlekot ha-kokabim (Libro del cálculo de los movimientos de los astros) de R. Abraham Bar Ḥiyya ha-Bargelonī*, Barcelona: Consejo Superior de Investigaciones Científicas, 1959.

Millás Vallicrosa, José María, *Las tablas astronómicas del rey Don Pedro el Ceremonioso. Edición crítica de los textos hebraico, catalán y latino, con estudio y notas*, Madrid/Barcelona: Consejo Superior de Investigaciones Científicas, 1962.

Nallino, Carlo Alfonso, *Al-Battānī sive Albatenii Opus astronomicum. Ad fidem codicis Escurialensis arabice editum. Latine versum, adnotationibus instructum*, 3 vols, Milan: Ulrico Hoepli, 1899–1907.

Neubauer, Adolf, *Catalogue of the Hebrew Manuscripts in the Bodleian Library*, Oxford: Clarendon Press, 1886.

Neugebauer, Otto, 'Thâbit ben Qurra "On the Solar Year" and "On the Motion of the Eighth Sphere"', *Proceedings of the American Philosophical Society* 106 (1962), pp. 264–99.

Neugebauer, Otto, *The Astronomical Tables of al-Khwārizmī*, Copenhagen: Det Kongelige Danske Videnskabernes Selskab/Munksgaard, 1962.

Pedersen, Fritz S., *The Toledan Tables. A Review of the Manuscripts and the Textual Versions with an Edition*, 4 vols, Copenhagen: Reitzels, for Det Kongelige Danske Videnskabernes Selskab, 2002.

Ragep, F. Jamil, 'Al-Battānī, Cosmology and the History of Trepidation in Islam', in Josep Casulleras and Julio Samsó (eds), *From Baghdad to Barcelona. Studies in the Islamic Exact Sciences in Honour of Prof. Juan Vernet*, Barcelona: Instituto Millás Vallicrosa de Historia de la Ciencia Árabe, 1996, Vol. I, pp. 267–98.

Ratdolt, Erhard, *Tabule astronomice illustrissimi Alfontij regis castelle*, Venice, 1483.

Reif, Stefan C., *Hebrew Manuscripts at Cambridge University Library. A Description and Introduction*, Cambridge: Cambridge University Press, 1997.

Renan, Ernst, and Adolf Neubauer. *Les rabbins français du commencement du quatorzième siècle*, Paris: Imprimerie nationale, 1877.

Richler, Benjamin, and Malachi Beit-Arié, *Hebrew Manuscripts in the Biblioteca Palatina in Parma. Catalogue*, Jerusalem: Jewish National and University Library, 2001.

Steinschneider, Moritz, 'Prophatii Judaei Montepessulani Massiliensis (a. 1300). Prœmium in almanach', *Bullettino di bibliografia e di storia delle scienze matematiche e fisiche* 9 (1876), pp. 595–614.

Steinschneider, Moritz, *Die hebraeischen Übersetzungen des Mittelalters und die Juden als Dolmetscher*, Berlin: Kommisionsverlag des Bibliographischen Bureaus, 1893.

Toomer, Gerald J., 'A Survey of the Toledan Tables', *Osiris* 15 (1968), pp. 5–174.

Toomer, Gerald J., 'Prophatius Judaeus and the Toledan Tables', *Isis* 64 (1973), pp. 351–55.

Toomer, Gerald J., *Ptolemy's Almagest*, London: Duckworth / New York: Springer, 1984.

Copying and Computing Tables in Late Medieval Monasteries*

Seb Falk

Richard of Wallingford, abbot of St Albans (1327–36), was perhaps the greatest astronomer of medieval England.[1] His legacy encapsulates a problem facing historians of medieval astronomy: how can we analyse technical and mathematical practices in their proper contexts? Whilst Richard's most notable works were composed at the University of Oxford, he continued to study astronomy after moving to St Albans, and devised a complex astronomical clock for his abbey church. His contemporaries and successors seem to have been as proud of his astronomic achievements as of his devotional writings or work restoring the abbey's lands; they worked hard to cement the reputation for learning he brought to the abbey.[2] Richard's most notable work was his treatise on the 'Albion' instrument he had invented, an astronomical compendium of great complexity and ingenuity. At least three of the surviving manuscripts of the *Tractatus Albionis* (1326–27) were produced at St Albans. In addition to the usual spiritual benefits arising from the monastic labour of reading, copying and correcting, the monks who produced these manuscripts were showing respect for their predecessor and demonstrating their own humility, qualities that were both central to the Rule of St Benedict.[3]

It is clear that the motivations for producing and studying astronomical works, the techniques required to compute and use them, and the networks of

* I am grateful to Matthieu Husson, Clemency Montelle, Glen Van Brummelen and Benno van Dalen for their advice on earlier drafts of this article. The research underlying it was funded by the Arts and Humanities Research Council, and supervised by Liba Taub. My arguments were refined at a workshop hosted by the TAMAS research project in January 2017, and I would like to thank the participants in that workshop for their constructive feedback.

[1] This was the judgement of Price, 'Review of J. D. North', p. 219.

[2] As narrated by Thomas Walsingham (*c.* 1390); see Riley, *Gesta abbatum monasterii Sancti Albani*, vol. II, pp. 182, 201, 207. See also Falk, 'I Found This Written', pp. 133–34.

[3] Oxford, Bodleian Library, MS Laud Misc. 657, fols 2r–45r; Oxford, Bodleian Library, MS Ashmole 1796, fols 118r–159v; Oxford, Corpus Christi College, MS 144 fols 44r–78v. Corpus Christi MS 144 is usually identified as a Tynemouth manuscript (see Thomson, *A Descriptive Catalogue*, pp. 72–73), but its tables for latitude 51;50°, the high quality of its parchment, and the fact that its chart of saints' days (fol. 59v) includes St Alban but not Tynemouth's patron St Oswine, all point to a southern production. *The Rule of St Benedict* is edited in Fry, *RB 1980*, chs 5, 7, pp. 29–38.

Editing and Analysing Numerical Tables: Towards a Digital Information System for the History of Astral Sciences, ed. by Matthieu Husson, Clemency Montelle and Benno van Dalen, PALS 2 (Turnhout, 2021), pp. 79–105
© BREPOLS PUBLISHERS 10.1484/M.PALS-EB.5.127696

scientific communication that made astronomy possible, varied between different contexts. Yet it is equally clear that such contexts are hard to define: texts circulated widely, and astronomers too could move between different settings. However, tables can help historians study these contexts. Where we can identify the algorithms and sources behind them, they can reveal not only astronomers' techniques, but also their wider practices and purposes.

Monks certainly had distinctive reasons to study the science: it was essential for the regulation of the ecclesiastical calendar; it supported monasteries' function as local centres of (astrological) medicine; and it was part of a long-established monastic culture of learning that monks worked hard to perpetuate.[4] Nevertheless, it is debatable how far these priorities resulted in distinctive ways of practising astronomy.[5] Put bluntly: how much did monks really *do* astronomy? In order to answer such questions, a case study approach, paying close attention to the individual contexts of production and transmission of its individual sources, may be helpful. Tables, whose astronomical content often allows them to be dated or geographically located with greater precision than other written sources, can be a valuable source for such case studies. Understanding whether – and how – a particular set of tables was copied or calculated can add some depth to analyses of monastic activity.

A manuscript well suited to this kind of case study is Oxford, Bodleian Library, MS Laud Misc. 657. Written almost entirely in a single hand around 1380, it collates two versions of the *Albion* treatise, critically copying and adding text and tables. It begins as follows:

> It should be known that Master Richard, abbot of the monastery of St Albans, first composed this book; and through it he devised and made that marvellous instrument which is called 'Albion'. But later a certain Simon Tunsted, professor of sacred theology, changed certain things not only in the book but also in the instrument, as will be clear to scholars in this book. Also, he added certain things.
>
> Master John of Westwyke gave this book to [the priory of] God and the blessed Mary and St Oswin, king and martyr, at Tynemouth; and to the monks serving the same God there. May the soul of the said John and the souls of all the faithful, through the mercy of God, rest in peace. Amen.[6]

[4] The links between astrology and medicine are well established. Evidence that this applied in a monastic context as much as elsewhere is provided by the existence of medico-astrological books in monastic libraries. See, for example, Cambridge, University Library, MS Gg.6.3; Oxford, Bodleian Library, MS Rawlinson D.238.

[5] The contexts of astronomy were particularly blurred because monks were encouraged to enhance their learning by studying at universities. See Pantin, 'The General and Provincial Chapters', pp. 209–10; Pope Benedict XII, 'Summi magistri' (1336), edited in Wilkins, *Concilia Magnae Britanniae*, vol. II, pp. 588–613, here p. 594.

[6] 'Sciendum est quod Dominus Ricardus Abbas monasterii sancti Albani primo composuit istum librum; Et per eum excogitavit & fecit instrumentum illud mirificum quod dicitur Albeon. Sed postea quidam Symon tounstede sacre theologie professor quedam mutavit tam in

This John was most likely from the manor of Westwick two miles west of St Albans, and probably moved to St Albans's dependent priory of Tynemouth around 1380.[7] The quality of parchment makes it likely that MS 657 was produced at the wealthy mother house, but the manuscript was always intended for its daughter, an outpost far to the north.[8] The clearest evidence of this is a table Westwyk added to the treatise, giving the oblique ascensions for 55°, the latitude of Tynemouth. Analysis of this and related tables can allow us to uncover the methods Westwyk used to produce MS 657; such analysis can supplement study of other features of the manuscript, in order to build a clearer picture of the ways – and perhaps the reasons – monks made astronomical books.

Westwyk evidently modelled his table for 55° on a table of oblique ascensions for 51;50° (the latitude of Oxford), which was already present in the treatise. This essay begins by evaluating Westwyk's copying of that earlier table and the *Tractatus Albionis* as a whole, as a way of exploring his scholarly competence and purposes. Following an explanation of the function of oblique ascensions, I then analyse the new table of oblique ascensions Westwyk added to Oxford, Bodleian Library, MS Laud Misc. 657, separating his processes of computation from copying. The *Albion* text which Westwyk copied hints at how this was done, alluding to Ptolemaic techniques, but this cannot be relied on as an account of his practices. However, we can – within certain limits – reconstruct his practices, and the rest of this essay attempts that reconstruction. The analysis is supported by some statistical tables, as well as by editions of Westwyk's tables of oblique ascensions for 51;50° and 55°. Neither has previously been edited. The *Albion* table for 51;50° has previously been published in a complete edition of the writings of Richard of Wallingford (which drew on Westwyk's manuscript among others).[9] However, its editor, John North, gave a corrected version of the table: an internally consistent table that reproduced what North judged as Richard's intention, with errors removed.[10] Since my focus is the contextualised practices of historical actors, my tables are correct

libro quam in instrumento, sicut patet studentibus in libro isto. Quedam eciam superaddidit. / Hunc librum dedit Dompnus Iohannes de Westwyke deo & beate marie & sancto Oswyno regi et martiri de tynemuth. Et monachis ibidem deo servientibus. Anima dicti Johannis & omnium fidelium anime per dei misericordiam requiescant in pace. Amen', MS Laud Misc. 657, fol. 1v.

[7] Rand, 'The Authorship of *The Equatorie*', p. 21.

[8] Falk, 'I Found This Written', pp. 134–36.

[9] North, *Richard of Wallingford*, vol. III, pp. 96–97.

[10] North, *Richard of Wallingford*, vol. II, pp. 238–39, 247–48. North changed thirteen values where the table was not internally symmetrical (that is, where $\rho(\lambda) \neq 360 - \rho(360 - \lambda)$). He substituted values from the table of oblique ascensions attributed to John Maudith in MS Laud Misc. 674, fols 72r–v.

in a philological sense: reproducing what appears in the manuscripts, including any errors and noting differences between six different manuscripts. It is hoped that the inclusion of blemishes and vestiges of production improves our understanding of such tables, and the contexts in which they were made.

John Westwyk, copying and compilation

In his prefatory remarks (quoted above), John Westwyk highlighted the work he had done to collate and compare two versions of the *Albion*: one apparently as written by Richard of Wallingford; the other adapted by Simon Tunsted. This was an explicit act of *compilatio*, not unusual in the later Middle Ages.[11] Throughout his copy of the treatise and accompanying tables Westwyk notes differences between the versions of 'the lord Abbot' and 'master Simon', and also compares them with an albion instrument ('instrumento nostro') which must have been available to him at the monastery.[12] The areas where he adds to the text reveal something of his interests. Chief among these were the practical aspects of instruments. He notes discrepancies between the differing instrument dimensions given in his source texts, and the dimensions of his own instrument; furthermore, in his most extended original contribution to the treatise, he adds commentary comparing the features of the albion with the saphea of al-Zarqālī (Arzachel) and the astrolabe (Wallingford had himself stated that his invention provided the functions of those and other instruments, but without giving details).[13] By contrast, Westwyk gave less attention to the diagrams that accompany part II of the treatise, illustrating the construction of the instrument. His diagrams are superficially acceptable, appearing in the same places as, and looking fairly similar to, those in other copies of the treatise; but closer inspection reveals that they do not accurately represent the instrument markings carefully described in the text.[14] Given the abilities Westwyk showed elsewhere in his compilation, the errors in his diagrams are unlikely to have been caused by imperfect understanding; rather, he may have realised that the diagrams were simply illustrations for processes that were sufficiently explained in Richard of Wallingford's text, and therefore chose to focus his efforts on making the compilation most useful to its readers.

[11] Hathaway, 'Compilatio', pp. 19–44.

[12] 'dominus abbas ... magister Symon'; MS Laud Misc. 657, fols 45r, 22v. For detailed analysis of Westwyk's comparisons, see Falk, 'I Found This Written', pp. 133–40.

[13] MS Laud Misc. 657, fols 10v, 43r–44r; *Tractatus Albionis* III, in North, *Richard of Wallingford*, vol. I, p. 340.

[14] The best copy of the diagrams is in Corpus Christi MS 144. See the comparison in Falk, 'I Found This Written', pp. 137–39. It is just possible that the diagrams were drawn by another person, but they were certainly labelled by Westwyk so it seems more likely he drew them too.

32r	'True motus of the sphere of Saturn' (IV.1) [true centre: the arc at earth between a planet's *aux* and epicycle centre][15]
32v	'True motus of the sphere of Jupiter' (IV.2)
33r	'True motus of the sphere of Mars' (IV.3)
33v	'True motus of the sphere of the Sun and Venus' (IV.4)
34r	'True motus of the sphere of Mercury' (IV.5)
34v	'True motus of the sphere of the Moon' (IV.6)
35r–35v	'True motus of the Moon and of the equation of the argument for the hour of conjunction' (IV.7) [true argument: the arc at the epicycle centre between the Moon and the true epicyclic apogee]
36r	'Table of the equation of iomyn, that is, of the natural day' (IV.8) [normed equation of days + longitude]
36v	Latitude of the Moon (IV.9) [as a function of longitude measured from the node] Table of longitude with its twelfth part; table of twelve conjunctions (IV.10) [twelve equal steps of 1^{s}2;30° and 11^{s}19;17°]
37r	Motion of the Moon in one hour at aux, mean distance, and opposite aux (IV.11)
37v	Table of fixed stars (IV.12)
38r–38v	Mean motus of Mercury (IV.13)
39r–39v	Mean motus of the Moon (IV.14)
40r–40v	Mean argument of the Moon (IV.15)
41r	Right ascensions (starting at Capricorn) (IV.16)
41v	Right ascensions (starting at vernal equinox) (IV.16)
42r	Oblique ascensions at latitude 51;50° (Oxford) (IV.17)
42v	Oblique ascensions at latitude 55° (Tynemouth)
44v–45r	Lunar elongations (to be inserted after table of mean motus of the Moon (39v))

Table 1: Tables in *Tractatus Albionis*, in Oxford, Bodleian Library, MS Laud Misc. 657.

This meant accurately copying, and where necessary updating, the tables in part IV of the treatise (see Table 1). Westwyk's copies of the standard sequence of tables of the *Tractatus Albionis* are exemplary.[16] This is demonstrated by examination of copies of four tables – IV.12, IV.16 (two versions) and IV.17

[15] These tables, explicitly drafted to aid in the construction of the Albion instrument, are standard in copies of the *Tractatus Albionis* (apart from the one for Tynemouth); their functions are fully explained in North, *Richard of Wallingford*, vol. II, pp. 237–48.

[16] The standard set is edited in North, *Richard of Wallingford*, vol. III, pp. 76–107. The chapter numbers given in brackets are used in some manuscripts (though not MS Laud Misc. 657) and in North's edition.

– in five manuscripts of the Albion, including the three St Albans copies.[17] The fact that tables in these five manuscripts were copied as part of the complete treatise can be established by some consistently occurring errors in the table of oblique ascensions for 51;50°, which can be observed by comparison with other contemporary tables of the same function.[18] Although it is hard to know whether unique errors in any one manuscript are the fault of the scribe of that manuscript, or arise from faithful copying of a faulty source, it is clear from the critical edition in Appendix 1a (see pp. 95–98) that the St Albans manuscripts exhibit greater consistency than the others, and have none of the obvious mistakes (such as in the degrees column) found elsewhere.[19] Two manuscripts are identical throughout table IV.17: Oxford, Corpus Christi College, MS 144 and Oxford, Bodleian Library, MS Laud Misc. 657. Thus John Westwyk may have copied MS Laud Misc. 657 in part from the earlier Corpus Christi MS 144 (while also using the copy ascribed to Simon Tunsted). The copy of the *Albion* in Corpus Christi MS 144 is 'by far the best', according to John North; like Laud Misc. 657 it was later to be taken to Tynemouth.[20] Westwyk's copy contains a diagram, in which the limb of the first face of the second disc is divided into 18 days, which is only found in MS 144, and Westwyk's error in naming a star 'Altayn' in table IV.12 might be traced to the slightly unclear way that the scribe of MS 144 formed the final letters of 'Altayir' (see Plates 4a and 4b).[21] All three St Albans manuscripts contain three identical discrepancies between the two versions of the table of right ascensions (IV.16), which do not appear in the other *Albion* manuscripts; that is, the values for right ascension for longitudes $11^s22°$, $8^s12°$ and $8^s25°$ in the table starting at Capricorn are the same across the three manuscripts, as are the values for those longitudes in the table starting at the vernal equinox, but within each manuscript the two tables disagree. It is clear, then, that although errors exist in the manuscripts, the overall standard of copying was high. Across the four tables examined, MS 657 differs from MS 144 in only two values, one in each table of right ascensions. One of Westwyk's values is unique in copies of this table, so the error was most likely introduced by Westwyk himself. However, the other difference appears at a place where the two tables within MS 144 do not match; Westwyk's change makes his two tables match at that point. This may be coincidental, but perhaps he had noticed the discrepancy and con-

[17] MS Laud Misc. 657, Corpus Christi MS 144 and MS Ashmole 1796, plus London, British Library, Harley MSS 80 and 625.

[18] Corpus Christi MS 144, fol. 122r, and MS Laud Misc. 674, fol. 72r, were used for this purpose.

[19] Such high standards are emphasised by the fact that North's edition of table IV.17 (North, *Richard of Wallingford*, vol. III, pp. 96–97) contains five errors.

[20] North, *Richard of Wallingford*, vol. II, p. 127; Thomson, *A Descriptive Catalogue*, p. 73.

[21] MS Laud Misc. 657, fol. 37v; Corpus Christi College, MS 144, fol. 76v.

sciously corrected it. Either way, it is safe to conclude that he was a particularly accurate copyist.[22]

Computation

If Westwyk's copying was impressively accurate, his computation was equally so. We see this in the table of oblique ascensions for latitude 55°, a table which is not found in contemporary manuscripts and which Westwyk surely computed for this manuscript. The heading of the table, which Westwyk adapted from Wallingford's heading for the table of oblique ascensions for latitude 51;50°, reads:

> Table of ascensions of signs on the oblique circle at latitude 55°. It was calculated and composed as explained in the canons in the second book of the Almagest; and with it the second circle on the second limb of the second face of the instrument should be divided, as is explained in chapter 18 of the second part of this [treatise]. // Tynemouth.[23]

The table gives oblique ascensions as a function of ecliptic longitude. Section II.18 of Richard of Wallingford's treatise explains that the limb (rim) of the albion is to be engraved with three scales: the ecliptic, right ascensions and, in the innermost circle, the oblique ascensions, calculated for the latitude of a place 'where we intend to stay for a long time and make many observations'.[24] This innermost scale was designed to allow the user to easily find the ascendant degree and divide the astrological houses. It was to be marked with degrees of ecliptic longitude ('gradus zodiaci', the argument of the table) corresponding to degrees of oblique ascension read on the 'ecliptic' scale. Thus on

[22] We should perhaps note a contrast with the other table Westwyk added to his copy of the *Tractatus Albionis*: a table of lunar elongations (MS Laud Misc. 657, fols 44v–45r) that does not appear in other copies of the treatise. The presence in the table of obvious copying errors indicates that Westwyk did not compute it himself, but transcribed it from another source. The large number of these errors (20 in a table with 366 values in signs, degrees and minutes) could be deemed a stain on Westwyk's otherwise impressive copying record, but it is quite possible that he made an accurate copy of a corrupt exemplar. The twenty errors do include some that are more likely to be computational, such as 20° instead of 19°; the nature of these, and the fact that the table does not follow a consistent arithmetical progression, suggest that it was computed by subtracting values for solar mean motion from an existing table of lunar mean motions. But that need not have been the table in the *Albion*; tables of mean motions and lunar elongations were sufficiently common to make it most likely that Westwyk copied the table, and did so from a source which itself probably had nothing to do with the *Albion*.

[23] 'Tabula ascencionum signorum in circulo obliquo in latitudine .55. gra. calculata est et composita sicut docent canones in secundo libro Almagesti; et debet per eam dividi circulus secundus in limbo secundo secunde faciei instrumenti sicut docetur capitulo 18° secunde partis huius'. 'tynemuth' is added as a gloss beneath the heading. MS Laud Misc. 657, fol. 42v.

[24] *Tractatus Albionis* II.18, in North, *Richard of Wallingford*, vol. I, p. 325.

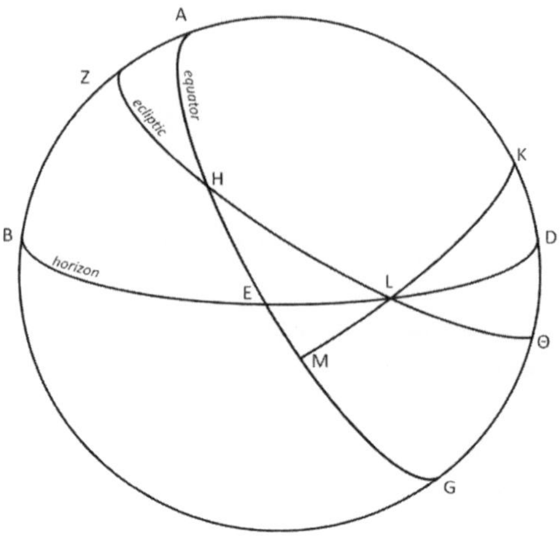

Figure 1: Diagram of ascensions on oblique circle, adapted from *Almagest* II.7. H is the vernal equinox and K the north pole of the equator. The table of oblique ascensions gives the arc of the equator (HE, ρ) rising at the same time as a given arc of the ecliptic (HL, λ). HE is computed by subtraction of the ascensional difference (EM, γ) from the right ascension (HM, α). ∠LEM is 90°–ϕ, where ϕ is the latitude of the observer.

the instrument longitude was 'tabulated' as a function of oblique ascension: the reverse of what we find in the tables.[25]

The table and scale of oblique ascension track the rising and setting of points on the celestial equator and ecliptic. Celestial longitude and latitude were defined as positions along and above or below the ecliptic, and these coordinates were the most common way of measuring the positions of stars, including the Sun, Moon and planets; the ecliptic is the path of the Sun (eclipses always take place at 0° latitude), and of course the best-known constellations were those of the zodiac, the band around the ecliptic. However, the rising and setting of signs on the ecliptic does not occur in equal times; rather, it is the equator, set at an angle to the ecliptic, on which points rise and set in equal times (see Figure 1). So the facility to read oblique ascensions – to convert between the rising of the ecliptic and equator – allowed the albion's user to find the ascendant degree at any given time, as Wallingford explains in Section III.18 of the *Albion* treatise.

The table heading cites *Almagest* Book II, where Ptolemy provides tables of rising times (equivalent to oblique ascensions) for a range of latitudes, as defined by the length of the longest day, from 12 hours (0°) to 17 hours (54;1°). Ptolemy explains how these may be computed from the right ascensions, by

[25] Owing to complications in the multiple uses of the ecliptic scale, Richard of Wallingford also instructed that the ascensions scales be graduated in the opposite direction to the ecliptic scale. See North, *Richard of Wallingford*, vol. II, pp. 177–78, 226–32.

calculating the ascensional difference (EM in the right spherical triangle ELM, in which the angle at E is 90°−ϕ, where ϕ is the latitude of the observer). In Figure 1, the oblique ascension (ρ) is the arc (HE) of the equator which rises in the same time as a given arc (HL) of the ecliptic; to find this, the ascensional difference (EM or γ) can be subtracted from the right ascension (HM or α). The right ascensions (rising times at *sphaera recta*) are explained in *Almagest* I.16, where Ptolemy notes that *sphaera recta* is a special case in which the horizon passes through the poles of the equator; an observer at the equator sees the stars ascending at right angles to the horizon. Ptolemy applied the spherical theorem he had proved in *Almagest* I.13 (known as Menelaus' Theorem, though Ptolemy only mentions Menelaus in the context of observations) to find the right ascension.[26] His method used the table of chords which he had provided in I.11, and is mathematically equivalent to the modern formula:

$$\sin \alpha = \tan \delta \cdot \cot \varepsilon, \tag{1}$$

where δ is the declination (arc LM in Figure 1) and ε is the obliquity of the ecliptic.[27] In *Almagest* I.12 he showed how the obliquity can be found by observation, and stated its extent as one half of approximately 11:83 of a complete circle. This matches the maximal value he gives in his table of declinations (I.15), which tabulates the length of the arc of a meridian between the equator and the ecliptic for longitudes (λ) from 0° to 90°. At λ = 0° the declination is 0, since that is the vernal equinox, where the equator and ecliptic intersect; its maximal value at λ = 90° is equal to the obliquity of the ecliptic, which Ptolemy specifies as 23;51,20°. The formula underlying the table of declinations is equivalent to:

$$\sin \delta = \sin \lambda \cdot \sin \varepsilon, \tag{2}$$

Thus obliquity is used twice in the process of computing right ascension, and so two different values of ε could, in principle, be used, though this would upset the clear symmetry of the right ascensions and prevent the right ascension at λ = 90° being equal to 90°, which would be a starkly unacceptable result.

The size of the obliquity also underlies the ascensional difference (γ), for which Ptolemy outlines a method equivalent to the modern formula:

$$\sin \gamma = \tan \delta \cdot \tan \phi. \tag{3}$$

The oblique ascension HE can then be found by a simple subtraction.[28]

[26] Toomer, *Ptolemy's Almagest*, ch. VII.3, pp. 336, 338.

[27] Pedersen, *A Survey of the Almagest*, pp. 96–97.

[28] Pedersen, *A Survey of the Almagest*, pp. 110–13. It is possible to compute the oblique ascension directly, by a single formula in which λ, ϕ and ε are the only variables, but such

Analysis of tables of oblique ascensions for 51;50° and 55°

I analysed the oblique ascensions tables for 51;50° and 55° in MS Laud Misc. 657. It was established above that John Westwyk copied the former, perhaps from Corpus Christi MS 144, but that the latter involved at least some computation on his part. This analysis allows us to understand the extent of this computation, and some of the methods used. John Westwyk's heading for his table repeats Richard of Wallingford's statement that it has been computed with reference to *Almagest* Book II. This cannot be taken completely at face value, but we must start from the assumption that a method like Ptolemy's, which computes the oblique ascension by subtracting the ascensional difference from the right ascension, was used.

The first step in our analysis of the table is to check both its overall symmetry and that of its presumed underlying functions of right ascension (α) and ascensional difference (γ). These two functions have different symmetries: the former is symmetrical such that $\alpha(180 - \lambda) = 180 - \alpha(\lambda)$, while the latter is symmetrical such that $\gamma(180 - \lambda) = \gamma(\lambda)$. Both α and γ can therefore be extracted from oblique ascension ρ by the following formulae:

$$\alpha(\lambda) = 90 + \tfrac{1}{2}(\rho(\lambda) - \rho(180-\lambda)) \quad (4)$$

$$\gamma(\lambda) = 90 - \tfrac{1}{2}(\rho(\lambda) + \rho(180-\lambda)) \quad (5)$$

It follows from the above that the entire oblique ascension function is symmetrical such that $\rho(\lambda) = 360 - \rho(360 - \lambda)$.[29] Therefore we begin our analysis of Westwyk's tables by using this overall symmetry to check for large errors in the tables. Once any such errors have been isolated, we can then proceed to separate the functions of right ascension and ascensional difference in order to identify the underlying parameters.

This first check – whether pairs of values on either side of 180° add up to 360° – revealed 13 asymmetries (occasions where $\rho(\lambda) + \rho(360 - \lambda) \neq 360$) in Westwyk's 51;50° table, and 18 in his 55° table. All those in the table for 51;50° occurred in places where the *Albion* copies of this table are consistent but do not match the one in in the non-*Albion* MS Laud Misc. 674 (see Appendix 1a); the latter manuscript's values in those places are symmetrical (though it has mistakes of its own elsewhere).[30] This suggests that these asymmetries arose in an early copy of the *Albion* tables. Of the 18 asymmetries in Westwyk's 55° table, 14 were exactly 1′ in size, which suggests they may have arisen in

a formula does not exist in any medieval source. (It would also preclude the combination of more than one value for the obliquity, which does occur in medieval tables).

[29] van Dalen, *Ancient and Mediaeval Astronomical Tables*, pp. 67, 185.

[30] At λ = 0s19°, 2s8°, 2s29°, 4s18°, 6s2°, 6s18°, 7s3°, 7s6°, 7s29°, 8s0°, 9s12°, 10s29° and 11s28°.

the process of calculation, and may be analysed further.[31] The remaining four discrepancies (noted in Appendix 1b on pp. 99–101) were 1°, 30′, 1° and 10′. The large size and roundness of these suggests they came about through the misplacement of digits (whether in calculation or copying cannot be known), and so they may be corrected – made symmetrical – in order to produce an idealised (mathematically consistent) table for use in further analyses. Finally for this stage, a similar check for symmetry can be made in the table of right ascensions, where the values for $\alpha(\lambda)$ should be symmetrical with three other values: those for $180 - \lambda$, $180 + \lambda$ and $360 - \lambda$. Comparison of these sets of four values revealed seven asymmetries, of which five occurred in the same rows (same values of λ) as asymmetries in Westwyk's 55° table.[32] Since in all seven cases of asymmetry three of the four values remained in agreement, we may correct all seven to obtain a mathematically consistent right ascension table. The results of these checks are summarised in the table in Appendix 2 on pp. 102–05 (columns B to D); in most of the columns of this table results are only included where they are significant (i.e. where there are asymmetries).

Next we may derive values for right ascensions from the tables of oblique ascensions using formula 4, and compare these with both the manuscript tables of right ascensions and values computed using formula 1. First, the oblique ascensions table for 51;50° yields values that match the table of right ascensions exactly, except in places where asymmetries have already been noted in this oblique ascensions table, and at $\lambda = 15°$, where an asymmetry was noted in the right ascensions table but none in the oblique ascensions. In all of these cases the correction of the asymmetries as indicated above also removes the mismatch, thus further supporting the plausibility of the corrections.[33] The table for 55° yields values that also match the table of right ascensions, except in seven places where there are discrepancies of 30″.[34] Such discrepancies could, in principle, arise from the use of a table showing more sexagesimal places – a precision to seconds – to calculate the ascensional differences and/or the right ascensions; but the consistent use of such a table would have led to 30″ discrepancies in around half the extracted values – rather more than the seven found. In fact, as Appendix 2 shows, some of these 30″ discrepancies (namely those for arguments 15 and 85) arise from the asymmetries of 1′ in the oblique ascensions; others remain unaccounted for (they correspond to 30″ discrepan-

[31] At λ or $360 - \lambda$ = 15°, 20°, 34°, 35°, 50°, 56°, 85°, 87°, 92°, 136°, 137°, 165°, 168° and 169°. Some of these are apparent confusions of 1/2 and 2/3, which could arise from scribal misreading, but others, such as 3/4 and 7/8, are highly unlikely to arise from misreading.

[32] Asymmetries in the right ascensions table were found at λ (or equivalents) = 15°, 34°, 35°, 71°, 85°, 86° and 87°.

[33] Discrepancies at λ = 0s15°, 0s19°, 2s8°, 2s29° and 4s18° (the procedure only uses values from 0 to 180°).

[34] At λ = 0s11°, 0s15°, 1s13°, 1s14°, 1s20°, 2s25° and 2s28°.

cies in the derived right ascension values in column E). Comparing the corrected mathematically consistent values with values computed using formulas 1 and 2, using a least-squares fit (the sum of the squares of all differences, so that a smaller number indicates a closer match), the following results were obtained (Table 2; see also Appendix 2, columns D and E).[35] The numbers *n* in parentheses are the number of individual discrepancies.[36]

ε	Σ (51;50°)	Σ (55°)	Notes
23;33,30°	19 (n=19)	20 (n=20)	Value of ε used in Toledan Tables
23;34,45	8 (n=8)	9 (n=9)	Value of ε producing lowest Σ (but not attested in any medieval source)
23;35°	9 (n=9)	10 (n=10)	Value of ε attributed to al-Battānī.
23;51°	589 (n=80)	586 (n=80)	Value of ε used in Ptolemy's *Handy Tables*
23;51,20	611 (n=80)	608 (n=80)	Value of ε used in Ptolemy's *Almagest*

Table 2: Least squares fit for obliquities underlying right ascensions tables used to compute oblique ascensions.

These results suggest that the tables of oblique ascensions were based on a right ascensions table with obliquity 23;35° (the same conclusion applies irrespective of whether we use a least-squares fit or the simple number of discrepancies). It may be noted that an (insignificantly) closer match with the manuscript tables occurs with an obliquity of 24;34,45°, but this is not sufficient grounds to claim that that was the parameter used by medieval astronomers. We must maintain cautious of the spurious precision offered by the spreadsheet techniques used here. These techniques allow us to compare and choose from a limited range of discrete values attested in surviving manuscripts. The match will inevitably be imperfect, owing to the vagaries of calculation techniques and the imperfections of medieval reference tables. However, even a somewhat anachronistic technique, if used consistently, allows the degree of closeness to be measured so that different values for the obliquity can be compared. The use of squared residuals (Σ) is a fairly crude method of statistical analysis, but it is precise enough to allow us to compare a selection of historically attested parameters. Whatever the results obtained by such techniques of recomputation and statistical analysis, they can only ever be an adjunct to the examination of tangible manuscript evidence.[37]

[35] Values for Σ were rounded to integers. It is, as noted above, theoretically possible to use two different values of ε in formulas 1 and 2, but even small differences yield strikingly discrepant results.

[36] In principle this is to a maximum of 90 comparisons, but in practice four comparisons of the 51;50° table were excluded owing to flaws in the manuscripts.

[37] On the use of these and more complex statistical techniques in history of astronomy, see van Dalen, 'A Statistical Method'; Van Brummelen and Butler, 'Determining the Interdependence'.

Having analysed the right ascensions, we may proceed to derive values for the ascensional difference from the corrected oblique ascensions tables, using formula 5. These may be compared with values computed using formula 3, with the following results (Table 3). As above, the numbers in parentheses are the number of individual discrepancies.

ε	Σ (51;50°)	Σ (55°)	Notes
23;32,30	105 (n=51)	56291 (n=90)	Value of ε used in the Maghribian *zīj* of Ibn Isḥāq (*c.* 1300, value not attested in Latin sources)
23;33,0°	37 (*n*=37)	53507 (*n*=90)	Value of ε used in the *Mumtaḥan Zīj* (Baghdad, *c.* 830), which was also known in Muslim Spain (value not attested in Latin sources)
23;33,15°	23 (*n*=23)	52062 (*n*=90)	Value of ε producing lowest Σ for 51;50° table (but not attested in any medieval source)
23;33,30°	28 (*n*=28)	50594 (*n*=90)	Value of ε used in Toledan Tables
23;35°	367 (*n*=83)	42035 (*n*=89)	Value of ε attributed to al-Battānī.
23;50,44°	36850 (*n*=90)	17 (*n*=14)	Value of ε producing lowest Σ for 55° table (but not attested in any medieval source)
23;51°	38210 (*n*=90)	26 (*n*=23)	Value of ε used in Ptolemy's *Handy Tables*
23;51,20	39651 (*n*=90)	70 (*n*=64)	Value of ε used in Ptolemy's *Almagest*

Table 3: Least squares fit for obliquities underlying ascensional differences used to compute oblique ascensions.

In his 1976 edition of Richard of Wallingford's writings, John North noted that the table of oblique ascensions could incorporate two different obliquities; he remarked that 'there are too many possibilities for it to be profitable to investigate them all', but did test some values using λ = 45° and suggested obliquities of 23;35° and 23;33,30° for the two stages of computing the table for latitude 51;50°.[38] The extraction of the underlying ascensional difference function, greater computing power available nowadays, and the development of statistical estimators for these purposes now allow us to state with some confidence that North's suggestion was correct. Although Richard of Wallingford had stated that his table was computed as explained in Book II of the *Almagest*, he clearly did not use Ptolemaic values for the obliquity of the ecliptic.

John Westwyk, on the other hand, updating Richard of Wallingford's treatise for use at a new latitude, did use a Ptolemaic value: the table above shows

[38] North, *Richard of Wallingford*, vol. II, pp. 247–48.

that his oblique ascensions match those computed with an obliquity of around 23;51°. It may not be possible to be certain that Westwyk used precisely 23;51° rather than 23;51,20°. The formula for the ascensional difference (formula 3) incorporates a value for the tangent of latitude (tan ϕ); this would surely have been rounded for the purposes of computation. Reverse-engineering the function will only yield this rounded value of tan 55°, which would produce a value of the latitude ϕ that is not precisely 55°, thus introducing an element of uncertainty into estimates at a precision of sexagesimal seconds. Nevertheless, we can certainly be confident that the new table was computed using a fresh set of ascensional differences with an obliquity different from that used by Richard of Wallingford.

There are two ways John Westwyk could have computed his full table of oblique ascensions: either (1) he could have subtracted ascensional differences from the existing table of right ascensions right across the table from 0–360°; or (2) exploiting the symmetry of the oblique ascensions function, he could have carried out the subtraction of ascensional difference only for $\lambda = 0$ to 180°, and then completed the second half of the table by subtracting the first-half values from 360°. The small asymmetries of Westwyk's table of oblique ascensions suggest that he used the first method; this is supported by the fact that the differences between the manuscript and a computed version are not symmetrically arranged (unlike the table for 51;50°). This hypothesis can be tested by comparing tables computed by both methods, using the manuscript right ascensions and an idealised ascensional difference; the two tables will be identical from 0 to 180° but slightly different from 180 to 360°.[39] Since the ascensional difference is itself derived from the table of oblique ascensions, this is simply a way of correlating errors in the oblique ascension tables with non-symmetric errors in the right ascensions.

As the parameters used for this comparison are merely standardised values of those within the tables themselves, it is not surprising that there are few discrepancies (see Appendix 2, columns H and J). However, the values computed using the symmetry of the oblique ascension function contained more discrepancies with Westwyk's 55° table than the one computed by subtracting ascensional differences right across the table (10, as opposed to 6). On the other hand, when the same comparison was carried out on the table of oblique ascensions for 51;50°, it was found to be a better match with a table computed using the symmetry of the oblique ascension function (4 discrepancies, as opposed to 10). These numbers are small in every case, so it is hard to be sure, but the

[39] The idealised ascensional difference was the value that can be most consistently derived from the manuscript tables of oblique and right ascensions. Since the ascensional difference function is symmetrical such that $\gamma(180 - \lambda) = \gamma(\lambda)$, tables from 0 to 360° will contain 90 values of γ, each repeated four times. The manuscript values were always consistent in at least 3 of the 4 repetitions, so it was easy to identify the idealised ascensional difference.

evidence suggests that whilst Richard of Wallingford computed the second half of his table of oblique ascensions by subtracting the first 180 values from 360°, John Westwyk did so by subtracting ascensional differences from the entirety of Wallingford's table of right ascensions.

Conclusion

In 1326–27, in the heading to his table of oblique ascensions for the latitude of Oxford, where he was then a scholar, Richard of Wallingford wrote that it had been 'calculated and composed as explained in the canons in the second book of the *Almagest*'. This analysis has shown that statement to be only half-true, since Richard used parameters significantly different from those of Ptolemy. Yet half a century later, when the monk John Westwyk came to adapt Richard of Wallingford's tables for the latitude of Tynemouth, he seems to have taken Wallingford's table heading at face value. It is likely that, for the latitude of 55°, he worked through a process starting from his own accurate copy of the table of right ascensions (computed using an obliquity of 23;35°), adapting them by subtraction right across the table of an ascensional difference computed using an ecliptic obliquity of 23;51°.[40] This was done with only four clear errors. In other words, his claim to be following Ptolemy's method was truer than the source from which he copied that claim. It is hard to be certain to what extent either astronomer's choice of obliquity was a deliberate one, but in another (later) manuscript Westwyk used two different obliquities, so it is possible that he made a conscious choice to follow Ptolemy's method faithfully, in contrast to Wallingford's more flexible approach.[41]

Westwyk was not the only person to adapt the *Albion* tables to a new latitude. In three fifteenth-century copies of the version adapted by John of Gmunden in or around 1430, the Oxford table of oblique ascensions is followed by one for Nuremberg.[42] But we cannot know whether the monks of Tynemouth took advantage of Westwyk's efforts: his table was not annotated, and no instrument survives that draws on the table data in the way the *Albion* treatise instructs. On the other hand, it seems that the astrological subject of house divisions associated with Westwyk's table did excite the monks' interest, as blank pages Westwyk left in the manuscript were filled with tables of houses by a near-contemporary hand.[43] These tables are apparently unique in

[40] In the one place where John Westwyk's table of right ascensions starting at the vernal equinox does not match the table in Corpus Christi MS 144, Westwyk's value (which is correct) matches his table of oblique ascensions.

[41] Cambridge, University Library, Peterhouse MS 75.I, fols 63v (23;35°) and 64 (23;33,30°).

[42] Vienna, Österreichische Nationalbibliothek, MSS 5412, 5415; Munich, Bayerische Staatsbibliothek, Clm. 10662.

[43] Falk, *Improving Instruments*, pp. 34–36.

combining a layout starting at the tenth house (midheaven) with a time column enabling the user to adjust the noon values to other times of day.[44] But the copy is a poor one, and a user with moderate astronomical expertise would surely not have been satisfied with its obvious errors and omissions. The values in these tables accord best with an obliquity of 23;33,30° and latitude of 51;50°, but of course this does not preclude their having been copied at Tynemouth, since tables for the Oxford latitude were widespread.[45]

In a quest to understand the monastic context for astronomy, analysis of tables can only provide a small part of the picture, but it can make an important contribution. The mere existence of the collation of Richard of Wallingford's *Albion* treatise told us that John Westwyk, and the monks who followed him, were interested in instruments and astrology. Study of his tables has confirmed that this monk, about whose education little is known, was not only a painstaking copyist, but a careful and competent calculator, capable of using the existing tables available to him to extend his source materials and make them useful in new locations. In this way he played his part in venerating his predecessor Richard of Wallingford and perpetuating the legacy of monastic astronomy, and broader scholarship, which the abbot had left at St Albans and its network of daughter houses.

Bibliography

Dalen, Benno van, 'A Statistical Method for Recovering Unknown Parameters from Medieval Astronomical Tables', *Centaurus* 32 (1989), pp. 85–145.

Dalen, Benno van, *Ancient and Mediaeval Astronomical Tables. Mathematical Structure and Parameter Values*, PhD dissertation, Utrecht: University of Utrecht, 1993.

Falk, Seb, *Improving Instruments. Equatoria, Astrolabes, and the Practices of Monastic Astronomy in Late Medieval England*, PhD dissertation, Cambridge: University of Cambridge, 2016; https://www.repository.cam.ac.uk/handle/1810/256996.

Falk, Seb, '"I Found This Written in the Other Book": Learning Astronomy in Late Medieval Monasteries', in Morwenna Ludlow, Charlotte Methuen and Andrew Spicer (eds), *Churches and Education*, Cambridge: Cambridge University Press, 2019, pp. 129–44.

Fry, Timothy, *RB 1980: The Rule of St Benedict in Latin and English with Notes*, Collegeville: The Liturgical Press, 1982.

Hathaway, Neil, 'Compilatio. From Plagiarism to Compiling', *Viator* 20 (1989), pp. 19–44.

[44] They bear some resemblance to the tables of John Walter, devised at Oxford in the late 1380s. See North, *Horoscopes and History*, pp. 126–30.

[45] Since the tables give only degrees, not minutes, we cannot be certain about this obliquity, and an obliquity of 23;35° may well have been used at some stage in their production.

North, John D., *Richard of Wallingford. An Edition of his Writings with Introduction, English Translation and Commentary*, 3 vols, Oxford: Clarendon Press, 1976.

North, John D., *Horoscopes and History*, London: Warburg Institute, 1986.

Pantin, William Abel, 'The General and Provincial Chapters of the English Black Monks, 1215–1540', *Transactions of the Royal Historical Society* 10 (1927), pp. 195–263.

Pedersen, Olaf, *A Survey of the Almagest. With Annotation and New Commentary by Alexander Jones*, New York: Springer, 2011.

Price, Derek de Solla, 'Review of J. D. North (ed.), *Richard of Wallingford. An Edition of his Writings*', *Journal for the History of Astronomy* 9 (1978), pp. 219–20.

Rand, Kari Anne, 'The Authorship of *The Equatorie of the Planetis* Revisited', *Studia Neophilologica* 87 (2015), pp. 15–35.

Riley, Henry T., *Gesta abbatum monasterii Sancti Albani a Thoma Walsingham*, 3 vols, Cambridge: Cambridge University Press, 1867–1869. Reprinted Cambridge: Cambridge University Press, 2012.

Thomson, R. M., *A Descriptive Catalogue of the Medieval Manuscripts of Corpus Christi College Oxford*, Cambridge: D. S. Brewer, 2011.

Toomer, Gerald J., *Ptolemy's Almagest*, London: Duckworth / New York: Springer, 1984.

Van Brummelen, Glen, and Kenneth Butler, 'Determining the Interdependence of Historical Astronomical Tables', *Journal of the American Statistical Association* 92 (1997), pp. 41–48.

Wilkins, David, *Concilia Magnae Britanniae et Hiberniae*, 4 vols, London: Gosling, 1737.

Appendix 1a: Edition of the table of oblique ascensions for 51;50°

Manuscripts

A = Oxford, Bodleian Library, MS Ashmole 1796, fol. 159r (s. xiv^{med}): *Tractatus Albionis* (St Albans)
C = Oxford, Corpus Christi College, MS 144, fol. 78v (s. xiv^{med}): *Tractatus Albionis* (St Albans)

$\mathbf{H_1}$ = London, British Library, Harley MS 80, fol. 54r (s. xiv[2]): *Tractatus Albionis*
$\mathbf{H_2}$ = London, British Library, Harley MS 625, fol. 164r (s. xiv[ex]): *Tractatus Albionis* (Merton College, Oxford)
L = Oxford, Bodleian Library, MS Laud Misc. 657, fol. 42r (*c.* 1380): *Tractatus Albionis* (St Albans, written by John Westwyk)
M = Oxford, Bodleian Library, MS Laud Misc. 674, fols 72r–v (s. xv): table attributed to John Maudith (d. *c.* 1343)

All but one of these copies of the table are taken from copies of the *Tractatus Albionis*. The last, **M**, is a near contemporary table of the same function. It was used by John North as a source of variant readings in his edition, and has been collated here for comparison purposes, especially to highlight the unity among the *Albion* manuscripts.

L has been used as the copy-text for this table. I have preserved the layout of the original table as far as possible, but had to split it into two parts (with the argument column repeated) due to space limitations.

The St Albans manuscripts (**A**, **C** and **L**) contain fewer errors, and no large errors (mistakes in the degree column or multiple adjacent cells), in contrast with the other copies ($\mathbf{H_1}$, $\mathbf{H_2}$ and **M**). All the *Albion* manuscripts give degrees up to 30, noting the signs where they change (**M** gives degrees to 360, which should be borne in mind when examining the variant readings below the table). Most signs start with 0°, which allowed scribes to write the number of signs in the column usually used for degrees. These sign numbers are highlighted in various ways: **A**, **C** and $\mathbf{H_1}$ write the number larger and usually in a different colour; **L** (John Westwyk) draws a red box around the number. Where there is an additional 1° at the start of a new sign (4, 7 and 9 signs), **C** and $\mathbf{H_1}$ note this by writing the 1 alongside the sign number; **A**, $\mathbf{H_2}$ and **L** do not.

Gloss

[Hec tabula que intitulatur] tabula ascensionum signorum in circulo obliquo, ubi videlicet est elevatio poli 51 gra. et 50 mi. cuiusmodi est latitudo civitatis Oxonie, calculata est et composita sicut docent canones in 2° libro *Almagesti*; et debet per eam dividi circulus 3[us] in limbo secundo secunde faciei huius instrumenti, sicut docetur capitulo 18° secunde partis huius. [Et hec est forma tabule.]

Gloss: Hec tabula que intitulatur] *om.* **L** signorum] *om.* $\mathbf{CH_1H_2}$ est latitudo] altitudo est $\mathbf{H_1}$ circulus 3[us]] primus circulus $\mathbf{H_1}$ secundus circulus $\mathbf{H_2}$ circulus **C** secundo] primo $\mathbf{CH_1H_2}$ huius$_1$] *om.* $\mathbf{CH_1H_2}$ Et hec est forma tabule] *om.* **L**

The gloss is not present in **A** or **M**.

Gradus zodiaci	**Tabula [ascensionum] signorum in circulo obliquo in latitudine 51 g et 50 m** (first half)											
	0		1		2		3		4		5	
	g	m	g	m	g	m	g	m	g	m	g	m
1	0	25	13	19	**1[,0]**	30	27	25	5	37	18	31
2	0	49	13	48	1	13	28	31	7	0	19	58
3	1	14	14	17	1	56	29	39	8	24	21	24
4	1	38	14	46	2	40	**2[,0]**	47	9	49	22	50
5	2	3	15	16	3	24	1	56	11	14	24	17
6	2	27	15	46	4	9	3	7	12	38	25	42
7	2	52	16	16	4	55	4	17	14	3	27	9
8	3	16	16	46	5	45	5	28	15	29	28	35
9	3	42	17	17	6	30	6	40	16	54	**5[,0]**	1
10	4	6	17	48	7	19	7	52	18	20	1	26
11	4	31	18	20	8	8	9	6	19	46	2	53
12	4	56	18	52	8	58	10	20	21	12	4	19
13	5	22	19	24	9	48	11	35	22	38	5	45
14	5	47	19	56	10	39	12	51	24	4	7	11
15	6	12	20	29	11	32	14	8	25	29	8	36
16	6	37	21	4	12	25	15	25	26	56	10	3
17	7	3	21	38	13	19	16	42	28	22	11	28
18	7	29	22	12	14	14	18	0	29	38	12	54
19	7	59	22	48	15	10	19	18	**4[,1]**	14	14	19
20	8	20	23	22	16	6	20	37	2	40	15	44
21	8	47	23	58	17	4	21	56	4	7	17	10
22	9	13	24	35	18	2	23	17	5	34	18	36
23	9	39	25	11	19	1	24	37	7	0	20	2
24	10	6	25	50	20	1	25	59	8	26	21	27
25	10	33	26	28	21	2	27	20	9	52	22	53
26	11	0	27	7	22	3	28	42	11	18	24	18
27	11	28	27	46	23	5	**3[,0]**	4	12	45	25	44
28	11	56	28	26	24	9	1	27	14	12	27	9
29	12	23	29	7	25	14	2	50	15	39	28	35
30	12	51	29	48	26	18	4	14	17	5	**6[,0]**	0

Title *adds* Maudith **M** ascensionum] ascensionis **L** in latitudine 15 g et 50 m] *om.* **CAH**$_1$ oxonie cuius latitude est 51 g et 50 m verificata oxonia anno domini 1310 **M**

0,4 1;38] 1;28 **H**$_1$ **0,19** 7;59] 7;55 **M** **0,25** 10;33] 10;31 **H**$_1$ **0,27** 11;28] 11;18 **H**$_1$ **2,8** 5;45] 35;43 **M** **2,27** 23;5] 23;11 **H**$_1$ **2,29** 25;14] 55;13 **M** **4,1** 5;37] 5;27 **H**$_1$ **4,17** 28;22] 119;22 **M** **4,18** 29;38] 119;48 **M** **4,25** 9;52] 7;52 **H**$_2$ **4,27** 12;45] 12;25 **H**$_1$ **4,29** 15;39] 15;30 **H**$_1$ **5,20** 15;44] 15;24 **H**$_1$ **5,28** 27;9] 27;0 **H**$_1$

Gradus zodiaci	**Tabula [ascensionum] signorum in circulo obliquo in latitudine 51 g et 50 m** (second half)											
	6		7		8		9		10		11	
	g	m	g	m	g	m	g	m	g	m	g	m
1	1	25	14	21	27	10	4	47	0	53	17	37
2	2	21	15	48	28	33	5	51	1	34	18	4
3	4	16	17	18	29	56	6	55	2	14	18	32
4	5	42	18	42	**9[,1]**	18	7	57	2	53	19	0
5	7	7	20	8	2	40	8	58	3	32	19	27
6	8	33	21	32	4	1	9	59	4	10	19	54
7	9	58	23	0	5	23	10	59	4	49	20	21
8	11	24	24	26	6	43	11	58	5	25	20	47
9	12	50	25	53	8	4	12	56	6	2	21	13
10	14	16	27	20	9	23	13	54	6	38	21	40
11	15	41	28	46	10	42	14	50	7	12	22	5
12	17	6	**8[,0]**	12	12	0	15	44	7	48	22	31
13	18	32	1	38	13	18	16	41	8	22	22	57
14	19	57	3	4	14	35	17	35	8	56	23	23
15	21	24	4	31	15	52	18	28	9	31	23	48
16	22	49	5	56	17	9	19	21	10	4	24	13
17	24	15	7	22	18	25	20	12	10	36	24	38
18	25	51	8	48	19	40	21	2	11	8	25	4
19	27	7	10	14	20	54	21	52	11	40	25	29
20	28	34	11	40	22	8	22	41	12	12	25	54
21	29	59	13	6	23	20	23	30	12	43	26	18
22	**7[,1]**	25	14	31	24	32	24	17	13	14	26	44
23	2	51	15	57	25	43	25	5	13	44	27	8
24	4	18	17	22	26	53	25	51	14	14	27	33
25	5	43	18	46	28	4	26	36	14	44	27	57
26	7	10	20	11	29	13	27	20	15	14	28	22
27	8	36	21	36	**10[,0]**	21	28	4	15	43	28	46
28	10	2	23	0	1	29	28	47	16	12	29	2
29	11	29	24	22	2	35	29	30	16	42	29	35
30	12	55	25	4	3	42	**11[,0]**	12	17	9	**12[,0]**	0

Title *adds* Maudith **M** ascensionum] ascensionis **L** in latitudine 15 g et 50 m] *om.* **CAH_1** oxonie cuius latitude est 51 g et 50 m verificata oxonia anno domini 1310 **M**

6,2 2;21] 2;31 **H_1** 182;51 **M** **6,6** 8;33] 8;23 **H_1** **6,18** 25;51] 205;41 **M** **6,22** 1;25] 1;27 **H_1** **6,29** 11;29] 11;19 **H_1** **6,30** 12;55] 22;55 (should be 222;55) **M** **7,3** 17;18] 227;15 **M** **7,6** 21;32] 231;34 **M** **7,18** 8;48] 8;58 **H_2** **7,21** 13;6] 13;8 **H_1** **7,29** 24;22] 264;23 **M** **7,30** 25;4] 25;46 **H_1** 265;46 **M** **8,4** 9,...18] 9,...58 **H_2** **8,28** 1;29] 1;39 **H_1** **9,12** 15,44] 315;46 **M** **9,14** 17;35] 318;35 **M** **9,15** 18;28] 319;28 **M** **9,16** 19;21] 320;21 **M** **9,17** 20;12] 321;12 **M** **10,5** 3;32] 3;22 **H_1** **10,7** 4;49] 4;40 **H_1** **10,12** 7;48] 7;44 **H_1** **10,13** 8;22] 8;28 **H_1** **10,18** 11;8] 11;12 **H_1** **10,29** 16;42] 16;43 **H_1** 346;41 **M** **10,30** 17;9] 17;0 **H_1** **11,1** 17;37] 13;37 **H_1** **11,6** 19;54] 20;54 **H_1** **11,13** 22;57] *om.* 57 **H_1** **11,12** 22;31] 22;21 **A** **11;15** 23;48] 23;28 **H_1** **11.19** 25;29] 25;20 **H_1** **11,22** 26;44] *om.* 44 **H_1** **11,28** 29;2] 359;11 **M** 29;4 **H_2**

Appendix 1b : Edition of the table of oblique ascensions for 55°

Manuscript

L = Oxford, Bodleian Library, MS Laud Misc. 657, fol. 42v (*c.* 1380): *Tractatus Albionis*, written by John Westwyk. See Plate 5.

This table is unique to **L**. I have preserved the layout of the original table as far as possible (see Plate 5), but had to split it into two parts (with the argument column repeated) due to space limitations.

As with the table for 51;50°, John Westwyk gives degrees up to 30, noting the signs where they change. Most signs start with 0°, so that Westwyk could write the new number of signs in the degrees column. He highlighted these sign numbers with a red box around the number. However, in some cases the new sign does not start with 0° (in other words, the oblique ascension jumps from 29;...° to 31;...°). Where this occurs (at the start of 3, 4, 7 and 10 signs), Westwyk did not note the additional 1°. So, for example, 2,0;6° and 3,1;24° are written as **2** 6 and **3** 24.

Gloss

Tabula ascencionum signorum in circulo obliquo in latitudine .55. gra. calculata est et composita sicut docent canones in secundo libro Almagesti; et debet per eam dividi circulus secundus in limbo secundo secunde faciei instrumenti sicut docetur capitulo 18° secunde partis huius.

// tynemuth

Gradus zodiaci	**Tabula ascensionum signorum in circulo obliquo in latitudo .55. gra.** (first half)											
	0		1		2		3		4		5	
	g	m	g	m	g	m	g	m	g	m	g	m
1	0	20	11	8	26	10	21	58	**3[,1]**	24	16	29
2	0	41	11	33	26	49	23	5	2	52	17	59
3	1	1	11	57	27	28	24	13	4	21	19	30
4	1	22	12	23	28	8	25	21	5	49	21	1
5	1	42	12	48	28	49	26	32	7	18	22	32
6	2	2	13	13	29	30	27	42	8	47	24	1
7	2	23	13	39	**1[,0]**	13	28	54	10	17	25	33
8	2	43	14	4	1	57	**2[,0]**	6	11	46	27	3
9	3	4	14	31	1	41	1	20	13	16	28	33
10	3	25	14	58	2	26	2	34	14	46	**5[,0]**	3
11	3	45	15	25	3	12	3	50	16	15	1	33
12	4	6	15	53	3	59	5	6	17	46	3	4
13	4	27	16	20	4	46	6	23	19	16	4	34
14	4	48	16	48	5	35	7	42	20	48	6	4
15	5	8	17	17	6	25	9	1	22	17	7	33
16	5	30	17	48	7	16	10	21	23	47	9	4
17	5	52	18	16	8	7	11	40	25	17	10	33
18	6	14	18	46	9	0	13	1	26	49	12	4
19	6	35	19	17	9	54	14	22	28	19	13	34
20	6	57	19	47	10	48	15	44	29	50	15	3
21	7	19	20	20	11	44	17	7	**4[,1]**	21	16	32
22	7	41	20	52	12	40	18	31	2	52	18	3
23	8	3	21	25	13	38	19	55	4	23	19	33
24	8	25	21	59	14	36	21	20	5	53	21	2
25	8	48	22	32	15	37	22	45	7	24	22	32
26	9	11	23	7	16	37	24	10	8	55	24	2
27	9	34	23	43	17	39	25	36	10	25	25	31
28	9	57	24	18	18	42	27	3	11	57	27	1
29	10	21	24	54	19	46	28	30	13	28	28	30
30	10	45	25	31	20	51	29	57	14	59	**6[,0]**	0

Errors noted (with suggested correction based on the internal symmetry of the oblique ascension function)

2,8 1;57] 0;57

Gradus zodiaci	Tabula ascensionum signorum in circulo obliquo in latitudo .55. gra. (second half)											
	6		7		8		9		10		11	
	g	m	g	m	g	m	g	m	g	m	g	m
1	1	30	16	32	1	30	10	14	5	6	19	39
2	2	59	18	3	2	57	11	18	5	42	20	3
3	4	29	19	35	4	24	12	20	6	17	20	26
4	5	58	21	5	5	50	13	23	6	54	20	49
5	7	28	22	36	7	15	14	24	7	28	21	12
6	8	58	24	7	8	40	15	24	8	1	21	35
7	10	27	25	37	10	5	16	22	8	35	21	57
8	11	57	27	8	11	29	17	20	9	8	22	19
9	13	28	28	39	12	53	18	16	9	40	22	41
10	14	57	**8[,0]**	10	14	16	19	12	10	12	23	2
11	16	27	1	41	15	38	20	6	10	43	23	25
12	17	55	3	11	16	59	20	0	11	14	23	46
13	19	27	4	42	18	20	21	53	11	44	24	8
14	20	56	6	12	19	39	22	44	12	12	24	30
15	22	26	7	43	20	59	23	35	12	43	24	51
16	23	56	9	12	22	18	24	25	13	12	25	12
17	25	26	10	44	23	37	25	14	13	40	25	33
18	26	56	12	14	24	54	26	1	14	17	25	54
19	28	27	13	45	26	10	26	48	14	35	26	15
20	29	57	15	14	27	26	27	34	15	2	26	35
21	**7[,1]**	27	16	44	28	40	28	19	15	29	26	56
22	2	57	18	14	29	54	29	3	15	56	27	17
23	4	27	19	43	**10[,1]**	6	29	47	16	21	27	37
24	5	59	21	13	2	18	**11[,0]**	30	16	47	27	58
25	7	28	22	42	3	28	1	11	17	13	28	18
26	8	59	24	11	4	39	1	52	17	38	28	38
27	10	0	25	39	5	47	2	32	18	3	28	59
28	12	1	27	8	6	56	3	11	18	27	29	19
29	13	31	28	36	8	2	3	50	18	52	29	40
30	15	1	**9[,0]**	3	9	9	4	29	19	15	**12**	0

Errors noted (with suggested correction based on the internal symmetry of the oblique ascension function)

6,27 10;0] 10;30 **9,12** 20;0] 21;0 **10,18** 14;17] 14;07

Appendix 2

Comparison of values in John Westwyk's table of oblique ascensions for 55° (Oxford, Bodleian Library, MS Laud Misc. 657, fol. 42v) and related tables, showing significant results and deviations from symmetry

Columns

- A: Longitude (λ), 0–360°.
- B: Manuscript values of oblique ascension (ρ) as a function of λ, and as a function of 360°–λ.
- C: Deviation from symmetry (in minutes): where columns B and C do not add up to 360°.
- D: Manuscript values of right ascension (α) as a function of λ, of 180°–λ of, 180°+λ, and of 360°–λ. Values in columns D2 to D4 only supplied where these are not symmetrical with D1.
- E: Right ascensions derived from symmetrically arranged values from John Westwyk's oblique ascensions table (see formula 4 above). Values supplied where these do not match D1.
- F: Ascensional difference (γ) derived by subtracting the manuscript values of oblique ascension from manuscript values of right ascension (α – ρ; D1 – B1).
- G: Ascensional difference derived from symmetrically arranged values from John Westwyk's oblique ascensions table (see formula 5 above). Values supplied where these do not match F.
- H: Idealised manuscript oblique ascensions 360–180°, computed by subtracting most consistent value of derived ascensional difference from manuscript right ascension for λ = 180–360°. Values only supplied where these do not match column B2.
- J: Idealised manuscript oblique ascensions 360–180°, computed by subtracting most consistent value of derived ascensional difference from manuscript right ascension for λ = 0–180° and subtracting the result from 360°. Values only supplied where these do not match column B2.

A	B1	B2	C	D1	D2	D3	D4	E	F	G	H	J
λ	ρ_{λ}	$\rho_{360-\lambda}$		α_{λ}	$\alpha_{180-\lambda}$	$\alpha_{180+\lambda}$	$\alpha_{360-\lambda}$	$\alpha_{derived}$	γ_1	γ_2	ρ_1	ρ_2
1	0;20	359;40		0;55					0;35			
2	0;41	359;19		1;50					1; 9			
3	1; 1	358;59		2;45					1;44			
4	1;22	358;38		3;40					2;18			
5	1;42	358;18		4;35					2;53			
6	2;02	357;58		5;30					3;28			
7	2;23	357;37		6;25					4; 2			
8	2;43	357;17		7;20					4;37			
9	3; 4	356;56		8;16					5;12			
10	3;25	356;35		9;11					5;46			
11	3;45	356;15		10; 6				10;5½	6;21	6;20½		
12	4; 6	355;54		11; 1					6;55			
13	4;27	355;33		11;57					7;30			
14	4;48	355;12		12;52					8; 4			
15	5; 8	354;51	−1	13;47	166;12	193;48	346;12	13;47½	8;39	8;39½		354;52
16	5;30	354;30		14;43					9;13			
17	5;52	354;08		15;39					9;47			
18	6;14	353;46		16;35					10;21			
19	6;35	353;25		17;31					10;56			
20	6;57	353;02	−1	18;27					11;30		353; 3	353; 3
21	7;19	352;41		19;23					12; 4			
22	7;41	352;19		20;19					12;38			
23	8; 3	351;57		21;15					13;12			
24	8;25	351;35		22;12					13;47			
25	8;48	351;12		23; 8					14;20			
26	9;11	350;49		24; 5					14;54			
27	9;34	350;26		25; 2					15;28			
28	9;57	350;03		25;59					16; 2			
29	10;21	349;39		26;56					16;35			
30	10;45	349;15		27;53					17; 8			
31	11; 8	348;52		28;50					17;42			
32	11;33	348;27		29;48					18;15			
33	11;57	348;03		30;46					18;49			
34	12;23	347;38	1	31;44			328;17		19;21			347;37
35	12;48	347;13	1	32;42			327;19		19;54			347;12
36	13;13	346;47		33;40					20;27			
37	13;39	346;21		34;38					20;59			
38	14; 4	345;56		35;36					21;32			
39	14;31	345;29		36;35					22; 4			
40	14;58	345;02		37;34					22;36			

A	B1	B2	C	D1	D2	D3	D4	E	F	G	H	J
λ	ρ_{λ}	$\rho_{360-\lambda}$		α_{λ}	$\alpha_{180-\lambda}$	$\alpha_{180+\lambda}$	$\alpha_{360-\lambda}$	$\alpha_{derived}$	γ_1	γ_2	ρ_1	ρ_2
41	15;25	344;35		38;33					23; 8			
42	15;53	344;17	10	39;32					23;39			
43	16;20	343;40		40;31				40;31½	24;11	24;11½		
44	16;48	343;12		41;30				41;30½	24;42	24;42½		
45	17;17	342;43		42;30					25;13			342;42
46	17;48	342;12		43;30					25;42			
47	18;16	341;44		44;30					26;14			
48	18;46	341;14		45;30					26;44			
49	19;17	340;43		46;31					27;14			
50	19;47	340;12	−1	47;31				47;30½	27;44	27;43½		
51	20;20	339;40		48;32					28;12			
52	20;52	339; 8		49;33					28;41			
53	21;25	338;35		50;34					29; 9			
54	21;59	338; 1		51;36					29;37			
55	22;32	337;28		52;37					30; 5			
56	23; 7	336;54	1	53;39					30;32		336;53	336;53
57	23;43	336;17		54;41					30;58			
58	24;18	335;42		55;43					31;25			
59	24;54	335; 6		56;45					31;51			
60	25;31	334;29		57;47					32;16			
61	26;10	333;50		58;50					32;40			
62	26;49	333;11		59;53					33; 4			
63	27;28	332;32		60;56					33;28			
64	28; 8	331;52		61;59					33;51			
65	28;49	331;11		63; 2					34;13			
66	29;30	330;30		64; 5					34;35			
67	30;13	329;47		65; 9					34;56			
68	31;57	329;03	60	66;13					35;16			
69	31;41	328;19		67;17					35;36			
70	32;26	327;34		68;21					35;55			
71	33;12	326;48		69;25		249;24			36;13			
72	33;59	326;01		70;29					36;30			
73	34;46	325;14		71;33					36;47			
74	35;35	324;25		72;37					37;02			
75	36;25	323;35		73;42					37;17			
76	37;16	322;44		74;47					37;31			
77	38; 7	321;53		75;52					37;45			
78	39; 0	320; 0	−60	76;57					37;57			
79	39;54	320; 6		78; 2					38; 8			
80	40;48	319;12		79; 7					38;19			

A	B1	B2	C	D1	D2	D3	D4	E	F	G	H	J
λ	ρ_λ	$\rho_{360-\lambda}$		α_λ	$\alpha_{180-\lambda}$	$\alpha_{180+\lambda}$	$\alpha_{360-\lambda}$	α_{derived}	γ_1	γ_2	ρ_1	ρ_2
81	41;44	318;16		80;12					38;28			
82	42;40	317;20		81;17					38;37			
83	43;38	316;22		82;22					38;44			
84	44;36	315;24		83;27					38;51			
85	45;37	314;24	1	84;33	95;28	264;32	275;28	84;32½	38;56	38;55½		314;23
86	46;37	313;23		85;38			274;21		39; 1		313;22	
87	47;39	312;20	−1	86;43			273;16		39; 4			312;21
88	48;42	311;18		87;49				87;48½	39; 7	39; 6½		
89	49;46	310;14		88;54					39; 8			
90	50;51	309; 9		90;00					39; 9			
91	51;58	308; 2										
92	53; 5	306;56	1									
93	54;13	305;47										
94	55;21	304;39										
⋮												
108	73; 1	286;59										
109	74;22	285;38				249;24					285;37	
110	75;44	284;16										
⋮												
135	112;17	247;43										
136	113;47	246;12	−1									
137	115;17	244;42	−1									
138	116;49	243;11										
⋮												
152	137;59	222; 1										
153	139;30	220; 0	−30									
154	141; 1	218;59										
⋮												
164	156; 4	203;56										
165	157;33	202;26	−1	13;47	166;12	193;48	346;12				202;27	202;27
166	159; 4	200;56										
167	160;33	199;27										
168	162; 4	197;55	−1								197;56	197;56
169	163;34	196;27	1									
170	165; 3	194;57										
⋮												
179	178;30	181;30										
180	180; 0											

Determining the Sine Tables Underlying Early European Tangent Tables

Kailyn Pritchard

Introduction

Trigonometry originated in ancient Greece, when Hipparchus of Rhodes began using a method which related the length of an arc in a circle to the length of the chord it subtends to predict the motions of the celestial bodies.[1] Later, Indian astronomers based their trigonometry on the Sine,[2] or half-Chord, a tradition that was transmitted into Arabic astronomy. The history of the Tangent function is more complicated. The first known manifestations of tables containing a function resembling the modern Tangent were known as 'shadow tables', and appeared within the context of sundial theory in early Arabic *zīj*es (astronomical handbooks) around the 9th century.[3] Shadow tables appeared in separate sections of astronomical handbooks, and did not form part of trigonometry in the way that Sines and Cosines did.

Shadow tables were transmitted from Arabic scholars into Europe through Muslim Spain, in the form of works such as the 11th century *Toledan Tables*, but remained firmly within the realm of sundial theory. Though Arabic astronomers integrated the shadow function into their trigonometry as early as the 10th century, the function we now know as the Tangent didn't appear in European astronomy until the 15th century. Manuscripts by Levi ben Gerson (1288–1344) demonstrate significant influence from Arabic sources, as well as from the works of Ptolemy, al-Battānī, and Jābir ibn Aflaḥ. Though the shadow function had made its way into eastern Arabic trigonometry by this time, the same was not true in Europe. Levi ben Gerson's work was based on Sines and Versed Sines, notably omitting the shadow function. When Levi might have made use of this function in his work, he resorted instead to a more complex procedure based on Sines.[4]

[1] Van Brummelen, *The Mathematics of the Heavens*, pp. 37–46.

[2] Capitalized functions, such as Sine, indicate the use of a base circle of radius $R \neq 1$, distinguishing them from modern sines.

[3] Van Brummelen, *The Mathematics of the Heavens*, pp. 149–55.

[4] Van Brummelen, *The Mathematics of the Heavens*, p. 231.

Editing and Analysing Numerical Tables: Towards a Digital Information System for the History of Astral Sciences, ed. by Matthieu Husson, Clemency Montelle and Benno van Dalen, PALS 2 (Turnhout, 2021), pp. 107–142
© BREPOLS PUBLISHERS 10.1484/M.PALS-EB.5.127697

Tables resembling those of the shadow function, separate from sundial theory, first appeared in Europe in the 15th century. The earliest such table appeared as an auxiliary table in Giovanni Bianchini's *Tabulae primi mobilis*.[5] Auxiliary functions first appeared in Arabic astronomy. These functions contain no astronomical meaning on their own, but appear frequently as units in solutions to a variety of astronomical problems. This particular unit appeared in Bianchini's solution of the problem of converting stellar positions from ecliptic to equatorial coordinates. This function, now known as the Tangent, took a while to become a trigonometric function.[6] This is illustrated by its omission from Regiomontanus' *De triangulis omnimodis*, a comprehensive overview of both planar and spherical trigonometry that uses only the Sine and Versed Sine. Regiomontanus did, however, include a Tangent table in his *Tabulae directionum* (an idea borrowed from Bianchini) which spread to his successors. By the end of the 16th century, the Tangent function was an integral part of trigonometry.

In contrast with Arabic astronomy, which mostly used a circle with base radius $R = 60$, European tables used a variety of different R values, such as 10 000, 60 000, and 100 000. The use of very large R values allowed European astronomers to give their Sine values as integers, whereas the Arabic Sine tables with $R = 60$ always had fractional sexagesimal digits.

Tangents can be computed in two different ways. The first simply divides the Sine by the Cosine and multiplies the result by R:[7]

$$\operatorname{Tan}\vartheta = \frac{\operatorname{Sin}\vartheta}{\operatorname{Cos}\vartheta}\cdot R.$$

The second method uses a variant of the Pythagorean Theorem, which was well known by astronomers and mathematicians in this time period:[8]

$$\operatorname{Tan}^2\vartheta + R^2 = \operatorname{Sec}^2\vartheta.$$

While the secant was not conceived of as a trigonometric function until the 1550s, the identity can also be expressed as $\operatorname{Tan}\vartheta = \sqrt{(\frac{R^2}{\operatorname{Cos}\vartheta})^2 - R^2}$ and thus provides another way to compute the Tangent.

[5] Van Brummelen, 'The End of an Error', pp. 553, 558.

[6] Van Brummelen, *The Mathematics of the Heavens*, pp. 261–63.

[7] Note that the R in this formula is not necessarily the R of the base circle for the Sine and Cosine values. It is the length of the gnomon in sundial theory, and the length of the radius of the base circle of the Tangent when doing trigonometry.

[8] The first secant tables to be published (other than those included in Georg Rheticus' six-function conception of trigonometry) were in Francesco Maurolico's *Theodosii sphaericorum elementorum* in 1558. For more information, see Van Brummelen and Byrne, 'Maurolico, Rheticus'.

Unfortunately, whatever method they chose, the astronomers computing Tangent tables were in trouble, and many of them knew it. Both of these methods are problematic for arguments near 90°. For these arcs, the Cosine values are small. A minor deviation in these values is magnified significantly by the division process.[9] The Tangent table in the *Opus palatinum* by Rheticus, for example, contains large errors that are symptomatic of this numerical sensitivity. Less than a year after having obtained a copy of these tables, astronomer Adrianus Romanus (1561–1615) noticed that they were in error in the last 3, 4, 5 or more places. He argued (correctly) that the use of the first method, in conjunction with Sines and Cosines accurate to only as many places as the Tangents, was the source of the 'inexcusable' errors in Rheticus' table.[10]

This paper examines four of the earliest Tangent tables to appear in Europe in order to determine how they were computed. We begin by examining the 'tabulae magistralis quarta', found in Giovanni Bianchini's *Tabulae primi mobilis*.[11] We then examine Regiomontanus' 'tabula fecunda', published a few decades later in his *Tabulae directionum*. The structure of Regiomontanus' *Tabulae directionum*, including this new auxiliary table, was found recently to copy that of Bianchini's *Tabulae primi mobilis*.[12] Finally, we examine two sets of trigonometric tables by Georg Rheticus in his *Canon doctrinae triangulorum* and *Opus palatinum*. The tables found in the *Opus palatinum* are perhaps the most significant set of trigonometric tables in European history. A corrected version of these tables was the best available in Europe for over 300 years, replaced only in the early 20th century.

Historians of mathematics have developed some understanding of how Sines and Cosines were computed in several different cultures. In contrast, historical practices for computing the Tangent function have been explored very little. This paper addresses this gap, by establishing how one can determine the method by which a given historical Tangent table was computed, as well as the radius used by the underlying Sine and Cosine table. After determining the radii of the Sine and Cosine tables underlying each Tangent table, we will reconstruct the Sine and Cosine values that the historical authors must have used to compute the Tangent values found in their tables. We will compare those reconstructed tables to known Sine and Cosine tables

[9] For a discussion on how errors found in underlying Sine and Cosine tables can impact Tangent values computed from them, see the Appendix of van Dalen, 'Islamic and Chinese Astronomy'.

[10] Bockstaele, 'Adrianus Romanus'.

[11] Specifically, we refer to the manuscript Cracow, Biblioteca Jagiellonska, MS 556, referred to as Tabulae primi mobilis B in Van Brummelen, 'Before the End of an Error'.

[12] Van Brummelen, 'The End of an Error', pp. 560–62.

by the same authors, allowing us to better understand how the astronomers interacted with their Co/Sine tables as they constructed these early Tangent tables. Very little direct evidence is available regarding how historical authors completed their computations behind the scenes. The results of this project provide us with unique insight into the working practises of these historical mathematicians.

Methods

Both methods for computing Tangents rely on Sines and Cosines taken from another table. Our objective is to determine the method by which the Tangent tables were computed, the radii of the underlying tables, and the Sine and Cosine values that were used. Given either of the two possible methods for computing a Tangent, each entry depends on at most three choices: a choice of R, a choice of $\operatorname{Sin}\vartheta$, and a choice of $\operatorname{Cos}\vartheta$. This means that there are a limited number of recomputational possibilities for the historical table.

Existing methods for determining the interdependence of historical tables are insufficient for answering the relevant questions of this project. The method developed by Glen Van Brummelen and Kenneth Butler determines whether a given table (of Tangents, for example) is likely to have been computed from another particular table containing potential underlying values (Co/Sines). It is only effective in cases where a method is assumed, and there is a single value upon which the Tangent value relies.[13] As the first method of computing the Tangent requires two underlying values (Sine and Cosine), this technique cannot be used. When using it to examine the second method, the results were inconclusive. Another method, developed by Benno van Dalen, uses statistics to perform a parameter estimation.[14] In this case, the parameter that we seek is R, which is easily found without statistics. Statistical methods like this are more useful when analyzing complex computations. Our situation is so computationally simple that direct recomputation is preferable to statistical analysis. The objective at hand is to determine the mathematical procedures which produced the values in the historical Tangent tables we are examining, rather than verifying a specific interdependence relationship or estimating a particular parameter. Thus, a method which reproduces the methods used by historical mathematicians when creating their tables is better suited to the purpose.

In the following work, 'correct value' or 'accurate value' refers to the value obtained by modern computation, 'reconstructed value' or 'underlying value'

[13] Van Brummelen and Butler, 'Determining the Interdependence', p. 43.

[14] van Dalen, 'A Statistical Method'.

refers to that determined by this method to have been used to compute the historical Tangent value, and 'historical value' refers to a Co/Sine printed in a known, published Co/Sine table.

In order to illustrate our method for analyzing the tables studied in this paper, we take as a case study the Tangent table contained in Regiomontanus' *Tabulae directionum*, which uses an R value of 100 000. In Regiomontanus' table, $\mathrm{Tan}(87^\circ) = 1\,908\,217$, as opposed to the correct value, 1 908 114. The Sine and Cosine tables underlying Regiomontanus' Tangent table can reasonably be assumed to have used an R value similar to either the Tangent table itself, or to Regiomontanus' known Sine tables (60 000, 6 000 000, and 10 000 000). Recall that both the Sine tables and Tangent tables contained only whole number values. The underlying Sine and Cosine values can be assumed to be within two units of the accurate values for whichever radius is used, since Regiomontanus' published Sine tables are at least this accurate.[15] This provides narrow constraints for Regiomontanus' possible computations.

Next, we generate a variety of fictional underlying tables, compute Tangents from them, and compare the resulting values to those found in Regiomontanus' Tangent table. While a single match of a recomputed value to a historical value could be coincidental, consistent preference of one R value over the others being examined would suggest Regiomontanus used that R value. For each given table, we examined only the Tangents with the largest arguments,[16] as both methods become unstable in this section of the table. They require division by Cosine values, which are relatively small for arguments near 90°. Slight inaccuracies in those Cosine values can therefore create significant differences in the Tangent values resultant from those computations, making this area of the tables the most fruitful for our study.[17] The usefulness of this method is limited to examining only the latter, more unstable portion of the tables, and cannot confirm whether the method used in those sections was used throughout the table with certainty. The underlying Co/Sine values could not be reliably reconstructed in the more stable sections of the table using this method. It is, however, highly likely that the table compiler would have used the same method to compute the entirety of the table and a hypothesis of a different method for the early part of the table would require some strong evidence that is not there by current analysis.

[15] Regiomontanus, *Tabulae directionum*.

[16] We examined the arguments for the 10° interval between 80° and 90° for all tables except for that found in Rheticus, *Opus palatinum*, for which we examined the 1° interval between 89° and 90°.

[17] For more details on the numerical sensitivity involved when computing Tangents, see the earlier discussion in the Introduction.

Considering Regiomontanus' Tangent table, we posit $R = 60\,000$.[18] We select five possible Sines and five possible Cosines: the mathematically correct value for $R = 60\,000$, rounded to a whole number, as well as the two values greater and the two values less than that value. In the case of Tan(87°), we choose the Sines 59916, 59917, 59918, 59919, and 59920, and the Cosines 3138, 3139, 3140, 3141, and 3142.

We examine the first method by calculating the 25 values of Tan(87°) generated by all of the possible combinations of these Sines and Cosines and compare the results with the author's entry Tan(87°) = 1908217. The twenty-five entries generated in such tables will later be referred to the entries generated by the best-fit Co/Sine tables.

	59916	59917	59918	59919	59920
3138	1909369	1909401	1909433	1909465	1909496
3139	1908761	1908793	1908824	1908856	1908888
3140	1908153	1908185	1908217	1908248	1908280
3141	1907545	1907577	1907609	1907641	1907673
3142	1906938	1906970	1907002	1907034	1907066

Table 1: 25 possible Tangent values, generated from combinations of five possible Co/Sine values.

We examine the second method by calculating the five values of Tan(87°) generated by using these Cosines and compare the results with the author's entry Tan(87°) = 1908217.

3138	3139	3140	3141	3142
1909429	1908819	1908210	1907600	1906992

Table 2: Five possible Tangent values, generated from five possible Cosine values.

Using the first method, the historical value of Tan(87°) = 1908217 is generated when Sin(87°) = 59918, and Cos(87°) = 3140 (Table 1). The closest that the second method is able to get to this value is 1908210: 7 units less. While this alone might just be a fortuitous match, if the same method and value of R produces mostly 'fortuitous' matches while other methods and values of R produce hardly any matches, we will have found with near certainty both the historical value of R and the Sine and Cosine values that were used. Once we have reconstructed the Sine and Cosine tables that must have been used to compute the historical Tangent values, we can compare those tables to known Co/Sine tables by the same author and based on the same R value

[18] We repeat the analysis for all possible attested and plausible R values, examining eight possibilities for each author.

to determine whether or not it is likely that those exact tables were used. The method itself does not directly make use of any known Co/Sine tables; it simply compares the reconstructed tables to known tables in order to establish a possible source.

To compare the results obtained by using each of these different *R* values and different methods, we examine the *mean error*: the sum of the differences between the most accurate of the 25 generated entries and the historical entry for each argument, divided by the total number of entries examined. This metric was selected from the many available because of its ability to quickly and simply give perspective on the magnitude of the errors created by selecting the various *R* values considered. Other options achieved this less effectively and elegantly. We also record the number of *exact matches*: the number of times that the generated entry closest to the entry found in the historical table is in fact identical to the historical entry. While these two metrics are not guaranteed to agree, since every exact match correlates with a zero in the sum contributing to the mean, a larger number of exact matches often correlates with a lower mean error.

For each of the Tangent tables discussed in this paper, our examination of these metrics conclusively determined that the second method was not used. This is unsurprising, given that the second method involves computations that are significantly more complex than the first, without resolving the issues of instability created by having a Cosine in the denominator of the relevant fraction. In every case, however, the *R* values correlating with the least mean errors and greatest number of exact matches were the same when using both the first and second methods. The second method will not be discussed in any of the analysis sections following. Tables containing the mean errors and exact matches for each Tangent table using the second method can be found at www.kailynpritchard.com.

Giovanni Bianchini

Astronomer Giovanni Bianchini (*c.* 1410–1469) was a Venetian merchant until 1427, when he began three decades of work as an administrator for the ruling d'Este family. Bianchini is also known to have taught at the University of Ferrara. He corresponded with Regiomontanus in 1463 and 1464, discussing mostly astronomical and mathematical problems.[19] It was recently discovered that the structure of Regiomontanus' *Tabulae directionum* directly copies that of Bianchini's *Tabulae primi mobilis*.[20] It is unknown whether the two ever met in person.

[19] Chabás and Goldstein, *The Astronomical Tables*, pp. 13–14, 19–20.

[20] cf. the reference in footnote 11.

Bianchini's scientific works were produced between 1440 and 1460. His master work is his *Flores Almagesti*. The *Flores Almagesti* is the theoretical backbone of most of Bianchini's works. The first three treatises of this work (which contains 8 to 10 treatises, depending on the manuscript tradition) provide a mathematical introduction. The remainder of the work deals with astronomical matters directly, following Ptolemy's *Almagest* up to Book VI.[21] Other titles include his *Tabulae astronomae*, a series of tables, and instructions for their use,[22] and his eclipse tables *Tabulae de eclypsibus*.

Tabulae primi mobilis: Background

Giovanni Bianchini's *Tabulae primi mobilis* is believed to have been written towards the end of his academic career, perhaps between 1455 and 1460.[23] It consists of approximately 40 pages of canons, which contain instructions for solving problems in spherical astronomy and mathematical astrology, and 100 pages of astronomical tables.[24] One of its tables, entitled 'tabula magistralis quarta', is the first appearance of what would come to be known as a Tangent table. An excerpt may be found in Appendix 1a. In Bianchini's work, it is represented and used as an auxiliary function, to aid in the computations necessary to solve the problem of the conversion of stellar coordinates. It is also found in a collection of auxiliary functions, Bianchini's *Tabulae magistralis*.

Tabulae primi mobilis: Analysis

The Tangent table in the *Tabulae primi mobilis*,[25] with radius $R = 10\,000$, is amenable to the analysis described in the previous section. We consider the last 60 entries of the table, associated with the arguments for each $10'$ of arc between 80° and 90°. Many of these values are erroneous, to some extent.

Exact reconstructions of 24 of these printed entries were generated by using the most accurate possible Sine and Cosine values with $R = 60\,000$. Using this radius, we generate a perfect match for another 25 entries with one accurate Co/Sine and the other in error by one in the last place. The number of matches achieved by using various Co/Sine tables with different R values is summarized below.

The greatest argument for which we were able to obtain an exact match was $89^\circ 30'$. Table 4 depicts the differences between each of the 25 Tangent

[21] Chabás and Goldstein, *The Astronomical Tables*, p. 19.

[22] Chabás and Goldstein, *The Astronomical Tables*, p. 14.

[23] Chabás and Goldstein, *The Astronomical Tables*, p. 19. Note that the *Tabulae primi mobilis* is referred to as *Canones tabularum super primo mobile* in this text.

[24] Van Brummelen, 'The End of an Error', p. 547.

[25] Cracow, BJ, MS 556, f. 52r-52v.

Radius (R)	Exact matches
6000	2
10000	5
60000	49
100000	27
600000	19
1000000	14
6000000	6
10000000	4

Table 3: Exact matches obtained using best-fit Co/Sine tables to recompute the last 60 entries from the Tangent table in Bianchini's *Tabulae primi mobilis*.

values generated by our method, and the Tangent value printed by Bianchini in his table. The column headers are the Sines, divided by the Cosines labelling each row. Note that the zero in the fourth column from the left, and fourth row down signifies that the exact Tangent value printed by Bianchini was obtained by dividing 59998 by 524, and multiplying by 10000.

	59996	59997	59998	59999	60000
522	4349	4368	4387	4406	4425
523	2151	2170	2189	2208	2228
524	−38	−19	0	19	38
525	−2219	−2200	−2181	−2162	−2143
526	−4392	−4373	−4354	−4335	−4316

Table 4: The differences between the 25 Tangent values generated by our method and Bianchini's printed Tangent value for argument $89°30'$.

The mean error achieved using Co/Sine tables for all of the different candidate R values is summarized in Table 5.

Radius (R)	Mean error
6000	388.2
10000	292.2
60000	0.8
100000	31.8
600000	18.1
1000000	24.7
6000000	40
10000000	41.8

Table 5: Mean error obtained when using best-fit Co/Sine tables to recompute the Tangent table in Bianchini's *Tabulae primi mobilis*. Note that the error associated with the entry for argument $89°50'$ was removed, as it could not be reliably reconstructed.

Based on these results, we conclude that Giovanni Bianchini used $\mathrm{Tan}\,\vartheta = \frac{\mathrm{Sin}\,\vartheta}{\mathrm{Cos}\,\vartheta} \cdot 10\,000$ in conjunction with a Sine table with a radius of $R = 60\,000$ in order to generate his Tangent values.

Of the 60 Tangent values examined, only one entry (for argument 89°50′) could not be reliably reconstructed. The least error for that entry ranged from 3738 to 86407, depending on the method and value of R. The error for this entry is two orders of magnitude greater than any other entry, and no change of a single digit creates a value that can be reliably reconstructed using this method. For these reasons, we believe this entry to be the result of a computational error.

A comparison of the trios of Sine values in the reconstruction tables (Appendix 1b) presents the following results:

- The correct values (according to modern recomputation) exactly match both the reconstructed value and the value printed by Bianchini in his tables in 41 of the 59 reliably reconstructed entries.
- Of the remaining 18 entries, in 11 cases the reconstructed value matches the value printed by Bianchini in his Sine tables, but not the correct value.
- For 5 of the 7 remaining entries, Bianchini's printed values match the correct values, but the reconstructed value is different.
- For the final 2 entries, the correct value, the reconstructed value, and the value printed in Bianchini's table are all different.

These results suggest that the underlying table is an old version of Bianchini's Sine table, which the printed table improves upon. Whether or not this is the case, the reconstructed table is a variant of the Sine table that Bianchini published in his *Tabulae primi mobilis*.

We turn next to the Cosine table (Appendix 1c).

- The three columns match each other for 42 out of 59 reliably reconstructed entries.[26]
- For 14 of the remaining 17 entries, the reconstructed value matches the value printed by Bianchini in his Cosine table, but not the correct value.
- For 2 of the 3 remaining entries, the reconstruction matches the computed values, but not the values printed in Bianchini's Cosine table.
- For the final entry, the computed value matches the value printed in Bianchini's table, but it does not match the reconstruction.

[26] As before, the entry for argument 89°50′ reveals a significant computational error.

This evidence suggests that the reconstructed Cosine table is an improved version of the printed table, but not strongly enough to reach a firm conclusion. The reconstructed table is clearly a variant of the published version, but not identical to it.

The reconstructed underlying Sine table matches Bianchini's published Sine table with $R = 60\,000$ for 52 of the 60 entries, and the reconstructed underlying Cosine table matches Bianchini's printed Cosine table with the same radius for 56 of 60 entries. As only 41 and 42 of these entries, respectively, are correct, this suggests that Bianchini's Tangent tables were computed based on a version of his Sine table with $R = 60\,000$.

Of the 10 entries that are reliable reconstructions and do not match, for 6 entries the value printed by Bianchini is more accurate than the reconstruction, for 2 entries the reconstruction is more accurate than Bianchini's printed values, and for 2 entries neither the printed values nor the reconstruction are more accurate. We therefore cannot conclude whether or not the Co/Sine tables Bianchini used to compute his Tangents were a revised version of those seen in his *Tabulae primi mobilis.* One could reasonably conclude that they were either, given that Bianchini would not have known which entries were more accurate when he computed them.

Regiomontanus

Born Johannes Müller (1436–1476), the astronomer and mathematician better known as Regiomontanus was, and remains, easily the most recognized European astronomer of the 15th century.[27] Much of his career was spent at the University of Vienna, studying trigonometry and astronomy under his mentor Georg Peurbach.[28] Though his life and career were both very short, Regiomontanus' work went on to influence scientists like Copernicus for over a century.[29]

Regiomontanus' most significant books included the *Epytoma in Almagestum Ptolemei, Tabulae primi mobilis, Tabulae directionum* and *De triangulis omnimodis.* The *Epytoma in Almagestum Ptolemei*, completing Georg Peurbach's work, provided a streamlined commentary on the mathematics contained in the *Almagest.* His *Tabulae primi mobilis*, reflecting Arabic astronomical works on auxiliary functions, contains a table that could solve a variety of problems encountered in spherical astronomy. *De triangulis omnimodis* is

[27] For additional information on Regiomontanus, see Hughes, *Regiomontanus* and Zinner, *Regiomontanus.* For discussion on Regiomontanus' trigonometry, see Van Brummelen, *The Mathematics of the Heavens*, pp. 251–263.

[28] Zinner, *Regiomontanus*, p. 17–30.

[29] Van Brummelen, *The Mathematics of the Heavens*, pp. 251 and 261; Zinner, 'Regiomontanus', p. 93.

a comprehensive treatment of both plane and spherical triangles, modeled on the style of reasoning in Euclid's *Elements*. Though Regiomontanus already may have begun working with the Tangent function in an astronomical context, the Tangent is notably absent from *De triangulis omnimodis*.[30]

Tabulae directionum: Background

One of Regiomontanus' most famous books, and undoubtedly his most popular set of astronomical tables, is his *Tabulae directionum*. It was written in 1467, though it remained unpublished until 1490, 14 years after his death.[31] As has been mentioned above, it has been demonstrated recently that the structure of Regiomontanus' tables in the *Tabulae directionum* was taken directly from the structure of Bianchini's *Tabulae primi mobilis*. In the *Tabulae directionum* we find a table called the 'tabula fecunda', or 'fruitful table'. It is believed to be the second Tangent table to appear in Europe, after the table by Bianchini that we treated in the previous section.[32] Regiomontanus' table is more precise than Bianchini's, using a radius of $R = 100\,000$ rather than $R = 10\,000$, but it contains far fewer entries, giving the Tangent for every degree of arc, as opposed to Bianchini's every $10'$ of arc.

Tabulae directionum: Analysis

For the Tangent table in the *Tabulae directionum*, we apply our analysis to the entries for 80°, 81°, ..., 89°. Exact reconstructions of eight of these printed entries were generated by using the most accurate possible Sine and Cosine values with $R = 60\,000$. Table 6 displays the number of exact matches achieved by using best-fit Co/Sine tables of all the historically plausible R values we examined.

The greatest argument for which we were able to obtain an exact match was 88°. Table 7 depicts the differences between each of the 25 Tangent values generated by our method, and the Tangent value printed by Regiomontanus in his table for this argument. The column labels are the Sines, divided by the Cosines labelling each row. Note that the zero in the fourth column from the left, and fourth row down signifies that the exact Tangent value printed by Regiomontanus was obtained by dividing 59963 by 2094, and multiplying by 100000.

[30] Van Brummelen, *The Mathematics of the Heavens*, pp. 261.

[31] Van Brummelen, *The Mathematics of the Heavens*, pp. 251, 261 and 263.

[32] Rosińska, 'Tables trigonométriques', p. 49. Rosińska incorrectly identifies the radius of Bianchini's table as 1000, rather than 10000.

Radius (R)	Exact matches
6000	0
10000	0
60000	8
100000	2
600000	3
1000000	1
6000000	1
10000000	0

Table 6: Exact matches obtained using best-fit Co/Sine tables to recompute the last ten entries from the Tangent table in Regiomontanus' *Tabulae directionum*. Note: 600000 is 10 times as large as 60000. Thus, supposing that $R = 60000$ was used with method 1, one should expect an occasional accidental match when using $R = 600000$: namely, when the underlying Sine/Cosine value happens to end in 0.

	59961	59962	59963	59964	59965
2092	2642	2689	2737	2785	2833
2093	1272	1320	1368	1415	1463
2094	−96	−48	0	47	95
2095	−1463	−1415	−1367	−1320	−1272
2096	−2828	−2781	−2733	−2685	−2637

Table 7: The differences between the 25 Tangent values generated by our method and Regiomontanus' printed Tangent value for argument 88°.

Table 8 gives the mean error achieved by best-fit Co/Sine tables of all the historically plausible R values we examined.

Radius (R)	Mean error
6000	2044.5
10000	1718.5
60000	0.9
100000	15.3
600000	18
1000000	27.9
6000000	114.5
10000000	123.5

Table 8: Mean error obtained when using best-fit Co/Sine tables to recompute Regiomontanus' Tangent table in the *Tabulae directionum*.

The fit for $R = 60000$ is well over an order of magnitude better than any other R value. We conclude that Regiomontanus used $\text{Tan}\,\vartheta = \frac{\text{Sin}\,\vartheta}{\text{Cos}\,\vartheta} \cdot 100000$ in conjunction with Co/Sine values with a radius of $R = 60000$ in order to generate his Tangent values. The only entries for which calculations based

Argument	Calculated	Reconstruction	Regiomontanus	Bianchini
80°	59088	59088	59088	59088
81°	59261	59261	59261*	59261
82°	59416	59416	59416	59416
83°	59553	59553	59552	59554
84°	59671	59671	59671	59671
85°	59772	59774*	59771	59772
86°	59854	59854	59853	59854
87°	59918	59918	59917	59918
88°	59963	59963	59963	59964
89°	59991	59991	59990	59990

Table 9: Reconstruction of the Sine table used to compute Regiomontanus' Tangent table, found in his *Tabulae directionum*. Asterisks denote entries for which the reconstruction is unreliable. Note: The entry denoted with an asterisk was published as 59161 in Regiomontanus, *Tabula directionum* (1504), due to a typographical error, and we have corrected it here.

Argument	Calculated	Reconstruction	Regiomontanus	Bianchini
80°	10419	10419	10418	10419
81°	9386	9386	9386	9386
82°	8350	8350	8350	8350
83°	7312	7312	7312	7312
84°	6272	6272	6271	6272
85°	5229	5229*	5229	5229
86°	4185	4185	4185	4185
87°	3140	3140	3140	3140
88°	2094	2094	2093	2094
89°	1047	1047	1047	1048

Table 10: Reconstruction of the Cosine table used to compute Regiomontanus' Tangent table, found in his *Tabulae directionum*. Regiomontanus' Cosine values were obtained by reading the values for $\mathrm{Sin}(90° - \vartheta)$ (Sine of the complement) from his Sine table. Asterisks denote entries for which the reconstruction is unreliable.

on Sine and Cosine tables using $R = 60000$ do not match exactly those found in the Tangent table are those for $\vartheta = 85°$ and $89°$. But even for these arguments, using the first method with $R = 60000$ produces a much better fit to Regiomontanus' Tangent value than any other choice of method and R.

Our reconstruction of the Sine and Cosine tables underlying Regiomontanus' printed Tangent tables is found in Tables 9 and 10, respectively. Since the structure of Regiomontanus' *Tabulae directionum*, including this table, was taken directly from the structure of Bianchini's *Tabulae primi mobilis*, we shall consider the possibility that Regiomontanus' Tangent table may depend on one of Bianchini's Sine tables, rather than his own. Thus, we will compare the reconstructed Co/Sine table to both Regiomontanus' and Bianchini's Co/Sine tables of the appropriate R value. The last two columns are therefore Regiomontanus and Bianchini's published Sine tables that use $R = 60000$.

The reconstructed table does not match perfectly either Regiomontanus' nor Bianchini's Sine tables.

- Comparing the reconstructed Sine values to both Regiomontanus' and Bianchini's Sine tables, we find a fit with Regiomontanus in 5 of 9 entries and with Bianchini in 6 of 9 entries.[33]
- All 4 of the remaining entries in the case of Regiomontanus, and all 3 in the case of Bianchini, are exactly correct at the level of precision of the table. This suggests that the Sine table underlying Regiomontanus' Tangent table is an improved version of either Regiomontanus' or Bianchini's Sine table.

We turn next to the Cosine table.

- Comparing the reconstructed Cosine values to both Regiomontanus' and Bianchini's Cosine tables, we find a fit with Regiomontanus in 7 of 10 entries and with Bianchini in 9 of 10 entries.[33]
- All 3 of the remaining entries in the case of Regiomontanus, and the final entry in the case of Bianchini, are exactly correct at the level of precision of the table.

This suggests that the Sine and Cosine tables underlying Regiomontanus' Tangent table is an improved version of either Regiomontanus' or Bianchini's Sine and Cosine tables. As Bianchini's Co/Sine table is slightly better than Regiomontanus' table, it is difficult to determine whether the underlying Co/Sine table is a slightly improved version of Bianchini's table, or a significantly improved version of Regiomontanus' table.

Georg Rheticus

Georg Rheticus (1514–1574) is most famous for his work under Nicolaus Copernicus (1473–1543). Despite initial resistance, Rheticus eventually received permission from Copernicus to publish the *Narratio prima*, an announcement of the heliocentric theory, within a year of his arrival in 1539. Over the next four years, Rheticus also convinced Copernicus to publish his *De revolutionibus orbium coelestium*.[34]

After Copernicus' death, Rheticus began producing his own astronomical work, primarily on trigonometry, as a scholar at the University of Leipzig.

[33] The entry associated with 85° has been excluded, since it was not possible to reconstruct it reliably.

[34] Rosen, 'Rheticus', p. 396.

In 1551 he published his first set of astronomical tables, the *Canon doctrinae triangulorum*, introducing Europe to a revolutionary new perspective on trigonometry. However, this work had little impact since it was banned by the Catholic church. Rheticus spent the better part of his remaining years practicing medicine in Cracow, and developing his masterwork, the *Opus palatinum*. This comprehensive work is over 1400 pages long, and contains extremely large tables with the same structure as the *Canon doctrinae triangulorum*. It was finally completed and published in 1596 by his student Lucius Valentin Otho.[35]

Rheticus' conception of trigonometry relied on the definition of three different 'species' of triangles. Each of these species is defined by setting one side length (the base, perpendicular or hypotenuse) equal to R; with this construct, each of the remaining six side lengths defines a trigonometric function.[36] Rheticus used his own unique terminology, eschewing even the word 'Sine'. Instead, he referred to each of his trigonometric functions as the base, perpendicular, or hypotenuse of one of his species of triangles.

Each of the six functions can be paired with another, such that the column containing one function read forwards from 0° to 90° is the same as the column containing another function read backwards from 90° to 0°. Rheticus exploits this symmetry in order to avoid redundancy in his tables by including arguments only up to 45°.

Canon doctrinae triangulorum: Background

Rheticus' *Canon doctrinae triangulorum* (1551) is a 24-page pamphlet, consisting of a 17-page trigonometric table containing all six trigonometric functions and concluding with a short dialogue extolling the virtues of Rheticus' new trigonometric system.[37] This extraordinary table presents Rheticus' revolutionary reformulation of trigonometry in Europe for the first time.[38] It contains entries for every $10'$ of arc, spanning the entire width of each pair of pages, using a radius of $R = 10\,000\,000$. The Tangents in this table are therefore significantly more precise than Regiomontanus' or Bianchini's, containing 2 and 3 more significant figures respectively.

Canon doctrinae triangulorum: Analysis

For the Tangent table in the *Canon doctrinae triangulorum*, we apply our analysis to the sixty entries for every $10'$ of arc between 80° and 90°. For

[35] Van Brummelen, *The Mathematics of the Heavens*, pp. 273–75.

[36] Van Brummelen, *The Mathematics of the Heavens*, p. 275.

[37] Rheticus, *Canon doctrinae triangulorum* and Archibald, '*Canon Doctrinae Triangvlorvm*', p. 131.

[38] Van Brummelen, *The Mathematics of the Heavens*, p. 273.

28 of the 60 printed values, a perfect match with the table's value is generated by using accurate Sine and Cosine values with $R = 10\,000\,000$. Using this same radius, another 15 matches were obtained by using Co/Sines in error by at most one in the last place. Table 11 displays the number of exact matches achieved by best-fit Sine tables of all the historically plausible R values examined.

Radius (R)	Exact matches
60 000	0
100 000	0
600 000	0
1 000 000	1
6 000 000	1
10 000 000	43
60 000 000	0
100 000 000	9

Table 11: Exact matches obtained using best-fit Sine tables to recompute Georg Rheticus' Tangent table in the *Canon doctrinae triangulorum*. Note: 100 000 000 is 10 times as large as 10 000 000. Thus, supposing that $R = 10\,000\,000$ was used, one should expect an occasional accidental match when using $R = 100\,000\,000$: namely, when the underlying Sine/Cosine value happens to end in 0.

We were able to obtain an exact match for the value associated with the greatest argument in this table: 89°50′. The following table depicts the differences between each of the 25 Tangent values generated by our method, and the Tangent value printed by Rheticus in his table for this argument. The column labels are the Sines, divided by the Cosines labelling each row. Note that the zero in the third column from the left, and third row down signifies that the exact Tangent value printed by Rheticus was obtained by dividing 9 999 957 by 29 088, and multiplying by 10 000 000.

	9 999 956	9 999 957	9 999 958	9 999 959	9 999 960
29 087	117 847	118 191	118 535	118 878	119 222
29 088	−344	0	343	687	1 031
29 089	−118 527	−118 183	−117 840	−117 496	−117 152
29 090	−236 702	−236 358	−236 015	−235 671	−235 327
29 091	−354 869	−354 525	−354 182	−353 838	−353 494

Table 12: The differences between the 25 Tangent values generated by our method and Rheticus' printed Tangent value for argument 89°50′.

Table 13 gives the mean error achieved by best-fit Sine tables for all the historically plausible R values we examined. The fit for $R = 10\,000\,000$ is approximately two orders of magnitude better than any other R value. We

Radius (R)	Mean error
60000	185352.6
100000	53708.2
600000	14335.1
1000000	7295
6000000	985.8
10000000	4.1
60000000	1013.6
100000000	1336.4

Table 13: Mean error achieved by using best-fit Co/Sine tables to recompute the Tangent table found in Georg Rheticus' *Canon doctrinae triangulorum*. The error associated with the entry for argument 81°30′ was removed, since it could not be reliably reconstructed using any historically plausible R value examined.

conclude that Georg Rheticus used $\mathrm{Tan}\,\vartheta = \frac{\mathrm{Sin}\,\vartheta}{\mathrm{Cos}\,\vartheta} \cdot 10\,000\,000$ in conjunction with Co/Sine values with a radius of $R = 10\,000\,000$. Our reconstructions of the Sine and Cosine tables underlying Rheticus' printed Tangent table appear in Appendices 3b and 3c, respectively.

A comparison of the trios of Sine values in the reconstructed table presents the following results:

- The accurate values (rounded to the nearest whole number) match both the reconstructed value and the historical value exactly in 39 of the 58 reliably reconstructed entries.[39]
- Of the remaining 19 entries, in 13 cases the reconstructed value matches the value printed by Rheticus in his Sine table, but not the accurate value.
- For 3 of the remaining 6 arguments, reconstruction matches the computed values, they do not match the values printed in Rheticus' Sine table.
- For the other three the opposite is true: Rheticus' printed values match the accurate values, but the reconstructed value is different.

Therefore, the underlying table is not clearly either an old version of Rheticus' Sine table or an improvement on the printed version.

We turn now to the reconstructed Cosine table.

- The computed values match both the reconstruction and Rheticus' printed Cosine values for 45 out of the 58 reliably reconstructed entries.[40]

[39] The entries for arguments 81°30′ and 86°20′ could not be reconstructed reliably.

[40] As before, the entries for arguments 81°30′ and 86°20′ reveal significant computational errors, making reconstruction unreliable.

- For 11 of the remaining 13 entries, the reconstructed value matches the value printed by Rheticus in his Cosine table, but not the accurate value.
- For the two remaining entries, the reconstruction matches the computed values, but not the values printed in Rheticus' Cosine table.

This suggests that the reconstructed Cosine table is an improved version of the printed table, but the evidence is not strong enough to reach a firm conclusion.

For 52 of the 58 entries the reconstructed underlying Sine table is identical to Rheticus' printed Sine table with $R = 10\,000\,000$, and for 56 of 58 entries the reconstructed underlying Cosine table is identical to Rheticus' printed Cosine table with the same radius. As only 39 and 45 of these entries, respectively, are correct, this suggests that Rheticus' Tangent tables were computed based on a version of his Sine table with $R = 10\,000\,000$, which also appears in the *Canon doctrinae triangulorum*.

Opus palatinum: Background

The *Opus palatinum* comprehensively covers every topic in trigonometry. It is divided approximately evenly between text and tables, with the text divided into four parts. The first three sections, written by Rheticus, describe (i) the methods used to compute the tables, (ii) trigonometry with plane triangles, and (iii) trigonometry with right spherical triangles. The fourth section, written by Otho, discusses oblique spherical triangles.[41] The content of this material awaits scholarly analysis.

In these tables, Rheticus again employs his unique conception of trigonometry.[42] This set of tables, however, is far larger than the *Canon doctrinae triangulorum*. The entries are given for every $10''$ of arc, using radius $R = 10\,000\,000\,000$, and span the entire width of each pair of open pages. The Tangents found in this table are by far the most precise[43] of any we have encountered in this paper, containing 5 and 6 more significant digits than Regiomontanus' and Bianchini's tables, respectively.

While the entries in the *Opus palatinum* table are extremely precise, they are also inaccurate.[44] In the last few entries of this table, up to 9 of the 15 places are in error. These flaws were first noticed by Adrianus Romanus

[41] Van Brummelen, *The Mathematics of the Heavens*, pp. 273–75.

[42] Van Brummelen, *The Mathematics of the Heavens*, p. 273.

[43] Precision is evaluated based on the number of significant digits given.

[44] Accuracy is evaluated by the number of significant digits that are incorrect, compared to the number of significant digits given.

(1561–1615), who used the identity

$$\sec\vartheta + \tan\vartheta = \tan\left[\vartheta + \frac{1}{2}(90^\circ - \vartheta)\right]$$

to check the accuracy of the entries.[45] The tables were later corrected by Bartholomeus Pitiscus, and these revised versions went on to provide computational foundations for a variety of scientific fields until the early 1900s.[46]

Opus palatinum: Analysis

For the Tangent table in the *Opus palatinum*, we analyze the 360 entries for arguments between 89° and 90°. Exact reconstructions (±1) of 287 of these printed entries were generated by using accurate Sine and Cosine values (rounded to the degree of precision of the table) with $R = 10\,000\,000\,000$. Using this same radius, another 71 matches were obtained by using Co/Sines in error by at most one in the last place. Table 14 displays the number of exact matches and almost exact matches (exact ±1) achieved by using best-fit Co/Sine tables for all the historically plausible R values examined.

Radius (R)	Exact matches	Almost exact matches (±1)
6×10^8	1	0
10^9	3	1
6×10^9	6	8
10^{10}	188	170
6×10^{10}	128	110
10^{11}	48	41
6×10^{11}	0	2
10^{12}	0	0

Table 14: Exact matches and almost exact matches (exact ±1) obtained using best-fit tables to recompute Rheticus' Tangent table in the *Opus palatinum*.[48]

[45] For more information, see Bockstaele, 'Adrianus Romanus'.

[46] Van Brummelen, *The Mathematics of the Heavens*, pp. 280–82; Archibald, 'Rheticus', p. 588, and Andoyer, *Nouvelles tables trigonométriques*.

[48] For each argument, when $R = 6 \times 10^{10}$, there is a 5 in 6 probability that one of the Sines (or Cosines) will be a multiple of 6. There is therefore a 25/36 probability that one of the best possible Sines and one of the best possible Cosines will both be multiples of 6. As the Co/Sines used to produce the exact, or near-exact error matches when $R = 10^{10}$ were never in error by more than one in the last place, it is highly likely that these values, multiplied by 6, will be very accurate Co/Sines for $R = 6 \times 10^{10}$. Thus, it makes sense that

We were able to obtain an exact match for the value associated with the greatest argument in this table: 89°59′50″. Table 15 depicts the differences between each of the 25 Tangent values generated by our method, and the Tangent value printed by Rheticus in his table for this argument. The column labels are the Sines, divided by the Cosines labelling each row. Note that the zero in the fourth column from the left, and fourth row down signifies that the exact Tangent value printed by Rheticus was obtained by dividing 9999999988 by 484814, and multiplying by 10000000000.

	9999999986	9999999987	9999999988	9999999989	9999999990
484812	850864543	850885169	850905796	850926422	850947049
484813	425410767	425431394	425452020	425472647	425493273
484814	−41253	−20626	0	20626	41253
484815	−425491518	−425470892	−425450265	−425429639	−425409012
484816	−850940028	−850919402	−850898775	−850878149	−850857523

Table 15: The differences between the 25 Tangent values generated by our method and Rheticus' printed Tangent value for argument 89°50′.

Table 16 gives the mean error achieved by best-fit Sine tables for all the historically plausible R values examined.

Radius (R)	Mean error
6×10^8	7290212.3
10^9	6729995.4
6×10^9	1080471.6
10^{10}	0.875
6×10^{10}	4012.1
10^{11}	213854.7
6×10^{11}	496032.4
10^{12}	534242.9

Table 16: Mean error achieved by using best-fit Sine tables to recompute the Tangent table found in Georg Rheticus' *Opus palatinum*.

The fit for $R = 10000000000$ is well over three orders of magnitude better than any other R value. We conclude that Rheticus computed his Tangents by using $\mathrm{Tan}\,\vartheta = \frac{\mathrm{Sin}\,\vartheta}{\mathrm{Cos}\,\vartheta} \cdot 10000000000$ in conjunction with Co/Sine values with

we see so many error matches for this value of R, even though we can determine from the additional evidence that it is not the value that Rheticus used.

10^{11} is 10 times as large as 10^{10}. Thus, supposing that $R = 10000000000$ was used, one should expect an occasional accidental match when using $R = 100000000000$: namely, when the underlying Sine/Cosine value happens to end in 0.

a radius of $R = 10\,000\,000\,000$. Our reconstructions of the Sine and Cosine tables underlying Rheticus' printed Tangent table for arguments between 89° and 90° appear in Appendices 4b and 4c, respectively.

A comparison of the trios of Sine values in the reconstructed table (Appendix 4b) presents the following:

- The correct values (rounded to the nearest whole number) exactly match both the reconstructed value and the value printed by Rheticus in his tables in 300 of the 358 reliably reconstructed entries.[49]
- Of the remaining 58 entries, in 39 cases the reconstructed value matches the value printed by Rheticus in his Sine tables, but not the correct value.
- For the 19 remaining entries, while the reconstruction matches the computed values, they do not match the values printed in Rheticus' Sine table.

We turn next to the reconstructed Cosine table (Appendix 4c).

- Here, the computed values match both the reconstruction and Rheticus' printed Cosine values for 308 out of the 358 reliably reconstructed entries.[50]
- For 33 of the remaining 50 entries, the reconstructed value matches the value printed by Rheticus in his Cosine table, but not the correct value.
- For 16 of the 17 remaining entries, the reconstruction matches the computed values, but not the values printed in Rheticus' Cosine table.
- For only one entry, the value printed by Rheticus matches the computed value, but this value does not appear in the reconstructed table.

We conclude that the underlying Sine and Cosine tables are improved versions of Rheticus' printed Sine and Cosine tables.

The reconstructed underlying Sine table is identical to Rheticus' printed Sine table with $R = 10\,000\,000\,000$ for 339 of the 358 reliably reconstructed entries, and the reconstructed underlying Cosine table is identical to Rheticus' printed Cosine table with the same radius for 341 of 358 reliably reconstructed entries. Of the remaining entries, all 19 of the reconstructed Sine values and 16 of the reconstructed Cosine values appear to be improved versions of the values that appear in the Sine table printed in the *Opus palatinum*.

[49] The entries for arguments 89°16′0″ and 89°28′50″ could not be reconstructed reliably.

[50] As with the reconstruction of the Sine table, two of the reconstructions were unreliable, likely due to computational error.

Conclusion

We have determined how four early European Tangent tables were computed. Unsurprisingly, all four use the relationship $\text{Tan}\,\vartheta = \frac{\text{Sin}\,\vartheta}{\text{Cos}\,\vartheta} \cdot R$. The Tangent tables in Giovanni Bianchini's *Tabulae primi mobilis* and Georg Rheticus' *Canon doctrinae triangulorum* were computed using versions of the Co/Sine tables printed alongside the Tangent tables in those same books with radii $R = 60\,000$ and $R = 10\,000\,000$, respectively. While they were not identical to the tables in the manuscripts, they were very similar. The reconstructed entries which were different from those found in the manuscript were sometimes more accurate, and other times less accurate, than those found in the manuscript. In contrast, the Tangent tables in Regiomontanus' *Tabulae directionum* and Georg Rheticus' *Opus palatinum* were computed using Co/Sine tables that were significantly better than those in the manuscripts. The underlying tables use the radii $R = 60\,000$ and $R = 10\,000\,000\,000$, respectively.

Our results help us to form a clearer picture of how these scholars thought and worked behind the scenes, as they constructed early Tangent tables. Their choices of R values, compared to the R values of the underlying Sine and Cosine tables, demonstrate differences in their conceptions of what the Tangent function was. While Bianchini and Regiomontanus use different R values for their Tangent tables and for their Sine and Cosine tables, Georg Rheticus uses the same R value for both sets of tables. Regiomontanus and Bianchini understood the Tangent as an auxiliary function without astronomical meaning, which allowed them to choose any R value they saw fit as the base for their table. In contrast, Rheticus used the same R value for all of his tables, since his understanding of the six trigonometric functions was based upon the construction of three similar triangles, each with one length equal to the R he had chosen.

The differences between our reconstructed Co/Sine tables and the Co/Sine tables in the texts themselves provide valuable insight into the authors' relationships with their tables. For the *Canon doctrinae triangulorum* and the *Tabulae primi mobilis*, the two sets of Co/Sine tables are similar. In the cases of the *Tabulae directionum* and *Opus palatinum*, however, the underlying Co/Sine tables were found to be significantly more accurate than the published Co/Sine tables. This indicates the authors' continuing commitment to improving the accuracy of their tables, even if those revised versions would never be published.

As committed as these authors were, a comparison of their Tangent tables to Arabic Tangent tables shows that these European table makers were at an early stage, not as aware of issues of error propagation as the Arabic scholars or those who came after them. Tables and their underlying computations grew dramatically in sophistication in the 17th century.

Acknowledgements

I would like to express my immense appreciation to my mentor and research supervisor, Dr. Glen Van Brummelen. My greatest thanks also to the referees who improved this paper immeasurably. All remaining shortcomings are the responsibility of the author.

Bibliography

Andoyer, Henri, *Nouvelles tables trigonométriques fondamentales*, 3 vols, Paris: A. Hermann et fils, 1915–1918.

Archibald, Raymond Claire, 'The *Canon Doctrinae Triangvlorvm* (1551) of Rheticus (1514-1576)', in *Mathematical Tables and Other Aids to Computation* 7 (1953), p. 131.

Archibald, Raymond Claire, 'Rheticus, with Special Reference to his Opus Palatinum', in *Mathematical Tables and Other Aids to Computation* 3 (1949), pp. 552–61.

Bockstaele, Paul, 'Adrianus Romanus and the Trigonometric Tables of Georg Joachim Rheticus', in Sergei S. Demidov, Menso Folkerts, David E. Rowe and Christoph J. Scriba (eds), *Amphora: Festschrift für Hans Wussing zu seinem 65. Geburtstag*, Basel/Boston/Berlin: Birkhäuser, 1992, pp. 55–66.

Chabás, José, and Bernard Goldstein, *The Astronomical Tables of Giovanni Bianchini*, Leiden/Boston: Brill, 2009.

Dalen, Benno van, 'A Statistical Method for Recovering Unknown Parameters from Medieval Astronomical Tables', *Centaurus* 32 (1989), pp. 85–145.

Dalen, Benno van, 'Islamic and Chinese Astronomy under the Mongols: A Little-Known Case of Transmission', in Yvonne Dold-Samplonius, Joseph W. Dauben, Menso Folkerts and Benno van Dalen (eds), *From China to Paris: 2000 Years Transmission of Mathematical Ideas*, Stuttgart: Steiner, 2002, pp. 327--56.

Rosen, Edward, 'Rheticus, George Joachim', in Charles Gillispie (ed.), *Dictionary of Scientific Biography*, vol. XI, New York: Scribner's Sons, 1981, pp. 395–98.

Regiomontanus, Johannes, *Tabula directionum*, Augsburg: Erhard Ratdolt, 1490.

Regiomontanus, Johannes, *Tabula directionum*, Venice: Petri Liechtensteyn, 1504.

Hughes, Barnabas, *Regiomontanus. On Triangles*, Madison: University of Wisconson Press, 1967.

Rheticus, Georg Joachim, *Canon doctrinae triangulorum*, Leipzig: Wolfgang Gunter, 1551.

Rheticus, Georg Joachim. *Opus palatinum de triangulis*, Neostadii in Palatinatu (Neustadt an der Haardt): Matthaeus Harnisius, 1596.

Rosińska, Grazyna, 'Tables trigonométriques de Giovanni Bianchini', *Historia Mathematica* 8 (1981), pp. 46–55.

Van Brummelen, Glen, and Kenneth Butler, 'Determining the Interdependence of Historical Astronomical Tables', *Journal of the American Statistical Association* 92 (1997), pp. 41–48.

Van Brummelen, Glen, *The Mathematics of the Heavens and the Earth*, Princeton: Princeton University Press, 2009.

Van Brummelen, Glen, 'The End of an Error: Bianchini, Regiomontanus, and the Tabulation of Stellar Coordinates', *Archive for History of Exact Sciences* 72 (2018), pp. 547–63.

Van Brummelen, Glen, and James Byrne, 'Maurolico, Rheticus, and the Birth of the Secant Function', *Journal for History of Astronomy*, to appear.

Van Brummelen, Glen, 'Before the End of an Error: Giovanni Bianchini's Original Flawed Treatise on the Conversion of Stellar Coordinates', *Archive for History of Exact Sciences*, to appear.

Zinner, Ernst, *Regiomontanus: His Life and Work*, tr. Ezra Brown, Amsterdam: North Holland, 1990.

Appendix 1a

Excerpt from Giovanni Bianchini's 'tabula magistralis quarta', found in his *Tabulae primi mobilis*, containing the entries for each 5° of arc from 0° to 80°, and every 10′ of arc from 80° to 90°. The complete table can be found at www.kailynpritchard.com.

Argument	Tangent	Error
0°	0	
5°	875	
10°	1 763	
15°	2 679	
20°	3 640	
25°	4 663	
30°	5 773	−1
35°	7 002	
40°	8 391	
45°	10 000	
50°	11 918	
55°	14 282	+1
60°	17 321	
65°	21 445	
70°	27 474	−1
75°	37 321	
80°	56 712	−1
80°10′	57 693	−1
80°20′	58 708	
80°30′	59 757	−1
80°40′	60 842	−2
80°50′	61 972	+2
81°	63 138	
81°10′	64 346	−2
81°20′	65 607	+1
81°30′	66 908	−4
81°40′	68 268	−1
81°50′	69 683	+1
82°	71 158	+4
82°10′	72 692	+5
82°20′	74 293	+6
82°30′	75 964	+6
82°40′	77 910	+206
82°50′	79 535	+5
83°	81 447	+4
83°10′	83 449	−1
83°20′	85 562	+7
83°30′	87 771	+2

Argument	Tangent	Error
83°40′	90 044	−54
83°50′	92 555	+2
84°	95 139	−5
84°10′	97 867	−15
84°20′	100 771	−9
84°30′	103 850	−4
84°40′	107 119	
84°50′	110 598	+4
85°	114 309	+8
85°10′	118 250	−12
85°20′	122 491	−14
85°30′	127 050	−12
85°40′	131 983	+14
85°50′	137 280	+13
86°	143 023	+16
86°10′	149 293	+49
86°20′	156 092	+44
86°30′	163 539	+40
86°40′	171 726	+33
86°50′	180 773	+23
87°	190 822	+11
87°10′	202 046	−10
87°20′	214 744	+40
87°30′	229 052	+14
87°40′	245 434	+16
87°50′	264 361	+45
88°	286 342	−21
88°10′	312 511	+95
88°20′	343 691[51]	+13
88°30′	381 789	−96
88°40′	429 678	+37
88°50′	490 892	−147
89°	572 980	+80
89°10′	687 216	−285
89°20′	859 542	+144
89°30′	1 145 000	−887
89°40′	1 719 170	+316
89°50′	3 420 851	−16 886
90°	Infinitum	

[51] Entry corrected from 383 691.

Appendix 1b

Reconstruction of the Sine table underlying Giovanni Bianchini's 'tabula magistralis quarta', found in his *Tabulae primi mobilis*. The computed values and values from the *Tabulae primi mobilis* are given as differences between those values and the reconstructed values. Asterisks denote entries for which the reconstruction is unreliable.

Argument	Recon-structed	Com-puted	Bianchini
80°	59 088		
80°10′	59 118	+1	
80°20′	59 148		
80°30′	59 177		
80°40′	59 205	+1	
80°50′	59 233	+1	
81°	59 261		
81°10′	59 288		
81°20′	59 315		
81°30′	59 341		
81°40′	59 366		
81°50′	59 391	+1	
82°	59 417	−1	−1
82°10′	59 440		
82°20′	59 464		
82°30′	59 487		
82°40′	59 510	−1	
82°50′	59 532	−1	
83°	59 554	−1	
83°10′	59 574		
83°20′	59 594		
83°30′	59 614		
83°40′	59 636	−2	−3
83°50′	59 652	+1	
84°	59 671		
84°10′	59 689		−10
84°20′	59 707		
84°30′	59 724		
84°40′	59 740		
84°50′	59 756		

Argument	Recon-structed	Com-puted	Bianchini
85°	59 772		
85°10′	59 787		
85°20′	59 800	+1	+1
85°30′	59 815		
85°40′	59 828		
85°50′	59 841		
86°	59 855	−1	−1
86°10′	59 866		
86°20′	59 877		
86°30′	59 888		
86°40′	59 898		
86°50′	59 908		
87°	59 918		
87°10′	59 927		
87°20′	59 935		
87°30′	59 943		
87°40′	59 952	−2	−2
87°50′	59 957		
88°	59 961	+2	+3
88°10′	59 971	−2	−2
88°20′	59 974	+1	
88°30′	59 979		
88°40′	59 983	+1	
88°50′	59 987	+1	
89°	59 991		
89°10′	59 994		
89°20′	59 996		
89°30′	59 998		
89°40′	59 999		
89°50′	60 000 *		

Appendix 1c

Reconstruction of the Cosine table underlying Giovanni Bianchini's 'tabula magistralis quarta', found in his *Tabulae primi mobilis*. The computed values and values from the *Tabulae primi mobilis* are given as differences between those values and the reconstructed values. Bianchini's Cosine values were obtained by reading the values for $\mathrm{Sin}(90° - \vartheta)$ (Sine of the complement) from his Sine table. He did not publish a separate Cosine table in the *Tabulae primi mobilis*. Asterisks denote entries for which the reconstruction is unreliable.

Argument	Reconstructed	Computed	Bianchini
80°	10 419		
80°10′	10 247		
80°20′	10 075		
80°30′	9 903		
80°40′	9 731		
80°50′	9 558		
81°	9 386		
81°10′	9 214		
81°20′	9 041		
81°30′	8 869		
81°40′	8 696		
81°50′	8 523		
82°	8 350		
82°10′	8 177	+1	
82°20′	8 004	+1	
82°30′	7 831	+1	
82°40′	7 658		
82°50′	7 485		
83°	7 312		
83°10′	7 139		
83°20′	6 965	+1	
83°30′	6 792		
83°40′	6 621	−2	−2
83°50′	6 445		
84°	6 272		
84°10′	6 099	−1	
84°20′	5 925	−1	
84°30′	5 751		
84°40′	5 577		
84°50′	5 403		
85°	5 229		
85°10′	5 056	−1	
85°20′	4 882		
85°30′	4 708		
85°40′	4 533	+1	
85°50′	4 359		
86°	4 185		
86°10′	4 010	+1	
86°20′	3 836	+1	
86°30′	3 662	+1	
86°40′	3 488	+1	
86°50′	3 314		
87°	3 140		
87°10′	2 966		
87°20′	2 791	+1	
87°30′	2 617		
87°40′	2 443		−1
87°50′	2 268		
88°	2 094		
88°10′	1 919	+1	
88°20′	1 745		
88°30′	1 571		
88°40′	1 396		
88°50′	1 222		
89°	1 047		+1
89°10′	873		
89°20′	698		
89°30′	524		
89°40′	349		
89°50′	175 *		

Appendix 2

Excerpt from Regiomontanus' 'tabula fecunda', found in his *Tabulae directionum*, containing the entries for each 5° of arc from 0° to 80°, and every degree of arc from 80° to 90°. The complete table can be found at www.kailynpritchard.com.

Argument	Tangent	Error
0°	0	
5°	8 748	−1
10°	17 633	
15°	26 794	−1
20°	36 396	−1
25°	46 631	
30°	57 734	−1
35°	70 022	+1
40°	83 909	−1
45°	100 000	
50°	119 197	+22
55°	142 813	−2
60°	173 207	+2
65°	214 450	−1
70°	274 753	+5
75°	373 211	+6
80°	567 118	−10
81°	631 377	+2
82°	711 569	+32
83°	814 456	+21
84°	951 387	−49
85°	1 143 131	+126
86°	1 430 203	+136
87°	1 908 217	+103
88°	2 863 563	−62
89°	5 729 796	+800
90°	Infinitum	

Appendix 3a

Excerpt from Georg Rheticus' Tangent table, found in his *Canon doctrinae triangulorum*, containing the entries for each 5° of arc from 5° to 80°, and every 10′ of arc from 80° to 90°. The complete table can be found at www.kailynpritchard.com.

Argument	Tangent	Error
5°	874 886	−1
10°	1 763 269	−1
15°	2 679 491	−1
20°	3 639 702	
25°	4 663 077	
30°	5 773 503	
35°	7 002 075	
40°	8 390 996	
45°	10 000 000	
50°	11 917 537	+1
55°	14 281 480	
60°	17 320 508	
65°	21 445 068	−1
70°	27 474 777	+3
75°	37 320 514	+6
80°	56 712 813	−5
80°10′	57 693 673	−15
80°20′	58 708 044	+2
80°30′	59 757 645	+1
80°40′	60 844 394	+13
80°50′	61 970 266	−13
81°	63 137 498	−17
81°10′	64 348 408	−20
81°20′	65 605 540	+2
81°30′	66 916 224	+4 662
81°40′	68 269 413	−24
81°50′	69 682 330	−5
82°	71 153 699	+2
82°10′	72 687 230	−25
82°20′	74 287 083	+19
82°30′	75 957 539	−2
82°40′	77 703 478	−28
82°50′	79 530 235	+11
83°	81 443 497	+33
83°10′	83 449 584	+26
83°20′	85 555 482	+14
83°30′	87 768 888	+14
83°40′	90 098 230	−31
83°50′	92 553 002	−33
84°	95 143 612	−33
84°10′	97 881 716	−16
84°20′	100 780 357	+46
84°30′	103 853 920	−51
84°40′	107 119 198	+72
84°50′	110 594 305	−5
85°	114 300 579	+56
85°10′	118 261 757	+90
85°20′	122 505 018	−37
85°30′	127 062 036	−11
85°40′	131 968 930	+100
85°50′	137 267 523	+145
86°	143 006 616	−47
86°10′	149 244 149	−21
86°20′	156 048 656	+815
86°30′	163 498 661	+106
86°40′	171 693 462	+93
86°50′	180 749 538	−236
87°	190 811 200	−167
87°10′	202 055 702	+167
87°20′	214 704 086	+76
87°30′	229 037 584	−71
87°40′	245 417 544	−34
87°50′	264 316 359	+363
88°	286 362 496	−37
88°10′	312 416 183	+416
88°20′	343 677 947	+238
88°30′	381 885 288	+695
88°40′	429 641 796	+1 023
88°50′	491 038 024	−782
89°	572 899 830	+214
89°10′	687 500 739	−133
89°20′	859 395 374	−2 533
89°30′	1 145 891 136	+4 635
89°40′	1 718 863 108	+9 115
89°50′	3 437 829 002	+91 927
90°	Infinitum	

Appendix 3b

Reconstruction of the Sine table underlying Georg Rheticus' Tangent table found in his *Canon doctrinae triangulorum*, containing entries for every 10′ of arc from 89° to 89°50′. The computed values and values from the *Canon doctrinae triangulorum* are given as differences between those values and the reconstructed values. Asterisks denote entries for which the reconstruction is unreliable.

Argument	Recon-structed	Com-puted	Rheticus
80°	9 848 078		
80°10′	9 853 087		
80°20′	9 858 014	−1	
80°30′	9 862 856		
80°40′	9 867 616	−1	
80°50′	9 872 291		
81°	9 876 883		
81°10′	9 881 393	−1	−1
81°20′	9 885 817		
81°30′	9 890 161*	−2	−2
81°40′	9 894 416		
81°50′	9 898 591	−1	
82°	9 902 681		
82°10′	9 906 688	−1	
82°20′	9 910 610		
82°30′	9 914 449		
82°40′	9 918 204		
82°50′	9 921 874		
83°	9 925 462		−1
83°10′	9 928 965		
83°20′	9 932 384		
83°30′	9 935 719		
83°40′	9 938 970	−1	
83°50′	9 942 136		
84°	9 945 219		
84°10′	9 948 218	−1	
84°20′	9 951 133	−1	−1
84°30′	9 953 962		
84°40′	9 956 708		
84°50′	9 959 371	−1	−1

Argument	Recon-structed	Com-puted	Rheticus
85°	9 961 947		
85°10′	9 964 440		
85°20′	9 966 849		
85°30′	9 969 173		
85°40′	9 971 414	−1	
85°50′	9 973 570	−1	
86°	9 975 641		−1
86°10′	9 977 628	−1	
86°20′	9 979 532*	−2	−2
86°30′	9 981 348		
86°40′	9 983 082		
86°50′	9 984 731		
87°	9 986 295		
87°10′	9 987 775		
87°20′	9 989 172	−1	
87°30′	9 990 482		
87°40′	9 991 709		+1
87°50′	9 992 850	+1	
88°	9 993 908		
88°10′	9 994 881		
88°20′	9 995 770		
88°30′	9 996 573		
88°40′	9 997 292		
88°50′	9 997 927		
89°	9 998 477		
89°10′	9 998 942		
89°20′	9 999 323		
89°30′	9 999 619		
89°40′	9 999 830	+1	
89°50′	9 999 957	+1	

Appendix 3c

Reconstruction of the Cosine table underlying Georg Rheticus' Tangent table found in his *Canon doctrinae triangulorum*, containing entries for every 10′ of arc from 89° to 89°50′. The computed values and values from the *Canon doctrinae triangulorum* are given as differences between those values and the reconstructed values. Asterisks denote entries for which the reconstruction is unreliable.

Argument	Recon-structed	Com-puted	Rheticus
80°	1 736 482		
80°10′	1 707 828		
80°20′	1 679 159		
80°30′	1 650 476		
80°40′	1 621 779		
80°50′	1 593 069		
81°	1 564 345		
81°10′	1 535 608	−1	
81°20′	1 506 857		
81°30′	1 478 092*	+2	+2
81°40′	1 449 319		
81°50′	1 420 531		
82°	1 391 731		
82°10′	1 362 920	−1	
82°20′	1 334 096		
82°30′	1 305 262		
82°40′	1 276 417	−1	
82°50′	1 247 560		
83°	1 218 693		
83°10′	1 189 816		
83°20′	1 160 929		
83°30′	1 132 032		
83°40′	1 103 126		
83°50′	1 074 210		
84°	1 045 285		
84°10′	1 016 351		
84°20′	987 408		
84°30′	958 458		
84°40′	929 498	+1	
84°50′	900 532		−1

Argument	Recon-structed	Com-puted	Rheticus
85°	871 557		
85°10′	842 575	+1	
85°20′	813 587		
85°30′	784 591		
85°40′	755 588	+1	
85°50′	726 579	+1	
86°	697 565		
86°10′	668 544		
86°20′	639 515*	+2	+2
86°30′	610 485		
86°40′	581 448		
86°50′	552 407	−1	
87°	523 360		
87°10′	494 308		
87°20′	465 253		
87°30′	436 194		
87°40′	407 131		
87°50′	378 064	+1	
88°	348 995		
88°10′	319 922		
88°20′	290 847		
88°30′	261 769		
88°40′	232 689	+1	
88°50′	203 608		
89°	174 524		+1
89°10′	145 439		
89°20′	116 353		
89°30′	87 265		
89°40′	58 177		
89°50′	29 088	+1	

Appendix 4a

Excerpt from Georg Rheticus' Tangent table, found in his *Opus palatinum*, containing the entries for each 5° of arc from 5° to 80°, every degree from 80° to 89°, every 5′ from 89° to 89°50′, every minute from 89°51′ to 89°53′, and every 10″ from 89°53′ to 90°. The complete table can be found at www.kailynpritchard.com. The asterisk denotes the obviously incorrect value for 45° in the source.

Argument	Tangent	Error
5°	874 886 635	
10°	1 763 269 808	+1
15°	2 679 491 924	
20°	3 639 702 343	
25°	4 663 076 581	−1
30°	5 773 502 692	
35°	7 002 075 382	
40°	8 390 996 312	
45°	1 000 000 000 *	
50°	11 917 535 925	−1
55°	14 281 480 068	+1
60°	17 320 508 076	
65°	21 445 069 206	+1
70°	27 474 774 197	+2
75°	37 320 508 076	
80°	56 712 818 196	
81°	63 137 515 147	
82°	71 153 697 224	
83°	81 443 464 279	−1
84°	95 143 644 515	−27
85°	114 300 523 091	+63
86°	143 006 662 649	+82
87°	190 811 367 023	+146
88°	286 362 532 844	+15
89°	572 899 617 499	+1 191
89°5′	624 991 535 656	−1 344
89°10′	687 500 874 524	+2 419
89°15′	763 900 091 458	−1 653
89°20′	859 397 907 351	+99
89°25′	982 179 427 668	+1 779
89°30′	1 145 886 501 120	−173
89°35′	1 375 074 470 983	+3 439
89°40′	1 718 854 008 950	+16 149
89°45′	2 291 816 628 035	−8 059
89°50′	3 437 737 056 005	−18 501
89°51′	3 819 709 888 983	−18 572
89°52′	4 297 175 649 163	−57 296
89°53′	4 911 060 157 173	+129 159

Continued on the next page

Excerpt from Georg Rheticus' Tangent table (continued)

Argument	Tangent	Error
89°53′	4 911 060 157 173	+129 159
89°53′10″	5 030 842 265 276	−41 791
89°53′20″	5 156 613 715 535	+23 542
89°53′30″	5 288 834 966 475	+83 231
89°53′40″	5 428 015 211 929	+135 875
89°53′50″	5 574 718 694 224	+180 429
89°54′	5 729 572 021 517	−112 026
89°54′10″	5 893 274 416 416	−105 912
89°54′20″	6 066 606 340 655	−113 232
89°54′30″	6 250 443 205 118	−136 085
89°54′40″	6 445 770 261 771	+237 895
89°54′50″	6 653 698 619 582	+201 999
89°55′	6 875 488 837 803	+144 371
89°55′10″	7 112 574 899 755	+60 525
89°55′20″	7 366 595 644 761	−53 420
89°55′30″	7 639 432 702 116	−202 971
89°55′40″	7 933 257 809 981	+232 965
89°55′50″	8 250 588 250 873	+41 105
89°56′	8 594 362 842 728	−205 725
89°56′10″	8 968 031 622 554	+285 237
89°56′20″	9 375 669 422 708	−33 222
89°56′30″	9 822 130 762 094	+524 977
89°56′40″	10 313 237 189 509	+109 246
89°56′50″	10 856 038 936 937	−426 636
89°57′	11 459 153 193 466	+199 732
89°57′10″	12 133 220 628 617	−520 994
89°57′20″	12 891 548 003 369	+198 602
89°57′30″	13 750 985 747 924	+1 088 856
89°57′40″	14 733 198 214 159	+30 400
89°57′50″	15 866 522 557 360	+1 100 753
89°58′	17 188 731 457 652	−457 015
89°58′10″	18 751 345 086 678	+841 860
89°58′20″	20 626 477 397 360	−1 611 311
89°58′30″	22 918 310 306 158	−44 632
89°58′40″	25 783 101 825 985	+2 337 935
89°58′50″	29 466 397 178 769	−2 582 423
89°59′	34 377 467 277 806	+539 580
89°59′10″	41 252 966 938 221	+6 496 816
89°59′20″	51 566 193 264 939	−7 650 378
89°59′30″	68 754 936 735 144	+1 804 319
89°59′40″	103 132 441 165 520	+38 365 292
89°59′50″	206 264 670 327 177	−135 758 907
90°	Infinitum	

Appendix 4b

Partial reconstruction of the Sine table underlying Georg Rheticus' Tangent table found in his *Opus palatinum*, containing entries for every 1′ of arc from 89° to 89°59′, after which point it contains entries for every 10″ of arc. The complete reconstruction contains entries for every 10″ of arc from 89° to 89°59′50″, and can be found at www.kailynpritchard.com. The computed values and values from the *Opus palatinum* are given as differences between those values and the reconstructed values. Rheticus' Sine values were obtained by reading the values for Cos(90° − ϑ) (Cosine of the complement) from his Cosine table. While he published both a Sine and Cosine table in his *Opus palatinum*, he only included these for arguments up to 45°. Asterisks denote entries for which the reconstruction is unreliable.

Argument	Reconstructed	Computed	Rheticus
89°0′	9 998 476 951	+1	
89°1′	9 998 527 295	+1	
89°2′	9 998 576 793		
89°3′	9 998 625 445		
89°4′	9 998 673 251		
89°5′	9 998 720 211		
89°6′	9 998 766 325		
89°7′	9 998 811 592	+1	
89°8′	9 998 856 014		
89°9′	9 998 899 590		
89°10′	9 998 942 319		
89°11′	9 998 984 203		
89°12′	9 999 025 240		
89°13′	9 999 065 431		
89°14′	9 999 104 777		
89°15′	9 999 143 276		
89°16′	9 999 180 928*	+1	+1
89°17′	9 999 217 736		
89°18′	9 999 253 697	−1	
89°19′	9 999 288 811		
89°20′	9 999 323 080		
89°21′	9 999 356 503		
89°22′	9 999 389 079		
89°23′	9 999 420 809		
89°24′	9 999 451 694		
89°25′	9 999 481 732		
89°26′	9 999 510 924		
89°27′	9 999 539 270		
89°28′	9 999 566 769		
89°29′	9 999 593 423		
89°30′	9 999 619 231		
89°31′	9 999 644 192		
89°32′	9 999 668 307		
89°33′	9 999 691 576		
89°34′	9 999 713 999		
89°35′	9 999 735 576		
89°36′	9 999 756 307		
89°37′	9 999 776 192		
89°38′	9 999 795 230		
89°39′	9 999 813 422		
89°40′	9 999 830 768	+1	
89°41′	9 999 847 268	+1	
89°42′	9 999 862 922		
89°43′	9 999 877 730		
89°44′	9 999 891 692		
89°45′	9 999 904 807		
89°46′	9 999 917 076		
89°47′	9 999 928 500	−1	
89°48′	9 999 939 076	+1	
89°49′	9 999 948 807		
89°50′	9 999 957 692		
89°51′	9 999 965 730	+1	
89°52′	9 999 972 923		
89°53′	9 999 979 269		
89°54′	9 999 984 769		
89°55′	9 999 989 423		
89°56′	9 999 993 231		
89°57′	9 999 996 192		
89°58′	9 999 998 308		
89°59′	9 999 999 577		
89°59′10″	9 999 999 706		
89°59′20″	9 999 999 812		
89°59′30″	9 999 999 894		
89°59′40″	9 999 999 953		
89°59′50″	9 999 999 988		

Appendix 4c

Partial reconstruction of the Cosine table underlying Georg Rheticus' Tangent table found in his *Opus palatinum*, containing entries for every 1′ of arc from 89° to 89°59′, after which point it contains entries for every 10″ of arc. The complete reconstruction contains entries for every 10″ of arc from 89° to 89°59′50″, and can be found at www.kailynpritchard.com. The computed values and values from the *Opus palatinum* are given as differences between those values and the reconstructed values. Rheticus' Cosine values were obtained by reading the values for $\mathrm{Sin}(90^\circ - \vartheta)$ (Sine of the complement) from his Sine table. While he published both a Sine and Cosine table in his *Opus palatinum*, he only included these for arguments up to 45°.

Argument	Reconstructed	Computed	Rheticus
89°0′	174 524 064		
89°1′	171 615 618		
89°2′	168 707 157		
89°3′	165 798 682		
89°4′	162 890 193		
89°5′	159 981 690		
89°6′	157 073 173		
89°7′	154 164 643		
89°8′	151 256 100		
89°9′	148 347 545		
89°10′	145 438 976	+1	
89°11′	142 530 396		
89°12′	139 621 803		
89°13′	136 713 199		
89°14′	133 804 583		
89°15′	130 895 956		
89°16′	127 987 317*		
89°17′	125 078 668		
89°18′	122 170 008		
89°19′	119 261 338		
89°20′	116 352 658		
89°21′	113 443 968		
89°22′	110 535 268		
89°23′	107 626 559		
89°24′	104 717 841		
89°25′	101 809 114		
89°26′	98 900 378	+1	
89°27′	95 991 635		
89°28′	93 082 882	+1	
89°29′	90 174 122	+1	
89°30′	87 265 355		
89°31′	84 356 580		
89°32′	81 447 798		
89°33′	78 539 009		
89°34′	75 630 213		
89°35′	72 721 411		
89°36′	69 812 603		
89°37′	66 903 789		
89°38′	63 994 969		
89°39′	61 086 144		
89°40′	58 177 313	+1	
89°41′	55 268 478		
89°42′	52 359 638		
89°43′	49 450 794		
89°44′	46 541 945		
89°45′	43 633 093		
89°46′	40 724 237		
89°47′	37 815 377		
89°48′	34 906 514		
89°49′	31 997 648		
89°50′	29 088 780		−1
89°51′	26 179 909		
89°52′	23 271 036		−1
89°53′	20 362 160	+1	
89°54′	17 453 284		
89°55′	14 544 405		
89°56′	11 635 526		
89°57′	8 726 645		
89°58′	5 817 764		
89°59′	2 908 882		
89°59′10″	2 424 068		
89°59′20″	1 939 255		
89°59′30″	1 454 441		
89°59′40″	969 627		
89°59′50″	484 814		

Part 2

Editing and Analysing Astronomical Tables

Editing Sanskrit Astronomical Tables: The *Candrārkī* of Dinakara (1578 CE)

Clemency MONTELLE

1. Introduction

Numerical tables were a popular medium for mathematical astronomy in second millennium India. Initial cataloguing efforts suggest there may be hundreds of thousands, if not millions of folia containing tables still extant today. From astronomical data to astrological correspondences, from horoscopes to calendars, from star-lists to sine tables, such repositories of data, often referred to by the Sanskrit word *koṣṭhaka* ('granary') or *sāraṇī* ('stream'), became widespread and had a notable impact on cultures of scientific practice. Given the scale and scope of such a genre, the situation for the historian of astronomy is thus as daunting as it is thrilling.[1]

The critical editor of numerical tables has a special set of challenges above and beyond the regular editor of textual material.[2] For one, the notion of variant reading takes on new significance when dealing with computed data. Discrepancies in the numerical data may arise from simple unintentional copying slips, but they also might be systematic and reveal a computational preference on the part of the compiler. Modern recomputations based on reconstructed algorithms may offer guidance to the editor as to how to emend erroneous values, however this must be done with some caution, as recomputed data brings with it its own set of implicit biases. The editor has to make some

[1] Despite the extent of the corpus, the study of numerical tables in India is just beginning. Initial cataloguing efforts began with David Pingree's publications 'Sanskrit Astronomical Tables in the United States' (*SATIUS*) and 'Sanskrit Astronomical Tables in England' (*SATE*). Over the last few decades, with increased international scholarly interest in tables as a genre, studies dedicated to numerical tables are becoming more numerous. See, for instance, Neugebauer and Pingree, 'The Astronomical Tables of Mahādeva'; Pingree, 'On the Classification'; Pingree, *Śrīdhara's Laghukhecarasiddhi*; Ikeyama and Plofker, 'The *Tithicintāmaṇi* of Gaṇeśa'; Montelle and Plofker, '*Karaṇakesari* of Bhāskara II'; Montelle and Plofker, 'The Transformation of a Handbook'; Montelle and Plofker, *Sanskrit Astronomical Tables*.

[2] Despite numerous publications dealing with critical editing, few studies have focussed on the issues faced by editing numerical data. See Montelle and Plofker, '*Karaṇakesari* of Bhāskara II', esp. pp. 32–34; Montelle and Plofker, 'The Transformation of a Handbook', and Misra et al., *The Sanskrit Astronomical Table Text Brahmatulyasāraṇī*.

Editing and Analysing Numerical Tables: Towards a Digital Information System for the History of Astral Sciences, ed. by Matthieu Husson, Clemency Montelle and Benno van Dalen, PALS 2 (Turnhout, 2021), pp. 145–185
© BREPOLS PUBLISHERS 10.1484/M.PALS-EB.5.127698

difficult decisions in light of the variety of discrepancies and whether or not they should be emended or modified.[3]

Paratextual features including titles, column/row headings and marginal notes pose interesting nuances to the task of editing numerical data. Furthermore, the spatial arrangement of data on a page is ever more relevant. All these subtleties are important, as they give us insight into the practices surrounding table preparation, use, and transmission. They simultaneously provide insight into individual compilers and copiers as well as a glimpse of some of the more general trends and priorities related to this milieu.

In order to explore these issues and some of the ways in which they may be embraced, we consider here the task of editing the numerical data related to a sixteenth century Sanskrit table-text concerning solar and lunar phenomena, the *Candrārkī* by the astronomer Dinakara. In addition to a critical edition of the tables, we offer an exploration of the specific challenges that arose when preparing this edition and the ways in which they were resolved. It is hoped that this specific case study can contribute more generally to much needed discussion on the resolution of the many issues surrounding editing numerical data, both in the Indian context and more broadly with numerical tables from different cultures of inquiry.

2. The *Candrārkī*

The *Candrārkī*, aptly named 'Pertaining to the Moon and the Sun', is a short text of around 30 verses with accompanying numerical tables concerning lunar and solar phenomena. This table-text was composed by Dinakara, an astronomer who flourished in the latter half of the sixteenth century, about whom we know very little.

Judging from the number of extant manuscripts—over 150 have been identified[4]—the *Candrārkī* was an extremely popular work. The majority of extant manuscripts were copied in the eighteenth century; the earliest extant copy was produced in 1624. A critical edition of the versified text has been prepared on the basis of around a dozen manuscripts,[5] and a translation and comprehensive astronomical analysis of the text has also been completed.[6] An edition of the tables still remains to be done, however.

[3] An excellent description of various senses of tabular errors is given in van Dalen, *Ancient and Mediaeval Astronomical Tables*, pp. 12–19.

[4] See Pingree, *Census*, vol. A3, pp. 102–104; vol. A4, p. 109; vol. A5, pp. 138–139.

[5] See Kolachana et al., 'A Critical Edition of the *Candrārkī*'.

[6] See Kolachana et al., 'The *Candrārkī* of Dinakara'.

The text reveals that Dinakara's intention is that the reader be able to complete a customised set of tables for any desired year. His opening verse underscores this aim:[7]

सूर्यं चन्द्रं सद्गुरुं भक्तिपूर्वं
नत्वा वक्ष्ये सूर्यचन्द्रोद्भवं च ।
पत्रं पञ्चाङ्गाभिधं बुद्धिवृद्ध्यै
ग्राह्यं तज्ज्ञैर्युक्तिमत् तन्मयोक्तम् ॥ १ ॥

sūryaṃ candraṃ sadguruṃ bhaktipūrvaṃ
natvā vakṣye sūryacandrodbhavaṃ ca |
patraṃ pañcāṅgābhidhaṃ buddhivṛddhyai
grāhyaṃ tajjñair yuktimat tanmayoktam || 1 ||

Having paid homage to the Sun, the Moon, and the great *guru* with complete devotion, I [am going to] narrate tables (*patra*), generated from the positions of the sun and the moon, called *pañcāṅga*. This rationale-based [*pañcāṅga*] stated by me may be received by the specialists of that for expanding [the horizons of] their knowledge.

Indeed, the *Candrārkī* includes a set of tables from which to produce a *pañcāṅga* ('five-components') or calendar for the year in question that provides details on the five key aspects: the weekday, *tithi*, *yoga*, *nakṣatra*, and *karaṇa*, time units which are based on critical instants involving the moon and the sun.[8] The tables provide general solar and lunar phenomena which can be customised by incorporating corrections for time and local circumstances for any year and locality the table user wants. This ambition is emphasised later in the work as well, as Dinakara includes specific instructions in his text for filling in table values.[9] In this sense it is a table-text for the purpose of creating calendars.

An overview of the verses included in the *Candrārkī* is given in Table 1. The table-text nexus is further complicated by the fact that some manuscripts include the tables only, some the text only, and yet others still both the text and the tables. Further studies and acquisition of more manuscripts will help us understand better the trajectory, transformation, and use of this table-text.

2.1. A description of the sources

Hand made paper was a common medium for written documents in India. Unfortunately it is rather fragile; largely because of the ambient climate con-

[7] As edited and translated by Kolachana et al., 'The *Candrārkī* of Dinakara', pp. 9–10.

[8] For more details on the Indian calendar, see Montelle and Plofker, *Sanskrit Astronomical Tables*, Section 1.4.3, and Plofker and Knudsen, 'Calendars in India'.

[9] See verses 10 and 15–17 in Kolachana et al., 'The *Candrārkī* of Dinakara'.

Verse number(s)	Topics
1	Invocation
2–5	Key annual and epoch parameters concerning lord of the year, epact, and anomaly determination
6	*Rāmabīja* corrections
7	Determining local time
8–9	Longitude of sun at local sunrise
10	Solar true daily motion
11–13	Conversions between lunar and solar days
14	Rates of motion of moon and anomaly
15–17	Making one's own customised annual table
18	Lunar true daily motion
19	Inaccuracy of previous authors
20–22	Determining *tithis* and *karaṇas*
23	Determining *nakṣatras* and *yogas*
24	Adding an intercalary month
25–26	Determining omitted *tithis*, *nakṣatras*, and *yogas*
27	Closing verse

Table 1: Overview of the contents of the *Candrārkī* table-text.

ditions, the paper doesn't survive more than two to three centuries.[10] The corpus that remains today, conservatively estimated to be around seven million, is thus generally made up of copies (of copies) no earlier than the eighteenth century. Sanskrit scribal convention is to number folia on the verso. Text and tables generally begin on the second page (the verso of the first folio). The first page of a manuscript is usually left blank; occasionally the title of the text is written on this.

Securing copies of manuscripts is not always easy, and we were able to gain access to 12 manuscripts of the *Candrārkī*. These manuscripts fell into three groups: those which contained the text and tables, those which contained the text only, and those which contained the tables only.[11] Five manuscripts fell into the latter two categories which we used when preparing our critical

[10] For an excellent survey of Indian manuscripts, their extent, and characteristics, see Wujastyk, 'Indian Manuscripts'.

[11] Nine manuscripts include a copy of the text, and the sigla we gave these along with the library which carries them and their shelf mark details are (see Kolachana et al., 'A Critical Edition of the *Candrārkī*', Table 1, p. 3): B_1 Baroda, Central Library, 3119; BO_1 Pune, Bhandarkar Oriental Research Institute (BORI), 315/Viśrāmbag (i); BO_3 Pune, BORI, 308/1882-83; J_1 Jaipur, Palace Library, Khasmohor 5015; R_2 Jodhpur, Rajasthan Oriental Research Institute (RORI), 10180; R_4 Jodhpur, RORI, 5482; R_5 Jodhpur, RORI, 11633; R_6 Jodhpur, RORI, 9026; and O_1 Oxford, Bodleian Library, Walker 208b. Two additional manuscripts, which were originally given the sigla J_2 and BO_2, turned out to be unrelated to the *Candrārkī*.

edition of the tables. These are

Siglum	Library and shelfmark	No. of folia
J_1	Jaipur, Palace Library, Khasmohor 5015	15 ff.
J_2	Jaipur, Palace Library, Khasmohor 5081	10 ff.
R_2	Jodhpur, Rajasthan Oriental Research Institute (RORI), 10180	12 ff.
R_3	Jodhpur, RORI, 20220	11 ff.
R_7	Jodhpur, RORI, 7752	8 ff.

The manuscripts J_1 and R_2 include both the text and the tables. The remaining include the tables only.

Digital colour copies of the manuscripts were obtained from the Palace Library in Jaipur and the Rajasthan Oriental Research Institute in Jodhpur in India. All manuscripts are written in Nāgarī script on hand-made paper and are generally stored together as separate leaves. Brief descriptions of the manuscripts are given below.

2.1.1. J_1: *Jaipur, Palace Library, Khasmohor 5015, ff. 1–10 and ff. 1–2*

This MS contains tables (first, ff. 1–10) and text (last, ff. 1–2). The foliation appears to have been made by the scribe who copied the manuscript, so the two sets of foliation may simply reflect their decision to recommence numbering after the tables were complete. The MS is 12 × 28 cm.[12] It is neatly written using black ink for the entries and red ink for the double margin lines and table grids. It includes table titles and row headers identifying the units or contents of the table cells on every page. Occasional errors have been corrected using whitish paste. On the very first page (f. 1r) there is written *sāraṇī caṃdrārkī* ('Tables *Candrārkī*'). See Figures 4 and 11 for examples of this manuscript.

2.1.2. J_2: *Jaipur, Palace Library, Khasmohor 5081, ff. 1–11*

This MS contains tables only. It measures 11 × 25.5 cm.[13] It is legibly written using black ink for the entries and paratext, and red ink for the double margin lines and table grids. It includes row headers identifying the contents of the table cells. On the very first page (f. 1r) there is written *caṃdrārkī sāraṇī patra 10*. See Figures 5 and 12 for examples of this manuscript.

[12] Pingree, *A Descriptive Catalogue*, p. 66.

[13] Pingree, *A Descriptive Catalogue*, p. 67.

2.1.3. R_2: *Jodhpur, Rajasthan Oriental Research Institute (RORI), 10180, ff. 1–12*

This MS contains the text (first) and the tables (last), the latter of which begins on f. 2v. It is neatly written using black ink for the entries and paratext, and red ink for the double margin lines and table grids. It includes table titles and very occasionally row headings. See Figures 6 and 9 for examples of this manuscript.

2.1.4. R_3: *Jodhpur, Rajasthan Oriental Research Institute (RORI), 20220, ff. 1–11*

This MS contains the tables only. It is described in the catalogue as having 16 lines per page and 27 characters per line which appears to correspond roughly to the rows of digits and columns respectively. The catalogue notes it is in good condition. The MS is 11×25.4 cm.[14] It is clearly written using black ink for the entries and paratext, and red ink for the double margin lines and table grids. It includes occasional paratext and row headings most often along the right hand side of the table. See Figures 7 and 10 and Plate 6 for examples of this manuscript.

2.1.5. R_7: *Jodhpur, Rajasthan Oriental Research Institute (RORI), 7752, ff. 1–7*

This MS contains the tables only. It appears to have been somewhat rapidly written using black ink for the entries and red ink for the double margin lines and table grids. There is occasional smudging which may compromise legibility in places. See Figure 8 and Plate 7 for examples of this manuscript.

3. Technical summary of the tables

As far as we know, the *Candrārkī* is the only text in the Indian astronomical tradition dedicated exclusively to solar and lunar phenomena. In the text that accompanies the tables, Dinakara reveals that the parameters he has based his solar and lunar phenomena on derive from the Brāhmapakṣa school to which many other famous authors, notably Brahmagupta and Bhāskara II, adhered too.[15] Furthermore, Dinakara's choice of year length reveals he follows various modifications made by Bhojarāja (fl. 11th century) and noted table text composer Māhadeva (fl. 14th century). A fuller discussion of the relation of this text and its parameters to others can be found in recent studies.[16] There are four distinct tables in the *Candrārkī*. These are listed in Table 2.[17] These four tables and the data they contain provide the user with enough

[14] Jinavijaya, *A Catalogue of Sanskrit and Prakrit Manuscripts*, pp. 320–21.

[15] For a detailed discussion of the *pakṣa* tradition in Indian astronomy, see Montelle and Plofker, *Sanskrit Astronomical Tables*, Section 1.2.3, pp. 23–24.

[16] Kolachana et al., 'The *Candrārkī* of Dinakara', pp. 6–9.

[17] Preliminary identifications of these tables have been outlined in *SATIUS*, pp. 52–53.

Number	Table type	Manuscripts (folio numbers)				
		J_1	J_2	R_2	R_3	R_7
I	True solar longitude and corresponding velocity; argument 0 to 365 days	1v–4v	1r–4r	2v–6r	1v–4v	1r–2v
II	Mean motion of moon and anomaly; argument 0 to 365 days	4v–7v	4v–6v	6r–9r	4v–7v	3r–5v
III	Mean motion of moon and anomaly; argument 1 to 60 *ghaṭīs*	10r–10v	7r–7v	11v–12v	8r	5v
IV	Lunar *manda*-equation; argument 0 to 360 degrees	8r–10r	8r–10v	9r–11v	8v–11r	6r–7v

Table 2: Overview of the contents and organisation of the *Candrārkī* tables.

information to construct the annual calendar with its various details, all of which ultimately rely on knowing the true positions and motions of the sun and the moon. The annual calendar, or *pañcāṅga*, includes the arrangement of five divisions of time in India, which were fundamental for timekeeping.

These five units were all based on solar or lunar motion, or a combination of both and included the weekday (*vāra*), the *tithi*, or lunar day, namely the time interval over which the elongation of the moon and the sun increases by 12°, the *karaṇa*, which is half of a *tithi*, the *nakṣatra*, namely the time it takes the moon to cover an arc of 13°20′, the *yoga*, here the interval of time taken for the sum of the sun's and the moon's motion to increase by 13°20′. While *tithis*, *karaṇas*, and *yogas* depend on both the sun and the moon, *nakṣatras* were an interval of time based solely on the moon. Thus determining the positions of the sun and the moon along with their motions were fundamental for determining these five elements.

Computing lunar and solar positions in Indian astronomy is based on adjusting a mean longitude, measured from the vernal equinox, with a correction term, known as the *manda* ('slow') correction. This correction, akin to the equation of center, is a function of the angular separation of the mean position of the planet from its apogee.[18] In essence, the mean anomaly $\varkappa$ is the difference between the mean longitude of the planet $\bar{\lambda}$ and its apogee A, the point where it reaches its largest distance from the earth:

$$\varkappa = \bar{\lambda} - A.$$

[18] Details of this correction can be found in Montelle and Plofker, *Sanskrit Astronomical Tables*, Sections 2.1.1 and 2.1.2, as well as Montelle and Plofker, 'The Transformation of a Handbook', pp. 12–14.

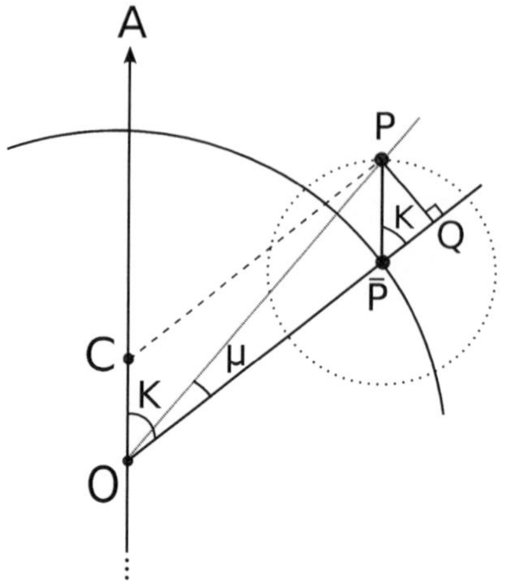

Figure 1: Diagram of the *manda*-correction μ as a function of the anomaly $\varkappa = AO\overline{P}$ producing the position of the true planet P from its mean position $\overline{P}$. Here $O\overline{P} = R$ is the radius of the deferent circle and PQ and $\overline{P}Q$ are $r \sin \varkappa$ and $r \cos \varkappa$ respectively, where r is the radius of the epicycle, or equivalently, the eccentricity.

The *manda*-equation μ is then computed via the trigonometric relations:

$$\sin \mu = \frac{r \cdot \sin \varkappa}{H},$$

where r $(= P\overline{P})$ is the radius of the epicycle and H is the distance between the earth and the true (corrected) planet (OP in Figure 1). This hypotenuse can be geometrically determined (using the Pythagorean theorem) as

$$H = \sqrt{(r \sin \varkappa)^2 + (R + r \cos \varkappa)^2}.$$

However, in the Indian tradition H here is usually replaced with R, the trigonometric radius of the circle, to simplify the computation, without much loss of accuracy. Therefore, with this simplifying assumption, the *manda*-equation is computed via:

$$\sin \mu = \frac{r \cdot \sin \varkappa}{R}$$

or, in other words, the *manda*-equation is simply a scaled function of $\sin \varkappa$. Because of this substitution, the values of the *manda*-equation will be symmetric about the quadrants, being a minimum when the mean planet is at the apogee or perigee, and a maximum when the anomaly is 90° and 270°. For this reason tables of the *manda*-equation typically contain only values for the first quadrant, as all other values can be supplied by symmetry.

In the *Candrārkī*, there is a single table for the sun which incorporates the *manda*-correction in its values, thus giving the true position of the sun directly. However, the moon requires three tables which must be used in conjunction with each other to produce its true position. This is because, unlike the solar apogee, the lunar apogee is moving and must be computed independently.

While these tables are referred to often in the text, the references are generally focused on table use and very little detail is given as to how the numerical data was generated. Using basic assumptions and knowledge about how such phenomena were computed in other treatises,[19] we tentatively propose algorithms and techniques that account for the general quantitative or qualitative trends of the numerical data. We leave the finer details of reconstructing the numerical data for future studies.

In the ensuing discussion, the table entries are identified as follows: argument, row 1 containing the numerical entries associated with that argument (usually a string of successive significant places), row 2 containing the numerical entries of a different relation associated with that argument, and so on.

Although the *Candrārkī* has an epoch (1578 CE) and various epoch positions are discussed in the text, the tables do not contain any epoch offsets in their entries. Their purpose is for the user to manipulate the data contained therein for their own circumstances.

Table I: True solar longitude and corresponding velocity (0 to 365 days)

Argument 0 to 365 days

Row 1 True longitude of the sun. The first value (day 0) is computed for when the mean sun is at Aries 0°. The procedure for determining the true longitude of the sun involves computing the mean position and applying the so-called *manda*-correction. This solar *manda*-equation μ is computed via:

$$\sin \mu = \frac{\sin \varkappa_S \cdot r_S}{R}, \tag{1}$$

where $\varkappa_S$ is the angular displacement of the sun from the apogee, or anomaly, r_S is the radius of the solar epicycle, or eccentricity, and R the trigonometric radius. The conventional value for the solar apogee in the Indian tradition is 78°.[20] Despite the fact that this value is not given anywhere by Dinakara, we assume it is the one he used.

[19] All the algorithms proposed here can be found in Bhāskara II's canonical astronomical handbook, the *Karaṇakutūhala* (epoch 1183 CE). See, for instance, Mishra, *Karaṇakutūhala*, and Balachandra Rao and Uma, *Karaṇakutūhalam*.

[20] See, for instance, Bhāskara II's *Karaṇakutūhala*, Chapter 2, verse 1.

Given the data in the table, we can back-compute to generate approximations for some of the parameters, most notably the ratio of the radius of the epicycle r_S to the radius of the base circle R. For instance, from the very first value, we can get a good approximation to this parameter. That is, when $\overline{\lambda_S} = 0°$, the solar anomaly is $\varkappa_S = |0 - 78| = 78$, and from the tables, the corresponding entry can be read off as $\mu = 2;9,13$. We can then use this data to approximate the ratio of the radius of the epicycle to the radius using equation 1, namely:

$$\frac{r_S}{R} = \frac{\sin\mu}{\sin\varkappa_S} \approx \frac{\sin(2;9,13)}{\sin 78} \approx 0.03841828,$$

where modern sines have been used. In comparison, the *Siddhāntaśiromaṇi* of Bhāskara II (Chapter 2.22) states parameters which produce $\frac{r_S}{R} = \frac{41}{1080} = 0.0379629$ for the above ratio.[21]

Using this parameter to reconstruct the tabular values via equation 1, one produces a reasonable fit.[22] Further table cracking efforts to refine this ratio would need to take account of the way of computing sines in line with early techniques (i.e., an appropriate Indian sine table using linear interpolation for non-tabulated values).

Row 2 True daily velocity of the sun. This appears to be the true angular velocity, rather than the difference in successive true solar positions.[23] That it is not the latter can be quickly confirmed by taking successive differences in solar positions, and comparing these with the tabular entries, which are close, but distinctly different.

In the table, the maximum value 61;23 minutes occurs at argument values 260 to 266 (around the perigee). The minimum value 56;54 minutes occurs at argument values 71 to 86 (around the apogee).

A recomputation of this can be made using the standard algorithm for true velocity v given in the *Siddhāntaśiromaṇi* of Bhāskara II (2.36–38):

$$v = \overline{v} - \frac{\overline{v} \cdot \cos\varkappa_S \cdot r_S}{R},$$

[21] Furthermore, using Bhāskara II's accurate sine table with $R = 3438$ produces a value of $\mu_{max} = 2;10,31$.

[22] In general, the recomputed values differ from the tabulated values by no more than 25 seconds in extreme cases without an apparent pattern. This can be explained by the effects of intermediate rounding, using a modern sine table, and interpolation.

[23] Indeed, true angular velocity is simply the rate of change in true longitude. Called *tātkālikā* ('at that time'), precise formulations for true angular velocity were given by Sanskrit authors. See, for instance, Bhāskara II, *Siddhāntaśiromaṇi* 2.36–38.

taking mean daily solar velocity[24] to be $\bar{v} = 59;8,10,12'$ and the above ratio for $\frac{r_S}{R}$, and computing the trigonometric ratio using the modern cosine. The resulting recomputation fits the tabular data quite well with differences between the recomputed values and the tabulated ones generally remaining within 5 seconds. In the recomputed data, the maximum value of $61;24'$ occurs at argument values 254 to 262. The minimum value of $56;51'$ occurs at argument values 76 to 80. This shift in extremal values in the recomputations await further exploration with more detailed table cracking efforts.[25]

Table II: Mean motion and mean anomaly of the moon (0 to 365 days)

Argument 0 to 365 days

Row 1 Mean position of the moon at sunrise. Initial value is $11^s 29;54,17°$ at argument value 0. Differences between subsequent values are not constant but range from around $13;9,54°$ to $13;11,0°$. Mean daily motion of the moon is given by Dinakara as $13;10,35°$ (verse 14).

Row 2 Mean lunar anomaly at sunrise. Initial value is $11^s 29;54,20°$ at argument value 0. Differences between subsequent values are not constant but range from around $13;3,25°$ to $13;4,5°$. The daily mean motion of the moon's anomaly is given by Dinakara as $13;3,54°$ (verse 14).

Notably, these two tabulated quantities do not begin at zero for the first entry (argument 0). This can be explained as follows: The daily mean motion of the moon is adjusted to account for the fact that the instant of sunrise is constantly changing. This correction is computed by multiplying the lunar mean daily motion by the sum of the equation of time and the equation of daylight and incorporating it positively or negatively as appropriate to the mean daily displacement in anomaly at mean sunrise. This produces the mean lunar anomaly at true sunrise per day, i.e., accounting for the fact that the interval from one sunrise to the next is not constant.

This tabulation of lunar motion and anomaly is identical in its mathematical basis to that found in Haridatta's *Jagadbhūṣana*, a table-text written over 50 years later in *Śaka* 1560 (1638 CE).[26] This similarity was crucial to us for understanding Dinakara's tables. However there are some differences in the two. While the *Candrārkī* tabulates lunar motion and anomaly for every day, the *Jagadbhūṣana* has only given values for every 14 day (*avadhi*) period (see

[24] This parameter has been reconstructed from the so-called 'lord of the year' parameter. See Kolachana et al., 'The *Candrārkī* of Dinakara'.

[25] In particular, we have not yet explained to our satisfaction a lack of symmetry with respect to values around the perigee and the apogee, i.e., why the minimum value occurs for more values around the apogee than the maximum around the perigee.

[26] See *SATIUS*, Tables 10 and 16, pp. 56–58.

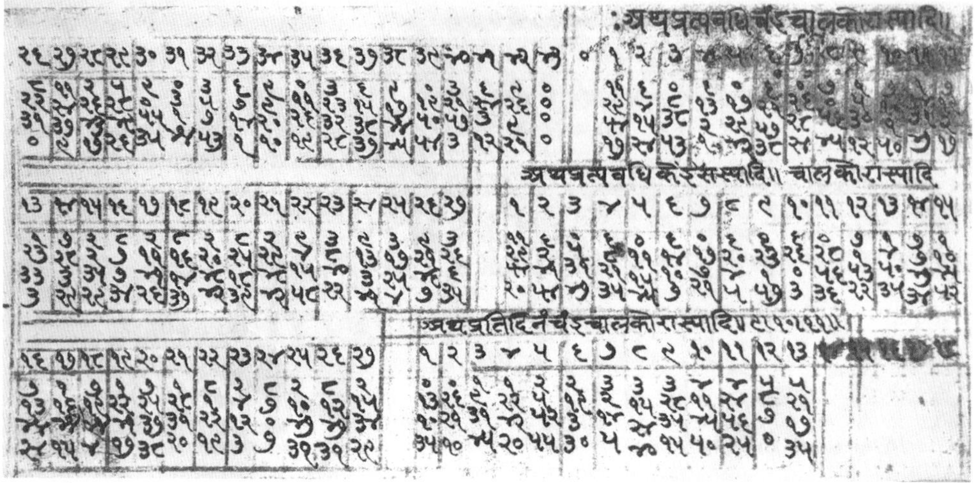

Figure 2: Table of lunar mean motion for true sunrise every 14 days (begins top right) from a manuscript of the *Jagadbhūṣana* of Haridatta (Smith Indic 146, f. 98r).

Figure 3: Table of lunar mean motion for true sunrise for 1 to 60 *ghaṭikās* (begins top right) from a manuscript of the *Jagadbhūṣana* of Haridatta (Smith Indic 146, f. 98v).

Figures 2 and 3). Comparing every 14th value in the *Candrārkī* with the data in the *Jagadbhūṣana*, however, produces an exact match. Furthermore, the *Jagadbhūṣana* includes a table giving the equation of time and equation of daylight that is not included in the *Candrārkī*.[27] This allowed us to apply the data as per the method described above to confirm the resulting entries. Further reconstruction efforts may reveal more accurate values of the underlying parameters on which Dinakara's table is based, especially the obliquity

[27] The data in this table is reproduced in *SATIUS*, Table 16, pp. 57–58.

and solar eccentricity. And indeed, further study into the relations between Dinakara's table corpus, and those produced by Haridatta half a century later, will shed more light on the development of lunar tables of this nature.

One significant point that immediately arises is that these tables are specific to a certain geographical location as terrestial latitude underpins the length of daylight. From related data in Haridatta's tables, the terrestrial latitude can be reconstructed.[28] Doing this reveals that these tables are geographically determined for $\varphi = 24°$.[29] This is more or less consistent with Dinakara's location of Bariya, Gujarat.[30] This implies that this particular lunar table is geographically determined and some modifications must be made by the *pañcāṅga* makers to compute values which are specific to their locality if it differs from Gujarat.

Table III: Mean motion of the moon and anomaly per *ghaṭikā* (1 to 60)

Argument 1 to 60 *ghaṭikās* (sixtieths of a day)

Row 1 Mean motion of the moon. First value 0;13,11° at argument value 1 and last value 13;10,35° at argument value 60. This table is constructed using constant differences of 0;13,10,35°, and entries are rounded to seconds.

Row 2 Mean motion of the lunar anomaly. First value 0;13,4° at argument value 1 and last value 13;3,54° at argument value 60. This table is constructed using constant differences of 0;13,3,54°, and entries are rounded to seconds.

This table is also identical to the one given by Haridatta in his *Jagadbhūṣana* (see Figure 3).

Table IV: Lunar *manda*-equation (0 to 360 degrees)

Argument 0 to 360 degrees

Row 1 Lunar *manda*-equation for 0 to 360 degrees of lunar anomaly. The maximum equation is 5;2,35° at argument values 90 and 270. We assume the lunar equation was generated via the standard relation:

$$\sin \mu = \frac{\sin \varkappa_M \cdot r_M}{R}$$

[28] Haridatta's tables contain information that the length of the longest day is 33;48 *ghaṭīs*, or equivalently around 13 hours 31 minutes. Using the relation $\sin \omega = \tan \varphi \tan \delta$ where ω is the half equation of daylight, φ is the terrestrial latitude, and δ the declination of the sun (at the instant of longest day, this is equal to the obliquity of the ecliptic, or $\varepsilon = 24°$), the terrestial latitude can be determined.

[29] As stated in *SATIUS*, p. 57.

[30] For further discussion on this geographical location, see Kolachana et al., 'The *Candrārkī* of Dinakara', p. 2.

where $\varkappa_M$ is the angular displacement of the moon from the apogee, r_M is the radius of the lunar epicycle, and R the trigonometric radius.

Back-computing using the data $\mu = 5;2,35°$ at $\varkappa_M = 90°$, we can determine the ratio of the radius of the lunar epicycle to the radius, namely:

$$\frac{r_M}{R} = \frac{\sin\mu}{\sin\varkappa_S} = \frac{\sin(5;2,35)}{\sin 90} \approx 0.08790432,$$

where modern sines have been used. Again, this approximation fits the data quite well when this parameter is used to recompute the data for other arguments but further investigation needs to be carried out to determine the precise parameters and method of computation.

Row 2 True lunar velocity. The maximum daily velocity is 14;19,24° at argument value 180. The minimum daily velocity is 12;1,46° at argument value 0. We assume the algorithm used to compute true velocity v was

$$v = \bar{v} - \frac{v_A \cdot \cos\varkappa_M \cdot r_M}{R},$$

where $\bar{v}$ is the mean lunar velocity and v_A is the velocity of the lunar anomaly.

A recomputation of this numerical data can be made, using the mean daily lunar velocity $\bar{v}_M = 13;10,35°$, mean lunar anomaly $v_A = 13;3,53°$, the above ratio for $\frac{r_S}{R}$, and computing the trigonometric ratio using the modern cosine. The resulting recomputation fits the tabular data very closely. In the recomputed data, the minimum velocity 12;1,40° occurs at argument value 0. The maximum value of 14;19,29° occurs at argument value 180.

4. Table Edition

The Appendices contain a critical edition of Tables I–IV along with an *apparatus criticus* given under the respective parts of the tables. In addition, we outline the issues arising from the process of editing these tables, as well as give an account of the editorial resolutions that we followed when preparing the edition, the software we developed to assist the editing process and how to read the apparatus criticus.

4.1. Preliminaries

4.1.1. *Issues raised in editing the Candrārkī tables*

Creating an edition of the numerical tables presented many issues which needed to be resolved. In addition to general problems of editing sources, numerical data in the *Candrārkī* raised the following specific issues:

Layout

Page orientation The arrangement of tabular data in Sanskrit numerical tables generally is in landscape format, reflecting the traditional orientation of manuscripts in India. Numerical data is tabulated horizontally: the argument and entries run from left to right across the page. When the available space on the page is filled, the table breaks off and continues underneath in a similar manner. Long tables may extend across a number of pages. See any of the images throughout the paper for this stacking effect.

Row breaks The length of tables in the *Candrārkī* range from 60 entries to 365 entries. They are thus broken up into stacks of roughly equal numbers of columns and spread over several pages. Scribes make individual decisions about how many stacks to fit on a page. For instance, MSS J_1 and R_3 (see Figures 4 and 7) have two stacks on the page. MS R_7 (see Figure 8) has managed to fit three. The number of columns per stack is variable, but is usually around 30. MS R_3 for instance (see Figure 7) has 24 columns per stack for this particular page. Later on R_3 has 30 (see Figure 10). MS J_1 (see Figure 11) has a page with 33 columns, MS R_7 (see Figure 8) has 31.

Beginning a new table Some scribes start new tables on new stacks: see for instance MS R_3 (Figure 7) in which Table I ends (argument 365) at the end of the first stack, and Table II begins at the beginning of the second stack (argument 0). Other scribes continue with a new table on the same stack, taking advantage of the preruled grid. For instance MS R_2 (Figure 9) finished Table I in the middle of the first stack and begins Table II several columns later on the same stack. Some scribes are happy to leave tabular cells blank at the end of a table, as in MSS J_2 or R_7 (Figures 5 and 8). Other scribes occasionally leave empty cells for no apparent reason: see MS R_2 (see Figure 6) with its three empty cells at the end of the first stack.

Paratext Table titles and row/column headers are handled differently from scribe to scribe.

Row headings There is no consistency between manuscripts regarding row headers. In Table I, MS J_1 (see Figure 4) has labelled each of the rows, in order: *gaṇa* ('number') for the argument row, and *rāśi, aṃśa, kalā, vikalā* ('sign', 'degree', 'minute', 'second' respectively) for the successive units of mean motion and *gatiḥ* ('velocity') for the solar true velocity. MS J_2 (see Figure 5) in contrast labels the mean motion rows simply with *spaṣṭo'rka*, literally 'accurate sun'. Similarly MS R_3 (see Figure 7) labels these rows *raviḥ*, literally 'sun'. In Table II, MS R_7 (see Figure 8) has somewhat haphazardly added row identifications to one stack (top right hand corner) but left others blank. The scribe has added *dina, caṃdra, keṃdra* ('day', 'moon', 'anomaly') for the argument, lunar mean motion, and anomaly respectively. In Table IV, MS R_3 (see Figure 10) adds several extra details in their row headers concerning the mathematical way in which the data is to be applied. The scribe often places row headers on the right hand side. The argument is labelled *keṃdrarāśi 'ṃśādi* ('anomaly [in] signs and degrees etc.'). The lunar equation row is labelled *maṃdaphala* × ('*manda*-equation, negative') to indicate at that particular point in the function the values are to be applied negatively. The difference row is labelled *aṃtara* | *dhana* ('Difference. Positive'), to indicate that the differences are positive. The final row, the true lunar velocity is labelled *gati* ('velocity') or *gatayaḥ* ('velocities'). In a similar spirit, but with slightly less information, MS J_1 gives the labels *keṃdra, maṃdaphala, 'ṃtara,* and *gati* respectively.

Column headings Very occasionally, scribes will tag some paratext to a particular column. For instance, in Table IV, the scribe of MS R_2 (see bottom stack of Figure 12) has written *tulaḥ* ('Libra') next to the column which contains the argument 6,0 (i.e., the beginning of the sixth zodiacal sign, Libra). Likewise, they have written *vṛściḥ*, short for *vṛścikaḥ*, ('Scorpio') close to the column which contains the argument 7,0 (the beginning of Scorpio).

Table titles Tables may or may not be given titles by scribes. MS J_1 has labelled Table I (see Figure 4) *sūryapaṃkti* ('solar result-line') and repeats this on each new page. MS R_2 (see Figure 6) has labelled Table I *ravikoṣṭhakā* ('solar table entries') and repeats this (or a close version of it) at the top of each stack. MSS J_2, R_3 and R_7 do not include table titles.

Table colophons MS R_2 (see Figure 9) includes some paratext at the end of Table I announcing that the table is finished *iti ravikoṣṭhakā saṃpūrṇo yaṃ samāptaḥ* ('Thus, the solar tabular entries are finished. This is completed.') and announces the beginning of Table II *atha caṃdrakoṣṭhakāy[a]ṃarabhyate* ('Now, the lunar tabular entries. This begins').

Morphology Morphology of Sanskrit words or phrases in row/column headings and general paratext varies inconsistently between abbreviated form, stem form, and full inflected form. For instance, Table I in MS J_1 reveals various states of abbreviation of row headers. Rows in the first stack are labelled with their stem forms: *gaṇa, rāśi, 'ṃśa, kalā, vika·*, and *gati*. The second stack are labelled with abbreviated forms: *gaṇa, rā·, 'ṃśa, ka·, vi·*, and *ga·*.

Table ordering Tables do not always appear to be in the same order across manuscripts. For instance, Table III in MS R_7 appears as the last table. In the other four manuscripts it is the third table.

Empty cells Some cells were not filled in for some inexplicable reason; other entries may have been illegible. For instance, Table IV in MS R_3 (see last column 6th row in Figure 10) has a missing value.

Abbreviating numerical data Scribes occasionally adopt a short-hand for representing numbers, particularly when they are several significant figures in length and one of the digits is repeated for long stretches. This occurs often in arguments once they hit three significant figures. For instance, in Table I in MS R_3 (see the first stack in Figure 7) the first argument begins with 342 (days). It then continues 43, 44, 45, and so on, until the penultimate value 64, and the ultimate value 365. The first and last values in this row therefore

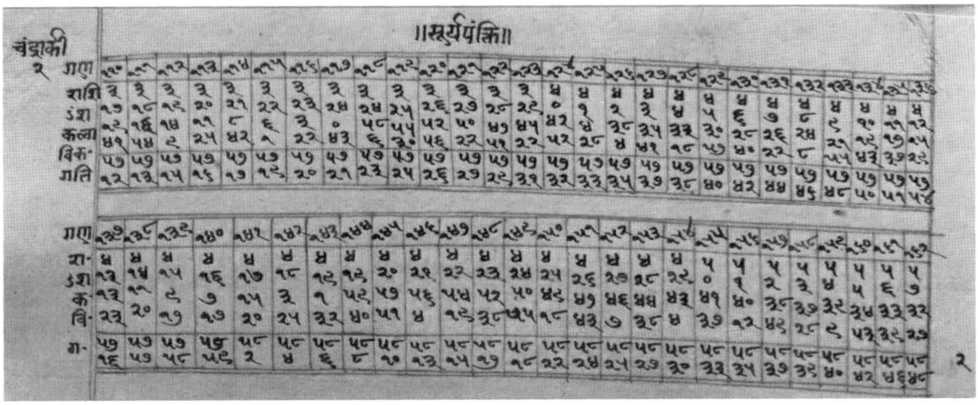

Figure 4: An excerpt from the solar longitude and velocity table from J_1 (MS Khasmohor 5015, f. 2v) with two distinct row stacks and row headers repeated for each row stack.

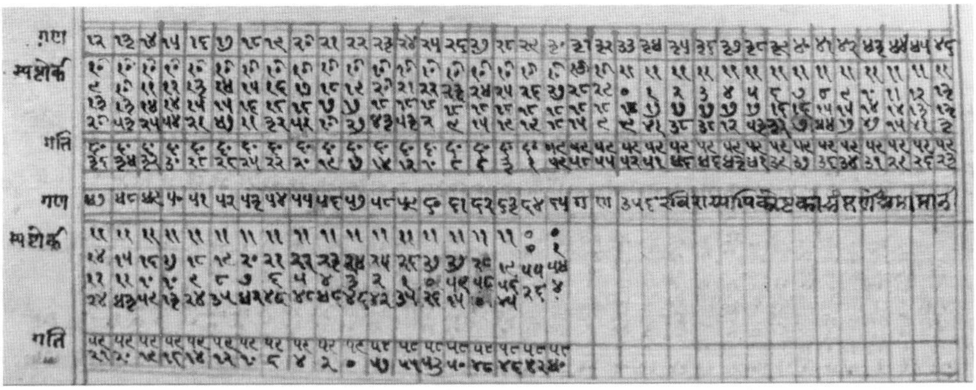

Figure 5: An excerpt from the solar longitude and velocity table from J_2 (MS Khasmohor 5081, f. 4r) with rows headers in the left-hand margin and a group of values corrected with white paste. There are many pre-ruled cells which have been left empty.

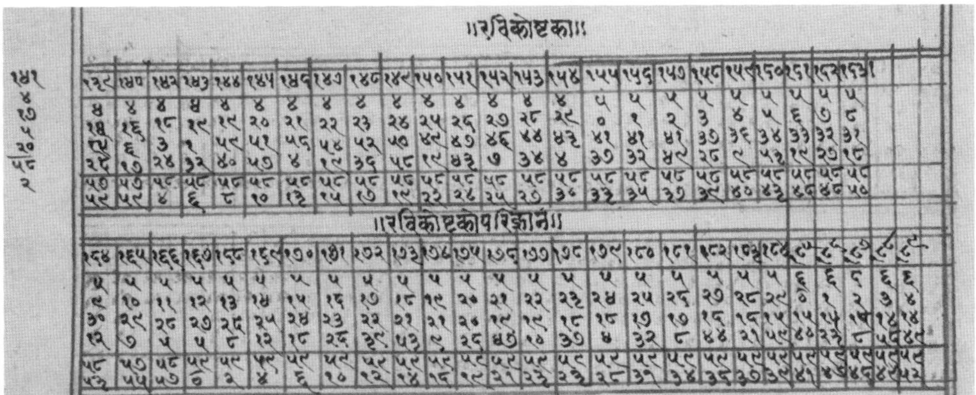

Figure 6: An excerpt from the solar longitude and velocity table from R_2 (MS RORI 10180, f. 4r) with a missing entry inserted in the left-hand margin and three empty cells on the right-hand side.

Figure 7: An excerpt from the solar longitude and velocity table from R_3 (MS RORI 20220, f. 4v) with row headers in the left-hand margin. A new table begins on the second stack without a title.

Figure 8: An excerpt from the solar longitude and velocity table from R_7 (MS RORI 7752, f. 5v) with three table stacks arranged on one page and extra cells added in the right-hand margin.

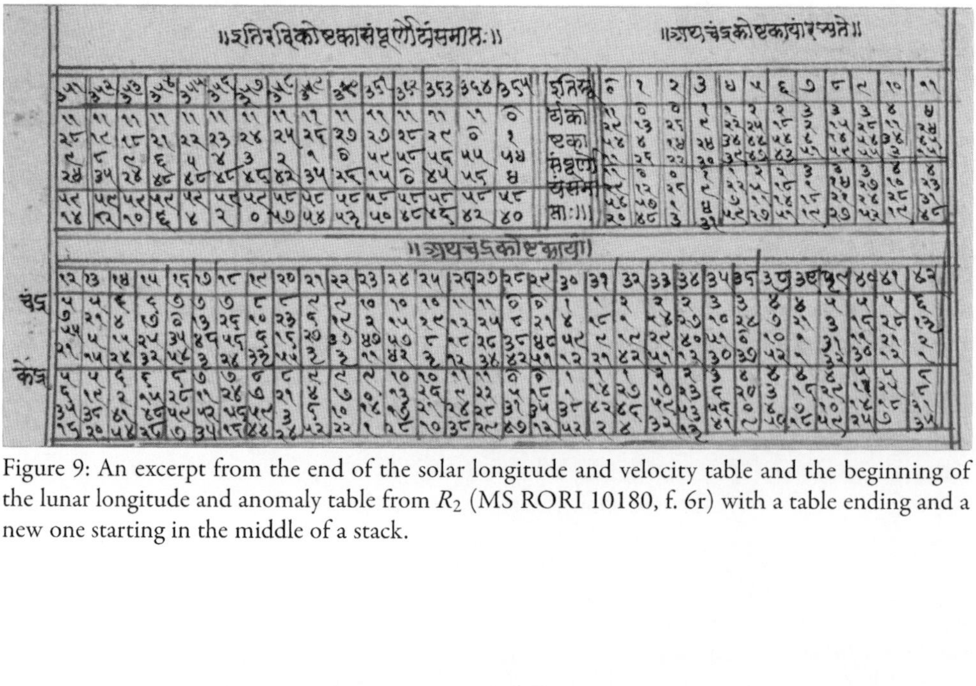

Figure 9: An excerpt from the end of the solar longitude and velocity table and the beginning of the lunar longitude and anomaly table from R_2 (MS RORI 10180, f. 6r) with a table ending and a new one starting in the middle of a stack.

Figure 10: An excerpt from the lunar equation table from R_3 (MS RORI 20220, f. 9r) with additional abbreviated paratext on the right-hand side explaining how to apply the correction.

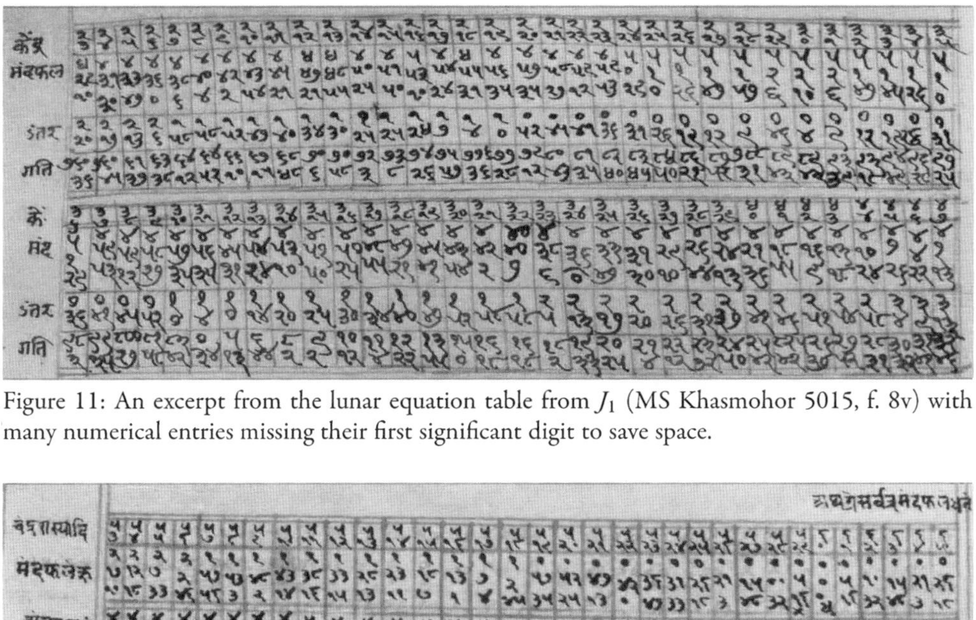

Figure 11: An excerpt from the lunar equation table from J_1 (MS Khasmohor 5015, f. 8v) with many numerical entries missing their first significant digit to save space.

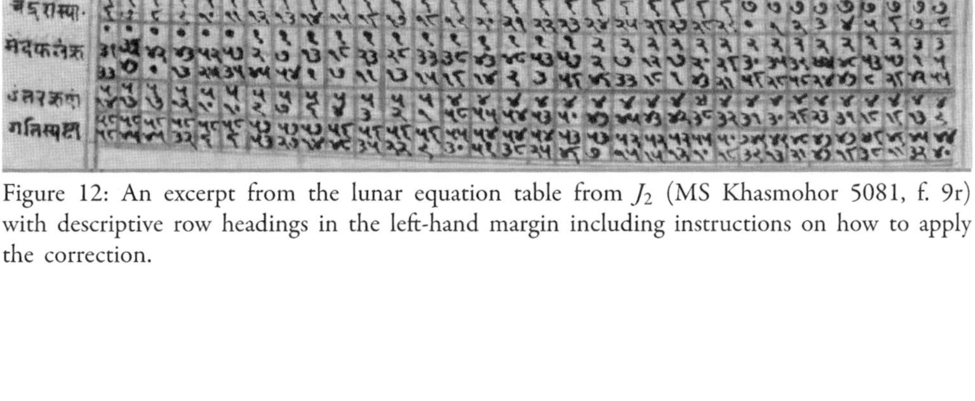

Figure 12: An excerpt from the lunar equation table from J_2 (MS Khasmohor 5081, f. 9r) with descriptive row headings in the left-hand margin including instructions on how to apply the correction.

include the correct rendering of the number with its three significant figures, the intermediate arguments drop the three; it is to be assumed by the reader from context. Scribes do not always put this helpful reminder at the beginning and end of a stack however. For instance, in Table I in MS J_2 (see Figure 5) the argument starts at 12 (for 312) and continues in double digits until 65 (for 365). The reader must therefore be very mindful of where they are in the table to correctly reconstruct the correct argument. This practice occurs in table entries as well. For instance in Table IV in MS J_1 (see Figure 11), the true lunar velocity is given at the bottom of the stack. This is given in minutes and seconds and varies from minute amounts from the 700s to the 800s. The first stack starts by giving these as 760;36, 760;45, after which point it drops the 7, and continues 61;37 (for 761;37), 63;38 (for 763;38), only occasionally writing out the number in full (see entry under argument 2,18: 776;36). In the bottom stack, at argument value 3,8 the lunar velocity enters the 800s and this is indicated as 800;27. However, dropping digits leads to some scribal carelessness as the next few numbers read: 81;58 (for 801;58), 83;42 (for 803;42), 0;34 (for 804;34), 5;13 (for 805;13), and so on. The table reader therefore must have an idea of the broader context of the entry with respect to the others to ascertain the correct value.

Additional features Some table cells include the symbols × ('negative')[31] and *dha* ('positive') to indicate where values are subtractive/negative or additive/positive. This is done somewhat inconsistently across manuscripts. For instance, Table IV in MS J_2 (see Figure 12) includes a cross in the *manda*-equation value (argument 5,29) and a *dha* (an abbreviation for *dhanaṃ*, 'positive') in argument 6,0 to indicate the equation is to henceforth be applied positively to the mean longitude.

Colour For the most part, the table data and paratext is written in black ink and the table rulings in red. However, occasionally red ink is used for some table headings. See for instance the table heading in Table II in MS R_2 (see bottom stack in Figure 9) in which the table heading *atha caṃdrakoṣṭhakāyāṃ* ('Now, in the lunar tabular cell[s]') is rendered in red ink.

Corrections

Overwriting Table errors that are noticed by the scribe are generally corrected by the application of white/yellow paste and the correct values are written on top. This can be seen in entries 55–62 of Table I in MS J_2 (see Figure 5) in which the degrees, minutes and seconds components of mean motion have been corrected.

[31] This symbol derives from a short-hand for the *nāgarī* character ऋ or *ṛ*, an abbreviation of *ṛṇaṃ* meaning 'negative' in Sanskrit. For further details and instances, see Montelle and Plofker, *Sanskrit Astronomical Tables*, p. 88 *et passim*.

Inserting a table entry An omitted table entry is sometimes inserted in the margin. For instance, in the left hand margin of Table I in MS R_2 (see Figure 6), the entry for argument 141 is written as it has been left out of the main numerical data.

4.1.2. *Editorial resolutions*

Taking note of every single variant that tables contain, with respect to the variety of features described in Section 4.1.1, is impossible. Some variants are more meaningful than others, especially given the basic assumptions of the critical edition. In light of this and given the various issues raised by the numerical data and its arrangement on the page by scribes, we observed the following resolutions in producing the critical edition:

1. We consider the table is the fundamental object of interest and not the manuscript page.
2. We preserve salient aspects of the layout in the edition, including the horizontal format placed on the landscape orientation of the page. For consistency we default to row breaks of 30 columns per stack and two stacks per page. We argue that variations to this are insignificant in terms of their ability to highlight relationships between the manuscripts. We also preserve the vertical orientation of the numbers in which successive components of the number are placed one below the other.
3. We preserve multiple functions that are tabulated with respect to a single argument in the same table, delimiting them with a single ruled line, as has been done by the scribes.
4. We have transcribed all Nāgarī numerals into their modern Indo-Arabic equivalents.
5. We silently emend stem form or abbreviated forms of Sanskrit words to their correct inflected equivalent.
6. We note an empty cell by *x*.
7. We note an illegible number by ?.
8. We silently emend the first significant digit which is inconsistently omitted by all scribes in the case of three digit numbers (Table IV, lunar velocity; the arguments for Tables I and II).
9. We have not attempted to regularise numerical content in any way. Discrepancies between attested and recomputed tabular values can arise from a variety of factors, including different computation techniques, uncertainty surrounding precision, intermediate rounding, truncation, copying errors, calculating errors, inappropriate modern assumptions and

the like. Recomputing historical tabular data is a study in and of itself. The edition thus leaves the data in a state as close as possible to manuscripts so that table cracking efforts can proceed from this.

4.1.3. *How to read the apparatus criticus*

The critical edition was created using the specially designed online software package called CATE (Computer-Assisted Tables Editor). This was designed and implemented by Paul Brouwers and is available on the online platform found at `http://uc.hamsi.org.nz/cate/#`. CATE provides some relief to the laborious task of critically editing tables by automating parts of the process. With two or more electronic copies of manuscript versions of a table as input, this package produces a base text and critical apparatus listing the numerical variants according to their argument and row number.

CATE generates the base table according to either of two editing strategy options. The first strategy is 'nominate the base table'. This is where one manuscript version is selected by the user as the base table, against which all others are compared. The second editing strategy is 'majority rules'. This is where the base table is formed from the most common entry for that cell among all table versions. The second option defaults to the first option in the case of an absence of a majority. In this particular version, we found MS R_3 to be the most reliable. Thus we chose the 'majority rules' editing strategy with the default option being MS R_3 where there was no majority consensus.

The apparatus criticus is placed at the bottom of the pages covered by each table. Variant readings are noted first with respect to the argument the variant falls under and secondly by the row under this in which it occurs. The argument is given in bold for ease of reading. Multiple variant readings are listed in alphabetical order according to the manuscript siglum. Variant readings are listed in the apparatus criticus with respect to each row in ascending order associated to an argument in ascending order.

Bibliography

Balachandra Rao, S., and S. K. Uma, *Karaṇakutūhalam of Bhāskarācārya II: An English Translation with Notes and Appendices*, New Delhi: Indian National Science Academy, 2008. [Reprinted from *Indian Journal for the History of Science*, 42/1–2 (2007) and 43/1, 3 (2008).]

Dalen, Benno van, *Ancient and Mediaeval Astronomical Tables: Mathematical Structure and Parameter Values*, PhD dissertation, Utrecht University, 1993.

Ikeyama, Setsuro, and Kim Plofker, 'The *Tithicintāmaṇi* of Gaṇeśa, a Medieval Indian Treatise on Astronomical Tables', *SCIAMVS* 2 (2001), pp. 251–89.

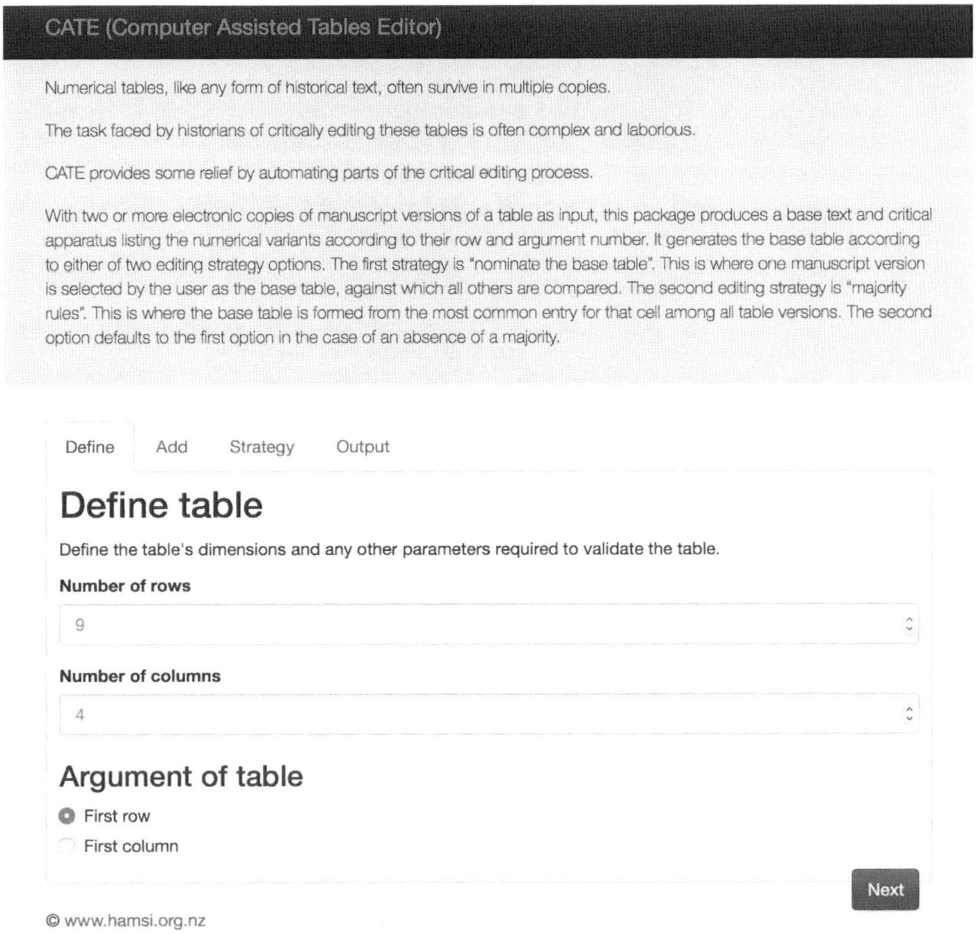

Figure 13: A screen-shot of the automated critical editing package for numerical tables CATE.

Jinavijaya, Padmashri Muni (ed.), *A Catalogue of Sanskrit and Prakrit Manuscripts in the Rajasthan Oriental Research Institute. Part IV*, Jodhpur: Rajasthan Oriental Research Institute, 1976.

Kolachana, Aditya, Clemency Montelle, Jambugahapitiye Dhammaloka, Keshav Melnad, K. Mahesh, Pravesh Vyas, K. Ramasubramanian, M. S. Sriram and Venketeswara Pai, 'The *Candrārkī* of Dinakara: A Table-text for the Moon and the Sun', *Journal for the History of Astronomy*, 49 (2018), pp. 306–44.

Kolachana, Aditya, Clemency Montelle, Jambugahapitiye Dhammaloka, Keshav Melnad, K. Mahesh, Pravesh Vyas, K. Ramasubramanian, M. S. Sriram and Venketeswara Pai, 'A Critical Edition of the *Candrārkī* of Dinakara: A Text Concerning Solar and Lunar Tables', *History of Science in South Asia* 6 (2018), pp. 127–62.

Mishra, Satyendra, *Karaṇakutūhala*, Varanasi: Krishnadas Academy, 1991. [Krishnadas Sanskrit Series, 129.]

Misra, Anuj, Clemency Montelle, and Kim Plofker, *The Sanskrit astronomical table text Brahmatulyasāraṇī: Numerical tables in textual scholarship*, Brill, Expected release 2020.

Montelle, Clemency, and Kim Plofker, '*Karaṇakesari* of Bhāskara II: a 17th-century table text for computing eclipses', *History of Science in South Asia* 1 (2013), pp. 1–63.

Montelle, Clemency, and Kim Plofker, 'The Transformation of a Handbook into Tables: The *Brahmatulyasāraṇī* and the *Karaṇakutūhala* of Bhāskara II', *SCIAMVS* 16 (2015), pp. 1–34.

Montelle, Clemency, and Kim Plofker, *Sanskrit Astronomical Tables*, Cham: Springer, 2018.

Neugebauer, O., and David Pingree, 'The Astronomical Tables of Mahādeva', *Proceedings of the American Philosophical Society* 111 (1967), pp. 69–92.

Pingree, David, *Śrīdhara's Laghukhecarasiddhi*, Baroda: Oriental Institute, 1976.

Pingree, David, *A Descriptive Catalogue of the Sanskrit Astronomical Manuscripts preserved at the Maharaja Man Singh II Museum in Jaipur, India*, Philadelphia: American Philosophical Society, 2003.

Pingree, David, 'History of Mathematical Astronomy in India', in *Dictionary of Scientific Biography*, Detroit: Charles Scribner's Sons, vol. XV (1978), pp. 533–633.

Pingree, David, 'Bīja-corrections in Indian astronomy', *Journal for the History of Astronomy* 27 (1996), pp. 161–72.

Pingree, David, *Census of the Exact Sciences in Sanskrit*, Series A, vols. I–V, Philadelphia: American Philosophical Society, 1970–1994.

Pingree, D. E., *Jyotiḥśāstra*, Wiesbaden: Otto Harrassowitz, 1995.

Pingree, David, 'On the Classification of Indian Planetary Tables', *Journal for the History of Astronomy* 1 (1970), pp. 95–108.

Plofker, Kim, and Toke L. Knudsen, 'Calendars in India', in John M. Steele (ed.), *Calendars and Years II: Astronomy and Time in the Ancient and Medieval World*, Oxford: Oxbow Books, 2011, pp. 53–68.

SATE= Pingree, David, *Sanskrit Astronomical Tables in England*, Madras: The Kuppuswami Sastri Research Institute, 1973.

SATIUS= Pingree, David, 'Sanskrit Astronomical Tables in the United States', *Transactions of the American Philosophical Society* New Series 58/3 (1968), pp. 1–77.

Wujastyk, Dominik, 'Indian manuscripts' in Jörg Quenzer, Dmitry Bondarev and Jan-Ulrich Sobisch (eds.), *Manuscript Cultures: Mapping the Field*, Berlin: De Gruyter, 2011, pp. 159–82.

Appendix A1. True solar longitude and corresponding velocity (0 to 365 days)

0	**1**	**2**	**3**	**4**	**5**	**6**	**7**	**8**	**9**	**10**	**11**	**12**	**13**	**14**	**15**	**16**	**17**	**18**	**19**	**20**	**21**	**22**	**23**	**24**	**25**	**26**	**27**	**28**	**29**
0	0	0	0	0	0	0	0	0	0	0	0	0	0	0	0	0	0	0	0	0	0	0	0	0	0	0	0	0	1
2	3	4	5	6	7	8	8	9	10	11	12	13	14	15	16	17	18	19	20	21	22	23	24	25	26	27	28	29	0
9	7	6	4	3	1	0	58	57	55	53	51	50	48	46	44	42	40	38	36	34	32	30	28	26	23	21	19	17	14
13	50	24	56	28	58	20	42	3	21	38	53	7	16	25	31	37	40	41	39	36	30	24	14	2	50	35	19	1	40
58	58	58	58	58	58	58	58	58	58	58	58	58	58	58	58	58	58	58	57	57	57	57	57	57	57	57	57	57	57
40	39	37	35	33	30	27	24	22	20	18	17	15	13	10	8	6	4	2	59	57	56	54	51	50	48	46	44	42	40

30	**31**	**32**	**33**	**34**	**35**	**36**	**37**	**38**	**39**	**40**	**41**	**42**	**43**	**44**	**45**	**46**	**47**	**48**	**49**	**50**	**51**	**52**	**53**	**54**	**55**	**56**	**57**	**58**	**59**
1	1	1	1	1	1	1	1	1	1	1	1	1	1	1	1	1	1	1	1	1	1	1	1	1	1	1	1	1	1
1	2	3	4	5	6	6	7	8	9	10	11	12	13	14	15	16	17	18	19	20	21	22	23	24	25	26	26	27	28
12	9	7	5	2	0	57	54	52	49	47	44	41	39	36	33	30	28	25	22	19	16	14	11	7	5	2	59	56	53
18	54	29	1	34	3	31	57	22	46	9	28	46	6	24	40	55	9	22	33	44	55	3	9	16	21	25	28	32	34
57	57	57	57	57	57	57	57	57	57	57	57	57	57	57	57	57	57	57	57	57	57	57	57	57	57	57	57	57	57
38	36	35	33	32	31	29	27	26	25	23	21	20	19	17	16	15	13	12	11	10	9	8	7	6	5	4	3	2	1

60	**61**	**62**	**63**	**64**	**65**	**66**	**67**	**68**	**69**	**70**	**71**	**72**	**73**	**74**	**75**	**76**	**77**	**78**	**79**	**80**	**81**	**82**	**83**	**84**	**85**	**86**	**87**	**88**	**89**
1	2	2	2	2	2	2	2	2	2	2	2	2	2	2	2	2	2	2	2	2	2	2	2	2	2	2	2	2	2
29	0	1	2	3	4	5	6	7	8	9	10	11	12	13	14	15	15	16	17	18	19	20	21	22	23	24	25	26	27
50	47	44	41	38	35	32	29	26	23	20	17	14	10	7	4	1	58	55	52	49	46	43	40	36	33	30	27	24	21
34	34	33	32	31	28	25	21	17	15	9	5	1	56	50	44	40	34	29	23	18	11	6	1	55	49	43	39	36	32
57	56	56	56	56	56	56	56	56	56	56	56	56	56	56	56	56	56	56	56	56	56	56	56	56	56	56	56	56	56
0	59	58	58	58	57	56	56	55	55	55	55	54	54	54	54	54	54	54	54	54	54	54	54	54	54	54	55	55	55

[**1**:3] 3 R2, [**1**:6] 37 R2, [**3**:4] 54 R2, [**4**:4] 38 J2, 26 R7, [**5**:4] 46 J1, 46 J2, 56 R2, 56 R7, [**6**:3] 8 R7, [**6**:6] 24 R2, [**7**:3] 6 R2, [**7**:4] 24 J1, 24 J2, [**8**:6] 23 R2, [**14**:3] 45 J2, [**14**:4] 20 J1, [**16**:4] 27 J2, 34 R2, [**19**:6] 58 R2, [**21**:4] 32 J1, [**23**:6] 57 J1, [**28**:6] 43 J1, 43 J2, [**29**:6] 42 J1, 42 J2, [**30**:4] 28 R2, [**31**:6] 37 R2, 37 R3, [**32**:4] 19 J2, [**32**:6] 33 R2, [**34**:3] 1 R2, [**34**:6] 30 R2, [**35**:4] 31 J1, 31 J2, [**35**:6] 30 R2, [**36**:4] 37 R2, 35 R7, [**36**:6] 20 R2, [**37**:6] 20 R2, 26 R7, [**38**:6] 20 R2, [**39**:6] 20 R2, [**42**:6] 19 R2, [**43**:4] 46 R2, [**43**:6] 59 R2, [**44**:4] 14 J1, [**44**:6] 16 J2, 57 R2, [**45**:6] 17 J2, [**46**:6] 10 R2, [**47**:3] 29 J1, 39 J2, [**49**:3] 23 R2, [**51**:4] 35 R3, [**51**:6] 6 R2, [**54**:3] 8 J1, 8 J2, [**55**:4] 20 R3, [**56**:6] 3 R2, [**57**:4] 58 J1, 38 J2, 39 R2, [**60**:4] 35 R2, [**63**:4] 31 R2, [**64**:3] 48 J2, [**64**:4] 30 R2, 41 R3, [**64**:6] 57 R2, [**65**:4] 38 R2, [**66**:3] 39 R2, 35 R7, [**66**:4] 5 R2, [**66**:6] 57 R2, [**67**:4] 31 R2, [**67**:6] 57 R2, [**68**:3] 27 J1, 27 J2, [**68**:6] 57 R2, 56 R3, [**69**:3] 22 R2, 22 R3, [**69**:4] 13 J1, 13 J2, [**70**:4] 10 R7, [**71**:6] 54 J2, [**72**:6] 55 R3, [**77**:2] 16 R2, [**77**:3] 28 J2, [**77**:4] 38 J1, 38 J2, 35 R2, [**78**:4] 19 J1, 59 J2, [**79**:4] 24 J2, 24 R7, [**82**:3] 46 J2, [**86**:4] 4 J2, [**87**:4] 29 J1, [**88**:3] 34 J1, [**89**:4] 23 R7.

Appendix A1: True solar longitude and velocity (continued)

90	**91**	**92**	**93**	**94**	**95**	**96**	**97**	**98**	**99**	**100**	**101**	**102**	**103**	**104**	**105**	**106**	**107**	**108**	**109**	**110**	**111**	**112**	**113**	**114**	**115**	**116**	**117**	**118**	**119**
2	2	3	3	3	3	3	3	3	3	3	3	3	3	3	3	3	3	3	3	3	3	3	3	3	3	3	3	3	3
28	29	0	1	2	3	4	4	5	6	7	8	9	10	11	12	13	14	15	16	17	18	19	20	21	22	23	24	24	25
18	15	12	9	6	3	0	57	54	51	48	45	42	39	35	33	30	28	25	22	19	16	14	11	8	6	3	0	58	55
28	25	22	18	18	17	16	16	16	18	21	25	29	34	40	47	58	6	16	28	41	54	9	25	42	1	22	43	6	30
56	56	56	56	56	56	56	56	57	57	57	57	57	57	57	57	57	57	57	57	57	57	57	57	57	57	57	57	57	57
56	56	56	57	57	58	58	59	0	1	2	3	4	5	6	7	8	9	10	11	12	13	14	16	17	19	20	21	23	25

120	**121**	**122**	**123**	**124**	**125**	**126**	**127**	**128**	**129**	**130**	**131**	**132**	**133**	**134**	**135**	**136**	**137**	**138**	**139**	**140**	**141**	**142**	**143**	**144**	**145**	**146**	**147**	**148**	**149**
3	3	3	3	4	4	4	4	4	4	4	4	4	4	4	4	4	4	4	4	4	4	4	4	4	4	4	4	4	4
26	27	28	29	0	1	2	3	4	5	6	7	8	9	10	11	12	13	14	15	16	17	18	19	19	20	21	22	23	24
52	50	47	45	42	40	38	35	33	30	28	26	24	21	19	17	15	13	11	9	7	5	3	1	59	57	56	54	52	50
56	22	51	22	52	28	4	41	18	57	40	22	8	55	43	37	29	23	20	17	17	20	25	32	40	51	4	19	38	55
57	57	57	57	57	57	57	57	57	57	57	57	57	57	57	57	57	57	57	57	57	58	58	58	58	58	58	58	58	58
26	27	29	31	32	33	35	37	38	40	42	44	46	48	50	51	54	56	57	58	59	2	4	6	8	10	13	15	17	18

150	**151**	**152**	**153**	**154**	**155**	**156**	**157**	**158**	**159**	**160**	**161**	**162**	**163**	**164**	**165**	**166**	**167**	**168**	**169**	**170**	**171**	**172**	**173**	**174**	**175**	**176**	**177**	**178**	**179**
4	4	4	4	4	5	5	5	5	5	5	5	5	5	5	5	5	5	5	5	5	5	5	5	5	5	5	5	5	5
25	26	27	28	29	0	1	2	3	4	5	6	7	8	9	10	11	12	13	14	15	16	17	18	19	20	21	22	23	24
49	47	46	44	43	41	40	38	37	39	34	33	32	31	30	29	28	27	26	25	24	23	22	21	21	20	19	19	18	18
18	43	7	38	4	37	12	49	28	9	53	39	27	18	12	7	5	5	8	12	18	26	39	53	9	26	47	10	37	4
58	58	58	58	58	58	58	58	58	58	58	58	58	58	58	58	58	59	59	59	59	59	59	59	59	59	59	59	59	59
22	24	25	27	30	33	35	37	39	40	42	46	48	50	53	55	57	0	2	4	6	10	12	14	16	19	21	23	26	28

[**91**:3] 18 R2, [**91**:4] 28 R2, [**94**:4] 16 R2, [**94**:6] 58 R2, 58 R3, [**99**:4] 15 R2, [**103**:4] 39 J1, 24 J2, [**104**:3] 38 J2, 46 R2, 36 R7, [**104**:4] 43 J1, 43 R3, [**105**:4] 27 J1, [**106**:4] 56 R2, [**111**:3] 15 J1, [**111**:4] 56 R2, [**112**:6] 15 J1, 15 R3, [**113**:6] 17 R2, [**116**:2] 22 R7, [**116**:4] 23 J2, 43 R2, [**117**:4] 41 R7, [**117**:6] 22 J2, [**118**:6] 22 R2, [**119**:6] 24 R2, [**120**:6] 56 R7, [**121**:6] 26 R2, [**122**:3] 58 R2, 51 R7, [**122**:4] 41 R2, 47 R7, [**124**:4] 57 J2, 54 R2, 55 R7, [**125**:3] 4 J1, [**125**:6] 35 R2, [**126**:4] 24 J2, [**129**:4] 17 J2, [**129**:6] 39 R2, [**130**:6] 41 J2, [**131**:6] 40 J2, [**132**:4] 5 J2, [**133**:4] 56 R2, [**133**:6] 50 R2, [**134**:2] 11 R2, [**134**:3] 17 R2, [**134**:4] 37 R2, [**135**:6] 52 R2, [**136**:4] 19 R3, 19 R7, [**136**:6] 55 J2, 56 R2, [**137**:4] 25 R2, [**137**:6] 16 J1, 57 J2, [**138**:4] 10 J2, 21 R2, [**138**:6] 58 J2, [**139**:4] 28 R2, [**139**:6] 59 R2, [**140**:3] 6 R2, [**140**:6] 58 J2, [**141**:3] 15 J1, [**142**:4] 24 R2, [**145**:3] 51 R2, [**145**:4] 57 R2, [**148**:4] 36 R2, [**149**:4] 58 R2, [**149**:6] 19 R2, [**150**:4] 19 R2, [**153**:4] 26 J2, 34 R2, [**154**:6] 31 R7, [**156**:3] 41 R2, [**156**:4] 32 R2, [**157**:3] 41 R2, [**159**:3] 36 J2, 36 R2, [**160**:4] 54 J2, [**161**:4] 19 R2, [**161**:6] 43 J2, [**162**:6] 46 J2, [**163**:6] 57 J1, [**169**:3] 24 R3, [**174**:4] 26 J2, [**175**:3] 19 J2, [**175**:4] 47 J2, [**176**:4] 10 J2, [**177**:3] 18 J2, [**177**:4] 37 J2, [**178**:4] 4 J2, [**178**:6] 23 R2, [**179**:3] 17 J2, [**179**:4] 5 J2, [**179**:6] 29 R7.

180	**181**	**182**	**183**	**184**	**185**	**186**	**187**	**188**	**189**	**190**	**191**	**192**	**193**	**194**	**195**	**196**	**197**	**198**	**199**	**200**	**201**	**202**	**203**	**204**	**205**	**206**	**207**	**208**	**209**
5	5	5	5	5	6	6	6	6	6	6	6	6	6	6	6	6	6	6	6	6	6	6	6	6	6	6	6	6	6
25	26	27	28	29	0	1	2	3	4	5	6	7	8	9	10	11	12	13	14	15	16	17	18	19	20	21	22	23	24
17	17	16	16	15	15	15	15	14	14	14	14	14	14	14	14	14	14	15	15	15	15	16	16	16	17	17	18	18	19
33	5	35	18	57	40	23	8	56	49	43	38	34	33	35	38	43	51	1	13	26	41	2	23	44	8	35	2	32	3
59	59	59	59	59	59	59	59	59	59	59	59	59	59	60	60	60	60	60	60	60	60	60	60	60	60	60	60	60	60
30	34	36	37	39	41	43	46	49	52	53	54	57	59	1	3	6	8	10	12	14	17	19	20	22	25	26	28	30	31

210	**211**	**212**	**213**	**214**	**215**	**216**	**217**	**218**	**219**	**220**	**221**	**222**	**223**	**224**	**225**	**226**	**227**	**228**	**229**	**230**	**231**	**232**	**233**	**234**	**235**	**236**	**237**	**238**	**239**
6	6	6	6	6	7	7	7	7	7	7	7	7	7	7	7	7	7	7	7	7	7	7	7	7	7	7	7	7	7
25	26	27	28	29	0	1	2	3	4	5	6	7	8	9	10	11	12	13	14	15	16	17	18	19	20	21	22	23	24
19	20	20	21	22	22	23	24	25	25	26	27	28	29	30	31	32	33	34	35	36	37	38	39	41	42	43	44	45	47
36	11	49	29	10	52	35	22	8	58	49	41	34	29	28	27	26	27	29	31	36	42	48	55	3	14	26	37	51	4
60	60	60	60	60	60	60	60	60	60	60	60	60	60	60	60	60	60	61	61	61	61	61	61	61	61	61	61	61	61
32	34	36	38	39	41	43	45	46	47	49	50	51	53	55	56	57	59	0	1	3	4	5	5	6	7	8	9	10	11

240	**241**	**242**	**243**	**244**	**245**	**246**	**247**	**248**	**249**	**250**	**251**	**252**	**253**	**254**	**255**	**256**	**257**	**258**	**259**	**260**	**261**	**262**	**263**	**264**	**265**	**266**	**267**	**268**	**269**
7	7	7	7	7	8	8	8	8	8	8	8	8	8	8	8	8	8	8	8	8	8	8	8	8	8	8	8	8	8
25	26	27	28	29	0	1	2	3	5	6	7	8	9	10	11	12	13	14	15	16	17	18	19	20	21	22	23	24	25
48	49	50	52	53	54	56	57	58	0	1	2	4	5	6	8	9	11	12	13	15	16	18	19	20	22	23	25	26	27
18	34	51	7	26	55	4	24	54	6	28	50	12	34	55	19	42	5	39	51	16	39	3	27	50	14	37	1	24	47
61	61	61	61	61	61	61	61	61	61	61	61	61	61	61	61	61	61	61	61	61	61	61	61	61	61	61	61	61	61
12	13	14	15	16	17	18	18	19	20	20	20	21	21	21	22	22	22	22	23	23	23	23	23	23	23	23	22	22	21

[**180**:4] 5 J2, 32 R2, [**180**:6] 31 R2, [**181**:3] 16 J2, [**181**:4] 41 J2, 8 R2, [**182**:4] 18 J2, 44 R2, [**183**:3] 15 J2, [**183**:4] 57 J2, 21 R2, [**184**:4] 55 J1, 59 R2, [**186**:6] 47 R2, [**187**:4] 16 J2, [**187**:6] 49 R3, [**188**:4] 54 J2, [**191**:6] 55 R2, [**192**:6] 56 J2, 56 R7, [**193**:4] 34 R2, [**193**:6] 0 R2, [**195**:4] 43 R2, [**196**:4] 44 J1, [**197**:6] 10 J1, [**198**:4] 0 R2, [**200**:4] 31 J2, [**201**:4] 47 J2, 47 R7, [**203**:4] 33 R2, [**203**:6] 21 R2, [**204**:2] 29 R2, [**204**:4] 54 J2, [**204**:6] 23 R2, [**205**:6] 24 R2, [**206**:4] 38 J2, [**208**:4] 2 J2, 33 R7, [**209**:3] 18 J2, [**210**:6] 33 R7, [**211**:4] 38 J2, 9 R2, [**212**:4] 48 J2, 44 R2, [**213**:4] 22 R2, [**214**:4] 14 J2, 2 R2, [**214**:6] 40 R2, [**215**:3] 23 J2, [**215**:4] 40 R2, [**216**:2] 23 R7, [**216**:3] 24 J1, 35 R7, [**216**:4] 23 R2, 2 R7, [**217**:2] 24 R7, [**217**:4] 8 R2, [**218**:3] 24 R2, [**218**:4] 54 R2, [**219**:4] 48 R7, [**220**:6] 48 R2, [**221**:4] 34 J2, [**221**:6] 5 J1, [**222**:4] 39 J2, 31 R2, [**224**:4] 27 J1, [**227**:4] 26 J1, [**230**:6] 2 J2, [**231**:6] 3 J2, [**232**:4] 49 J2, [**232**:6] 4 J2, [**233**:6] 6 J1, 6 R2, [**234**:6] 7 J1, 7 R2, [**235**:6] 8 J1, 8 R2, [**236**:4] 23 J2, [**236**:6] 9 J1, 9 R2, [**237**:4] 36 R2, [**237**:6] 10 J1, 10 R2, [**238**:6] 11 J1, 11 R2, [**239**:6] 12 J1, 12 R2, [**240**:6] 13 J1, 13 R2, [**241**:3] 50 J2, [**241**:4] 56 J2, [**241**:6] 14 J1, 14 R2, [**242**:3] 52 J2, [**242**:4] 7 J2, 21 R2, [**242**:6] 15 J1, 15 R2, [**243**:3] 53 J2, [**243**:4] 21 J2, [**243**:6] 16 J1, 16 R2, [**244**:3] 54 J2, 59 R2, [**244**:4] 55 J2, [**244**:6] 17 J1, 17 R2, [**245**:4] 45 J1, 57 R2, [**245**:6] 18 J1, 18 R2, [**248**:4] 24 J1, 44 J2, 44 R7, [**250**:6] 21 J2, [**251**:4] 55 J2, [**252**:3] 3 R2, [**252**:4] 52 J2, [**252**:6] 22 R2, [**253**:6] 22 J2, [**254**:4] 34 J2, [**257**:3] 51 J1, [**258**:4] 29 J2, [**262**:4] 6 R2, [**263**:4] 37 J2, [**265**:4] 19 R2, [**266**:4] 27 R3, [**267**:6] 23 J1, 21 R2, [**268**:4] 25 R7, [**268**:6] 21 R2, [**269**:3] 47 R2.

Appendix A1: True solar longitude and velocity (continued)

270	**271**	**272**	**273**	**274**	**275**	**276**	**277**	**278**	**279**	**280**	**281**	**282**	**283**	**284**	**285**	**286**	**287**	**288**	**289**	**290**	**291**	**292**	**293**	**294**	**295**	**296**	**297**	**298**	**299**
8	8	8	8	9	9	9	9	9	9	9	9	9	9	9	9	9	9	9	9	9	9	9	9	9	9	9	9	9	9
26	27	28	29	0	1	2	3	4	5	6	7	8	9	10	11	12	13	14	15	16	17	18	19	20	21	23	24	25	26
29	30	31	33	34	35	37	38	39	41	42	43	44	46	47	48	49	50	52	53	54	55	56	57	58	59	0	1	2	3
8	30	51	12	34	54	14	32	52	9	27	43	57	12	25	37	48	58	9	17	25	31	37	40	45	45	44	42	40	35
61	61	61	61	61	61	61	61	61	61	61	61	61	61	61	61	61	61	61	61	61	61	61	61	61	61	61	60	60	60
21	21	20	20	20	19	18	18	18	17	16	16	15	14	13	12	11	10	9	8	7	6	5	4	3	2	1	59	57	56

300	**301**	**302**	**303**	**304**	**305**	**306**	**307**	**308**	**309**	**310**	**311**	**312**	**313**	**314**	**315**	**316**	**317**	**318**	**319**	**320**	**321**	**322**	**323**	**324**	**325**	**326**	**327**	**328**	**329**
9	9	9	10	10	10	10	10	10	10	10	10	10	10	10	10	10	10	10	10	10	10	10	10	10	10	10	10	10	10
27	28	29	0	1	2	3	4	5	6	7	8	9	10	11	12	13	14	15	16	17	18	19	20	21	22	23	24	25	26
4	5	6	6	7	8	9	10	10	11	12	12	13	13	14	14	15	15	16	16	16	17	17	17	17	18	18	18	18	18
30	22	13	58	52	40	23	7	39	29	8	44	20	53	25	54	21	47	11	33	51	10	27	41	53	2	9	15	19	19
60	60	60	60	60	60	60	60	60	60	60	60	60	60	60	60	60	60	60	60	60	60	60	60	60	60	60	60	60	59
55	53	51	50	49	47	46	45	43	41	39	38	36	34	32	30	28	26	25	22	20	19	17	14	12	10	8	6	3	1

330	**331**	**332**	**333**	**334**	**335**	**336**	**337**	**338**	**339**	**340**	**341**	**342**	**343**	**344**	**345**	**346**	**347**	**348**	**349**	**350**	**351**	**352**	**353**	**354**	**355**	**356**	**357**	**358**	**359**
10	10	10	11	11	11	11	11	11	11	11	11	11	11	11	11	11	11	11	11	11	11	11	11	11	11	11	11	11	11
27	28	29	0	1	2	3	4	5	6	7	8	9	10	11	12	13	14	15	16	17	18	19	20	21	22	23	24	25	26
18	18	18	18	17	17	17	17	16	16	16	15	15	14	14	13	13	12	11	10	10	9	8	7	6	5	4	3	2	1
18	15	9	0	52	41	38	12	53	32	9	44	17	47	15	41	3	24	43	59	13	24	35	42	48	48	48	46	42	35
59	59	59	59	59	59	59	59	59	59	59	59	59	59	59	59	59	59	59	59	59	59	59	59	59	59	59	59	58	58
59	58	54	52	51	49	46	43	41	39	37	36	34	30	28	26	23	21	20	19	16	14	12	10	6	4	2	0	57	55

360	**361**	**362**	**363**	**364**	**365**
11	11	11	11	0	0
27	27	28	29	0	1
0	59	58	56	55	54
26	15	0	45	26	4
58	58	58	58	58	58
53	50	48	46	42	40

[**270**:4] 0 J2, [**272**:4] 15 J2, 56 R2, [**272**:6] 22 J1, 21 R2, [**274**:4] 38 J2, [**275**:4] 34 R2, [**276**:3] 36 R2, [**276**:4] 57 R2, [**276**:5] x R2, [**276**:6] x R2, [**277**:6] 17 J2, [**278**:6] 17 J2, [**279**:4] 5 J2, 10 R2, [**279**:6] 16 J2, [**285**:4] 47 J2, [**286**:3] 48 J2, [**286**:4] 49 J2, 49 R7, [**287**:4] 48 J2, [**290**:4] 27 R2, [**290**:6] 9 J2, [**291**:4] 51 J2, [**294**:4] 25 J2, [**296**:4] 54 J2, [**296**:6] 29 J2, [**300**:6] 53 J2, [**301**:4] 33 R2, 32 R3, [**301**:6] 54 R2, [**302**:4] 12 J1, 30 R2, [**303**:2] 29 R7, [**303**:3] 7 J2, 7 R7, [**303**:4] 3 J2, 8 R2, 3 R7, [**304**:3] 8 J2, [**305**:6] 49 R2, [**306**:4] 33 R2, [**307**:3] 15 R2, [**310**:4] 44 J1, [**311**:3] 13 R2, [**314**:6] 22 J1, [**319**:4] 32 J2, 32 R7, [**320**:4] 52 J2, 50 R3, [**323**:3] 18 J2, [**323**:4] 43 J2, 28 R2, [**324**:3] 18 J2, [**324**:6] 14 R2, [**325**:2] 21 R2, [**325**:3] 17 R2, [**325**:4] 53 R2, 53 R3, [**325**:6] 12 R2, [**326**:2] 33 R2, [**326**:6] 10 R2, [**327**:6] 8 R2, [**328**:4] 14 J1, 3 R2, [**328**:6] 6 R2, [**329**:5] 60 R2, [**329**:6] 3 R2, [**330**:5] 60 R2, [**330**:6] 1 R2, [**331**:4] 25 R2, [**331**:6] 59 R2, [**332**:6] 55 J2, 58 R2, 55 R7, [**333**:4] 9 J2, [**333**:6] 54 R2, [**334**:4] 51 J1, 41 J2, [**334**:6] 50 J1, 52 R2, [**335**:4] 42 J1, 38 J2, [**335**:6] 46 J2, 51 R2, [**336**:4] 28 R7, [**337**:4] 38 J1, [**337**:6] 45 R2, [**338**:3] 17 J2, [**338**:6] 40 J1, [**339**:4] 33 J1, [**340**:4] 7 J2, [**341**:4] 45 J1, [**343**:6] 31 J2, 31 R7, [**344**:6] 29 J2, 29 R7, [**345**:4] 43 J2, [**350**:4] 53 J1, [**351**:2] 28 R2, [**353**:2] 18 R2, [**353**:3] 9 R2, [**353**:4] 24 R2, [**354**:4] 46 J2, 50 R7, [**355**:4] 51 R7, [**355**:6] 6 J1, [**356**:4] 52 R7, [**356**:6] 3 J1, [**357**:4] 59 R7, [**359**:6] 54 R2, [**361**:4] 19 J1, [**362**:4] 7 R7, [**363**:2] 19 J2, [**364**:4] 56 R2, [**365**:6] 43 J1.

Appendix A2. Mean motion of moon and anomaly (0 to 365 days)

0	1	2	3	4	5	6	7	8	9	10	11	12	13	14	15	16	17	18	19	20	21	22	23	24	25	26	27	28	29
11	0	0	1	1	2	2	3	3	3	4	4	5	5	6	6	7	7	7	8	8	9	9	10	10	10	11	11	0	0
29	13	26	9	22	5	18	2	15	28	11	24	7	21	4	17	0	13	26	10	23	6	19	2	15	29	12	25	8	21
54	4	14	24	34	44	54	4	14	24	35	45	55	5	15	24	35	46	56	6	16	27	37	47	57	8	18	28	38	48
17	26	22	30	39	47	42	51	59	55	3	12	21	15	24	32	54	3	24	33	55	3	11	33	42	3	12	34	43	51
11	0	0	1	1	2	2	3	3	3	4	4	5	5	6	6	6	7	7	8	8	9	9	10	10	10	11	11	0	0
29	12	26	9	22	5	18	1	14	27	10	23	6	19	2	15	28	11	24	7	21	4	17	0	13	26	9	22	5	18
54	57	1	4	7	11	14	18	21	24	28	31	35	38	41	45	59	52	56	59	3	6	10	14	17	21	24	28	31	35
20	48	3	31	59	27	51	19	27	52	19	48	16	20	54	28	7	35	16	44	24	52	20	1	28	10	38	19	47	12

30	31	32	33	34	35	36	37	38	39	40	41	42	43	44	45	46	47	48	49	50	51	52	53	54	55	56	57	58	59
1	1	2	2	2	3	3	4	4	5	5	5	6	6	7	7	8	8	9	9	9	10	10	11	11	0	0	1	1	1
4	18	1	14	27	10	24	7	20	3	16	29	13	26	9	22	5	18	2	15	28	11	24	7	21	4	17	0	13	27
59	9	19	29	40	51	0	10	21	31	41	51	2	12	23	33	44	54	5	15	26	37	47	57	8	19	29	40	50	1
12	21	42	51	13	30	30	52	1	22	30	52	1	25	11	46	7	42	16	51	26	1	36	58	33	8	42	17	52	27
1	1	1	2	2	3	3	4	4	4	5	5	6	6	7	7	8	8	8	9	9	10	10	11	11	11	0	0	1	1
1	14	27	10	23	6	20	3	16	29	12	25	8	21	4	17	0	13	26	9	22	5	19	2	15	28	11	24	7	20
38	42	46	59	53	56	0	3	7	10	14	18	21	25	29	33	36	40	44	48	52	56	0	4	7	11	15	19	23	27
52	2	4	32	13	41	9	50	18	59	25	17	35	29	23	17	58	52	46	46	33	28	22	3	57	51	45	39	33	27

[**0**:2] 19 J1, [**0**:4] 11 R2, 11 R3, [**3**:8] 34 R7, [**4**:7] 34 J2, [**4**:8] 39 J2, [**5**:2] 25 R2, [**5**:4] 43 R7, [**5**:7] 44 J2, [**5**:8] 45 J2, [**6**:4] 43 R2, [**7**:8] 9 R3, 9 R7, [**8**:7] 24 J2, [**8**:8] 59 J2, 28 R7, [**9**:3] 54 R2, [**9**:4] 54 J1, 54 J2, 45 R7, [**9**:5] 4 J1, 4 J2, [**9**:6] 10 J2, [**9**:7] 28 J2, [**9**:8] 19 J2, [**10**:3] 34 R2, [**10**:7] 18 J1, [**11**:3] 15 J1, [**11**:4] 15 R2, [**11**:7] 21 J1, [**11**:8] 28 R7, [**14**:2] 14 J2, [**14**:4] 23 R7, [**14**:8] 45 J1, 45 J2, [**15**:2] 7 R7, [**15**:3] 25 R2, [**15**:4] 54 R7, [**15**:7] 48 R2, [**15**:8] 25 J2, [**16**:3] 25 J1, [**16**:7] 49 R3, 49 R7, [**17**:8] 34 R7, [**19**:4] 26 R7, [**20**:4] 54 J1, 15 R7, [**22**:1] 10 J1, 10 J2, [**22**:4] 3 R2, 10 R7, [**22**:8] 22 R2, [**23**:4] 11 R2, [**23**:5] 9 R2, [**23**:8] 42 R7, [**24**:1] x R7, [**24**:2] x R7, [**24**:4] 43 R7, [**24**:8] 26 R7, [**25**:3] x R7, [**26**:6] 8 R7, [**27**:4] 38 J2, [**27**:8] 29 J2, 29 R2, [**28**:4] 42 R2, [**29**:3] 41 R3, [**29**:7] 34 R7, [**31**:2] 17 J1, [**32**:4] 41 R7, [**32**:7] 47 R3, [**33**:7] 49 R3, 49 R7, [**34**:4] 12 R2, [**34**:8] 23 J2, [**35**:4] 31 R7, [**36**:4] 37 R2, [**36**:5] 4 R2, [**36**:7] 7 R7, [**37**:7] 4 R2, [**38**:2] 21 R2, [**38**:3] 1 R2, [**39**:4] 32 J1, 32 R2, [**39**:6] 19 R3, [**40**:3] 11 R2, [**40**:4] x R7, [**41**:2] 28 R2, [**41**:3] 21 R2, [**41**:4] 12 R2, 56 R7, [**41**:8] 7 R2, 7 R3, [**42**:4] 35 R7, [**42**:8] 3 R7, [**43**:1] 8 R2, [**43**:2] 27 R2, [**43**:3] 56 J2, 51 R2, [**43**:4] 36 J1, 19 J2, 36 R7, [**43**:8] 39 J2, [**44**:3] 13 R2, [**44**:4] 13 R2, [**45**:3] 13 R2, [**48**:2] 18 J2, [**48**:8] 16 J2, 26 R2, [**49**:8] 36 R3, [**50**:2] 26 R7, [**50**:8] 34 J1, [**51**:3] 57 R2, [**51**:7] 58 J2, 26 R7, [**51**:8] 48 R3, [**52**:1] 11 J1, 11 J2, [**52**:4] 38 J1, [**52**:8] 23 R2, [**53**:3] 58 J1, [**53**:4] 8 R2, [**54**:4] 31 J1, [**55**:4] 4 J2, [**56**:3] 9 R7, [**56**:8] 16 J1, 16 J2, [**57**:8] 35 J1, 35 J2, [**58**:3] 54 J1, 54 J2, [**58**:8] 23 J1, [**59**:6] 2 J1, [**59**:8] 37 R2.

Appendix A2. Mean motion of moon and anomaly (continued)

60	**61**	**62**	**63**	**64**	**65**	**66**	**67**	**68**	**69**	**70**	**71**	**72**	**73**	**74**	**75**	**76**	**77**	**78**	**79**	**80**	**81**	**82**	**83**	**84**	**85**	**86**	**87**	**88**	**89**
2	2	3	3	4	4	4	5	5	6	6	7	7	8	8	8	9	9	10	10	11	11	0	0	0	1	1	2	2	3
10	23	6	19	2	16	29	12	25	8	21	5	18	1	14	27	11	24	7	20	3	16	0	13	26	9	22	6	19	2
12	22	32	43	54	4	15	25	36	47	57	8	19	30	40	51	2	13	23	34	45	56	6	17	28	39	49	0	11	22
2	37	59	34	9	43	18	53	28	3	38	26	14	2	37	25	13	1	49	38	25	1	48	36	24	12	47	36	24	12
2	2	2	3	3	4	4	5	5	6	6	6	7	7	8	8	9	9	9	10	10	11	11	0	0	1	1	1	2	2
3	16	29	12	25	8	21	4	18	1	14	27	10	23	6	19	2	15	28	11	24	7	20	4	17	0	13	26	9	22
31	35	38	42	46	50	54	58	2	6	10	14	18	22	26	30	34	38	41	46	51	55	59	3	7	11	15	19	23	27
20	15	56	50	44	38	32	26	20	14	7	15	22	29	23	30	37	44	51	59	5	0	7	14	21	28	22	29	36	43

90	**91**	**92**	**93**	**94**	**95**	**96**	**97**	**98**	**99**	**100**	**101**	**102**	**103**	**104**	**105**	**106**	**107**	**108**	**109**	**110**	**111**	**112**	**113**	**114**	**115**	**116**	**117**	**118**	**119**
3	3	4	4	5	5	6	6	7	7	7	8	8	9	9	10	10	10	11	11	0	0	1	1	2	2	2	3	3	4
15	28	11	25	8	21	4	17	0	14	27	10	23	6	20	3	16	29	12	25	9	22	5	18	1	15	28	11	24	7
32	43	54	4	15	26	37	53	58	9	20	31	42	52	3	14	25	36	47	57	8	19	30	41	51	2	13	24	35	46
46	34	22	57	45	34	22	57	45	32	20	11	3	47	35	23	24	12	0	47	35	23	12	0	47	35	35	24	12	0
3	3	4	4	4	5	5	6	6	7	7	7	8	8	9	9	10	10	11	11	11	0	0	1	1	2	2	2	3	3
5	18	1	14	27	10	23	6	20	3	16	29	12	25	8	21	4	17	0	13	26	9	23	6	19	2	15	28	11	24
31	35	39	43	47	51	56	59	4	8	12	16	20	23	28	32	37	41	45	49	53	57	1	6	10	14	18	22	26	31
35	42	49	43	50	57	4	58	5	13	20	27	34	51	48	55	15	22	29	36	43	50	57	5	12	19	39	46	53	0

[**61**:2] 22 R2, [**61**:3] 27 R2, [**61**:4] 27 J2, 32 R2, [**61**:8] 16 R2, [**62**:6] 38 R2, [**62**:7] 29 R2, 58 R3, 39 R7, [**66**:1] x J2, [**66**:2] x J2, [**66**:3] x J2, [**66**:4] x J2, [**66**:5] x J1, x J2, [**66**:6] x J2, [**66**:7] 45 J1, x J2, [**66**:8] x J2, 35 R7, [**67**:4] 54 J1, 54 J2, 13 R2, [**67**:6] 8 R3, [**67**:7] 28 R2, 50 R3, [**67**:8] 38 R3, [**68**:3] 38 R2, [**68**:6] 28 R2, [**69**:6] 2 J2, [**69**:8] 10 R2, [**73**:2] 14 J2, [**73**:4] 42 R2, [**74**:6] 8 R2, [**74**:8] 3 J1, 3 J2, [**75**:4] 27 R7, [**76**:4] 23 J1, [**76**:8] 17 J2, [**78**:3] 3 J1, [**78**:4] 89 R7, [**78**:7] 42 R3, 42 R7, [**80**:4] 22 J1, 22 J2, [**83**:4] 26 J2, 46 R2, [**85**:4] 47 J2, [**86**:1] 2 R2, [**88**:1] 3 R2, [**88**:7] 33 J1, 33 J2, [**89**:6] 12 R3, [**90**:3] 35 R2, 31 R7, [**91**:2] 27 R2, [**91**:3] 42 R7, [**91**:4] 24 R2, [**91**:8] 43 J1, 43 J2, [**92**:2] 10 R2, [**92**:4] 52 J1, 2 R7, [**93**:4] 7 R2, [**95**:2] 20 R2, [**95**:3] 16 R7, [**95**:4] 36 R2, [**97**:2] 7 J2, [**97**:3] 48 J1, 48 J2, 47 R7, [**97**:4] 50 R2, [**100**:2] 7 J2, [**102**:3] 41 R7, [**102**:6] 22 R2, [**103**:7] 24 R2, 24 R7, [**103**:8] 42 J1, 42 J2, 41 R2, [**105**:3] 24 J1, [**105**:8] 52 R2, [**106**:2] 18 R7, [**107**:7] 42 J1, 43 J2, [**108**:7] 49 R2, 49 R3, [**109**:8] 39 J1, 49 J2, [**111**:3] 9 R7, [**111**:7] 59 J2, [**111**:8] 57 J2, 17 R7, [**112**:4] 17 R7, [**113**:6] 9 J2, [**114**:3] 57 R2, [**116**:2] 38 J2, [**116**:8] 29 R2, [**117**:3] 14 R7, [**117**:6] 18 R7, [**117**:7] 21 J2, 42 R2, 12 R7, [**118**:7] 46 J1, 46 J2, [**118**:8] 4 R7.

120	**121**	**122**	**123**	**124**	**125**	**126**	**127**	**128**	**129**	**130**	**131**	**132**	**133**	**134**	**135**	**136**	**137**	**138**	**139**	**140**	**141**	**142**	**143**	**144**	**145**	**146**	**147**	**148**	**149**
4	5	5	6	6	6	7	7	8	8	9	9	9	10	10	11	11	0	0	1	1	1	2	2	3	3	4	4	4	5
20	4	17	0	13	26	10	23	6	19	2	15	29	12	25	8	21	4	18	1	14	27	10	24	7	20	3	16	29	13
56	7	18	29	40	51	1	12	23	34	44	55	6	17	27	38	48	59	10	21	31	42	53	4	14	25	36	47	57	8
49	37	38	26	14	2	50	38	13	1	49	24	13	0	35	23	58	46	34	10	57	36	20	8	56	44	19	7	56	31
4	4	5	5	5	6	6	7	7	8	8	9	9	9	10	10	11	11	0	0	0	1	1	2	2	3	3	4	4	4
7	20	3	16	29	12	26	9	22	5	18	1	14	27	10	23	6	19	2	15	28	12	25	8	21	4	17	0	13	26
35	39	43	47	51	55	0	4	8	12	16	20	24	28	32	36	40	44	48	52	56	0	4	8	12	16	20	25	29	33
7	14	34	41	49	56	3	10	4	11	18	12	19	26	20	27	21	28	35	29	36	43	37	44	51	58	52	1	7	1

150	**151**	**152**	**153**	**154**	**155**	**156**	**157**	**158**	**159**	**160**	**161**	**162**	**163**	**164**	**165**	**166**	**167**	**168**	**169**	**170**	**171**	**172**	**173**	**174**	**175**	**176**	**177**	**178**	**179**
5	6	6	7	7	8	8	8	9	9	10	10	11	11	0	0	0	1	1	2	2	3	3	3	4	4	5	5	6	6
26	9	22	5	19	2	15	28	11	24	8	21	4	17	0	14	27	10	23	6	19	3	16	29	12	25	8	22	5	18
19	30	40	51	2	13	23	34	45	56	6	17	28	39	50	0	11	22	33	43	54	5	16	26	37	48	29	9	20	31
18	6	54	29	17	5	54	42	16	4	52	40	28	17	5	40	28	16	3	54	42	17	5	53	41	16	3	51	39	26
5	5	6	6	7	7	7	8	8	9	9	10	10	10	11	11	0	0	1	1	2	2	2	3	3	4	4	5	5	5
9	22	5	18	1	14	28	11	24	7	20	3	16	29	12	25	8	21	4	17	0	14	27	10	23	6	19	2	15	28
37	41	45	49	53	57	1	5	9	13	17	21	26	30	34	38	42	46	50	54	58	2	6	10	15	18	23	27	31	35
7	14	21	15	22	29	36	43	37	44	51	58	5	12	19	13	40	28	35	42	49	43	50	57	4	58	5	12	19	26

[**121**:3] 47 R7, [**121**:4] 35 J1, [**122**:3] 28 R2, [**122**:8] 37 J1, [**123**:4] 36 R2, [**123**:8] 51 J2, [**124**:4] 15 R3, [**124**:7] 50 J1, [**124**:8] 25 R2, [**125**:4] 9 J2, [**125**:7] 0 R2, [**127**:3] 22 R7, [**128**:4] 12 J1, [**129**:4] 18 R2, [**129**:8] 1 J1, [**130**:4] 48 R2, [**131**:4] 29 J2, [**131**:8] 15 R7, [**132**:3] 4 R2, [**132**:4] 19 R2, [**132**:6] 15 R7, [**133**:7] 38 J2, [**135**:7] 38 J2, [**136**:4] 51 R2, [**137**:3] 49 R7, [**138**:2] 17 R2, [**138**:3] 20 R2, [**138**:4] 36 J1, [**138**:8] 36 R7, [**139**:2] 0 R2, 0 R3, [**139**:3] 11 J2, [**140**:7] 59 J1, [**141**:2] 29 J2, [**141**:3] 43 J2, [**141**:4] 30 J2, [**141**:8] 46 R7, [**142**:3] 13 J2, [**142**:6] 24 J2, [**142**:8] 36 J2, [**143**:4] 36 J2, 18 R2, [**144**:2] 0 J2, [**144**:8] 58 J2, [**145**:3] 30 J2, [**145**:4] 4 R7, [**145**:8] 56 J2, [**146**:4] 9 R2, [**146**:7] 30 J1, [**146**:8] 59 J2, [**147**:2] 19 J2, [**147**:4] 17 R2, [**147**:8] 7 J2, [**148**:3] 5 R3, [**148**:6] 23 J2, [**149**:4] 33 J2, [**149**:8] 2 J1, 10 J2, [**150**:3] 29 J2, [**151**:8] 26 J2, [**152**:6] x R3, [**152**:8] 20 J1, 20 R2, [**153**:4] 22 J2, [**153**:7] 59 R2, [**153**:8] 22 J2, [**154**:4] 57 R2, [**154**:6] 12 J2, [**155**:3] 53 R2, [**156**:3] 24 R2, [**156**:7] 23 J2, [**156**:8] 54 J2, [**157**:7] 4 J1, 4 J2, [**158**:4] 19 J2, [**158**:8] 27 J2, [**159**:7] 12 R3, [**159**:8] 51 R3, [**160**:2] 50 R2, [**160**:3] 60 R2, [**160**:7] 57 J2, [**161**:3] 10 R2, [**161**:4] 47 R2, [**164**:4] 0 R2, [**164**:8] 29 R2, [**165**:4] 44 J1, [**165**:6] 2 R3, [**166**:3] 14 J2, [**166**:8] 20 R3, 20 R7, [**167**:3] 23 J2, [**168**:7] 5 R7, [**168**:8] 30 R7, [**169**:3] 42 R2, [**169**:7] 44 R7, [**170**:2] 29 J2, [**171**:6] 17 R7, [**171**:7] 3 R2, [**173**:3] 54 J2, [**173**:4] 42 J2, [**175**:7] 19 R2, [**176**:3] 56 J1, 56 J2, 59 R2, 59 R7, [**178**:3] 30 J2, [**178**:4] 29 R2, [**179**:7] 36 J2, [**179**:8] 36 J1.

Appendix A2. Mean motion of moon and anomaly (continued)

180	**181**	**182**	**183**	**184**	**185**	**186**	**187**	**188**	**189**	**190**	**191**	**192**	**193**	**194**	**195**	**196**	**197**	**198**	**199**	**200**	**201**	**202**	**203**	**204**	**205**	**206**	**207**	**208**	**209**
7	7	7	8	8	9	9	10	10	11	11	11	0	0	1	1	2	2	2	3	3	4	4	5	5	6	6	6	7	7
1	14	28	11	24	7	20	3	17	0	13	26	9	23	6	19	2	15	28	12	25	8	21	4	18	1	14	27	10	23
42	52	3	14	25	36	47	57	8	19	30	41	52	2	13	24	35	46	47	8	18	29	40	51	2	13	24	34	45	56
1	59	37	25	14	24	2	50	38	4	28	16	4	52	53	41	29	17	19	7	55	43	44	32	20	21	9	58	46	47
6	6	7	7	8	8	9	9	9	10	10	11	11	0	0	0	1	1	2	2	3	3	3	4	4	5	5	6	6	7
11	24	7	20	3	17	0	13	26	9	22	5	18	1	14	27	10	23	6	19	3	16	29	12	25	8	21	4	17	0
39	43	47	51	55	0	4	8	12	16	20	25	29	33	37	41	45	50	54	58	2	6	11	15	19	23	27	31	35	40
20	27	34	41	48	9	16	23	30	42	57	4	11	18	38	45	52	0	20	27	24	41	1	8	15	35	42	49	36	17

210	**211**	**212**	**213**	**214**	**215**	**216**	**217**	**218**	**219**	**220**	**221**	**222**	**223**	**224**	**225**	**226**	**227**	**228**	**229**	**230**	**231**	**232**	**233**	**234**	**235**	**236**	**237**	**238**	**239**
8	8	9	9	9	10	10	11	11	0	0	1	1	1	2	2	3	3	4	4	5	5	5	6	6	7	7	8	8	8
7	20	3	16	29	13	26	9	22	5	18	2	15	28	11	24	8	21	4	17	0	13	27	10	23	6	19	3	16	29
7	18	29	40	51	2	13	24	35	46	57	8	19	30	41	52	3	14	25	36	47	58	8	19	30	41	52	3	14	25
34	36	37	38	27	28	16	17	19	19	7	9	10	11	26	27	15	17	17	5	7	9	57	59	46	48	48	36	37	39
7	7	8	8	9	9	10	10	10	11	11	0	0	1	1	1	2	2	3	3	4	4	5	5	5	6	6	7	7	8
13	26	9	22	6	19	2	15	28	11	24	7	20	3	16	29	12	25	9	22	5	18	1	14	27	10	23	6	19	2
44	48	53	57	1	5	9	14	18	22	27	31	35	40	44	48	53	57	1	5	10	14	18	23	27	31	35	39	44	48
24	44	14	24	31	51	58	18	39	59	6	26	46	6	39	59	6	27	47	54	13	33	40	0	7	28	48	56	15	35

[**180**:3] 43 J2, 41 R2, 41 R7, [**181**:4] 49 J1, 49 J2, [**182**:2] 29 J2, [**182**:3] 13 J1, 33 J2, 5 R7, [**182**:8] 44 R7, [**183**:8] 40 R7, [**184**:4] 13 R7, [**184**:7] 54 J2, [**184**:8] 49 R3, 49 R7, [**185**:8] 2 J2, [**186**:7] 40 R7, [**187**:3] 58 J1, 58 J2, [**187**:4] 10 J1, 10 J2, 5 R7, [**187**:8] 13 R2, [**189**:2] 10 R2, [**189**:4] 40 J1, 40 J2, [**190**:5] 11 J1, 11 J2, [**190**:7] 30 R7, [**190**:8] 50 J2, 28 R2, [**191**:4] 46 R2, [**192**:2] 2 J2, 8 R2, [**192**:3] 42 R2, 42 R3, [**192**:5] 12 J1, 12 J2, [**192**:7] 42 J2, 18 R7, [**192**:8] 29 R7, [**193**:3] 12 R2, [**193**:8] 22 R7, [**194**:4] 3 R7, [**194**:8] 18 J1, [**195**:8] 41 R2, [**197**:3] 47 J1, 47 J2, [**197**:8] 4 J2, [**198**:3] 50 J1, 57 J2, 57 R7, [**198**:7] 24 J1, [**198**:8] 10 J2, [**199**:2] 13 J1, [**199**:8] 17 R7, [**200**:1] 4 R2, [**200**:2] 35 J1, 8 R2, [**200**:3] 29 R2, [**200**:4] 43 R2, [**200**:8] 34 R3, 34 R7, [**201**:2] 21 R2, [**202**:3] 29 R2, [**202**:4] 43 R2, [**202**:6] 19 J1, [**203**:3] 21 J2, 59 R3, [**203**:4] 30 J2, [**204**:4] 30 J2, [**204**:6] 5 R2, [**204**:8] 25 J1, [**207**:4] 48 J2, 28 R7, [**207**:7] 30 R2, 1 R3, [**207**:8] 29 J1, 29 J2, 39 R2, [**208**:8] 56 J1, 46 J2, 56 R2, [**209**:4] 57 J1, 57 J2, [**211**:8] 45 R2, [**212**:4] 38 R2, [**212**:7] 52 R2, [**212**:8] 24 R2, [**213**:4] 37 R2, [**213**:7] 56 R2, [**214**:3] 57 J1, [**214**:4] 21 J2, [**215**:3] 10 R2, [**215**:8] 15 R2, [**216**:4] 17 R2, [**216**:6] 20 J1, [**216**:8] 8 R2, [**218**:5] 11 J1, 11 R2, [**218**:7] 19 J2, [**219**:4] 16 R7, [**219**:6] 12 J1, [**219**:8] 19 J1, [**220**:2] 5 R2, [**220**:3] 47 R2, [**220**:7] 37 R7, [**221**:2] 1 R2, [**223**:3] 37 R2, [**223**:7] 39 R2, [**225**:3] 42 R2, [**225**:8] 39 J1, 39 J2, [**226**:2] 4 J1, [**226**:7] 57 J2, 52 R2, 52 R7, [**226**:8] 53 R7, [**227**:1] 2 J1, [**227**:7] 7 R2, 47 R3, [**227**:8] 57 R7, [**228**:8] 46 R2, [**229**:8] 24 R2, 24 R3, [**230**:7] 14 R2, [**231**:3] 57 R2, [**231**:7] 24 J1, 24 J2, 18 R2, [**232**:7] 33 R2, 33 R3, [**232**:8] 4 R7, [**233**:4] 58 J1, 58 J2, [**234**:4] 48 R2, [**234**:8] 0 J1, 0 J2, [**235**:3] 48 R2, [**235**:4] 41 R2, [**235**:8] 38 R2, [**237**:2] 16 R2, [**237**:7] 41 R2, [**237**:8] 48 R2, 46 R3, [**238**:3] 34 J1.

240	**241**	**242**	**243**	**244**	**245**	**246**	**247**	**248**	**249**	**250**	**251**	**252**	**253**	**254**	**255**	**256**	**257**	**258**	**259**	**260**	**261**	**262**	**263**	**264**	**265**	**266**	**267**	**268**	**269**
9	9	10	10	11	11	0	0	0	1	1	2	2	3	3	4	4	4	5	5	6	6	7	7	7	8	8	9	9	10
12	25	8	22	5	18	1	14	28	11	24	7	20	3	17	0	13	26	9	23	6	19	2	15	28	12	25	8	21	4
36	47	58	9	20	31	42	53	4	15	26	37	48	48	9	20	31	41	52	3	14	25	35	46	57	8	18	29	40	51
27	28	30	18	19	20	21	3	10	12	0	1	3	50	38	26	1	49	38	26	14	2	49	24	12	1	39	37	12	0
8	8	9	9	10	10	11	11	0	0	0	1	1	2	2	3	3	3	4	4	5	5	6	6	6	7	7	8	8	9
15	28	12	25	8	21	4	17	0	13	26	9	22	5	18	1	15	28	11	24	7	20	3	16	29	12	25	8	21	4
52	57	1	5	9	14	18	22	26	31	35	39	44	48	52	56	0	4	8	12	16	20	25	28	33	37	41	45	49	53
42	2	22	29	50	10	30	27	57	17	24	44	4	12	19	26	20	27	34	41	48	55	2	56	3	10	17	24	18	24

270	**271**	**272**	**273**	**274**	**275**	**276**	**277**	**278**	**279**	**280**	**281**	**282**	**283**	**284**	**285**	**286**	**287**	**288**	**289**	**290**	**291**	**292**	**293**	**294**	**295**	**296**	**297**	**298**	**299**
10	11	11	11	0	0	1	1	2	2	2	3	3	4	4	5	5	6	6	6	7	7	8	8	9	9	10	10	10	11
18	1	14	27	10	23	7	20	3	16	29	12	26	9	21	5	18	2	15	28	11	24	7	21	4	17	0	13	26	10
1	12	22	33	44	55	5	16	27	37	48	59	9	20	30	41	51	2	12	23	33	44	55	5	15	26	36	46	57	7
34	26	57	45	34	9	57	32	20	54	42	27	39	14	49	24	46	21	55	17	52	27	2	23	58	20	41	50	13	33
9	10	10	10	11	11	0	0	1	1	1	2	2	3	3	4	4	5	5	5	6	6	7	7	8	8	8	9	9	10
17	1	14	27	10	23	6	19	2	15	28	11	24	7	20	3	17	0	13	26	9	22	5	18	1	14	27	10	23	6
57	1	5	9	13	17	21	25	29	33	37	41	45	49	53	56	0	4	8	12	15	19	23	27	31	35	38	42	45	49
19	26	20	2	35	29	36	30	37	31	38	32	13	7	1	55	36	30	24	5	55	52	46	26	20	1	42	10	51	32

[**240**:8] 41 R7, [**241**:6] 27 J2, [**242**:4] 3 R7, [**244**:7] 10 R7, [**245**:4] 21 R7, [**247**:4] 21 R7, [**248**:3] 14 R2, [**248**:8] 27 J1, 27 R2, [**250**:6] 2 R7, [**252**:7] 43 R2, 43 R3, [**252**:8] 7 R2, 7 R3, [**253**:3] 58 J1, 58 J2, [**253**:7] 41 R7, [**253**:8] 7 J2, 42 R7, [**254**:5] 5 R3, 5 R7, [**254**:7] 58 J1, [**255**:7] 6 R2, 46 R3, [**255**:8] 23 R7, [**256**:8] 28 J1, 28 J2, [**257**:3] 42 R2, [**258**:3] 53 R2, [**259**:3] 4 R2, [**259**:6] 14 R3, [**260**:2] 9 J2, [**260**:8] 49 R7, [**263**:4] 34 R2, [**265**:4] 12 R2, [**265**:7] 27 J1, [**265**:8] 50 R7, [**266**:3] 57 J2, [**266**:4] 12 J2, [**266**:7] 18 J1, 15 J2, [**266**:8] 39 J1, 39 J2, [**267**:3] 22 J2, [**269**:4] 1 R2, [**270**:7] 17 R2, [**271**:3] 11 R2, [**271**:4] 56 J2, [**272**:2] 24 R2, [**273**:2] 17 R7, [**273**:8] 21 R2, [**274**:7] 23 R7, [**275**:2] 27 J2, [**275**:7] 7 R7, [**276**:2] 6 R2, [**276**:4] 51 J1, 51 J2, [**276**:7] 31 J1, 25 J2, [**277**:2] 19 R2, [**278**:2] 2 R2, [**279**:2] 15 R2, 11 R3, [**279**:8] 41 R2, [**280**:4] 22 J1, 22 J2, [**280**:7] 33 J2, [**281**:3] 49 J1, 58 R7, [**281**:8] 31 R2, [**282**:3] 2 J2, [**282**:4] 29 J1, [**282**:6] 14 R3, [**283**:1] 9 J1, [**283**:7] 56 J2, 48 R7, [**283**:8] 55 J2, [**284**:2] 22 J1, 22 J2, [**284**:3] 40 J1, 40 J2, [**284**:4] 46 J1, 1 J2, [**284**:7] 35 R7, [**285**:7] 57 R2, [**286**:3] 52 R2, [**286**:4] 49 J2, [**286**:6] 27 J2, [**286**:8] 56 R7, [**287**:8] 38 J1, 38 R2, 3 R7, [**288**:8] 29 R2, [**289**:3] 33 J2, [**289**:4] 27 J1, 57 R2, [**289**:7] 13 J2, [**291**:4] 37 J1, 37 J2, [**293**:8] 36 R2, [**294**:4] 18 J1, 48 J2, 18 R2, [**294**:6] 0 J2, [**295**:3] 25 R2, [**295**:8] 23 J2, 58 R7, [**296**:2] 1 R2, [**296**:6] 28 J2, [**297**:4] 5 R2, [**297**:7] 41 R7, [**297**:8] 12 J1, 17 R2, 20 R3, [**298**:4] 33 R7, [**298**:8] 41 J1, 41 J2, [**299**:8] 33 J1, 50 J2.

Appendix A2. Mean motion of moon and anomaly (continued)

300	**301**	**302**	**303**	**304**	**305**	**306**	**307**	**308**	**309**	**310**	**311**	**312**	**313**	**314**	**315**	**316**	**317**	**318**	**319**	**320**	**321**	**322**	**323**	**324**	**325**	**326**	**327**	**328**	**329**
11	0	0	1	1	1	2	2	3	3	4	4	5	5	5	6	6	7	7	8	8	8	9	9	10	10	11	11	0	0
23	6	19	2	15	29	12	25	8	21	5	18	1	14	27	10	24	7	20	3	16	29	13	26	9	22	5	18	2	15
17	28	38	48	59	9	19	30	40	50	0	11	21	31	41	52	2	12	22	33	43	53	3	13	23	34	44	54	4	14
55	4	36	48	9	30	39	1	22	32	53	2	23	32	53	2	10	32	41	2	11	33	41	49	58	7	2	11	19	28
10	11	11	11	0	0	1	1	2	2	3	3	3	4	4	5	5	6	6	6	7	7	8	8	8	9	9	10	10	11
19	2	16	29	12	25	8	21	4	17	0	13	26	9	22	5	18	1	14	28	11	24	7	20	3	16	29	12	25	8
53	56	0	4	7	11	14	18	22	25	29	32	36	40	43	47	50	54	57	1	4	8	12	15	18	22	25	29	32	34
13	41	22	3	44	25	57	38	19	42	23	51	32	0	4	9	37	18	45	26	54	35	3	31	59	27	41	9	37	5

330	**331**	**332**	**333**	**334**	**335**	**336**	**337**	**338**	**339**	**340**	**341**	**342**	**343**	**344**	**345**	**346**	**347**	**348**	**349**	**350**	**351**	**352**	**353**	**354**	**355**	**356**	**357**	**358**	**359**
0	1	1	2	2	3	3	4	4	4	5	5	6	6	7	7	7	8	8	9	9	10	10	11	11	11	0	0	1	1
28	11	24	7	21	4	17	0	13	26	10	23	6	19	2	15	29	12	25	8	21	4	18	1	14	27	10	23	7	20
24	34	44	54	4	15	25	35	45	55	5	15	25	35	45	55	5	15	26	35	46	56	6	16	26	36	46	56	6	16
35	44	39	48	57	5	14	23	18	26	21	20	35	34	43	51	47	55	3	58	7	3	11	7	15	24	17	27	23	30
11	0	0	1	1	1	2	2	3	3	4	4	4	5	5	6	6	7	7	7	8	8	9	9	10	10	11	11	11	0
21	4	17	0	13	26	10	23	6	19	2	15	28	11	24	7	20	3	16	29	12	25	8	21	5	18	1	14	27	10
39	43	46	49	53	56	0	3	6	10	13	17	20	23	17	30	33	37	40	44	47	50	54	57	0	4	7	11	14	17
33	1	15	41	11	39	7	35	50	17	32	0	28	43	11	39	53	21	49	4	31	46	13	28	56	24	39	7	21	39

360	**361**	**362**	**363**	**364**	**365**
2	2	2	3	3	4
3	16	29	12	26	9
26	36	46	56	6	16
27	42	31	39	35	29
0	1	1	2	2	2
23	6	19	2	15	28
21	24	27	31	34	37
4	32	47	15	29	44

[**300**:3] 7 J2, [**300**:6] 26 J2, [**300**:7] 57 J2, [**300**:8] 53 R2, [**301**:3] 20 J1, [**301**:8] 43 J2, [**302**:4] 26 J1, 26 J2, [**302**:6] 26 R3, [**302**:7] 17 J2, [**302**:8] 55 J2, [**303**:4] 58 J1, [**303**:5] 1 R7, [**303**:8] 4 J2, [**304**:7] 37 J2, [**304**:8] 42 J2, [**305**:4] 34 R2, 3 R7, [**305**:8] 50 J2, [**306**:3] 18 R2, 9 R7, [**306**:4] 48 R2, [**306**:6] 18 J2, [**306**:7] 24 J1, 24 R2, [**306**:8] 59 J2, [**307**:2] 15 J1, [**307**:6] 20 J2, [**308**:8] 29 J2, [**309**:4] 31 J1, 31 J2, 30 R2, [**311**:6] 12 R2, [**313**:4] 29 J2, 22 R7, [**316**:3] 22 R7, [**316**:8] 30 R2, [**318**:4] 42 R2, [**319**:4] 3 R2, [**320**:2] 15 R2, [**320**:4] 10 J2, [**321**:6] 28 R2, [**322**:7] 13 J1, 13 J2, [**324**:4] 18 R2, [**324**:5] 9 R3, 9 R7, [**325**:7] 12 R3, [**326**:7] 26 R2, [**326**:8] 9 J2, [**327**:4] 18 R2, [**329**:7] 35 R2, 36 R7, [**329**:8] 35 R2, [**330**:7] 29 J2, [**332**:7] 49 J2, [**333**:8] 51 R3, [**334**:3] 41 R2, [**334**:4] 47 R2, [**337**:4] 33 J2, [**337**:5] 3 J2, [**338**:2] 12 R7, [**338**:7] 3 J1, 3 J2, [**340**:4] 27 J1, 27 J2, [**341**:4] 38 J1, 38 J2, 30 R7, [**341**:7] 7 R7, [**342**:6] 29 J1, 19 J2, [**344**:7] 47 J1, 47 J2, 27 R2, 27 R7, [**345**:7] 38 R3, [**345**:8] 29 J2, [**346**:8] 11 R2, [**347**:8] 11 J2, 11 R2, [**348**:3] 25 R2, [**348**:8] 46 J2, [**349**:6] 19 J1, [**350**:2] 51 R2, [**352**:2] 4 R2, [**352**:3] 56 R2, [**352**:4] 13 R2, [**352**:8] 53 R2, [**353**:2] 4 R2, [**353**:3] 56 R2, [**354**:4] 25 J1, 25 J2, [**355**:3] 26 R7, [**355**:4] 34 J1, [**356**:3] 47 R7, [**357**:4] 57 R3, [**358**:8] 15 R7, [**359**:4] 31 R3, [**359**:8] 49 R3, 34 R7, [**360**:5] 20 R7, [**360**:7] 11 R2, [**360**:8] 14 R7, [**361**:2] 26 J1, [**361**:4] 43 J1, 43 J2, 31 R2, [**361**:6] 19 R2, [**361**:7] 27 R2, [**361**:8] 47 R2, [**362**:1] 3 R2, [**362**:2] 12 R2, [**362**:3] 56 R2, [**362**:4] 39 R2, [**362**:5] 2 R2, [**362**:6] 2 R2, [**362**:7] 31 R2, [**362**:8] 15 R2, [**363**:2] 13 J1, [**364**:4] 25 R2, [**365**:4] 19 R2, [**365**:7] 33 R2.

Appendix A3. Mean motion of moon and anomaly per *ghaṭikā* (0 to 59)

1	2	3	4	5	6	7	8	9	10	11	12	13	14	15	16	17	18	19	20	21	22	23	24	25	26	27	28	29	30
0	0	0	0	0	0	0	0	0	0	0	0	0	0	0	0	0	0	0	0	0	0	0	0	0	0	0	0	0	0
0	0	0	0	1	1	1	1	1	2	2	2	2	3	3	3	3	3	4	4	4	4	5	5	5	5	5	6	6	6
13	26	39	52	5	19	32	45	58	11	24	38	51	4	17	30	44	57	10	23	36	49	3	16	29	42	55	8	22	35
11	21	32	42	53	3	14	25	35	46	56	7	18	28	39	49	0	10	21	32	42	53	3	14	35	35	46	56	7	17
0	0	0	0	0	0	0	0	0	0	0	0	0	0	0	0	0	0	0	0	0	0	0	0	0	0	0	0	0	0
0	0	0	0	1	1	1	1	1	2	2	2	2	3	3	3	3	3	4	4	4	4	5	5	5	5	5	6	6	6
13	26	39	52	5	18	31	44	57	10	23	36	49	2	15	29	42	55	8	21	34	47	0	13	26	39	52	5	18	31
4	8	12	15	20	23	27	32	35	39	42	47	51	54	59	2	6	10	14	18	22	26	29	34	38	41	46	49	53	57

31	32	33	34	35	36	37	38	39	40	41	42	43	44	45	46	47	48	49	50	51	52	53	54	55	56	57	58	59	60
0	0	0	0	0	0	0	0	0	0	0	0	0	0	0	0	0	0	0	0	0	0	0	0	0	0	0	0	0	0
6	7	7	7	7	7	8	8	8	8	9	9	9	9	9	10	10	10	10	10	11	11	11	11	12	12	12	12	12	13
48	1	14	28	41	54	7	20	33	47	0	13	26	39	52	6	19	32	45	58	12	25	38	51	4	17	31	44	57	10
28	39	49	0	10	21	32	42	53	3	14	24	35	46	56	7	17	28	39	49	0	10	21	31	42	53	3	14	24	35
0	0	0	0	0	0	0	0	0	0	0	0	0	0	0	0	0	0	0	0	0	0	0	0	0	0	0	0	0	0
6	6	7	7	7	7	8	8	8	8	8	9	9	9	9	10	10	10	10	10	11	11	11	11	11	12	12	12	12	13
45	48	11	24	37	50	3	16	29	42	55	8	21	34	47	1	14	27	40	53	6	19	32	45	58	11	24	37	50	3
1	5	8	13	16	20	25	28	32	36	40	43	48	52	55	0	3	7	12	15	19	22	27	30	34	39	42	46	50	54

[**1**:4] 10 R7, [**1**:8] 41 J2, 0 R2, [**2**:4] 22 R3, [**3**:4] 33 R3, [**5**:4] 56 R7, [**7**:3] 34 R7, [**7**:8] 17 J2, [**13**:4] 58 R3, [**15**:4] 49 J2, [**15**:8] 52 J2, 49 R7, [**16**:3] 17 J2, [**16**:4] 39 R7, [**18**:4] 18 R3, [**20**:8] 19 R7, [**22**:4] 54 R7, [**23**:7] 1 R3, 1 R7, [**24**:4] 39 J1, 34 J2, 39 R2, [**24**:7] 0 J2, [**24**:8] 29 J2, [**25**:3] 39 J1, 39 R2, [**25**:4] 25 R3, 25 R7, [**25**:7] 13 J2, [**25**:8] 34 J2, 36 R3, [**26**:4] 36 J1, [**26**:8] 40 J2, [**27**:6] 6 R2, [**27**:7] 55 J2, [**28**:3] 5 J2, [**28**:4] 49 J2, [**29**:4] 57 J2, [**32**:7] 58 R3, 58 R7, [**32**:8] 7 J2, [**35**:7] 36 J1, [**37**:7] 7 R3, [**39**:7] 26 J1, [**40**:7] 43 J2, 41 R7, [**41**:2] 8 J2, [**42**:4] 35 J2, [**43**:3] 36 J2, [**43**:4] 44 J2, [**44**:3] 32 J2, [**44**:4] 36 R2, [**46**:3] 46 R2, [**48**:4] 38 R2, 29 R7, [**50**:7] 15 R7, [**52**:7] 16 J2, [**52**:8] 52 J2, [**53**:3] 25 J1, [**53**:4] 10 J1, [**53**:7] 9 J1, [**53**:8] 22 J1, [**54**:7] 32 J1, [**54**:8] 27 J1, [**55**:2] 11 R7, [**55**:4] 43 J1, [**55**:6] 12 J2, [**55**:7] 5 J2, [**55**:8] 49 J2, 55 R7, [**56**:3] 1 J1, [**56**:4] 52 R7, [**57**:4] 2 R7, [**57**:7] 11 J2, [**57**:8] 39 J2, [**59**:7] 52 R7. R2 and R3 have not included intial row of zeros.

Appendix A4. Lunar equation (0 to 360 degrees)

0	**0**	**0**	**0**	**0**	**0**	**0**	**0**	**0**	**0**	**0**	**0**	**0**	**0**	**0**	**0**	**0**	**0**	**0**	**0**	**0**	**0**	**0**	**0**	**0**	**0**	**0**	**0**	**0**	**0**
0	**1**	**2**	**3**	**4**	**5**	**6**	**7**	**8**	**9**	**10**	**11**	**12**	**13**	**14**	**15**	**16**	**17**	**18**	**19**	**20**	**21**	**22**	**23**	**24**	**25**	**26**	**27**	**28**	**29**
0	0	0	0	0	0	0	0	0	0	0	0	1	1	1	1	1	1	1	1	1	1	1	1	2	2	2	2	2	2
0	5	10	15	21	26	31	36	42	47	52	57	2	7	13	18	23	28	33	38	43	48	53	57	2	7	12	27	21	26
0	16	32	48	3	18	33	47	0	13	25	35	45	54	1	7	11	13	15	16	14	9	3	56	46	33	18	1	43	21
5	5	5	5	5	5	5	5	5	5	5	5	5	5	5	5	5	5	5	4	4	4	4	4	4	4	4	4	4	4
16	16	16	15	15	15	14	13	13	12	10	10	9	7	6	4	2	2	1	58	55	54	53	50	47	43	43	42	38	35
721	721	721	721	721	721	722	722	722	722	723	723	723	723	723	724	724	724	725	725	726	726	726	727	728	728	728	730	730	731
46	46	46	59	59	59	12	12	25	38	4	4	27	43	56	22	48	48	1	40	19	32	45	24	3	55	55	55	32	31

1	**1**	**1**	**1**	**1**	**1**	**1**	**1**	**1**	**1**	**1**	**1**	**1**	**1**	**1**	**1**	**1**	**1**	**1**	**1**	**1**	**1**	**1**	**1**	**1**	**1**	**1**	**1**	**1**	**1**
0	**1**	**2**	**3**	**4**	**5**	**6**	**7**	**8**	**9**	**10**	**11**	**12**	**13**	**14**	**15**	**16**	**17**	**18**	**19**	**20**	**21**	**22**	**23**	**24**	**25**	**26**	**27**	**28**	**29**
2	2	2	2	2	2	2	3	3	3	3	3	3	3	3	3	3	3	3	3	3	3	3	4	4	4	4	4	4	4
30	35	39	44	48	53	57	1	5	10	14	18	22	25	28	33	37	40	44	47	51	54	58	1	4	7	10	13	16	18
56	28	48	24	47	8	26	42	55	4	8	7	3	56	47	31	12	51	25	55	22	43	0	13	22	26	24	18	9	55
4	4	4	4	4	4	4	4	4	4	3	3	3	3	3	3	3	3	3	3	3	3	3	3	3	2	2	2	2	2
32	30	26	23	21	18	16	13	9	4	59	56	53	51	44	41	38	34	30	27	21	17	13	9	4	58	54	51	46	41
731	732	733	733	734	734	735	736	737	738	739	739	740	741	742	742	743	743	744	745	746	747	748	749	750	751	752	753	754	755
49	49	28	54	33	59	38	30	30	40	40	58	37	48	34	13	13	59	51	30	48	40	32	24	29	47	34	18	23	28

[**0·0**:5] 6 R3, [**0·3**:3] 18 J1, [**0·5**:6] 722 R7, [**0·5**:7] 12 R7, [**0·6**:7] 11 R3, 22 R7, [**0·7**:7] 12 J1, 12 R2, [**0·8**:2] 41 R7, [**0·8**:5] 12 R2, [**0·9**:5] 11 R2, [**0·11**:2] 52 J2, [**0·11**:3] 36 J2, [**0·11**:5] 9 R2, [**0·12**:2] 12 R2, [**0·12**:5] 7 R2, 7 R3, [**0·12**:7] 17 R3, 17 R7, [**0·13**:5] 6 R2, [**0·13**:7] 46 J1, 46 J2, [**0·14**:2] 12 R7, [**0·14**:5] 8 J1, 3 R2, [**0·16**:5] 3 J2, 3 R2, [**0·16**:7] 22 R2, [**0·19**:7] 49 R2, [**0·20**:7] 32 J2, 39 R2, [**0·22**:3] 13 R2, [**0·22**:6] 727 R2, [**0·22**:7] 44 J2, 32 R2, [**0·23**:3] 16 R7, [**0·23**:6] 728 J1, 728 J2, [**0·24**:7] 23 R2, [**0·25**:7] 33 R2, [**0·26**:7] 25 R2, [**0·27**:2] 27 J1, 27 R2, [**0·27**:5] 41 R2, [**0·27**:6] 728 R3, [**0·27**:7] 8 J1, 8 J2, 0 R2, 0 R7, [**0·28**:5] 35 R2, [**0·28**:7] 51 R2, 0 R3, 52 R7, [**0·29**:2] 16 R7, [**0·29**:3] 1 R7, [**0·29**:6] 733 R3, [**0·29**:7] 52 R3, [**1·0**:5] 30 J1, 30 J2, 22 R7, [**1·0**:6] 733 J1, 733 J2, [**1·0**:7] 51 R2, 31 R3, 57 R7, [**1·1**:3] 38 J2, [**1·1**:5] 26 R7, [**1·1**:6] 731 R3, [**1·1**:7] 57 R3, [**1·2**:3] 58 J2, 54 R7, [**1·2**:6] 732 R3, [**1·2**:7] 54 J2, 49 R3, [**1·3**:3] 32 J2, [**1·3**:7] 59 J2, 20 R3, [**1·4**:6] 733 R3, [**1·4**:7] 59 J2, 54 R3, 13 R7, [**1·5**:7] 48 R2, 33 R3, [**1·6**:3] 16 J1, [**1·6**:6] 734 R3, [**1·6**:7] 35 R2, 59 R3, [**1·7**:6] 735 R3, [**1·7**:7] 38 R3, [**1·8**:3] 15 R3, [**1·8**:6] 736 R3, [**1·8**:7] 25 J1, 25 J2, 35 R2, 35 R7, [**1·9**:6] 737 R3, [**1·9**:7] 35 R3, 4 R7, [**1·10**:6] 738 R3, [**1·10**:7] 39 J1, 40 R3, [**1·11**:5] 53 R2, 53 R3, [**1·11**:6] 740 R2, [**1·11**:7] 18 J2, 37 R2, 19 R3, [**1·12**:3] 53 J2, 30 R2, [**1·12**:6] 739 R3, [**1·12**:7] 58 R3, [**1·13**:5] 50 R2, [**1·13**:6] 740 R3, [**1·13**:7] 46 R2, 30 R3, 34 R7, [**1·14**:2] 28 J1, 28 R2, [**1·14**:6] 741 R3, [**1·14**:7] 48 R3, [**1·15**:5] 38 J2, [**1·15**:6] 743 R7, [**1·15**:7] 34 R3, [**1·16**:2] 31 J1, [**1·16**:3] 2 J1, 2 J2, [**1·16**:5] 34 J2, [**1·17**:2] 41 R2, [**1·17**:5] 30 J2, [**1·17**:7] 58 R2, [**1·18**:5] 27 J2, 31 R7, [**1·18**:7] 56 J1, 56 J2, [**1·19**:5] 21 J2, [**1·20**:5] 17 J2, 17 R2, [**1·20**:6] 747 R2, 747 R3, [**1·21**:5] 13 J2, [**1·21**:6] 748 R7, [**1·21**:7] 32 R7, [**1·22**:5] 9 J2, [**1·22**:7] 38 R7, [**1·23**:5] 4 J2, [**1·24**:5] 58 J2, [**1·24**:7] 38 R2, [**1·25**:5] 54 J2, [**1·26**:5] 52 J2, [**1·26**:7] 29 R3, 39 R7, [**1·27**:2] 3 R2, [**1·27**:5] 51 J1, 46 J2, 51 R7, [**1·27**:7] 23 J2, [**1·28**:5] 41 J2, [**1·28**:7] 43 R2, [**1·29**:5] 37 J2, [**1·29**:7] 29 R7.

2	**2**	**2**	**2**	**2**	**2**	**2**	**2**	**2**	**2**	**2**	**2**	**2**	**2**	**2**	**2**	**2**	**2**	**2**	**2**	**2**	**2**	**2**	**2**	**2**	**2**	**2**	**2**	**2**	**2**
0	**1**	**2**	**3**	**4**	**5**	**6**	**7**	**8**	**9**	**10**	**11**	**12**	**13**	**14**	**15**	**16**	**17**	**18**	**19**	**20**	**21**	**22**	**23**	**24**	**25**	**26**	**27**	**28**	**29**
4	4	4	4	4	4	4	4	4	4	4	4	4	4	4	4	5	4	4	4	4	4	4	4	5	5	5	5	5	5
21	24	26	29	31	33	36	38	40	42	43	45	47	48	50	51	53	54	55	56	57	58	59	59	0	1	1	1	1	2
36	13	44	10	30	47	0	6	4	2	54	41	21	55	25	50	10	24	31	35	35	27	12	53	29	0	26	45	57	6
2	2	2	2	2	2	2	1	1	1	1	1	1	1	1	1	1	1	1	1	0	0	0	0	0	0	0	0	0	0
37	31	26	20	17	13	6	58	58	52	47	40	34	30	25	20	14	7	4	0	52	45	41	36	31	26	19	12	9	4
756	757	758	760	760	761	763	764	764	766	767	768	770	770	772	773	774	775	776	777	779	780	781	782	783	784	786	787	788	789
20	38	48	6	45	37	8	12	52	10	15	48	6	58	3	8	26	57	36	28	12	43	35	40	45	50	21	53	31	43

3	**3**	**3**	**3**	**3**	**3**	**3**	**3**	**3**	**3**	**3**	**3**	**3**	**3**	**3**	**3**	**3**	**3**	**3**	**3**	**3**	**3**	**3**	**3**	**3**	**3**	**3**	**3**	**3**	**3**
0	**1**	**2**	**3**	**4**	**5**	**6**	**7**	**8**	**9**	**10**	**11**	**12**	**13**	**14**	**15**	**16**	**17**	**18**	**19**	**20**	**21**	**22**	**23**	**24**	**25**	**26**	**27**	**28**	**29**
5	5	5	5	5	5	5	4	4	4	4	4	4	4	4	4	4	4	4	4	4	4	4	4	4	4	4	4	4	4
2	2	1	1	1	1	0	59	59	58	57	56	55	54	53	51	50	48	47	45	43	42	40	38	36	33	31	29	26	24
10	6	57	45	26	0	29	53	12	27	35	35	31	24	10	50	25	55	21	41	54	2	7	6	0	47	30	10	44	13
0	0	0	0	0	0	0	0	0	0	1	1	1	1	1	1	1	1	1	1	1	1	1	2	2	2	2	2	2	2
4	9	12	19	26	31	36	41	45	52	0	4	7	14	20	25	30	34	40	47	52	58	58	6	13	17	20	26	31	37
789	793	793	794	796	797	798	799	800	801	803	804	805	806	808	809	810	811	812	813	815	816	816	818	819	820	821	822	823	824
43	39	18	49	20	25	30	35	27	48	42	34	13	44	2	7	12	4	22	55	0	18	18	2	33	25	4	12	32	50

[**2·1**:7] 48 R3, 37 R7, [**2·2**:2] 29 J1, 29 J2, 24 R2, [**2·3**:7] 36 J1, 36 J2, [**2·6**:7] 38 J1, [**2·7**:3] 4 R7, [**2·7**:4] 2 R2, [**2·7**:5] 1 R2, [**2·7**:7] 14 R2, [**2·8**:3] 7 R2, [**2·8**:5] 55 R2, [**2·8**:7] 33 R2, [**2·11**:3] 21 J1, 21 J2, [**2·11**:7] 40 J2, [**2·13**:6] 771 R2, [**2·14**:7] 43 R7, [**2·15**:5] 25 J1, [**2·15**:7] 16 R7, [**2·16**:1] 4 R3, 4 R7, [**2·16**:5] 24 J1, [**2·16**:7] 36 R7, [**2·22**:7] 43 R7, [**2·23**:7] 4 R2, [**2·24**:3] 26 R7, [**2·24**:5] 36 R3, [**2·25**:5] 31 R3, [**2·25**:7] 45 J2, [**2·26**:5] 26 R3, [**2·26**:7] 31 R7, [**2·27**:3] 47 J1, [**2·27**:5] 19 R3, [**2·27**:7] 52 J1, 52 J2, 41 R7, [**2·28**:5] 12 R3, [**2·29**:5] 46 J1, 45 J2, [**2·29**:6] x R3, [**3·1**:6] 792 R7, [**3·2**:3] 47 J1, [**3·5**:7] 50 R7, [**3·6**:2] 1 J1, 1 J2, [**3·6**:3] 19 R7, [**3·6**:7] 3 J1, 3 J2, [**3·8**:5] 40 J2, [**3·8**:7] 37 R7, [**3·9**:6] 81 J1, 81 J2, [**3·9**:7] 58 J1, 58 J2, 50 R2, [**3·10**:6] 83 J1, 83 J2, [**3·10**:7] 43 R2, [**3·11**:6] 0 J1, 0 J2, [**3·12**:3] 35 R2, [**3·13**:2] 58 J2, [**3·13**:3] 55 J2, [**3·15**:5] 30 J2, [**3·15**:6] 810 J2, [**3·15**:7] 2 J1, 12 J2, [**3·16**:5] 34 J2, 31 R2, [**3·16**:6] 811 J2, [**3·16**:7] 4 J2, 7 R2, [**3·17**:5] 40 J2, [**3·17**:6] 812 J2, [**3·17**:7] 22 J2, [**3·18**:5] 47 J2, [**3·18**:6] 813 J2, [**3·18**:7] 55 J2, 25 R7, [**3·19**:5] 52 J2, [**3·19**:6] 815 J2, [**3·19**:7] 0 J2, 45 R2, [**3·20**:5] 58 J2, [**3·20**:6] 816 J2, [**3·20**:7] 18 J2, 10 R7, [**3·21**:5] 55 R2, [**3·21**:6] 815 R2, [**3·21**:7] 37 R2, 48 R3, [**3·22**:3] 4 R3, 4 R7, [**3·22**:4] 2 R2, [**3·22**:5] 6 J2, 1 R2, [**3·22**:6] 818 J2, [**3·22**:7] 18 J1, 2 J2, 56 R2, 42 R7, [**3·23**:2] 28 R2, [**3·23**:5] 5 J1, 13 J2, [**3·23**:6] 819 J2, [**3·23**:7] 33 J2, [**3·24**:5] 17 J2, [**3·24**:6] 820 J2, [**3·24**:7] 25 J2, [**3·25**:2] 37 J2, [**3·25**:5] 20 J2, 12 R7, [**3·25**:6] 821 J2, [**3·25**:7] 4 J2, [**3·26**:5] 26 J2, [**3·26**:6] 822 J2, [**3·26**:7] 12 J2, [**3·27**:4] 26 R2, [**3·27**:5] 31 J2, 21 R2, [**3·27**:7] 22 R2, 23 R3, 32 R7, [**3·28**:5] 32 J2, [**3·29**:7] 0 R3.

Appendix A4. Lunar equation (continued)

4	**4**	**4**	**4**	**4**	**4**	**4**	**4**	**4**	**4**	**4**	**4**	**4**	**4**	**4**	**4**	**4**	**4**	**4**	**4**	**4**	**4**	**4**	**4**	**4**	**4**	**4**	**4**	**4**	**4**
0	**1**	**2**	**3**	**4**	**5**	**6**	**7**	**8**	**9**	**10**	**11**	**12**	**13**	**14**	**15**	**16**	**17**	**18**	**19**	**20**	**21**	**22**	**23**	**24**	**25**	**26**	**27**	**28**	**29**
4	4	4	4	4	4	4	4	3	3	3	3	3	3	3	3	3	3	3	3	3	3	3	3	2	2	2	2	2	2
21	18	16	13	10	7	4	1	58	54	51	47	44	40	37	33	29	25	22	18	14	10	5	1	57	53	48	44	39	34
36	55	9	18	24	26	22	13	0	43	22	55	25	51	12	31	47	56	3	7	8	4	55	42	26	8	27	24	58	28
2	2	2	2	2	3	3	3	3	3	3	3	3	3	3	3	3	3	3	3	4	4	4	4	4	4	4	4	4	4
41	46	51	54	58	4	9	13	17	21	27	30	34	38	42	44	51	53	56	59	4	9	13	16	18	21	23	26	30	32
825	825	826	828	829	830	831	832	833	834	835	836	837	837	838	839	840	841	841	842	843	844	845	846	846	843	847	848	849	849
42	47	52	31	23	41	46	32	30	18	40	19	11	57	36	21	33	12	51	30	35	40	32	11	37	16	42	21	36	39

5	**5**	**5**	**5**	**5**	**5**	**5**	**5**	**5**	**5**	**5**	**5**	**5**	**5**	**5**	**5**	**5**	**5**	**5**	**5**	**5**	**5**	**5**	**5**	**5**	**5**	**5**	**5**	**5**	**5**
0	**1**	**2**	**3**	**4**	**5**	**6**	**7**	**8**	**9**	**10**	**11**	**12**	**13**	**14**	**15**	**16**	**17**	**18**	**19**	**20**	**21**	**22**	**23**	**24**	**25**	**26**	**27**	**28**	**29**
2	2	2	2	2	2	2	1	1	1	1	1	1	1	1	1	1	1	1	0	0	0	0	0	0	0	0	0	0	0
30	36	21	17	12	7	2	57	53	48	43	38	33	28	23	18	13	7	2	57	52	47	42	36	31	26	21	15	10	5
56	21	43	1	18	33	46	56	3	9	14	16	15	13	11	7	1	54	45	35	25	13	0	47	33	18	3	48	32	16
4	4	4	4	4	4	4	4	4	4	4	5	5	5	5	5	5	5	5	5	5	5	5	5	5	5	5	5	5	5
35	38	42	43	43	47	50	53	54	55	58	1	2	3	4	6	7	9	10	10	12	13	13	14	15	15	15	16	16	16
850	851	852	852	852	853	853	854	854	854	855	856	856	856	856	857	857	857	858	858	858	858	858	858	859	859	859	859	859	859
18	10	12	15	15	7	46	25	38	51	30	9	22	35	48	14	27	53	6	19	32	45	45	58	11	11	11	24	24	24

[**4·1**:5] 48 J2, [**4·1**:6] 825 J1, 825 R2, [**4·1**:7] 42 J1, 23 J2, [**4·2**:6] 826 J1, 826 R2, [**4·2**:7] 27 J1, [**4·3**:6] 827 J1, [**4·3**:7] 52 J1, [**4·4**:6] 828 J1, [**4·4**:7] 31 J1, 32 J2, [**4·5**:3] 36 R2, 36 R3, [**4·5**:7] 32 J1, [**4·6**:7] 41 J1, 41 J2, [**4·7**:6] 831 J2, 833 R2, [**4·7**:7] 46 J1, 46 J2, [**4·8**:7] 32 J1, [**4·9**:7] 30 J1, [**4·10**:7] 18 J1, [**4·11**:2] 48 J1, 48 J2, [**4·11**:7] 40 J1, [**4·12**:3] 55 J2, [**4·12**:7] 19 J1, [**4·13**:7] 11 J1, [**4·14**:2] 27 R2, [**4·14**:5] 42 J1, 42 R2, [**4·14**:7] 57 J1, 41 J2, [**4·15**:7] 36 J1, 20 R7, [**4·16**:5] 50 J1, [**4·16**:7] 21 J1, 30 J2, 32 R7, [**4·17**:7] 33 J1, 42 R7, [**4·18**:7] 12 J1, [**4·19**:6] 841 J2, [**4·20**:6] 842 J2, [**4·21**:6] 843 J2, [**4·22**:5] 12 R2, [**4·22**:6] 844 J2, [**4·24**:2] 17 R2, [**4·25**:6] 847 R2, 4 R3, 846 R7, [**4·25**:7] 46 R7, [**4·26**:3] 27 J1, 27 R2, 48 R7, [**4·26**:7] 12 J1, 12 J2, 22 R7, [**4·28**:2] 29 J2, [**4·28**:5] 28 J1, 28 J2, [**4·28**:7] 13 R3, 30 R7, [**5·1**:2] 26 R3, 26 R7, [**5·3**:3] 10 J1, 10 J2, [**5·4**:7] 45 R7, [**5·5**:3] 37 R2, [**5·5**:5] 48 J1, 48 J2, [**5·7**:2] 54 J1, 54 J2, [**5·9**:7] 22 J1, 38 J2, [**5·10**:5] 18 R7, [**5·11**:5] 10 R7, [**5·12**:5] 1 R7, [**5·12**:7] 2 J1, 2 J2, 12 R7, [**5·13**:3] 23 R2, [**5·13**:5] 2 R3, 2 R7, [**5·13**:7] 22 R3, 22 R7, [**5·15**:2] 13 R7, [**5·15**:3] 16 R7, [**5·16**:2] 7 R7, [**5·16**:3] 14 R7, [**5·17**:3] 4 J1, 4 J2, 14 R7, [**5·19**:5] 11 J1, 11 J2, [**5·19**:7] 6 R3, 6 R7, [**5·20**:7] 45 J1, [**5·21**:5] 14 J1, 14 J2, [**5·22**:7] 58 J1, [**5·23**:2] 26 R2, [**5·23**:7] 11 J1, [**5·24**:7] 10 R7, [**5·26**:5] 16 R7, [**5·28**:2] 20 J1, 20 J2.

6	**6**	**6**	**6**	**6**	**6**	**6**	**6**	**6**	**6**	**6**	**6**	**6**	**6**	**6**	**6**	**6**	**6**	**6**	**6**	**6**	**6**	**6**	**6**	**6**	**6**	**6**	**6**	**6**	**6**
0	**1**	**2**	**3**	**4**	**5**	**6**	**7**	**8**	**9**	**10**	**11**	**12**	**13**	**14**	**15**	**16**	**17**	**18**	**19**	**20**	**21**	**22**	**23**	**24**	**25**	**26**	**27**	**28**	**29**
0	0	0	0	0	0	0	0	0	0	0	0	1	1	1	1	1	1	1	1	1	1	1	1	2	2	2	2	2	2
0	5	10	15	21	26	31	36	42	47	52	57	2	7	13	18	23	28	33	38	43	48	53	57	2	7	12	17	21	26
0	16	32	48	3	18	33	47	0	13	25	25	45	54	1	7	12	13	15	16	14	9	3	56	46	33	18	1	43	21
5	5	5	5	5	5	5	5	5	5	5	5	5	5	5	5	5	5	5	4	4	4	4	4	4	4	4	4	4	4
16	16	16	15	15	15	14	13	13	12	10	10	9	7	6	4	2	2	1	58	55	54	53	50	47	43	43	42	38	35
859	859	859	859	859	859	858	858	858	858	858	858	857	857	857	856	856	856	856	855	854	854	854	853	853	852	852	852	851	850
24	24	24	11	11	11	58	45	45	32	6	6	53	27	14	48	35	22	9	30	51	38	25	46	7	15	15	12	10	18

7	7	7	7	7	7	7	7	7	7	7	7	7	7	7	7	7	7	7	7	7	7	7	7	7	7	7	7	7	7
0	**1**	**2**	**3**	**4**	**5**	**6**	**7**	**8**	**9**	**10**	**11**	**12**	**13**	**14**	**15**	**16**	**17**	**18**	**19**	**20**	**21**	**22**	**23**	**24**	**25**	**26**	**27**	**28**	**29**
2	2	2	2	2	2	2	3	3	3	3	3	3	3	3	3	3	3	3	3	3	3	3	4	4	4	4	4	4	4
30	35	39	44	48	53	57	1	5	10	14	18	22	25	29	33	37	40	44	47	51	54	58	1	4	7	10	13	16	18
56	28	58	24	47	8	26	42	55	4	8	7	3	56	47	31	12	51	25	55	22	43	0	13	22	26	24	18	9	55
4	4	4	4	4	4	4	4	4	4	3	3	3	3	3	3	3	3	3	3	3	3	3	3	3	2	2	2	2	2
32	30	26	23	21	18	16	13	9	4	59	56	53	51	44	41	38	34	30	27	21	17	13	9	4	58	54	51	46	41
859	849	848	847	847	846	846	845	844	844	842	841	841	840	839	838	837	837	836	835	834	833	832	831	830	829	828	827	826	825
39	13	21	42	16	37	11	32	40	35	30	51	12	33	21	36	57	11	19	40	18	30	32	1	41	23	31	52	47	42

[**6·0**:3] 42 J1, [**6·6**:5] 13 J1, [**6·6**:6] 859 R7, [**6·6**:7] 10 R7, [**6·8**:2] 41 R7, [**6·9**:2] 43 J2, [**6·9**:5] 13 R2, [**6·10**:3] 24 J1, 24 J2, [**6·10**:5] 12 R2, [**6·10**:7] 19 R7, [**6·11**:3] 35 J1, 35 J2, [**6·11**:5] 11 R2, [**6·12**:5] 10 R2, [**6·13**:5] 9 R2, [**6·14**:5] 7 R2, [**6·15**:5] 6 R2, [**6·16**:3] 12 J1, 12 R2, [**6·16**:5] 3 J1, 3 J2, 4 R2, [**6·16**:7] 22 R3, 32 R7, [**6·17**:5] 3 R2, [**6·18**:5] 2 R2, [**6·19**:5] 1 R2, [**6·20**:5] 58 R2, [**6·21**:5] 55 R2, [**6·21**:7] 28 R7, [**6·22**:5] 54 R2, [**6·23**:5] 53 R2, [**6·24**:5] 50 R2, [**6·25**:5] 45 J1, 45 J2, 47 R2, [**6·25**:7] 37 R7, [**6·26**:2] 18 R3, [**6·27**:5] 43 R2, [**6·28**:5] 43 R2, [**6·29**:5] 32 J2, 38 R2, [**7·0**:5] 31 J2, [**7·1**:5] 37 J1, [**7·2**:5] 28 J1, [**7·3**:7] 47 J2, [**7·4**:5] 31 J2, [**7·4**:6] 846 R3, [**7·5**:7] 38 J1, 38 J2, 36 R7, [**7·6**:7] 12 R7, [**7·7**:7] 31 R2, [**7·8**:6] 4 R7, [**7·9**:6] 843 R2, 843 R3, [**7·9**:7] 34 J2, [**7·13**:7] 30 J2, [**7·14**:4] 5 R7, [**7·15**:7] 21 R2, [**7·17**:3] 41 R2, [**7·19**:3] 51 R3, [**7·20**:7] 28 J2, [**7·22**:2] 38 R3, [**7·23**:7] 46 J2, 46 R7, [**7·25**:4] 4 R7, [**7·27**:6] 857 R2, [**7·28**:3] 18 R2, [**7·28**:6] 824 R2, 824 R7, [**7·28**:7] 42 R2, [**7·29**:2] 19 R2, [**7·29**:3] 52 R2, [**7·29**:5] 46 R2, 45 R7, [**7·29**:6] 826 R2, [**7·29**:7] 45 J2.

Appendix A4. Lunar equation (continued)

8	8	8	8	8	8	8	8	8	8	8	8	8	8	8	8	8	8	8	8	8	8	8	8	8	8	8	8	8	8
0	**1**	**2**	**3**	**4**	**5**	**6**	**7**	**8**	**9**	**10**	**11**	**12**	**13**	**14**	**15**	**16**	**17**	**18**	**19**	**20**	**21**	**22**	**23**	**24**	**25**	**26**	**27**	**28**	**29**
4	4	4	4	4	4	4	4	4	4	4	4	4	4	4	4	4	4	4	4	4	4	4	4	4	5	5	5	5	5
21	24	26	29	31	33	36	38	40	42	43	45	47	48	50	51	53	54	55	56	57	58	59	59	0	1	1	1	1	2
36	13	44	10	30	47	0	6	7	2	54	41	21	55	25	50	10	24	31	35	35	27	12	53	29	0	26	45	57	6
2	2	2	2	2	2	2	2	1	1	1	1	1	1	1	1	1	1	1	1	0	0	0	0	0	0	0	0	0	0
37	31	26	20	17	13	6	1	55	52	47	40	34	30	25	20	14	7	4	0	52	45	41	36	31	26	19	12	9	4
824	823	822	821	820	819	818	816	815	815	813	812	811	810	809	808	806	805	804	803	801	880	799	798	797	796	794	793	792	791
50	20	22	20	25	33	2	56	57	0	55	22	4	12	7	2	44	13	34	42	58	27	35	30	25	20	49	18	39	27

9	9	9	9	9	9	9	9	9	9	9	9	9	9	9	9	9	9	9	9	9	9	9	9	9	9	9	9	9	9
0	**1**	**2**	**3**	**4**	**5**	**6**	**7**	**8**	**9**	**10**	**11**	**12**	**13**	**14**	**15**	**16**	**17**	**18**	**19**	**20**	**21**	**22**	**23**	**24**	**25**	**26**	**27**	**28**	**29**
5	5	5	5	5	5	5	4	4	4	4	4	4	4	4	4	4	4	4	4	4	4	4	4	4	4	4	4	4	4
2	2	1	1	1	1	0	59	59	58	57	56	55	54	53	51	50	48	47	45	43	42	40	38	36	33	31	29	26	24
10	6	57	45	26	0	29	53	12	27	35	35	31	24	10	50	25	55	11	41	54	2	7	6	0	47	30	10	44	13
0	0	0	0	0	0	0	0	0	0	1	1	1	1	1	1	1	1	1	1	1	1	2	2	2	2	2	2	2	2
4	9	12	19	26	31	36	41	45	52	0	4	7	14	20	25	30	34	40	47	52	55	1	6	13	17	20	26	31	37
798	788	787	786	784	783	782	781	780	779	776	776	775	774	773	772	770	770	768	767	766	765	764	763	761	760	760	758	757	756
43	31	52	21	50	45	40	35	43	12	36	36	57	26	8	3	58	6	48	15	10	33	14	8	37	45	9	48	38	20

[**8·0**:1] 22 R2, [**8·0**:2] 31 R3, [**8·1**:2] 34 J2, [**8·1**:7] 20 J1, 22 J2, 20 R2, [**8·2**:7] 32 R7, [**8·3**:7] 4 R2, 4 R3, [**8·5**:3] 7 R7, [**8·6**:3] 7 R2, [**8·7**:4] 1 R3, [**8·7**:5] 58 R3, [**8·7**:7] 48 R3, 48 R7, [**8·8**:3] 40 R3, [**8·8**:5] 58 R3, [**8·8**:6] 816 R3, [**8·8**:7] 37 R2, 18 R3, [**8·11**:2] 41 R7, [**8·11**:3] 51 J1, 51 J2, [**8·14**:2] 48 R7, [**8·14**:3] 55 R7, [**8·15**:2] 50 R7, [**8·15**:3] 25 R7, [**8·18**:6] 803 R2, [**8·19**:7] 43 R7, [**8·20**:3] 37 J2, [**8·21**:5] 25 R2, [**8·21**:6] 800 R3, 800 R7, [**8·22**:6] 779 R2, [**8·23**:3] 3 R7, [**8·23**:7] 2 R7, [**8·24**:1] 5 R2, 5 R3, [**8·24**:2] 59 R7, [**8·24**:3] 39 R7, [**8·27**:5] 9 J2, [**8·27**:6] 794 R3, [**8·27**:7] 3 R2, [**8·28**:7] 30 J1, [**8·29**:6] 789 R3, 790 R7, [**8·29**:7] 47 R3, 20 R7, [**9·0**:3] 70 R2, 70 R3, [**9·0**:6] 789 R3, 789 R7, [**9·1**:7] 12 R7, [**9·2**:7] 12 J2, [**9·3**:3] 49 J1, 26 J2, [**9·3**:5] 18 R7, [**9·4**:7] 60 J1, [**9·5**:7] 47 J2, [**9·6**:5] 26 R2, [**9·7**:6] 780 R2, [**9·8**:5] 47 R7, [**9·9**:2] 46 R2, [**9·9**:3] 35 R2, [**9·9**:4] 1 R2, [**9·9**:5] 4 R2, [**9·9**:6] 778 R2, 777 R7, [**9·9**:7] 36 R2, [**9·10**:2] 46 R7, [**9·10**:5] 7 R7, [**9·10**:6] 777 R2, 777 R3, [**9·10**:7] 28 R2, 28 R3, [**9·11**:3] 36 R7, [**9·11**:5] 7 R7, [**9·11**:6] 777 J1, 777 J2, 774 R2, [**9·12**:5] 0 R2, [**9·12**:7] 47 J1, 27 J2, [**9·13**:6] 776 R2, [**9·16**:7] 50 J1, 6 J2, [**9·17**:7] 48 J2, 50 R7, [**9·18**:3] 21 R3, 21 R7, [**9·19**:6] 768 R7, [**9·19**:7] 10 R7, [**9·20**:5] 2 R3, 51 R7, [**9·20**:7] 3 R7, [**9·21**:2] 43 R7, [**9·21**:5] 56 R3, [**9·21**:7] 14 R7, [**9·22**:3] 4 R3, [**9·22**:4] 1 R3, [**9·22**:5] 56 R3, [**9·22**:7] 12 R3, 10 R7, [**9·23**:1] 48 R7, [**9·23**:2] 36 R7, [**9·23**:4] 8 J1, 8 J2, [**9·24**:7] 27 R7, [**9·25**:5] 13 R7, [**9·25**:7] 25 R7, [**9·26**:5] 24 R7, [**9·26**:6] 759 R7, [**9·26**:7] 6 R2, 6 R3, [**9·27**:7] 49 R7, [**9·29**:5] 7 R3, [**9·29**:6] 745 R7, [**9·29**:7] 40 R7.

10	**10**	**10**	**10**	**10**	**10**	**10**	**10**	**10**	**10**	**10**	**10**	**10**	**10**	**10**	**10**	**10**	**10**	**10**	**10**	**10**	**10**	**10**	**10**	**10**	**10**	**10**	**10**	**10**	**10**
0	**1**	**2**	**3**	**4**	**5**	**6**	**7**	**8**	**9**	**10**	**11**	**12**	**13**	**14**	**15**	**16**	**17**	**18**	**19**	**20**	**21**	**22**	**23**	**24**	**25**	**26**	**27**	**28**	**29**
4	4	4	4	4	4	4	4	3	3	3	3	3	3	3	3	3	3	3	3	3	3	3	3	2	2	2	2	2	2
21	18	16	13	10	7	4	1	58	54	51	47	44	40	37	33	29	25	22	18	14	10	5	1	57	53	48	44	39	35
36	55	9	18	24	26	22	13	0	43	22	55	25	51	12	31	47	56	3	7	8	4	55	42	26	8	47	24	58	28
2	2	2	2	2	3	3	3	3	3	3	3	3	3	3	3	3	3	3	3	4	4	4	4	4	4	4	4	4	4
41	46	51	54	58	4	9	13	17	21	27	30	34	38	41	44	51	53	56	59	4	9	13	16	18	21	23	26	30	32
755	754	753	752	751	750	749	748	747	746	745	744	743	743	742	741	740	739	739	738	737	736	735	734	734	733	733	732	731	731
28	23	18	39	40	29	24	32	40	48	10	51	59	13	34	48	37	58	19	40	35	30	38	59	33	54	28	49	57	31

11	**11**	**11**	**11**	**11**	**11**	**11**	**11**	**11**	**11**	**11**	**11**	**11**	**11**	**11**	**11**	**11**	**11**	**11**	**11**	**11**	**11**	**11**	**11**	**11**	**11**	**11**	**11**	**11**	**11**
0	**1**	**2**	**3**	**4**	**5**	**6**	**7**	**8**	**9**	**10**	**11**	**12**	**13**	**14**	**15**	**16**	**17**	**18**	**19**	**20**	**21**	**22**	**23**	**24**	**25**	**26**	**27**	**28**	**29**
2	2	2	2	2	2	2	1	1	1	1	1	1	1	1	1	1	1	1	0	0	0	0	0	0	0	0	0	0	0
30	26	21	17	12	7	2	57	53	48	43	38	33	28	23	18	13	7	2	57	52	47	42	36	31	26	21	15	10	5
56	21	43	1	18	33	46	56	3	9	14	16	15	13	11	7	1	54	45	35	25	13	0	47	33	18	3	48	32	16
4	4	4	4	4	4	4	4	4	4	4	5	5	5	5	5	5	5	5	5	5	5	5	5	5	5	5	5	5	5
35	38	42	43	43	47	50	53	54	55	58	1	2	2	4	6	7	9	10	12	12	13	14	14	15	15	15	16	16	16
730	730	728	728	728	728	727	726	726	726	725	725	724	724	724	723	723	723	723	722	722	722	722	722	721	720	721	721	721	721
52	0	55	55	55	3	24	45	32	19	4	1	44	35	22	56	43	17	4	51	38	25	25	12	59	59	59	46	46	46

[**10·1**:3] 15 R7, [**10·3**:3] 19 J1, [**10·4**:5] 28 R2, [**10·4**:7] 47 R3, 47 R7, [**10·5**:4] 2 R7, [**10·6**:3] 21 R3, [**10·6**:5] 8 R7, [**10·6**:7] 44 J2, [**10·8**:3] 7 R2, [**10·9**:3] 53 R3, [**10·9**:5] 51 R2, [**10·10**:5] 30 J2, [**10·10**:7] 30 R2, 30 R3, [**10·14**:7] 43 J2, [**10·15**:2] 23 R7, [**10·16**:6] 4 R7, [**10·16**:7] 27 R3, [**10·17**:5] 52 R7, [**10·19**:6] 739 R7, [**10·22**:7] 48 R7, [**10·23**:2] 0 R7, [**10·23**:5] 18 J2, [**10·23**:7] 33 J2, [**10·24**:6] 738 R2, 733 R7, [**10·24**:7] 24 J2, 54 R7, [**10·25**:7] 24 J1, 24 J2, [**10·26**:5] 13 J2, 13 R2, [**10·27**:5] 28 R7, [**10·27**:7] 59 R3, [**10·28**:2] 38 J1, 38 J2, [**11·2**:2] 2 R3, [**11·4**:5] 47 J1, 45 R7, [**11·4**:7] 50 J1, [**11·5**:5] 50 J1, [**11·5**:7] 2 J1, 2 J2, 14 R7, [**11·6**:5] 53 J1, [**11·7**:5] 54 J1, [**11·8**:3] 7 J1, [**11·8**:5] 55 J1, [**11·8**:7] 39 R2, [**11·9**:5] 58 J1, [**11·10**:5] 1 J1, [**11·10**:7] 40 R3, [**11·11**:5] 2 J1, [**11·12**:5] 3 J1, [**11·12**:7] 48 R3, [**11·13**:5] 4 J1, 2 R3, [**11·13**:7] 44 R2, 48 R3, [**11·14**:5] 6 J1, [**11·15**:5] 7 J1, [**11·16**:5] 9 J1, [**11·17**:3] 14 R7, [**11·17**:5] 10 J1, [**11·17**:7] 43 R2, [**11·18**:5] 11 J1, [**11·19**:5] 12 J1, 11 J2, 12 R2, 11 R7, [**11·19**:6] 723 R3, [**11·19**:7] 38 R2, 4 R3, [**11·20**:5] 13 J1, [**11·20**:6] 723 R3, [**11·20**:7] 51 R2, 28 R3, [**11·21**:6] 723 R3, [**11·22**:5] 14 J1, 14 R7, [**11·22**:6] 723 R3, [**11·23**:2] 38 J1, 38 J2, 34 R7, [**11·24**:3] 32 R2, [**11·24**:5] 14 J1, 14 J2, [**11·25**:3] 3 R7, [**11·25**:6] 721 R3, 721 R7, [**11·26**:7] 46 J2, [**11·28**:2] 15 R2, [**11·28**:3] 35 R7.

Recomputing Sanskrit Astronomical Tables: The *Amṛtalaharī* of Nityānanda (*c.* 1649/50 CE)

Anuj Misra*

1. Introduction

Astronomical tables (*koṣṭhakas* or *sāraṇīs*) begin to appear in Sanskrit astral sciences from around the twelfth century CE. These tables described different calendrical quantities (like the division of synodic lunar months or the lunar mansions), a variety of mathematical and trigonometric relations, and the planetary positions and motions. By the early modern period of Indian history, the corpus of Sanskrit astronomical tables had grown to reflect incredible ingenuity in the way complex calendrical and planetary elements were calculated and represented. In Mughal India,[1] as medieval Islamicate astronomy began interacting with Sanskrit mathematical astronomy, the computational practices of Sanskrit astronomers started to reflect this exchange of ideas. It is in this historical context that we find the *Amṛtalaharī* of Nityānanda.

Nityānanda was a seventeenth-century Sanskrit astronomer at the court of the Mughal emperor Shāh Jāhān (r. 1592 to 1666 CE). He was commissioned by Āsaf Khān, the emperor's chief minister (*vazīr*), to translate into Sanskrit the *Zīj-i Shāh-Jahānī*, an enormous compilation of Persian astronomical tables prepared by Mullā Farīd al-Dīn Masʿūd al-Dihlavī in October 1629 CE. Nityānanda dedicated himself to the task and in the early 1630s, he completed his translation the *Siddhāntasindhu* 'Ocean of Siddhāntas'.[2] Around a decade later, in 1639 CE, Nityānanda published his canonical treatise (*siddhānta*)

* Preliminary numerical computations were done with the assistance of Zachary Hynd (Seequent, New Zealand).

[1] Mughal India refers to the cosmopolitan society under the rule of the Mughal emperors (1526 to 1857 CE) where artistic, scientific, and linguistic exchanges between Islamicate (Arabic and Persian) and Sanskrit scholars flourished for over three hundred years, see Truschke, *Culture of Encounters*.

[2] The four complete extant manuscripts of the *Siddhāntasindhu*, one bearing the seal of Emperor Shāh Jāhān himself, are currently held at the City Palace Museum Library in Jaipur, India. These manuscripts are over 450 folia each and contain vast numbers of mathematical, astronomical, and astrological tables of different kinds, see Pingree, *A Descriptive Catalogue*, pp. 138–43.

Editing and Analysing Numerical Tables: Towards a Digital Information System for the History of Astral Sciences, ed. by Matthieu Husson, Clemency Montelle and Benno van Dalen, PALS 2 (Turnhout, 2021), pp. 187–249
© BREPOLS PUBLISHERS 10.1484/M.PALS-EB.5.127699

the *Sarvasiddhāntarāja* 'King of all Siddhāntas' as an attempt to explain Islamicate (Ptolemaic) astronomical models and parameters in the language of a traditional Sanskrit *siddhānta*.[3] Misra, *The Golādhyāya*, pp. 12–17, discusses the scientific milieu of Mughal India in which Nityānanda lived and worked.

A short paper by David Pingree brought Nityānanda's *Amṛtalaharī* to my attention.[4] The *Amṛtalaharī* is a collection of astronomical tables for computing Indian calendrical elements, planetary positions, and ascensions of zodiacal signs. Pingree made some insightful observations on how the *Amṛtalaharī* was an experiment in bringing elements of Islamicate and Sanskrit astronomy together. The list below summarises some of his main remarks on the tables of the *Amṛtalaharī* (based on MS Sanskrit 19 from the collection of the University of Tokyo).

1. The name *Amṛtalaharī* is reconstructed. As the incomplete incipit on f. 1v indicates, Nityānanda may have called his work *Kheṭakṛti*. However, the manuscript catalogue of the collection of the University of Tokyo identifies this work as the *Amṛtalaharī*, and accordingly, I follow Pingree in referring to this work with its catalogued name.
2. Brief notes (in the paratext surrounding the tables) refer to earlier Sanskrit works, e.g. Makaranda's *Makaranda* (1428 CE) is mentioned in the paratext surrounding the *tithi* tables on f. 2r.[5] There are also certain calendrical elements that, according to Pingree, are Nityānanda's own inventions. For instance, the mean motion tables employ a lunar-solar calendar equivalent to three Metonic cycles of 57 solar years found in Jewish calendars (and explained in Islamicate *zījes*).[6]
3. Pingree conjectures the epoch of the *Amṛtalaharī* as 21 February Julian in 1593 CE. According to him, the epoch year 1593 is the beginning of the 57-year long period within which the work was composed. This puts the *terminus ante quem* of the work around 1649/50 agreeing with Nityānanda's floruit in the early parts of the seventeenth century.
4. Certain features of the lunar and planetary tables of the *Amṛtalaharī* closely resemble those seen in similar tables from Islamicate and Ptolemaic traditions, and mostly absent in Sanskrit astronomical works, e.g. tabulating the mean motions of the anomalies of the Moon, Venus,

[3] See Pingree, 'Indian Reception', pp. 476–80; Pingree, 'The Sarvasiddhāntarāja'; Montelle et al., 'Computation of Sines', and Montelle and Ramasubramanian, 'Determining the Sine'.

[4] Pingree, 'Amṛtalaharī'. In Misra, *The Golādhyāya* only two works are credited to Nityānanda, the *Sarvasiddhāntarāja* and the *Siddhāntasindhu*. The existence of the *Amṛtalaharī* was unknown to me at the time.

[5] See Pingree, 'Amṛtalaharī', footnote 9 on p. 210.

[6] ibid., pp. 211–12.

and Mercury, or the positive norming in the tables of planetary equations.

Pingree concludes his paper with the remark:[7]

> 'It remains unclear why Nityānanda wrote it [the *Amṛtalaharī*]; indeed, it is indeed [*sic*] astonishing that even one copy of this unusual attempt to reform siddhāntic astronomy has survived. It is a curiosity, but perhaps it played some role in history by suggesting to Jayasiṃha's astronomers how they might express de La Hire's Latin tables, which use the Julian and Gregorian calendars, in the form of an adjusted Indian calendar.'

To understand better the implication of Nityānanda's 'attempt to reform siddhāntic astronomy', I recompute and analyse a set of astronomical tables from the *Amṛtalaharī* in this study. My goal is to recompute the *attested* values seen in the manuscript (MS Sanskrit 19) instead of suggesting the *correct* values derived from historically apposite procedures. By identifying the computational methods (including irregularities) and analysing the differences between the attested values and our recomputed results, we can gain an insight into the subtle mathematical practices of *table authors*.[8] The analytical and historical methods applied in this study demonstrate how numerical tables can be seen as mathematical artefacts in the transmission of scientific knowledge between cultures.

Section 2 begins with a description of the source manuscript and a general overview of the tables of the *Amṛtalaharī*. Following this, I describe the set of six tables selected for this study (hereafter referred to as the 'selected corpus') and provide an English translation of the Sanskrit text associated with these tables. Towards the end of the section, I discuss my methodological framework to study the selected corpus, and also describe the mathematical standards adopted in this study. In Section 3, for each table, I first outline my recomputation strategies (including any irregular recomputations that exactly reproduce an attested value), and then analyse the differences between the attested values and my recomputed results individually. These discussions also include a few proposed emendations to the attested values based on my

[7] ibid., p. 213.

[8] In this study, I use the term *table authors* to indicate those historical actors who change certain numerical values based on their own computational decisions as they recopy a table. Other actors, like scribes, who copy the tables without making any computational changes are set apart. This separation is made for expedient reasons; it is not an attempt to divide them into mutually exclusive categories. In fact, both kinds of actors modify a table as they copy it (e.g. through their inadvertent oversights in copying); however, what sets them apart in this study are recomputational interventions. Scribes and table authors may both intentionally intervene to rectify a corrupted/illegible/missing entry, but table authors (often) do so by applying a mathematical algorithm (e.g. interpolating) whereas scribes may simply fill in the numbers by observing a pattern. More on this in Section 2.3.

analysis, particularly, when an inadvertent or intentional scribal effect is evident. Finally, in Section 4, I summarise the main observations of this study, and discuss the methodological questions that arise when recomputing historical tables using modern computational tools.

2. The *Amṛtalaharī* of Nityānanda

2.1. Description of the digitised microfilm

For this study, I have used a digital copy of the only verified manuscript of the *Amṛtalaharī* currently known to be extant.[9] MS Sanskrit 19 (henceforth identified with the siglum 'Tk') is a part of the Sanskrit manuscript collections of the University of Tokyo and contains the tables of the *Amṛtalaharī*.[10] The digitised microfilm of MS Tk[11] contains the (catalogue?) reference numbers 547 (old) and 19 (new) at the very beginning of the reel. The second frame captures an image of the cover page of the manuscript with the word '*Amṛtalaharī*' (in the centre) and number '13' (at the top-left corner) written in Sanskrit. The handwriting on the cover page is notably different from that of the scribe who copied this manuscript. I suspect an earlier cataloguer, or perhaps Prof. Junjirō Takakusu, who brought the manuscript from Nepal to Japan in 1913,[12] wrote this on the cover page of the manuscript. The reel number of the microfilm (MF_13) and the catalogued name *Amṛtalaharī*[13] appear to be based on this writing on the cover page. All remaining frames contain images of two folia of the manuscript, one above the other, with the digital stamp General Library. University of Tokyo at the bottom-right corner of each frame. Plates 8 and 9 show ff. 1v–2r from MS Tk, photographed by Taro Mimura in January 2021, as examples.

2.1.1. *Manuscript description from surrogate*

According to Matsunami's catalogue,[14] MS Tk is a paper manuscript with 51 folia of dimensions 11 × 5 inches that contain a collection of Sanskrit

[9] An unconfirmed manuscript of a work called *Amṛtalaharīsāraṇī* (of unknown authorship) is catalogued in the collection of the Nepalese-German Manuscript Cataloguing Project maintained by the University of Hamburg (https://catalogue.ngmcp.uni-hamburg.de/receive/aaingmcp_ngmcpdocument_00002491). At the time of writing this chapter, I have been unable to independently verify the authenticity or the contents of this manuscript.

[10] MS Tk is referenced in Matsunami, *A Catalogue of the Sanskrit Manuscripts*, pp. 8–9.

[11] This is available at http://picservice.ioc.u-tokyo.ac.jp/03_150219~UT-library_sanskrit_ms/MF13_03_004~MF13_03_004/?pageId=001.

[12] See Pingree, 'Amṛtalaharī', p. 210.

[13] See Matsunami, *A Catalogue of the Sanskrit Manuscripts*, pp. 8–9.

[14] ibid.

astronomical tables in Devanāgarī. I list below some additional features of MS Tk from its digital surrogate.

1. The folio edges of the manuscript are frayed. There are no visible binding marks or string holes suggesting, perhaps, that the stacked folia were merely wrapped in cloth and held together between (wooden?) cover boards (resembling loose-leaf, unbound books called *pothīs*). The handwriting is in black ink, legible, free of any corrections, and produced by a single scribal hand. The tables themselves appear to be written between (faint) double-ruled margins and, in a few instances, the paratext and numbers extend into the margins. The folio numbers are written on verso pages, towards the middle of the page in the right margin.
2. On f. 1v, an incomplete incipit verse is partially visible along the frayed top edge of the folio. It contains the last three quatrains (*pādas*) (of an incomplete verse in the *indravaṃśā* meter):

   yā paṇḍitair indrapurī virājate |
   śrīdevadattasya suto dvijānugaḥ
   tasyāṃ vasan khetakṛtiṃ cikīrṣati ||

   The [city of] Indrapurī that appears beautiful with [the presence] of scholars (*paṇḍita*), the 'twice-born' (*dvija*) [i.e. Brahmin] son of Śrī Devadatta resident in that city desires to complete [this work called] Kheṭakṛti.
3. Towards the top-left corner of f. 1v, the title ⟨*A*⟩*mṛtalaharī* (the initial *a* is lost to the frayed edge of the folio) appears in the left margin. According to Pingree, it is written by a different hand compared to the copyist of the manuscript.[15] The digital surrogate makes it difficult to validate this claim with surety; however, I believe this was written by the same hand as the main copyist. The shape of the remaining letters in the word matches the chirography of the primary scribe.
4. The Sanskrit numerals, along with the paratext, table titles, and row headings in Sanskrit, are written in the Nepālī (Pracalita Lipi, Newar, or Nepāla Lipi) and Devanāgarī scripts, occasionally, conflating the two scripts together. For example, Table 1 shows samples of Sanskrit numbers in MS Tk written in Pracalita Lipi and Devanāgarī.
5. The paratext surrounding the tables also use the two scripts in an intermixed manner. On the top-left corner of f. 1v (below the frayed top edge), the words of the incipit ...यापंडितैरिंद्रपूरी... (...*yā paṃḍitair iṃdrapūrī*...) are in Pracalita Lipi. In other places, identical Sanskrit words are written variously in Pracalita Lipi or Devanāgarī. For example, the compounded words ...धनमृणं... (...*dhanam ṛṇaṃ*...) on lines 3–4

[15] See Pingree, 'Amṛtalaharī', p. 210.

Script	Sample
Hindu-Arabic Numerals	0 1 2 3 4 5 6 7 8 9
Devanāgarī Numerals	० १ २ ३ ४ ५ ६ ७ ८ ९
Pracalita Lipi Numerals	𑑐 𑑑 𑑒 𑑓 𑑔 𑑕 𑑖 𑑗 𑑘 𑑙
Digits '0–9' in Devanāgarī (from f. 14r)	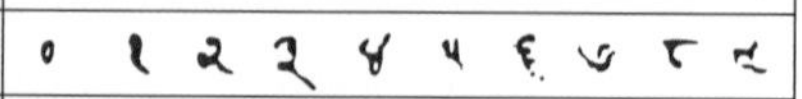
Digits '0–9' in Pracalita Lipi (from f. 1v)	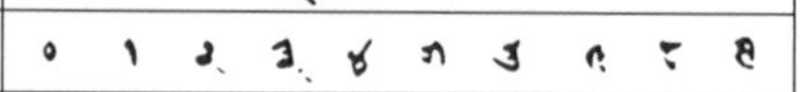
Mixed scripts (from f. 1v): Numbers '25–29' in Pracalita Lipi (top line) Numbers '35–39' in Devanāgarī (bottom line)	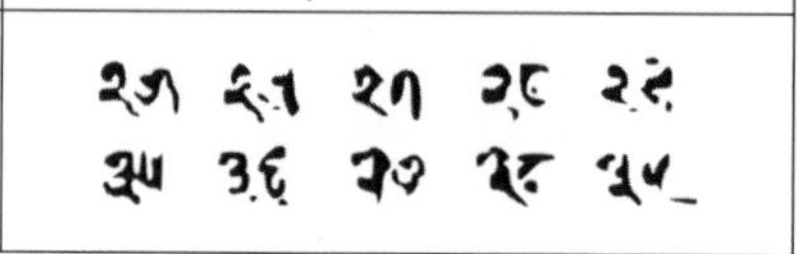

Table 1: Samples from MS Tk showing Sanskrit numbers written in different scripts.

of the text block to the right of f. 1v are in Pracalita Lipi, while the same set of words ...धनमृणं... (...*dhanam ṛṇaṃ*...) on line 4 of the text block to the right on f. 2r are in Devanāgarī.

6. When letters in the two scripts are homoglyphic (in handwritten Sanskrit), the scribe appears to write the letters using Pracalita Lipi, e.g. the letter *la*, seen as ल in MS Tk, is closer in appearance to the letter 𑐮 in Pracalita Lipi than the letter ऌ in Devanāgarī.
7. Finally, the number '1' is written at the beginning of every table title. In Prachalita Lipi, the number 𑑑 stands for the Sanskrit invocation *siddhir astu* 'may there be success' as a benedictory supplication.[16]

2.2. Overview of the tables of the *Amṛtalaharī*

Table 2 includes an overview of the types and foliation of the tables of the *Amṛtalaharī* in MS Tk. A more detailed description of these tables, and the different table parameters in each instance, can be found (in the Appendix) in Pingree, 'Amṛtalaharī', pp. 214–17.

My study focuses on the collection of tables seen in row VI of Table 2. The selected corpus includes the table of Sines (*kramajyā*);[17] the table of solar declinations (*krānti*); the three tables of shadow lengths for gnomons (*śaṅku-chāyā*) of heights 60 digits, 12 digits, and 7 digits; and the table of lunar latitudes (*śara*).

For each of these six different tables, the arguments range from 1° to 90° in one-degree steps, and their corresponding values are expressed in sexages-

[16] See Sircar, *Indian Epigraphy*, pp. 92–97 for a discussion on auspicious marks in Indian texts and epigraphs.

[17] Throughout this chapter, I use capitalised initials for trigonometric functions 'Sine', 'Cosine', 'Chord', etc. to indicate a non-unitary radius $\mathcal{R}$ (*sinus totus*). Mathematically, $\mathrm{Sin}\,\theta \equiv \mathcal{R}\sin\theta$, $\mathrm{Cos}\,\theta \equiv \mathcal{R}\cos\theta$, $\mathrm{Crd}\,\theta \equiv \mathcal{R}\,\mathrm{crd}\,\theta$, etc.

Number	Table types	Foliation
I.A	Tables of *tithis*.[18]	Ff. 1v–6v
I.B	Tables of *nakṣatras*.[19]	Ff. 7r–11v
I.C	Tables of *yogas*.[20]	Ff. 11v–17v
II	Tables of *abdapas* and *saṅkrāntīs*.[21]	Ff. 17v–18r
III	Tables of planetary mean motions of the Sun, the Moon, Lunar anomaly, Lunar node, Mars, Mercury's anomaly, Jupiter, Venus' anomaly and Saturn.	Ff. 18v–27r
IV	Tables of planetary equations: (a) *manda* equations for the Sun and the Moon; and (b) the set of first *śīghra*, *manda*, and second *śīghra* equations for the five star-planets.[22]	Ff. 27v–44r
V	Tables of rising times of zodiacal signs (right and oblique ascensions).	Ff. 44v–49r
VI	Tables of (a) Sines (here: Table VI.A); (b) solar declinations (Table VI.B); (c) shadow lengths of gnomons of heights 60 digits (Table VI.C_1), 12 digits (Table VI.C_2), and 7 digits (Table VI.C_3); and (d) lunar latitudes (Table VI.D).	Ff. 49v–50v
VII	Tables of adjustments for the five star-planets.	F. 51r

Table 2: An overview of the tables of the *Amṛtalaharī* in MS Tk.

[18] A *tithi* is the thirtieth part of a synodic lunar month, or the time interval during which the longitudinal difference between the Sun and the Moon increases by 12°.

[19] A *nakṣatra* (or lunar mansion) is the constellation in which the Moon is located. Typically, Sanskrit astronomy lists 27 *nakṣatras* each spanning 13°20′ along the 360° orbit of revolution of the Moon.

[20] A *yoga* (or *nityayoga* 'daily yoga') is the duration in which the combined motions of the Sun and the Moon amount to 1 *nakṣatra* or 13°20′. There are 27 identified *yogas* corresponding to the 27 *nakṣatras*.

[21] The *abdapas* are the weekdays on which particular years commence, and *saṅkrāntīs* refer to the solar ingress (*saṅkramaṇa*) into the 12 zodiacal signs (*rasis*) and 27 lunar mansions (*nakṣatras*).

[22] In Indian astronomy, the *manda-saṃskāras* are the equation-of-centre corrections applied to the mean longitude of the planets (*madhyama-grahas*) to produce the *manda*-corrected longitudes or *manda-sphuṭa-grahas*. In case of the Sun and the Moon, this is the only correction required to obtain their true longitudes (*sphuṭa-grahas*). However, for the other five star-planets—the two interior planets Mercury and Venus and the three exterior planets Mars, Jupiter, and Saturn—an additional *śīghra-saṃskāra* (correction due to the anomaly of conjunction) is applied to their *manda-sphuṭa-grahas* to obtain their true longitudes. For exterior planets, the *manda-sphuṭa-grahas* are their true heliocentric longitudes and the *śīghra-saṃskāra* converts these values to their true geocentric longitudes. For interior planets, the *manda-sphuṭa-graha* is the *manda*-corrected mean Sun that gets *śīghra*-corrected to produce their true geocentric longitudes.

imal numbers (up to a fractional precision of seconds). The six tables are identically arranged over three folia (ff. 49v–50v) of MS Tk. Each folio has thirty arguments in the first row, followed by six successive rows listing the corresponding six function values (i.e. the attested values of each table) in individual rows. Appendix A (pp. 226–31) includes the images of ff. 49v–50v from MS Tk and a diplomatic transcription of the six tables on these folia.

2.2.1. *Translation of the table titles*

The three table titles (seen at the top of ff. 49r–50v of MS Tk respectively) are presented below. The Sanskrit text is transliterated with Latin characters and also translated into English.

|| 1 *atha kramajyā-krānti-ṣaṣṭyaṅgula-śaṅku-dvādaśāṅgula-saptāṅgula-śaṅku-chāyā-candra-śarāṃśāḥ* ||

Now, the Sines (*kramajyā*); the solar declinations (*krānti*); the shadow lengths (*chāyā*) [of] 60-digit gnomon (*ṣaṣṭi-aṅgula-śaṅku*), 12-digit (*dvādaśa-aṅgula*) [gnomon], 7-digit gnomon (*sapta-aṅgula-śaṅku*); [and] the degrees of lunar latitudes (*candra-śara-aṃśa*).

|| 1 *pratyaṃśa-kramajyā-krānti-chāyāḥ śarāśca* ||

For every degree (*aṃśa*), the Sines (*kramajyā*), the solar declinations (*krānti*), the shadow lengths (*chāyā*), and the [lunar] latitudes (*śara*).

|| 1 *iti pratyaṃśaka-kramajyā-kranti-chāyāḥ śarāśca samāptaḥ* ||

Thus, the Sines (*kramajyā*), the solar declinations (*krānti*), the shadow lengths (*chāyā*), and the [lunar] latitudes (*śara*) for every degree (*aṃśaka*) ends.

At the bottom of f. 50v of MS Tk, we find the following text:

|| *pātonacandro bhujā kārye bhujyaṃśebhyaḥ śaro grāhyāḥ yadi pātonacandraḥ ṣaḍbhonas tadā śaraḥ saumyaḥ yadādhikas tadā yāmyaḥ* ||[23]

In [taking] the longitude (*bhujā*) of the Moon (*candra*) minus the node (*pāta*), the lunar latitude (*śara*) is to be understood from the degrees fulfilling it [i.e. calculated according to the degrees of lunar elongation]. If the [longitude of] the Moon minus the node (*pāta*) is less than six signs (*ṣaḍ-bha*) [i.e. less than 180°] then the lunar latitude (*śara*) is in the northern direction (*saumya*) [i.e. north of the ecliptic plane]; if it is more [i.e. greater than 180°] then [the lunar latitude] is in the southern direction (*yāmya*) [i.e. south of the ecliptic plane].

[23] This sentence is grammatically ill-formed; for example, the attested words *bhujyaṃśebhyaḥ* (instead of *bhujāṃśebhyaḥ*) and *grāhyāḥ* (instead of *grāhyaḥ*) have orthographic defects.

2.3. Methodology of recomputation and analysis

Before describing my general methodology for recomputing and analysing individual tables, I note the following remarks on the selected corpus, and on my mathematical practice of recomputing numerical tables.

1. Ff. 49v–50v of MS Tk do not contain any instructions to compute the attested values of the six functions (Tables VI.A–VI.D). As the translations of the table titles show, the titular text merely identifies the types of tables written on a particular folio.[24] The other table titles throughout the rest of this manuscript (as well as the paratext surrounding those tables) also lack any computational instructions. Hence, the recomputation strategies used in this study are derived from other apposite Sanskrit and Islamicate sources.

   Recent studies on Nityānanda's texts,[25] and more generally, the culture of science that thrived at the Mughal courts of early seventeenth century India,[26] suggest that he was well acquainted with Islamicate (Persianate) theories in addition to Sanskrit siddhāntic astronomy.[27] His *Amṛtalaharī* uses certain parameters that are distinctly Islamicate, e.g. a *sinus totus* of 60, as well as those that are traditionally siddhāntic, e.g. an ecliptic obliquity of 24°. In fact, the *Amṛtalaharī* contains several instances that testify to Nityānanda's familiarity with (and acceptance of) both traditions of knowledge. It is, therefore, reasonable to choose recomputational methods from the Sanskrit texts (e.g. *siddhāntas*, *karaṇas*, or *koṣṭhakas*) or the Islamicate *zījes* that were in circulation in Mughal India during his time.[28]

2. Establishing an absolute agreement between the attested and recomputed values is extremely difficult, if not nearly impossible. While some differences can be explained computationally, there are other unknown factors that lead to differences between the attested and recomputed values.[29] In fact, even at the level of recomputations, the arithmetical practices of table authors (e.g. dividing mixed fractions, rounding/truncating

[24] The text at the bottom of f. 50v of MS Tk explains certain aspects of the lunar latitude (*śara*); however, it does not describe a computational procedure or algorithm.

[25] For example, Misra, *The Golādhyāya*; Montelle et al., 'Computation of Sines', and Montelle and Ramasubramanian, 'Determining the Sine'.

[26] For example, Minkowski, 'Astronomers and Their Reasons' and Truschke, *Culture of Encounters*.

[27] For example, Misra, 'Persian Astronomy in Sanskrit' and Misra, 'Sanskrit Recension of Persian Astronomy'.

[28] See Ansari, 'On the Transmission' and Ansari, 'Survey of Zījes'.

[29] See Appendix A2 *Tabular errors* in van Dalen, 'A Statistical Method', pp. 116–19 for a statistical description of the errors in numerical tables.

the fractions, approximating/interpolating between fractions, etc.) affect our own calculations at every step. The cumulative effect of these decisions create an uncertainty in precisely reproducing the attested value. In this study, all recomputed values are presented up to a level of computational efficacy that retains a residual arithmetical noise.

3. In some instances, the differences between the attested and recomputed values can indicate scribal discrepancies. Typically, these include
   (a) *inadvertent copying oversights* in the digits of an entry (or the whole entry), e.g.
      - permutation or transposition of digits/entries,
      - unwitting alteration of homoglyphic digits (due to misreading);
      - dittography, i.e. copying a sequence of digits/entries twice;
      - haplography, i.e. omitting a sequence of identical digits/entries while copying; or
      - mistranscription, i.e. a general non-purposive mistake in reading and copying an individual digit, a whole entry, or a sequence of digits/entries; and
   (b) *intentional interventions* by historical actors (scribes/table authors) to rectify corrupted/illegible/missing digits of an entry (or the whole entry), e.g.
      - ad hoc substitution, i.e. replacing illegible digits by other digits;
      - assimilation, i.e. merging digits of adjacent entries to create new entries;
      - insertion, i.e. filling missing digits or whole entries by inspecting the sequence; or
      - contamination, i.e. inserting digits/entries from elsewhere (on the folio) to fill missing entries.

   It is worth noting that the lists above are neither exhaustive nor mutually exclusive. It is often the case that distinguishing between inadvertent or intentional actions is simply not possible. Moreover, even in the clearest of examples, any emendations to the attested value (that are meant to correct/rectify these actions) remain conjectural. With these caveats, it is nevertheless useful to analyse the differences between the attested and recomputed values. If a difference can be justifiably explained as the result of an inadvertent copying mistake or an intentional (but inaccurate) intervention, the attested value can be emended to a recomputed result as a *proposed emendation*.

   For example, in Table VI.A, the attested digits (in the minutes places) for $\mathrm{Sin}\,16°$, $\mathrm{Sin}\,17°$, $\mathrm{Sin}\,18°$, and $\mathrm{Sin}\,19°$ are 32, 33, 32, and 32 respectively. The recomputed Sines for these arguments suggest that digits in the minutes places for each of these arguments should be 32. The abrupt increase of $+1^{m}$ (for Sin 17°) in an overall monotonic sequence suggests a plausible error in coping '33' instead of '32'. The digits २ and ३ in handwritten Devanāgarī are often homoglyphic, and hence, an unwitting alteration of these digits is not uncommon. Accordingly, I emend the digits in the minutes place of Sin 17° from '33' to the recomputed result '32' in Table VI.A.

4. A particular class of intentional actions, different from the ones listed in the previous remark, are *recomputational interventions*. While scribes may intervene to correct a corrupted/illegible/missing entry following some rudimentary logic, table authors do the same but they recalculate (or estimate) the values using more elaborate mathematical procedures. Sometimes, table authors apply these mathematical procedures to intentionally intervene, but do so inattentively which leads to an erroneous result. By retracing their calculations (using historically apposite procedures instead of modern ones), we can detect the irregularities along the way that lead to the errant result. The goal of this study is to recompute the attested results, and therefore, identifying *recomputational irregularities* is an important part of the process. In my study of the selected corpus, I have identified the following kinds of recomputational irregularities:
   (a) instances where table authors (unwittingly) err in applying a mathematical procedure, e.g. misidentifying an appropriate interval when interpolating;
   (b) instances where table authors perpetuate an erroneous calculation, e.g. using an erroneous Sine to compute the solar declination; and
   (c) instances where table authors round/truncate the sexagesimal digits in a calculation inconsistently.

The six individual tables from the selected corpus are recomputed and analysed following a common methodological routine:

Routine of recomputation

1. Recompute the values of the table for the entire range of arguments using apposite historical procedures.
2. Compare the attested values (in MS Tk) and the results of the first recomputation, and note the differences (between the digits in corresponding sexagesimal places).
3. Inspect all non-zero differences, and where possible, identify any irregular recomputations that reproduce the attested values (and thereby, eliminate these differences).

4. Reassess the revised differences between the attested values and the results of the second recomputation (i.e. recomputations including irregular ones).

Routine of analysis

5. Re-examine the attested values in (the diplomatic transcription and the digital surrogate of) MS Tk for those arguments that still have large (revised) differences.
6. Identify, if possible, any copying oversights or intentional (non-recomputational) interventions in the attested value, and propose emendations or corrections to those values with justifications.

2.3.1. *Mathematical standards*

1. I follow two main mathematical standards to recompute the individual tables in this study:
   (a) recomputed sexagesimal values are reduced to the second fractional place by systematically rounding the digits in the final result instead of truncating them (at the seconds place),[30] and
   (b) recomputed Sines are chosen over attested Sines (in MS Tk) for all calculations.[31]

   Appendices C.1–2 include my statistical justifications for choosing these mathematical standards in this study.
2. When the division of sexagesimal numbers is an intermediate part of a computation, the result of the division is rounded to seconds before proceeding further. Effectively, this implies that,
   - while calculating the solar declination in Section 3.3, $\mathrm{Sin}\,\delta = \mathrm{Sin}\,\lambda \times {}^{\mathrm{Sin}\,24^\circ}/_{60}$ is computed as a sexagesimal number (rounded to seconds) before proceeding to find δ as the inverse arc of $\mathrm{Sin}\,\delta$;
   - while calculating the shadow lengths in Section 3.5, ${}^{\mathrm{Cos}\,a^\circ}/_{\mathrm{Sin}\,a^\circ}$ is computed as a sexagesimal number (rounded to seconds) before multiplying it by the different gnomon heights h to determine the value of their respective shadow lengths; and

[30] For a sexagesimal number $a; b, c, d$ with $a, b, c, d \in [0, 59]$, *systematic rounding* results in either $a; b, c$ for $d < 29$ or $a; b, c+1$ for $d \geq 30$. All calculations in this study follow this standard of systematic rounding. In contrast, *truncation* ignores the final (third) sexagesimal digit d and simply takes the result as $a; b, c$ for any value of d.

[31] Sines are required for recomputing the solar declinations (in Table VI.B), the shadow lengths for gnomons of various heights (in Tables VI.C_1–VI.C_3), and lunar latitudes (in Table VI.D).

- while calculating the lunar latitude in Section 3.7, $\text{Sin}\,\beta = 4;42,25 \times {}^{\text{Sin}\,\omega}/_{60}$ is computed as a sexagesimal number (rounded to seconds) before proceeding to find β as the inverse arc of $\text{Sin}\,\beta$.

3. In addition to this:
   (a) the lunar latitudes (in Table VI.D) are recomputed using an exact expression in lieu of an approximate one, and
   (b) the lunar latitude recomputations use $\text{Sin}\,4°30' = 4;45,25$.

   Appendices C.3–4 include my statistical justifications for these choices.

3. Recomputation strategies and analyses of differences for Tables VI.A–D

Following the general methodology described above, my recomputation strategies for each of the six tables from the selected corpus, along with an analyses of the differences between the attested values and my recomputed results, are presented below in separate subsections.

3.1. Table of Sines (*kramajyā*): Recomputation strategy

The Sine table of the *Amṛtalaharī* (in MS Tk) is computed for every degree of arc from 1° to 90° and has a maximum value (*sinus totus* $\mathcal{R}$) of 60;0,0. I recompute the Sines following a sequence of interdependent mathematical operations based on arithmetical, geometrical, and trigonometric arguments. My recomputed table of Sines for the first ninety degrees of arc is presented in Table VI.A on page 233.

The *Amṛtalaharī* (in particular, MS Tk) does not describe any method to compute the Sines; however, Nityānanda's *Sarvasiddhāntarāja* (1639 CE) includes a detailed discussion on Sine computations (sixty verses including several diagrams in six sections) in the *spaṣṭādhikāra* of the *gaṇitādhyāya*, I.3: 19–85. A critical edition, English translation, and technical commentary of the verses from the first five sections can be found in Montelle et al., 'Computation of Sines', and those from the sixth section can be found in Montelle and Ramasubramanian, 'Determining the Sine'. Considering the *Amṛtalaharī* was composed almost contemporaneously with the *Sarvasiddhāntarāja* (i.e. around the first half of the seventeenth century), it is reasonable to assume that Nityānanda used analogous geometrical arguments and trigonometric formulae (including the iterative algorithm for calculating the Sine of 1°) to construct the Sine tables of the *Sarvasiddhāntarāja* and the *Amṛtalaharī*.[32]

[32] See Van Brummelen, *The Mathematics of the Heavens* and Plofker, *Mathematics in India* for a more detailed discussion on the history and development of trigonometry in India.

3.1.1. *Recomputing the Sines of the elementary arcs based on geometrical arguments*

I first compute the Sines of 90°, 72°, 60°, 54°, 45°, 36°, 30°, and 18°. In the *Sarvasiddhāntarāja* I.3: 24–54, Nityānanda computes these Sines using (**a**) geometrical arguments in a circle of radius 60, (**b**) the half-arc and double-arc formulae for Sines, and (**c**) the sum and difference laws for Sines. I list below the different expressions for calculating these Sines. All of these expressions can be derived using simple geometrical arguments; readers may refer to Montelle et al., 'Computation of Sines' where Nityānanda's derivations from the *Sarvasiddhāntarāja* are described in greater detail.

1. Sin 90° corresponds to the radius (*vyāsa-khaṇḍa*) of a circle, i.e. we have $\mathrm{Sin}\,90^\circ \equiv \mathcal{R} = 60;0,0,0$ (*Sarvasiddhāntarāja* I.3: 24).[33]
2. Sin 45° can be expressed as $\frac{1}{\sqrt{2}}\sqrt{\mathcal{R}^2} \approx 42;25,35,3$ (*Sarvasiddhāntarāja* I.3: 28). This expression is derived using the Pythagorean theorem in an inscribed right triangle at the centre of a circle of radius $\mathcal{R}$.[34]
3. Sin 30° can be expressed as $\frac{1}{2}\mathcal{R} = 30;0,0,0$ (*Sarvasiddhāntarāja* I.3: 24). An equilateral triangle subtended at the centre of a circle of radius $\mathcal{R}$ has sides measuring $\mathrm{Crd}\,60^\circ \equiv \mathcal{R}$. The Sine (*jyārdha* 'half the chord') corresponding to an arc (*cāpa*) of 30° is 'half the chord of double the arc', i.e. $\frac{1}{2}\,\mathrm{Crd}\,60^\circ$.[35]
4. Sin 60° is approximately 51;57,41,29. This value is computed using Nityānanda's procedure for the Sine of double the arc (*Sarvasiddhānta-rāja* I.3: 37) for an arc of 30°. Montelle et al., 'Computation of Sines', pp. 28–29 discuss the two-step procedure for this calculation as well as Nityānanda's Sanskrit expressions of the formula for the Sine of double the arc.
5. Sin 18° can be expressed as $\sqrt{\left(\frac{\mathcal{D}}{4}\right)^2 + \frac{1}{4}\left(\frac{\mathcal{D}}{4}\right)^2} - \frac{1}{2}\left(\frac{\mathcal{D}}{4}\right) \approx 18;32,27,40$ (*Sarvasiddhāntarāja* I.3: 24), where the diameter $\mathcal{D} \equiv 2\mathcal{R} = 120$. Nityānanda's geometrical demonstration for this expression (in the *Sarvasiddhāntarāja* I.3: 25–27), and its equivalence to Bhāskara II's expression $\frac{1}{4}(\sqrt{5\mathcal{R}^2} - \mathcal{R})$ stated in terms of the radius $\mathcal{R}$ (in his *Jyotpatti*: 9, 1150 CE) is discussed in Montelle et al., 'Computation of Sines', pp. 18–22.[36]

[33] Montelle et al., 'Computation of Sines', p. 18.

[34] ibid., pp. 22–23.

[35] ibid., p. 18.

[36] Bhāskara II does not derive this equation; Munīśvara (*fl.* 1638 CE), in his commentary *Marīci-ṭīkā* on the *Jyotpatti*, offers a geometrical explanation for it. In fact, Munīśvara proposes the lemma *daśāśra-bhujā-vargo'yaṃ bhuja-trijyā-vadhena yuk trijyāvargo bhavet* 'The square of a side of a regular decagon together with the product of the side and the radius (of the

6. Sin 36° is approximately 35;16,1,36. Like Sin 60°, this value is also computed using Nityānanda's procedure for the Sine of double the arc (*Sarvasiddhāntarāja* I.3: 37) for an arc of 18°.[37]
7. Sin 54° is approximately 48;32,27,40. This value is computed using Nityānanda's procedure for the Sine of the difference of two arcs (*Sarvasiddhāntarāja* I.3: 49) for two arcs measuring 90° and 36°, with Sin 90° = 60 and Sin 36° = 35;16,1,36. Nityānanda's geometrical demonstration of this expression (in the *Sarvasiddhāntarāja* I.3: 50–54) is discussed in Montelle et al., 'Computation of Sines', pp. 38–41.
8. And finally, Sin 72° is approximately 57;3,48,12, also using Nityānanda's procedure for the Sine of the difference of two arcs (*Sarvasiddhāntarāja* I.3: 49) for two arcs measuring 90° and 18°, with Sin 90° = 60 and Sin 18° = 18;32,27,40.[38]

3.1.2. *Recomputing the Sines for multiples of 3° of arc*

Next, I compute the Sines of multiples of 3° of arc (in a circle of radius 60). These values are calculated by successively applying the trigonometric formulae for (**a**) the Sine of half the arc and (**b**) the Sine of the sum and differences of arcs.

In his *Sarvasiddhāntarāja* I.3: 31–32 and 36, Nityānanda gives two expressions to determine the Sine of half the arc. The first method calculates the Sine in terms of the Versine (*utkramajyā*), while the second method computes it iteratively. See Montelle et al., 'Computation of Sines', pp. 23–27 for a more detailed description of these methods, including their derivations and equivalence.

As an example, Sin 27° is calculated from Sin 54° ≈ 48;32,27,40 (with the first method) as

$$\text{Sin}\,27^\circ = \text{Sin}\left(\frac{54^\circ}{2}\right) = \sqrt{\left(\frac{\text{Vers}\,54^\circ}{2}\right)^2 + \left(\frac{\text{Sin}\,54^\circ}{2}\right)^2}$$
$$(\text{where Vers}\,54^\circ = \mathcal{R} - \text{Cos}\,54^\circ)$$
$$\Rightarrow \text{Sin}\,27^\circ \approx 27;14,21,56.$$

In the *Sarvasiddhāntarāja* I.3: 41 and 49, Nityānanda also gives the expressions for the Sine of the addition of (or the subtraction between) two arcs; see Montelle et al., 'Computation of Sines', pp. 29–46 for Nityānanda's

circumscribing circle) is equal to the square of the radius' to derive an expression for Sin 18°, see Gupta, 'Sine of Eighteen Degrees'.

[37] Montelle et al., 'Computation of Sines', pp. 28–29.

[38] ibid., pp. 38–41.

geometrical arguments to derive these expressions. Essentially, these formulae help calculate new Sines using previously determined Sines (and corresponding Cosines). For example, $\mathrm{Sin}\,48^\circ$ is calculated from $\mathrm{Sin}\,30^\circ = 30;0,0,0$ and $\mathrm{Sin}\,18^\circ \approx 18;32,27,40$ as

$$\begin{aligned}\mathrm{Sin}\,48^\circ &= \tfrac{1}{60}\,\mathrm{Sin}\,(30^\circ + 18^\circ)\\ &= \tfrac{1}{60}(\mathrm{Sin}\,30^\circ\,\mathrm{Cos}\,18^\circ + \mathrm{Cos}\,30^\circ\,\mathrm{Sin}\,18^\circ) \approx 44;35,19,16.\end{aligned}$$

Similarly, the Sine of the difference between two arcs is calculated using the Sines (and corresponding Cosines) of the two arcs. For example, $\mathrm{Sin}\,6^\circ$ is calculated from $\mathrm{Sin}\,36^\circ = 35;16,1,36$ and $\mathrm{Sin}\,30^\circ = 30;0,0,0$ as

$$\begin{aligned}\mathrm{Sin}\,6^\circ &\equiv \tfrac{1}{60}\,\mathrm{Sin}\,(36^\circ - 30^\circ)\\ &= \tfrac{1}{60}(\mathrm{Sin}\,36^\circ\,\mathrm{Cos}\,30^\circ - \mathrm{Cos}\,36^\circ\,\mathrm{Sin}\,30^\circ) \approx 6;16,18,8.\end{aligned}$$

I calculate the Sines for the thirty arguments that are multiples of 3° of arc by successively applying the formulae for the Sine of half the arc and the Sine of the sums and differences of arcs.

3.1.3. *Recomputing the Sine of 1° of arc*

To calculate the remaining Sines, in particular, the Sines for arguments that are multiples of 2° (distinct from the multiples of 3°), e.g. $\mathrm{Sin}\,4^\circ$ or $\mathrm{Sin}\,56^\circ$, the value of Sine of 1° is essential.

Typically, in the Indian tradition, the Sines were tabulated in 24 blocks of 3°45′ (or 225′) for the first 90° (the first quadrant) of a circle of specified radius (identified as the *trijyā* or *sinus totus*).[39] The Sine of a non-tabulated argument was calculated by interpolating between appropriate (successive) values using different interpolation (and iterative) schemes.[40]

[39] Bag, 'Sine Table' describes the different sine tables in the Indian tradition. Also, Subbarayappa and Sarma, *Indian Astronomy*, pp. 62–73 present translations and analyses of the verses (from primary sources) that discuss Sine computations from major Sanskrit texts.

[40] For example, see Hayashi, 'Āryabhaṭa's Rule' for Āryabhaṭa's rule of differences for computing Sines in his *Āryabhaṭīya* (*c.* 499 CE); Gupta, 'Second Order Interpolation', p. 88 for Brahmagupta's second-order finite-difference interpolation scheme for approximating Sines in his *Dhyānagraha* (*c.* early 7th century)—a technique also repeated in his later and more famous work *Khaṇḍakhādyaka* (665 CE); Plofker, 'An Example of the Secant Method' for Parameśvara's fixed-point iterations to compute Sines in his *Siddhāntadīpikā* (*c.* 14th century); Ramasubramanian and Sriram, *Tantrasaṅgraha*, pp. 52–68 for Āryabhaṭa's commentator Nīlakaṇṭha Somayājī's interpolation techniques to compute desired Sines in his *Tantrasaṅgraha* (1501 CE); and Sarma et al., *Gaṇita-yukti-bhāṣā*, pp. 90–102 for Jyeṣṭhadeva's demonstrations of the sine and cosine series approximations—attributed to the famous Kerala astronomer Mādhava of Saṅgamagrāma (fl. *c.* 1380/1420 CE)—in his *Gaṇitayuktibhāṣā* (*c.* 16th century).

In his *Sarvasiddhāntarāja* I.3: 60–66, Nityānanda gives three different iterative algorithms to determine the Sine of one degree as a solution to a cubic equation. Montelle and Ramasubramanian, 'Determining the Sine' discuss, in detail, Nityānanda's algebraic and geometrical rationales in using a cubic equation, his derivation of the Sine of one degree as a recursive solution of a cubic equation, as well as the historical and technical context of this derivation—including its origin in al-Kāshī's method from the 15th century.

I describe below the main steps in calculating Sin 1° following Nityānanda's first iterative method described in his *Sarvasiddhāntarāja* I.3: 60–63.[41]

1. Calculate Sin 3°. In the *Sarvasiddhāntarāja* I.3: 66, Nityānanda expressly mentions the value of Sin 3° as 3;8,24,33,59,34,28,14,50; however, for the present purpose, a recomputed value (to thirds) provides an identical estimate of Sin 1° up to the fourth fractional place in this algorithm. The formula for the Sine of half the arc for an arc of 6° (with Sin 6° $\approx$ 6;16,18,8) gives Sin 3° $\approx$ 3;8,24,33.

2. Solve a cubic equation in X (with $X \equiv 2 \times \sin 1^\circ$) of the form (in modern notation)

$$X = \frac{2 \times \sin 3^\circ}{3} + \frac{X^3}{3R^2} \quad \text{or} \quad \sin 1^\circ = \frac{\sin 3^\circ}{3} + \frac{(\sin 1^\circ)^3}{3R^2}.$$

   By treating the number X as a sequence of successive sexagesimal digits $p_0, p_1, p_2, \ldots, p_n$ (up to the nth level of precision), Nityānanda's first iterative method (*Sarvasiddhāntarāja* I.3: 60–63) generates the individual digits p_i for $i \in \mathbb{N}_n$ recursively in n iterations. Essentially, this method uses successive divisions of remainders to determine a progressively more accurate root of the cubic equation in Sin 1°. According to Montelle and Ramasubramanian, 'Determining the Sine', pp. 15–16, this algorithm gives Sin 1° = 1;2;49,43,11 (calculated up to the fourth fractional place).

3.1.4. *Recomputing the Sines of the remaining arcs*

The Sines for all remaining integer-valued arcs between 1° and 90° can be easily recomputed with Sin 1° and the formulae for Sines of half the arc and the sums and differences of arcs. For example, Sin 2° is calculated from Sin 3° $\approx$ 3;8,24,33 and Sin 1° $\approx$ 1;2;49,43 as

$$\sin 2^\circ \equiv \sin(3^\circ - 1^\circ) = \tfrac{1}{60}(\sin 3^\circ \cos 1^\circ - \cos 3^\circ \sin 1^\circ) \approx 2;5,38,17.$$

[41] See Montelle and Ramasubramanian, 'Determining the Sine', pp. 13–14.

3.2. Table of Sines (*kramajyā*): Analysis of differences

List of proposed emendations to the attested Sines in MS Tk:[42]

Based on inadvertent copying oversights

1. Sin $1^\circ{}_m$: 5 → 2 and Sin $2^\circ{}_m$: 0 → 5. The value of Sin 1° is an important part of the recomputation of Sines, and hence, an error in the minutes place of Sin 1° suggests an unintentional copying mistake rather than an irregular recomputation. The digits '5' and '0' in Sin $1^\circ{}_m$ and Sin $2^\circ{}_m$ could have been mistakenly transposed during copying; however, Sin $1^\circ{}_m = 0$ is still a significant error.
2. Sin $17^\circ{}_m$: 33 → 32. Suspected alteration of homoglyphic digits '2' and '3' in handwritten Devanāgarī. Also, Sin $16^\circ{}_m$, Sin $17^\circ{}_m$, Sin $18^\circ{}_m$, and Sin $19^\circ{}_m$ appear in the sequence '32', '33', '32', and '32' respectively so a mistranscription is just as likely.
3. Sin $37^\circ{}_m$: 16 → 6. Suspected dittography. Sin $36^\circ{}_m$ and Sin $37^\circ{}_m$ appear in the sequence '16' and '16' respectively.
4. Sin $50^\circ{}_u$: 46 → 45. Suspected mistranscription. Sin $49^\circ{}_u$, Sin $50^\circ{}_u$, and Sin $51^\circ{}_u$ appear in the sequence '45', '46', and '46' respectively.
5. Sin $75^\circ{}_s$: 30 → 20. Suspected alteration of homoglyphic digits '2' and '3' in handwritten Devanāgarī.

Based on intentional interventions

6. Sin $88^\circ{}_{m,s}$: 59,27 → 57,48. Suspected contamination. Adjacent entries Sin 88° and Sin 89° are both 59;59,27. This could also suggest a dittography; however, the entries for *all* six functions corresponding to the 88th and 89th arguments are identical in MS Tk. (Pages 230–31 show the printed reproduction and a diplomatic transcription of f. 50v from MS Tk.) I suspect a table author intentionally copied the entire column of (correct) entries corresponding to the 89th argument (from a parent manuscript) to replace a corrupted/illegible/missing column of entries for the 88th argument.

Remarks on Table VI.A

1. On f. 49v of MS Tk, the digits '2' and '0' (of the number 20) in Sin $57^\circ{}_m$ have overhead marks: , 2̌0̈. This could suggest a correction

[42] The subscripts 'u', 'd', 'm', and 's' are used to indicate digits in the units, degrees, minutes, and seconds place respectively. I use '→' to represent a change between digits, in other words, the digits to the left of '→' are emended to the ones on its right. I follow these conventions to indicate my proposed emendations for the rest of this chapter.

to the (digits in the) number 20; however, there are no marginal corrections visible on the folio and hence I simply record this entry as 20 in my transcription.

2. The attested and recomputed Sin 46°, Sin 49°, Sin 50°, Sin 52°, Sin 53°, Sin 54°, and Sin 57° differ as $\pm 1'$. My recomputations (including irregular ones) have been unsuccessful in removing this difference, and there are no discernible copying mistakes or scribal corrections in any of these instances. Therefore, I present the attested digits (in the minutes place) of these Sines in Table VI.A without suggesting any emendations.

   However, looking at Nityānanda's Sine table from his *Sarvasiddhāntarāja*,[43] we find:[44]

$$\mathrm{Sin}\,46^\circ = 43;\underline{9},37,23,49$$

$$\mathrm{Sin}\,49^\circ = 45;\underline{16},57,16,10 \qquad \mathrm{Sin}\,50^\circ = 45;\underline{57},45,35,59$$

$$\mathrm{Sin}\,52^\circ = 47;\underline{16},50,19,22 \qquad \mathrm{Sin}\,53^\circ = 47;\underline{55},\ 5,16,13$$

$$\mathrm{Sin}\,54^\circ = 48;\underline{32},27,40,15 \qquad \mathrm{Sin}\,57^\circ = 50;\underline{19},12,50,34$$

   The underlined digits (in the minutes place) of these values are identical to the corresponding digits of my recomputed Sines in Table VI.A. The similarity between these Sines in the *Sarvasiddhāntarāja* and the *Amṛtalaharī* alludes to a common computational nuance, or perhaps a common textual ancestor.

3. Sin 14°_s, Sin 21°_s, Sin 45°_s, and Sin 65°_s have a difference of $+1$ between the attested values (from MS Tk) and the recomputed results. This difference appears to be the result of an unknown (and possibly, irregular) arithmetical calculation by a table author. I leave the digits (in the seconds place) of these Sines unchanged in Table VI.A.

3.3. Table of solar declinations (*krānti*): Recomputation strategy

The table of solar declination (*krānti*) of the *Amṛtalaharī* (in MS Tk) is computed for every degree of celestial (tropical or *sāyana*) longitude λ from 1° to 90° and has a maximum value (equal to the obliquity of the ecliptic ε) of 24°0′0″. The solar declination δ is related to the celestial longitude λ with the expression

$$\mathrm{Sin}\,\delta = \mathrm{Sin}\,\lambda \times \frac{\mathrm{Sin}\,\varepsilon}{\mathcal{R}} \equiv \mathrm{Sin}\,\lambda \times \frac{\mathrm{Sin}\,24^\circ}{60} \qquad \because\ \mathrm{Sin}\,90^\circ = \mathcal{R} = 60.$$

[43] MS Sans γ550, f. 19r, from the Wellcome Institute for the History of Medicine and MS Reel No. B 354/15, f. 15r, from the National Archive Kathmandu.

[44] The fifth digit '49' of Sin 46° is illegible in MS B 354/15. Also, Sin 54°_s resembles '20' in MS B 354/15.

This expression is commonly found in most Sanskrit *siddhāntas* from very early times, e.g. Brahmagupta's *Brāhmasphuṭasiddhānta* (628 CE): II.55. See Plofker, 'An Example of the Secant Method', pp. 91–92 for a simple geometric derivation of this expression applying the 'rule of three' to similar right triangles inscribed between the ecliptic and the equator. Having calculated the Sine of the declination, the method to find the arc of declination corresponding to it involves estimating the inverse arc of Sine. Several Sanskrit texts describe the method to find the inverse Sine, i.e. the arc measure (*cāpa* or *dhanus*, 'bow') corresponding to a particular Sine (*kramajyā*) value.[45]

Typically, the unknown arc θ for a given $\text{Sin}\,\theta$ is linearly interpolated using localised Sine differences. The general algorithm of this method (in modern notation) is as follows:

1. Identify the interval $\text{Sin}\,\theta_i < \text{Sin}\,\theta < \text{Sin}\,\theta_{i+1}$ for $i \in \mathbb{Z}^+_{90}$ in the table of Sines. The Sine function is a monotonic function that increases from 0 to $\mathcal{R}$ in the interval $[0°, 90°]$, and therefore, the corresponding interval of the argument θ can be identified as $\theta_i < \theta < \theta_{i+1}$ for $i \in \mathbb{Z}^+_{90}$.
2. Compute $\delta\theta$, where $\delta\theta \stackrel{\text{def}}{=} \theta - \theta_i$ and hence $\theta = \theta_i + \delta\theta$. The increment $\delta\theta$ can be computed from a linear incremental ratio in the unit interval $[\theta_i, \theta_{i+1}]$ as
$$\frac{\text{Sin}\,\theta - \text{Sin}\,\theta_i}{\theta - \theta_i} = \frac{\text{Sin}\,\theta_{i+1} - \text{Sin}\,\theta_i}{1}$$
$$\Rightarrow \theta - \theta_i = \frac{\text{Sin}\,\theta - \text{Sin}\,\theta_i}{\text{Sin}\,\theta_{i+1} - \text{Sin}\,\theta_i} \Rightarrow \delta\theta = \frac{\delta\text{Sin}\,\theta}{\Delta_i\text{Sin}\,\theta}.$$
3. Calculate θ from θ_i and $\delta\theta$ with $\theta = \theta_i + \delta\theta$.

It is worth noting that table authors are not as systematic in linearly interpolating between successive values as described above. Sometimes, certain (re)computational irregularities are easy to identify, e.g. choosing $\text{Sin}\,\theta_{i+2} - \text{Sin}\,\theta_{i+1}$ instead of $\text{Sin}\,\theta_{i+1} - \text{Sin}\,\theta_i$ in calculating $\delta\theta$. However, in other instances, table authors make intuitive choices like approximating the argument instead of interpolating it (for smaller values), making it difficult to explain an anomalous entry. My recomputations of the solar declinations attested in MS Tk admit to this level of uncertainty in a few instances.

3.3.1. *Worked example*

Calculating the solar declination δ corresponding to a celestial longitude λ of 52°:

[45] For example, see Bhāskara II's *Karaṇakutūhala* (1183 CE): II.8 (Rao and Uma, *Karaṇakutūhalam*, p. S19) or Nilakaṇṭha's *Tantrasaṅgraha* (1501 CE): II.7 (Ramasubramanian and Sriram, *Tantrasaṅgraha*, pp. 68–70).

1. For the celestial longitude $\lambda = 52°$, using the recomputed results $\text{Sin}\,52° = 47;16,50$ and $\text{Sin}\,24° = 24;24,15$ from Table VI.A,

$$\text{Sin}\,\delta(52°) = \text{Sin}\,52° \times \frac{\text{Sin}\,24°}{60} \approx 19;13,50,33.$$

2. To determine the arc $\delta(52°)$ corresponding to a Sine of 19;13,51 (rounded to seconds), observe from Table VI.A that $\text{Sin}\,18° \equiv 18;32,28 < \text{Sin}\,\delta(52°) < \text{Sin}\,19° \equiv 19;32,3$. Therefore,

$$\begin{aligned}\delta(52°) &= 18° + \frac{\text{Sin}\,\delta(52°) - \text{Sin}\,18°}{\text{Sin}\,19° - \text{Sin}\,18°}\\ &= 18° + \left[\frac{19;13,51 - 18;32,28}{19;32,3 - 18;32,28}\right]_{\text{in degrees}} = 18° + \left[\frac{0;41,23}{0;59,35}\right]_{\text{in degrees}}\\ &= 18° + 0°41'40'' \text{ (rounded to seconds)} \approx 18°41'40''\end{aligned}$$

The recomputed solar declination corresponding to a celestial longitude of 52° is 18°41′40″.

Table VI.B on page 234 presents the recomputed solar declinations for every degree of celestial longitude from 1° to 90°. Most of these recomputations follow the algorithm described above; however, a few entries are calculated irregularly as described below.

3.3.2. *Recomputational irregularities in solar declination calculations*

1. Recomputing the solar declination for a celestial longitude of 28°. For $\lambda = 28°$, $\text{Sin}\,\delta(28°) = \text{Sin}\,28° \times {}^{\text{Sin}\,24°}/_{60}$. With $\text{Sin}\,28° = 28;10,6$ and $\text{Sin}\,24° = 24;24,15$, $\text{Sin}\,\delta(28°) = 11.45707836 \approx 11;27,25$. A regular interval to determine the inverse arc of this Sine (by interpolation) is

$$\begin{aligned}&\text{Sin}\,11° < \text{Sin}\,\delta(28°) < \text{Sin}\,12°\\ \Rightarrow\ &11;26,55 < \text{Sin}\,\delta(28°) \approx 11;27,25 < 12;28,29,\end{aligned}$$

which gives $\delta \approx 11;0,30$ (rounded to seconds). However, the irregular interval

$$\begin{aligned}&\text{Sin}\,10° < \text{Sin}\,\delta(28°) < \text{Sin}\,12°\\ \Rightarrow\ &10;25,8 < \text{Sin}\,\delta(28°) \approx 11;27,25 < 12;28,29\end{aligned}$$

gives

$$\begin{aligned}\delta(28°) &= 10° + \frac{\text{Sin}\,\delta(28°) - \text{Sin}\,10°}{\text{Sin}\,12° - \text{Sin}\,11°}\\ &\approx 10° + 1°0'41''54''' \approx 11°0'41''54'''.\end{aligned}$$

The truncated value $\delta(28°) = 11°0'41''$ is identical to the attested value in MS Tk.

2. Recomputing the solar declinations for the celestial longitudes 2°, 7°, 12°, 15°, 18°, 37°, 43°, 45°, 49°, 53°, 55°, 61°, 64°, 72°, 74°, 78°, 80°, and 82°. The recomputed values match the attested values in MS Tk when the final results are truncated to seconds (instead of systematically rounding them to seconds), e.g.

δ (recomputed, up to thirds)		δ (attested in MS Tk)
$\delta(2°) = 0°48'47''44'''$	$\longleftrightarrow$	$\delta(2°) = 0°48'47''$
$\delta(18°) = 7°13'14''45'''$	$\longleftrightarrow$	$\delta(18°) = 7°13'14''$

The truncated versions of these recomputed values are not suggested as emendations; instead, the third row of differences between the attested and recomputed values (corresponding to these arguments) in Table VI.B registers them as a difference of '−1'.

3.4. Table of solar declinations (*krānti*): Analysis of differences

List of proposed emendations to the attested solar declinations in MS Tk:

Based on inadvertent copying oversights

1. $\delta(23°)_d$: 8 → 9. Suspected mistranscription. $\delta(22°)_d$, $\delta(23°)_d$, and $\delta(24°)_d$ appear in the sequence '8', '<u>8</u>', and '9' respectively.
2. $\delta(67°)_s$: 25 → 15. Suspected alteration of homoglyphic digits '1' and '2' in handwritten Devanāgarī.
3. $\delta(70°)_m$: 29 → 28. Suspected mistranscription. $\delta(68°)_m$, $\delta(69°)_m$, and $\delta(70°)_m$ appear in the sequence '9', '19', and '<u>29</u>' respectively.
4. $\delta(77°)_s$: 44 → 55. Suspected alteration of homoglyphic digits '4' and '5' in handwritten Devanāgarī.

Based on intentional interventions

5. $\delta(88°)_s$: 46 → 4. Suspected contamination. Adjacent entries $\delta(88°)$ and $\delta(89°)$ are both 23°59′46″. All six functions corresponding to the 88th and 89th arguments are identical in MS Tk. See note 6 on page 204.

Remarks on Table VI.B

The digits in the seconds place of the attested and recomputed solar declinations for several degrees of celestial longitudes vary by ±1. A few entries differ by up to ±4″, with one instance of a +5″ variation. I suspect these differences are a result of irregular arithmetic calculations, or selecting incorrect interpolation intervals. However, I have not been able to explain these differences mathematically (or justify them as interventions/oversights), and therefore, I do not emend the attested digits (in the seconds places) of the solar declinations corresponding to these longitudes in Table VI.B.

3.5. Tables of shadow lengths (*śaṅkuchāyā*): Recomputation strategy

The tables of lengths of shadows (*chāyā*) of gnomons (*śaṅku*) of the *Amṛtalaharī* (in MS Tk) are computed for every degree of solar altitude (*lambaka*)[46] from 1° to 90° for gnomons of heights 60 digits, 12 digits, and 7 digits.[47] The shadow length (*śaṅkuchāyā*) of a gnomon of height h digits, hereafter abbreviated as $Ch\bar{a}y\bar{a}_h$, is related to the solar altitude a with the expression

$$Ch\bar{a}y\bar{a}_h\, a = h \times \frac{\operatorname{Cos} a}{\operatorname{Sin} a},$$

where $a \equiv$ solar altitude and $h \equiv$ gnomon height (in digits). A simple geometric derivation for this expression (for a 12 digit gnomon, a typical measure in Indian astronomy) is described in Ramasubramanian and Sriram, *Tantrasaṅgraha*, p. 135. Another way to interpret the shadow length is to consider the argument as the terrestrial latitude φ of an observer. The tabulated shadow lengths then represent the length of the equinoctial noon shadow cast by the gnomon (of a particular height h). On the day of the equinox, the declination of the Sun is zero and hence the diurnal path of the Sun (almost) traces the celestial equator in the sky. At midday on this day, the local zenith crossing of the Sun corresponds to the local terrestrial latitude (measured from the local zenith). Thus, the equinoctial noon shadow of the gnomon (*viṣuvatchāyā*) can be expressed as a function of the local terrestrial latitude, i.e. $h \times {}^{\operatorname{Sin}\varphi}/_{\operatorname{Cos}\varphi}$.[48] Several Sanskrit texts, beginning from very early times, describe how the shadow lengths of gnomons (for known heights) are computed, e.g. Kauṭilīya's *Arthaśāstra* (2nd century–3rd century CE) or Āryabhaṭa's *Āryabhaṭīya* (*c.* 499 CE).[49]

3.5.1. *Worked example*

Calculating the shadow length $Ch\bar{a}y\bar{a}_h$ corresponding to a solar altitude a of 52° for gnomons of heights $h = 60$ digits, 12 digits, and 7 digits:

1. For the solar altitude $a = 52°$ and gnomon height $h = 60$, $\operatorname{Sin} a \equiv \operatorname{Sin} 52° \approx 47;16,50$ and $\operatorname{Cos} a \equiv \operatorname{Cos} 52° = \operatorname{Sin}(38°) \approx 36;56,23$ (using recomputed Sines from Table VI.A). Thus,

$$Ch\bar{a}y\bar{a}_{60}\, 52° = h \times \frac{\operatorname{Cos} 52°}{\operatorname{Sin} 52°} \equiv 60 \times \frac{36;56,23}{47;16,50} \approx 46;52,38$$

[46] The solar altitude (*lambaka*) (above the horizon) is the complement of the zenith distance (*natāṃśa*) of the Sun.

[47] A digit or *aṅgula* is a unit of linear measure of a finger breadth, approximately, $(1/24)^{th}$ part of a cubit (*hasta*).

[48] Ramasubramanian and Sriram, *Tantrasaṅgraha*, p. 140.

[49] See, respectively, Abraham, 'The Gnomon' and Shukla and Sarma, *Āryabhaṭīya*.

(rounded to seconds). The recomputed shadow length for a gnomon of height 60 digits and corresponding to a solar altitude of 52° is 46;52,38.

2. Similarly, for $h = 12$ and $h = 7$: *Chāyā*$_{12}$ $52^\circ \equiv 12 \times \frac{36;56,23}{47;16,50} \approx 9;22,32$ and *Chāyā*$_7$ $52^\circ \equiv 7 \times \frac{36;56,23}{47;16,50} \approx 5;28,8$. The recomputed shadow lengths for gnomons of height 12 digits and 7 digits corresponding to a solar altitude of 52° are 9;22,32 and 5;28,8 respectively. Both values are rounded to the second fractional place.

Tables VI.C$_1$ (page 235), VI.C$_2$ (page 236), and VI.C$_3$ (page 237) present the recomputed shadow lengths for every degree of solar altitude from 1° to 90° for gnomon lengths 60 digits, 12 digits, and 7 digits respectively. Most of these recomputations follow the algorithm described above; however, a few entries are calculated irregularly as described below.

3.5.2. *Recomputational irregularities in shadow-length calculations: 60-digit gnomon (Table VI.C$_1$)*

1. Recomputing the shadow length for a solar altitude of 14°. Using the attested value Sin 14° = 14;30,56 from MS Tk (see Table VI.A) gives *Chāyā*$_{60}$ $(14^\circ) = 60 \times \frac{\text{Cos } 14^\circ}{\text{Sin } 14^\circ} = 60 \times \frac{58;13,4}{14;30,56} \approx 60 \times 4;0,38,34,45 \approx$ 240;38,34,45 (truncated to seconds). This value is identical to the attested value in MS Tk. The shadow length value with the recomputed Sin 14° as 14;30,55 is 240;38,51 (rounded to seconds).
2. The recomputations of the shadow lengths for the following arguments agree with their attested values in MS Tk if irregular Sine (or Cosine) values are considered. These recomputational scenarios are seemingly random; nevertheless, I list them below for completeness. The attested or recomputed Sines stated below can be found in Table VI.A.
   (a) With the recomputed Sin 11° = 11;26,55 and an *arbitrary* Cos 11° = Sin 79° = 58;53,53, *Chāyā*$_{60}$ (11°) ≈ 308;40,25 (rounded to seconds), which agrees with the attested value in MS Tk. The shadow length with the attested/recomputed Cos 11° as 58;53,51 is 308;40,14 (rounded to seconds).
   (b) With the *attested* Sin 21° = 21;30,8 and the recomputed Cos 21° = Sin 69° = 56;0,53, *Chāyā*$_{60}$ (21°) ≈ 156;18,14 (rounded to seconds), which agrees with the attested value in MS Tk. The shadow length with the recomputed Sin 21° as 21;30,6,59 is 156;18,22 (rounded to seconds).
   (c) With an *arbitrary* Sin 42° = 40;8,50 and the recomputed Cos 42° = Sin 48° = 44;35,19, *Chāyā*$_{60}$ (42°) ≈ 66;38,16 (rounded to sec-

onds), which agrees with the attested value in MS Tk. The shadow length with the attested/recomputed Sin 42° as 40;8,52 is 66;38,12 (rounded to seconds).

(d) With an *arbitrary* Sin 43° = 40;55,16 and the recomputed Cos 43° = Sin 47° = 43;52,52, $\textit{Chāyā}_{60}(43°) \approx 64;20,24$ (rounded to seconds), which agrees with the attested value in MS Tk. The shadow length with the attested/recomputed Sin 43° as 40;55,12 is 64;20,30 (rounded to seconds).

(e) With an *arbitrary* Sin 50° = 45;57,47 and the recomputed Cos 50° = Sin 40° = 38;34 (rounded to minutes), $\textit{Chāyā}_{60}(50°) \approx 50;20,41$ (rounded to seconds), which agrees with the attested value in MS Tk. The shadow length with the attested Sin 50° as 46;56,46 and the recomputed Cos 50° as 38;34,2 (up to the seconds) is 49;17,29 (rounded to seconds), whereas the shadow length with the recomputed Sin 50° as 45;57,46,0 and the recomputed Cos 50° as 38;34,2 (up to the seconds) is 50;20,45 (rounded to seconds).

3. Recomputing the shadow lengths of a 60-digit gnomon for the solar altitudes 3°, 5°, 23°, 27°, 40°, 41°, 49°, 51°, 59°, 61°, 64°, 68°, 80°, 83°, 86°, and 87°. The recomputed values match the attested values in MS Tk when the final results are truncated to seconds instead of systematically rounding them to seconds), e.g.

$\textit{Chāyā}_{60}$ (recomputed to thirds)		$\textit{Chāyā}_{60}$ (attested value)
$\textit{Chāyā}_{60}(3°) = 1114;49,27,53$	$\longleftrightarrow$	$\textit{Chāyā}_{60}(3°) = 1114;49,27$
$\textit{Chāyā}_{60}(49°) = 52;9,26,35$	$\longleftrightarrow$	$\textit{Chāyā}_{60}(49°) = 52;9,26$

The truncated versions of these recomputed values are not suggested as emendations; instead, the third row of differences between the attested and recomputed values (corresponding to these arguments) in Table VI.C_1 registers them as a difference of '−1'.

3.5.3. *Recomputational irregularities in shadow length calculations: 12-digit gnomon (Table VI.C_2)*

1. Recomputing the attested shadow length for a solar altitude of 4°. The shadow-lengths of gnomons of heights 60 and 12 digits are related by $\textit{Chāyā}_{12} = \frac{1}{5}\,\textit{Chāyā}_{60}$. For a solar altitude of 4°, using the attested value of $\textit{Chāyā}_{60}(4°) = 859;3,48$ from MS Tk (see Table VI.A) gives $\textit{Chāyā}_{12}(4°) = \frac{859;3,48}{5} \approx 171;48,45,36 \approx 171;48,46$ (rounded to seconds). This value agrees with the attested value 172; 48,46 in MS Tk

if the digits '172' in the units place are considered a copying oversight for '171'. (The digits '1' and '2' are homoglyphic in handwritten Devanāgarī.) Using the recomputed value $Ch\bar{a}y\bar{a}_{60}(4^\circ)$ as 858;3,48, $Ch\bar{a}y\bar{a}_{12}(4^\circ) \approx 171;36,46$ (rounded to seconds).

2. Recomputing the shadow length for a solar altitude of 24°. With the *arbitrary* shadow length $Ch\bar{a}y\bar{a}_{60}(24^\circ) = 134;45,5$, $Ch\bar{a}y\bar{a}_{12}(24^\circ) = \frac{1}{5} \times 134;45,5 = 26;57,1$, which agrees with the attested value in MS Tk. The shadow length $Ch\bar{a}y\bar{a}_{12}(24^\circ)$ with the attested/recomputed $Ch\bar{a}y\bar{a}_{60}(24^\circ)$ as 134;45,45 is 26;57,9.
3. Recomputing the shadow length for a solar altitude of 87°. The regular expression for $Ch\bar{a}y\bar{a}_{12}\, 87^\circ$ is $12 \times \frac{\text{Cos } 87^\circ}{\text{Sin } 87^\circ}$. However, using $\text{Cos } 88^\circ = \text{Sin } 2^\circ = 2;0,38$ (the attested value in MS Tk) instead of $\text{Cos}(87^\circ)$ gives $Ch\bar{a}y\bar{a}_{12}(87^\circ) = 12 \times \frac{\text{Cos } 88^\circ}{\text{Sin } 87^\circ} = 12 \times \frac{2;0,38}{59;55,4} = 0;24,9,35 \approx 0;24,9$ (truncated to seconds), which agrees with the attested value in MS Tk. A regular recomputation of $Ch\bar{a}y\bar{a}_{12}(87^\circ)$ (using recomputed Sin 87° and $\text{Cos } 87^\circ = \text{Sin } 3^\circ$) gives 0;37,44 (rounded to seconds).
4. Recomputing the shadow lengths of a 12-digit gnomon for the solar altitudes 3°, 17°, 20°, 39°, 46°, 47°, 56°, 58°, 65°, and 86°. The recomputed values match the attested values in MS Tk when final results are truncated to seconds (instead of systematically rounding them to seconds), e.g.

$Ch\bar{a}y\bar{a}_{12}$ (recomputed to thirds)		$Ch\bar{a}y\bar{a}_{12}$ (attested value)
$Ch\bar{a}y\bar{a}_{12}(3^\circ) = 228;57,53,34$	$\longleftrightarrow$	$Ch\bar{a}y\bar{a}_{12}(3^\circ) = 228;57,53$
$Ch\bar{a}y\bar{a}_{12}(39^\circ) = 14;49,7,47$	$\longleftrightarrow$	$Ch\bar{a}y\bar{a}_{12}(39^\circ) = 14;49,7$

The truncated versions of these recomputed values are not suggested as emendations; instead, the third row of differences between the attested and recomputed values (corresponding to these arguments) in Table VI.C_2 registers them as a difference of '−1'.

3.5.4. *Recomputational irregularities in shadow length calculations: 7 digits (Table VI.C_3)*

1. Recomputing the shadow length for a solar altitude of 4°. The shadow-lengths of gnomons of heights 60 and 7 digits are related by $Ch\bar{a}y\bar{a}_7 = \frac{7}{60}\, Ch\bar{a}y\bar{a}_{60}$. For a solar altitude of 4°, using the attested value $Ch\bar{a}y\bar{a}_{60}(4^\circ) = 859;3,48$ from MS Tk (see Table VI.A) gives $Ch\bar{a}y\bar{a}_7(4^\circ) = \frac{7 \times 859;3,48}{60} \approx 100;13,26,36 \approx 100;13,27$ (rounded to seconds). This value agrees with the attested value in MS Tk. Using the re-

computed value *Chāyā*$_{60}$ (4°) as 858;3,48 gives *Chāyā*$_7$ (4°) ≈ 100;6,27 (rounded to seconds).

2. Recomputing the shadow length for a solar altitude of 12°. For an *arbitrary* value of Sin 12° = 12;28,30, *Chāyā*$_7$ (12°) = 32;55,54, which agrees with the attested value in MS Tk. The recomputed Sin 12° as 12;28,28,55 gives 32;55,57 (rounded to seconds).
3. Recomputing the shadow length for a solar altitude of 37°. With the recomputed Sin 37° = 36;6,32 and an *arbitrary* Cos 37° = Sin 53° = 47;56,50, *Chāyā*$_7$ (37°) ≈ 9;17,42 (rounded to seconds), which agrees with the attested value in MS Tk. The shadow length with the recomputed Sin 37° as 36;6,32 and the attested Cos 37° as 47;56,5 is 9;17,33 (rounded to seconds), whereas the shadow length with the recomputed Sin 37° as 36;6,32 and the recomputed Cos 37° as 47;55,5 is 9;17,21 (rounded to seconds).
4. Recomputing the shadow lengths of a 7-digit gnomon for the solar altitudes 39°, 42°, and 59°. The recomputed values match the attested values in MS Tk when final results are truncated to seconds (instead of systematically rounding them to seconds), e.g.

Chāyā$_7$ (recomputed to thirds)		*Chāyā*$_7$ (attested value)
Chāyā$_7$ (39°) = 8;38,39,40	⟷	*Chāyā*$_7$ (39°) = 8;38,39
Chāyā$_7$ (42°) = 7;46,27,48	⟷	*Chāyā*$_7$ (42°) = 7;46,27

The truncated versions of these recomputed values are not suggested as emendations; instead, the third row of differences between the attested and recomputed values (corresponding to these arguments) in Table VI.C3 registers them as a difference of '−1'.

3.6. Table of shadow lengths (*śaṅkuchāyā*): Analysis of differences

In the following subsections, I present a list of proposed emendations to the attested values of shadow lengths for gnomons of heights 60 digits, 12 digits, and 7 digits respectively.

3.6.1. *Shadow length for gnomon of height 60 digits (Table VI.C$_1$)*

Based on inadvertent copying oversights

1. *Chāyā*$_{60}$ (10°)$_s$: 24 → 34. Suspected alteration of homoglyphic digits '2' and '3' in handwritten Devanāgarī.
2. *Chāyā*$_{60}$ (12°)$_s$: 19 → 39. Suspected alteration of homoglyphic digits '1' and '3' in handwritten Devanāgarī.

3. *Chāyā*$_{60}$ $(67^\circ)_s$: 0 $\rightarrow$ 7. Suspected alteration of homoglyphic digits '0' and '7' in handwritten Devanāgarī.

Based on intentional interventions

5. *Chāyā*$_{60}$ (88°): 1;2,50 $\rightarrow$ 2;5,43. Suspected contamination. Adjacent entries *Chāyā*$_{60}$ (88°) and *Chāyā*$_{60}$ (89°) are both 1;2,50. All six functions corresponding to the 88th and 89th arguments are identical in MS Tk. See note 6 on page 204.

Remarks on Table VI.C$_1$

1. The attested entry '859' for *Chāyā*$_{60}$ $(4^\circ)_u$ could be emended to '858' as a suspected mistranscription by a table author (or scribe). This emendation would agree with the recomputed result, and also avoid the difference of 1 integer unit between the attested and recomputed entries (a significant statistical anomaly). However, the attested shadow lengths of the 60-digit and 12-digit gnomons corresponding to 4° of solar altitude in MS Tk are computationally interrelated. The irregular recomputation of *Chāyā*$_{12}$ (4°) uses 859;3,48 as the attested value of *Chāyā*$_{60}$ (4°) (see note 1 in Section 3.5.3).
2. On f. 49v of MS Tk, the digit '0' (of the number 30) in *Chāyā*$_{60}$ $(23^\circ)_s$ had a dot under it: ३०̣, 30̣. An underdot is sometimes used as a *signe de renvoi* (cancellation mark) in Sanskrit, and the recomputational evidence also suggests *Chāyā* $(23^\circ)_s = 3$. Hence, I record the value of *Chāyā*$_{60}$ $(23^\circ)_s$ as 3 in my transcription.
3. The digits in the seconds place of the attested and recomputed shadow lengths of a 60-digit gnomon for certain degrees of solar altitudes (e.g. 38°, 56°, or 65°) vary by up to ± 3. I have not been able to justify these differences mathematically (or as obvious interventions/oversights), and therefore, I do not propose any emendations in Table VI.C$_1$ to change the attested digits (in the seconds place) of the shadow lengths corresponding to these arguments.

3.6.2. *Shadow length for gnomon of height 12 digits (Table VI.C$_2$)*

Based on inadvertent copying oversights

1. *Chāyā*$_{12}$ $(4^\circ)_u$: 172 $\rightarrow$ 171. Suspected alteration of homoglyphic digits '1' and '2' in handwritten Devanāgarī.
2. *Chāyā*$_{12}$ $(19^\circ)_m$: 1 $\rightarrow$ 51. Suspected mistranscription. *Chāyā*$_{12}$ $(18^\circ)_m$, *Chāyā*$_{12}$ $(19^\circ)_m$, and *Chāyā*$_{12}$ $(20^\circ)_m$ appear in the sequence '55', '1', and '58' respectively.

3. *Chāyā* $_{12}$ $(23°)_s$: 23 → 13. Suspected alteration of homoglyphic digits '1' and '2' in handwritten Devanāgarī.

4. *Chāyā* $_{12}$ $(25°)_s$: 12 → 2. Suspected mistranscription. *Chāyā* $_{12}$ $(24°)_s$, *Chāyā* $_{12}$ $(25°)_s$ and *Chāyā* $_{12}$ $(26°)_s$ appear in the sequence '1', '<u>12</u>', and '14' respectively.

5. *Chāyā* $_{12}$ $(33°)_m$: 48 → 28. Suspected mistranscription. *Chāyā* $_{12}$ $(32°)_m$, *Chāyā* $_{12}$ $(33°)_m$ and *Chāyā* $_{12}$ $(34°)_m$ appear in the sequence '12', '<u>48</u>', and '47' respectively.

6. *Chāyā* $_{12}$ $(43°)_m$: 55 → 52. Suspected mistranscription. *Chāyā* $_{12}$ $(42°)_m$, *Chāyā* $_{12}$ 43°, and *Chāyā* $(44°)_m$ appear in the sequence '19', '<u>55</u>', and '25' respectively.

7. *Chāyā* $_{12}$ $(53°)_s$: 34 → 14. Suspected alteration of homoglyphic digits '1' and '3' in handwritten Devanāgarī.

8. *Chāyā* $_{12}$ $(68°)_s$: 4 → 54. Suspected mistranscription (perhaps, an inadvertent omission of the digit '5' in '54').

9. *Chāyā* $_{12}$ $(86°)_s$: 29 → 21. Suspected alteration of homoglyphic digits '1' and '9' in handwritten Devanāgarī.

10. *Chāyā* $_{12}$ $89°_m$: 32 → 12 and *Chāyā* $_{12}$ $89°_s$: 24 → 34. Suspected alteration of homoglyphic digits '1', '2', and '3' in handwritten Devanāgarī.

Based on intentional interventions

11. *Chāyā* $_{12}$ (88°) : 0;32,24 → 0;25,9. Suspected contamination. Adjacent entries *Chāyā* $_{12}$ (88°) and *Chāyā* $_{12}$ (89°) are both 0;32,24. All six functions corresponding to the 88th and 89th arguments are identical in MS Tk. See note 6 on page 204.

Remarks on Table VI.C₂

The digits in the seconds place of the attested and recomputed shadow lengths of a 12-digit gnomon for several degrees of solar altitudes vary by ±1. For a few entries, the values differ by up to $+3''$ or $-4''$. I suspect these differences are a result of irregular sexagesimal divisions. However, I have not been able to justify these differences mathematically (or observe inadvertent or intentional scribal discrepancies). Therefore, I present the attested digits (in the seconds place) of the shadow lengths corresponding to these arguments in Table VI.C_2 without suggesting any emendations.

3.6.3. *Shadow length for gnomon of height 7 digits (Table VI.C$_3$)*

Based on inadvertent copying oversights

1. *Chāyā*$_7$ $(1°)_u$: 410 → 400. Suspected mistranscription. Also, the digits '0' and '1 can (sometimes) appear homoglyphic in handwritten Devanāgarī suggesting a possible unwitting alteration.
2. *Chāyā*$_7$ $(67°)_m$: 59 → 58. Suspected mistranscription. *Chāyā*$_7$ $(67°)_m$ and *Chāyā*$_7$ $(68°)_m$ appear in the sequence '<u>59</u>' and '49' respectively. Also, the digits '8' and '9' can (sometimes) appear homoglyphic in handwritten Devanāgarī suggesting a possible alteration.
3. *Chāyā*$_7$ $(81°)_u$: 16 → 6. Suspected mistranscription. *Chāyā*$_7$ $(80°)_u$ and *Chāyā*$_7$ $(81°)_u$ appear in the sequence '14' and '<u>16</u>' respectively.
4. *Chāyā*$_7$ $(82°)_u$: 34 → 2. Suspected mistranscription. *Chāyā*$_7$ $(82°)$, *Chāyā*$_7$ $(83°)$, and *Chāyā*$_7$ $(84°)$ appear in the sequence '34', '<u>34</u>', and '34' respectively.
5. *Chāyā*$_7$ $(84°)_u$: 34 → 9. Suspected mistranscription. *Chāyā*$_7$ $(82°)$, *Chāyā*$_7$ $(83°)$, and *Chāyā*$_7$ $(84°)$ appear in the sequence '34', '34', and '<u>34</u>' respectively.
6. *Chāyā*$_7$ $(89°)_s$: 10 → 20. Suspected alteration of homoglyphic digits '1' and '2' in handwritten Devanāgarī.

Based on intentional interventions

7. *Chāyā*$_7$ $(87°)$: 0;14,40 → 0;22,1. Suspected contamination. The recomputed value of *Chāyā*$_7$ $(88°)$ is 0;14,40; this value appears under the 87th argument as a dislocated or displaced entry (perhaps, to replace a corrupted/illegible/missing entry; however, this could also be an unintentional mistranscription).
8. *Chāyā*$_7$ $(88°)$: 0;7,20 → 0;14,40. Suspected contamination. Adjacent entries *Chāyā*$_7$ $(88°)$ and *Chāyā*$_7$ $(89°)$ are both 0;7,20. All six functions corresponding to the 88th and 89th arguments are identical in MS Tk. See note 6 on page 204.

Remarks on Table VI.C$_3$

The digits in the seconds place of the attested and recomputed shadow lengths of a 7-digit gnomon for the solar altitudes of 18°, 25°, 34°, 38°, 43°, 51°, and 77° vary by +1. Without any mathematical justification for these differences (or any evidence to suggest scribal interventions/oversights), I leave digits (in the seconds place) of these shadow lengths in Table VI.C$_3$ unemended.

3.7. Table of lunar latitudes (*śara*): Recomputation strategy

The table of lunar latitude (*śara*) of the *Amṛtalaharī* (in MS Tk) is computed for every degree of the lunar-nodal elongation[50] from 1° to 90° and has a maximum value (equal to the inclination i of the lunar orbit) of 4°30′. The lunar latitude β is related to the lunar-nodal elongation (also known as the argument of lunar latitude) ω with the expression

$$\operatorname{Sin}\beta = \operatorname{Sin} i \times \frac{\operatorname{Sin}\omega}{\mathcal{R}} \equiv \operatorname{Sin} 4^\circ 30' \times \frac{\operatorname{Sin}\omega}{60} \quad \because\ i = 4^\circ 30' \text{ and } \mathcal{R} = 60.$$

1. Most Sanskrit *siddhāntas* approximate the lunar latitude β as $4;30 \times {}^{\operatorname{Sin}\omega}/_{\mathcal{R}}$ (in degrees), e.g. Lalla's *Śiṣyadhīvṛddhidatantra* (*c.* early 9th century): V.11.[51] However, MS Tk uses the exact form of the expression to calculate the lunar latitude.[52] Appendix C.3 includes a statistical analysis of the differences between the attested lunar latitudes (from MS Tk) and the recomputed results when the approximate expression ($4;30 \times \operatorname{Sin}\omega/\mathcal{R}$) or the exact equation ($\operatorname{Sin} 4^\circ 30' \times \operatorname{Sin}\omega/\mathcal{R}$) are used separately.
2. The value of the parameter Sin 4°30′ can be calculated in two different ways:
   (a) by linear interpolation using the recomputed values of Sin 4° and Sin 5° as 4;11,17 and 5;13,46 (from Table VI.A) respectively, or
   (b) by using the formula for the Sine of half the arc for an arc of 9° and the recomputed value Cos 9° = Sin 81° = 59;15,41 (from Table VI.A).

   The method of linear interpolation gives 4;42,26,29,59 (with all subsequent fractions greater than 30), or 4;42,27 (successively rounded to seconds). Using the trigonometric formula gives 4;42,26,8,59, or approximately 4;42,26 (rounded to seconds). My recomputations, however, indicate that the lunar latitude calculations in MS Tk use Sin 4°30′ = 4;42,25. I select the value 4;42,25 by statistically testing the differences between the attested values in MS Tk and my recomputed results (using all three values of the parameter Sin 4°30′ separately) to find the parametric estimate that minimises these differences, see Appendix C.4.

[50] The lunar-nodal elongation is the difference between the celestial longitude of the orbital lunar node (☊ or ☋) and the orb of the Moon, i.e. $\omega = \lambda_{\text{Moon}} - \lambda_{☊\text{ or }☋}$. The lunar-nodal elongation ranges from 0° to ±180° depending on the position of the Moon (along its orbit) and the lunar node.

[51] Chatterjee, *Śiṣyadhīvṛddhida Tantra*, pp. 113–14, includes a derivation of Lalla's method to compute the lunar latitude using the approximate expression.

[52] The maximum value of β (at $\omega = 90^\circ$) is equal to the inclination of the lunar orbit, i.e. 4°30′. As $\operatorname{Sin} 4^\circ 30' \approx 4;30$, most Sanskrit texts take $\operatorname{Sin}\beta \approx \beta$ for all $0^\circ \leq \beta \leq 4^\circ 30'$.

The method of determining the lunar latitude (from its Sine) is similar to that of the solar declination. Having calculated the Sine of the lunar latitude, the corresponding latitude (in degrees) is determined by finding the inverse arc of Sine. See Section 3.3 for the algorithm to inversely interpolate the measure of arc corresponding to a particular Sine.

3.7.1. *Worked example*

Calculating the lunar latitude β corresponding to a lunar-nodal elongation ω of 52°:

1. For a lunar-nodal elongation $\omega = 52^\circ$, using the recomputed $\mathrm{Sin}\,52^\circ = 47;16,50$ from Table VI.A, $\mathrm{Sin}\,\beta(52^\circ) = 4;42,25 \times {}^{\mathrm{Sin}\,52^\circ}/_{60} \approx 3;42,33$ (rounded to seconds).
2. To determine the lunar latitude $\beta(52^\circ)$ corresponding to a Sine of 3;42,33, observe from Table VI.A that $\mathrm{Sin}\,3^\circ \equiv 3;8,25 < \mathrm{Sin}\,\beta(52^\circ) < \mathrm{Sin}\,4^\circ \equiv 4;11,7$. Therefore,

$$\begin{aligned}\beta(52^\circ) &= 3^\circ + \frac{\mathrm{Sin}\,\beta(52^\circ) - \mathrm{Sin}\,3^\circ}{\mathrm{Sin}\,4^\circ - \mathrm{Sin}\,3^\circ}\\ &= 3^\circ + \left[\frac{3;42,33 - 3;8,25}{4;11,7 - 3;8,25}\right]_{\text{in degrees}} = 3^\circ + \left[\frac{0;34,9}{1;2,42}\right]_{\text{in degrees}}\\ &= 3^\circ + 0^\circ 32' 40'' \text{ (rounded to seconds)} \approx 3^\circ 32' 40''.\end{aligned}$$

   The recomputed lunar latitude corresponding to a lunar-nodal elongation of 52° is 3°32′40″.

Table VI.D on page 238 presents the recomputed lunar latitudes for every degree of lunar-nodal elongation from 1° to 90°. Most of these recomputations follow the algorithm described above; however, a few entries are calculated irregularly as described below.

3.7.2. *Recomputational irregularities in lunar latitude calculations*

1. Recomputing the lunar latitude β for a lunar-nodal elongation of $\omega = 90^\circ$. For $\omega = 90^\circ$, $\mathrm{Sin}\,\beta(90^\circ) = \mathrm{Sin}\,4^\circ 30'$ as $\mathrm{Sin}\,\omega = \mathrm{Sin}\,90^\circ = \mathcal{R}$. Hence, $\beta(90^\circ)$ is simply 4°30′ (the inclination of the lunar orbit). Alternatively, with $\mathrm{Sin}\,(4^\circ 30') \approx 4;42,25$, $\beta(90^\circ) \equiv \mathrm{arcSin}\,(4;42,25) \approx 4^\circ 29' 59''$ (rounded to seconds). This value is inversely interpolated using the recomputed values $\mathrm{Sin}\,4^\circ = 4;11,7$ and $\mathrm{Sin}\,5^\circ = 5;13,46$ from Table VI.A. The attested value of 4;30′ in MS Tk agrees with this interpolated value (rounded to minutes).
2. Recomputing the lunar latitudes for the lunar-nodal elongations 4°, 7°, 24°, 25°, 26°, 42°, 48°, 50°, 62°, and 79°. The recomputed values

match the attested values in MS Tk when the final results are truncated to seconds (instead of systematically rounding them to seconds), e.g.

β (recomputed, up to thirds)		β (attested in MS Tk)
$\beta(4^\circ) = 0^\circ 18' 48'' 42'''$	$\longleftrightarrow$	$\beta(4^\circ) = 0^\circ 18' 48''$
$\beta(24^\circ) = 1^\circ 49' 42'' 48'''$	$\longleftrightarrow$	$\beta(24^\circ) = 1^\circ 49' 42''$

The truncated versions of these recomputed values are not suggested as emendations; instead, the third row of differences between the attested and recomputed values (corresponding to these arguments) in Table VI.D registers them as a difference of '-1'.

3.8. Table of lunar latitudes (*śara*): Analysis of differences

List of proposed emendations to the attested lunar latitudes in MS Tk:

Based on inadvertent copying oversights

1. $\beta(12^\circ)_d$: $1 \rightarrow 0$. Suspected mistranscription. $\beta(10^\circ)_d$, $\beta(11^\circ)_d$, $\beta(12^\circ)_d$, and $\beta(13^\circ)_d$ appear in the sequence '0', '0', '$\underline{1}$', and '1' respectively.
2. $\beta(44^\circ)_s$: $36 \rightarrow 26$. Suspected alteration of homoglyphic digits '2' and '3' in handwritten Devanāgarī.

Based on intentional interventions

3. $\beta(85^\circ)_s$: $20 \rightarrow 57$. Suspected contamination. The recomputed value of $\beta(86^\circ)_s$ is '19'; the number '20' ($\sim$ '19' at the level of arithmetical noise) appears under the 85th argument as a dislocated or displaced entry, perhaps, to replace a corrupted/illegible/missing entry. However, this could also be an unintentional mistranscription by a scribe/table author.
4. $\beta(86^\circ)_s$: $37 \rightarrow 20$. Suspected contamination. The recomputed value of $\beta(87^\circ)_s$ is '37'; this value appears under the 86th argument as a dislocated or displaced entry (perhaps, to replace a corrupted/illegible/missing entry or a perpetuated mistranscription).
5. $\beta(87^\circ)_s$: $50 \rightarrow 37$ Suspected contamination. The recomputed value of $\beta(88^\circ)_s$ is '50'; the number '49' ($\sim$ '50' at the level of arithmetical noise) appears under the 87th argument as a dislocated or displaced entry (perhaps, to replace a corrupted/illegible/missing entry or a perpetuated mistranscription).
6. $\beta(88^\circ)_s$: $57 \rightarrow 50$. Suspected contamination. Adjacent entries $\beta(88^\circ)_s$ and $\beta(89^\circ)_s$ are '57'. All six functions corresponding to the 88th and 89th arguments are identical in MS Tk. See note 6 on page 204.

Remarks on Table VI.D

1. The lunar latitudes for 57° to 60° of lunar-nodal elongation are illegible in the minutes and seconds places in MS Tk. I represent these illegible entries, the differences between the corresponding sexagesimal digits, and their proposed emendations as '[--]' in Table VI.D.
2. The attested and recomputed lunar latitudes for 70°, 71°, 74°, 75°, 76°, 77°, 81°, and 82° of lunar-nodal elongation differ by $\pm 1'$. My recomputations (including irregular ones) have been unsuccessful in removing this difference, and there are no discernible copying mistakes or scribal corrections in any of these instances. Therefore, I present the attested digits (in the minutes place) of these lunar latitudes in Table VI.D without suggesting any emendations.
3. Also, the digits in the seconds place of the attested and recomputed lunar latitudes for several degrees of lunar-nodal elongations vary by up to ± 3. I suspect these differences are a result of irregular arithmetic calculations or selecting incorrect interpolation intervals. However, I have not been able to explain these differences mathematically (or justify them as interventions/oversights), and therefore, I do not emend the attested digits (in the seconds place) for these arguments in Table VI.D.

4. Conclusion and Discussion

In this study, I recomputed a selection of six tables from Nityānanda's *Amṛtalaharī* to understand the algorithms, the irregularities, and the interdependencies that capture the mathematics of these tables. I also analysed the differences between the attested values (in a single witness MS Tk) and my recomputed results to identify plausible scribal discrepancies (inadvertent copying oversights or intentional interventions), which then allowed to propose a few emendations to the attested values. The process of recomputing attested tables not only reveals the subtle mathematical decisions that table authors make as they recalculate or rectify entries, but also indicates patterns of errors and oversights that get transmitted as the tables are recopied over time. This study brings to light the challenges in applying this process when working with a single manuscript witness. I summarise below the main observations of my study, and the ensuing questions they pose as we begin to build modern digital tools to understand better the historical process of computing astronomical tables.

1. The attested values corresponding to the 88th and 89th arguments are identical for all six functions tabulated on the manuscript. The digital surrogates (of ff. 49v–50v) of MS Tk show faint vertical rules separating thirty columns of arguments on each folio, with corresponding six sets

of functions vertically stacked below them and mutually separated by horizontal rules. This formatted (grid-like) presentation of the six tables on MS Tk suggests that a professional scribe could have copied the entries from a parent manuscript, column by column, and while doing so, inadvertently duplicated all six sets of values for the 88th and 89th arguments as they populated the grid.

However, there are other instances where individual digits (in the sexagesimal places of the value of a function) appear to be shifted horizontally into adjacent cells, e.g. the leftwards displacement of the digits (in the seconds place) for lunar latitudes corresponding to the 86th, 87th, and 88th arguments. These horizontal shifts suggest that the tables (or certain parts of the tables) were perhaps copied cell by cell along each row. Certain mathematical aspects of a function (e.g. monotonicity) become evident when copying the values progressively, and hence, table authors may have found it intuitive to copy the sexagesimal digits (of the value of a function) row-wise. The various patterns of computational irregularities or scribal discrepancies noted in this study suggest different directions in which the tables were possibly copied. The extent to which anomalous entries can expose the direction of copying, and perhaps, the intention of the copyist themselves, is a challenging question that requires more advanced methods of analysis applied to larger selections of tables from a manuscript.

2. While the identical sets of values for the 88th and 89th arguments on f. 50v of MS Tk could be the result of an inadvertent copying oversight, it is just as likely the result of an intentional change. At some point in the transmission of the tables, a diligent scribe (or a table author) may have simply copied the six sets of values for the 89th argument into the column of the 88th argument to rectify a corrupted, illegible, or missing column in a parent manuscript (perhaps, treating the small differences between these values to be mathematically insignificant). These speculations indicate how inadvertent or intentional choices of successive historical actors (scribes or table authors) modify a particular table, and separate each subsequent copy from the previous one (and the original) by an added degree of uncertainty.

3. In this study, there are some cases where irregular recomputations eliminate the differences between the attested values (in MS Tk) and my recomputed results. In other instances, inadvertent or intentional scribal changes are evident enough to justify emending the attested values, and by doing so, reduce or remove the differences. Nevertheless, there are still several (small) differences between the attested and recomputed val-

ues in every table that cannot be justified as anomalous calculations or scribal discrepancies. Perhaps, in some measure, these differences are the result of historical actors making tacit decisions *ad libitum*. Most historical recomputations of astronomical tables, including those presented here, admit to this level of residual noise.

4. In my study of the selected corpus, I found a single instance where an attested Sine from MS Tk (different from my recomputed Sine) reproduces an attested value (of another function) identically and exclusively. With $\text{Cos}\,88° = \text{Sin}\,2° = 2;0,38$, the recomputed shadow length of a 12-digit gnomon for a solar altitude of 87° is identical to its attested value in MS Tk. Mathematically, this recomputation is highly irregular as it not only enters a wrong Cosine in the algorithm ($\text{Cos}\,88°$ instead of the regular $\text{Cos}\,87° = \text{Sin}\,3°$), but also uses an inaccurate Sine ($\text{Sin}\,2°$ should be 2;5,38) in the calculation that follows. Accordingly, this attested (or irregularly recomputed) shadow length for the 87th argument makes the sequence $\textit{Chāyā}_{12}\,(86°) = 0;50,20$, $\textit{Chāyā}_{12}\,(87°) = 0;24,9$, and $\textit{Chāyā}_{12}\,(88°) = 0;32,24$ in MS Tk mathematically inconsistent. (The shadow length is a monotonically decreasing function for the first ninety degrees of the argument.) The recomputational irregularities that involve interdependencies between attested values from different tables are a strong indication of secondary interventions. In this case, it is very likely that a (later) table author (mis)calculated the shadow length for a corrupted/illegible/missing entry corresponding to the 87th argument by simply using the attested value of Sine (in the parent manuscript).
5. The three tables of shadow lengths in MS Tk reveal further interdependencies between their entries, e.g. the shadow lengths $\textit{Chāyā}_{7}\,(4°)$ and $\textit{Chāyā}_{60}\,(4°)$, $\textit{Chāyā}_{12}\,(4°)$ and $\textit{Chāyā}_{60}\,(4°)$, or $\textit{Chāyā}_{12}\,(24°)$ and $\textit{Chāyā}_{60}\,(24°)$. These computational interdependencies also indicate that historical actors (presumably, different from the original author) regularly modified tables by recomputing certain entries using attested values from a parent manuscript.

The observations of this study show how historical actors carelessly or consciously modify a table as they copy it. Their modifications increasingly distance earlier versions of the table from what is attested in a present witness. Essentially, each witness is a mathematical artefact from a particular time that contains an aggregated picture of the changes made (and unmade) by previous actors. Our modern recomputations simulate historical procedures, identify computational irregularities, and analyse scribal discrepancies to help us trace the mathematical practices of these actors. As more advanced tools from data sciences (in particular, knowledge discovery processes and machine learning) are adapted to analyse and predict patterns in these table entries, methodological questions become important for designing meaningful algorithms. For instance, how do table authors modify theoretical (canonical) formulae for practical computations? What combinations of arithmetical operations reproduce the anomalous values attested in a table? Do residual differences follow a behavioural trend for a selected corpus? What is a sensible taxonomy of recomputational irregularities and scribal discrepancies? How can competing recomputational strategies be statistically chosen? This study addresses some of these questions by examining a few selected tables of the *Amṛtalaharī*.

Acknowledgements

I gratefully acknowledge the University of Tokyo Library for hosting open-source digital images of the manuscript of the *Amṛtalaharī* on their online platform. I am particularly grateful to Professor Taro Mimura (Tokyo University) for making the photographs from MS Tk that are included with this article. I would like to thank Matthieu Husson, Clemency Montelle, and Benno van Dalen for their patience in reading draft versions of this work and offering critical remarks. I also express my gratitude to Glen Van Brummelen, Richard Kremer, Agathe Keller, Karine Chemla, Krishnamurthi Ramasubramanian, Christophe Schmit, Galla Topalian, and Zachary Hynd for their insightful comments and technical expertise in shaping this work.

This work was generously supported by the 'Tables Analysis Method for the history of Astral Sciences' project funded by the PSL grant '*Jeunes équipes/nouvelles équipes*' and hosted by the Systèmes de Référence Temps Espace Laboratory (UMR 8630) at the Observatoire de Paris.

Bibliography

Abraham, George, 'The Gnomon in Early Indian Astronomy', *Indian Journal of History of Science* 16 (1981), pp. 215–18.

Ansari, S. M. Razaullah, 'On the Transmission of Arabic-Islamic Astronomy to Medieval India', *Archives Internationales d'Histoire des Sciences* 45 (1995), pp. 273–97.

Ansari, S. M. Razaullah, 'Survey of Zījes Written in the Sub-Continent', *Indian Journal of History of Science* 50 (2015), pp. 573–601.

Bag, A. K., 'Sine Table in Ancient India', *Indian Journal of History of Science* 4 (1969), pp. 79–85.

Chatterjee, Bina, *Śiṣyadhīvṛddhida Tantra of Lalla*. Vol II: *Translation and Mathematical Notes*, New Delhi: Indian National Science Academy, 1981.

van Dalen, Benno, 'A Statistical Method for Recovering Unknown Parameters from Medieval Astronomical Tables', *Centaurus* 32 (1989), pp. 85–145.

Gupta, Radha Charan, 'Second Order Interpolation in Indian Mathematics up to the Fifteenth Century', *Indian Journal of History of Science* 4 (1969), pp. 1–10.

Gupta, Radha Charan, 'Sine of Eighteen Degrees in India up to the Eighteenth Century', *Indian Journal of History of Science* 11 (1976), pp. 86–98.

Hayashi, Takao, 'Āryabhaṭa's Rule and Table for Sine-Differences', *Historia Mathematica* 24 (1997), pp. 396–406.

Matsunami, Seiren, *A Catalogue of the Sanskrit Manuscripts in the Tokyo University Library*, Tokyo: University of Tokyo, 1965 [http://utlsktms.ioc.u-tokyo.ac.jp/detail.php?key=20].

Minkowski, Christopher, 'Astronomers and Their Reasons: Working Paper on Jyotiḥśāstra', *Journal of Indian Philosophy* 30 (2002), pp. 495–514.

Misra, Anuj, *The Golādhyāya of Nityānanda's Sarvasiddhāntarāja: An Examination of 'The Chapter on Spheres' in a Seventeenth-Century Text on Mathematical Astronomy*, PhD dissertation, Christchurch: University of Canterbury, 2016 [http://hdl.handle.net/10092/12246].

Misra, Anuj, 'Persian Astronomy in Sanskrit: A Comparative Study of Mullā Farīd's *Zīj-i Shāh Jahānī* and its Sanskrit Translation in Nityānanda's *Siddhāntasindhu*', *History of Science in South Asia* 9 (2021), pp. 30–127.

Misra, Anuj, 'Sanskrit Recension of Persian Astronomy: The Computation of True Declination in Nityānanda's *Sarvasiddhāntarāja*', *History of Science in South Asia*, forthcoming.

Montelle, Clemency, and K. Ramasubramanian, 'Determining the Sine of One Degree in the *Sarvasiddhāntarāja* of Nityānanda', *SCIAMVS* 19 (2018), pp. 1–52.

Montelle, Clemency, K. Ramasubramanian, and J. Dhammaloka, 'Computation of Sines in Nityānanda's *Sarvasiddhāntarāja*', *SCIAMVS* 17 (2016), pp. 1–53.

Pingree, David, 'Indian Reception of Muslim Versions of Ptolemaic Astronomy', in: F. Jamil Ragep and Sally P. Ragep (eds), *Tradition, Transmission, Transformation. Proceedings of Two Conferences on Pre-Modern Science Held at the University of Oklahoma*, Leiden: Brill, 1996, pp. 471–85.

Pingree, David, 'Amṛtalaharī of Nityānanda', *SCIAMVS* 1 (2000), pp. 209–17.

Pingree, David, *A Descriptive Catalogue of the Sanskrit Astronomical Manuscripts Preserved at the Maharaja Man Singh II Museum in Jaipur, India*, Philadelphia: American Philosophical Society, 2003. [Memoirs of the American Philosophical Society, 250.].

Pingree, David, 'The Sarvasiddhāntarāja of Nityānanda', in: Jan P. Hogendijk and Abdelhamid I. Sabra (eds), *The Enterprise of Science in Islam*, Cambridge (Mass.): MIT Press, 2003, pp. 269–84.

Plofker, Kim, 'An Example of the Secant Method of Iterative Approximation in a Fifteenth-Century Sanskrit Text', *Historia Mathematica* 23 (1996), pp. 246–56.

Plofker, Kim, *Mathematics in India*, Princeton: Princeton University Press, 2009.

Ramasubramanian, K., and M. S. Sriram, *Tantrasaṅgraha of Nīlakaṇṭha Somayājī*, London: Springer, 2011.

Rao, S. B., and S. K. Uma, *Karaṇakutūhalam of Bhāskarācārya II: An English Translation with Notes and Appendices*, New Delhi: Indian National Science Academy, 2008.

Sarma, K. V., K. Ramasubramanian, M. D. Srinivas, and M. S. Sriram, *Gaṇita-yuktibhāṣā of Jyeṣṭhadeva (Rationales in Mathematical Astronomy)*, London: Springer, 2008.

Shukla, K. S., and K. V. Sarma, *Āryabhaṭīya of Āryabhaṭa: Critically Edited with Introduction, English Translation, Notes, Comments, and Indexes*, New Delhi: Indian National Science Academy, 1976.

Sircar, Dinesh Chandra, *Indian Epigraphy*, New Delhi: Motilal Banarasidass, 1996. [Epigraphy, Palaeography, Numismatics, 10.]

Subbarayappa, B. V., and K. V. Sarma, *Indian Astronomy: A Source Book*, Mumbai: Nehru Centre, 1985.

Truschke, Audrey, *Culture of Encounters: Sanskrit at the Mughal Court*, New Delhi: Penguin Books, 2016.

Van Brummelen, Glen, *The Mathematics of the Heavens and the Earth: The Early History of Trigonometry*, Princeton: Princeton University Press, 2009.

Appendix A: Diplomatic transcription of the table on ff. 49v–50v of MS Tk

Pages 226–31 include the printed reproductions and diplomatic transcriptions of the six tables on ff. 49v–50v of MS Tk. The diplomatic transcriptions preserve the attested (landscape) layout of the tables. The orthography of the attested text is transcribed without modifying any erroneous or missing letters. Illegible letters are indicated as '[-*x*-]' (where each '*x*' indicates an individual letter) or '[-*x*?-]' (where '*x*?' indicates an unknown number of missing letters). The paratext and table titles are transliterated into the Latin script. Sanskrit euphonic rules (*saṃdhi*) are silently applied to make the transliterated words morphologically regular, e.g. the solar declination or *krānti* is correctly transliterated with internal /*n*/ instead of *krāṃti* (with the allophonic /*ṃ*/). The numbers in the tables are copied as they appear in the manuscript. Any missing or illegible numbers are indicated as '[--]' with the number of dashes (within the square brackets) indicating, tentatively, the maximum size of the missing number. Any unclear or dysmorphic numbers are disambiguated by inspecting the handwriting for consistency, or by identifying local or global trends in the sequence in which these numbers appear.

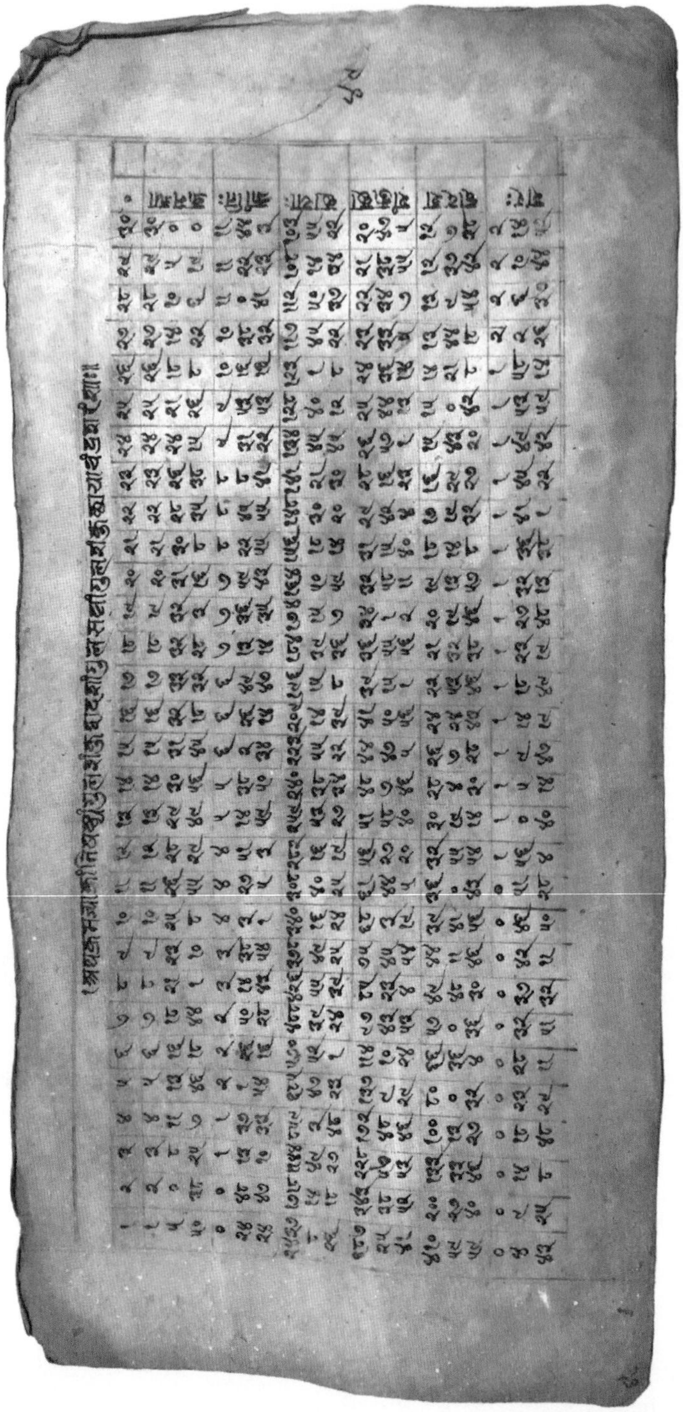

F. 49v of MS Tk

Diplomatic transcription of the tables on f. 49v of MS Tk

\|\| atha kramajyākrāntiṣaṣṭyaṅgulaśaṅkudvādaśāṅgulasaptāṅgulaśaṅkuchāyācandraśaraṃśāḥ \|\|

1	2	3	4	5	6	7	8	9	10	11	12	13	14	15	16	17	18	19	20	21	22	23	24	25	26	27	28	29	30	o	
1	2	3	4	5	6	7	8	9	10	11	12	13	14	15	16	17	18	19	20	21	22	23	24	25	26	27	28	29	30	*kramajyā*	VI.A
5	0	8	11	13	16	18	21	23	25	26	28	29	30	31	32	33	32	32	31	30	28	26	24	21	18	14	10	5	0		
50	38	25	7	46	18	44	1	10	8	55	29	49	56	45	18	32	28	3	16	8	35	38	15	26	8	22	6	19	0		
0	0	1	1	2	2	2	3	3	4	4	4	5	5	6	6	6	7	7	7	8	8	8	9	9	10	10	11	11	11	*krāntiḥ*	VI.B
24	48	13	37	1	26	50	14	38	3	27	51	14	38	2	26	49	13	36	59	22	42	8	31	53	16	38	0	22	44		
24	47	10	33	54	16	28	43	54	1	1	3	59	50	34	14	47	14	35	43	52	52	41	22	53	16	32	41	23	3		
3437	1718	1144	859	685	570	488	426	378	340	308	282	259	240	223	209	196	184	174	164	156	148	141	134	128	123	117	112	108	103	*chāyāḥ*	VI.C_1
8	14	49	2	47	52	39	55	49	16	40	16	53	38	55	14	15	39	15	50	18	30	21	45	40	1	45	50	14	55		
26	18	27	48	23	1	24	39	25	34	25	19	27	34	22	39	8	36	7	59	14	20	30	45	12	8	22	37	34	22		
687	343	228	172	137	114	97	85	75	68	61	56	51	48	44	41	39	36	34	32	31	29	28	26	25	24	23	22	21	20	*śaṅkuchā*	VI.C_2
25	38	57	48	9	10	43	23	45	3	44	27	58	7	47	50	15	55	1	58	15	42	16	57	44	36	33	34	38	47		
41	52	53	46	29	24	52	4	54	19	5	20	40	46	5	56	1	56	2	11	40	4	23	1	2	14	5	7	55	5		
410	200	133	100	80	66	57	49	44	39	36	32	30	28	26	24	22	21	20	19	18	17	16	15	15	14	13	13	12	12	*dvādaśa*	VI.C_3
59	27	33	12	0	36	0	48	11	41	0	55	19	4	7	24	53	32	19	13	14	19	29	43	0	21	44	9	37	7		
59	40	46	27	32	4	36	30	46	56	43	54	14	30	28	43	46	38	46	57	8	32	27	20	42	8	18	54	42	28		
0	0	0	0	0	0	0	0	0	0	0	1	1	1	1	1	1	1	1	1	1	1	1	1	1	1	2	2	2	2	*śaraḥ*	VI.D
4	9	14	18	23	28	32	37	42	46	51	56	0	5	9	14	18	23	27	32	36	41	45	49	53	58	2	6	10	14		
43	25	8	48	29	11	51	32	11	50	28	4	40	14	47	19	49	19	46	13	38	1	22	42	59	14	26	30	44	52		

49

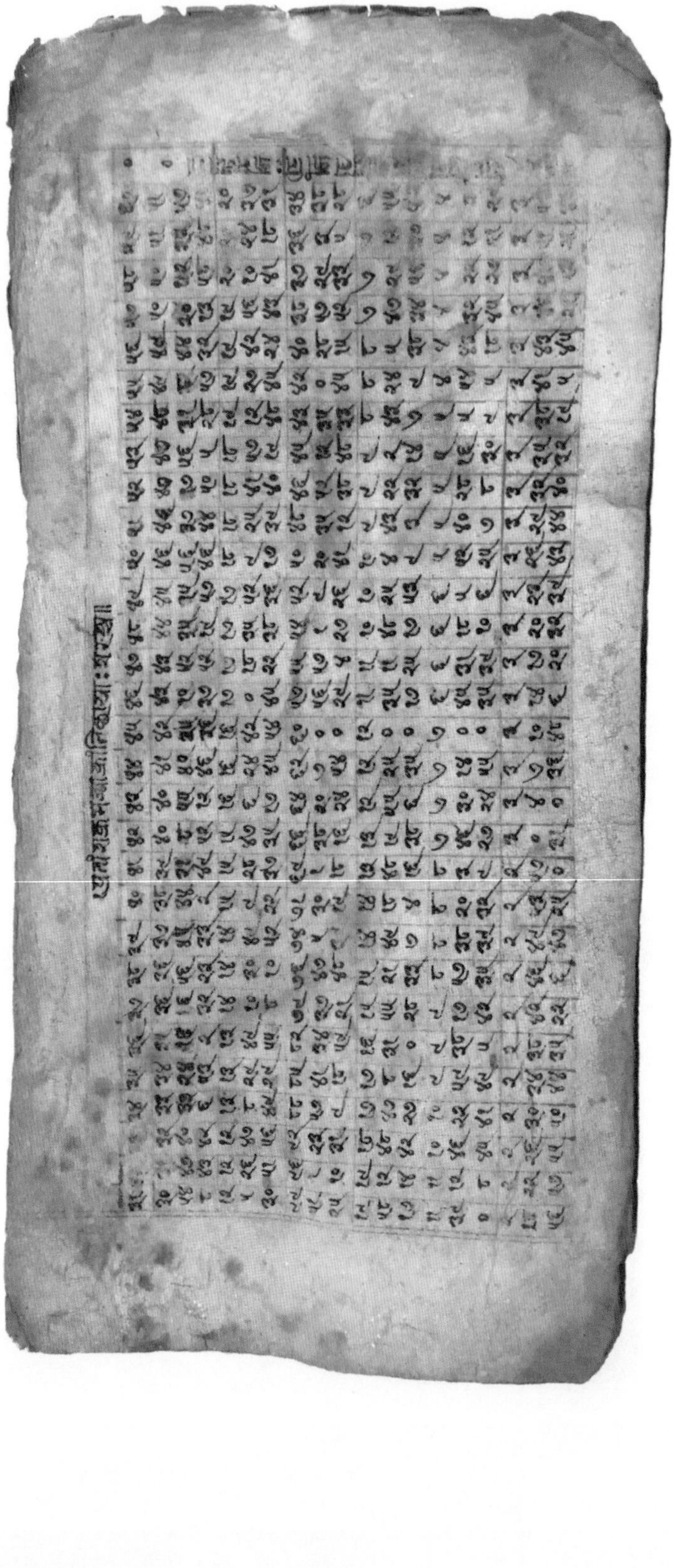

F. 50r of MS Tk

Diplomatic transcription of the tables on f. 50r of MS Tk

|| pratyaṃśakramajyākrāntichāyāḥ śaraśca ||																															
31	32	33	34	35	36	37	38	39	40	41	42	43	44	45	46	47	48	49	50	51	52	53	54	55	56	57	58	59	60	o	
30	31	32	33	34	35	36	36	37	38	39	40	40	41	42	43	43	44	45	46	46	47	47	48	49	49	50	50	51	51		
54	47	40	33	24	16	6	56	45	34	21	8	55	40	25	10	52	35	15	56	37	17	56	31	8	44	20	52	22	57	o	VI.A
8	43	42	6	53	2	32	23	33	2	49	52	12	46	36	37	52	19	57	46	44	50	5	28	57	32	13	58	48	41		
12	12	12	13	13	13	14	14	14	15	15	15	16	16	16	17	17	17	17	18	18	18	18	19	19	19	19	20	20	20		
5	26	47	8	29	49	10	30	49	9	28	47	6	24	42	0	18	35	52	9	25	41	57	12	27	42	56	10	24	37	*kramajyā*	VI.B
30	51	56	49	29	55	8	10	52	22	37	35	17	45	54	45	22	38	36	17	35	40	19	48	45	24	43	41	18	31		
99	96	92	88	85	82	79	76	74	71	69	66	64	62	60	57	55	54	52	50	48	46	45	43	42	40	38	37	36	34		
51	1	23	57	41	34	37	47	5	30	1	38	20	7	0	56	57	1	9	20	35	52	12	35	0	28	57	29	3	38	[-x-]*āṅgula krāntiḥ*	VI.C$_1$
25	10	31	9	18	59	21	48	37	19	18	16	24	54	0	29	4	27	26	41	12	38	48	33	45	15	52	33	5	28		
19	19	18	17	17	16	15	15	14	14	13	12	13	12	12	11	11	10	10	10	9	9	9	8	8	8	7	7	7	6		
58	12	48	47	8	31	55	21	49	18	48	19	55	25	0	35	11	48	25	4	43	22	2	43	24	5	47	29	12	55	[-*xxx*-]	VI.C$_2$
17	14	42	27	16	0	28	33	7	4	16	38	6	35	0	17	25	17	53	9	3	32	14	7	9	38	34	54	37	42		
11	11	10	10	9	9	9	8	8	8	8	7	7	7	7	6	6	6	6	5	5	5	5	5	4	4	4	4	4	4		
39	12	46	22	59	38	17	57	38	20	3	46	30	14	0	45	31	18	5	52	40	28	16	5	54	43	32	22	12	2	[-*x*?-]	VI.C$_3$
0	8	40	41	49	5	42	35	39	32	9	27	24	55	0	35	39	10	6	25	7	8	30	8	5	18	45	27	21	29		
2	2	2	2	2	2	2	2	2	2	2	3	3	3	3	3	3	3	3	3	3	3	3	3	3	3	3	3	3	3		
18	22	26	30	34	38	42	46	49	53	57	0	4	7	10	14	17	20	23	26	29	32	35	38	41	43	[--]	[--]	[--]	[--]	[-*x*?-]	VI.D
56	57	55	50	44	32	22	6	47	25	0	31	0	36	48	6	25	32	39	42	44	40	32	59	5	45	[--]	[--]	[--]	[--]		

F. 50v of MS Tk

Diplomatic transcription of the tables on f. 50v of MS Tk

|| iti pratyaṃśakakramajyākrantichāyāḥ śaraścasamāptaḥ ||

61	62	63	64	65	66	67	68	69	70	71	72	73	74	75	76	77	78	79	80	81	82	83	84	85	86	87	88	89	90	o		
52	52	53	53	54	54	55	55	56	56	56	57	57	57	57	58	58	58	58	59	59	59	59	59	59	59	59	59	59	60			
28	58	27	55	22	48	13	37	0	22	43	3	22	40	57	13	27	41	53	5	15	24	33	40	46	51	55	59	59	0	o		VI.A
38	37	37	40	42	46	49	52	53	54	52	48	42	33	30	4	44	20	51	18	41	58	10	17	18	14	4	27	27	0			
20	21	21	21	21	21	21	22	22	22	22	22	22	23	23	23	23	23	23	23	23	23	23	23	23	23	23	23	23	24			
50	2	14	26	37	48	59	9	19	29	37	45	53	0	8	14	20	26	31	36	41	45	48	51	54	56	57	59	59	0	o		VI.B
21	48	53	37	54	46	25	21	2	17	6	28	26	55	2	42	54	40	59	49	13	9	40	38	11	17	54	46	46	0			
33	31	30	29	27	26	25	24	23	21	20	19	18	17	16	14	13	12	11	10	9	8	7	6	5	4	3	2	1	0			
15	54	34	15	58	42	28	14	1	50	39	29	20	12	4	57	51	45	39	34	30	26	22	18	14	11	8	2	2	0	o	50	VI.C₁
30	8	18	49	42	49	0	29	55	18	35	41	38	17	37	35	7	12	46	46	11	57	1	22	58	43	40	50	50	0			
6	6	6	5	5	5	5	4	4	4	4	3	3	3	3	2	2	2	2	2	1	1	1	1	1	0	0	0	0	0			
39	22	6	51	35	20	5	50	36	22	7	53	40	26	12	59	46	33	19	6	54	41	28	15	3	50	24	32	32	0	o		VI.C₂
6	50	52	10	44	34	37	4	23	4	55	57	8	27	55	31	14	2	57	57	2	11	25	41	0	20	9	24	24	0			
3	3	3	3	3	3	2	2	2	2	2	2	2	2	1	1	1	1	1	1	1	0	0	0	0	0	0	0	0	0			
52	43	34	24	15	7	59	49	41	32	24	16	8	0	52	44	36	29	21	14	16	59	51	44	36	29	14	7	7	0	o		VI.C₃
49	19	0	51	51	0	17	41	13	52	37	28	24	26	32	42	59	16	38	3	31	34	34	34	45	22	40	10	10	0			
3	3	4	4	4	4	4	4	4	4	4	4	4	4	4	4	4	4	4	4	4	4	4	4	4	4	4	4	4	4			
56	58	0	2	4	6	8	10	12	14	16	16	18	22	24	22	24	24	25	25	25	26	27	28	28	29	29	29	29	30	o		VI.D
4	19	31	27	38	36	29	17	1	40	13	45	10	31	46	57	2	5	0	52	39	21	59	32	20	37	50	57	57	0			

pātonacandro bhujā kārye bhujyaṃśedbhyaḥ śaro grāhyāḥ yadi pāto nacandraḥ ṣadbhonas tadā śaraḥ saumyaḥ yadādhikas tadā yāmya ||

Appendix B: Recomputation and analysis of Tables VI.A–D

Conventions for representing the tables

The six tables (Tables VI.A, VI.B, VI.C_{1-3}, and VI.D) from the selected corpus are presented on pp. 233–38.

1. Each table has four separate rows of (sexagesimal) entries, placed one below the other, in three argument blocks 1° to 30°, 31° to 60°, and 61° to 90°. The arguments (in degrees) represent the following different quantities for the respective tables:
   (a) Table of Sines (VI.A): measure of arc;
   (b) Table of solar declinations (VI.B): celestial longitude;
   (c) Table of shadow lengths of gnomons of 60-digit (VI.C_1), 12-digit (VI.C_2), and 7-digit heights (VI.C_3): solar altitude; and
   (d) Table of lunar latitudes (VI.D): lunar-nodal elongation.
2. The sexagesimal values of the table entries are written vertically. The digits at the top of a vertical stack represent the integer part (i.e. units or degrees) of the number, those in the middle indicate the first fractional part (i.e. minutes), and the digits at the bottom of a stack signify the second fractional part (i.e. seconds).
3. In each argument block of thirty degrees,
   (a) the first row lists the attested values from MS Tk;
   (b) the second row presents the recomputed values with
      – the digits (in individual sexagesimal places) that result from irregular recomputations enclosed in a rectangular box;
   (c) the third row shows the difference in digits between corresponding sexagesimal places of the attested and recomputed values (from the previous two rows) with
      – all non-zero differences enclosed in shaded grey boxes; and
   (d) the fourth row lists the proposed emendation to the attested values where
      – any modified entries (in individual sexagesimal places) are enclosed in circles.

These conventions allow (**a**) recomputational irregularities (digits in rectangular boxes), (**b**) non-zero revised differences (digits in grey cells), and (**c**) proposed emendations (encircled digits) to be clearly identified. For a collection of tables from a single manuscript, this visual representation allows the recomputational and the text-critical versions of individual tables to be seen concurrently.

Table VI.A: Recomputations and analyses of the Table of Sines (*kramajyā*) discussed in Sections 5.1–5.2

Measure of arc		1	2	3	4	5	6	7	8	9	10	11	12	13	14	15	16	17	18	19	20	21	22	23	24	25	26	27	28	29	30
Attested Sine in MS Tk	UNITS	1	2	3	4	5	6	7	8	9	10	11	12	13	14	15	16	17	18	19	20	21	22	23	24	25	26	27	28	29	30
	MINUTES	5	0	8	11	13	16	18	21	23	25	26	28	29	30	31	32	33	32	32	31	30	28	26	24	21	18	14	10	5	0
	SECONDS	50	38	25	7	46	18	44	1	10	8	55	29	49	56	45	18	32	28	3	16	8	35	38	15	26	8	22	6	19	0
Recomputed Sine (rounded to seconds)	UNITS	1	2	3	4	5	6	7	8	9	10	11	12	13	14	15	16	17	18	19	20	21	22	23	24	25	26	27	28	29	30
	MINUTES	2	5	8	11	13	16	18	21	23	25	26	28	29	30	31	32	32	32	32	31	30	28	26	24	21	18	14	10	5	0
	SECONDS	50	38	25	7	46	18	44	1	10	8	55	29	49	55	45	18	32	28	3	16	7	35	38	15	26	8	22	6	19	0
Difference between the digits in corresponding sexagesimal places	UNITS	0	0	0	0	0	0	0	0	0	0	0	0	0	0	0	0	0	0	0	0	0	0	0	0	0	0	0	0	0	0
	MINUTES	+3	−5	0	0	0	0	0	0	0	0	0	0	0	0	0	0	+1	0	0	0	0	0	0	0	0	0	0	0	0	0
	SECONDS	0	0	0	0	0	0	0	0	0	0	0	0	0	+1	0	0	0	0	0	0	+1	0	0	0	0	0	0	0	0	0
Proposed emendation to the attested value in MS Tk	UNITS	1	2	3	4	5	6	7	8	9	10	11	12	13	14	15	16	17	18	19	20	21	22	23	24	25	26	27	28	29	30
	MINUTES	(2)	(5)	8	11	13	16	18	21	23	25	26	28	29	30	31	32	(32)	32	32	31	30	28	26	24	21	18	14	10	5	0
	SECONDS	50	38	25	7	46	18	44	1	10	8	55	29	49	56	45	18	32	28	3	16	8	35	38	15	26	8	22	6	19	0

Measure of arc		31	32	33	34	35	36	37	38	39	40	41	42	43	44	45	46	47	48	49	50	51	52	53	54	55	56	57	58	59	60
Attested Sine in MS Tk	UNITS	30	31	32	33	34	35	36	36	37	38	39	40	40	41	42	43	43	44	45	46	46	47	47	48	49	49	50	50	51	51
	MINUTES	54	47	40	33	24	16	16	56	45	34	21	8	55	40	25	10	52	35	15	56	37	17	56	31	8	44	20	52	25	57
	SECONDS	8	43	42	6	53	2	32	23	33	2	49	52	12	46	36	37	52	19	57	46	44	50	5	28	57	32	13	58	48	41
Recomputed Sine (rounded to seconds)	UNITS	30	31	32	33	34	35	36	36	37	38	39	40	40	41	42	43	43	44	45	45	46	47	47	48	49	49	50	50	51	51
	MINUTES	54	47	40	33	24	16	6	56	45	34	21	8	55	40	25	9	52	35	16	57	37	16	55	32	8	44	19	52	25	57
	SECONDS	8	43	42	6	53	2	32	23	33	2	49	52	12	46	35	37	52	19	57	46	44	50	5	28	57	32	13	58	48	41
Difference between the digits in corresponding sexagesimal places	UNITS	0	0	0	0	0	0	0	0	0	0	0	0	0	0	0	0	0	0	0	+1	0	0	0	0	0	0	0	0	0	0
	MINUTES	0	0	0	0	0	0	+10	0	0	0	0	0	0	0	0	+1	0	0	−1	−1	0	+1	+1	−1	0	0	+1	0	0	0
	SECONDS	0	0	0	0	0	0	0	0	0	0	0	0	0	0	+1	0	0	0	0	0	0	0	0	0	0	0	0	0	0	0
Proposed emendation to the attested value in MS Tk	UNITS	30	31	32	33	34	35	36	36	37	38	39	40	40	41	42	43	43	44	45	(45)	46	47	47	48	49	49	50	50	51	51
	MINUTES	54	47	40	33	24	16	(6)	56	45	34	21	8	55	40	25	10	52	35	15	56	37	17	56	31	8	44	20	52	25	57
	SECONDS	8	43	42	6	53	2	32	23	33	2	49	52	12	46	36	37	52	19	57	46	44	50	5	28	57	32	13	58	48	41

Measure of arc		61	62	63	64	65	66	67	68	69	70	71	72	73	74	75	76	77	78	79	80	81	82	83	84	85	86	87	88	89	90
Attested Sine in MS Tk	UNITS	52	52	53	53	54	54	55	55	56	56	56	57	57	57	57	58	58	58	58	59	59	59	59	59	59	59	59	59	59	60
	MINUTES	28	58	27	55	22	48	13	37	0	22	43	3	22	40	57	13	27	41	53	5	15	24	33	40	46	51	55	59	59	0
	SECONDS	38	37	37	40	43	46	49	52	53	54	52	48	42	33	30	4	44	20	51	18	41	58	10	17	18	14	4	27	27	0
Recomputed Sine (rounded to seconds)	UNITS	52	52	53	53	54	54	55	55	56	56	56	57	57	57	57	58	58	58	58	59	59	59	59	59	59	59	59	59	59	60
	MINUTES	28	58	27	55	22	48	13	37	0	22	43	3	22	40	57	13	27	41	53	5	15	24	33	40	46	51	55	57	59	0
	SECONDS	38	37	37	40	42	46	49	52	53	54	52	48	42	33	20	4	44	20	51	18	41	58	10	17	18	14	4	48	27	0
Difference between the digits in corresponding sexagesimal places	UNITS	0	0	0	0	0	0	0	0	0	0	0	0	0	0	0	0	0	0	0	0	0	0	0	0	0	0	0	0	0	0
	MINUTES	0	0	0	0	0	0	0	0	0	0	0	0	0	0	0	0	0	0	0	0	0	0	0	0	0	0	0	+2	0	0
	SECONDS	0	0	0	0	+1	0	0	0	0	0	0	0	0	0	+10	0	0	0	0	0	0	0	0	0	0	0	0	−21	0	0
Proposed emendation to the attested value in MS Tk	UNITS	52	52	53	53	54	54	55	55	56	56	56	57	57	57	57	58	58	58	58	59	59	59	59	59	59	59	59	59	59	60
	MINUTES	28	58	27	55	22	48	13	37	0	22	43	3	22	40	57	13	27	41	53	5	15	24	33	40	46	51	55	(57)	59	0
	SECONDS	38	37	37	40	43	46	49	52	53	54	52	48	42	33	(20)	4	44	20	51	18	41	58	10	17	18	14	4	(48)	27	0

Table VI.B: Recomputations and analyses of the Table of solar declinations (*krānti*) discussed in Sections 3.3–3.4

Celestial longitude		**1**	**2**	**3**	**4**	**5**	**6**	**7**	**8**	**9**	**10**	**11**	**12**	**13**	**14**	**15**	**16**	**17**	**18**	**19**	**20**	**21**	**22**	**23**	**24**	**25**	**26**	**27**	**28**	**29**	**30**
Attested solar declination in MS Tk	DEGREES	0	0	1	1	2	2	2	3	3	4	4	4	5	5	6	6	6	7	7	7	8	8	8	9	9	10	10	11	11	11
	MINUTES	24	48	13	37	1	26	50	14	38	3	27	51	14	38	2	26	49	13	36	59	22	45	8	31	53	16	38	0	22	44
	SECONDS	24	47	10	33	54	16	28	43	54	1	5	3	59	50	34	14	47	14	35	43	55	55	41	22	53	16	32	41	23	3
Recomputed solar declination (rounded to seconds)	DEGREES	0	0	1	1	2	2	2	3	3	4	4	4	5	5	6	6	6	7	7	7	8	8	9	9	9	10	10	11	11	11
	MINUTES	24	48	13	37	1	26	50	14	38	3	27	51	14	38	2	26	49	13	36	59	22	45	8	31	53	16	38	0	22	44
	SECONDS	24	48	11	33	54	12	29	42	54	1	5	4	59	50	35	14	47	15	35	47	54	51	41	22	53	16	28	[41]	23	3
Difference between the digits in corresponding sexagesimal places	DEGREES	0	0	0	0	0	0	0	0	0	0	0	0	0	0	0	0	0	0	0	0	0	0	−1	0	0	0	0	0	0	0
	MINUTES	0	0	0	0	0	0	0	0	0	0	0	0	0	0	0	0	0	0	0	0	0	0	0	0	0	0	0	0	0	0
	SECONDS	0	−1	−1	0	0	+4	−1	+1	0	0	0	−1	0	0	−1	0	0	−1	0	−4	+1	+4	0	0	0	0	+4	0	0	0
Proposed emendation to the attested value in MS Tk	DEGREES	0	0	1	1	2	2	2	3	3	4	4	4	5	5	6	6	6	7	7	7	8	8	(9)	9	9	10	10	11	11	11
	MINUTES	24	48	13	37	1	26	50	14	38	3	27	51	14	38	2	26	49	13	36	59	22	45	8	31	53	16	38	0	22	44
	SECONDS	24	47	10	33	54	16	28	43	54	1	5	3	59	50	34	14	47	14	35	43	52	52	41	22	53	16	32	41	23	3

Celestial longitude		**31**	**32**	**33**	**34**	**35**	**36**	**37**	**38**	**39**	**40**	**41**	**42**	**43**	**44**	**45**	**46**	**47**	**48**	**49**	**50**	**51**	**52**	**53**	**54**	**55**	**56**	**57**	**58**	**59**	**60**
Attested solar declination in MS Tk	DEGREES	12	12	12	13	13	13	14	14	14	15	15	15	16	16	16	17	17	17	17	18	18	18	18	19	19	19	19	20	20	20
	MINUTES	5	26	47	8	29	49	10	30	49	9	28	47	6	24	42	0	18	35	52	9	25	41	57	12	27	42	56	10	24	37
	SECONDS	30	51	56	49	29	55	8	10	52	22	37	35	17	45	54	45	22	38	36	17	39	40	19	48	45	24	43	41	18	31
Recomputed solar declination (rounded to seconds)	DEGREES	12	12	12	13	13	13	14	14	14	15	15	15	16	16	16	17	17	17	17	18	18	18	18	19	19	19	19	20	20	20
	MINUTES	5	26	47	8	29	49	10	30	49	9	28	47	6	24	42	0	18	35	52	9	25	41	57	12	27	42	56	10	24	37
	SECONDS	33	51	56	49	31	56	9	9	52	22	37	35	18	45	55	45	22	39	37	17	39	40	20	43	46	25	42	41	18	31
Difference between the digits in corresponding sexagesimal places	DEGREES	0	0	0	0	0	0	0	0	0	0	0	0	0	0	0	0	0	0	0	0	0	0	0	0	0	0	0	0	0	0
	MINUTES	0	0	0	0	0	0	0	0	0	0	0	0	0	0	0	0	0	0	0	0	0	0	0	0	0	0	0	0	0	0
	SECONDS	−3	0	0	0	−2	−1	−1	+1	0	0	0	0	−1	0	−1	0	0	−1	−1	0	0	0	−1	+5	−1	−1	+1	0	0	0
Proposed emendation to the attested value in MS Tk	DEGREES	12	12	12	13	13	13	14	14	14	15	15	15	16	16	16	17	17	17	17	18	18	18	18	19	19	19	19	20	20	20
	MINUTES	5	26	47	8	29	49	10	30	49	9	28	47	6	24	42	0	18	35	52	9	25	41	57	12	27	42	56	10	24	37
	SECONDS	30	51	56	49	29	55	8	10	52	22	37	35	17	45	54	45	22	38	36	17	39	40	19	48	45	24	43	41	18	31

Celestial longitude		**61**	**62**	**63**	**64**	**65**	**66**	**67**	**68**	**69**	**70**	**71**	**72**	**73**	**74**	**75**	**76**	**77**	**78**	**79**	**80**	**81**	**82**	**83**	**84**	**85**	**86**	**87**	**88**	**89**	**90**
Attested solar declination in MS Tk	DEGREES	20	21	21	21	21	21	21	22	22	22	22	22	22	23	23	23	23	23	23	23	23	23	23	23	23	23	23	23	23	24
	MINUTES	50	2	14	26	37	48	59	9	19	29	37	45	53	0	8	14	20	26	31	36	41	45	48	51	54	56	57	59	59	0
	SECONDS	21	48	53	37	54	46	25	21	2	17	6	28	26	55	2	42	44	40	59	49	13	9	40	38	11	17	54	46	46	0
Recomputed solar declination (rounded to seconds)	DEGREES	20	21	21	21	21	21	21	22	22	22	22	22	22	23	23	23	23	23	23	23	23	23	23	23	23	23	23	23	23	24
	MINUTES	50	2	14	26	37	48	59	9	19	28	37	45	53	0	8	14	20	26	31	36	41	45	48	51	54	56	57	59	59	0
	SECONDS	22	48	55	38	54	48	15	21	2	17	6	29	25	56	2	42	55	41	59	50	13	10	38	39	12	17	55	4	46	0
Difference between the digits in corresponding sexagesimal places	DEGREES	0	0	0	0	0	0	0	0	0	0	0	0	0	0	0	0	0	0	0	0	0	0	0	0	0	0	0	0	0	0
	MINUTES	0	0	0	0	0	0	0	0	0	+1	0	0	0	0	0	0	0	0	0	0	0	0	0	0	0	0	0	0	0	0
	SECONDS	−1	0	−2	−1	0	−2	+10	0	0	0	0	−1	+1	−1	0	0	−11	−1	0	−1	0	−1	+2	−1	−1	0	−1	+42	0	0
Proposed emendation to the attested value in MS Tk	DEGREES	0	21	21	21	21	21	21	22	22	22	22	22	22	23	23	23	23	23	23	23	23	23	23	23	23	23	23	23	23	24
	MINUTES	50	2	14	26	37	48	59	9	19	(28)	37	45	53	0	8	14	20	26	31	36	41	45	48	51	54	56	57	59	59	0
	SECONDS	21	48	53	37	54	46	(15)	21	2	17	6	28	26	55	2	42	(55)	40	59	49	13	9	40	38	11	17	54	(4)	46	0

Table VI.C$_1$: Recomputations and analyses of the Table of shadow lengths (*śaṅkuchāyā*) of a 60-digit gnomon discussed in Sections 3.5–3.6

Solar altitude	1	2	3	4	5	6	7	8	9	10	11	12	13	14	15	16	17	18	19	20	21	22	23	24	25	26	27	28	29	30	
Attested shadow length of a 60-digit gnomon in MS Tk	3437	1718	1144	859	685	570	488	426	378	340	308	282	259	240	223	209	196	184	174	164	156	148	141	134	128	123	117	112	108	103	UNITS
	8	14	49	3	47	52	39	55	49	16	40	16	53	38	55	14	15	39	15	50	18	30	21	45	40	1	45	50	14	55	MINUTES
	26	18	27	48	23	1	24	39	25	24	25	19	27	34	22	39	8	36	7	59	14	20	3	45	12	8	22	37	34	22	SECONDS
Recomputed shadow length of a 60-digit gnomon (rounded to seconds)	3437	1718	1144	858	685	570	488	426	378	340	308	282	259	240	223	209	196	184	174	164	156	148	141	134	128	123	117	112	108	103	
	8	14	49	3	47	51	39	55	49	16	40	16	53	38	55	14	15	39	15	50	18	30	21	45	40	1	45	50	14	55	
	26	18	28	48	24	59	24	40	26	34	25	39	27	34	22	39	8	36	7	59	14	20	4	45	10	8	23	37	34	22	
Difference between the digits in corresponding sexagesimal places	0	0	0	+1	0	0	0	0	0	0	0	0	0	0	0	0	0	0	0	0	0	0	0	0	0	0	0	0	0	0	
	0	0	0	0	0	+1	0	0	0	0	0	0	0	0	0	0	0	0	0	0	0	0	0	0	0	0	0	0	0	0	
	0	0	−1	0	−1	−58	0	−1	−1	−10	0	−20	0	0	0	0	0	0	0	0	0	0	−1	0	+2	0	−1	0	0	0	
Proposed emendation to the attested value in MS Tk	3437	1718	1144	859	685	570	488	426	378	340	308	282	259	240	223	209	196	184	174	164	156	148	141	134	128	123	117	112	108	103	
	8	14	49	3	47	52	39	55	49	16	40	16	53	38	55	14	15	39	15	50	18	30	21	45	40	1	45	50	14	55	
	26	18	27	48	23	1	24	39	25	34	25	39	27	34	22	39	8	36	7	59	0	20	3	45	12	8	22	37	34	22	

Solar altitude	31	32	33	34	35	36	37	38	39	40	41	42	43	44	45	46	47	48	49	50	51	52	53	54	55	56	57	58	59	60	
Attested shadow length of a 60-digit gnomon in MS Tk	99	96	92	88	85	82	79	76	74	71	69	66	64	62	60	57	55	54	52	50	48	46	45	43	42	40	38	37	36	34	UNITS
	51	1	23	57	41	34	37	47	5	30	1	38	20	7	0	56	57	1	9	20	35	52	12	35	0	28	57	29	3	38	MINUTES
	25	10	31	9	18	59	21	48	39	19	18	16	24	54	0	29	4	27	26	41	12	38	48	33	45	15	52	33	5	28	SECONDS
Recomputed shadow length of a 60-digit gnomon (rounded to seconds)	99	96	92	88	85	82	79	76	74	71	69	66	64	62	60	57	55	54	52	50	48	46	45	43	42	40	38	37	36	34	
	51	1	23	57	41	34	37	47	5	30	1	38	20	7	0	56	57	1	9	20	35	52	12	35	0	28	57	29	3	38	
	25	10	31	12	17	56	21	47	39	20	19	16	24	54	0	29	4	27	27	41	13	38	48	33	45	14	52	33	6	28	
Difference between the digits in corresponding sexagesimal places	0	0	0	0	0	0	0	0	0	0	0	0	0	0	0	0	0	0	0	0	0	0	0	0	0	0	0	0	0	0	
	0	0	0	0	0	0	0	0	0	0	0	0	0	0	0	0	0	0	0	0	0	0	0	0	0	0	0	0	0	0	
	0	0	0	−3	+1	+3	0	+1	0	−1	−1	0	0	0	0	0	0	0	−1	0	−1	0	0	0	0	+1	0	0	−1	0	
Proposed emendation to the attested value in MS Tk	99	96	92	88	85	82	79	76	74	71	69	66	64	62	60	57	55	54	52	50	48	46	45	43	42	40	38	37	36	34	
	51	1	23	57	41	34	37	47	5	30	1	38	20	7	0	56	57	1	9	20	35	52	12	35	0	28	57	29	3	38	
	25	10	31	9	18	59	21	48	37	19	18	16	24	54	0	29	4	27	26	41	12	38	48	33	45	15	52	33	5	28	

Solar altitude	61	62	63	64	65	66	67	68	69	70	71	72	73	74	75	76	77	78	79	80	81	82	83	84	85	86	87	88	89	90	
Attested shadow length of a 60-digit gnomon in MS Tk	33	31	30	29	27	26	25	24	23	21	20	19	18	17	16	14	13	12	11	10	9	8	7	6	5	4	3	1	1	0	UNITS
	15	54	34	15	58	42	28	14	1	50	39	29	20	12	4	57	51	45	39	34	30	26	22	18	14	11	8	2	2	0	MINUTES
	30	9	18	49	42	49	0	29	55	18	35	41	38	17	37	35	7	12	46	46	11	57	1	22	58	43	40	50	50	0	SECONDS
Recomputed shadow length of a 60-digit gnomon (rounded to seconds)	33	31	30	29	27	26	25	24	23	21	20	19	18	17	16	14	13	12	11	10	9	8	7	6	5	4	3	2	1	0	
	15	54	34	15	58	42	28	14	1	50	39	29	20	12	4	57	51	45	39	34	30	25	22	18	14	11	8	5	2	0	
	31	9	18	50	43	49	7	30	54	17	35	43	37	17	37	35	7	12	46	47	11	56	2	22	58	44	41	43	50	0	
Difference between the digits in corresponding sexagesimal places	0	0	0	0	0	0	0	0	0	0	0	0	0	0	0	0	0	0	0	0	0	0	0	0	0	0	0	−1	0	0	
	0	0	0	0	0	0	0	0	0	0	0	0	0	0	0	0	0	0	0	0	0	+1	0	0	0	0	0	−3	0	0	
	−1	0	0	−1	−1	0	−7	−1	+1	+1	0	−2	+1	0	0	0	0	0	0	−1	0	+1	−1	0	0	−1	−1	−7	0	0	
Proposed emendation to the attested value in MS Tk	33	31	30	29	27	26	25	24	23	21	20	19	18	17	16	14	13	12	11	10	9	8	7	6	5	4	3	2	1	0	
	15	54	34	15	58	42	28	14	1	50	39	29	20	12	4	57	51	45	39	34	30	26	22	18	14	11	8	5	2	0	
	30	8	18	49	42	49	7	29	55	18	35	41	38	17	37	35	7	12	46	47	11	57	1	22	58	43	40	43	50	0	

Table VI.C$_2$: Recomputations and analyses of the Table of shadow lengths (*śaṅkuchāyā*) of a 12-digit gnomon discussed in Sections 3.5–3.6

Solar altitude	1	2	3	4	5	6	7	8	9	10	11	12	13	14	15	16	17	18	19	20	21	22	23	24	25	26	27	28	29	30	
Attested shadow length of a 60-digit gnomon in MS Tk	687	343	228	172	137	114	97	85	75	68	61	56	51	48	44	41	39	36	34	32	31	29	28	26	25	24	23	22	21	20	UNITS
	25	38	57	48	9	10	43	23	45	3	44	27	58	7	47	50	15	55	1	58	15	42	16	57	44	36	33	34	38	47	MINUTES
	41	52	53	46	29	24	53	4	54	19	5	20	40	46	5	56	1	56	2	11	40	4	23	1	12	14	5	7	55	5	SECONDS
Recomputed shadow length of a 12-digit gnomon (rounded to seconds)	687	343	228	171	137	114	97	85	75	68	61	56	51	48	44	41	39	36	34	32	31	29	28	26	25	24	23	22	21	20	
	25	38	57	48	9	10	43	23	45	3	44	27	58	7	47	50	15	55	51	58	15	42	16	57	44	36	33	34	38	47	
	41	52	54	46	29	24	53	8	53	19	3	20	41	43	4	56	2	55	1	12	39	4	13	1	2	14	5	7	55	4	
Difference between the digits in corresponding sexagesimal places	0	0	0	+1	0	0	0	0	0	0	0	0	0	0	0	0	0	0	0	0	0	0	0	0	0	0	0	0	0	0	
	0	0	0	0	0	0	0	0	0	0	0	0	0	0	0	0	0	0	−50	0	0	0	0	0	0	0	0	0	0	0	
	0	0	−1	0	0	0	0	−4	+1	0	+2	0	−1	+3	+1	0	−1	+1	+1	−1	+1	0	+10	0	+10	0	0	0	0	+1	
Proposed emendation to the attested value in MS Tk	687	343	228	171	137	114	97	85	75	68	61	56	51	48	44	41	39	36	34	32	31	29	28	26	25	24	23	22	21	20	
	25	38	57	48	9	10	43	23	45	3	44	27	58	7	47	50	15	55	51	58	15	42	16	57	44	36	33	34	38	47	
	41	52	53	46	29	24	53	4	54	19	5	20	40	46	5	56	1	56	2	11	40	4	13	1	2	14	5	7	55	5	

Solar altitude	31	32	33	34	35	36	37	38	39	40	41	42	43	44	45	46	47	48	49	50	51	52	53	54	55	56	57	58	59	60	
Attested shadow length of a 12-digit gnomon in MS Tk	19	19	18	17	17	16	15	15	14	14	13	13	12	12	12	11	11	10	10	10	9	9	9	8	8	8	7	7	7	6	UNITS
	58	12	48	47	8	31	55	21	49	18	48	19	55	25	0	35	11	48	25	4	43	22	2	43	24	5	47	29	12	55	MINUTES
	17	14	42	27	16	0	28	33	7	4	16	38	6	35	0	17	25	17	53	9	3	32	14	7	9	38	34	54	37	42	SECONDS
Recomputed shadow length of a 12-digit gnomon (rounded to seconds)	19	19	18	17	17	16	15	15	14	14	13	13	12	12	12	11	11	10	10	10	9	9	9	8	8	8	7	7	7	6	
	58	12	28	47	8	31	55	21	49	18	48	19	52	25	0	35	11	48	25	4	43	22	2	43	24	5	47	29	12	55	
	17	14	42	26	16	0	28	33	8	4	16	39	5	35	0	18	26	17	53	9	3	32	34	7	9	39	34	55	37	42	
Difference between the digits in corresponding sexagesimal places	0	0	0	0	0	0	0	0	0	0	0	0	0	0	0	0	0	0	0	0	0	0	0	0	0	0	0	0	0	0	
	0	0	+20	0	0	0	0	0	0	0	0	0	+3	0	0	0	0	0	0	0	0	0	0	0	0	0	0	0	0	0	
	0	0	0	+1	0	0	0	0	−1	0	0	−1	+1	0	0	−1	−1	0	0	0	0	0	−20	0	0	−1	0	−1	0	0	
Proposed emendation to the attested value in MS Tk	19	19	18	17	17	16	15	15	14	14	13	13	12	12	12	11	11	10	10	10	9	9	9	8	8	8	7	7	7	6	
	58	12	28	47	8	31	55	21	49	18	48	19	52	25	0	35	11	48	25	4	43	22	2	43	24	5	47	29	12	55	
	17	14	42	27	16	0	28	33	7	4	16	38	6	35	0	17	25	17	53	9	3	32	34	7	9	38	34	54	37	42	

Solar altitude	61	62	63	64	65	66	67	68	69	70	71	72	73	74	75	76	77	78	79	80	81	82	83	84	85	86	87	88	89	90	
Attested shadow length of a 12-digit gnomon in MS Tk	6	6	6	5	5	5	5	4	4	4	4	3	3	3	3	2	2	2	2	2	1	1	1	1	1	0	0	0	0	0	UNITS
	39	22	6	51	35	20	5	50	36	22	7	53	40	26	12	59	46	33	19	6	54	41	28	15	3	50	24	32	32	0	MINUTES
	6	50	52	10	44	34	37	4	23	4	55	57	8	27	55	31	14	2	57	57	2	11	25	41	0	29	9	24	24	0	SECONDS
Recomputed shadow length of a 12-digit gnomon (rounded to seconds)	6	6	6	5	5	5	5	4	4	4	4	3	3	3	3	2	2	2	2	2	1	1	1	1	1	0	0	0	0	0	
	39	22	6	51	35	20	5	50	36	22	7	53	40	26	12	59	46	33	19	6	54	41	28	15	3	50	24	25	12	0	
	6	50	52	10	45	34	37	54	23	3	55	57	7	27	55	31	13	2	57	57	2	11	24	40	0	21	9	9	34	0	
Difference between the digits in corresponding sexagesimal places	0	0	0	0	0	0	0	0	0	0	0	0	0	0	0	0	0	0	0	0	0	0	0	0	0	0	0	0	0	0	
	0	0	0	0	0	0	0	0	0	0	0	0	0	0	0	0	0	0	0	0	0	0	0	0	0	0	0	+7	+20	0	
	0	0	0	0	−1	0	0	−50	0	+1	0	0	+1	0	0	0	+1	0	0	0	0	0	+1	+1	0	+8	0	+15	−10	0	
Proposed emendation to the attested value in MS Tk	6	6	6	5	5	5	5	4	4	4	4	3	3	3	3	2	2	2	2	2	1	1	1	1	1	0	0	0	0	0	
	39	22	6	51	35	20	5	50	36	22	7	53	40	26	12	59	46	33	19	6	54	41	28	15	3	50	24	25	12	0	
	6	50	52	10	44	34	37	54	23	4	55	57	8	27	55	31	14	2	57	57	2	11	25	41	0	21	9	9	34	0	

Table VI.C$_3$: Recomputations and analyses of the Table of shadow lengths (*śaṅkuchāyā*) of a 7-digit gnomon discussed in Sections 3.5–3.6

Solar altitude	1	2	3	4	5	6	7	8	9	10	11	12	13	14	15	16	17	18	19	20	21	22	23	24	25	26	27	28	29	30	
Attested shadow length of a 7-digit gnomon in MS Tk	410	200	133	100	80	66	57	49	44	39	36	32	30	28	26	24	22	21	20	19	18	17	16	15	15	14	13	13	12	12	UNITS
	59	27	33	13	0	36	0	48	11	41	0	55	19	4	7	24	53	32	19	13	14	19	29	43	0	21	44	9	37	7	MINUTES
	59	40	46	27	32	4	36	30	46	56	43	54	14	30	28	43	46	38	46	57	8	32	27	20	42	8	18	54	42	28	SECONDS
Recomputed shadow length of a 7-digit gnomon (rounded to seconds)	400	200	133	100	80	66	57	49	44	39	36	32	30	28	26	24	22	21	20	19	18	17	16	15	15	14	13	13	12	12	
	59	27	33	13	0	36	0	48	11	41	0	55	19	4	7	24	53	32	19	13	14	19	29	43	0	21	44	9	37	7	
	59	40	46	27	32	4	36	30	46	56	43	54	14	30	28	43	46	37	46	57	8	32	27	20	41	8	18	54	42	28	
Difference between the digits in corresponding sexagesimal places	+10	0	0	0	0	0	0	0	0	0	0	0	0	0	0	0	0	0	0	0	0	0	0	0	0	0	0	0	0	0	
	0	0	0	0	0	0	0	0	0	0	0	0	0	0	0	0	0	0	0	0	0	0	0	0	0	0	0	0	0	0	
	0	0	0	0	0	0	0	0	0	0	0	0	0	0	0	0	0	+1	0	0	0	0	0	0	+1	0	0	0	0	0	
Proposed emendation to the attested value in MS Tk	400	200	133	100	80	66	57	49	44	39	36	32	30	28	26	24	22	21	20	19	18	17	16	15	15	14	13	13	12	12	
	59	27	33	13	0	36	0	48	11	41	0	55	19	4	7	24	53	32	19	13	14	19	29	43	0	21	44	9	37	7	
	59	40	46	27	32	4	36	30	46	56	43	54	14	30	28	43	46	38	46	57	8	32	27	20	42	8	18	54	42	28	

Solar altitude	31	32	33	34	35	36	37	38	39	40	41	42	43	44	45	46	47	48	49	50	51	52	53	54	55	56	57	58	59	60	
Attested shadow length of a 7-digit gnomon in MS Tk	11	11	10	10	9	9	9	8	8	8	8	7	7	7	7	6	6	6	6	5	5	5	5	5	4	4	4	4	4	4	UNITS
	39	12	46	22	59	38	17	57	38	20	3	46	30	14	0	45	31	18	5	52	40	28	16	5	54	43	32	22	12	2	MINUTES
	0	8	45	41	49	5	42	35	39	32	9	27	24	55	0	35	39	10	6	25	7	8	30	9	5	18	45	27	21	29	SECONDS
Recomputed shadow length of a 7-digit gnomon (rounded to seconds)	11	11	10	10	9	9	9	8	8	8	8	7	7	7	7	6	6	6	6	5	5	5	5	5	4	4	4	4	4	4	
	39	12	46	22	59	38	17	57	38	20	3	46	30	14	0	45	31	18	5	52	40	28	16	5	54	43	32	22	12	2	
	0	8	45	40	49	5	42	34	40	32	9	28	23	55	0	35	39	10	6	25	6	8	30	9	5	18	45	27	22	29	
Difference between the digits in corresponding sexagesimal places	0	0	0	0	0	0	0	0	0	0	0	0	0	0	0	0	0	0	0	0	0	0	0	0	0	0	0	0	0	0	
	0	0	0	0	0	0	0	0	0	0	0	0	0	0	0	0	0	0	0	0	0	0	0	0	0	0	0	0	0	0	
	0	0	0	+1	0	0	0	+1	−1	0	0	−1	+1	0	0	0	0	0	0	0	+1	0	0	0	0	0	0	0	−1	0	
Proposed emendation to the attested value in MS Tk	11	11	10	10	9	9	9	8	8	8	8	7	7	7	7	6	6	6	6	5	5	5	5	5	4	4	4	4	4	4	
	39	12	46	22	59	38	17	57	38	20	3	46	30	14	0	45	31	18	5	52	40	28	16	5	54	43	32	22	12	2	
	0	8	45	41	49	5	42	35	39	32	9	27	24	55	0	35	39	10	6	25	7	8	30	9	5	18	45	27	21	29	

Solar altitude	61	62	63	64	65	66	67	68	69	70	71	72	73	74	75	76	77	78	79	80	81	82	83	84	85	86	87	88	89	90	
Attested shadow length of a 7-digit gnomon in MS Tk	3	3	3	3	3	3	2	2	2	2	2	2	2	2	1	1	1	1	1	1	1	0	0	0	0	0	0	0	0	0	UNITS
	52	43	34	24	15	7	59	49	41	32	24	16	8	0	52	44	36	29	21	14	16	59	51	44	36	29	14	7	7	0	MINUTES
	49	19	0	51	51	0	17	41	13	52	37	28	24	26	32	43	59	16	38	3	31	34	34	34	45	22	40	10	10	0	SECONDS
Recomputed shadow length of a 7-digit gnomon (rounded to seconds)	3	3	3	3	3	3	2	2	2	2	2	2	2	2	1	1	1	1	1	1	1	0	0	0	0	0	0	0	0	0	
	52	43	34	24	15	7	58	49	41	32	24	16	8	0	52	44	36	29	21	14	6	59	51	44	36	29	22	14	7	0	
	49	19	0	51	51	0	17	41	13	52	37	28	24	26	32	43	58	16	38	3	31	2	34	9	45	22	1	40	20	0	
Difference between the digits in corresponding sexagesimal places	0	0	0	0	0	0	0	0	0	0	0	0	0	0	0	0	0	0	0	0	0	0	0	0	0	0	0	0	0	0	
	0	0	0	0	0	0	+1	0	0	0	0	0	0	0	0	0	0	0	0	0	+10	0	0	0	0	0	−8	−7	0	0	
	0	0	0	0	0	0	0	0	0	0	0	0	0	0	0	0	+1	0	0	0	0	+32	0	+25	0	0	+39	−30	−10	0	
Proposed emendation to the attested value in MS Tk	3	3	3	3	3	3	2	2	2	2	2	2	2	2	1	1	1	1	1	1	1	0	0	0	0	0	0	0	0	0	
	52	43	34	24	15	7	58	49	41	32	24	16	8	0	52	44	36	29	21	14	6	59	51	44	36	29	22	14	7	0	
	49	19	0	51	51	0	17	41	13	52	37	28	24	26	32	43	59	16	38	3	31	2	34	9	45	22	1	40	20	0	

Table VI.D: Recomputations and analyses of the Table of lunar latitudes (*śara*) discussed in Sections 3.7–3.8

Lunar-nodal elongation	**1**	**2**	**3**	**4**	**5**	**6**	**7**	**8**	**9**	**10**	**11**	**12**	**13**	**14**	**15**	**16**	**17**	**18**	**19**	**20**	**21**	**22**	**23**	**24**	**25**	**26**	**27**	**28**	**29**	**30**	
Attested lunar latitude in MS Tk	0	0	0	0	0	0	0	0	0	0	0	1	1	1	1	1	1	1	1	1	1	1	1	1	1	1	2	2	2	2	DEGREES
	4	9	14	18	23	28	32	37	42	46	51	56	0	5	9	14	18	23	27	32	36	41	45	49	53	58	2	6	10	14	MINUTES
	43	25	8	48	29	11	51	32	11	50	28	4	40	14	47	19	49	19	48	13	38	1	22	42	59	14	26	37	44	52	SECONDS
Recomputed lunar latitude (rounded to seconds)	0	0	0	0	0	0	0	0	0	0	0	0	1	1	1	1	1	1	1	1	1	1	1	1	1	1	2	2	2	2	
	4	9	14	18	23	28	32	37	42	46	51	56	0	5	9	14	18	23	27	32	36	41	45	49	54	58	2	6	10	14	
	43	24	7	49	30	11	52	32	11	49	27	4	40	14	49	21	51	21	49	16	40	3	24	43	0	15	28	39	47	54	
Difference between the digits in corresponding sexagesimal places	0	0	0	0	0	0	0	0	0	0	0	+1	0	0	0	0	0	0	0	0	0	0	0	0	0	0	0	0	0	0	
	0	0	0	0	0	0	0	0	0	0	0	0	0	0	0	0	0	0	0	0	0	0	0	0	−1	0	0	0	0	0	
	0	+1	+1	−1	−1	0	−1	0	0	+1	+1	0	0	0	−2	−2	−2	−2	−1	−3	−2	−2	−2	−1	+59	−1	−2	−2	−3	−2	
Proposed emendation to the attested value in MS Tk	0	0	0	0	0	0	0	0	0	0	0	(0)	1	1	1	1	1	1	1	1	1	1	1	1	1	1	2	2	2	2	
	4	9	14	18	23	28	32	37	42	46	51	56	0	5	9	14	18	23	27	32	36	41	45	49	53	58	2	6	10	14	
	43	25	8	48	29	11	51	32	11	50	28	4	40	14	47	19	49	19	48	13	38	1	22	42	59	14	26	37	44	52	

Lunar-nodal elongation	**31**	**32**	**33**	**34**	**35**	**36**	**37**	**38**	**39**	**40**	**41**	**42**	**43**	**44**	**45**	**46**	**47**	**48**	**49**	**50**	**51**	**52**	**53**	**54**	**55**	**56**	**57**	**58**	**59**	**60**	
Attested lunar latitude in MS Tk	2	2	2	2	2	2	2	2	2	2	2	3	3	3	3	3	3	3	3	3	3	3	3	3	3	3	3	3	3	3	DEGREES
	18	22	26	30	34	38	42	46	49	53	57	0	4	7	10	14	17	20	23	26	29	32	35	38	41	43	[--]	[--]	[--]	[--]	MINUTES
	56	57	55	50	44	35	22	6	47	25	0	31	0	36	48	6	20	32	39	43	44	40	32	19	5	45	[--]	[--]	[--]	[--]	SECONDS
Recomputed lunar latitude (rounded to seconds)	2	2	2	2	2	2	2	2	2	2	2	3	3	3	3	3	3	3	3	3	3	3	3	3	3	3	3	3	3	3	
	18	22	26	30	34	38	42	46	49	53	57	0	4	7	10	14	17	20	23	26	29	32	35	38	41	43	46	48	51	53	
	56	58	56	52	44	35	22	6	47	25	0	32	0	26	48	6	21	33	40	44	44	40	32	20	5	45	21	53	21	45	
Difference between the digits in corresponding sexagesimal places	0	0	0	0	0	0	0	0	0	0	0	0	0	0	0	0	0	0	0	0	0	0	0	0	0	0	0	0	0	0	
	0	0	0	0	0	0	0	0	0	0	0	0	0	0	0	0	0	0	0	0	0	0	0	0	0	0	[--]	[--]	[--]	[--]	
	0	−1	−1	−2	0	0	0	0	0	0	0	−1	0	+10	0	0	−1	−1	−1	−1	0	0	0	−1	0	0	[--]	[--]	[--]	[--]	
Proposed emendation to the attested value in MS Tk	2	2	2	2	2	2	2	2	2	2	2	3	3	3	3	3	3	3	3	3	3	3	3	3	3	3	3	3	3	3	
	18	22	26	30	34	38	42	46	49	53	57	0	4	7	10	14	17	20	23	26	29	32	35	38	41	43	[--]	[--]	[--]	[--]	
	56	57	55	50	44	35	22	6	47	25	0	31	0	(26)	48	6	20	32	39	43	44	40	32	19	5	45	[--]	[--]	[--]	[--]	

Lunar-nodal elongation	**61**	**62**	**63**	**64**	**65**	**66**	**67**	**68**	**69**	**70**	**71**	**72**	**73**	**74**	**75**	**76**	**77**	**78**	**79**	**80**	**81**	**82**	**83**	**84**	**85**	**86**	**87**	**88**	**89**	**90**	
Attested lunar latitude in MS Tk	3	3	4	4	4	4	4	4	4	4	4	4	4	4	4	4	4	4	4	4	4	4	4	4	4	4	4	4	4	4	DEGREES
	56	58	0	2	4	6	8	10	12	14	16	16	18	20	21	22	24	24	25	25	25	26	27	28	28	29	29	29	29	30	MINUTES
	4	19	31	37	38	36	29	17	1	40	13	45	10	31	46	57	2	5	0	52	39	21	59	32	20	37	50	57	57	0	SECONDS
Recomputed lunar latitude (rounded to seconds)	3	3	4	4	4	4	4	4	4	4	4	4	4	4	4	4	4	4	4	4	4	4	4	4	4	4	4	4	4	4	
	56	58	0	2	4	6	8	10	12	13	15	16	18	19	20	21	23	24	25	25	26	27	27	28	28	29	29	29	29	30	
	4	20	30	36	38	36	29	17	0	40	15	45	10	30	46	57	3	4	1	52	38	21	58	29	57	19	37	49	56	0	
Difference between the digits in corresponding sexagesimal places	0	0	0	0	0	0	0	0	0	0	0	0	0	0	0	0	0	0	0	0	0	0	0	0	0	0	0	0	0	0	
	0	0	0	0	0	0	0	0	0	+1	+1	0	0	+1	+1	+1	+1	0	0	0	−1	−1	0	0	0	0	0	0	0	0	
	0	−1	+1	+1	0	0	0	0	+1	0	−2	0	0	+1	0	0	−1	+1	−1	0	+1	0	+1	+3	−37	+18	+13	+8	+1	0	
Proposed emendation to the attested value in MS Tk	3	3	4	4	4	4	4	4	4	4	4	4	4	4	4	4	4	4	4	4	4	4	4	4	4	4	4	4	4	4	
	56	58	0	2	4	6	8	10	12	14	16	16	18	20	21	22	24	24	25	25	25	26	27	28	28	29	29	29	29	30	
	4	19	31	37	38	36	29	17	1	40	13	45	10	31	46	57	2	5	0	52	39	21	59	32	(57)	(20)	(37)	(50)	57	0	

Appendix C: Statistical Analysis

C.1. Choosing systematic rounding over truncation

All regular recomputations in this study express sexagesimal numbers up to the second fractional place. To reduce a sexagesimal number in the final result of a recomputation, I chose to *systematically round* the number to the seconds place instead of *truncating* it; in other words, for a number of the form $a; b, c, d$, I round the number to $a; b, c$ when $d < 30$ or $a; b, c + 1$ when $d \geq 30$ (instead of truncating it to $a; b, c$ for any value of d). To validate this choice, I statistically test the proportion of differences between the attested and recomputed values when two mutually independent strategies are used to reduce the final result, namely, systematic rounding and truncation. In both reduction strategies, computing the differences between the attested values and the recomputed results are considered as binary events, i.e. they generate zero (0-state) or non-zero values (1-state) of the differences. The z-test for two population proportions is then used to test the efficacy of these two strategies in minimising the proportion of the differences for every table from the selected corpus. The parameters, hypotheses, and test statistic in implementing this test are described below.

1. The ninety reduced entries (i.e. the final results) using systematic rounding and those using truncation are considered as two independent populations with a common size. The total number of determinate events n_{det} is selected as the common sample size from both populations. The determinate events are those instances where a clear distinction can be made between the choice of sexagesimal reduction.[53] For every table, the reduced sample size n_{det} is large enough (i.e. greater than thirty) to assume normality, and the individual events (in the 0-state or 1-state) in the sample are mutually independent.

2. In the two samples of size n_{det}, $x_{\text{det}}^{\text{sys.rnd}}$ indicates the number of 1-state events (i.e. those producing non-zero differences between the attested and recomputed values) generated by the first population (systematic rounding) and $x_{\text{det}}^{\text{trunc}}$ indicates the 1-state events generated by the second population (truncation). With these values, the sample proportions for the two populations are computed as

[53] For a sexagesimal result $a; b, c, d$ with $d \leq 29$, systematic rounding or truncation reduce the number to $a; b, c$ identically. Such instances are called *indeterminate* events n_{indet} as the two reduction strategies are indistinguishable. The present analysis only includes *determinate* events n_{det} where the reduction strategies can be clearly identified from one another; in other words, cases where the recomputed results are $a; b, c, d$ with $d \geq 30$ (and hence reduced to $a; b, c + 1$ by systematic rounding or $a; b, c$ by truncation). For every table, $n_{\text{det}} + n_{\text{indet}} = 90$.

$$\hat{p}^{\text{sys.rnd}}_{\text{det}} = \frac{x^{\text{sys.rnd}}_{\text{det}}}{n_{\text{det}}} \quad \text{and} \quad \hat{p}^{\text{trunc}}_{\text{det}} = \frac{x^{\text{trunc}}_{\text{det}}}{n_{\text{det}}}.$$

3. To statistically test:

   the null hypothesis H_0: $\hat{p}^{\text{sys.rnd}}_{\text{det}} \leq \hat{p}^{\text{trunc}}_{\text{det}}$ against
   the alternative hypothesis H_a: $\hat{p}^{\text{sys.rnd}}_{\text{det}} > \hat{p}^{\text{trunc}}_{\text{det}}$.

   The null hypothesis maintains that the proportion of 1-state events in the first population is lower or equal to those in the second population, whereas the alternative hypothesis claims the converse. In other words, the null hypothesis expresses the belief that systematic rounding is statistically better (or at least, equivalent to) truncating the digits when the two reduction strategies are compared. The alternative hypothesis, if true, shows that truncating the digits, instead of systematically rounding them, is significantly better at minimising the non-zero differences between the attested and recomputed results.
4. The test statistic based on the pooled sample proportion is:

$$z\text{-statistic: } z = \frac{\hat{p}^{\text{sys.rnd}}_{\text{det}} - \hat{p}^{\text{trunc}}_{\text{det}}}{\sqrt{\hat{p}_{\text{det}} \times (1 - \hat{p}_{\text{det}}) \times \left(\frac{2}{n_{\text{det}}}\right)}},$$

   where

$$\hat{p}_{\text{det}} \equiv \frac{\hat{p}^{\text{sys.rnd}}_{\text{det}} \times n_{\text{det}} + \hat{p}^{\text{trunc}}_{\text{det}} \times n_{\text{det}}}{n_{\text{det}} + n_{\text{det}}} = \frac{x^{\text{sys.rnd}}_{\text{det}} + x^{\text{trunc}}_{\text{det}}}{2n_{\text{det}}}$$

   is the pooled proportion. For every table, n_{det} is large enough to ensure $\hat{p}_{\text{det}} \times n_{\text{det}} \geq 5$ and $(1 - \hat{p}_{\text{det}}) \times n_{\text{det}} \geq 5$. This allows the z-statistic to be validly applied.
5. The hypothesis is tested at a 5% level of significance α using the right-tailed z-test for two population proportions. For $\alpha = 0.05$, the decision rule is:

   Reject H_0 $\forall$ $z \in R$, where the rejection region $R := \{z : z > 1.645\}$.

   The critical value of the right-tailed z-test is taken as $z_c \equiv z_\alpha = 1.645$.

As Table 3 shows, the calculated z-statistic lies outside the rejection region for all six tables of the selected corpus, and therefore, the null hypothesis H_0 is retained and the alternative H_a is rejected. At a 5% level of significance, the proportion of non-zero differences between the attested and recomputed values using systematic rounding is lower (or at the very least, equal to) the proportion when truncation is used. The recomputations in this study, in particular, the final results of a calculation, are reduced to seconds by systematically rounding the digits based on this statistical inference.

Type of recomputation	Sexagesimal reduction strategies	
	Systematic rounding	Truncation
Sines Table VI.A	$n_{\text{det}} = 46$ and $n_{\text{indet}} = 44$	
	$x_{\text{det}}^{\text{sys.rnd}} = 3, \quad \hat{p}_{\text{det}}^{\text{sys.rnd}} \approx 0.065$	$x_{\text{det}}^{\text{trunc}} = 46, \quad \hat{p}_{\text{det}}^{\text{trunc}} = 1$
	$\hat{p}_{\text{det}} = {}^{49}/_{92} \approx 0.533$ and $z \approx {}^{-0.935}/_{0.104} \approx -8.985 < z_c = 1.645$	
	$\because z \notin R \Rightarrow$ Accept H_0 and reject H_a	
	statistically preferred	statistically rejected
Solar declinations Table VI.B	$n_{\text{det}} = 45$ and $n_{\text{indet}} = 45$	
	$x_{\text{det}}^{\text{sys.rnd}} = 26, \quad \hat{p}_{\text{det}}^{\text{sys.rnd}} = 0.578$	$x_{\text{det}}^{\text{trunc}} = 26, \quad \hat{p}_{\text{det}}^{\text{trunc}} \approx 0.578$
	$\hat{p}_{\text{det}} = {}^{52}/_{90} \approx 0.578$ and $z \approx {}^{0}/_{0.104} = 0 < z_c = 1.645$	
	$\because z \notin R \Rightarrow$ Accept H_0 and reject H_a	
	statistically preferred	statistically rejected
Shadow lengths: 60-digit gnomon Table VI.C_1	$n_{\text{det}} = 42$ and $n_{\text{indet}} = 48$	
	$x_{\text{det}}^{\text{sys.rnd}} = 19, \quad \hat{p}_{\text{det}}^{\text{sys.rnd}} \approx 0.452$	$x_{\text{det}}^{\text{trunc}} = 23, \quad \hat{p}_{\text{det}}^{\text{trunc}} \approx 0.548$
	$\hat{p}_{\text{det}} = {}^{42}/_{84} = 0.5$ and $z \approx {}^{-0.096}/_{0.109} \approx -0.881 < z_c = 1.645$	
	$\because z \notin R \Rightarrow$ Accept H_0 and reject H_a	
	statistically preferred	statistically rejected
Shadow lengths: 12-digit gnomon Table VI.C_2	$n_{\text{det}} = 48$ and $n_{\text{indet}} = 42$	
	$x_{\text{det}}^{\text{sys.rnd}} = 13, \quad \hat{p}_{\text{det}}^{\text{sys.rnd}} \approx 0.271$	$x_{\text{det}}^{\text{trunc}} = 37, \quad \hat{p}_{\text{det}}^{\text{trunc}} = 0.771$
	$\hat{p}_{\text{det}} = {}^{50}/_{96} \approx 0.521$ and $z \approx {}^{-0.5}/_{0.102} \approx -4.903 < z_c = 1.645$	
	$\because z \notin R \Rightarrow$ Accept H_0 and reject H_a	
	statistically preferred	statistically rejected
Shadow lengths: 7-digit gnomon Table VI.C_3	$n_{\text{det}} = 41$ and $n_{\text{indet}} = 49$	
	$x_{\text{det}}^{\text{sys.rnd}} = 5, \quad \hat{p}_{\text{det}}^{\text{sys.rnd}} = 0.122$	$x_{\text{det}}^{\text{trunc}} = 38, \quad \hat{p}_{\text{det}}^{\text{trunc}} = 0.927$
	$\hat{p}_{\text{det}} = {}^{43}/_{82} \approx 0.524$ and $z \approx {}^{-0.805}/_{0.110} \approx -7.297 < z_c = 1.645$	
	$\because z \notin R \Rightarrow$ Accept H_0 and reject H_a	
	statistically preferred	statistically rejected
Lunar latitudes Table VI.D	$n_{\text{det}} = 49$ and $n_{\text{indet}} = 41$	
	$x_{\text{det}}^{\text{sys.rnd}} = 29, \quad \hat{p}_{\text{det}}^{\text{sys.rnd}} \approx 0.592$	$x_{\text{det}}^{\text{trunc}} = 36, \quad \hat{p}_{\text{det}}^{\text{trunc}} \approx 0.735$
	$\hat{p}_{\text{det}} = {}^{65}/_{98} \approx 0.663$ and $z \approx {}^{-0.143}/_{0.095} \approx -1.496 < z_c = 1.645$	
	$\because z \notin R \Rightarrow$ Accept H_0 and reject H_a	
	statistically preferred	statistically rejected

Table 3: Statistical test (right-tailed z-test for two population proportions) to select between systematic rounding or truncation (two mutually independent reduction strategies) to reduce the final results of the recomputations to the second fractional place for the six tables from MS Tk.

C.2. Choosing recomputed Sines over the attested Sines in MS Tk

The solar declinations, shadow lengths of gnomons of various heights, and lunar latitudes in this study are calculated using the recomputed Sines (Sin_r) instead of the attested Sines (Sin_a) in MS Tk.[54] I justify this choice on the basis of the following two statistical measures:

1. The first measure compares the differences d_is between the attested values of these functions (from MS Tk) and their recomputed values using Sin_r and Sin_a separately, i.e.

$$d_i^{\text{Sin}_a} = |\text{Value}_i^{\text{recomp}}[\text{Sin}_a] - \text{Value}_i^{\text{attest}}| \qquad \text{and}$$
$$d_i^{\text{Sin}_r} = |\text{Value}_i^{\text{recomp}}[\text{Sin}_r] - \text{Value}_i^{\text{attest}}| \forall\ i \in \mathbb{N}_{90}$$

   Similar to the 0-state and 1-state described in Appendix C.1, these differences (i.e. $d_i^{\text{Sin}_a}$ and $d_i^{\text{Sin}_r}$) are considered as binary events. Accordingly, I consider

   - $x_{\text{sim}}^{\text{Sin}_a}$ and $x_{\text{sim}}^{\text{Sin}_r}$ as the number of 0-states (i.e. instances when the differences d_is are similar or zero) using the attested and recomputed Sines respectively, and
   - $x_{\text{diss}}^{\text{Sin}_a}$ and $x_{\text{diss}}^{\text{Sin}_r}$ as the number of 1-states (i.e. instances when the differences d_is are dissimilar or non-zero) using the attested and recomputed Sines respectively.

   For a total of $n = 90$ entries for each function, the proportion of 0 and 1 states using Sin_r and Sin_a separately can be expressed as

$$p_{\text{sim}}^{Sin_a} = \frac{x_{\text{sim}}^{Sin_a}}{n}, \quad p_{\text{sim}}^{Sin_r} = \frac{x_{\text{sim}}^{Sin_r}}{n}, \quad p_{\text{diss}}^{Sin_a} = \frac{x_{\text{diss}}^{Sin_a}}{n}, \quad \text{and} \quad p_{\text{diss}}^{Sin_r} = \frac{x_{\text{diss}}^{Sin_r}}{n}.$$

   Table 4 presents these four proportions (in percentages) for the recomputations of the solar declinations, shadow lengths of gnomons of various heights, and lunar latitudes in 2×2 contingency tables. For lunar latitudes, the attested values for 57°, 58°, 59 °, and 60° are illegible in MS Tk, and accordingly, $n = 86$ for calculating these proportions. The percentage proportion of dissimilar (non-zero) differences between the attested and recomputed function values are typically lower (or, at the very least, comparably equal) when recomputed Sines are used instead of the attested Sines from MS Tk. Equivalently, the percentage

[54] I use the attested Sine values with my proposed emendations (to correct for scribal discrepancies) in this analysis. For example, the attested $\text{Sin}_a 1°$ is taken as 1;2,50 instead of 1;5,50 (seen in MS Tk). Without these emendation, the recomputed function values based on the attested Sines become highly irregular and statistically superfluous. Also, for all calculations in this analysis, the final sexagesimal results are systematically rounded to the second fractional place.

Solar declinations		
Differences / Sines	0-state (similar or zero)	1-state (dissimilar or non-zero)
Attested	$p^{Sin_a}_{\text{sim}} = {}^{50}/_{90} \approx 55.56\%$	$p^{Sin_a}_{\text{diss}} = {}^{40}/_{90} \approx 44.44\%$
Recomputed	$p^{Sin_r}_{\text{sim}} = {}^{50}/_{90} \approx 55.56\%$	$p^{Sin_r}_{\text{diss}} = {}^{40}/_{90} \approx 44.44\%$

Shadow lengths: 60-digit gnomon		
Differences / Sines	0-state (similar or zero)	1-state (dissimilar or non-zero)
Attested	$p^{Sin_a}_{\text{sim}} = {}^{49}/_{90} \approx 54.44\%$	$p^{Sin_a}_{\text{diss}} = {}^{41}/_{90} \approx 45.56\%$
Recomputed	$p^{Sin_r}_{\text{sim}} = {}^{58}/_{90} \approx 64.44\%$	$p^{Sin_r}_{\text{diss}} = {}^{32}/_{90} \approx 35.56\%$

Shadow lengths: 12-digit gnomon		
Differences / Sines	0-state (similar or zero)	1-state (dissimilar or non-zero)
Attested	$p^{Sin_a}_{\text{sim}} = {}^{52}/_{90} \approx 57.78\%$	$p^{Sin_a}_{\text{diss}} = {}^{38}/_{90} \approx 42.22\%$
Recomputed	$p^{Sin_r}_{\text{sim}} = {}^{63}/_{90} = 70\%$	$p^{Sin_r}_{\text{diss}} = {}^{27}/_{90} = 30\%$

Shadow lengths: 7-digit gnomon		
Differences / Sines	0-state (similar or zero)	1-state (dissimilar or non-zero)
Attested	$p^{Sin_a}_{\text{sim}} = {}^{67}/_{90} \approx 74.44\%$	$p^{Sin_a}_{\text{diss}} = {}^{23}/_{90} \approx 25.56\%$
Recomputed	$p^{Sin_r}_{\text{sim}} = {}^{80}/_{90} \approx 88.89\%$	$p^{Sin_r}_{\text{diss}} = {}^{10}/_{90} \approx 11.11\%$

Lunar latitudes		
Differences / Sines	0-state (similar or zero)	1-state (dissimilar or non-zero)
Attested	$p^{Sin_a}_{\text{sim}} = {}^{14}/_{86} \approx 16.28\%$	$p^{Sin_a}_{\text{diss}} = {}^{72}/_{86} \approx 83.72\%$
Recomputed	$p^{Sin_r}_{\text{sim}} = {}^{34}/_{86} \approx 39.53\%$	$p^{Sin_r}_{\text{diss}} = {}^{52}/_{86} \approx 60.47\%$

Table 4: 2×2 contingency tables showing the proportions of differences (in percentages) between the attested and recomputed values of solar declinations, shadow lengths for gnomons of various heights, and lunar latitudes calculated using the attested Sines (in MS Tk) and the recomputed Sines separately.

Type of Recomputation	Sines	
	Attested	Recomputed
Solar declinations	RMSD $\approx 7.258^s$ AAD $\approx 2.779^s$	RMSD $\approx 1.265^s$ AAD $\approx 0.689^s$
Shadow lengths: 60-digit gnomon	RMSD $\approx 380.840^s$ AAD $= 52.7^s$	RMSD $\approx 379.528^s$ AAD $\approx 41.067^s$
Shadow lengths: 12-digit gnomon	RMSD $\approx 6.536^s$ AAD $\approx 2.811^s$	RMSD $\approx 0.767^s$ AAD $= 0.367^s$
Shadow lengths: 7-digit gnomon	RMSD $\approx 3.485^s$ AAD $\approx 1.367^s$	RMSD $\approx 0.333^s$ AAD $\approx 0.111^s$
Lunar latitudes	RMSD $\approx 18.161^s$ AAD $\approx 6.779^s$	RMSD $\approx 18.219^s$ AAD $\approx 6.256^s$

Table 5: Table comparing the Root Mean Square Deviation (RMSD) and Average Absolute Deviation (AAD) (both measures in seconds) in recomputing the solar declinations, shadow lengths of gnomons of various heights, and lunar latitudes using the attested Sines (in MS Tk) and the recomputed Sines separately.

of similar (zero) differences between the attested and recomputed function values are typically higher (or comparably equal) when recomputed Sines are used. This provides the first measure of validation for using the recomputed Sines in calculating the other functions in this study.

2. In addition to the percentage proportions of differences, I calculate the Root Mean Square Deviation (RMSD) and the Average Absolute Deviation (AAD) for an Ordinary Least Squares (OLS) regression model as a second statistical measure to validate my choice. For each recomputed function, treating the attested $\text{Value}_i^{\text{attest}}$ as the predicted value y_i and the recomputed $\text{Value}_i^{\text{recomp}}[\text{Sin}_a]$ or $\text{Value}_i^{\text{recomp}}[\text{Sin}_r]$ as the observed value $\hat{y}_i^\alpha$ (α being Sin_a or Sin_r, and $i \in \mathbb{N}_{90}$), the ith residual is $e_i^\alpha = \hat{y}_i^\alpha - y_i$ (among the total $n = 90$ residuals). With this

$$\text{RMSD} = \sqrt{\frac{\sum_i^n (e_i^\alpha)^2}{n}} \equiv \sqrt{\frac{\sum_i^n (\hat{y}_i^\alpha - y_i)^2}{n}} \quad \text{and}$$

$$\text{AAD} = \sqrt{\frac{\sum_i^n |e_i^\alpha|}{n}} \equiv \sqrt{\frac{\sum_i^n |\hat{y}_i^\alpha - y_i|}{n}}.$$

The RMSD measures the square root of the variance of the residual; in other words, it indicates the standard deviation of the unexplained variance between the prediction and the observation. The AAD indicates the absolute average value of the residual, i.e. the average difference between the attested and recomputed values of the functions. Both measures of fit are absolute measures (in the units of the entries themselves) with lower values indicating a better fit. In OLS regression models, RMSD and AAD are used to indicate how accurately a model predicts the response. Table 5 lists the RMSD and AAD values (in seconds) for my recomputations of the solar declinations, shadow lengths of gnomons of various heights, and lunar latitudes using the attested and recomputed Sines. (The lunar latitude calculations use $n = 86$ as four attested entries in MS Tk are illegible.) The RMSD and AAD values are lower in most recomputations when recomputed Sines are used (instead of the attested Sines in MS Tk), and thus, provide a second reason to choose recomputed Sines to calculate the other functions in this study.[55]

C.3. Choosing the exact expression of lunar latitude over the approximate one

In this study, the lunar latitude β is recomputed for each degree of lunar-nodal elongation ω using the exact expression $\mathrm{Sin}\,\beta = \mathrm{Sin}\,4°30' \times {}^{\mathrm{Sin}\,\omega}/_{60}$ instead of the approximate expression $\beta \approx 4°30' \times {}^{\mathrm{Sin}\,\omega}/_{60}$. I justify this choice based on the following two statistical measures:

1. The first measure compares the proportion of differences between the attested and recomputed lunar latitudes when the two expressions are used separately. Similar to the first statistical measure in Appendix C.2 (note 1), the proportions of the 0-state (similar or zero) and 1-state (dissimilar or non-zero) differences using the exact and approximate expressions of lunar latitudes separately can be calculated as

$$p^{\text{exact}}_{\text{sim}} = \frac{x^{\text{exact}}_{\text{sim}}}{n}, \quad p^{\text{approx}}_{\text{sim}} = \frac{x^{\text{approx}}_{\text{sim}}}{n}, \quad p^{\text{exact}}_{\text{diss}} = \frac{x^{\text{exact}}_{\text{diss}}}{n}, \quad \text{and } p^{\text{approx}}_{\text{diss}} = \frac{x^{\text{approx}}_{\text{diss}}}{n}.$$

   where $x^{\text{exact}}_{\text{sim}}$ and $x^{\text{approx}}_{\text{sim}}$ are the number of 0-states using the respective expressions; $x^{\text{exact}}_{\text{diss}}$ and $x^{\text{approx}}_{\text{diss}}$ are the number of 1-states using the respective expressions; and $n = 86$ (since four entries corresponding to the arguments 57° to 60° are illegible in MS Tk). Table 6 presents these four proportions (in percentages) for the lunar latitude recomputations in a 2×2 contingency table. Following previous calculations, the final sexagesimal results are systematically rounded to the second fractional place, and recomputed Sines (instead of the attested Sines in MS Tk) are used. The percentage proportion of dissimilar (non-zero) differences between the attested and recomputed lunar latitudes is lower when the exact expression is used instead of the approximate one. Or equivalently,

[55] The RMSD is sensitive to outliers as the effect of each residual is proportional to the size of its squared value. On account of this, the RMSD value for the lunar latitude recomputations using recomputed Sines is slightly larger than the corresponding value using attested Sines in Table 5.

Expressions \ Differences	0 state (similar or zero)	1 state (dissimilar or non-zero)
Exact	$p^{\text{exact}}_{\text{sim}} = {}^{34}/_{86} \approx 39.53\%$	$p^{\text{exact}}_{\text{diss}} = {}^{52}/_{86} \approx 60.47\%$
Approximate	$p^{\text{approx}}_{\text{sim}} = {}^{10}/_{86} \approx 11.63\%$	$p^{\text{approx}}_{\text{diss}} = {}^{80}/_{86} \approx 93.02\%$

Table 6: 2×2 contingency table showing the proportions of differences (in percentages) between the attested and recomputed values of lunar latitudes calculated using the exact and approximate expressions separately.

the percentage of similar (zero) differences between the attested and recomputed lunar latitudes is higher when the exact expression is used. This provides the first measure of validation for using the exact expression to recompute lunar latitudes.

2. I calculate the Median Absolute Deviation (MAD) of the differences between the attested and recomputed lunar latitudes using the exact and approximate expressions separately to establish the second statistical measure. With the ith difference $d_i = \text{Value}_i^{\text{recomp}}[\tilde{\alpha}] - \text{Value}_i^{\text{attest}}$ where $\tilde{\alpha}$ is the exact or approximate expression and $i \in \mathbb{N}_{86}$,

$$\text{MAD} = \text{Median}\left(|d_i - \text{Median}(d_i)|\right).$$

MAD provides a robust measure of the variability of the differences with non-normal distributions.[56] With it, a median-centred interval $[\nu - 2\,\text{MAD}, \nu + 2\,\text{MAD}]$ can be constructed to identify statistical outliers that lie outside the limits. Table 7 provides the descriptive statistics for 86 entries of d_is using the exact and approximate expressions of lunar latitude. When the exact expression is used,

- 75 entries (out of 86) are within $\pm 2\,\text{MAD}$ of the median, in other words, a set of 75 differences $d_i^{\text{corrected}} \in [-2, 2]$ are statistically relevant; while
- 78 entries (out of 86) are within $\pm 2\,\text{MAD}$ of the median when the appropriate expression is used, i.e. 78 differences $d_i^{\text{corrected}} \in [-7, 17]$ are statistically relevant.

Among these outlier-corrected differences $d_i^{\text{corrected}}$,

- there are 41 dissimilar (non-zero) differences out of 75, i.e. around 54.67%, when the exact expression is used, and
- there are 72 dissimilar (non-zero) differences out of 78, i.e. around 92.03%, when the approximate expression is used.

[56] Typically, a normal distribution has skewness $\varsigma \sim 0$ and kurtosis $\kappa \sim 3$, with the mean $\mu \sim$ median ν. As Table 7 shows, the differences between the attested and recomputed lunar latitudes using the exact and approximate expressions are not normally distributed.

Type of Recomputation	Expressions of lunar latitude	
	Exact	Approximate
Median Absolute Deviation (MAD)	1	6
Median $\nu \equiv \text{Median}(d_i)$	0	5
Mean μ	≈ -2.442	≈ 1.953
Standard Deviation σ	≈ 18.055	≈ 18.578
Skewness $\varsigma(d_i)$	≈ -1.279	≈ -1.509
Kurtosis $\kappa(d_i)$	≈ 7.275	≈ 6.789

Table 7: Table showing the descriptive statistics, including the Median Absolute Deviation (MAD) of the differences between the attested and recomputed lunar latitudes calculated using the exact and approximate expressions separately.

The lower percentage of outlier-corrected dissimilar (non-zero) differences between the attested and recomputed lunar latitudes using the exact expression (compared to the approximate one) validates its choice in this study.

C.4. Choosing the parameter Sin $4°30' = 4;45,25$ for lunar latitude recomputations

In this study, the lunar latitude β is recomputed for each degree of lunar-nodal elongation ω using the exact expression with the parameter Sin $4°30' = 4;42,25$ instead of 4;42,26 or 4;42,27.[57] I justify this choice based on the following two statistical measures:

1. The first measure compares the proportion of differences between the attested and recomputed values when the three estimates of the parameter Sin $4°30'$ are used separately. Similar to the first statistical measures in Appendices C.2–3 (note 1), the proportion of the 0-state (similar or zero) and 1-state (dissimilar or non-zero) differences can be separately calculated using 4;42,25, 4;42,26, and 4;42,27 as

$$p^{25^s}_{\text{sim}} = \frac{x^{25^s}_{\text{sim}}}{n}, \quad p^{26^s}_{\text{sim}} = \frac{x^{26^s}_{\text{sim}}}{n}, \quad p^{26^s}_{\text{sim}} = \frac{x^{27^s}_{\text{sim}}}{n},$$

$$p^{25^s}_{\text{diss}} = \frac{x^{25^s}_{\text{diss}}}{n}, \quad p^{26^s}_{\text{diss}} = \frac{x^{26^s}_{\text{diss}}}{n}, \quad \text{and } p^{26^s}_{\text{diss}} = \frac{x^{27^s}_{\text{diss}}}{n},$$

where $x^{25^s}_{\text{sim}}$, $x^{26^s}_{\text{sim}}$, and $x^{27^s}_{\text{sim}}$ are the 0-states using 4;42,25, 4;42,26, and 4;42,27 respectively; $x^{25^s}_{\text{diss}}$, $x^{26^s}_{\text{diss}}$, and $x^{27^s}_{\text{diss}}$ are the 1-states using the same

[57] The different estimates of the parameter Sin $4°30'$ are derived using different methods, see Section 3.7 (note 2).

Differences / Sin 4°30′	0 state (similar or zero)	1 state (dissimilar or non-zero)
4;42,25	$p^{25^s}_{\text{sim}} = {}^{38}/_{86} \approx 39.53\%$	$p^{25^s}_{\text{diss}} = {}^{52}/_{86} \approx 60.47\%$
4;42,26	$p^{26^s}_{\text{sim}} = {}^{30}/_{86} \approx 34.88\%$	$p^{26^s}_{\text{diss}} = {}^{60}/_{86} \approx 69.77\%$
4;42,27	$p^{27^s}_{\text{sim}} = {}^{14}/_{86} \approx 16.28\%$	$p^{27^s}_{\text{diss}} = {}^{76}/_{86} \approx 88.37\%$

Table 8: 3×2 contingency table showing the proportions of differences (in percentages) between the attested and recomputed values of lunar latitudes calculated with the parametric estimates 4;42,25, 4;42,26, and 4;42,27 separately.

parametric estimates respectively; and $n = 86$ (since four entries corresponding to the arguments 57° to 60° are illegible in MS Tk). Table 8 presents these six proportions (in percentages) for the lunar latitude recomputations in a 3×2 contingency table. Like the previous calculations, the final sexagesimal results are systematically rounded to the second fractional place, and recomputed Sines (instead of those attested in MS Tk) are used. The percentage proportions of dissimilar (non-zero) differences between the attested and recomputed lunar latitudes is lower with the parametric estimate 4;42,25 instead of 4;42,26 or 4;42,27. Or equivalently, the percentage of similar (zero) differences between the attested and recomputed lunar latitudes is higher when 4;42,25 is used. This provides the first measure to statistically validate using $\text{Sin}\,4°30' = 4;42,25$ to recompute the lunar latitudes.

2. The second statistical measure uses the Median Absolute Deviation (MAD) calculated for the three parametric estimates separately. As described in note 2 of Appendix C.3, the MAD values determines a median-centred interval $[\nu - 2\,\text{MAD}, \nu + 2\,\text{MAD}]$ of differences d_i s between the attested and recomputed lunar latitudes for each of the three parametric estimates. Table 9 provides the descriptive statistics for 86 entries of d_i s calculated with the parametric estimates 4;42,25, 4;42,26, and 4;42,27 separately.
   - Using 4;42,25, 75 entries (out of 86) are within ±2 MAD of the median, i.e. 75 differences $d_i^{\text{corrected}} \in [-2, 2]$ are statistically relevant;
   - using 4;42,26, 77 entries (out of 86) are within ±2 MAD of the median, i.e. 77 differences $d_i^{\text{corrected}} \in [-1, 3]$ are statistically relevant; and
   - using 4;42,27, 78 entries (out of 86) are within ±2 MAD of the median, i.e. 78 differences $d_i^{\text{corrected}} \in [-2, 4]$ are statistically relevant.

Type of Recomputation	Sin $4°30'$		
	4;42,25	4;42,26	4;42,27
Median Absolute Deviation (MAD)	1	1	1.5
Median $\nu \equiv \text{Median}(d_i)$	0	1	1
Mean μ	≈ -2.442	≈ -1.930	≈ -1.291
Standard Deviation σ	≈ 18.055	≈ 18.086	≈ 18.072
Skewness $\varsigma(d_i)$	≈ -1.279	≈ -1.741	≈ -1.638
Kurtosis $\kappa(d_i)$	≈ 7.275	≈ 7.64	≈ 7.690

Table 9: Table showing the descriptive statistics, including the Median Absolute Deviation (MAD) of the differences between the attested and recomputed lunar latitudes calculated with the parametric estimates 4;42,25, 4;42,26, and 4;42,27 separately.

Among these outlier-corrected differences $d_i^{\text{corrected}}$,

- there are 41 dissimilar (non-zero) difference out of 75, i.e. around 54.67%, when $\text{Sin}\,4°30' = 4;42,25$;
- there are 50 dissimilar (non-zero) difference out of 77, i.e. around 64.94%, when $\text{Sin}\,4°30' = 4;42,26$; and
- there are 70 dissimilar (non-zero) difference out of 78, i.e. around 89.74%, when $\text{Sin}\,4°30' = 4;42,27$.

The lower percentage of outlier-corrected dissimilar (non-zero) differences between the attested and recomputed lunar latitudes calculated with the parameter $\text{Sin}\,4°30' = 4;42,25$ (compared to the estimates 4;42,26 and 4;42,27) validates its choice in this study.

Part 3

Computational Practices and Table Cracking

Tables of Sunrise and Sunset in Yuan and Ming China (1271–1644) and their Adoption in Korea*

Li Liang

Introduction

Tables that record times of sunrise and sunset were common both in the astral sciences and in daily life in ancient China, and sometimes they were presented in a different form as tables of daytime and nighttime. An early document named *Day Book* (*Rishu* 日書) contains an extant table of this kind. Written on bamboo slips, it was excavated from tomb M11 in Shuihudi in the province Hubei and is thought to have been sealed in 217 BCE. This type of table gives the durations of daytime and nighttime throughout the year and uses a division of the day into 16 parts. It is the type of astronomical table most frequently inserted in early manuscripts dealing with the astral sciences.[1] Tables of daytime and nighttime became widespread, although the division of a day was changed to 100 *ke* 刻 per day. Through the *Monographs on Harmonics and Astronomy* (*Lü-lizhi* 律曆志, further abbreviated as *Monographs*) in the *Standard Histories* (*Zhengshi* 正史) of the Chinese dynasties, we can find such tables in various calendrical systems[2] between the first and sixth centuries.[3] Generally, in these calendrical systems, authors would give the daytime and nighttime together with the length of the gnomon shadow for each of the twenty-four solar terms (*qi* 氣) during a year.

From the late sixth century, tables of sunrise and sunset were generally used in parallel with, or in place of, the tables of daytime and nighttime in the *Monographs*. For example, the *Great Patrimony System* (*Daye li* 大業曆),[4] used

* I would like to express my sincere gratitude to my colleagues Benno van Dalen, Clemency Montelle, Daniel Morgan, Matthieu Husson and Richard Kremer for their constructive suggestions and comments

[1] See Chemla and Li, 'Numerical Tables'.

[2] For the translation of *li* 曆 in this paper I use 'calendrical system' or 'system' for short. The character also stands for mathematical astronomy or calendrical astronomy in general.

[3] According to Chen Meidong, twenty-one Chinese calendrical systems recorded in the *Monographs* provide a method to find daytime and nighttime. Twelve of these rely on astronomical tables, six refer to algebraic formulae described in textual form, and three systems give both tables and a formula. See Chen and Li, 'Research of *lou-ke* Calculation'.

[4] In order to be consistent with previous studies, I borrow the translations of the names of calendrical systems and some technical terms from Sivin, *Granting the Seasons*.

Editing and Analysing Numerical Tables: Towards a Digital Information System for the History of Astral Sciences, ed. by Matthieu Husson, Clemency Montelle and Benno van Dalen, PALS 2 (Turnhout, 2021), pp. 253–285

10.1484/M.PALS-EB.5.127700

between 597 and 618, is the first calendrical system in the *Monographs* containing a table of sunrise and sunset instead of a table of daytime and nighttime. Following the table, the canon introduces a method named 'procedure to find sunrise and sunset' (*qiu richuru suozaishu* 求日出入所在術).[5] By means of this method the user may determine the times of sunrise and sunset for each day in a year using linear interpolation between the times of sunrise and sunset at the twenty-four solar terms given in this table. The lengths of daytime and nighttime depend on the times of dawn and dusk, which are also related to the times of sunrise and sunset.[6] In most ancient Chinese calendrical systems, regardless of the season, a constant value of 2.5 *ke* is assigned to twilight, the time between dawn and sunrise or between sunset and dusk. The table of daytime and nighttime and the table of sunrise and sunset can easily be converted to each other, which is why many calendrical systems in the *Monographs* present only one of them rather than both in order to save space.

In the Yuan period (1271–1368), the table of sunrise and sunset became more sophisticated and accurate.[7] The table attached to the *Season-Granting System* (*Shoushi li* 授時曆),[8] the official calendrical system of the Yuan dynasty finished by the astronomer Guo Shoujing 郭守敬 (1231–1316) and his colleagues and used between 1281 and 1384, is different from previous methods which were based on an algebraic expression. Guo Shoujing compiled this table on the basis of a geometrical model and relied on calculations similar to Western spherical trigonometry. Moreover, he used the 'altitude of the North Pole' (*beiji chudi* 北極出地, i.e. geographical latitude) of Dadu 大都 (nowadays Beijing), the capital city of the Yuan period.

[5] *Lidai tianwen lüli deng zhi huibian*, p. 1928.

[6] The *Book of Song* (*Songshu* 宋書) records 天之晝夜以日出入為分, 人之晝夜以昏明為限 'The daytime and nighttime for heaven are divided by sunrise and sunset; the daytime and nighttime for persons rely on dawn and dusk'. See *Lidai tianwen lüli deng zhi huibian*, p. 293.

[7] Chen Meidong analyzed the accuracy of the table of daytime and nighttime of the *Season-granting System*. The average difference between the table and modern theory is about 0.7 minute if the latitude is taken as 39.95 degrees. See Chen and Li, 'Research of *Lou-ke* Calculation'. Concerning further research of tables of sunrise and sunset in ancient China, Jin-Chyuan Lin discusses the record of sunrise and sunset in the *Concordant Epoch System* (*Tong Yuan li* 統元曆, used between 1135 and 1167); see Lin, 'The Pick-up Tables'. Mihn et al., 'Analysis of Interval Constants', provides a brief comparison of the nighttime lengths from the *Season-granting System* and the *Great Concordance System* with the results of modern calculations, but does not analyze the underlying methods.

[8] The *Season-granting System*, created for the Mongol emperor Kublai Khan 忽必烈 (r. 1260–1294), is considered to be the most sophisticated system of calendrical astronomy in ancient China. Its treatise records detailed instructions for computing eclipses and the motions of the planets, based on a rich archive of observations. See the Japanese translation Yabuuti and Nakayama, *Season-granting System*, and the English translation and study in Sivin, *Granting the Seasons*.

After the Ming dynasty replaced the Yuan in 1368, the director of the Astronomical Bureau, Yuan Tong 元統 (fl. 1384–1396)[9] attempted to revise the *Season-granting system*. With the assistance of Guo Boyu 郭伯玉 (a descendent of Guo Shoujing who flourished in the 1380s), Yuan Tong finished the *Great Concordance System* (*Datong li* 大統曆),[10] which was used between 1384 and 1644. In most parts, the *Great Concordance system* is adapted from the *Season-granting System* with only very few modifications. One of the distinct differences between the two systems is that the author of the former recalculated every value in the table of sunrise and sunset for the latitude of Nanjing, the new capital city of the Ming dynasty. The maximum deviation of the times of sunrise and sunset between Nanjing and Dadu reaches about three *ke* (almost 43 minutes).

Not long after these two calendrical systems were adopted in Yuan and Ming China, they were successively transmitted to Korea. In order to learn the methods of calendrical systems, in 1303 the Koryeo court (918–1392) sent the officer Choi Seongji 崔誠之 to the Yuan empire. Finally, the Koryeo Kingdom mastered computational techniques to adjust the *Season-granting System* for use in Korea after the year 1309.[11] Since the method of extracting square roots was not transmitted, the section on eclipses still needed to follow the old method of the *Extending Enlightenment System* (*Xuanming li* 宣明曆), used in China between 822 and 892.[12] As for the table of sunrise and sunset, it did not provide correct results because of the difference in latitude between Beijing and Koryeo. Even though the Korean astronomers perceived this problem, they were not able to revise the table accordingly. They could only duplicate the procedures of the *Season-granting System* from China and included these contents in the *History of Koryeo* (*Koryeo-Sa* 高麗史) without revision.

In 1392, the Koryeo dynasty was replaced by the Joseon dynasty, and Seoul was chosen as the new capital. To develop calendrical astronomy, King Sejong 世宗 (r. 1418–1450) directed his astronomers to undertake systematic research on the Chinese *Season-granting System* and the *Great Concordance System*. Their most significant achievement was the compilation of the Korean calendrical work *Inner Chapter of Computation of the Seven Regulators* (*Chiljeongsan Naepyon* 七政算内篇),[13] finished in 1442. The *Inner Chapter* mostly copied the

[9] Yuan Tong was originally a 'doctor of the clepsydra' (*louke boshi* 漏刻博士) at the Chinese Astronomical Bureau, but was promoted to director of the bureau in 1385 because of his contribution to the modification of the official calendrical system.

[10] The *Great Concordance System* is mostly identical to the *Season-granting System*, but has some small improvements such as several epoch constants.

[11] Lee, 'Korean Astronomical Calendar'.

[12] *Koryeo-Sa*, 50.1b.

[13] The *Chiljeongsan Naepyon* was considered to be the first original Korean calendrical system. However, its contents is in fact basically identical to its Chinese sources. This system used

Season-granting System and the *Great Concordance System*, but it provided a new table of sunrise and sunset for the city of Seoul.

As has been mentioned above, various versions of tables of sunrise and sunset were used in China and Korea during the Yuan and Ming periods; their underlying theories are not clear. This paper will detail the methods of construction and use of tables of sunrise and sunset in China and analyze how Korean astronomers updated these tables in Korea with their limited mathematical knowledge. Section 2 provides a survey of the sources containing the relevant tables, and Section 3 introduces the two types of tables of sunrise and sunset with different layouts and applications. The first type tabulates data required in some specific calendrical calculations, while the second type is more straightforward to use. Section 4 analyzes the algorithms and methods of calculation underlying these tables. Two methods to produce the tables of sunrise and sunset in the *Season-granting System* will be presented. One employs the arc-sagitta method of calculation (i.e. characteristic 'Chinese spherical trigonometry') and imposes a heavy calculational burden. The other method named 'Method of Nine Domains' is simple but approximative. In Section 5, it is discussed how Korean astronomers constructed their own tables without sufficient mathematical knowledge, and a possible method for the calculation of their tables is outlined.

Description of the Sources and Systems of Measuring Time

The documents included in the *Season-granting System* have four parts: *Evaluation* (*Liyi* 曆議),[14] *Canon* (*Lijing* 曆經), *Pick-up Tables* (*Licheng* 立成)[15] and *Detailed Procedures* (*Licao* 曆草).[16] The former two parts are the ones incorpo-

the value $38\frac{1}{4}$ *du* (corresponding to 37;41°) for the latitude of Hanyang 漢陽 (the old name for the Joseon capital and nowadays the northern part of Seoul). However, in this paper I find that the table of sunrise and sunset in the *Chiljeongsan Naepyon* does not fit this latitude value very well.

[14] The *Evaluation* sets out in detail how the Yuan astronomers used a remarkable archive of observations covering more than a thousand years to test the new astronomical system, proving that it would be more reliable than its predecessors. See Sivin, *Granting the Seasons*, p. 21.

[15] Nathan Sivin translates *licheng* 立成 as 'ready reckoner', but in this paper I use Karine Chemla's translation 'Pick-up tables'. See Chemla and Li, 'Numerical Tables'. The calculations required in ancient Chinese calendrical systems could be carried out by two different methods: by procedures described in texts, or using specific types of tables named *licheng*. This type of table seems to have come into use in the Sui period (581–618) and to have been widespread from the Tang period (618–907) onward. Its development appears to be correlated to the introduction of astronomical functions based on quadratic interpolation. Thus rather than having to perform calculations with numbers provided in the text, the results could be readily 'picked up' from a *licheng* table based on the values of its arguments. See Li, 'Astronomical Tables'.

[16] The term 'detailed procedures' (*cao* 草) initially came from ancient mathematical texts and referred to a kind of notebook used to explain algorithms or give detailed comments on

rated in the official *History of the Yuan Dynasty* (*Yuanshi* 元史), and both of them consist of two volumes. The *Pick-up Tables* were omitted from the Chinese official histories to save space, but they still exist as they were reprinted in Korea in the first half of the fifteenth century.[17] They also consist of two volumes: volume A contains the tables describing the motions of the sun, moon and planets, and the whole of volume B is devoted to the tables of sunrise and sunset.

In China, the *Detailed Procedures* were still available as late as the early eighteenth century. The mathematician Mei Wending 梅文鼎 (1633–1721) once referred to the *Detailed Procedures* and cited part of them in his contribution to the chapter *Monographs* in the *History of the Ming Dynasty*.

The *Season-granting System* contains two sets of tables of sunrise and sunset of different types, one in the *Canon* and the other in the *Pick-up Tables*. In order to distinguish them, we classify them as Type I and Type II respectively, and will explain their differences in the next section.

The corpus of tables of sunrise and sunset that we deal with in this paper includes general works on the astral sciences and the monographs on calendrical astronomy that are included in the official dynastical histories. This paper focuses on the contents of these tables but will not address the process of their compilation. Three-part abbreviations will be used to identify the characteristics of each table (see the column 'Type' in Table 1). The first letter (C or K) indicates whether it concerns a document from China or Korea. The second part (I or II) indicates the type of the table. The third letter (B, N or S, standing for Beijing, Nanjing and Seoul respectively) shows the specific location the table is computed for. For example, C-I-B indicates that the table is a table from China of Type I and is compiled for Beijing.

This article focuses on the practices surrounding the computation of various tables of sunrise and sunset. The main issues that have guided the inquiry include:

1. Algorithms and methods of calculation underlying these tables.
2. The differences between the two approaches in the *Season-granting System* to calculate the times of sunrise and sunset.
3. The actual latitude values these tables used.
4. The possible solutions the Korean astronomers may have sought to make the table suitable for the latitude of Seoul.

classic mathematical texts. Here, *licao* refers to a document explaining the mathematical algorithms behind the procedures that have been given in the canon.

[17] A copy of the *Licheng* is saved in the Gyujanggak Library in Seoul; see *Pick-Up Tables*.

<table>
<tr><th>Source</th><th>Date</th><th>Type</th></tr>
<tr><td>Canon of the Season-granting System in the History of the Yuan Dynasty
元史·授時曆經[18]</td><td>1369</td><td>C-I-B</td></tr>
<tr><td>Pick-up Tables (Licheng) in the Season-granting System
授時曆立成[19]</td><td>1280s</td><td>C-II-B</td></tr>
<tr><td>Times of Sunrise and Sunset in the Great Concordance System
大統日出入分[20]</td><td>1380s</td><td>C-II-N</td></tr>
<tr><td>Canon of the Season-granting System in the History of the Koryeo Dynasty
高麗史·授時曆經[21]</td><td>1451</td><td>K-I-B</td></tr>
<tr><td rowspan="2">Pick-up Tables (Licheng) in the Great Concordance System, History of the Ming Dynasty
明史·大统曆法立成[22]</td><td rowspan="2">final version finished in 1735</td><td>C-I-B</td></tr>
<tr><td>C-II-N</td></tr>
<tr><td rowspan="2">Inner Chapter of Computation of the Seven Regulators (Chiljeongsan Naepyon)
七政算内篇[23]</td><td rowspan="2">1442</td><td>K-I-S</td></tr>
<tr><td>K-II-S</td></tr>
</table>

Table 1: Tables of sunrise and sunset discussed in this paper

We will first present essential details of the ancient Chinese systems for measuring time. Three systems were most common throughout Chinese history.[24] The first of these is the 'double-hour system', in which the 'double-hour' (*shi* 時), a twelvefold division of the day, is employed as the basic unit.[25] The second system is the 'one-hundredth-of-a day system', which divides each day into 100 *ke* 刻 of 14.4 minutes (i.e. slightly less than a quarter of an hour). Each *ke* is further divided into 100 *fen* 分, so that one day equals ten thousand *fen*.

[18] *Lidai tianwen lüli deng zhi huibian*, pp. 3371–3441.

[19] See *Pick-up Tables*.

[20] See *Datong richu riru fen*.

[21] *Koryeo-Sa*, vol. 51. The *Canon for the Season-granting System* in the *History of the Koryeo Dynasty* is almost identical to its counterpart in the *History of the Yuan Dynasty*. The differences lie in that the former omitted the section 'To find clepsydra marks at [any] location within the nine domains' 求九服所在漏刻. See *Koryeo-Sa*, 51.33b.

[22] *Lidai tianwen lüli deng zhi huibian*, pp. 3633–94.

[23] See *Chiljeongsan Naepyon*.

[24] For a detailed discussion of ancient Chinese systems of measuring time, see Chen, 'Research on Chinese Ancient Time-Reckoning Systems'; Wang, 'Investigation into Systems of Reckoning Time', and Qu, 'Time-reckoning in Ancient Chinese Calendrical Systems'.

[25] In this system, the earthly branches (*ganzhi* 干支) are used to name the twelve *shi*, and each *shi* can also be divided into two parts named *chu* 初 ('beginning') and *zheng* 正 ('standard') respectively.

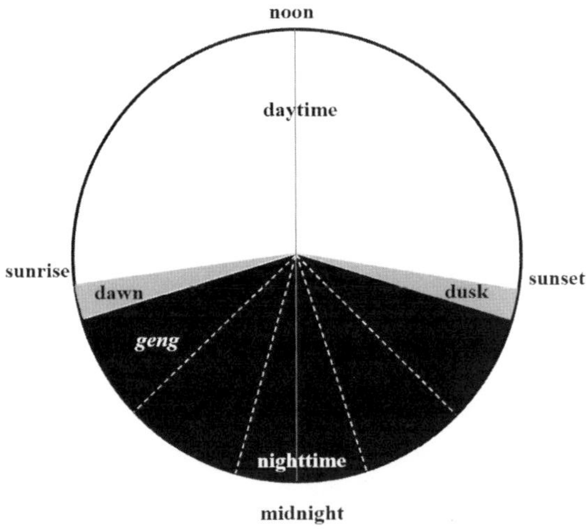

Figure 1: System of watches and points.

Different from the above two systems, a third system, named 'watches and points', is utilized specifically for nighttime. For the purpose of civil time-keeping, nighttime was the interval between dusk and dawn. Ancient Chinese astronomers divided it into five equal 'watches' (*geng* 更), and subdivided each *geng* into five equal 'points' (*dian* 點). In the course of a year, the number of watches and points remained constant, so the length of both units varied from day to day (Figure 1).[26] In order to provide the correct nighttime, most Chinese calendrical systems presented a table of daytime and nighttime according to this system.

In addition to keeping track of time the tables of daytime and nighttime or the tables of sunrise and sunset are necessary in some other calculations. For example, the times of sunrise and sunset are useful to determine whether a solar eclipse will be visible when it occurs. The times of dusk or dawn, on the other hand, can be used to determine which star is culminating (i.e. crossing the meridian) at these moments.

Two Types of Sunrise and Sunset Tables

As previously mentioned, the *Season-granting System* offers two types of tables related to sunrise and sunset for distinct purposes. The first type, which we classify as Type I, is a table entitled 'Declination and polar distance of the ecliptic, and half lengths of daytime and nighttime' (*Huangdao churu chidaoneiwai qujidu ji banzhouyefen* 黃道出入赤道內外去極度及半晝夜分).[27] This table

[26] Sivin, *Granting the Seasons*, p. 494.

[27] *Lidai tianwen lüli deng zhi huibian*, pp. 3405–14.

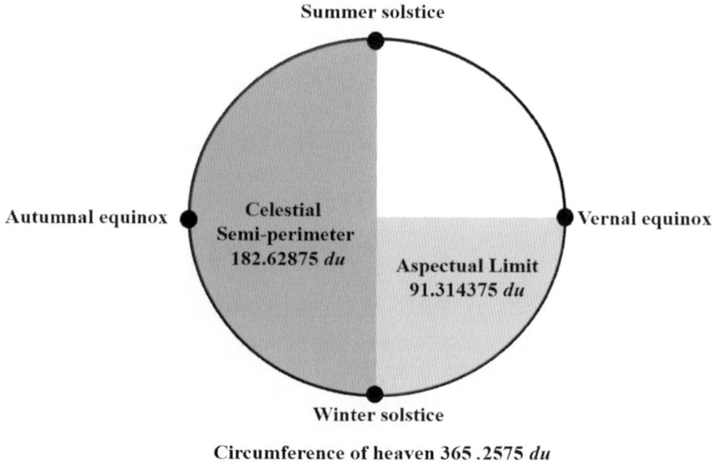

Figure 2: The Chinese measuring unit *du*.

belongs to the chapter 'The Pacing of the Centered Star' (*Buzhongxing* 步中星, the fifth section of the *Canon*), which concentrates on whatever star is culminating (crossing the meridian) at a given moment such as dusk or dawn (see Plate 9a).

Table 2 is a transcription of the table of Type I in the *Canon* of the *Season-granting System*, and its columns are arranged as follows:[28]

1. The argument, 'ecliptic accumulated degrees' (*huangdao jidu* 黄道積度), is degrees of the ecliptic measured from the solstice in 'Chinese degrees' (*du* 度).[29] In Chinese calendrical systems, the unit *du* corresponds to the mean solar motion in exactly one day, so that the number of *du* in a full circle is equal to the length of the solar year. In the *Season-granting System* this 'circumference of heaven' (*zhoutian* 周天, or 'circuit of heaven') amounts to 365.2575 *du* (see Figure 2).[30] The argument of the table runs from 0 to 91.31 *du* (i.e. a quarter of the circumference of heaven, which is 365.2575/4 = 91.314375 or 91.31 after rounding off), and the interval between consecutive arguments is one *du*, except for the last interval which is 0.31 *du*.
2. 'Inside/outside degrees' (*neiwai du* 內外度), the declination measured from the equator to the ecliptic.

[28] In all transcriptions of tables in this article, columns and rows have been transposed, because the original Chinese texts are read vertically and from right to left.

[29] Different from Western astronomical systems, which measure the degree of the ecliptic from the vernal equinox, almost all Chinese calendrical systems start from the winter solstice.

[30] In the *Season-granting System* the tropical year is 365.2425 days, the motion of the sun in a year is 365.2425 *du*, and the sidereal year is 365.2575 days (or 365.2575 *du*, the circumference of heaven, found as the sum of 365.2425 *du* and a precession of 0.0150 *du*).

(1)	(2)	(3)	(4)	(5)	(6)	(7)	(8)
0	23.9030	0.33	115.2173	67.4113	1907.96	3092.04	0.09
1	23.8997	0.99	115.2140	67.4146	1908.05	3091.95	0.29
2	23.8898	1.66	115.2041	67.4245	1908.34	3091.66	0.47
3	23.8732	2.31	115.1875	67.4411	1908.81	3091.19	0.66
4	23.8501	2.99	115.1644	67.4642	1909.47	3090.53	0.85
5	23.8202	3.65	115.1345	67.4941	1910.32	3089.68	1.04
6	23.7837	4.32	115.0980	67.5306	1911.36	3088.64	1.22
7	23.7405	4.98	115.0548	67.5738	1912.58	3087.42	1.42
8	23.6907	5.65	115.0050	67.6236	1914.00	3086.00	1.61
9	23.6342	6.36	114.9485	67.6801	1915.61	3084.39	1.79
10	23.5706	7.02	114.8849	67.7437	1917.40	3082.60	1.99
⋮	⋮	⋮	⋮	⋮	⋮	⋮	⋮
85	2.4583	38.93	93.7726	88.8560	2443.45	2556.55	8.97
86	2.0690	38.94	93.3833	89.2453	2452.42	2547.58	8.96
87	1.6796	38.94	92.9939	89.6347	2461.38	2538.62	8.96
88	1.2902	38.95	92.6045	90.0241	2470.34	2529.66	8.96
89	0.9007	38.95	92.2150	90.4136	2479.30	2520.70	8.96
90	0.5112	38.95	91.8255	90.8031	2488.26	2511.74	8.95
91	0.1217	12.17	91.4360	91.1926	2497.21	2502.79	2.79
91.31	0	0	91.3143	91.3143	2500	2500	0

Table 2: Transcription of the table of Type C-I-B (Type I for the site of Beijing, excerpt).

3. 'Inside/outside difference' (*neiwai cha* 內外差), the successive differences of *neiwai du*. It is used for carrying out linear interpolation in column 2.
4. 'Polar distance before/after the winter solstice' (*dongzhi qianhou quji* 冬至前後去極), a quarter cycle 91.314375 *du* plus the values from column 2.
5. 'Polar distance before/after the summer solstice' (*xiazhi qianhou quji* 夏至前後去極), a quarter cycle 91.314375 *du* minus the values from column 2.
6. 'Winter daytime/summer nighttime' (*dongzhou xiaye* 冬晝夏夜), half of the length of daytime or nighttime. It is the time between dawn and noon (or between noon and dusk) in the winter or the time between dusk and midnight (or between midnight and the next dawn) in the summer, since the duration of daytime or nighttime is symmetric about noon or midnight. The unit of this quantity is *fen* (1/10000 day),[31] and the times of sunrise and sunset can be obtained through the addition and subtraction of 250 *fen*, regardless of the season.
7. 'Summer daytime/winter nighttime' (*xiazhou dongye* 夏晝冬夜), the complement of column 6. It is the time between dawn and noon in summer or the time between dusk and midnight in winter.
8. 'Day-night difference' (*zhouyecha* 晝夜差), the successive differences of columns 6 or 7.

[31] Here it uses the 'one-hundredth-of-a-day system' (see above).

(1)	(2)	(3)	(4)	(5)	(6)
0	23.9030	0.33	1956.50	3043.50	0.08
1	23.8997	0.99	1956.58	3043.42	0.27
2	23.8898	1.66	1956.85	3043.15	0.43
3	23.8732	2.31	1957.28	3042.72	0.61
4	23.8501	2.99	1957.89	3042.11	0.78
5	23.8202	3.65	1958.67	3041.33	0.95
6	23.7837	4.32	1959.62	3040.38	1.12
7	23.7405	4.98	1960.74	3039.26	1.30
8	23.6907	5.65	1962.04	3037.96	1.48
9	23.6342	6.36	1963.52	3036.48	1.64
10	23.5706	7.02	1965.16	3034.84	1.83
⋮	⋮	⋮	⋮	⋮	⋮
85	2.4583	38.93	2448.06	2551.94	8.23
86	2.0690	38.94	2456.29	2543.71	8.23
87	1.6796	38.94	2464.52	2535.48	8.23
88	1.2902	38.95	2472.25	2527.25	8.23
89	0.9007	38.95	2480.98	2519.02	8.23
90	0.5112	38.95	2489.21	2510.79	8.22
91	0.1217	12.17	2497.43	2502.57	2.57
91.31	0	0	2500	2500	0

Table 3: Transcription of the table of Type K-I-S (Type I for the site of Seoul, excerpt).

A table of Type I (K-I-S) also exists in the Korean work *Inner Chapter of Computation of the Seven Regulators*. Different from Type C-I-B shown above, this table omits the two columns for 'polar distance', which can be calculated from the 'inside/outside degrees' in the second column. Type K-I-S also revised the values of column 6, 7 and 8 from Type C-I-B (and placed them in the columns 4, 5 and 6 respectively) to serve for the latitude of the Korean capital Seoul (see Table 3).

The lengths of half daytime or nighttime in tables C-I-B and K-I-S are displayed in Figure 3, in which the horizontal axis shows the 'accumulated degrees' after the solstices and the vertical axis depicts the length of time in the unit *fen*. It is noteworthy that the maximum difference between values of the two cities appears at the two solstices, where it reaches 48.54 *fen*. This means that the length of daytime at the summer solstice in Beijing is 97.08 *fen* (48.54 × 2, modern equivalent 13.98 minutes) longer than in Seoul.

The second type of table of sunrise and sunset, the one we classify as Type II, is more straightforward. It exists in the *Pick-up Tables* (*Licheng*) part of the *Season-granting System* and is titled 'Sunrise, sunset, dawn, dusk and half nighttime of the *Season-granting System*' (*Shoushi li richuru chenhun banzhoufen* 授時曆日出入晨昏半晝分).[32] The two sections of this table cover 'after

[32] *Pick-up Tables*, vol. 2, 1a–51b.

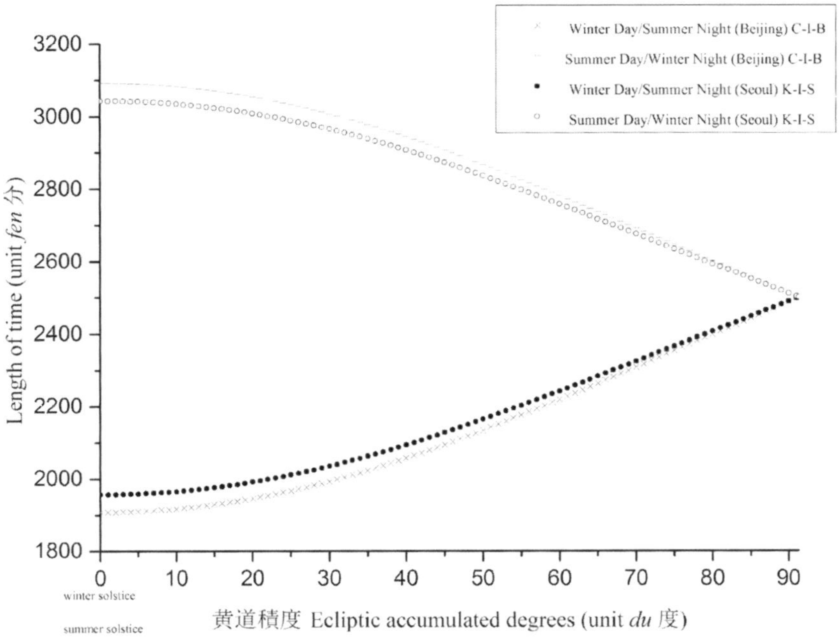

Figure 3: The half lengths of daytime and nighttime in C-I-B and K-I-S.

the winter solstice' and 'after the summer solstice', and both sections have as arguments days 0 to 182 after the solstice (see Plate 9b).

A transcription of this table is given in Table 4, and the columns of this table are arranged as follows:

1. The argument, 'accumulated days' (*jiri* 積日), the number of days after the solstices.
2. 'Time of dawn' (*chenfen* 晨分), the time from midnight (the beginning of a day) to dawn.
3. 'Time of sunrise' (*richufen* 日出分), the time from midnight to sunrise.
4. 'Half daytime' (*banzhoufen* 半晝分), the time from sunrise to midday or from midday to sunset.
5. 'Time of sunset' (*rirufen* 日入分), the time from midnight to sunset.
6. 'Time of dusk' (*hunfen* 昏分), the time from midnight to dusk.

In addition to 'C-II-B', two more documents that incorporated a table of Type II can be found in the *Great Concordance System* and the *Inner Chapter of Computation of the Seven Regulators*. They are here labelled Type C-II-N and Type K-II-S (see Table 5 and Table 6). Their values were revised for the sites of Nanjing and Seoul respectively. In order to clarify their differences, we display their times of sunrise in a diagram (see Figure 4).

(1)	(2)	(3)	(4)	(5)	(6)
0	2842.04	3092.04	1907.96	6907.96	7157.96
1	2841.94	3091.94	1908.06	6908.06	7158.06
2	2841.62	3091.62	1908.38	6908.38	7158.38
3	2841.09	3091.09	1908.91	6908.91	7158.91
4	2840.36	3090.36	1909.64	6909.64	7159.64
5	2839.42	3089.42	1910.58	6910.58	7160.58
6	2838.28	3088.28	1911.72	6911.72	7161.72
7	2836.93	3086.93	1913.07	6913.07	7163.07
8	2835.37	3085.37	1914.63	6914.63	7164.63
9	2833.60	3083.60	1916.40	6916.40	7166.40
10	2831.63	3081.63	1918.37	6918.37	7168.37
⋮	⋮	⋮	⋮	⋮	⋮
178	1659.83	1909.83	3090.17	8090.17	8340.17
179	1659.12	1909.12	3090.88	8090.88	8340.88
180	1658.58	1908.58	3091.42	8091.42	8341.42
181	1658.21	1908.21	3091.79	8091.79	8341.79
182	1658.01	1908.01	3091.99	8091.99	8341.99

(1)	(2)	(3)	(4)	(5)	(6)
0	1657.96	1907.96	3092.04	8092.04	8342.04
1	1658.04	1908.04	3091.96	8091.96	8341.96
2	1658.31	1908.31	3091.69	8091.69	8341.69
3	1658.74	1908.74	3091.26	8091.26	8341.26
4	1659.34	1909.34	3090.66	8090.66	8340.66
5	1660.11	1910.11	3089.89	8089.89	8339.89
6	1661.06	1911.06	3088.94	8088.94	8338.94
7	1662.17	1912.17	3087.83	8087.83	8337.83
8	1663.46	1913.46	3085.54	8086.54	8336.54
9	1664.93	1914.93	3085.07	8085.07	8335.07
10	1666.56	1916.56	3083.44	8083.44	8333.44
⋮	⋮	⋮	⋮	⋮	⋮
178	2840.34	3090.34	1909.34	6909.66	7159.66
179	2840.69	3090.69	1909.31	6909.31	7159.31
180	2841.33	3091.33	1908.67	6908.67	7158.67
181	2841.76	3091.76	1908.24	6908.246	7158.24
182	2841.99	3091.99	1908.01	6908.01	7158.01

Table 4: Transcription of the first half ('after the winter solstice', above) and second half ('after the summer solstice', below) of Type C-II-B (Type II for the site of Beijing, excerpt).

(1)	(2)	(3)	(4)	(5)	(6)
0	2681.70	2931.70	2068.30	7068.30	7318.30
1	2681.62	2931.62	2068.38	7068.38	7318.38
2	2681.39	2931.39	2068.61	7068.61	7318.61
3	2681.01	2931.01	2068.99	7068.99	7318.99
4	2680.48	2930.48	2069.52	7069.52	7319.52
5	2679.79	2929.79	2070.21	7070.21	7320.21
6	2678.96	2928.96	2071.04	7071.04	7321.04
7	2677.97	2927.97	2072.03	7072.03	7322.03
8	2676.83	2926.83	2073.17	7073.17	7323.17
9	2675.55	2925.55	2074.45	7074.45	7324.45
10	2674.11	2924.11	2075.89	7075.89	7325.89
⋮	⋮	⋮	⋮	⋮	⋮
178	1819.66	2069.66	2930.34	7930.34	8180.34
179	1819.14	2069.14	2930.86	7930.86	8180.86
180	1818.75	2068.75	2931.25	7931.25	8181.25
181	1818.49	2068.49	2931.51	7931.51	8181.51
182	1818.34	2068.34	2931.66	7931.66	8181.66

Table 5: Transcription of the first half ('after the winter solstice') of Type C-II-N (Type II for the site of Nanjing, excerpt)

(1)	(2)	(3)	(4)	(5)	(6)
0	2793.50	3043.50	1956.50	6956.50	7206.50
1	2793.41	3043.41	1956.59	6956.59	7206.59
2	2793.11	3043.11	1956.89	6956.89	7206.89
3	2792.63	3042.63	1957.37	6957.37	7207.37
4	2791.95	3041.95	1958.05	6958.05	7208.05
5	2791.09	3041.09	1958.91	6958.91	7208.91
6	2790.05	3040.05	1959.95	6959.95	7209.95
7	2788.81	3038.81	1961.19	6961.19	7211.19
8	2787.38	3037.38	1962.62	6962.62	7212.62
9	2785.76	3035.76	1964.24	6964.24	7214.24
10	2783.95	3033.95	1966.05	6966.05	7216.05
⋮	⋮	⋮	⋮	⋮	⋮
178	1708.20	1958.20	3041.80	8041.80	8291.80
179	1707.55	1957.55	3042.45	8042.45	8292.45
180	1707.06	1957.06	3042.94	8042.94	8292.94
181	1706.73	1956.73	3043.27	8043.27	8293.27
182	1706.55	1956.55	3043.45	8043.45	8293.45

Table 6: Transcription of the first half ('after the winter solstice') of Type K-II-S (Type II for the site of Seoul, excerpt).

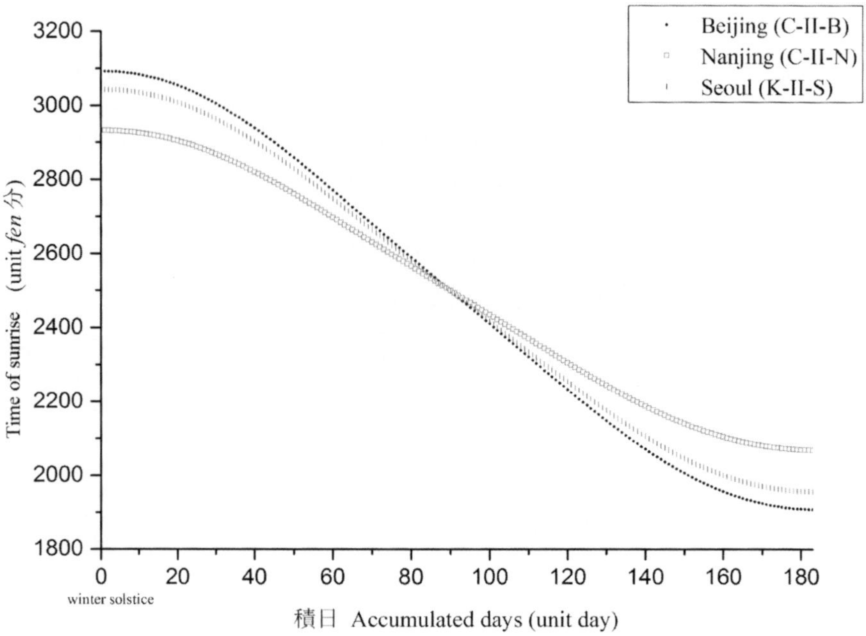

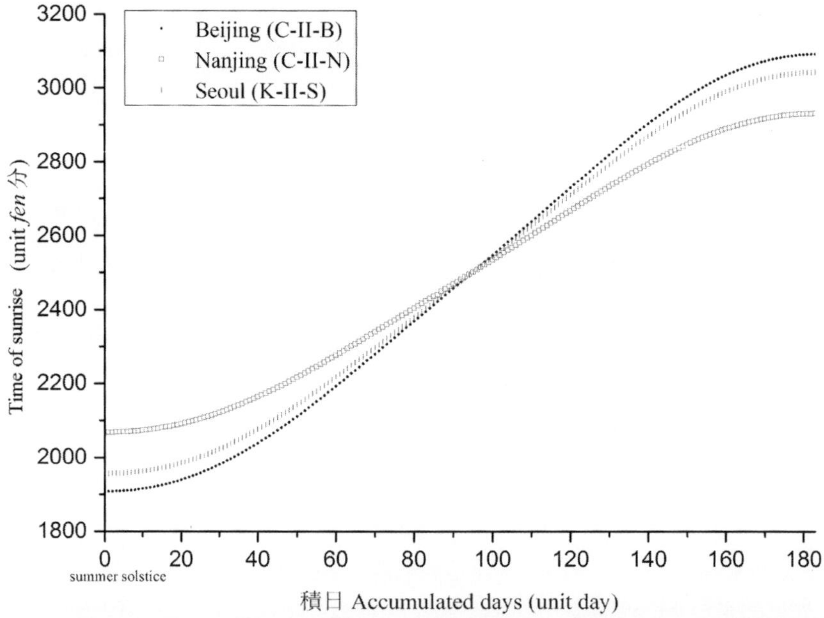

Figure 4: The times of sunrise in the first half (above) and second half (below) of the tables of types C-II-B, C-II-N and K-II-S.

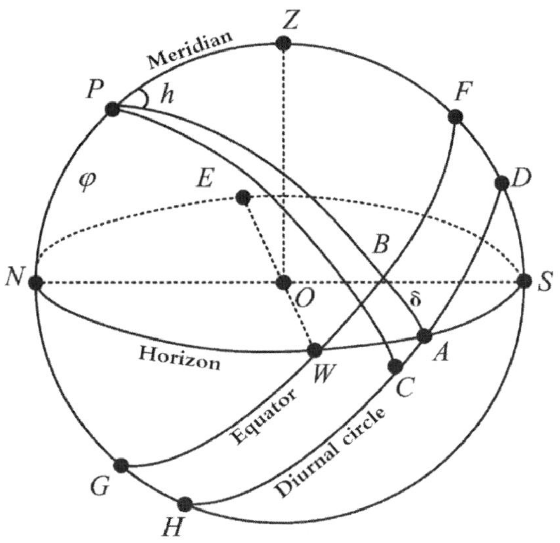

Figure 5: Modern geometrical model for calculating the hour angle (*h*).

4. The Underlying Mathematical Methods

4.1 Calculation of arc, chord, and sagitta

From a modern point of view, the determination of the times of sunrise and sunset is related to the calculation of the hour angle (*h*).

Figure 5 depicts the celestial sphere for an observer at *O*, whose geometrical latitude is φ (arc *NP*). Point *Z* is the zenith, point *P* the celestial north pole, and *N*, *W*, *S*, *E* the north, west, south and east points on the horizon. The meridian is circle *ZSNP*, the equator is circle *GWBF*, and the diurnal circle is circle *HCAD*. Arc *AB* is the declination of the sun (δ). Point *A* is the position of the sun at sunset, and point *C* is the position of the sun at dusk. To determine the hour angle *h*, we make use of the spherical triangle *PAS*.[33] Using modern spherical trigonometry, the hour angle can be found directly from the values of the latitude φ and the declination δ:

$$\begin{aligned}\cos h &= -\tan(\text{arc}\,PS)\cdot\cot(\text{arc}\,PA)\\ &= \tan(180^\circ-\varphi)\cdot\cot(90^\circ+\delta)\\ &= -\tan\varphi\cdot\tan\delta.\end{aligned}$$

However, it needs to be pointed out that early Chinese astronomy was a great deal less concerned with geometric deductions, so not much attention was paid to the relation between arcs and chords until the late eleventh century, when Shen Gua 沈括 (1031–1095) worked out a general formula for chords of two-dimensional arcs.[34] Shen Gua provided an approximate method which

[33] Qu, *Chinese Mathematical Astronomy*, p. 244.

[34] Sivin, *Granting the Seasons*, p. 66.

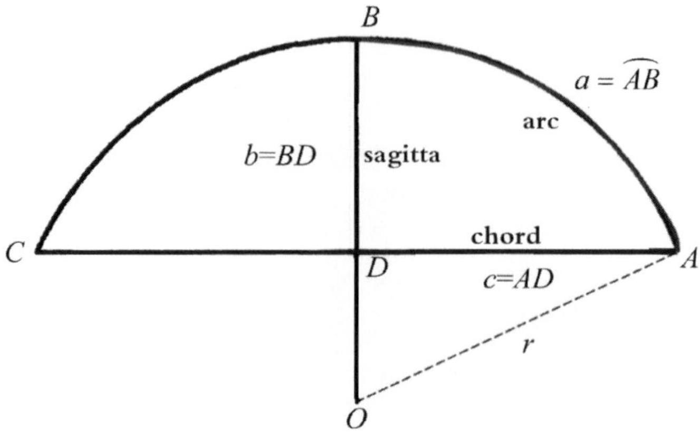

Figure 6: Schematic diagram of Shen Gua's formula.

originated from the Chinese classic mathematical document *The Nine Chapters of the Mathematical Art* (*Jiuzhang suanshu* 九章算術) (see Figure 6).[35] He considered the length of arc *AC* (2*a*) as a function of the chord *AC* (2*c*) subtending the arc, and the corresponding sagitta *BD* (*b*), as well as the diameter of the circle *d* (*d*=2*r*, where *r* represents radius *OB* or *OA*). For any value of arc, chord, and sagitta, the following relation holds:

$$a = c + \frac{b^2}{d}.$$

With the Theorem of Pythagoras, we can obtain:

$$c^2 = r^2 - (r - b)^2.$$

Eliminating the half chord c from the above two equations, we can achieve an equation of order four:

$$b^4 - (d^2 - 2da)b^2 - d^3b + d^2a^2 = 0.$$

This is one of the major formulae of the so-called 'Chinese spherical trigonometry' that Guo Shoujing used in the *Season-granting System* to compute many of the astronomical tables. However, as Martzloff points out, this type of trigonometry is based on arcs, chords and sagittas obtained from approximate formulae, instead of on angles and exact formulae. Moreover, it never involves triangles with three spherical sides, but only triangles with at most one spherical side, as well as plane triangles.[36] With this so-called 'method for determining the sagitta by the segment of the circle' (*hushi geyuan shu* 弧矢割圓術), a close approximation for a triangle with one spherical side, the authors of the *Season-granting System* made their own contribution to finding solutions for spherical problems.

[35] Liu, 'Summary of Season-granting System'.

[36] Martzloff, *A History of Chinese Mathematics*, p. 329.

4.2 Method for calculating tables of Type I

According to the *History of the Yuan Dynasty*, local latitude and solar declination are the two key values used to construct the table of sunrise and sunset in the *Season-granting System*. The text in *Biography of Guo Shoujing* (*Guo Shoujing zhuan* 郭守敬傳) states:

> 大明曆日出入晝夜刻，皆據汴京為準，其刻數與大都不同。今更以本方北極出地高下，黃道出入內外度，立術推求每日日出入晝夜刻。
>
> The [values of] sunrise and sunset, daytime and nighttime in the Revised Great Enlightenment System (used between 1181 and 1280) are all provided for [the site of] Bianjing (nowadays Kaifeng), and these times are different from [the values for] Dadu (nowadays Beijing). Now [we] change [the method], and use the 'altitude of the North Pole' (geographical latitude) and 'inside/outside degrees of the ecliptic' (declination measured from the equator to the ecliptic) to set up a method to calculate the times of sunrise and sunset and of daytime and nighttime for each day.[37]

Notably, we cannot find any specific discussion concerning the underlying method of Table C-I-B in the existing documents of the *Season-granting System* in the *History of the Yuan Dynasty*. Fortunately, the editors of the *History of the Ming Dynasty* realized the importance of the method, and discussed it when they re-edited the *Great Concordance System*, which is a revised version of the *Season-granting System*. The editor Mei Wending (1633–1721) pointed out that this method had been introduced in the book *Detailed Procedures* by Guo Shoujing; only very few copies of the book survived up to Mei's times. According to Mei Wending, the calculations of daytime and nighttime in Guo Shoujing's *Detailed Procedures* are based on the projection of the ecliptic, equator and diurnal circle into planes, and it uses the method of 'Chinese spherical trigonometry', to determine the declination δ.

4.2.1 Mei Wending's reasoning for tables of Type I

In the *History of the Ming Dynasty*, Mei Wending presented a method consisting of two procedures to calculate the daytime and nighttime for a desired position of the sun on the ecliptic based on Guo Shoujing's book *Detailed Procedures*. The first one is based on the difference in the times of sunrise and sunset at the solstices, and the second one is referred to as *qiu huangdao meidu zhouyeke* 求黃道每度晝夜刻, that is, 'to find the lengths of daytime and nighttime for each *du* of the ecliptic'.[38] Geometrical diagrams are not often included in Chinese calendrical works. However, Mei Wending emphasized that 'without a diagram [things] do not become clear' (*feitu buming* 非圖不明),[39] so to expound the theory of sunrise and sunset, he presented a diagram

[37] *Yuanshi*, p. 3850.

[38] *Lidai tianwen lüli deng zhi huibian*, p. 3621.

[39] *Lidai tianwen lüli deng zhi huibian*, p. 3583.

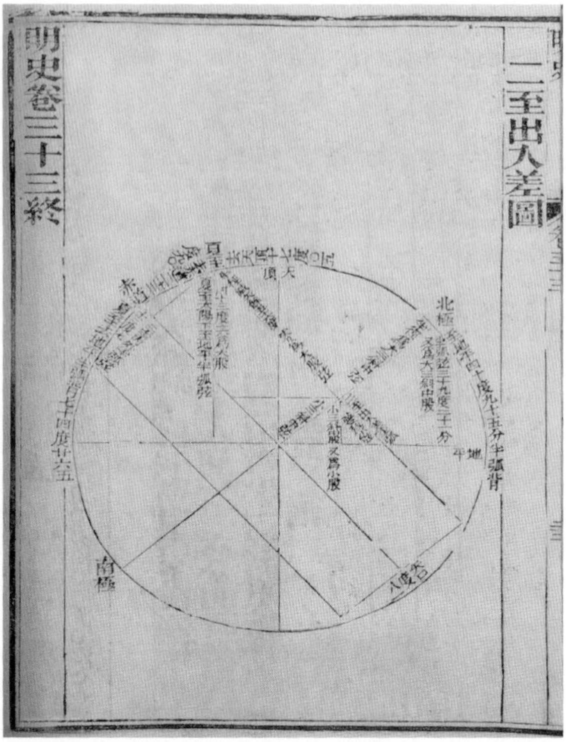

Figure 7: 'Diagram for the difference in the times of sunrise and sunset at the solstices' in the *History of the Ming Dynasty*. From the Qing copy of *Shanghai shenji shuzhuang* 上海慎記書莊刻本.

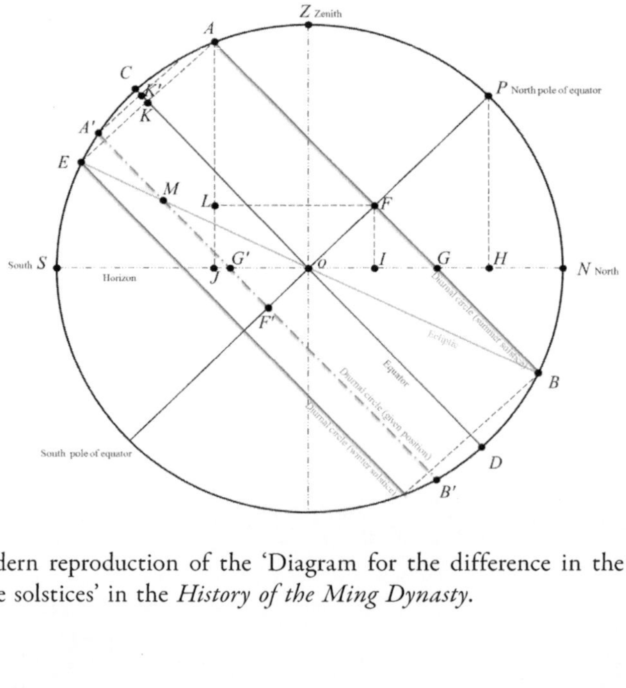

Figure 8: A modern reproduction of the 'Diagram for the difference in the times of sunrise and sunset at the solstices' in the *History of the Ming Dynasty*.

in which he projected all arcs onto the plane of the meridian, similar to the analemma known from Greek and Islamic sources (see Figure 7). To make this diagram easy to read for modern readers, we have redrawn it and added two auxiliary lines, namely the projections of the ecliptic and the diurnal circle of a given position (see Figure 8).

Because Mei Wending's reasoning is very long and not easy to understand for modern readers,[40] in the following section I will outline it using modern mathematical symbolism. In the reproduction of Mei Wending's projection (Figure 8), point P is the north pole of the equator, point Z is the zenith, and N and S are the north and south points of the horizon. Line CD is the projection of the equator; line BE is the projection of the ecliptic. Line AB represents the projection of the diurnal circle at the summer solstice and, at this time, point G is the position of the sun at the times of sunrise and sunset, and point A is the position of the sun at midday. Suppose M is the position of the sun on the ecliptic on a given day, then line $A'B'$ is the projection of the diurnal circle for this day and point G' is the corresponding position of the sun at the times of sunrise and sunset. The lengths of the arcs $A'G'$ and $B'G'$ are, respectively, half daytime (*banzhoufen* 半晝分) and half nighttime for this given day. Readers who are not interested in the details of Mei Wending's reasoning may pass over the following calculations and skip ahead to the final equation for the hour angle h.

In the *History of the Ming Dynasty*, the latitude of Beijing is taken as 40.95 *du*, and the obliquity of the ecliptic (ε) is 23.903 *du* (23.9 *du* after rounding). According to the method of 'Chinese spherical trigonometry' (*hushi geyuan shu* 弧矢割圓術), we can obtain $PH = 39.26$ *du* in $\triangle POH$, and $FO = AK = EK = 23.71$ *du*.

It is noteworthy that in this spherical system, the traditional Chinese rule 'the circumference of a circle is 3 if the diameter is 1' (*zhousan jingyi* 周三徑一) is applicable, which means that the value of π may be taken as 3. As Martzloff has demonstrated, this seemingly eccentric use of 3 for π made it possible to obtain considerably more accurate results than the use of the exact value of π.[41] The circumference of heaven is 365.25 *du* (truncated from its exact value 365.2575 *du*), so the diameter is 121.75 = 365.25/3 *du* and the radius OP or OC is 60.875 *du*. By means of the Pythagorean theorem, we get $CK = OC - OK = OC - \sqrt{r^2 - EK^2} = 4.81$ *du*. At noon of the summer solstice, the sun is at point A, so the arc $AS = 74.265$ *du*.[42] According to the rule of 'Chinese

[40] Qu Anjing provides annotations for Mei's reasoning in modern mathematical terms. See Qu, *Chinese Mathematical Astronomy*, pp. 279–83.

[41] Martzloff points out that $\pi = 3$ is not the value that gives the best results in combination with Guo Shoujing's approximate computations, but at any rate the best possible value is certainly very close to $\pi = 3$. See Martzloff, *A History of Chinese Mathematics*, p. 334.

[42] Here, arc AS = arc $ZS - \varphi + \varepsilon = 91.315 - 40.95 + 23.9 = 74.265$ *du*. Arc ZS is a quadrant of circle, equaling $365.2575/4 = 91.314375 \approx 91.315$ *du*.

spherical trigonometry', we find $AJ = 58.45$ *du*. The consequent steps can be transcribed as follows:

$$\triangle FOI \sim \triangle POH \rightarrow FI = \frac{FO \times PH}{r} = 15.29\ du$$

$$AL = AJ - LJ = AJ - FI = 43.16\ du$$

$$AF = OK = r - CK = 56.065\ du$$

$$\triangle FAL \sim \triangle GFI \rightarrow FG = \frac{FI \times AF}{AL} = 19.87\ du$$

With the help of 'Chinese spherical trigonometry', we now find:

$$\text{arc}\, FG = 19.9614\ du \qquad \text{and}$$

$$\triangle FGO \sim \triangle F'G'O \rightarrow \frac{\text{arc}\, F'G'}{F'O} = \frac{\text{arc}\, FG}{FO} = 0.8419.$$

This value 0.8419 is a ratio named '[latitude] difference variate' (*ducha fen* 度差分), which connects the values of arc $F'G'$ and $F'O$. Here $F'O = A'K'$, and $A'K'$ is the chord of the declination δ. The ratio of the equation of daylight, arc $F'G'$, to the apparent motion of the sun $6A'F'+1$ (line $A'F'$ is the radius of the diurnal circle, therefore $6A'F'$ is its circumference and $6A'F'+1$ the apparent motion of the sun) is:

$$\frac{\text{arc}\, F'G'}{6A'F' + 1} = \frac{0.8419 \times A'K'}{6A'F' + 1} \quad \text{(unit: day)}.$$

This ratio is called 'time deviation for sunrise and sunset' (*churu chake* 出入差刻), i.e. the length of time when the sun is displaced from position F' to position G'. Therefore, the length of half daytime (*banzhou* 半晝), i.e. from point G' (sunrise) to point A' (midday), can be obtained as follows:

$$h = 0.25 \pm \frac{0.8419 \times A'K'}{6A'F' + 1}$$

$$= 0.25 \pm \frac{0.8419 \times A'K'}{6\sqrt{r^2 - A'K'^2} + 1} \quad \text{(unit: day).}$$[43]

The above is the final equation for the hour angle, and the values of $A'K'$ and $A'F'$ can be picked up from the tables in the *History of the Ming Dynasty*. So the half daytime for each *du* along the ecliptic can be calculated by this method. With the value of the half daytime and the characteristics of symmetry, it is easy to obtain the times of dawn and dusk or the times of sunrise and sunset; thus the data in columns 6, 7 and 8 of the Table C-I-B can be determined. With the ratio '[latitude] difference variate' calculated according to the latitude of a particular place instead of Beijing's 0.8419 (corresponding to the latitude 40.95 *du*), we can compile the table for any geographical location.

Even though the underlying latitude of the Table K-I-S is unknown, the above method allows us to reproduce this type of table for any latitude, and thus to establish the latitude which leads to the closest result. With the assistance of a computer program, it can be found that the ratio '[latitude] differ-

[43] Here, '+' is used after the vernal equinox, and '–' is used after the autumnal equinox.

ence variate' of Table K-I-S is 0.7729, and therewith that a latitude of 38.6 *du* produces results which are almost identical to those in the table.

4.2.2 The 'Method of Nine Domains'

In addition to what we may recover from Mei Wending's narrative, the canon of the *Season-granting System* provides another concise method titled 'clepsydra time at [any location within] the nine domains' (*jiufu louke* 九服漏刻) to calculate the times of sunrise and sunset. The term 'nine domains' (*jiufu* 九服) is a literary allusion to an early theory of monarchy which already existed in China before the second or first century B.C. This theory divides the territory of the state into concentric squares, with the king's domain in the center and the residences of barbarians on the periphery.[44] The method in question is appropriate for sites outside of the capital city. The procedure states:

> 各於所在以儀測驗，或下水漏，以定其處冬至或夏至夜刻，與五十刻相減，餘為至差刻。置所求日黃道去赤道內外度及分，以至差刻乘之，進一位，如二百三十九而一，所得內減外加五十刻，即所求夜刻；以減百刻，餘為晝刻。
>
> Determine the length of local nighttime at the winter or summer solstice at each location through instrumental observations or with the help of the clepsydra. Subtract 50 *ke* [from the length of nighttime], the remainder is the 'solstice difference mark'. For the selected day, set out its declination of the ecliptic, inside or outside the equator, in units *du* and *fen*. Multiply it by the solstice difference mark, advance one column, and count 1 for each 239 in the result.[45] If on that day, the ecliptic is inside the equator, subtract the result from 50 *ke*; if it is outside, add 50 *ke* to the result. So the required length of nighttime is obtained. Subtract it from 100, the remainder is daytime.[46]

In this method, the 'solstice difference mark' (*zhicha* 至差刻), an empirical local measure of the difference between solstitial and equinoctial daylength, needs to be obtained in advance. Suppose H_W is the length of nighttime at the winter solstice and H_S is the length of nighttime at the summer solstice, then the 'solstice difference mark' (SD for short) is '$50 - H_W$' or '$50 - H_S$' in the unit *ke*. The nighttime of the desired day (H_N) is as follows:

$$H_N = 50 \pm SD \times \delta \times \frac{10}{239}.$$

In this equation, δ is the declination of the ecliptic inside or outside the equator, and '10/239' is one divided by the obliquity ε. Thus the equation is equivalent to

$$H_N = 50 \pm SD \times \frac{\delta}{\varepsilon}.$$

This procedure avoids the use of the complicated method of 'Chinese spherical trigonometry', and presumes that the changes in the lengths of daytime and nighttime are simply proportional to the change in solar declination. The result can thus be obtained as a simple linear function of the value δ. This

[44] Sivin, *Granting the Seasons*, p. 496.

[45] That is, divide the value by 239.

[46] *Lidai tianwen lüli deng zhi huibian*, p. 3416.

Ecliptic accumulated degrees	C-I-B[47] (Beijing)	Nine Domains (Beijing)	Differences (Beijing)	K-I-S[48] (Seoul)	Nine Domains (Seoul)	Differences (Seoul)
0	1907.96	1907.88569	0.07431	1956.5	1956.43178	0.06822
1	1908.05	1907.96743	0.08257	1956.58	1956.50682	0.07318
2	1908.34	1908.21267	0.12733	1956.85	1956.73195	0.11805
3	1908.81	1908.62388	0.18612	1957.28	1957.10945	0.17055
4	1909.47	1909.19610	0.2739	1957.89	1957.63476	0.25524
5	1910.32	1909.93677	0.38323	1958.67	1958.31470	0.3553
⋮	⋮	⋮	⋮	⋮	⋮	⋮
50	2133.83	2117.505	16.325	2163.86	2148.86489	14.99511
51	2142.09	2125.64246	16.44754	2171.44	2156.33517	15.10483
52	2150.41	2133.88147	16.52853	2179.08	2163.89869	15.18131
53	2158.81	2142.21462	16.59538	2186.79	2171.54862	15.24138
54	2167.27	2150.65429	16.61571	2194.56	2179.29634	15.26366
55	2175.81	2159.18809	16.62191	2202.4	2187.13047	15.26953
56	2184.4	2167.81106	16.58894	2210.29	2195.04647	15.24353
57	2193.04	2176.51826	16.52174	2218.22	2203.03979	15.18021
58	2201.73	2185.30473	16.42527	2226.2	2211.10587	15.09413
⋮	⋮	⋮	⋮	⋮	⋮	⋮
88	2470.34	2468.03975	2.30025	2472.25	2470.66010	1.5899
89	2479.3	2477.68827	1.61173	2480.98	2479.51755	1.46245
90	2488.26	2487.33678	0.92322	2489.21	2488.37501	0.83499
91	2497.21	2496.98530	0.2247	2497.43	2497.23247	0.19753
91.31	2500	2500	0	2500	2500	0

Table 7: Differences in 'winter daytime/summer nighttime' between Type I and the 'Method of Nine Domains'.

approximate 'Method of Nine Domains' is suitable for non-core sites outside of the capital. While this method is not particularly accurate, it helps to reduce the computational burden.

The canon of the *Season-granting System* gives Table C-I-B directly for the Yuan capital Beijing (geographical latitude 40.95 *du*) and provides the 'Method of Nine Domains' as an alternative method of calculation for sites outside of the capital. A comparison shows that the values obtained by the 'Method of Nine Domains' are smaller than the corresponding values in Table C-I-B, and the maximum difference 16.6219 *fen* (about 2.4 minutes in modern time) appears in the entry for 55 *du* after the solstices (see Table 7). In addition, if we apply

[47] Winter daytime/summer nighttime (*dongzhou xiaye* 冬晝夏夜) in Table C-I-B.

[48] Winter daytime/summer nighttime (*dongzhou xiaye* 冬晝夏夜) in Table K-I-S.

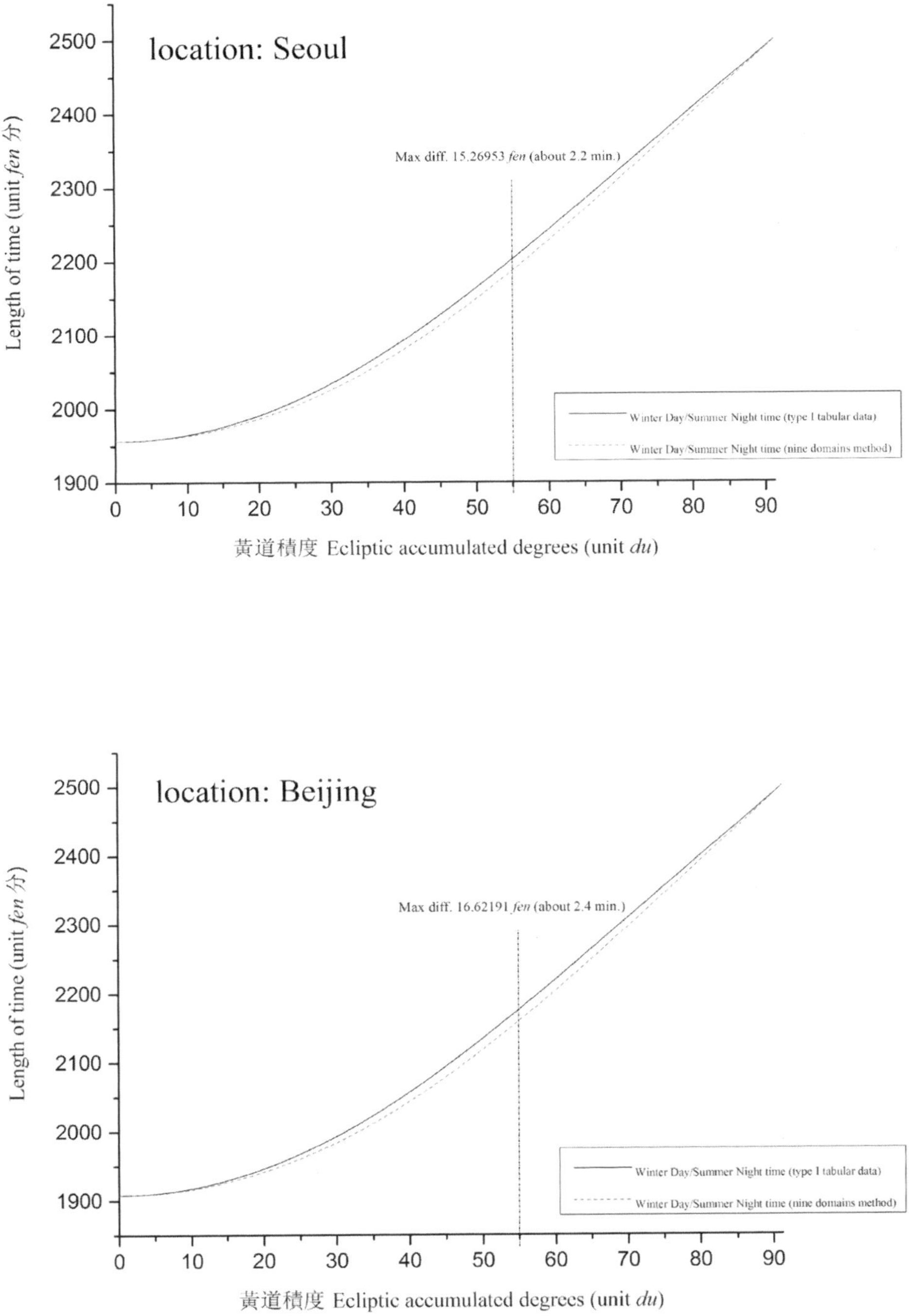

Figure 9: Comparison of two methods for the lengths of daytime and nighttime.

the 'Method of Nine Domains' to Seoul, the calculation results will be smaller than those in Table K-I-S, which is consistent with what happens for Beijing (see Figure 9).[49] This comparison indicates that neither Table C-I-B nor Table K-I-S is produced by the 'Method of Nine Domains'. A probable reason is that both tables were compiled for capital cities and thus had to be highly accurate.

4.3 Method of computation for tables of Type II

The difference between tables of Type I and of Type II is that they have different ranges of arguments. Both start from the solstices, but Type I has an argument range from 0 to 91.31 *du*, and Type II from 0 to 182 days. The interval of the former is one *du* in Chinese degrees, while the interval of the latter is one day in time. In the canon of the *Season-granting System*, the third section of the chapter 'The Pacing of the Centered Star' (*buzhongxing* 步中星, i.e. measuring the time of the meridian transit of stars) introduces a procedure titled 'Finding the half daytime/nighttime and the time of sunrise, sunset, dawn and dusk of each day' (*qiu meiri banzhouye ji richuru chenhunfen* 求每日半晝夜及日出入晨昏分).[50] This procedure begins by picking up data from the table and carrying out interpolation for the lengths of daytime and nighttime; the purpose is to transfer the argument from Type I to Type II. The text states:

> 置所求入初末限，滿積度，去之，餘以晝夜差乘之，百約之，所得，加減其段半晝夜分，為所求日半晝夜分；前多後少為減，前少後多為加。以半夜分便為日出分，用減日周，餘為日入分；以昏明分減日出分，餘為晨分；加日入分，為昏分。
>
> Set the Beginning/End Extent for the desired argument (the true solar position of the day, taken from the solar equation table) and cast the full accumulated *du* (counted along the ecliptic from the appropriate solstice).[51] Multiply its remainder (a value smaller than accumulated *du*) by the day/night difference (column 8 in Table C-I-B) and simplify the result by 100 and round it off. The result, when added to or subtracted from midday and midnight time for the corresponding increment, becomes midday and midnight time for the day. If it is greater before than after, subtract; if smaller before than after, add. Then take midnight time as sunrise time. Subtract from the day cycle (10000 *fen*) and the remainder is sunset time. Subtract dusk/daybreak time (twilight) from sunrise time and the remainder is dawn time. Add [dusk/daybreak time (250 *fen*)] to sunset time to produce dusk time.[52]

[49] The maximum difference of 15.26953 *fen* (about 2.2 minutes in modern time) appears for the entry 55 *du* after the solstices.

[50] *Lidai tianwen lüli deng zhi huibian*, p. 3414.

[51] In the *Season-granting System*, the sun's speed and the daily increment to the equation of center are identical in the four quadrants. Thus the Beginning/End Extent in quadrants I and IV is read from the winter solstice, and in quadrants II and III from the summer solstice. See Sivin, *Granting the Seasons*, p. 415.

[52] *Lidai tianwen lüli deng zhi huibian*, pp. 3414–15.

Solar equation table

Days (after solstice)	Solar correction (difference of true and mean position)
0	0
1	0.05108569
2	0.10167752
3	0.15177363
4	0.20137216
5	0.25047125
6	0.29906904
7	0.34716367
8	0.39475328
9	0.44183601
10	0.48841
11	0.53447339
12	0.58002432
13	0.62506093
14	0.66958136
…	…

⇨

True solar position

Days (after solstice)	True solar position (accumulated degrees from solstice)
0	0
1	1.05108569
2	2.10167752
3	3.15177363
4	4.20137216
5	5.25047125
6	6.29906904
7	7.34716367
8	8.39475328
9	9.44183601
10	10.48841
11	11.53447339
12	12.58002432
13	13.62506093
14	14.66958136
…	…

\+

Type I (Beijing)

Accumulated degrees (after solstice)	Half day time	Day/Night difference
0	1907.96	0.09
1	1908.05	0.29
2	1908.34	0.47
3	1908.81	0.66
4	1909.47	0.85
5	1910.32	1.04
6	1911.36	1.22
7	1912.58	1.42
8	1914	1.61
9	1915.61	1.79
10	1917.4	1.99
11	1919.39	2.18
12	1921.57	2.37
13	1923.94	2.56
14	1926.5	2.74
…	…	…

⇨

Type II (Beijing)

Days (after solstice)	Half day time
0	1907.96
1	1908.06
2	1908.38
3	1908.91
4	1909.64
5	1910.58
6	1911.72
7	1913.07
8	1914.63
9	1916.40
10	1918.37
11	1920.55
12	1922.94
13	1925.54
14	1928.33
…	…

Figure 10: Process of transforming from Type I to Type II.

In this procedure, the user needs to refer to the solar anomaly table and pick up the true position of the sun for the desired day. Because the *Season-granting System* chooses the winter solstice as the approximated solar apogee, and the mean motion of the sun is one *du* per day, the true solar position on day n after the solstice is n times 1 *du* plus or minus the correction of the solar equation of anomaly.

For example, for the tenth day after the winter solstice we have 10 *du* as the total motion of the mean sun and the solar equation of anomaly is 0.48841 *du*, thus the sum is 10.48841 *du*, which is the true position of the sun (measured from the winter solstice) on this day. With 10.48841 as the argument, we now cast out the whole ecliptical degrees, resulting in 0.48841. After multiplying the remainder by the corresponding day/night difference of 1.99 picked up from the table of Type I (column 8 in the table of Type C-I-B), we obtain 0.9719359. Adding this to 1917.4, the corresponding half daytime (column 10 in Table C-I-B), the half daytime for day 10 after the winter solstice can be obtained (see Figure 10). Finally, by symmetry, we can produce the times of sunrise, sunset, dawn and dusk on this day.

5. Tables revised in Korea

When the *Season-granting System* and the *Great Concordance System* were transmitted to Korea, the Korean astronomers were aware of the difference in latitude between the Chinese capital and the Korean capital, and tried to find a solution. A Korean commentary on the *Great Concordance System* mentions:

> 《授時曆》、《通軌》、《回回曆》日出入晝夜刻，各據所在推定，與本國不同。今更以本國漢都每日日出入晝夜刻錄於內、外篇中，永為定式。
>
> The times of sunrise, sunset, daytime and nighttime in [books such as] the Season-granting System, General Rules [of the Great Concordance System] and the

> Huihui Calendrical System (Chinese-Islamic System) are all based on calculation results of their corresponding sites, which are different from our own country. Now we replace these data with the times of sunrise and sunset, the lengths of daytime and nighttime for each day at our capital Hanyang (Seoul), and record them in the Inner Chapter and the Outer Chapter [of Computation of the Seven Regulators] as acknowledged facts that remain true forever.[53]

In addition to this text, the official record in the *The Veritable Records of the Joseon Dynasty* (*Joseon wangjo shillok* 朝鮮王朝實錄) also points out that 'the [Chinese] Great Ming dynasty refers to the book *General Rules* [*of the Great Concordance System*]' 大明用《通軌》日出分 to determine the time of sunrise, while 'our country uses the *Inner Chapter* [*of Computation of Seven Regulators*]' 本國用《內篇》日出分.[54] These records show that the Korean astronomers were aware that the differences in the times of sunrise and sunset were caused by the difference in latitude. Consequently, in order to compile a new table for Seoul, one had to possess not only astronomical knowledge but also enough computing skills.

Eun-Hee Lee points out that the half lengths of daytime and nighttime in Table K-I-S were calculated based on the 'inside/outside degrees of the ecliptic' (*huangdao churu chidao neiwaidu* 黃道出入赤道內外度, i.e., the declination measured from the equator to the ecliptic) and the 'solstice difference mark' (*zhichake* 至差刻). In addition, the solar declination was calculated by applying Shen Gua's formula from the late eleventh century. She also compares the values in Table K-I-S with a recalculation using modern methods, and says that the disagreement between Table K-I-S and modern methods is caused by the use of the approximate formula of Shen Gua, who adopted the value of $\pi = 3$.[55] However, we cannot find historical records to support the fact that the Korean astronomers had mastered the knowledge of determining the sagitta by the segment of the circle (by the rule of 'Chinese spherical trigonometry', *hushi geyuan shu* 弧矢割圓術). Shi Yunli criticizes Eun-Hee Lee's conclusion and points out that this table may have been produced by the 'Method of Nine Domains' with its algebraic formula, which is the concise method introduced in the canon of the *Season-granting System*. He supports his argument by pointing out that the Korean documents in their calculation emphasize the 'solstice difference mark' (*zhichake* 至差刻), which only appears in the procedure of the 'Method of Nine Domains'.[56] Indeed, our analysis has demonstrated that the procedure to construct this table is more complex than scholars have previously imagined.

[53] *Datong lifa tonggui*, epilogue 2a.
[54] *Joseon wangjo shillok*, Sejo annals 119.2b.
[55] See Lee, 'The Ch'iljŏngsan Naepiŏn'.
[56] Shi, 'The Study of *Shoushili*'.

As mentioned in the last section, according to the Table K-I-S in the *Inner Chapter of Computation of the Seven Regulators*, we find that the latitude adopted by the Korean astronomers is about 38.6 *du*. However, none of the historical documents mentions this value; even the *Inner Chapter of Computation of the Seven Regulators* avoids discussing the latitude value actually used.

The *History of the Yuan Dynasty* reports that as early as 1279 the Chinese astronomer Guo Shoujing carried out a survey to accurately measure the latitude at twenty-seven different sites including Koryeo (Korea). The latitude of Koryeo is recorded as $38\frac{1}{4}$ *du* (37°41′). This value is thought to belong to Gaegyeong, the capital of Koryeo and presently the North-Korean city Kaesong, 50 km north of Seoul.

In 1432, several Korean scholars including Jeong Inji 鄭麟趾 (1396–1478) were assigned by King Sejong to measure the latitude of Seoul. They established the latitude as 38 *du*, and came to the conclusion that this value is 'close to the measurement recorded in the *History of the Yuan Dynasty*' 少與《元史》所測合符.[57] Later, the Korean astronomer Yi Soonji 李純之 (fl. 1450s) recomputed the latitude and obtained $38\frac{1}{12}$ *du*. At first, King Sejong was not convinced, but in time, the Chinese Ming court issued an almanac to Korea, which says 'the latitude of Korea is $38\frac{1}{12}$ *du*' 高麗北極出地, 三十八度强.[58]

Because the *Inner Chapter of Computation of the Seven Regulators*, finished in 1442, only displays the tables and conceals the latitude and the method of calculation, the only clue we can find in this book is the following passage:

> 日出入隨處各異，諸曆不同。《內篇》據漢陽日至之晷，推求至差，得每日日出晝夜刻分，定為本國所用。
>
> Sunrise and sunset vary from place to place, and each calendrical system [has] different [values]. The Inner Chapter [of Computation of the Seven Regulators] fixes the hours of day and night appropriate for use in this country as calculated from the 'solstice difference [mark]' according to the solstitial gnomon [shadow] at Hanyang.[59]

In addition, the book Method for the Calculation of the Eclipses (Gyosik Chubobeob 交食推步法), finished in 1457, which is a revision of the *Inner Chapter of Computation of the Seven Regulators*, presents a similar expression:

> 日出入則以北極出地高下，隨處各異。而書雲觀只依中朝大統曆日出入用之，極爲踈闊。故軫慮精思，先測定我國漢陽北極出地三十八度少弱，參考其二至晷影，推求得日出入之分。
>
> The [times of] sunrise and sunset are related to the polar altitude (geographical latitude), and each site is different. Since the Hall of Heavenly Records (Korean Royal Observatory and Astronomical Bureau) merely relies on [the times of] sunrise and sunset mentioned in the Great Concordance System from China, data in this book

[57] *Joseon wangjo shillok*, Sejo annals 77.9b.

[58] *Joseon wangjo shillok*, Sejo annals 36.14a.

[59] *Chiljeongsan Naepyon*, vol. A, 38.b.

> are extremely inaccurate [for use in our country]. So [we are] anxious about this and have spent great efforts on it. Firstly, [we] determined the latitude of Hanyang (Seoul) to be $38\frac{1}{6}$ du. Then, referring to the shadow of the sundial at two solstices, [we] obtained the times of sunrise and sunset.[60]

In these two texts, we find that the Korean astronomers calculated the times of sunrise and sunset based on observations with a sundial in Hanyang (Seoul). The calculation of a table of Type I requires the exact local latitude and mastery of the method of 'Chinese spherical trigonometry'. The underlying latitude 38.6 *du* adopted by Table K-I-S is not consistent with other historical values such as $38\frac{1}{4}$ *du*, 38 *du*, $38\frac{1}{12}$ *du* or $38\frac{1}{6}$ *du*. Although the Korean astronomers received many astronomical and mathematical books from China in the fifteenth century, the catalogue of these books did not include *Detailed Procedures* which explains the underlying theory,[61] so probably they were not able to compile the tables of sunrise and sunset by the sophisticated method of 'Chinese spherical trigonometry'. Another relevant detail is that at that time, the Korean astronomers had limited mathematical skills in calendrical calculations. According to the official annals, many staff members of the Korean Royal Observatory and Astronomical Bureau only 'mastered preliminary arithmetic such as multiplication and division' 粗習乘除而已, and even 'the extraction of the cubic root was unknown' 立方開法尚未知也.[62] It is therefore very unlikely that the sophisticated theory was accessible to Korean astronomers and that their calculation ability was adequate to apply 'Chinese spherical trigonometry'.

In all historical texts mentioned above, we should pay special attention to the term 'solstice difference mark'. The 'Method of Nine Domains' provides us with a calculation procedure based on the 'solstice difference mark', but our analysis has shown that the data of Table K-I-S do not correspond with this method. The Korean astronomers may have found an ingenious way to obtain accurate data for Table K-I-S, that borrows data from the earlier Table C-I-B.

Column 8 of Table C-I-B gives the 'day-night difference' (*zhouye cha* 晝夜差), the increase or decrease of daytime for each day ranging from solstice to equinox, and the sum of the day-night differences is 592.04 *fen*. This is equivalent to the maximum difference of half daytime between the solstices and equinoxes and is half of the 'solstice difference mark'.[63]

The Korean astronomers must have carried out an empirical measurement of daylength at the two solstices at first, and found the 'solstice difference mark'

[60] *Gyosik Chubobeob*, vol. A, 1a.

[61] The book *Detailed Procedures* (*Licao*) was almost lost in China at the time, but we can access its contents through Mei Wending's record from the late seventeenth century.

[62] *Joseon wangjo shillok*, Sejo annals, 20.39a.

[63] The length of half daytime at the winter and summer solstices is 1907.96 *fen* and 3092.04 fen respectively. The length of half daytime at the two equinoxes is 2500 *fen*.

Ecliptic accumulated degrees	K-I-S (Seoul)	Recomputation (Seoul)	Differences (Seoul)
0	0.08	0.08232	0.00232
1	0.27	0.26524	−0.00476
2	0.43	0.42988	−0.00012
3	0.61	0.60366	−0.00634
4	0.78	0.77744	−0.00256
5	0.95	0.95122	0.00122
⋮	⋮	⋮	⋮
50	7.58	7.55488	−0.02512
51	7.64	7.60976	−0.03024
52	7.71	7.68293	−0.02707
53	7.77	7.7378	−0.0322
54	7.84	7.81098	−0.02902
55	7.89	7.85671	−0.03329
56	7.93	7.90244	−0.02756
57	7.98	7.94817	−0.03183
58	8.03	8.00305	−0.02695
⋮	⋮	⋮	⋮
88	8.23	8.19512	−0.03488
89	8.23	8.19512	−0.03488
90	8.22	8.18598	−0.03402
91	2.57	2.55183	−0.01817
91.31	0	0	0

Table 8: Differences of 'day-night difference' between K-I-S and my recomputation.

as 541.5 *fen*. Then they needed to find a way to distribute the 541.5 *fen* over 91.31 days. Table C-I-B gives the 'day-night difference' for Beijing in column 8. By scaling the data in this column, it would not be difficult to rebuild the 'day-night difference' (DNd for short) in Table K-I-S. The relationship between the two DNds is then as follows:

$$\frac{\text{DNd(n)\_KIS}}{541.5} = \frac{\text{DNd(n)\_CIB}}{592.04}$$

(here n is the day after the solstice from 0 to 91.31).

With the formula above, the average difference of the 'day-night difference' between K-I-S and my recomputation is about −0.02113 *fen* (0.18 seconds in modern time), and the maximum difference is −0.03488 *fen* (0.30 seconds in modern time). These small deviations may be caused by the used method of rounding or truncation (see Table 8).

With the 'day-night difference' of Table K-I-S in column 8 and the lengths of daytime and nighttime on the days of the summer and winter solstice, the lengths of daytime and nighttime of any given day listed in columns 4 and 5 in Table K-I-S can be obtained easily. In this way, the Korean astronomers could ignore the value of the geographical latitude, and avoid the complicated operations of 'Chinese spherical trigonometry' to produce a table practically as good as the one used in China. This reasoning is consistent with the record in the book *Inner Chapter of Computation of the Seven Regulators*. If true, it would be a typical case of drawing support from existing astronomical tables to bypass the theoretical and technological obstacles and compile a new one. As to the question of why the underlying latitude of table K-I-S is not given and its value is not consistent with actual historical observations, a reasonable explanation is that this table is probably made based on empirical measurement and Table C-I-B, rather than sticking to some fixed calculation procedure based on the latitude.

6. Concluding remarks

Firstly, this paper scrutinizes various tables of sunrise and sunset in the calendrical works of Yuan and Ming China, including the *Season-granting System* in the *History of the Yuan Dynasty* and the *Great Concordance System* in the *History of the Ming Dynasty*. These tables of sunrise and sunset belong to two types, and the distinct differences between them are their layouts and applications. The Type I table with the argument of 'ecliptic accumulated degrees' is a necessary tabular tool in some specific calendrical calculations, while the table of Type II is more straightforward and can be used by non-experts to calibrate time reckoning instruments such as the clepsydra.[64] These two types of tables can be converted to each other when required, and the canon of the *Season-granting System* includes the procedure to transform from Type I to Type II.

The *Season-granting System* provides two methods to produce the tables of sunrise and sunset. One employs arc and sagitta calculations and utilizes the characteristic 'Chinese spherical trigonometry' (*hushi geyuan shu* 弧矢割圓術). This approach is based on a specific geometrical model and can be clearly explained in theory, but as it imposes too much calculation burden, it is only used for the capital cities. This method is not recorded in the existing canon of the *Season-granting System* in the *History of the Yuan Dynasty*, but we can learn about it through Mei Wending's record from the late seventeenth century. The other method named 'Method of Nine Domains' is rough but simple, and it applies to districts outside of the capital. Based on our analysis, tables of sun-

[64] Sometimes it is necessary to convert the time from the 'one-hundredth-of-a-day system' to the 'watches-and-points system' and 'double-hour system'.

rise and sunset for Beijing, Nanjing, and Seoul in historical documents were all produced using the first method, because these three cities were the capitals of the Chinese Yuan and Ming dynasty and the Korean Joseon dynasty respectively. The historical document states the latitude used for table C-I-B as 40.95 *du*, which is consistent with the result of our recomputation. Tables C-I-N and K-I-S do not specify the latitude of their places of use; calculation finds that the most likely values of the two tables are 32.5 *du* and 38.6 *du*.

Secondly, there is no direct evidence telling us how the Korean astronomers constructed their own tables K-I-S. It is not clear whether they did calculations independently or adapted Chinese tables for Korea. The analysis shows that the underlying latitude of Tables K-I-S does not agree with any previous historical measurement. It is reasonable to deduce that the Korean astronomers probably compiled their tables on the basis of existing Chinese tables. Although the Korean astronomers were not familiar with the specific underlying theory, they managed to construct their own tables by referring to available Chinese tables when their mathematical and geometrical knowledge was inadequate.

Bibliography

Secondary literature

Chemla, Karine, and Li Liang, 'Numerical Tables in Chinese Documents Dealing with Mathematics or Astral Sciences', in Dominique Tournes (ed.), *History of Numerical Tables*, Berlin: Springer, forthcoming.

Chen Jiujin, 'Research on Chinese Ancient Time-reckoning Systems and Their Conversions' 中國古代時制研究及其換算, *Studies in the History of Natural Sciences* 自然科學史研究 2 (1983), pp. 118–32.

Chen Meidong and Li Dongsheng, 'Research of *lou-ke* Calculation in Ancient China' 中國古代晝夜漏刻長度的計算法, *Studies in the History of Natural Sciences* 自然科學史研究 9 (1990), pp. 47–61.

Lee Eun-Hee, 'The Ch'iljŏngsan Naepiŏn, an Adopted Version of the Shoushi-li and Datong-li', in Il-Seong Nha and F. Richard Stephenson (eds), *Oriental Astronomy from Guo Shoujing to King Sejong*, Seoul: Yonsei University Press, 1997, pp. 339–48.

Lee Eun-Hee, 'Korean Astronomical Calendar, *Chiljeongsan*', in Clive L. N. Ruggles (ed.), *Handbook of Archaeoastronomy and Ethnoastronomy*, New York: Springer, 2015, ch. 208, pp. 2157–62.

Li Liang, 'Astronomical Tables in the "*Lü-li zhi*". On the Characteristics and Adoption of *Licheng* Pick-up Tables', *Chinese Annals of History of Science and Technology* 2/2 (2018), pp. 41–67.

Lin Jin-Chyuan, 'The Pick-up Tables of Sunrise and Sunset in the Tong Yuan Calendar in the Song Dynasty' 宋《統元曆日出日入辰刻立成》, *Journal of Chinese Studies of National Chengkung University* 成大中文學報 18 (2007), pp. 1–38.

Liu Caonan, 'Summary of Season-granting System' (*Shoushi lishu shuyao* 授時曆術述要), in *Interpretation of Ancient Astronomy and Calendrical System* (*Gudai tianwen lifa shizheng* 古代天文曆法釋證), Hangzhou: Zhejiang daxue chubanshe, 2009, pp. 452–71.

Martzloff, Jean-Claude, *A History of Chinese Mathematics*, Berlin: Springer, 1997.

Mihn Byeong-Hee, Lee Ki-Won and Ahn Young-Sook, 'Analysis of Interval Constants in Shoushili-affiliated Calendars', *Research in Astronomy and Astrophysics* 14 (2014), pp. 485–96.

Qu Anjing, 'Time-reckoning in Ancient Chinese Calendrical Systems' 中國古代曆法中的計時制度, *Sinology Study Hanxue yanjiu* 漢學研究 12 (1994), pp. 157–72.

Qu Anjing, *Chinese Mathematical Astronomy Zhongguo shuli tianwenxue* 中國數理天文學, Beijing: Kexue chubanshe, 2008.

Shi Yunli, 'The Study of *Shoushili* by Ancient Korean Scholars' 古代朝鮮學者的《授時曆》研究, *Studies in the History of Natural Sciences* 自然科學史研究 17 (1998), pp. 312–21.

Sivin, Nathan, *Granting the Seasons. The Chinese Astronomical Reform of 1280, with a Study of its Many Dimensions and a Translation of its Records*, New York: Springer, 2009.

Wang Lixin, 'Investigation into Systems of Reckoning Time' 紀時制度考, in *Collected Works of Chinese History of Astronomy Zhongguo tianwenxueshi wenji* 中國天文學史文集, vol. IV, Beijing: Kexue chubanshe, 1986, pp. 1–47.

Yabuuti, Kiyosi, and Shigeru Nakayama, *Season-granting System: Annotated Translation and Study* (*Jujireki: Yakuchū to kenkyū* 授時曆：譯注と研究), Kawasaki: I. K. Corporation, 2006.

Traditional Works in Chinese

Chiljeongsan Naepyon = Jeong In-Ji 鄭麟趾 (1396–1478), Jeong Cho 鄭招 (?–1434) and Jeong Heum-ji 鄭欽之 (fl. 1430s), *Chiljeongsan Naepyon* 七政算內篇 (*Inner Chapter of Computation of Seven Regulators*), printed in 1444, kept in the Gyujanggak Library of Seoul National University, collection no. 48.

Datong lifa tonggui = Yuan Tong 元統 (fl. 1384–1396), *Datong lifa tonggui* 大統曆法通軌 (*General Rules of the Great Concordance System*), printed in 1444, kept in the Gyujanggak Library of Seoul National University, collection nos 12434–39.

Datong richu riru fen = Anonymous, *Datong richu riru fen* 大統日出日入分 (*Times of Sunrise and Sunset in the* Great Concordance System), in *Yuan guoli Beiping tushuguan jiaku shanben congshu* 原國立北平圖書館甲庫善本叢書 (第五〇六冊) (Rare Books Series in the Former National Library in Peking, vol. 506), Beijing: Guojia tushuguan chubanshe, 2013.

Gyosik Chubobeob = Yi Soonji 李純之 (fl. 1450s), *Gyosik Chubobeob* 交食推步法 (*Method for the Calculation of the Eclipses*), finished in 1457, kept in the Gyujanggak Library of Seoul National University, collection no. 67.

Joseon wangjo shillok = Anonymous, *Joseon wangjo shillok* 朝鮮王朝實錄 (*The Veritable Records of the Joseon Dynasty*), electronic texts and images at the official website of the National Institute of Korean History, http://sillok.history.go.kr.

Koryeo-Sa = Jeong In-ji 鄭麟趾 (1396–1478), *Koryeo-Sa* 高麗史 (*History of the Koryeo Dynasty*), finished in 1451, kept in the Gyujanggak Library of Seoul National University, collection no. 3539.

Monographs = *Lidai tianwen lüli deng zhi huibian* 歷代天文律曆等志彙編 (*Collection of Monographs on Heavenly Patterns, Harmonics and Calendrical Astronomy for Successive Dynasties*), modern edition Beijing: Zhonghua shuju, 1976.

Pick-up Tables = Guo Shoujing 郭守敬 (1231–1316) and Wang Xun 王恂 (1235–1281), *Shoushili licheng* 授時曆立成 (*Pick-up Tables of the Season-granting System*), printed in 1444, kept in the Gyujanggak Library of Seoul National University, collection no. 893.

Yuanshi = Song Lian 宋濂 (1310–1381) (ed.), *Yuanshi* 元史 (*History of the Yuan Dynasty*), modern edition Beijing: Zhonghua shuju, 1976.

The Tables of Planetary Latitudes in Jamshīd al-Kāshī's *Khāqānī Zīj*

Glen VAN BRUMMELEN

After the ninth century AD, the practice of Islamic mathematical astronomy was inspired mostly by the tradition initiated by Claudius Ptolemy's *Almagest*.[1] The determination of the positions of the celestial bodies as a function of time was at the heart of the project, with various applications (such as eclipse reckoning, religious observances such as prayer times, and astrological prognostications) following from them. Although there were several genres within Islamic astronomy, the largest, with more than 225 known representatives, was the *zīj*.[2] These treatises generally eschewed explicit representation of geometric theory in favour of application, containing astronomical tables and instructions for their use but comparatively little discussion of the underlying trigonometry or observations. In this respect they were closer in spirit to Ptolemy's *Handy Tables*; however, in their tabulations of the positions of the planets they frequently followed the methods of the *Almagest*.

The *Almagest* deals with the motions of the planets in Books IX–XI (and Books III and IV, if one includes the Sun and Moon). But in these books he deals only with the planets' *longitudes*, i.e., their positions as if they travel along the ecliptic (Figure 1) measured from the vernal equinox. However, the planets move above and below the ecliptic by up to several degrees, and the predictive power of the *Almagest* would be limited had he not dealt with *latitudes* as well. Ptolemy's approach to modeling latitudes in the XIIIth and last book involves somewhat complicated three-dimensional geometry. The mathematics that Ptolemy applies to the problem is filled with approximations, more than one finds within the treatment of planetary longitudes earlier in the *Almagest*.[3]

[1] Toomer, *Ptolemy's Almagest*.

[2] For surveys of the *zīj* literature and summaries of the most important *zīj*es (including those mentioned in this article), see Kennedy, 'A Survey'; and King et al., 'Astronomical Handbooks'.

[3] For overviews of Ptolemaic latitude theory see Neugebauer, *A History*, vol. I, pp. 208–16; and Pedersen, *A Survey*, pp. 355–86. See also Riddell, 'The Latitudes of Venus'; Swerdlow, 'Ptolemy's Theory of the Inferior Planets'; and Swerdlow, 'Ptolemy's Theories of the Latitude'. On planetary latitudes in medieval Islam (especially in Maragha and Samarqand) see Mozaffari, 'Planetary Latitudes'. For a summary of planetary latitude tables in medieval Islam see van Dalen, 'Tables of Planetary Latitude', pp. 325–28.

Editing and Analysing Numerical Tables: Towards a Digital Information System for the History of Astral Sciences, ed. by Matthieu Husson, Clemency Montelle and Benno van Dalen, PALS 2 (Turnhout, 2021), pp. 287–329
© BREPOLS PUBLISHERS 10.1484/M.PALS-EB.5.127702

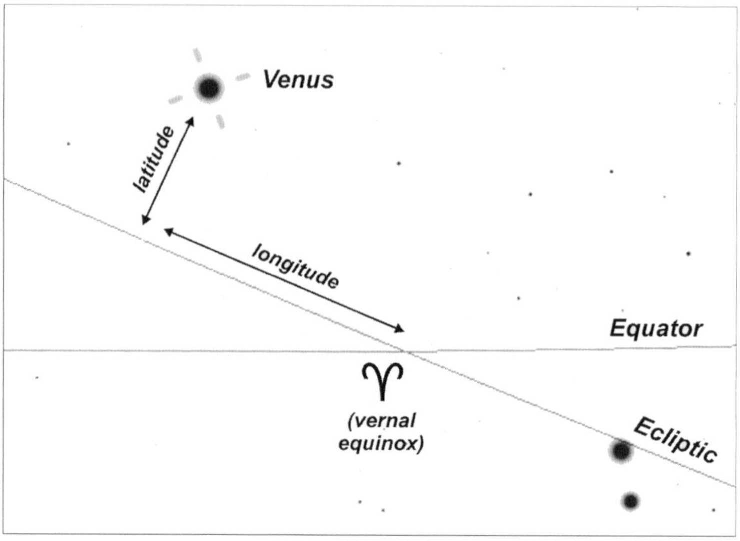

Figure 1: The longitude and latitude of Venus.

Due partly to the lesser significance of planetary latitudes (as opposed to longitudes) in astrology, they were often treated cursorily, and latitude tables were frequently taken directly from the *Almagest* or copied from one *zīj* to the next.

Two of the most sophisticated *zīj*es composed in the first half of the fifteenth century were related to each other. The first is the *Khāqānī Zīj* by Jamshīd Ghiyāth al-Dīn al-Kāshī of Kashan, Iran.[4] Written early in his working life, the *Khāqānī Zīj* still holds many secrets. Unusually for *zījes*, it contains extensive accounts of mathematical and astronomical innovations. Small sections of the work have been examined in the modern literature: for instance, the trigonometric tables,[5] its first determination of sin 1°,[6] its spherical astronomy,[7] aspects of its astrology,[8] its planetary longitude tables,[9] and its novel determination of planetary longitudes and latitudes that replaces Ptolemy's epicyclic circle with an epicyclic sphere.[10]

[4] For a survey of the contents of the entire *zīj*, see Kennedy, *On the Contents*.

[5] Hamadanizadeh, 'The Trigonometric Tables'.

[6] Van Brummelen, 'Crossing a Mathematical Rubicon'. This is not the well-known method that solves a cubic equation numerically; for that see Aaboe, 'Al-Kāshī's Iteration Method', or Rosenfeld and Hogendijk, 'A Mathematical Treatise'.

[7] Kennedy, 'Spherical Astronomy'.

[8] Kennedy, 'Treatise V of Kāshī's *Khāqānī Zīj*', and Kennedy, 'The Prime Vertical Method'.

[9] Tichenor, 'Late Medieval Two-Argument Tables'.

[10] Van Brummelen, 'Taking Latitude with Ptolemy'.

The *Khāqānī Zīj* was written as an upgrade to Naṣīr al-Dīn al-Ṭūsī's mid-thirteenth-century *Īlkhānī Zīj*.[11] Indeed, al-Kāshī's *Zīj* begins with a list of some 70 improvements of his work over its predecessor. It opens with an effusive passage praising his sponsor Ulugh Beg, sultan in Samarqand and an astronomer in his own right. Al-Kāshī would eventually end up in Ulugh Beg's scientific court,[12] and it is there that he would perform the computational feats for which he is best known today: the determinations of π and of sin 1° to the equivalent of 16 decimal places.[13]

Al-Kāshī died in 1429, but astronomical work continued at Ulugh Beg's court, culminating in the *Sulṭānī Zīj* written by 1447 – just two years before Ulugh Beg's own death. The *Sulṭānī Zīj* became perhaps the widest distributed *zīj* of all time, available in hundreds of manuscripts over a vast region and translated into Arabic and Turkish. Its tables are among the most impressive of their type; for instance, its sine table gives values of sines for each minute of arc (5400 entries altogether) to five sexagesimal (base 60) places, equivalent to roughly nine decimal places and generally accurate to within one unit of the last place. The text of the *zīj* was edited and translated to French over 150 years ago and recently the work was translated to Russian,[14] but an edition remains to be attempted.

This paper proposes to study al-Kāshī's planetary latitude tables, only partly completed in the *Khāqānī Zīj*. They consist of a set of single-argument tables mostly but not completely derived from their counterparts in the *Īlkhānī Zīj*; and a set of double-argument tables, one of only two such sets composed in medieval Islam. We shall study the relation between the tables in the *Īlkhānī Zīj* and the *Khāqānī Zīj*, paying special attention to the computational innovations proposed by al-Kāshī. Since the double-argument tables were not completed, we are afforded a brief glimpse into the activity of table production. Finally, we shall compare al-Kāshī's tables to the corresponding set in the *Sulṭānī Zīj*, providing another data point toward understanding the extent of al-Kāshī's influence on this latter great work.

[11] Although Naṣīr al-Dīn al-Ṭūsī is known for his *Tadhkira*, a work that replaces Ptolemy's geometric models of planetary motion with alternate models (see Ragep, *Naṣīr al-Dīn al-Ṭūsī's Memoir*), his highly popular *Īlkhānī Zīj*, extant in dozens of manuscripts, is firmly within the Ptolemaic tradition.

[12] We know something of the life of Ulugh Beg's court and observatory through several sources, including two letters al-Kāshī wrote to his father. See Kennedy, 'A Letter of Jamshid al-Kāshī'; Bagheri, 'A Newly Found Letter', and Giahi Yazdi and Rezvani, 'Chronology of the Events'.

[13] On sin 1° see note 6. On π see Luckey, *Der Lehrbrief über den Kreisumfang*, and Azarian, 'Al-Kāshī's Fundamental Theorem'.

[14] Sédillot, *Prolégoménes des Tables Astronomiques*.

Manuscripts

The *Khāqānī Zīj* is available in at least ten manuscripts, of which several are fragments. Our edition of the double argument tables for Mercury and Venus in the appendix is based on the following:[15]

- London, British Library, MS India Office 430 (Ethé 2232)[16]
- Istanbul, Süleymaniye Library, MS Ayasofya 2692
- Cairo, Dār al-Kutub, MS TR 149

The latitude tables are found in the following folios:

	India Office	*Ayasofya*	*Dār al-Kutub*
Lunar latitude	fol. 138v, 139r	fol. 100v	pp. 241, 242
Superior planets single-argument	139r, v	101r	242, 243
Venus single-argument	139v, 140r	101v	243, 244
Mercury single-argument	140v, 141r	102r	245
Saturn double-argument	152r – *empty*	113r – *empty*	268 – *empty*
Jupiter double-argument	152v – *empty*	*missing*	269 – *empty*
Mars double-argument	153r – *empty*	*missing*	270 – *empty*
Venus double-argument	153v; 154r, v *partly completed*	113v (*empty*); 114r, v (*partly completed*)	271–74 *partly completed*
Mercury double-argument	155r, v; 156r	115r, v, 116r *all empty*	276, 277, 278

As one can see, al-Kāshī apparently never began to fill in the double-argument tables for the superior planets; thus we shall focus on the inferior planets. Noting that the Ayasofya manuscript's double-argument table for Mercury and the first page of the Venus table are empty, Kennedy suggests either that a scribe left them out or that the Ayasofya manuscript was compiled before these tables were completed.[17] The Dār al-Kutub manuscript is missing one of four pages of the Mercury table, but has a few extra entries in the Venus table. The latitude tables are found along with other planetary tables (including his double-argument planetary latitude tables) at the end of Treatise III; al-Kāshī's text describing his mathematical innovations and criticisms of Ptolemy are in Treatise III, Chapter 2, Sections 6 and 7.

[15] Other complete manuscripts, not used for this study, include Hyderabad, OMLRI, MS Āṣafiyya 323; Jaipur, Maharaja Man Singh II Museum Library, MS 9, and Qum, Marʿashī Library, MS 8144.

[16] A rough translation of the instructions for the tables' use based on this manuscript, provided to the author by E. S. Kennedy for this purpose, was invaluable.

[17] Kennedy, *On the Contents*, p. 30.

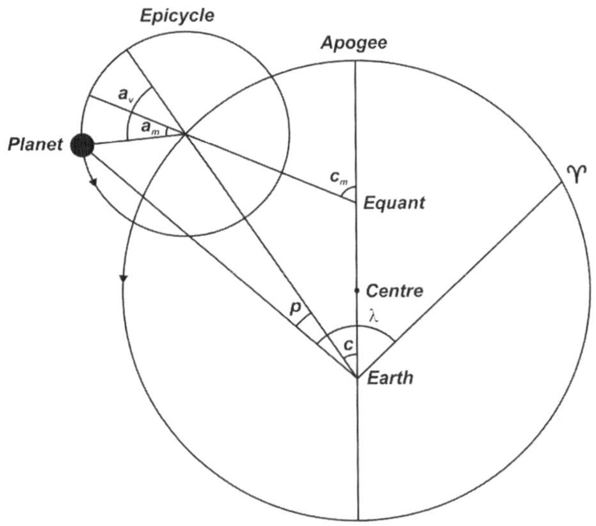

Figure 2: Ptolemy's model for the longitudes of the planets.

When the *Īlkhānī Zīj* and the *Sulṭānī Zīj* are cited below, folio numbers for the former refer to London, British Library, MS Or. 7464, and for the latter, Oxford, Bodleian Library, MS Marsh 396.

Ptolemy's latitude model

Ptolemy's complete set of models for the latitudes of the planets, as well as their motivations and derivations, are described in several places.[18] Various modern authors use different terms to describe the several planetary latitude effects; for the most part we base our language and notation on Pedersen's *A Survey of the Almagest*. In this paper we shall outline only enough to understand al-Kāshī's calculations. The model for the superior planets differs substantially from that of the inferior planets. Since al-Kāshī's single-argument tables for the superior planets are taken from the *Īlkhānī Zīj* and the double-argument tables were never computed, we shall focus entirely on the inferior planets Venus and Mercury.

Ptolemy's basic model for the motions of the planets (Figure 2) works as follows. The center of the *epicycle* travels along the edge of a larger circle, the *deferent*, while the planet revolves around the epicycle. The epicycle's motion along the deferent takes place so that the *mean centrum* (angle c_m – the angle at the *equant* point, corresponding to the arc from the apogee to the center of the epicycle) increases uniformly with time. The planet moves on the epicycle so that the *mean anomaly* (angle a_m) increases uniformly with time. In order to find the position of a planet as seen from the Earth, first the values of c_m and a_m are found using *mean motion* tables. Next, geometric arguments allow Ptol-

[18] See especially Pedersen, *A Survey*, Chapter 12, and Riddell, 'The Latitudes of Venus'.

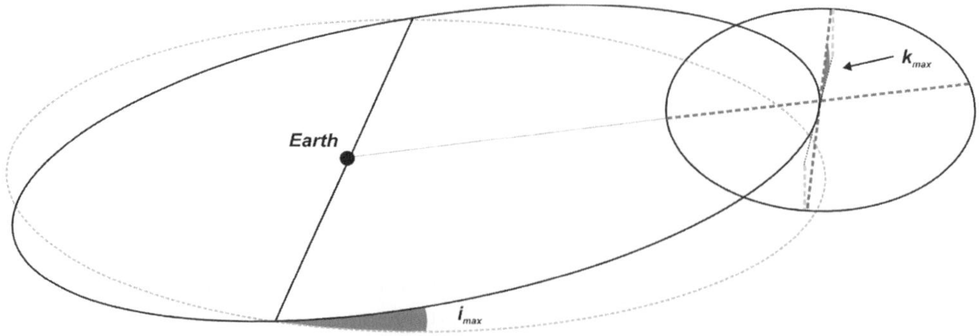

Figure 3(a): Ptolemy's model for the latitudes of the inferior planets, $c = 0°$.

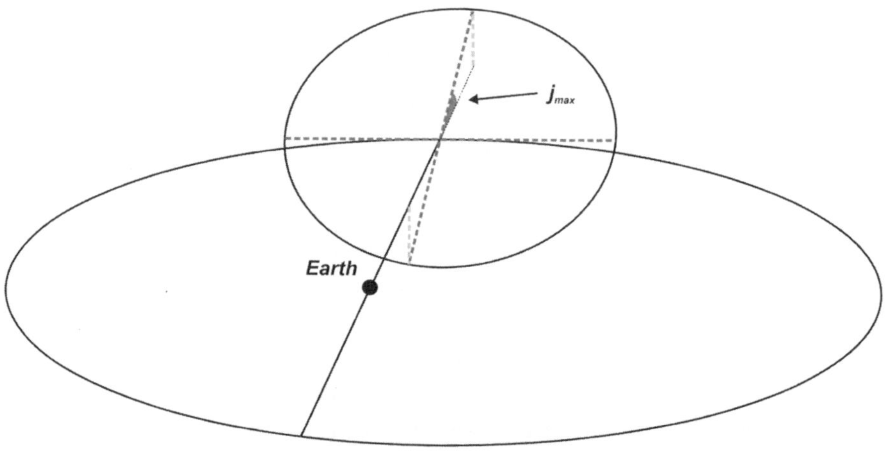

Figure 3(b): idem, $c = 90°$.

emy to find the value of the *true centrum* (angle c) from c_m, and the value of the *true anomaly* (angle a_v) from the values of a_m and c_m. From them Ptolemy can compute the *equation of anomaly p* (a function of both a_v and c), which when added to c gives the planet's position seen from the Earth with respect to the apogee. Finding the planet's *longitude* λ requires adding to this position the arc between the apogee and the zero point of longitude, the vernal equinox ♈.[19]

However, the entire model of Figure 2 takes place within the plane of the ecliptic, while the planets move up to several degrees or even more above and below it. Ptolemy handles these motions in latitude in Book XIII, at the end of the *Almagest*. He does so by causing the deferent and epicycle to move above (northward) and below (southward) the plane of the ecliptic. The model for the superior planets (Mars, Jupiter, Saturn) differs from the more complicated model for the inferior planets (Venus, Mercury). Although Ptolemy later simplified this

[19] Ptolemy's model for Mercury varies from this description. See Neugebauer, *A History*, pp. 158–69, and Pedersen, *A Survey*, pp. 309–28.

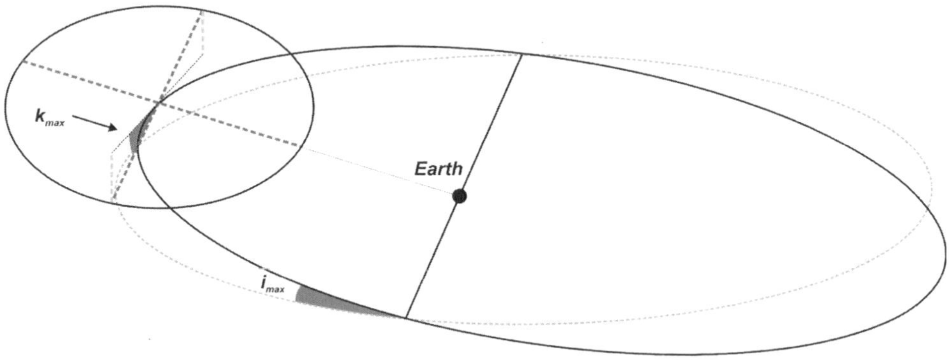

Figure 3(c): Ptolemy's model for the latitudes of the inferior planets, c = 180°.

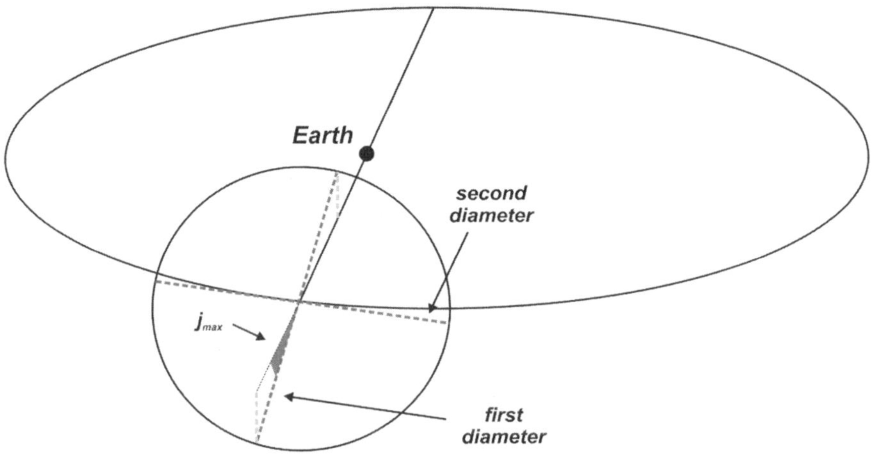

Figure 3(d): idem, c = 270°.

model and its tabulations in his *Handy Tables*,[20] the *zīj* tradition followed the *Almagest*. Since al-Kāshī seems never to have begun his double-argument tables for the superior planets, we describe here only the model for the inferior planets.

The latitude is a combination of three phenomena. The first, called the *inclination*, is a wobble of the deferent circle as the epicycle travels around the deferent. In the four diagrams comprising Figure 3, the rightmost point of the deferent corresponds to c = 0°. In Figure 3(a), c = 0° and the deferent is inclined to the ecliptic by its maximum value $i_{\max}$. For Venus we shall assume that the direction above the ecliptic plane is northward; for Mercury, that direction is southward. For Venus, $i_{\max}$ = 0;10°; for Mercury, $i_{\max}$ = 0;45°. In Figure 3(b), c has increased to 90° and the deferent has returned to the plane of the ecliptic.

[20] For the first of six planned volumes of the *Handy Tables* see Tihon and Mercier, *Πτολεμαίου Πρόχειροι Κανόνες*; see also Stahlman, *The Astronomical Tables*, and Swerdlow, 'Ptolemy's Theories of the Latitude'.

In Figure 3(c), c is now 180° and the deferent is again inclined by an angle of i_{max}, but in the other direction. In Figure 3(d) c has reached 270°; the deferent is again on the plane of the ecliptic. Finally, after c increases by another 90° we return to the situation of Figure 3(a). Thus, in Figures 3(b) and 3(d) the center of the epicycle is in the plane of the ecliptic; while at all other times, the center of the epicycle is above the ecliptic.

The second phenomenon, the *deviation*, is the first of two wobbles of the epicycle with respect to the deferent circle. The *first diameter* of the epicycle extends from the point nearest to the Earth to the point furthest from the Earth. In Figure 3(a) the first diameter is on the plane of the deferent. By Figure 3(b), its deviation from the plane of the deferent has gradually increased and is now at its maximal amount j_{max} (2;30° for Venus and 6;15° for Mercury). By Figure 3(c) the first diameter has returned to the deferent, and by Figure 3(d) it has reached its maximal deviation j_{max} in the opposite direction.

The third phenomenon, the *slant*, is a wobble of the *second diameter* of the epicycle, which is perpendicular to the first diameter. In Figure 3(a) the second diameter's leading edge slants by its maximal amount k_{max} (3;30° for Venus; 7° for Mercury). In Figure 3(b) it has returned to the plane of the deferent; in Figure 3(c) it now slants by k_{max} in the other direction; and in Figure 3(d) it has again returned to the plane of the deferent.

Calculating and tabulating the three latitude effects: The inclination

As we shall see, al-Kāshī follows the overall structure of Ptolemy's computation and tabulation of latitudes. However, in a number of instances he points out some approximation made by Ptolemy, either agreeing or disagreeing with Ptolemy's assumption that the error caused by the approximation is negligible and may be permitted.[21] This provides us an opportunity to gain insight into al-Kāshī's implied standards for the level of precision appropriate to planetary predictions; we shall also be able to judge whether or not al-Kāshī's standards are consistent from one such instance to the next.

The planet's latitude β is a function[22] of the epicycle's position on the deferent c, and of the planet's position on the epicycle a_v. In their tabulations Ptolemy

[21] It is worth noting here the assumption that numerical accuracy is a central criterion to measure the quality of a mathematical astronomical procedure. This is not always the case in al-Kāshī's work; he also concerns himself with the extent to which a theoretical approach conforms to geometric method (see, for instance, Van Brummelen, 'Crossing a Mathematical Rubicon'). However, occasionally he allows himself to violate geometric restrictions when the error is sufficiently small. If he had not, it seems certain he would never have accomplished anything on planetary latitudes.

[22] Of course the term 'function' is anachronistic in the context of ancient and medieval science. Our use of it here and elsewhere implies only the prescription of a procedure that takes a value of one quantity and transforms it into a value of another, dependent quantity.

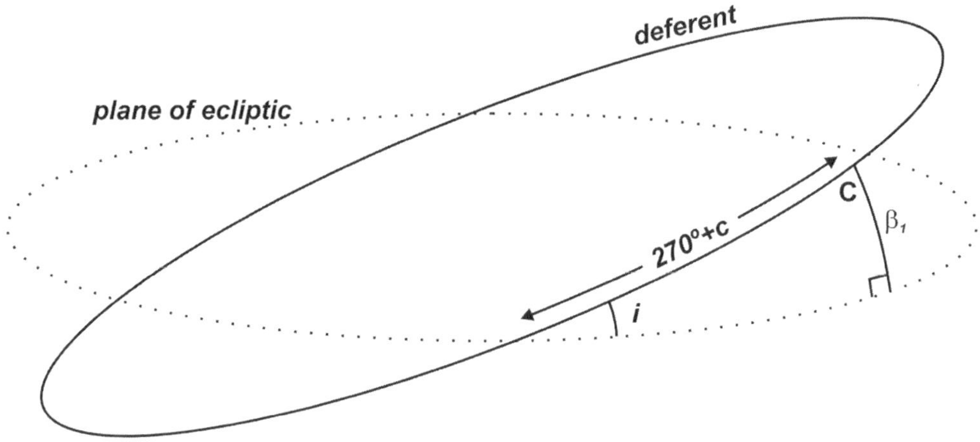

Figure 4: The inclination of an inferior planet.

and al-Kāshī deal with the inclination (β_1), deviation (β_2), and slant (β_3) separately and add these three effects together:

$$\beta(c, a_v) = \beta_1(c) + \beta_2(c, a_v) + \beta_3(c, a_v). \qquad (1)$$

Our first effect is the inclination, which is a function only of *c*. In Figure 4, the center of the epicycle *C* travels around the deferent (the epicycle itself is not drawn), with *c* measured from the rightmost position counterclockwise as seen in the figure. Then, where *i* is the current inclination of the deferent, spherical trigonometry gives β_1 according to[23]

$$\sin \beta_1 = \sin i \cdot \sin(270^\circ + c) = \sin i \cdot \cos c. \qquad (2)$$

However, *i* is a very small angle varying up to 0;10° (Venus) or 0;45° (Mercury). Ptolemy assumes then that $\beta_1 / i = \sin \beta_1 / \sin i$, so that

$$\beta_1 = i \cos c. \qquad (3)$$

Now, the value of *i* varies sinusoidally as the deferent oscillates, so

$$i = i_{\max} \cos c. \qquad (4)$$

[23] We have had to make some concessions to modern terminology to render the historical procedures comprehensible to the modern reader. For example: (a) Ptolemy does not have modern trigonometric functions, relying instead only on the chord; we convert to sines and cosines for ease of comparison with al-Kāshī. (b) Ancient and medieval authors defined their trigonometric functions not with a unit circle, but with a circle of radius 60. We convert to the modern functions here. (c) $\sin(270^\circ + c)$ is equal to $-\cos c$, but negative quantities did not exist at this time. Rather, quantities were added or subtracted and then given as northward/southward, or measured clockwise/counterclockwise from some reference point.

c	Venus	Mercury
0	0;10	0;45
1	0;10	0;45
2	0;10	0;45
3	0;10	0;45
⋮	⋮	⋮
10	0;10	0;44
⋮	⋮	⋮
20	0; 9	0;39
⋮	⋮	⋮
30	0; 8	0;34
⋮	⋮	⋮
40	0; 6	0;27
⋮	⋮	⋮
50	0; 4	0;19
⋮	⋮	⋮
60	0; 2	0;11
⋮	⋮	⋮
70	0; 1	0; 6
⋮	⋮	⋮
80	0; 0	0; 2
⋮	⋮	⋮
90	0; 0	0; 0

Table 1: Excerpts of al-Kāshī's inclination tables for Venus and Mercury.

Therefore, combining (3) and (4),[24]

$$\beta_1 = i_{\max} \cos^2 c. \tag{5}$$

Ptolemy does not tabulate $\beta_1(c)$ directly, but Islamic *zīj* authors did. Excerpts of al-Kāshī's inclination tables for Venus and Mercury are given in Table 1.[25] Here we arrive at the first opportunity to examine al-Kāshī's standards for approximation. In his text on the inclination,[26] he states that equation (3) should be replaced by

$$\sin \beta_1 = \sin i \cos c, \tag{6}$$

but he allows the approximation to stand. Our recomputations confirm that the errors involved here are never larger than 0.00013″ for Venus and 0.012″ for Mercury, which are indeed tiny.

A similar (although not quite identical) situation arises with respect to the Moon's latitude. In Figure 5 the Moon travels on a plane inclined to the ecliptic by 5°, and so

$$\sin \beta = \sin 5^\circ \sin \lambda_d. \tag{7}$$

[24] See Neugebauer, *A History*, p. 224; Toomer, *Ptolemy's Almagest*, pp. 631, 636; and Pedersen, *A Survey*, pp. 376–77.

[25] The arguments of the table for Mercury continue to 180°; since the function is symmetric around 90°, we do not include the second half of the table here.

[26] Treatise 3, Section II.7; fol. 102r in the India Office MS.

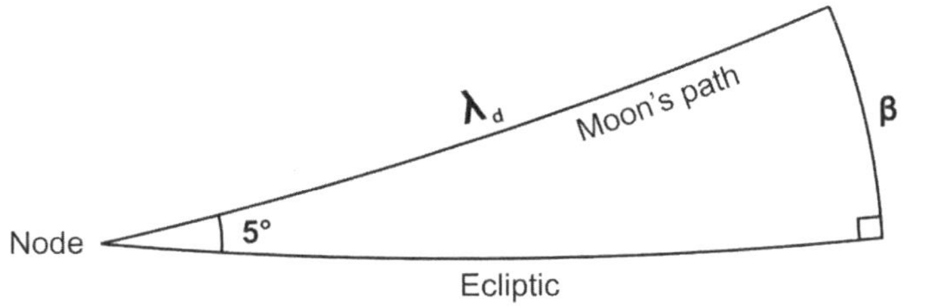

Figure 5: The Moon's latitude.

λ_d	Lunar latitude	Error vs (7)	Error vs (8)
0	0; 0, 0		
1	0; 5,14		
2	0;10,27		[−1]
3	0;15,42	[+1]	
⋮	⋮	⋮	⋮
10	0;52, 2		[−4]
⋮	⋮	⋮	⋮
20	1;42,28	[−1]	[−8]
⋮	⋮	⋮	⋮
30	2;29,51		[−9]
⋮	⋮	⋮	⋮
40	3;12,41	[−1]	[−9]
⋮	⋮	⋮	⋮
50	3;49,41	[−1]	[−8]
⋮	⋮	⋮	⋮
60	4;19,44		[−4]
⋮	⋮	⋮	⋮
70	4;41,52		[−2]
⋮	⋮	⋮	⋮
80	4;55,25	[−1]	[−2]
⋮	⋮	⋮	⋮
90	5; 0, 0		

Table 2: Excerpts from al-Kāshī's lunar latitude table, with errors displayed with respect to the precise formula (7) and the approximate formula (8). The table itself is given for every fifth of a degree. Errors are given in the usual way, as multiples of units of the last tabulated sexagesimal place. Blank entries in the error columns indicate that the historical table's entry is correct to all places.

However, Ptolemy seems to calculate using the approximation

$$\beta = 5^\circ \sin \lambda_d. \tag{8}$$

It is difficult to tell whether Ptolemy actually uses (8), since his lunar latitude table is given only to minutes and only one entry produces different values when computed with (7). Here al-Kāshī objects to the approximation, which he says is also used in other *zījes*.[27] Instead he calculates lunar latitudes to sec-

[27] Treatise 3, Section II.4; fol. 79r in the India Office MS.

onds rather than the usual minutes, and at this level of precision there is a significant difference between the two formulas. He gives his new, more accurate table as the 22nd of his 70 improvements over the *Īlkhānī Zīj*. Table 2 presents excerpts from this improved table, with errors given in comparison with both (7) and (8). Clearly al-Kāshī is correct that (8) does not produce sufficient accuracy on the order of seconds of arc.

As for the origins of the planetary inclination tables themselves, both tables (one each for Venus and Mercury) are almost identical to those in the *Īlkhānī Zīj*.[28] This sort of copying, especially for latitude tables, was common practice.[29]

Calculating and tabulating the three latitude effects: The deviation

The second latitude β_2, the effect of the deviation, may be seen from Figure 3 to be a function of two arguments. Clearly β_2 is affected by the planet's position on the epicycle a_v. But since the first diameter's inclination changes as the epicycle moves around the deferent, β_2 is also a function of c. In principle this would require that $\beta_2(a_v,c)$ be tabulated in a rectangular grid rather than a single column. But from Ptolemy onward, this had been considered to be a daunting computational prospect to be avoided at all costs. In the case of the deviation (and also the slant, as we shall see) the cost of avoiding a double-argument table is an error caused by approximation. Ptolemy and his successors reason as follows: imagine in Figure 3 that the planet is fixed in place on the epicycle, i.e., that a_v is constant. Then as the epicycle moves around the deferent, β_2 varies sinusoidally with extreme values in the positions of Figures 3(a) and 3(c), and with values of zero in the positions of Figures 3(b) and 3(d). In other words

$$\beta_2(c, a_v) \approx \beta_2(270^\circ, a_v) \cdot \sin c. \tag{9}$$

Thus astronomers need only tabulate two columns: one for the *maximal deviation* $d(a_v) = \beta_2(270°,a_v)$, and another for the *deviation interpolation function* $f_d(c) = \sin c$. To arrive at a value for β_2, the user simply looks up the values for $d(a_v)$ and $f_d(c)$ and multiplies them together.

The geometry behind $d(a_v)$ leads Ptolemy and al-Kāshī to a procedure equivalent to the following:

$$\sin d(a_v) = \frac{r \sin j_{\max} \cos a_v}{\sqrt{\rho^2 + 2\rho r \cos j_{\max} \cos a_v + r^2}}, \tag{10}$$

[28] Most of the 14 differences in the two Venus inclination tables are caused by a single error, a vertical shift in a string of entries caused by some scribe omitting an entry in the copying process. Three of the four differences in the Mercury table may have been caused by a similar copying error.

[29] See the discussion of possible reasons for this in van Dalen, 'Tables of Planetary Latitude', pp. 323–25.

where r is the radius of the epicycle (43;10 for Venus and 22;30 for Mercury), j_{max} is the maximum inclination of the first diameter, and ρ is the distance from the Earth to the center of the epicycle when $c = 270°$ (60 for Venus, 56;40 for Mercury). Ptolemy's values for j_{max} are 2;30 (Venus) and 6;15 (Mercury); al-Kāshī refers also to alternate 'modern' parameters 3;30 (Venus) and 7;0 (Mercury).[30]

Excerpts of al-Kāshī's tables for $d(a_v)$ and $f_d(c)$ for both Venus and Mercury are given in Table 3, with graphs of $d(a_v)$ in Figure 6. Repeating our finding for the inclination tables, the values are taken from the *Īlkhānī Zīj*.[31] In all but one case the parameter j_{max} implied by the entries in the table is the Ptolemaic one. A curiosity arises with respect to $d(a_v)$ for Venus: both al-Ṭūsī's and al-Kāshī's table use the Ptolemaic value for j_{max} (2;30) for the first 90 entries, and seem to use a value quite close to the modern one (3;30) for the last 90. This is the reason for the bend in the graph at the argument of 90°. Since the problem goes back at least as far as al-Ṭūsī, al-Kāshī's only involvement in the issue was to copy the table, and we leave the matter for future research.[32]

Calculating and tabulating the three latitude effects: The slant

The Ptolemaic approach to the third latitude β_3 parallels that of the second latitude β_2. As before, the slant is a function of both a_v and c and would therefore require the production of a double-argument table. Instead, as before imagining the planet to be fixed in place on the epicycle while the latter revolves around the deferent, β_3 reaches extreme values in Figures 3(b) and 3(d), and is zero in Figures 3(a) and 3(c). Thus once more we may split the calculation of the deviation into two parts:

$$\beta_3(c, a_v) \approx \beta_3(0, a_v) \cdot \cos c. \tag{11}$$

This again permits the tabulation of only two columns, one for the *maximal slant* $s(a_v) = \beta_3(0°, a_v)$, and the other the *slant interpolation function* $f_s(c) = \cos c$.

The computation of $s(a_v)$ is vexed already in the *Almagest*.[33] Although there is a geometric path to s similar to what we saw for the maximal deviation, in this case Ptolemy and successors reasoned differently. In Figure 7, looking more closely at the epicycle in Figure 3(a), E is the Earth, the dashed circle with center G is the epicycle before the rotation caused by the slant, FP is the epicycle after the slant has been applied, and P is the planet. In this configuration $c = 0$

[30] On the origins of these new parameters in the Maragha tradition see Mozaffari, 'Planetary Latitudes', pp. 520–22.

[31] All four pairs of tables match all 90 or 180 entries except for up to seven entries; where they do differ, it is always by one unit in the last place.

[32] On this anomalous table see Mozaffari, 'Planetary Latitudes', pp. 533–35.

[33] See Neugebauer, *A History*, vol. I, pp. 221–26; Pedersen, *A Survey*, pp. 379–85, and Van Brummelen, *Mathematical Tables*, pp. 367–72.

Argument	Venus deviation	Venus interpolation	Mercury deviation	Mercury interpolation
0	1; 2	0	1;45	0
1	1; 2	1	1;45	1
2	1; 2	2	1;45	2
3	1; 2	3	1;45	3
⋮	⋮	⋮	⋮	⋮
15	1; 2	15	1;43	15
⋮	⋮	⋮	⋮	⋮
30	0;57	30	1;35	30
⋮	⋮	⋮	⋮	⋮
45	0;50	42	1;20	42
⋮	⋮	⋮	⋮	⋮
60	0;35	52	1; 1	52
⋮	⋮	⋮	⋮	⋮
75	0;20	58	0;32	58
⋮	⋮	⋮	⋮	⋮
90	0; 0	60	0; 0	60
⋮	⋮		⋮	
105	−0;37		−0;40	
⋮	⋮		⋮	
120	−1;20		−1;25	
⋮	⋮		⋮	
135	−2;28		−2;16	
⋮	⋮		⋮	
150	−4;26		−3; 7	
⋮	⋮		⋮	
165	−6;53		−3;48	
⋮	⋮		⋮	
179	−8;40		−4; 4	

Table 3: The deviation tables; note that the two interpolation tables are identical. As elsewhere, we use the convention that positive values imply northward for Venus and southward for Mercury. The interpolation coefficients are expressed in minutes.

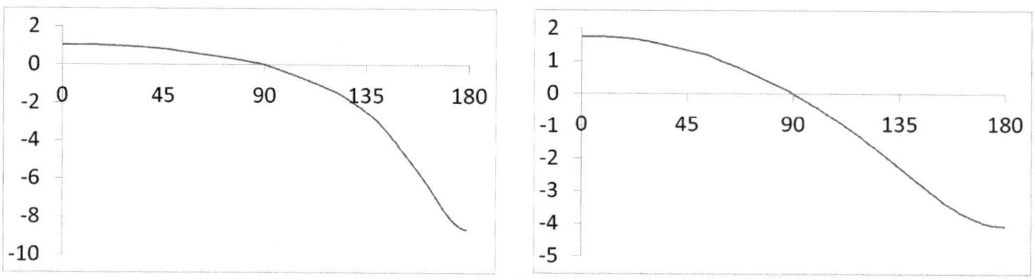

Figure 6: Graphs of al-Kāshī's tables of maximum deviation: Venus (left) and Mercury (right).

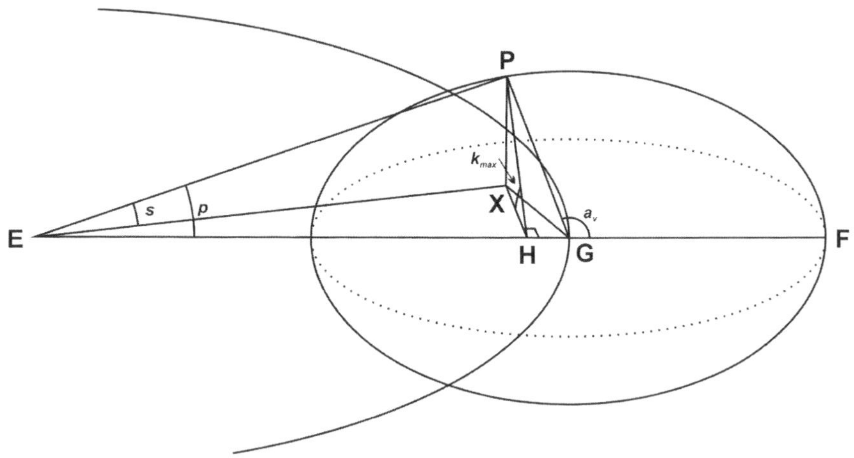

Figure. 7: Calculating the slant.

and so the epicycle's slant is at its maximum value $k_{\max} = \angle PHX$. In addition $a_v = \angle FGP$, $s(a_v) = \angle PEX$, and the equation of anomaly $p(a_v,0°) = \angle HEP$.[34] We begin the derivation of s by noting that

$$\sin s(a_v) = \sin k_{\max} \sin p(a_v, 0).^{35} \tag{12}$$

If we imagine as before the epicycle remaining fixed and the planet moving around the epicycle, we see that as a_v varies the angles s and p vary apparently in synchrony, seeming to reach their maxima, minima, and zero values for the same values of a_v. Assuming then that they reach their maxima $s_{\max}$ and $p_{\max}$ for the same value of a_v (a little more than 90°), we have

$$\frac{s(a_v)}{p(a_v,0)} \approx \frac{s_{\max}}{p_{\max}}. \tag{13}$$

If this approximation is to be trusted, we have now a method to tabulate $s(a_v)$ simply as a constant multiple of the table for $p(a_v,0)$. This is what Ptolemy did, with one minor caveat: he replaced $p(a_v,0)$ with $p(a_v,c_m{}^0)$, where $c_m{}^0$ is a value a bit larger than 90° chosen so that $\rho = 60$.[36] Thus for Ptolemy

$$\frac{s(a_v)}{p(a_v,c_m{}^0)} \approx \frac{s_{\max}}{p_{\max}}. \tag{14}$$

But al-Kāshī tabulated p in a different way. The only table that appears explicitly in his *zīj* is $p(a_v,0)$, so he prefers (13).

[34] Recall that p is a function of both a_v and c_m; in this configuration, $c_m = c = 0$. There is a question here concerning whether p is to be measured on the tilted epicycle or on the original epicycle, but this does not affect our study of the tables.

[35] Because $\sin s = PX/PE$, $\sin k_{\max} = PX/PG$, and $\sin p = PG/PE$.

[36] This is true for Venus. We shall deal with the variations caused by Mercury's different planetary model momentarily.

However, al-Kāshī has a further objection. Applying (12) at the situation where s and p are maximized, we have $\sin s_{\max} = \sin k_{\max} \sin p_{\max}$; and combining this relation with (12), instead of (14) we arrive at

$$\frac{\sin s(a_v)}{\sin p(a_v,0)} = \frac{\sin s_{\max}}{\sin p_{\max}}. \quad (15)$$

Al-Kāshī claims that the difference between (12) and (15) is significant enough to be taken into account,[37] which provides another opportunity to judge his standards for precision. For Mercury, the difference in the calculation of latitude is on average about 23″, with a maximum difference a bit more than 1′. For Venus we have differences on average around 2′. Again al-Kāshī's computational standards push the computations to the level of seconds rather than his predecessors' minutes.[38]

Al-Kāshī is now in a position to tabulate $s(a_v)$ easily as a multiple of the table for $p(a_v,0)$. For his Venus table the multiple is 2;30/45; for Mercury it is 2;30/19;1.[39] Excerpts from the Venus tables for $s(a_v)$ and the interpolation function $f_s(c)$ are given in Table 4. As with the other tables we have seen so far, the table for $s(a_v)$ is nearly identical to the corresponding *Īlkhānī Zīj* table, differing in only one entry. The *Īlkhānī Zīj* does not have a table for $f_s(c)$.

Turning to Mercury, recall that the Ptolemaic longitude model differs from that of the other planets. This difference has not affected the tables for inclination and deviation, but it does make a difference for the slant. The reason is that $\rho(c)$, the distance from the Earth to the center of the epicycle, varies by a large enough amount to affect the latitude calculation substantially (between 57 and 69 units, as opposed to 58;45 and 61;15 units for Venus). Ptolemy deals with this variation in ρ rather crudely. When $90° < c < 270°$ – that is, when the epicycle's center is on the left side of the deferent in Figure 3 – ρ is smaller than average, so the epicycle is closer to the Earth and the slant component of the latitude must be greater. Ptolemy increases it by a factor of one tenth:

$$\beta_3(c, a_v) = \tfrac{11}{10} s(a_v) \cdot \cos c \quad (16)$$

(compare with (11)). For the other values of c the slant component of the latitude is decreased by one tenth:

$$\beta_3(c, a_v) = \tfrac{9}{10} s(a_v) \cdot \cos c. \quad (17)$$

Now, *zīj*es sometimes made their users' work easier by incorporating these multiplicative factors into the tables.[40] So, rather than providing a table of $s(a_v)$,

[37] Treatise 3, Section II.7; fol. 102r in the India Office MS. This same approximation is questioned by a modern commentator on the *Almagest* in Pedersen, *A Survey*, p. 381.

[38] Toomer, 'Review of Olaf Pedersen', p. 145, notes that the errors in Ptolemy's approximations for the slant are small, i.e., for Venus less than 7 minutes. However, this level of error is clearly insufficient by al-Kāshī's standards.

[39] These values correspond to the maximum values in the tables of $s(a_v)$ and $p(a_v,0)$.

[40] See van Dalen, 'Tables of Planetary Latitude', pp. 323–25.

Argument	Slant	Interpolation
0	0; 0	60
1	0; 2	60
2	0; 3	60
3	0; 5	60
⋮	⋮	⋮
15	0;22	58
⋮	⋮	⋮
30	0;41	52
⋮	⋮	⋮
45	1; 2	42
⋮	⋮	⋮
60	1;20	30
⋮	⋮	⋮
75	1;39	15
⋮	⋮	⋮
90	1;57	0
⋮	⋮	
105	2;12	
⋮	⋮	
120	2;25	
⋮	⋮	
135	2;30	
⋮	⋮	
150	2;22	
⋮	⋮	
165	1;41	
⋮	⋮	
179	0;8	

Table 4: Slant tables for Venus. The interpolation coefficients are expressed in minutes.

they gave tables of both $\frac{11}{10}s(a_v)$ and $\frac{9}{10}s(a_v)$, one table or the other to be used depending on the value of c.

Al-Kāshī, however, objects to the use of one multiplicative constant for some of the values of c and another multiplicative constant for the others.[41] Instead, he argues, we must allow the factor multiplied by $s(a_v)$ to vary continuously, as follows. Let $s_0(a_v) = \frac{11}{10}s(a_v)$ be the *maximal slant at perigee*; it is the correct maximal slant when $c = 180°$. At apogee ($c = 0$), the correct maximal slant will be smaller than $s_0(a_v)$ by a factor of $\frac{2}{11}$.[42] For other values of c, this factor varies

[41] Treatise 3, Section II.7; fols 103v–104r in the India Office MS.

[42] Al-Kāshī calculates this factor as the ratio between (a) the difference between the greatest value of the maximal slant at perigee and the greatest value of the maximal slant at apogee (0;30), and (b) the greatest value of the maximal slant at perigee (2;45).

according to the value of $\rho(c)$. Since at Mercury's apogee ρ is equal to 69 while at perigee ρ is equal to 57,[43] al-Kāshī asserts that the factor should be

$$\frac{2}{11} \cdot \frac{\rho(c) - 57}{69 - 57} = \frac{2}{11} \cdot \frac{\rho(c) - 57}{12}. \tag{18}$$

This leads us to the formula

$$\beta_3(c, a_v) = s_0(a_v) \cdot \left[1 - \frac{2}{11} \cdot \frac{\rho(c) - 57}{12}\right] \cdot \cos c \tag{19}$$

(compare with (11)). Thus the new interpolation function for Mercury is

$$f_s(c) = \left[1 - \frac{2}{11} \cdot \frac{\rho(c) - 57}{12}\right] \cdot \cos c.^{44} \tag{20}$$

The tables for $s_0(a_v)$ and $f_s(c)$ are excerpted in Table 5. The table for $s_0(a_v)$ is identical in all but one entry with its equivalent in the *Īlkhānī Zīj*; of course, the latter *zīj* does not have a table for $f_s(c)$.

The revised process for computing Mercury's latitudes is the 28th of al-Kāshī's advertised improvements over the *Īlkhānī Zīj*. Our recomputations verify that al-Kāshī's method alters the latitude values by an average of about 2 minutes, and reaches a maximum difference of around 10 minutes. This is again in accord with our previous observations that al-Kāshī sought an improvement in the precision of calculations of planetary positions down from the level of minutes of arc to that of seconds.

Using the five tables described above, given values of c and a_v the user of al-Kāshī's single-argument tables may now determine a value for the planet's latitude as follows:

$$\beta(c, a_v) = \beta_1(c) + d(a_v) \cdot f_d(c) + s_0(a_v) \cdot f_s(c). \tag{21}$$

Avoiding approximations with a spherical epicycle

Throughout the process of computing the latitudes of the inferior planets, al-Kāshī has been following the essence of Ptolemy's procedures while altering them in several places to remove some of Ptolemy's more significant approximations. In the next section of his *zīj* he proposes a new approach to computing the positions of the inferior planets that takes a more dramatic stance, eliminating most of the approximations altogether and returning to something

[43] Strictly speaking 57 is the value of ρ at the point opposite the apogee; due to the peculiarities of Mercury's model the perigee is not directly opposite the apogee. See Pedersen, *A Survey*, p. 383, and Toomer, *Ptolemy's Almagest*, p. 630.

[44] Al-Kāshī illustrates the process for the example $c = 30°$. He states that $\rho(30°) = 66;20$ but does not offer a supporting calculation.

Argument	Slant	Interpolation
0	0; 0	49
1	0; 2	49
2	0; 4	49
3	0; 6	49
⋮	⋮	⋮
15	0;30	47
⋮	⋮	⋮
30	1; 0	44
⋮	⋮	⋮
45	1;30	38
⋮	⋮	⋮
60	1;55	26
⋮	⋮	⋮
75	2;16	13
⋮	⋮	⋮
90	2;34	0
⋮	⋮	⋮
105	2;44	−15
⋮	⋮	⋮
120	2;44	−31
⋮	⋮	⋮
135	2;30	−44
⋮	⋮	⋮
150	1;57	−53
⋮	⋮	⋮
165	1;5	−58
⋮	⋮	⋮
179	0;5	−60

Table 5: Slant tables for Mercury. The interpolation coefficients are expressed in minutes.

close to geometric argument.[45] He accomplishes this by replacing the epicyclic circle with an epicyclic sphere. In Figure 8, the original epicycle is enclosed as a great circle within a sphere. The deviation is modeled using a great circle on this sphere (the *circle of deviation*) that wobbles sinusoidally with respect to the epicycle. The slant is modeled via a great circle (the *circle of slant*) wobbling sinusoidally with respect to the circle of deviation. Finally, the planet travels on the circle of slant. The embedding of the deviation and slant on the surface of a sphere gives al-Kāshī the ability to apply spherical trigonometry to the problem. This allows him to avoid having to assume that the planetary longitude, the deviation, and the slant must be treated independently. In fact they do affect each other to some degree, and this new approach makes it possible

[45] Treatise 3, Section II.8; fol. 104v–108v in the India Office MS. We performed a detailed study of this section in Van Brummelen, 'Taking Latitude with Ptolemy'.

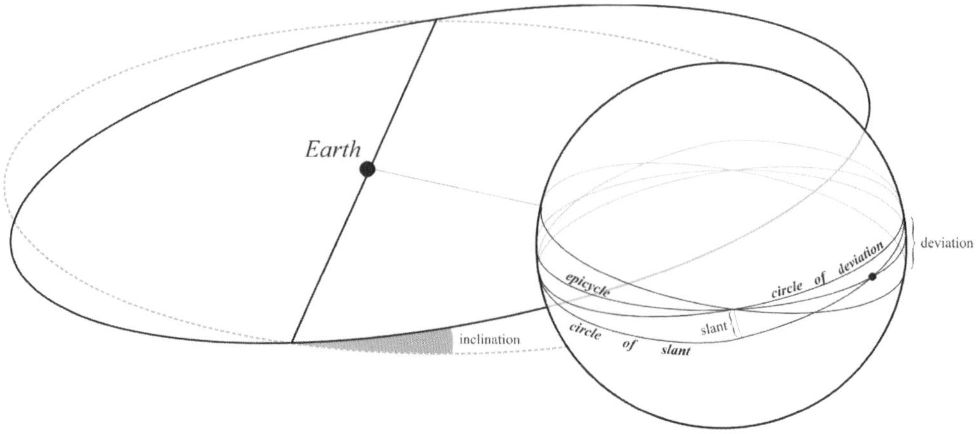

Figure 8: Al-Kashi's spherical epicycle for inferior planets.

for these interactions to be accounted for. He refers to this new method as the 29th on his list of 70 improvements over the *Īlkhānī Zīj*.

Al-Kāshī illustrates his approach with an example computation of the position of Venus, but he never claims that he used his new method in his tabulations. In fact, we shall see that he did not.

The double-argument latitude tables

The user's goal is to find β in terms of a_v and c, so the best possible solution is to give a double-argument table, rather than requiring the user to piece together the three effects from single-argument tables. However, Ptolemy and almost all of his successors avoided the mammoth task of tabulating this double-argument function. There were at least two exceptions, the first by thirteenth-century Moroccan astronomer Ibn Isḥāq in the *Tunisian Zīj*, the second by al-Kāshī.[46] There is no indication or reason to believe that al-Kāshī was in any way aware of Ibn Isḥāq or the *Tunisian Zīj*.[47] As we have seen, although al-Kāshī set out to provide double-argument latitude tables for all the planets, he completed only the Mercury table and part of the table for Venus (see Figure 9). The gridlines for the tables of the other planets are present in all manuscripts, but are blank. Perhaps assuming that he would complete the job some

[46] One finds another double-argument table of planetary latitudes in the *Huihui lifa*, a thirteenth-century Chinese table inspired by the Islamic tradition. See van Dalen, 'Tables of Planetary Latitude'.

[47] For discussions of double-argument planetary latitude tables in the European and Chinese traditions, see also Goldstein and Chabás, 'Ptolemy, Bianchini, and Copernicus', pp. 456–57; Chabás and Goldstein, 'Early Alfonsine Astronomy'; Husson, 'Remarks on Two Dimensional Array Tables', and Li Liang, 'Tables with "European" Layout'.

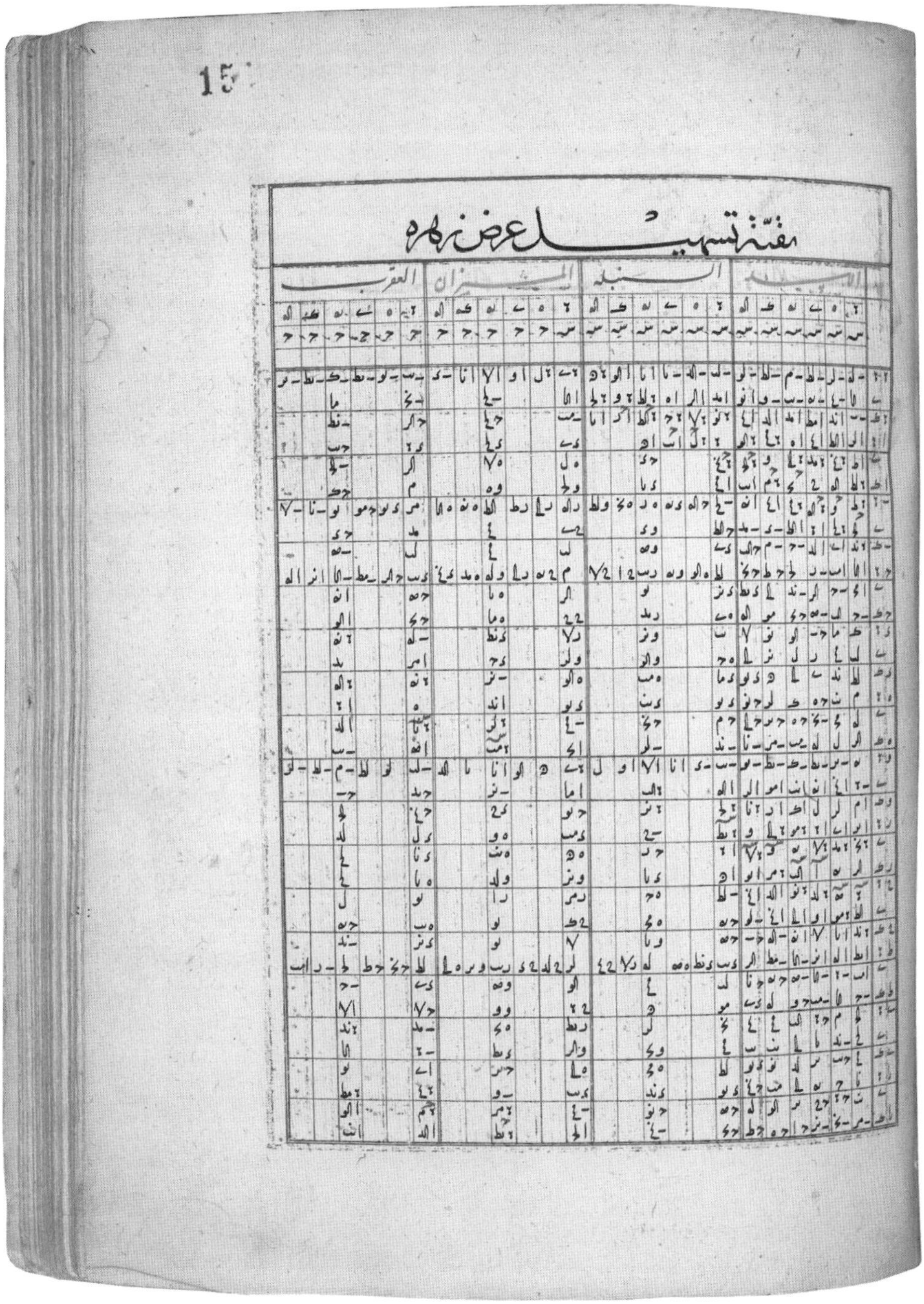

Figure 9: The second of the three pages of the Venus double-argument latitude table. For the first page, see Plate 11; for the third page, see p. 36. © The British Library Board, MS India Office 430, fol. 154r.

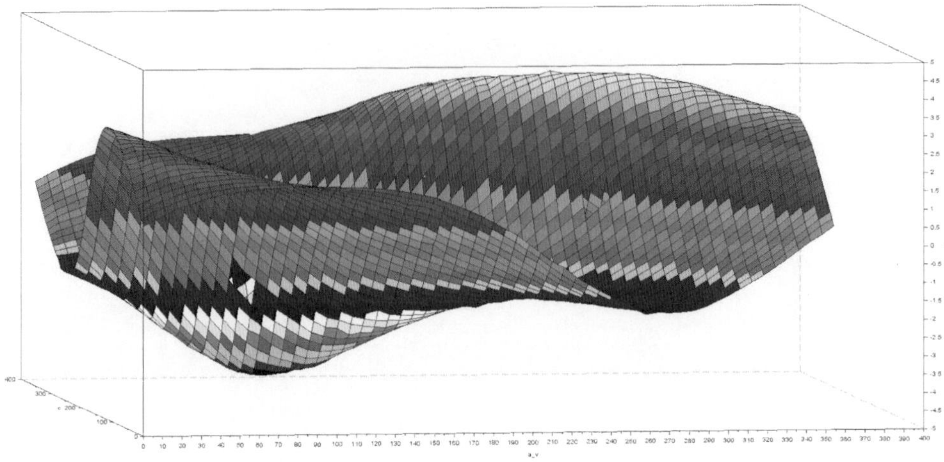

Figure 10: A graph of the double-argument table for Mercury.

day, he included his tables (which he calls 'simplified', thinking of the user) as the 33rd on his list of 70 improvements over the *Īlkhānī Zīj*.

The tables are edited in the appendix to this article. For Mercury the entries are tabulated with increments of 6° for both arguments, for a total of 3600 entries. For Venus the increment for c is 10°, and for a_v 5°; this would produce 2592 entries, but only about half of the table is completed. The tabulated function $\beta(a_v,c)$ is graphed for Mercury in Figure 10; the graph for Venus looks similar. In the case of Venus a positive value for β in our graphs and analyses indicates north of the ecliptic and a negative value indicates south; for Mercury a positive value indicates south and a negative value indicates north. In the tables themselves al-Kāshī signifies that a column of values implies the northerly direction by writing the letter ش (for *shimāl*, or 'northern') above it, while south is indicated by the letter ج (for *janūb*, or 'southern').

One confusion regarding al-Kāshī's description of his double-argument latitude tables is that he instructs the user to enter the table using the *mean* anomaly and centrum as arguments rather than the *true* anomaly and centrum, but in marginal notes on both tables in the India Office manuscript the user is instructed to use the *true* anomaly and centrum.[48] It is not difficult to see that the true, not the mean, arguments are intended.[49] The difference between entries in a table with mean versus true arguments is greatest for centrum values around 90° and 270°. If the true arguments are intended, the values in

[48] Treatise 3, Section I.6; fol. 80v in the India Office MS. The marginal notes are on fol. 153v and fol. 155r.

[49] Presumably a sophisticated historical user of the tables would have been able to make a similar deduction. It remains unclear (for this and other tables) how many users there were, what they used the tables for, and how familiar they were with the underlying theory.

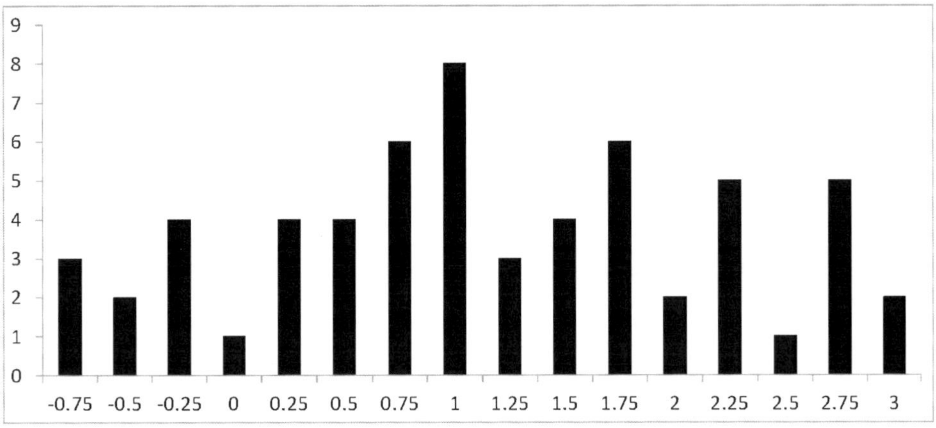

Figure 11: In the column for centrum 90° in the Mercury double-argument table, the difference between al-Kāshī's values and entries computed with mean arguments, minus the difference between al-Kāshī's values and entries computed with true arguments. The strong skew toward positive values in this histogram indicates that true arguments were used.

these two columns reduce to just $\pm d(a_v)$ (since when $c = 0$, $\beta_1 = \beta_2 = \beta_3 = 0$). This is not the case if mean arguments are intended.[50] For Mercury, al-Kāshī's entries in these columns fit the values with true centrum as argument much more closely than they fit the values with mean centrum as argument, and they reveal the symmetries that apply only when the true arguments are used (see Figure 11). Similar results apply for the other columns in the table. We shall see later that the Venus table even more clearly uses true arguments rather than mean.

Another curiosity of the two tables is that the two arguments are reversed. For Venus the anomaly is given along the columns while the centrum is given along the rows; for Mercury it is the other way around. Al-Kāshī gives no indication of this, either in the tables themselves or in the instructions for their use.[51] One wonders, then, how a user could possibly obtain correct values. This effect can be seen most easily by observing the table for the same arguments we considered in the previous paragraph: when $c = 90°$ or $270°$, the latitude function reduces to $\pm d(a_v)$. In the Venus table the two *rows* corresponding to arguments 90° and 270° are close matches to the table for $d(a_v)$ in Table 3; for Mercury the two *columns* for 90° and 270° are close matches to the table for $d(a_v)$.

[50] For instance, if the argument is the true anomaly, then $d(90°) = d(270°) = 0$; and indeed these entries are zero. The recomputation of the column described in this paragraph might be performed in several different ways, but they all produce similar results.

[51] The first page of the Venus table in the Cairo MS has the word 'center' written at the top of the column of arguments, but the arguments are not specified anywhere else in any of the manuscripts.

The entries in the double-argument tables provide some clues concerning how al-Kāshī calculated them. In the case of Venus, one fact is clear immediately. Recall that the table for $d(a_v)$, copied from al-Ṭūsī, contains a large anomaly: the entries for $a_v > 90°$ are much larger than they should be (up to more than 2°), as if they are calculated using a much larger parameter value, leading to a graph with a corner at $a_v = 90°$ (Figure 6, left). The double-argument table exhibits the same anomaly. This narrows the possibilities to two: either al-Kāshī used the table for $d(a_v)$ and therefore also the other single-argument tables according to (21), or his unknown computation process includes the same anomaly that led to al-Ṭūsī's table. The former would seem far more likely. However, one fact gives us pause: the entries in the rows for $a_v = 90°$ and 270°, although within a couple of minutes of the entries in the table for $d(a_v)$, are not identical.

In the case of Mercury, the possibilities fall into three categories:

1. The entries of $\beta(a_v,c)$ were computed, using a_v and c as arguments, from the geometric definitions of the three latitude effects in some way.
2. The entries were computed using al-Kāshī's spherical epicycle method described above.
3. The entries were computed using the single-argument tables, simply following the method that a user would employ to combine the three latitude effects.

We may eliminate option 2 immediately. For a given pair of arguments the spherical epicycle method produces the longitude and latitude simultaneously, using a method that al-Kāshī considered to be superior to any other. However, only the Venus and Jupiter tables of longitudes – not the Mercury table – have been completed. Also, al-Kāshī does not present anywhere a geometric method for the spherical epicycle that handles the peculiarities of the Mercury model.

Option 1 contains within it a number of possible choices that al-Kāshī might have made. These include:

- For the inclination β_1, he could choose either the correct formula (6) or its approximation (3).
- For the deviation β_2, he could use either the Ptolemaic value of the parameter $j_{max} = 6;15°$ or the 'modern' value $j_{max} = 7°$.
- For the slant β_3, he could use the table of $p(a_v,0)$ given earlier in the *zīj*, or he could compute the values of p from scratch.
- Also for β_3, he could choose either the more precise (15) or the less precise (13).
- Finally for β_3, he could use Ptolemy's (16) and (17), or his own improvement (19).

Other possible decisions may be considered, but these five are the most significant.

With five choices to make, we generated $2^5 = 32$ different tables, which produce several clear results. Firstly, it is evident that al-Kāshī used $j_{max} = 6;15°$ rather than $j_{max} = 7°$: the average difference between the entries in the sixteen recomputed tables that use $j_{max} = 6;15°$ and the entries in al-Kāshī's table is 2.91′, while the difference for the entries that use $j_{max} = 7°$ is 9.08′, more than three times larger.[52] So we may discard the sixteen tables that use $j_{max} = 7°$. Secondly, we find strong confirmation that al-Kāshī's improvement (19) and the associated interpolation function were used: the entries in the tables using the improvement differ from al-Kāshī's entries on average by 2.22′, as opposed to an average difference of 3.60′ with the entries in the tables that do not use it. After this, we find very little to choose between the various remaining options; these recomputations give results that are much closer to each other than they are to the tables' values themselves. Therefore, whatever the remaining discrepancies between table and recomputation are, they are not caused by the effects of the various options outlined here.

These conclusions are compatible with the hypothesis that al-Kāshī generated the double-argument table directly from his single-argument tables. Indeed, the average difference between the entries in the double-argument table and those that we compute directly from the single-argument tables is 2.40′, about the same as the best of the recomputations above. So in summary, we may assert that however he worked, al-Kāshī used $j_{max} = 6;15°$ and his new interpolation method for the slant. Beyond this he might have used the single-argument tables, but these data are not conclusive.

Indeed, there may be reasons to believe that he did not, at least, not entirely. Certain columns and rows of the double-argument table take on simplified values when the arguments a_v and c are multiples of 90°. For instance, as we saw with the Venus table, when $c = 90°$ or 270° the latitude function simplifies to $\pm d(a_v)$. Although all the entries in these two columns again come within a couple of minutes of the entries in al-Kāshī's table of $d(a_v)$, they do not match perfectly – a difficult phenomenon to explain if the single-argument tables had been used. Similar results are obtained for the other columns corresponding to arguments that are multiples of 90°.

Finally, a glance at the pattern of filled-in entries in the Venus table (Figure 9) suggests the possibility that interpolation was being used to complete the tables after certain rows and columns had been calculated. However, our analyses on both the Mercury and Venus tables reveals no evidence in favour of this hypothesis. It should be noted that if al-Kāshī used his single-argument tables, direct calculation would have been at least as easy as interpolation.

52 A statistical test determining parameters embedded within historical astronomical tables exists; see van Dalen, 'A Statistical Method'. However, the results here are so strong and the number of tabular entries so large that a statistical procedure is unnecessary.

The relation with the latitude tables in Ulugh Beg's *Sulṭānī Zīj*

Al-Kāshī likely was still in Kāshān when he completed the *Khāqānī Zīj*, but he spent the last part of his career in Samarqand at Ulugh Beg's observatory. The latter's *Sulṭānī Zīj* was completed after al-Kāshī's death, but traces of al-Kāshī's influence on the later *zīj* have already been found.[53] Thus comparing the respective latitude tables will be useful. Now, the *Sulṭānī Zīj* does not include double-argument tables, and its single-argument tables for the superior planets are based on newly determined parameter values, so they are not related to al-Kāshī's equivalents.[54]

The single-argument tables for the inferior planets, however, may be compared. Indeed, all ten tables (five for Venus, five for Mercury) are identical to al-Kāshī's, other than the odd scribal error in a couple of entries here and there. Recall that most of these tables in turn had been taken from al-Ṭūsī. However, we do find intact in Ulugh Beg's *zīj* an entirely new set of tables for computing the slant of Mercury, including the new interpolation function.

Conclusions

Our study of al-Kāshī's latitude tables for the inferior planets has yielded the following results:

- His single-argument tables (other than the tables for the slant of Mercury) were taken from al-Ṭūsī's *Īlkhānī Zīj*, and later found their way into Ulugh Beg's *Sulṭānī Zīj*.
- He innovated with a new scheme to calculate Mercury's slant. His improvement on Ptolemy was to allow a parameter relating to the distance between the Earth and the center of the epicycle, which for Ptolemy took on the two values 0.9 and 1.1, to vary continuously. The result was an improvement on the determination of Mercury's latitude by an average of 2′, and occasionally up to 10′. Al-Kāshī's table for this effect was adopted wholesale by Ulugh Beg in the *Sulṭānī Zīj*.
- The double-argument tables for Venus and Mercury reverse the roles of the arguments with respect to rows and columns, with no indication of this curious fact in the text or in the tables. This would have given an unsophisticated user only a 50% chance to extract a correct value.
- The Mercury double-argument table uses the Ptolemaic value of the parameter $j_{\max}$ rather than the modern value mentioned in the *zīj*; it

[53] For instance, Kim Plofker, Clemency Montelle, and I will soon publish a study of an early eighteenth-century Sanskrit text on various methods of calculating sin 1° inspired by al-Kāshī; in this work, Ulugh Beg's contributions are featured prominently.

[54] See Mozaffari, 'Planetary Latitudes', pp. 522, 525, 535–38.

also employs al-Kāshī's innovative scheme for computing the slant. This makes the double-argument table a close, but not perfect fit with computation from the single-argument tables.

- The Venus double-argument table exhibits the same anomalous pattern that has been noticed previously in the single-argument deviation table, a sudden change in the slope of the function when $d(a_v)$ changes sign.
- We have identified several instances where al-Kāshī points out approximations made in the *Almagest* and either accepts or rejects them for something better. In each case al-Kāshī makes a consistent and competent judgment: when he accepts an approximation, we find that the resulting error is smaller than the level of seconds of arc; and when he rejects an approximation, his replacement improves the fit between the geometric model and the computation on the level of seconds or minutes.

It will be an enormous task to identify and bring forward all the other novel aspects of the *Khāqānī Zīj*, which may be countless. Once this is done, I suspect it will be recognized as perhaps the most mathematically creative *zīj* ever written, and a significant source of inspiration for the astronomers of the Samarqand observatory and their successors. This paper is a small step in that direction.

Acknowledgments

My research on this set of tables began over fifteen years ago, at the former Institut für Geschichte der Naturwissenschaften in Frankfurt. So, my first thanks go to David King for his considerate hospitality as host, and Benno van Dalen, who provided assistance and friendship at every stage. Both of them, as well as Fuat Sezgin, graciously provided access to manuscripts. Edward Kennedy's early rough translation of sections of the *Khāqānī Zīj*, provided by him for the purpose of this project, proved to be immensely valuable. My work was supported by a grant from Vermont EPSCoR, which did not seem to mind when I became distracted with what became my 2006 paper in the bibliography of this paper. I have the TAMAS project, especially its principal investigator Matthieu Husson, to thank for the revival of my interest in tables in general and this project in particular. Every member of the TAMAS team has been collegial and supportive; it has been a great pleasure to work with each one of them. Finally, my thanks to the referees for their comments. As always, remaining inadequacies are my responsibility.

Bibliography

Aaboe, Asger, 'Al-Kāshī's Iteration Method for the Determination of Sin 1°', *Scripta Mathematica* 20 (1954), pp. 24–29.

Azarian, Mohammad, 'Al-Kāshī's Fundamental Theorem', *International Journal of Pure and Applied Mathematics* 14 (2004), pp. 493–503.

Bagheri, Mohammad, 'A Newly Found Letter of al-Kāshī on Scientific Life in Samarqand', *Historia Mathematica* 24 (1997), pp. 241–56.

Chabás, José, and Bernard R. Goldstein, 'Early Alfonsine Astronomy in Paris: The Tables of John Vimond (1320)', *Suhayl* 4 (2004), pp. 207–94.

van Dalen, Benno, 'A Statistical Method for Recovering Unknown Parameters from Medieval Astronomical Tables', *Centaurus* 32 (1989), 85–145.

van Dalen, Benno, 'Tables of Planetary Latitude in the *Huihui li* (II)', in Yung-Sik Kim and Francesca Bray (eds), *Current Perspectives in the History of Science in East Asia*, Seoul: Seoul National University Press, 1999, pp. 316–29.

Giahi Yazdi, Hamid-Reza, and Pouyan Rezvani, 'Chronology of the Events of the Samarqand "Observatory and School" Based on some Old Persian Texts: A Revision', *Suhayl* 14 (2015), pp. 145–65.

Goldstein, Bernard R., and José Chabás, 'Ptolemy, Bianchini, and Copernicus: Tables for Planetary Latitudes', *Archive for History of Exact Sciences* 58 (2004), pp. 453–73.

Hamadanizadeh, Javad, 'The Trigonometric Tables of al-Kāshī in his *Zīj-i Khāqānī*', *Historia Mathematica* 7 (1980), pp. 38–45.

Husson, Matthieu. 'Remarks on Two Dimensional Array Tables in Latin Astronomy: A Case Study in Layout Transmission', *Suhayl* 13 (2014), pp. 103–17.

Kennedy, Edward S., 'A Survey of Islamic Astronomical Tables', *Transactions of the American Philosophical Society*, new series 46 (1956), pp. 123–77.

Kennedy, Edward S., A Letter of Jamshīd al-Kāshī to his Father: Scientific Research and Personalities at a Fifteenth Century Court, *Orientalia* 29 (1960), pp. 191–213.

Kennedy, Edward S., 'Spherical Astronomy in Kāshī's *Khāqānī Zīj*', *Zeitschrift für Geschichte der Arabisch-Islamischen Wissenschaften* 2 (1985), pp. 1–46.

Kennedy, Edward S., 'The Prime Vertical Method for the Astrological Houses as Presented in Kāshī's *Khāqānī Zīj*', in Anton von Gotstedter (ed.), *Ad Radices: Festband zum fünfzigjährigen Bestehen des Instituts für Geschichte der Naturwissenschaften der Johann Wolfgang Goethe-Universität Frankfurt am Main*, Stuttgart: Steiner, 1994, pp. 95–97.

Kennedy, Edward S., 'Treatise V of Kāshī's *Khāqānī Zīj*: Determination of the Ascendent', *Zeitschrift für Geschichte der Arabisch-Islamischen Wissenschaften* 10 (1995/96), pp. 123–45.

Kennedy, Edward S., *On the Contents and Significance of the Khāqānī Zīj by Jamshīd Ghiyāth al-Dīn al-Kāshī*, Frankfurt: Institut für Geschichte der Arabisch-Islamischen Wissenschaften, 1998.

King, David, Julio Samsó and Bernard R. Goldstein, 'Astronomical Handbooks and Tables from the Islamic World (750–1900): An Interim Report', *Suhayl* 2 (2001), pp. 9–105.

Li Liang, 'Tables with "European" Layout in China: A Case Study in Tabular Layout Transmission', *Suhayl* 13 (2014), pp. 83–101.

Luckey, Paul, *Der Lehrbrief über den Kreisumfang (ar-Risāla al-muḥītīya) von Ǧamšīd b. Masʿūd al-Kāšī*, Berlin: Akademie Verlag, 1953. [Abhandlungen der Deutschen Akademie der Wissenschaften zu Berlin, 1950/6].

Mozaffari, S. Mohammad, 'Planetary Latitudes in Medieval Islamic Astronomy: An Analysis of the Non-Ptolemaic Latitude Parameter Values in the Maragha and Samarqand Astronomical Traditions', *Archive for History of Exact Sciences* 70 (2016), pp. 513–41.

Neugebauer, Otto. *A History of Ancient Mathematical Astronomy*, 3 vols, Berlin: Springer, 1975.

Pedersen, Olaf, *A Survey of the Almagest. With Annotation and New Commentary by Alexander Jones*, New York: Springer, 2011.

Ragep, F. Jamil, *Naṣīr al-Dīn al-Ṭūsī's* Memoir on Astronomy *(al-Tadhkira fī ʿilm al-hayʾa)*, 2 vols, New York: Springer, 1993.

Riddell, R. C., 'The Latitudes of Venus and Mercury in the *Almagest*', *Archive for History of Exact Sciences* 19 (1978), pp. 95–111.

Rosenfeld, Boris, and Jan Hogendijk, 'A Mathematical Treatise Written in the Samarqand Observatory of Ulugh Beg', *Zeitschrift für Geschichte der Arabisch-Islamischen Wissenschaften* 15 (2002/03), pp. 25–65.

Sédillot, Louis P. E. Amélie, *Prolégomènes des Tables Astronomiques d'Oloug-Beg*, 2 vols, Paris: L'Institut de France, 1847/1853.

Stahlman, William D., *The Astronomical Tables of Codex Vaticanus Graecus 1291*, PhD dissertation, Brown University, 1960.

Swerdlow, Noel, 'Ptolemy's Theory of the Inferior Planets', *Journal for the History of Astronomy* 20 (1989), pp. 29–59.

Swerdlow, Noel, 'Ptolemy's Theories of the Latitude of the Planets in the *Almagest*, *Handy Tables*, and *Planetary Hypotheses*', in Jed Buchwald and Allan Franklin (eds), *Wrong for the Right Reasons*, Dordrecht: Springer, 2005, pp. 41–71.

Tichenor, Mark J., 'Late Medieval Two-Argument Tables for Planetary Longitudes', *Journal of Near Eastern Studies* 26 (1967), pp. 126–28.

Tihon, Anne, and Raymond Mercier, *Πτολεμαίου Πρόχειροι Κανόνες: Les 'Tables Faciles' de Ptolémée / Ptolemy's Handy Tables*, vols 1a and 1b, Louvain-la-Neuve: Université Catholique de Louvain/Peeters, 2011.

Toomer, Gerald J., 'Review of Olaf Pedersen, *A Survey of the Almagest*', *Archives Internationales d'Histoire des Sciences* 27 (1977), pp. 137–50.

Toomer, Gerald J., *Ptolemy's Almagest*, New York: Springer and London: Duckworth, 1984. [Reprinted Princeton: Princeton University Press, 1998].

Van Brummelen, Glen, *Mathematical Tables in Ptolemy's Almagest*, PhD dissertation, Simon Fraser University, 1993.

Van Brummelen, Glen, 'Taking Latitude with Ptolemy: Jamshīd al-Kāshī's Novel Geometric Model of the Motions of the Inferior Planets', *Archive for History of Exact Sciences* 60 (2006), pp. 353–77.

Van Brummelen, Glen, 'Crossing a Mathematical Rubicon: Al-Kāshī's Two Methods of Computing Sin(1°)', to appear.

Appendix: Editions of Mercury and Venus Double-Argument Tables

Our edition is based on the three principal manuscripts:

IO: London, British Library, MS India Office 430 (Ethé 2232)

AS: Istanbul, Süleymaniye Library, MS Ayasofya 2692

C: Cairo, Dār al-Kutub, MS TR 149

In **C** the Venus table is rendered on four pages, with each page containing 90° intervals of the horizontal argument a_v; the other two manuscripts (and **C** for Mercury) represent the tables over three pages, each containing 120° sections of the horizontal argument. All three pages of the Mercury table in **AS** display the gridlines but the entries are blank. The first of the three pages in the **AS** table for Venus likewise displays gridlines but is otherwise blank. **C** contains an extra blank page of gridlines after the Venus table in the format appropriate for Venus; the first page of the Mercury table appears after pages two and three, not before. The other two manuscripts follow the same layout but occasionally leave certain cells blank.[55] All three manuscripts contain almost precisely the same pattern of incomplete entries in the Venus table. A few entries are blank in all manuscripts. The tables are represented here to fit within the page constraints of this volume, namely with 90° of the horizontal argument per page and the Mercury table furthermore split into their upper and bottom halves.

Entries are represented as northward or southward using the symbols ش and ج (the first letters of the Arabic words for 'northern' and 'southern', *shimāl* and *janūb*). A string of consecutive entries within a column represents southward if it has a ج written above it; similarly with ش for northward. Occasionally the letter is omitted, but it is usually easy to tell from context which entries indicate northward and which indicate southward. The Venus table has north/south indicators at the top of each column, which we translate here as 'N' and 'S';[56] the Mercury table does not. Here, for Venus we represent southward entries in italics; for Mercury we represent northward entries in italics. As is typical in *zīj*es, if a sequence of entries working down a column shares the same leading digit, that digit is usually written only in the first of these entries. Here we fill in the unwritten digits for clarity. Occasionally dots above letters in the manuscripts are accidentally added or omitted (again a common occurrence), for instance changing a '13' into a '53' or vice versa. These errors are silently corrected. Al-Kāshī usually groups the arguments in segments of

[55] In the **C** manuscript, the word 'centrum' appears in the cell above the first column in the first page of the Venus table.

[56] In the **C** and **AS** manuscripts, some of these indicators are missing.

30° and labels them by astrological signs (although in the Venus table the rows are indicated by sign numbers rather than names).

Manuscript variants of entries are given in the apparatus below each page of the table. Each variant is indicated first by the row argument, then by the column argument. Thus for the Mercury table '♈ 6°, ♉ 0°: **IO** 1;33' denotes that the entry in the **IO** manuscript corresponding to a_v = ♈ 6°, c = ♉ 0° is 1;33. For the Venus table '1,20°, ♋ 20°: **C** 0;53' denotes that the entry in the **C** manuscript corresponding to c = 1,20° (= 50°) , a_v = ♋ 20° is 0;53.

Simplification of the Latitude of Venus (first quarter)

	♈						♉						♊					
	0	5	10	15	20	25	0	5	10	15	20	25	0	5	10	15	20	25
	N	N	N	N	N	N	N	N	N	N	N	N	N	N	N	N	N	N
0, 0	0;10	0;17	0;23	0;30	0;37	0;45	0;51	0;58	1; 5	1;12	1;18	1;23	1;30	1;37	1;44	1;49	1;54	1;59
0,10	0;20	0;27	0;33	0;40	0;47	0;53	1; 0	1; 6	1;12	1;18	1;23	1;29	1;35	1;40	1;45	1;50	1;55	1;58
0,20	0;29	0;36	0;42	0;48	0;54	1; 0	1; 6	1;11	1;16	1;21	1;26	1;31	1;36	1;40	1;44	1;48	1;52	1;56
1, 0	0;38	0;44	0;49	0;55	1; 1	1; 6	1;11	1;15	1;19	1;23	1;27	1;31	1;34	1;37	1;40	1;43	1;45	1;47
1,10	0;45	0;50	0;55	1; 0	1; 5	1;10	1;14	1;17	1;20	1;23	1;26	1;28	1;30	1;32	1;33	1;34	1;35	1;36
1,20	0;51	0;55	0;59	1; 3	1; 7	1;11	1;14	1;16	1;18	1;20	1;22	1;23	1;24	1;24	1;23	1;23	1;22	1;22
2, 0	0;56	0;59	1; 2	1; 5	1; 8	1;11	1;13	1;14	1;14	1;15	1;15	1;16	1;16	1;15	1;13	1;11	1; 9	1; 7
2,10	1; 0	1; 2	1; 3	1; 5	1; 7	1; 8	1; 9	1; 8	1; 7	1; 6	1; 5	1; 4	1; 3	1; 1	0;59	0;57	0;53	0;49
2,20	1; 2	1; 2	1; 2	1; 3	1; 3	1; 4	1; 4	1; 2	1; 0	0;58	0;56	0;53	0;50	0;46	0;42	0;38	0;33	0;28
3, 0	1; 2	1; 2	1; 1	1; 1	1; 0	0;59	0;58	0;55	0;52	0;49	0;46	0;42	0;38	0;33	0;28	0;22	0;16	0; 9
3,10	1; 2	1; 1	0;59	0;57	0;55	0;53	0;51	0;47	0;43	0;39	0;34	0;29	0;24	0;18	0;12	0; 5	*0; 2*	*0; 9*
3,20	1; 1	0;58	0;55	0;52	0;49	0;46	0;43	0;38	0;33	0;28	0;22	0;16	0;10	0; 3	*0; 4*	*0;11*	*0;18*	*0;26*
4, 0	0;58	0;53	0;49	0;45	0;41	0;37	0;33	0;27	0;21	0;15	0; 9	0; 3	*0; 4*	*0;12*	*0;20*	*0;28*	*0;36*	*0;44*
4,10	0;53	0;48	0;43	0;38	0;33	0;28	0;23	0;17	0;10	0; 3	*0; 4*	*0;11*	*0;18*	*0;26*	*0;34*	*0;42*	*0;50*	*0;59*
4,20	0;47	0;41	0;35	0;29	0;23	0;17	0;11	0; 4	*0; 3*	*0;10*	*0;17*	*0;24*	*0;32*	*0;40*	*0;48*	*0;56*	*1; 5*	*1;14*
5, 0	0;40	0;33	0;27	0;20	0;14	0; 7	0; 1	*0; 6*	*0;13*	*0;20*	*0;27*	*0;35*	*0;43*	*0;51*	*0;59*	*1; 7*	*1;15*	*1;23*
5,10	0;31	0;25	0;19	0;12	0; 5	*0; 2*	*0; 9*	*0;16*	*0;23*	*0;30*	*0;38*	*0;46*	*0;54*	*1; 1*	*1; 8*	*1;16*	*1;24*	*1;32*
5,20	0;21	0;15	0; 8	0; 1	*0; 6*	*0;13*	*0;20*	*0;27*	*0;34*	*0;42*	*0;49*	*0;56*	*1; 3*	*1;10*	*1;17*	*1;24*	*1;31*	*1;38*
6, 0	0;10	0; 3	*0; 3*	*0;10*	*0;17*	*0;25*	*0;31*	*0;38*	*0;45*	*0;52*	*0;58*	*1; 3*	*1;10*	*1;17*	*1;24*	*1;29*	*1;34*	*1;39*
6,10	*0; 1*	0;6	*0;12*	*0;21*	*0;26*	*0;33*	*0;40*	*0;46*	*0;52*	*0;58*	*1; 4*	*1;10*	*1;15*	*1;20*	*1;25*	*1;30*	*1;35*	*1;40*
6,20	*0;13*	*0;19*	*0;25*	*0;31*	*0;37*	*0;43*	*0;49*	*0;54*	*0;59*	*1; 4*	*1; 9*	*1;14*	*1;18*	*1;22*	*1;26*	*1;30*	*1;34*	*1;36*
7, 0	*0;26*	*0;32*	*0;36*	*0;42*	*0;47*	*0;52*	*0;57*	*1; 1*	*1; 5*	*1; 8*	*1;11*	*1;15*	*1;19*	*1;22*	*1;25*	*1;27*	*1;29*	*1;31*
7,10	*0;37*	*0;41*	*0;45*	*0;50*	*0;55*	*0;59*	*1; 3*	*1; 6*	*1; 9*	*1;12*	*1;14*	*1;16*	*1;18*	*1;19*	*1;19*	*1;20*	*1;21*	*1;22*
7,20	*0;45*	*0;49*	*0;52*	*0;56*	*0;59*	*1; 3*	*1; 6*	*1; 7*	*1; 8*	*1; 9*	*1;10*	*1;11*	*1;12*	*1;13*	*1;14*	*1;14*	*1;12*	*1;10*
8, 0	*0;54*	*0;56*	*0;58*	*1; 0*	*1; 2*	*1; 4*	*1; 6*	*1; 6*	*1; 6*	*1; 5*	*1; 5*	*1; 4*	*1; 4*	*1; 3*	*1; 1*	*0;59*	*0;57*	*0;55*
8,10	*0;59*	*1; 0*	*1; 1*	*1; 2*	*1; 3*	*1; 4*	*1; 5*	*1; 4*	*1; 3*	*1; 1*	*0;59*	*0;57*	*0;55*	*0;53*	*0;50*	*0;47*	*0;43*	*0;39*
8,20	*1; 2*	*1; 2*	*1; 2*	*1; 1*	*1; 1*	*1; 1*	*1; 1*	*0;59*	*0;57*	*0;54*	*0;51*	*0;48*	*0;45*	*0;41*	*0;35*	*0;32*	*0;27*	*0;22*
9, 0	*1; 2*	*1; 1*	*1; 0*	*0;59*	*0;58*	*0;57*	*0;56*	*0;52*	*0;48*	*0;44*	*0;40*	*0;36*	*0;35*	*0;29*	*0;24*	*0;19*	*0;14*	*0; 7*
9,10	*1; 2*	*1; 0*	*0;58*	*0;56*	*0;54*	*0;51*	*0;48*	*0;43*	*0;38*	*0;33*	*0;28*	*0;23*	*0;18*	*0;13*	*0; 7*	0; 0	0; 7	0;15
9,20	*0;58*	*0;55*	*0;52*	*0;49*	*0;45*	*0;41*	*0;37*	*0;32*	*0;27*	*0;21*	*0;15*	*0; 9*	*0; 3*	0; 4	0;11	0;19	0;27	0;35
10, 0	*0;51*	*0;47*	*0;43*	*0;38*	*0;33*	*0;28*	*0;23*	*0;17*	*0;11*	*0; 5*	0; 2	0; 9	0;16	0;24	0;32	0;40	0;48	0;56
10,10	*0;43*	*0;38*	*0;33*	*0;28*	*0;23*	*0;18*	*0;12*	*0; 5*	0; 2	0; 9	0;16	0;23	0;30	0;38	0;46	0;54	1; 3	1;12
10,20	*0;33*	*0;28*	*0;22*	*0;16*	*0;10*	*0; 4*	0; 2	0; 9	0;16	0;23	0;30	0;37	0;45	0;53	1; 1	1;10	1;19	1;28
11, 0	*0;22*	*0;16*	*0;10*	*0; 4*	0; 2	0; 9	0;16	0;23	0;30	0;37	0;44	0;52	1; 0	1; 8	1;16	1;24	1;32	1;40
11,10	*0;11*	*0; 5*	0; 2	0; 9	0;16	0;23	0;30	0;37	0;44	0;51	0;58	1; 6	1;14	1;21	1;28	1;36	1;44	1;52
11,20	0; 0	0; 6	0;12	0;19	0;26	0;33	0;40	0;47	0;54	1; 1	1; 8	1;15	1;23	1;30	1;36	1;44	1;51	1;58

0,20°, ♈ 20°: **C** illegible. 2,20°, 5°: **C** 0;16. 5,10°, ♊ 20°: **C** *1;26*. 6,20°, ♊ 25°: **IO** *1;37*. 7,0°, ♈ 10°: **IO** *0;37*. 7,10°, ♈ 0°: **IO** *0;36*. 8,20°, ♉ 10°: **C** *0;52*. 11,20°, ♈ 5°: **IO** blank.

Continuation of the Simplification of the Latitude of Venus (second quarter)

	♋						♌						♍					
	0	5	10	15	20	25	0	5	0	0	0	0	0	0	10	15	20	25
	N	N	N	N	N	N	N	N	N	N	N	N	N	N	N	N	N	N
0, 0	2; 7	2;12	2;16	2;22	2;27	2;32	2;35	2;37	2;32	2;32	2;32	2;32	2;32	2;32	2;11	1;51	1;26	0;50
0,10	2; 5	2; 9	2;12	2;15	2;17	2;19	2;21	2;18	1;44	1;44	1;44	1;44	1;44	1;44	1;5	0;39	0; 6	*0;33*
0,20	2; 0	2; 3	2; 5	2; 5	2; 5	2; 4	2; 2	1;54	0;56	0;56	0;56	0;56	0;56	0;56	0; 3	*0;29*	*1; 7*	*1;51*
1, 0	1;49	1;49	1;48	1;46	1;44	1;41	1;37	1;29	0; 0	0; 0	0; 0	0; 0	0; 0	0; 0	*1;12*	*1;50*		
1,10	1;37	1;37	1;34	1;28	1;23	1;17	1; 9	0;58	*0;58*	*0;58*	*0;58*	*0;58*	*0;58*	*0;58*		*3; 4*		
1,20	1;21	1;19	1;11	1; 4	0;56	0;48	0;39	0;25	*1;48*	*1;48*	*1;48*	*1;48*	*1;48*	*1;48*		4;11		
2, 0	1; 5	1; 0	0;51	0;39	0;29	0;20	0; 9	*0; 6*	*2;38*	*2;38*	*2;38*	*2;38*	*2;38*	*2;38*	*4;15*	*5;7*	*5;53*	*6;39*
2,10	0;44	0;35	0;23	0;11	0; 0	*0;11*	*0;23*	*0;38*	*3;29*	*3;29*	*3;29*	*3;29*	*3;29*	*3;29*		*6; 4*		
2,20	0;23	0;13	0; 1	*0;14*	*0;27*	*0;40*	*0;54*	*1;10*	*4;10*	*4;10*	*4;10*	*4;10*	*4;10*	*4;10*		6;45		
3, 0	0; 2	*0;12*	*0;25*	*0;38*	*0;53*	*1; 6*	*1;21*	*1;42*	*4;39*	*4;39*	*4;39*	*4;39*	*4;39*	*4;39*	*6;15*	*7;12*	*8; 1*	*8;31*
3,10	*0;16*	*0;29*	*0;33*	*0;58*	*1;13*	*1;28*	*1;43*	*2; 3*	*4;57*	*4;57*	*4;57*	*4;57*	*4;57*	*4;57*		7;16		
3,20	*0;34*	*0;44*	*0;55*	*1;18*	*1;33*	*1;48*	*2; 3*	*2;22*	*5;10*	*5;10*	*5;10*	*5;10*	*5;10*	*5;10*		7;14		
4, 0	*0;53*	*1; 4*	*1;16*	*1;35*	*1;45*	*2; 2*	*2;20*	*2;41*	*5;12*	*5;12*	*5;12*	*5;12*	*5;12*	*5;12*	*6;22*	*6;57*		
4,10	*1; 8*	*1;19*	*1;32*	*1;48*	*2; 4*	*2;19*	*2;32*	*2;48*	*5; 3*	*5; 3*	*5; 3*	*5; 3*	*5; 3*	*5; 3*		6;26		
4,20	*1;23*	*1;33*	*1;45*	*2; 0*	*2;13*	*2;26*	*2;39*	*2;54*	*4;41*	*4;41*	*4;41*	*4;41*	*4;41*	*4;41*		5;42		
5, 0	*1;32*	*1;43*	*1;54*	*2; 5*	*2;17*	*2;29*	*2;40*	*2;52*	*4;16*	*4;16*	*4;16*	*4;16*	*4;16*	*4;16*		4;52		
5,10	*1;40*	*1;49*	*1;58*	*2; 7*	*2;14*	*2;26*	*2;35*	*2;43*	*3;40*	*3;40*	*3;40*	*3;40*	*3;40*	*3;40*		3;53		
5,20	*1;45*	*1;52*	*1;59*	*2; 6*	*2;13*	*2;20*	*2;27*	*2;30*	*2;54*	*2;54*	*2;54*	*2;54*	*2;54*	*2;54*		2;37		
6, 0	*1;47*	*1;52*	*1;57*	*2; 3*	*2; 7*	*2;12*	*2;15*	*2;17*	*2;12*	*2;12*	*2;12*	*2;12*	*2;12*	*2;12*	*1;51*	*1;31*	*1; 6*	*0;30*
6,10	*1;45*	*1;48*	*1;51*	*1;54*	*1;56*	*1;58*	*2; 0*	*1;58*	*1;25*	*1;25*	*1;25*	*1;25*	*1;25*	*1;25*		0;22		
6,20	*1;40*	*1;41*	*1;42*	*1;42*	*1;41*	*1;40*	*1;40*	*1;37*	*0;33*	*0;33*	*0;33*	*0;33*	*0;33*	*0;33*		0;57		
7, 0	*1;32*	*1;31*	*1;29*	*1;27*	*1;24*	*1;21*	*1;17*	*1;10*	0;19	0;19	0;19	0;19	0;19	0;19		2; 8		
7,10	*1;23*	*1;20*	*1;16*	*1;11*	*1; 5*	*0;59*	*0;53*	*0;44*	1; 0	1; 0	1; 0	1; 0	1; 0	1; 0		3; 7		
7,20	*1; 8*	*1; 4*	*0;58*	*0;50*	*0;43*	*0;35*	*0;27*	*0;15*	1;50	1;50	1;50	1;50	1;50	1;50		4;11		
8, 0	*0;53*	*0;49*	*0;41*	*0;30*	*0;21*	*0;11*	0; 0	0;15	2;39	2;39	2;39	2;39	2;39	2;39		5; 3		
8,10	*0;34*	*0;27*	*0;19*	*0; 8*	0; 2	0;15	0;29	0;46	3;15	3;15	3;15	3;15	3;15	3;15		5;43		
8,20	*0;16*	*0; 8*	0; 3	0;14	0;27	0;40	0;54	1;11	3;45	3;45	3;45	3;45	3;45	3;45		6;11		
9, 0	0; 1	0;11	0;24	0;36	0;49	1; 3	1;19	1;35	4;12	4;12	4;12	4;12	4;12	4;12	5;45	6;35	7;31	8;18
9,10	0;23	0;33	0;45	0;58	1;12	1;27	1;42	2; 0	4;32	4;32	4;32	4;32	4;32	4;32		6;48		
9,20	0;44	0;55	1; 7	1;20	1;34	1;48	2; 3	2;21	4;46	4;46	4;46	4;46	4;46	4;46		6;50		
10, 0	1; 5	1;16	1;28	1;41	1;55	2; 9	2;23	2;40	4;53	4;53	4;53	4;53	4;53	4;53		6;37	7;15	7;34
10,10	1;21	1;31	1;43	1;55	2; 9	2;23	2;38	2;54	4;48	4;48	4;48	4;48	4;48	4;48		6;13		
10,20	1;37	1;48	2; 0	2;12	2;24	2;36	2;48	3; 2	4;39	4;39	4;39	4;39	4;39	4;39		5;43		
11, 0	1;49	1;59	2; 9	2;19	2;29	2;40	2;51	3; 3	4;16	4;16	4;16	4;16	4;16	4;16		4;54		
11,10	2; 0	2; 8	2;17	2;25	2;34	2;43	2;52	3; 0	3;45	3;45	3;45	3;45	3;45	3;45		3;56		
11,20	2; 5	2;12	2;19	2;26	2;33	2;40	2;47	2;53	3;13	3;13	3;13	3;13	3;13	3;13		2;58		

0,10°, ♌ 10°: **C** 2;3. 1,0°, ♍ 0°: **AS** illegible. 1,0°, ♍ 5°: **IO** *0;30*. 1,20°, ♋ 20°: **C** 0;53. 2,0°, ♋ 25°: all MS's *0;20* (N/S marker above rather than below the entry). 2,20°, ♌ 10°: **IO** *1;34*. 2,20°, ♍ 15°: **AS** *6;47*. 3,0°, ♍ 0°: **C** *4;29*. 3,10°, ♌ 25°: **AS** *4;9*. 4,0°, ♍ 5°: **IO** blank. 4,0°, ♍ 10°: **IO** blank. 4,20°, ♋ 15°: **IO** *2;5*. 5,10°, ♋ 25°: **IO** *2;29*. 6,0°, ♋ 10°: **C** *1;52*. 6,0°, ♋ 15°: **C** *2;2*. 8,0°, ♌ 5°: **IO** 0;45. 10,0°, ♍ 20°: **IO** blank. 10,0°, ♍ 25°: **IO** blank. 10,20°, ♌ 5°: **IO** 3;12. 10,20°, ♌ 25°: **AS**, **IO** 4;16.

Continuation of the Simplification of the Latitude of Venus (third quarter)

	♎						♏						♐					
	0	5	10	15	20	25	0	5	10	15	20	25	0	5	10	15	20	25
	N	S	S	S	S	S	S	S	S	S	S	S	S	S	S	S	S	S
0, 0	0;10	*0;30*	*1; 6*	*1;31*	*1;51*	*2;4*	*2;12*	*2;16*	*2;19*	*2;20*	*2;19*	*2;17*	*2;15*	*2;12*	*2; 7*	*2; 2*	*1;57*	*1;52*
0,10	*1;21*			2;38			2;53			2;41			2;27					
0,20	*2;42*			3;38			3;27			2;59			2;34					
1, 0	*4;10*			4;38			*4; 0*			3;12			2;35					
1,10	*5;30*			5;31			4;27			3;23			2;36					
1,20	*6;33*			6;5			4;40			3;20			2;26					
2, 0	*7;25*	*7;28*	*7; 9*	*6;29*	*5;55*	*5;21*	4;47	4;16	3;46	3;16	2;51	2;31	2;17					
2,10	*8;10*			6;48			4;44			*3; 4*			*2; 1*					
2,20	8;32			6;48			4;32			2;45			1;42					
3, 0	*8;40*	*8;15*	*7;28*	*6;35*	*5;44*	*4;58*	*4;12*	*3;27*	*2;49*	*2;21*	*1;57*	*1;35*	*1;19*	*1; 3*	*0;50*	*0;36*	*0;23*	*0;11*
3,10	*8;27*			5;51			3;45			1;55			0;54					
3,20	*8; 8*			5;41			3;13			1;26			0;27					
4, 0	*7;31*			4;59			2;35			0;52			0; 5					
4,10	*6;37*			*4; 3*			1;47			0;14			0;33					
4,20	*5;26*			2;57			0;55			0;25			1; 4					
5, 0	*4;16*			1;54			*0; 5*			1; 0			1;30					
5,10	*2;48*			0;36			0;51			1;34			1;56					
5,20	*1;13*			0;42			1;45			2;12			2;20					
6, 0	0;10	0;50	1;26	1;51	2;11	2;24	2;32	2;36	2;39	2;40	2;39	2;37	2;35	2;32	2;27	2;22	2;17	2;12
6,10	1;41			2;57			3;14			3; 2			2;47					
6,20	3;16			4; 8			3;58			3;23			2;53					
7, 0	4;42			5; 6			4;30			3;34			2;54					
7,10	5;50			5;52			4;51			3;38			2;49					
7,20	6;57			6;34			5;11			3;38			2;40					
8, 0	7;47			7; 1			5;16			3;30			2;26					
8,10	8;20			7;16			5;12			3;15			2; 5					
8,20	8;31			7;16			4;57			2;54			1;43					
9, 0	8;37	8;34	8; 4	7;12	6;17	5;28	4;39	3;53	3;9	2;33	2;7	1;42	1;21	1; 7	0;53	0;39	0;25	0;13
9,10	8;26			6;45			4;10			2; 3			0;54					
9,20	8; 0			6; 6			3;31			1;31			0;25					
10, 0	7;19			5;13			2;44			0;54			*0; 3*					
10,10	6;27			4;19			2; 0			0;21			0;28					
10,20	5;28			3;17			1;10			0;16			0;57					
11, 0	4;12			2; 6			0;18			0;49			1;19					
11,10	2;48			0;47			0;40			1;26			1;44					
11,20	1;33			0;19			1;24			1;52			*2; 1*					

0,0°, ♐ 10°: **C** *1;56*. 1,10°, ♏ 15°: **IO** and **AS** *2;23*. 2,0°, ♏ 10°: **C** *3;16*. 3,0°, ♏ 10°: **C** *2;19*. 3,20°, ♐ 0°: **C** *0;26*. 4,0°, ♐ 0°: **C** 0;4. 4,20°, ♐ 0°: **C** 1;5. 10,0°, ♎ 5°: **AS** spurious entry 8;18.

Continuation of the Simplification of the Latitude of Venus (fourth quarter)

	♑						♒						♓					
	0	5	10	15	20	25	0	5	10	15	20	25	0	5	10	15	20	25
	S	S	S	S	S	S	S	S	S	S	S	S	S	S	S	S	S	N
0, 0	*1;47*	*1;39*	*1;34*	*1;29*	*1;24*	*1;17*	*1;10*	*1; 3*	*0;58*	*0;52*	*0;45*	*0;38*	*0;31*	*0;25*	*0;17*	*0;10*	*0; 3*	0; 3
0,10	*1;45*						1; 3						0;20					
0,20	*1;42*						0;56						0;12					
1, 0	*1;33*						0;44						0;0					
1,10	*1;26*						0;35						0;10					
1,20	*1;12*						0;22						0;20					
2, 0	*0;59*						0;10						0;29					
2,10	0;42						0; 5						0;39					
2,20	*0;23*						0;18						0;48					
3, 0	0; 0	0; 7	0;13	0;19	0;24	0;29	0;33	0;38	0;44	0;49	0;52	0;54	0;56	0;57	0;59	1; 0	1; 1	1; 2
3,10	0;16						0;45						1; 1					
3,20	0;36						0;58						1; 7					
4, 0	0;57						1;10						1;11					
4,10	1;16						1;20						1;14					
4,20	1;33						1;28						1;13					
5, 0	1;46						1;33						1;11					
5,10	1;58						1;36						1; 7					
5,20	2; 5						1;35						1; 0					
6, 0	2; 7	1;59	1;54	1;49	1;44	1;37	1;30	1;23	1;18	1;12	1; 5	0;58	0;51	0;45	0;37	0;30	0;23	0;17
6,10	2; 5						1;23						0;40					
6,20	1;58						1;12						0;27					
7, 0	1;46						0;57						0;13					
7,10	1;33						0;42						0; 1					
7,20	1;16						0;26						0;16					
8, 0	0;57						0; 7						0;29					
8,10	0;36						0; 8						0;41					
8,20	0;16						0;23						0;51					
9, 0	*0; 0*	*0; 8*	*0;14*	*0;21*	*0;26*	*0;31*	*0;36*	*0;41*	*0;46*	*0;50*	*0;53*	*0;56*	*0;58*	*0;59*	*1; 0*	*1; 1*	*1; 2*	*1; 2*
9,10	*0;23*						0;50						1; 4					
9,20	*0;42*						1; 1						1; 7					
10, 0	*0;59*						1;10						1; 7					
10,10	*1;12*						1;15						1; 6					
10,20	*1;26*						1;19						1; 2					
11, 0	*1;33*						1;18						0;56					
11,10	*1;42*						1;18						0;48					
11,20	*1;45*						1;15						0;40					

6,0°, ♑ 15°: **C** 1;19, **AS** 1;29. 6,20°, ♒ 0°: **IO** 1;2.

Table of the Simplification of the Latitude of Mercury
(first quarter, top half)

		♈					♉					♊				
		0	6	12	18	24	0	6	12	18	24	0	6	12	18	24
	0	0;45	0;55	1; 3	1;14	1;19	1;26	1;31	1;35	1;38	1;41	1;43	1;44	1;45	1;46	1;46
	6	0;55	1; 6	1;14	1;22	1;28	1;34	1;38	1;41	1;45	1;46	1;47	1;48	1;47	1;47	1;46
♈	12	1; 5	1;15	1;23	1;30	1;36	1;42	1;45	1;48	1;50	1;51	1;51	1;51	1;48	1;48	1;47
	18	1;15	1;24	1;32	1;39	1;45	1;50	1;52	1;54	1;56	1;57	1;55	1;53	1;51	1;48	1;45
	24	1;25	1;33	1;41	1;48	1;53	1;57	1;59	2; 0	2; 1	2; 0	1;58	1;55	1;51	1;46	1;44
	0	1;35	1;42	1;50	1;57	2; 1	2; 4	2; 5	2; 5	2; 3	2; 3	2; 0	1;56	1;50	1;45	1;41
	6	1;44	1;51	1;59	2; 5	2; 8	2; 9	2; 9	2; 8	2; 6	2; 4	1;59	1;55	1;48	1;42	1;36
♉	12	1;53	2; 0	2; 8	2;13	2;15	2;15	2;12	2;10	2; 7	2; 3	1;58	1;52	1;44	1;37	1;30
	18	2; 6	2; 9	2;16	2;20	2;21	2;19	2;15	2;12	2; 8	2; 2	1;56	1;48	1;39	1;31	1;24
	24	2;11	2;17	2;23	2;26	2;25	2;21	2;17	2;13	2; 9	2; 1	1;53	1;44	1;34	1;25	1;17
	0	2;19	2;25	2;28	2;30	2;28	2;23	2;18	2;14	2; 8	1;59	1;49	1;39	1;28	1;18	1; 8
	6	2;26	2;32	2;33	2;33	2;30	2;25	2;19	2;13	2; 5	1;56	1;45	1;33	1;21	1;10	0;59
♊	12	2;33	2;38	2;37	2;35	2;31	2;26	2;18	2;10	2; 1	1;51	1;39	1;26	1;13	1; 1	0;49
	18	2;40	2;42	2;40	2;37	2;32	2;25	2;16	2; 7	1;57	1;46	1;33	1;19	1; 4	0;51	0;38
	24	2;47	2;46	2;43	2;39	2;33	2;24	2;13	2; 3	1;52	1;40	1;26	1;10	0;53	0;39	0;26
	0	2;51	2;49	2;46	2;40	2;32	2;22	2;10	1;58	1;46	1;32	1;17	1; 0	0;44	0;26	0;12
	6	2;55	2;51	2;47	2;40	2;30	2;18	2; 5	1;51	1;37	1;25	1; 6	0;48	0;29	0;11	*0; 3*
♋	12	2;58	2;52	2;46	2;38	2;27	2;12	1;58	1;42	1;27	1;11	0;53	0;36	0;14	*0; 2*	*0;18*
	18	2;59	2;53	2;45	2;34	2;20	2; 4	1;48	1;32	1;14	0;58	0;39	0;19	*0; 1*	*0;16*	*0;33*
	24	3; 0	2;51	2;41	2;28	2;13	1;55	1;37	1;20	1; 2	0;43	0;24	0; 2	*0;16*	*0;34*	*0;50*
	0	2;59	2;48	2;36	2;21	2; 4	1;45	1;26	1; 7	0;47	0;28	0;6	*0;16*	*0;36*	*0;54*	*1;10*
	6	2;57	2;44	2;30	2;13	1;54	1;34	1;14	0;58	0;30	0;11	*0;12*	*0;35*	*1; 2*	*1;16*	*1;31*
♌	12	2;50	2;37	2;21	2; 2	1;42	1;21	0;59	0;36	0;12	*0; 9*	*0;32*	*0;57*		*1;38*	*1;52*
	18	2;42	2;27	2; 9	1;49	1;27	1; 4	0;41	0;16	*0; 6*	*0;30*	*0;54*	*1;18*	*1;39*	*2; 0*	*2;14*
	24	2;33	2;15	1;55	1;33	1; 9	0;43	0;18	*0; 6*	*0;29*	*0;53*	*1;18*	*1;42*	*2; 3*	*2;21*	*2;35*
	0	2;21	2; 0	1;39	1;15	0;50	0;21	*0; 3*	*0;29*	*0;53*	*1;16*	*1;41*	*2;5*	*2;26*	*2;43*	*2;52*
	6	2; 6	1;43	1;20	0;55	0;28	*0; 0*	*0;27*	*0;53*	*1;16*	*1;41*	*2; 4*	*2;27*	*2;48*	*3; 4*	*3;17*
♍	12	1;48	1;24	0;59	0;34	0; 6	*0;22*	*0;50*	*1;17*	*1;41*	*2;4*	*2;26*	*2;48*	*3; 8*	*3;25*	*3;36*
	18	1;28	1; 4	0;37	0;12	*0;17*	*0;45*	*1;13*	*1;39*	*2;3*	*2;26*	*2;46*	*3; 6*	*3;26*	*3;42*	*3;51*
	24	1; 7	0;42	0;15	*0;11*	*0;39*	*1; 7*	*1;35*	*2; 0*	*2;22*	*2;45*	*3; 4*	*3;23*	*3;40*	*3;53*	*4; 0*

♈ 6°, ♉ 0°: **IO** 1;33. ♌ 12°, ♊ 12°: all MS's *1;27*. ♍ 12°, ♈ 0°,: **IO** 1;38.

Table of the Simplification of the Latitude of Mercury
(first quarter, bottom half)

		♈					♉					♊				
		0	6	12	18	24	0	6	12	18	24	0	6	12	18	24
♎	0	0;45	0;19	*0;7*	*0;33*	*1; 0*	*1;28*	*1;54*	*2;17*	*2;40*	*3; 0*	*3;18*	*3;35*	*3;49*	*3;58*	*4; 3*
	6	0;21	*0; 3*	*0;28*	*0;53*	*1;19*	*1;45*	*2;10*	*2;33*	*2;52*	*3;10*	*3;27*	*3;40*	*3;52*	*4; 0*	*4; 3*
	12	0; 2	*0;23*	*0;47*	*1;11*	*1;35*	*1;59*	*2;22*	*2;44*	*3; 0*	*3;16*	*3;29*	*3;41*	*3;51*	*3;56*	*3;58*
	18	*0;18*	*0;41*	*1; 4*	*1;26*	*1;48*	*2;10*	*2;31*	*2;50*	*3; 4*	*3;19*	*3;30*	*3;39*	*3;46*	*3;49*	*3;47*
	24	*0;35*	*0;57*	*1;18*	*1;38*	*1;58*	*2;17*	*2;36*	*2;52*	*3; 3*	*3;18*	*3;27*	*3;33*	*3;37*	*3;37*	*3;32*
♏	0	*0;50*	*1; 9*	*1;29*	*1;47*	*2; 4*	*2;21*	*2;37*	*2;51*	*2;59*	*3;12*	*3;19*	*3;23*	*3;24*	*3;21*	*3;15*
	6	*1; 3*	*1;20*	*1;36*	*1;52*	*2; 7*	*2;22*	*2;35*	*2;46*	*2;53*	*3; 2*	*3; 6*	*3; 8*	*3; 7*	*3; 3*	*2;56*
	12	*1;15*	*1;28*	*1;41*	*1;55*	*2; 8*	*2;21*	*2;31*	*2;39*	*2;47*	*2;51*	*2;53*	*2;53*	*2;50*	*2;45*	*2;37*
	18	*1;20*	*1;34*	*1;45*	*1;56*	*2; 7*	*2;17*	*2;25*	*2;31*	*2;34*	*2;39*	*2;39*	*2;37*	*2;33*	*2;27*	*2;18*
	24	*1;26*	*1;37*	*1;47*	*1;56*	*2; 4*	*2;11*	*2;17*	*2;22*	*2;23*	*2;26*	*2;21*	*2;20*	*2;15*	*2; 8*	*1;58*
♐	0	*1;29*	*1;38*	*1;46*	*1;53*	*1;59*	*2; 4*	*2; 8*	*2;11*	*2;10*	*2;12*	*2; 9*	*2; 3*	*1;57*	*1;49*	*1;38*
	6	*1;30*	*1;37*	*1;43*	*1;48*	*1;52*	*1;55*	*1;58*	*1;59*	*1;56*	*1;58*	*1;54*	*1;47*	*1;39*	*1;30*	*1;19*
	12	*1;30*	*1;34*	*1;38*	*1;41*	*1;44*	*1;45*	*1;46*	*1;46*	*1;44*	*1;43*	*1;39*	*1;31*	*1;22*	*1;11*	*1; 0*
	18	*1;30*	*1;30*	*1;33*	*1;34*	*1;35*	*1;36*	*1;35*	*1;33*	*1;32*	*1;28*	*1;24*	*1;15*	*1; 5*	*0;54*	*0;42*
	24	*1;29*	*1;26*	*1;27*	*1;27*	*1;26*	*1;26*	*1;23*	*1;20*	*1;19*	*1;13*	*1; 8*	*0;59*	*0;49*	*0;38*	*0;26*
♑	0	*1;26*	*1;21*	*1;20*	*1;19*	*1;17*	*1;16*	*1;12*	*1; 8*	*1; 5*	*0;59*	*0;52*	*0;44*	*0;34*	*0;23*	*0;12*
	6	*1;15*	*1;15*	*1;13*	*1;10*	*1; 7*	*1; 4*	*1; 0*	*0;56*	*0;49*	*0;45*	*0;37*	*0;29*	*0;19*	*0;10*	0; 1
	12	*1; 9*	*1; 7*	*1; 4*	*1; 1*	*0;57*	*0;53*	*0;48*	*0;43*	*0;36*	*0;31*	*0;23*	*0;14*	*0; 5*	0; 3	0;14
	18	*1; 3*	*0;59*	*0;55*	*0;51*	*0;46*	*0;39*	*0;36*	*0;31*	*0;23*	*0;18*	*0;10*	*0; 1*	0; 8	0;16	0;27
	24	*0;56*	*0;51*	*0;46*	*0;40*	*0;35*	*0;28*	*0;24*	*0;19*	*0; 9*	*0; 5*	0; 3	0;12	0;20	0;29	0;39
♒	0	*0;49*	*0;42*	*0;37*	*0;30*	*0;25*	*0;18*	*0;12*	*0; 6*	0; 1	0; 8	0;15	0;24	0;32	0;41	0;50
	6	*0;41*	*0;33*	*0;27*	*0;21*	*0;14*	*0; 7*	*0; 0*	0; 7	0;16	0;20	0;27	0;35	0;43	0;51	1; 0
	12	*0;33*	*0;23*	*0;17*	*0;11*	*0; 3*	0; 4	0;12	0;19	0;25	0;32	0;39	0;46	0;53	1; 1	1; 9
	18	*0;25*	*0;14*	*0; 7*	*0; 0*	0; 7	0;15	0;23	0;30	0;35	0;43	0;50	0;56	1; 3	1;10	1;17
	24	*0;16*	*0; 5*	0; 3	0;11	0;19	0;27	0;34	0;41	0;45	0;53	1; 0	1; 6	1;12	1;18	1;24
♓	0	*0; 6*	0; 5	0;13	0;21	0;30	0;38	0;45	0;52	0;56	1; 3	1; 9	1;12	1;20	1;25	1;30
	6	0; 5	0;15	0;24	0;33	0;41	0;48	0;55	1; 2	1; 7	1;12	1;16	1;20	1;26	1;31	1;35
	12	0;15	0;25	0;34	0;43	0;51	0;58	1; 5	1;11	1;16	1;20	1;24	1;28	1;32	1;36	1;39
	18	0;25	0;35	0;44	0;53	1; 1	1; 8	1;14	1;20	1;24	1;28	1;31	1;37	1;37	1;42	1;42
	24	0;35	0;45	0;54	1; 2	1;10	1;17	1;23	1;28	1;31	1;35	1;38	1;40	1;42	1;43	1;44

♎ 18°, ♉ 24°: **IO** *3;39*. ♓ 12°, ♊ 0°: **IO** 1;27.

Continuation of the Simplification of the Latitude of Mercury
(second quarter, top half)

		♋					♌					♍				
		0	6	12	18	24	0	6	12	18	24	0	6	12	18	24
	0	1;46	1;46	1;45	1;44	1;44	1;42	1;41	1;38	1;35	1;31	1;27	1;19	1;12	1; 4	0;55
	6	1;45	1;44	1;42	1;40	1;38	1;36	1;33	1;29	1;25	1;21	1;16	1; 9	1; 1	0;52	0;43
♈	12	1;44	1;41	1;38	1;35	1;32	1;29	1;25	1;20	1;12	1;11	1; 6	0;58	0;49	0;40	0;31
	18	1;42	1;38	1;34	1;30	1;26	1;21	1;16	1;12	1; 6	1; 0	0;54	0;46	0;37	0;28	0;19
	24	1;40	1;35	1;29	1;25	1;19	1;13	1; 7	1; 1	0;55	0;49	0;42	0;36	0;25	0;16	0; 7
	0	1;36	1;29	1;23	1;17	1;10	1; 4	0;57	0;50	0;43	0;37	0;32	0;21	0;12	0; 3	*0; 5*
	6	1;30	1;23	1;16	1; 8	1; 0	0;53	0;45	0;38	0;31	0;24	0;17	0; 8	*0; 1*	*0;10*	*0;18*
♉	12	1;23	1;15	1; 7	0;58	0;50	0;42	0;33	0;25	0;18	0;11	0; 4	0; 0	*0;14*	*0;23*	*0;31*
	18	1;16	1; 7	0;57	0;48	0;39	0;30	0;21	0;13	0; 5	*0; 2*	*0; 9*	*0;18*	*0;26*	*0;35*	*0;43*
	24	1; 8	0;58	0;47	0;37	0;27	0;17	0; 8	0; 1	*0; 6*	*0;15*	*0;23*	*0;31*	*0;38*	*0;46*	*0;54*
	0	0;59	0;48	0;37	0;26	0;17	0; 4	*0; 4*	*0;12*	*0;20*	*0;28*	*0;36*	*0;43*	*0;50*	*0;56*	*1; 3*
	6	0;49	0;37	0;26	0;14	0; 3	*0; 9*	*0;18*	*0;26*	*0;34*	*0;41*	*0;49*	*0;56*	*1; 2*	*1; 8*	*1;13*
♊	12	0;38	0;25	0;13	0; 1	*0;11*	*0;22*	*0;32*	*0;40*	*0;48*	*0;55*	*1; 2*	*1; 8*	*1;14*	*1;19*	*1;23*
	18	0;27	0;13	*0; 0*	*0;13*	*0;25*	*0;36*	*0;46*	*0;54*	*1; 2*	*1; 8*	*1;14*	*1;20*	*1;25*	*1;29*	*1;32*
	24	0;13	0; 0	*0;13*	*0;27*	*0;40*	*0;51*	*1; 0*	*1; 8*	*1;15*	*1;21*	*1;27*	*1;32*	*1;36*	*1;39*	*1;41*
	0	*0; 0*	*0;13*	*0;27*	*0;42*	*0;55*	*1; 6*	*1;15*	*1;22*	*1;28*	*1;34*	*1;39*	*1;43*	*1;46*	*1;48*	*1;49*
	6	*0;14*	*0;28*	*0;42*	*0;58*	*1;10*	*1;21*	*1;30*	*1;37*	*1;42*	*1;47*	*1;50*	*1;53*	*1;55*	*1;56*	*1;54*
♋	12	*0;31*	*0;45*	*0;59*	*1;14*	*1;26*	*1;36*	*1;45*	*1;51*	*1;55*	*1;59*	*2; 1*	*2; 2*	*2; 3*	*2; 2*	*2; 0*
	18	*0;48*	*1; 3*	*1;18*	*1;31*	*1;42*	*1;51*	*1;59*	*2; 8*	*2; 3*	*2;10*	*2;11*	*2;11*	*2;10*	*2; 7*	*2; 3*
	24	*1; 6*	*1;21*	*1;37*	*1;49*	*1;59*	*2; 6*	*2;14*	*2;18*	*2;21*	*2;21*	*2;21*	*2;19*	*2;16*	*2;11*	*2; 5*
	0	*1;26*	*1;40*	*1;54*	*2; 6*	*2;17*	*2;21*	*2;28*	*2;31*	*2;33*	*2;34*	*2;29*	*2;25*	*2;21*	*2;14*	*2; 8*
	6	*1;46*	*1;59*	*2;12*	*2;24*	*2;34*	*2;36*	*2;47*	*2;43*	*2;43*	*2;41*	*2;36*	*2;31*	*2;24*	*2;16*	*2; 6*
♌	12	*2; 6*	*2;18*	*2;31*	*2;44*	*2;50*	*2;54*	*2;54*	*2;53*	*2;51*	*2;48*	*2;41*	*2;33*	*2;25*	*2;15*	*2; 2*
	18	*2;27*	*2;38*	*2;50*	*2;59*	*3; 5*	*3; 6*	*3; 5*	*3; 2*	*2;58*	*2;52*	*2;43*	*2;33*	*2;22*	*2; 9*	*1;54*
	24	*2;47*	*2;58*	*3; 7*	*3;15*	*3;18*	*3;17*	*3;14*	*3; 9*	*3; 3*	*2;54*	*2;43*	*2;30*	*2;15*	*2; 0*	*1;44*
	0	*3; 7*	*3;17*	*3;24*	*3;30*	*3;30*	*3;26*	*3;21*	*3;14*	*3; 5*	*2;53*	*2;41*	*2;24*	*2; 5*	*1;48*	*1;31*
	6	*3;26*	*3;33*	*3;38*	*3;42*	*3;40*	*3;34*	*3;25*	*3;15*	*3; 3*	*2;49*	*2;33*	*2;12*	*1;54*	*1;34*	*1;14*
♍	12	*3;42*	*3;46*	*3;49*	*3;50*	*3;47*	*3;39*	*3;26*	*3;14*	*2;59*	*2;42*	*2;22*	*2; 1*	*1;39*	*1;17*	*0;55*
	18	*3;54*	*3;56*	*3;54*	*3;53*	*3;47*	*3;40*	*3;22*	*3;8*	*2;50*	*2;32*	*2; 8*	*1;45*	*1;21*	*0;57*	*0;33*
	24	*4; 2*	*4; 1*	*3;59*	*3;52*	*3;44*	*3;30*	*3;15*	*2;57*	*2;37*	*2;13*	*1;48*	*1;26*	*0;59*	*0;33*	*0; 8*

♈ 24°, ♋ 6°: **IO** 1;39. ♊18°, ♋ 6°: **IO** 0;18.

Continuation of the Simplification of the Latitude of Mercury
(second quarter, bottom half)

		♋					♌					♍				
		0	6	12	18	24	0	6	12	18	24	0	6	12	18	24
♎	0	*4; 5*	*4; 3*	*4; 0*	*3;49*	*3;35*	*3;18*	*3; 0*	*2;40*	*2;18*	*1;54*	*1;28*	*1; 0*	*0;33*	*0; 7*	0;19
	6	*4; 2*	*3;57*	*3;50*	*3;40*	*3;24*	*3; 3*	*2;41*	*2;18*	*1;54*	*1;30*	*1; 3*	*0;34*	*0; 7*	0;21	0;47
	12	*3;54*	*3;46*	*3;38*	*3;25*	*3; 6*	*2;45*	*2;20*	*1;55*	*1;29*	*1; 4*	*0;36*	*0; 7*	0;21	0;48	1;14
	18	*3;34*	*3;32*	*3;19*	*3; 3*	*2;45*	*2;23*	*1;58*	*1;31*	*1; 3*	*0;37*	*0; 9*	0;19	0;47	1;14	1;39
	24		3;16	2;59	2;42	2;21	*1;59*	*1;33*	*1; 6*	*0;38*	*0;14*	0;16	0;44	1;12	1;38	2;1
♏	0	*3; 4*	*2;54*	*2;38*	*2;19*	*1;57*	*1;33*	*1; 7*	*0;40*	*0;13*	0;13	0;40	1; 8	1;35	2; 0	2;21
	6	*2;47*	*2;34*	*2;16*	*1;55*	*1;34*	*1; 7*	*0;41*	*0;15*	0;11	0;37	1; 3	1;31	1;56	2;19	2;39
	12	*2;27*	*2;13*	*1;54*	*1;31*	*1; 8*	*0;44*	*0;16*	0; 8	0;33	0;59	1;26	1;51	2;14	2;35	2;54
	18	*2; 6*	*1;51*	*1;34*	*1; 9*	*0;45*	*0;21*	0; 6	0;30	0;54	1;19	1;44	2; 8	2;29	2;49	3;6
	24	*1;46*	*1;29*	*1;10*	*0;48*	*0;24*	0; 0	0;25	0;50	1;13	1;36	1;59	2;21	2;42	3; 0	3;15
♐	0	*1;27*	*1; 9*	*0;50*	*0;28*	*0; 3*	0;19	0;43	1; 6	1;29	1;51	2;11	2;31	2;51	3; 8	3;21
	6	*1; 6*	*0;51*	*0;32*	*0;10*	0;14	0;37	0;59	1;21	1;42	2; 3	2;22	2;40	2;58	3;12	3;23
	12	*0;48*	*0;34*	*0;16*	0; 8	0;31	0;53	1;13	1;33	1;53	2;12	2;30	2;47	3; 2	3;14	3;23
	18	*0;28*	*0;17*	0; 0	0;23	0;45	1; 6	1;25	1;44	2; 3	2;20	2;37	2;52	3; 5	3;15	3;22
	24	*0;14*	0; 0	0;17	0;37	0;58	1;18	1;36	1;54	2;11	2;27	2;42	2;56	3; 7	3;15	3;20
♑	0	0; 0	0;15	0;32	0;50	0; 9	1;28	1;45	2; 1	2;17	2;32	2;46	2;58	3; 6	3;14	3;18
	6	0;13	0;28	0;45	1; 2	0;19	1;36	1;53	2; 8	2;22	2;36	2;48	2;58	3; 5	3;11	3;13
	12	0;26	0;40	0;56	1;12	0;28	1;44	1;59	2;13	2;25	2;37	2;47	2;56	3; 2	3; 6	3;7
	18	0;38	0;51	1; 5	1;20	0;36	1;49	2; 4	2;16	2;26	2;37	2;45	2;52	2;57	3; 0	2;59
	24	0;49	1; 1	1;14	1;27	0;41	1;54	2; 7	2;18	2;29	2;36	2;42	2;48	2;52	2;53	2;51
♒	0	0;59	1;10	1;22	1;34	0;46	1;58	2; 9	2;19	2;28	2;34	2;39	2;44	2;46	2;46	2;42
	6	1; 8	1;18	1;29	1;40	0;51	2; 1	2;10	2;19	2;26	2;32	2;36	2;39	2;40	2;33	2;35
	12	1;16	1;25	1;34	1;45	0;55	2; 3	2;11	2;18	2;25	2;29	2;32	2;33	2;33	2;29	2;26
	18	1;23	1;32	1;39	1;49	0;57	2; 4	2;11	2;17	2;22	2;25	2;27	2;27	2;25	2;20	2;16
	24	1;30	1;37	1;44	1;51	0;58	2; 4	2;10	2;15	2;18	2;20	2;21	2;19	2;16	2;11	2;5
♓	0	1;36	1;41	1;47	1;52	0;58	2; 3	2; 7	2;11	2;13	2;14	2;12	2;10	2; 7	2; 2	1;54
	6	1;40	1;44	1;49	1;53		2; 0	2; 3	2; 5	2; 6	2; 4	2; 5	2; 2	1;57	1;51	1;43
	12	1;42	1;46	1;49	1;52		1;57	1;58	1;59	1;59	1;58	1;56	1;51	1;46	1;39	1;31
	18	1;44	1;46	1;48	1;50		1;53	1;53	1;53	1;52	1;49	1;46	1;40	1;35	1;28	1;19
	24	1;45	1;46	1;47	1;48		1;48	1;47	1;46	1;44	1;40	1;37	1;30	1;24	1;16	1;7

♎ 24°, ♋ 0°: all MS's *3;26*.

Continuation of the Simplification of the Latitude of Mercury
(third quarter, top half)

		♎					♏					♐				
		0	6	12	18	24	0	6	12	18	24	0	6	12	18	24
	0	0;45	0;34	0;24	0; 9	*0; 5*	*0;20*	*0;30*	*0;46*	*0;58*	*1; 9*	*1;19*	*1;28*	*1;37*	*1;48*	*1;45*
	6	0;33	0;22	0;11	*0; 2*	*0;15*	*0;30*	*0;44*	*0;56*	*1; 7*	*1;16*	*1;25*	*1;33*	*1;40*	*1;49*	*1;46*
♈	12	0;21	0;10	*0; 0*	*0;13*	*0;23*	*0;40*	*0;53*	*1; 5*	*1;15*	*1;23*	*1;31*	*1;38*	*1;43*	*1;46*	*1;46*
	18	0; 9	*0; 1*	*0;12*	*0;24*	*0;37*	*0;49*	*1; 1*	*1;11*	*1;20*	*1;29*	*1;36*	*1;42*	*1;44*	*1;46*	*1;45*
	24	*0; 3*	*0;13*	*0;24*	*0;35*	*0;47*	*0;58*	*1; 8*	*1;17*	*1;25*	*1;33*	*1;39*	*1;43*	*1;45*	*1;44*	*1;44*
	0	*0;15*	*0;25*	*0;35*	*0;45*	*0;56*	*1; 6*	*1;15*	*1;23*	*1;30*	*1;36*	*1;41*	*1;44*	*1;44*	*1;41*	*1;41*
	6	*0;26*	*0;37*	*0;46*	*0;55*	*1; 4*	*1;13*	*1;21*	*1;28*	*1;34*	*1;38*	*1;42*	*1;43*	*1;42*	*1;41*	*1;37*
♉	12	*0;39*	*0;48*	*0;56*	*1; 4*	*1;12*	*1;19*	*1;26*	*1;30*	*1;37*	*1;40*	*1;42*	*1;41*	*1;40*	*1;37*	*1;31*
	18	*0;50*	*0;58*	*1; 5*	*1;12*	*1;19*	*1;25*	*1;30*	*1;35*	*1;39*	*1;41*	*1;41*	*1;39*	*1;36*	*1;32*	*1;24*
	24	*1; 0*	*1; 7*	*1;13*	*1;19*	*1;25*	*1;30*	*1;34*	*1;37*	*1;40*	*1;40*	*1;39*	*1;36*	*1;32*	*1;26*	*1;16*
	0	*1; 9*	*1;15*	*1;21*	*1;26*	*1;30*	*1;34*	*1;37*	*1;39*	*1;40*	*1;39*	*1;36*	*1;32*	*1;26*	*1;18*	*1; 9*
	6	*1;18*	*1;23*	*1;28*	*1;31*	*1;34*	*1;36*	*1;39*	*1;40*	*1;39*	*1;37*	*1;34*	*1;26*	*1;19*	*1;10*	*1; 0*
♊	12	*1;27*	*1;30*	*1;34*	*1;36*	*1;37*	*1;38*	*1;38*	*1;38*	*1;36*	*1;34*	*1;27*	*1;21*	*1;11*	*1; 1*	*0;50*
	18	*1;35*	*1;37*	*1;39*	*1;40*	*1;40*	*1;39*	*1;36*	*1;35*	*1;32*	*1;28*	*1;21*	*1;13*	*1; 3*	*0;51*	*0;39*
	24	*1;43*	*1;43*	*1;44*	*1;44*	*1;41*	*1;40*	*1;34*	*1;32*	*1;28*	*1;21*	*1;14*	*1; 2*	*0;53*	*0;40*	*0;24*
	0	*1;49*	*1;49*	*1;48*	*1;47*	*1;43*	*1;39*	*1;30*	*1;28*	*1;21*	*1;12*	*1; 5*	*0;54*	*0;42*	*0;28*	*0;14*
	6	*1;54*	*1;57*	*1;50*	*1;47*	*1;42*	*1;36*	*1;27*	*1;22*	*1;13*	*1; 5*	*0;54*	*0;42*	*0;29*	*0;15*	0; 1
♋	12	*1;57*	*1;55*	*1;50*	*1;45*	*1;38*	*1;31*	*1;23*	*1;14*	*1; 4*	*0;54*	*0;42*	*0;29*	*0;15*	0; 0	0;16
	18	*1;59*	*1;54*	*1;48*	*1;41*	*1;33*	*1;24*	*1;14*	*1; 4*	*0;53*	*0;42*	*0;29*	*0;15*	0; 0	0;17	0;33
	24	*2; 0*	*1;52*	*1;45*	*1;36*	*1;26*	*1;15*	*1; 3*	*0;52*	*0;40*	*0;28*	*0;14*	0; 1	0;17	0;34	0;50
	0	*1;59*	*1;49*	*1;39*	*1;29*	*1;13*	*1; 5*	*0;51*	*0;38*	*0;25*	*0;12*	0; 3	0;19	0;36	0;53	1; 9
	6	*1;56*	*1;45*	*1;33*	*1;21*	*1; 8*	*0;53*	*0;38*	*0;23*	*0; 8*	0; 6	0;23	0;39	0;56	1;13	1;33
♌	12	*1;49*	*1;38*	*1;24*	*1; 9*	*0;54*	*0;38*	*0;22*	*0; 5*	0;11	0;26	0;44	1; 1	1;18	1;36	1;51
	18	*1;39*	*1;26*	*1;11*	*0;54*	*0;37*	*0;19*	*0; 2*	0;17	0;33	0;50	1; 7	1;23	1;40	1;56	2;12
	24	*1;27*	*1;10*	*0;53*	*0;35*	*0;17*	0; 4	0;21	0;39	0;56	1;13	1;30	1;46	2; 3	2;19	2;34
	0	*1;12*	*0;52*	*0;33*	*0;12*	0; 5	0;23	0;44	1; 3	1;21	1;38	1;55	2;11	2;27	2;42	2;55
	6	*0;54*	*0;31*	*0;10*	0;11	0;28	0;46	1;10	1;28	1;46	2; 3	2;22	2;36	2;50	3; 4	3;15
♍	12	*0;32*	*0; 9*	0;20	0;34	0;53	1;10	1;37	1;54	2;11	2;28	2;43	2;59	3;12	3;23	3;33
	18	*0; 7*	0;16	0;38	0;59	1;21	1;38	2; 3	2;20	2;36	2;52	3; 4	3;19	3;31	3;41	3;49
	24	0;19	0;43	1; 5	1;26	1;48	2; 5	2;28	2;45	3; 0	3;13	3;24	3;37	3;46	3;54	3;59

♈ 0°, ♎ 6°: **C** 0;35. ♈ 0°, ♎ 24°: **IO** illegible. ♈ 6°, ♐ 18°: **IO** *1;44*. ♊ 6°, ♐ 6°: **C** 1;22. ♌ 6°, ♐ 12°: **IO** 1;16. ♌ 18°, ♏ 24°: **IO** 0;55.

Continuation of the Simplification of the Latitude of Mercury
(third quarter, bottom half)

		♎					♏					♐				
		0	6	12	18	24	0	6	12	18	24	0	6	12	18	24
	0	0;45	1; 9	1;31	1;53	2;14	2;34	2;52	3; 8	3;21	3;32	3;42	3;50	3;57	4; 1	4; 3
	6	1;11	1;34	1;56	2;18	2;38	2;56	3;12	3;25	3;36	3;46	3;54	4; 0	4; 3	4; 4	4; 3
♎	12	1;37	1;59	2;20	2;39	2;57	3;13	3;28	3;38	3;46	3;55	3;59	4; 2	4; 2	4; 1	3;59
	18	2; 2	2;22	2;42	3; 0	3;15	3;24	3;38	3;47	3;54	3;58	4; 5	4; 1	3;58	3;53	3;48
	24	2;24	2;45	3; 0	3;15	3;28	3;34	3;46	3;54	3;57	3;58	3;56	3;53	3;49	3;42	3;33
	0	2;42	3; 1	3;16	3;27	3;38	3;41	3;55	3;55	3;54	3;54	3;49	3;43	3;37	3;29	3;18
	6	2;57	3;13	3;26	3;36	3;44	3;49	3;52	3;53	3;51	3;46	3;38	3;31	3;22	3;11	2;59
♏	12	3; 9	3;22	3;33	3;42	3;48	3;52	3;50	3;48	3;44	3;36	3;28	3;18	3; 6	2;53	2;39
	18	3;19	3;30	3;38	3;45	3;48	3;49	3;45	3;41	3;35	3;25	3;15	3; 3	2;49	2;35	2;19
	24	3;27	3;35	3;41	3;45	3;46	3;44	3;39	3;31	3;23	3;13	3; 1	2;46	2;32	2;16	2; 0
	0	3;30	3;34	3;42	3;40	3;42	3;37	3;30	3;21	3;10	2;59	2;46	2;31	2;14	1;57	1;41
	6	3;31	3;33	3;38	3;37	3;35	3;29	3;20	3; 9	2;57	2;45	2;30	2;14	1;56	1;36	1;22
♐	12	3;29	3;34	3;33	3;31	3;27	3;19	3; 8	2;56	2;43	2;29	2;14	1;56	1;39	1;21	1; 4
	18	3;27	3;28	3;27	3;24	3;19	3; 8	2;56	2;43	2;29	2;14	1;58	1;40	1;22	1; 4	0;48
	24	3;24	3;28	3;21	3;16	3; 9	2;57	2;44	2;30	2;15	1;59	1;43	1;24	1; 6	0;48	0;31
	0	3;19	3;28	3;12	3; 4	2;58	2;46	2;32	2;17	2; 1	1;45	1;28	1; 9	0;50	0;32	0;15
	6	3;13	3;19	3; 6	2;56	2;47	2;34	2;19	2; 4	1;47	1;31	1;14	0;54	0;35	0;17	0; 0
♑	12	3; 5	3; 3	2;56	2;47	2;35	2;21	2; 6	1;51	1;34	1;16	0;59	0;40	0;21	0; 3	*0;13*
	18	2;57	2;53	2;45	2;34	2;22	2; 8	1;53	1;37	1;20	1; 2	0;44	0;26	0; 8	*0; 9*	*0;25*
	24	2;48	2;43	2;37	2;22	2; 9	1;54	1;40	1;22	1; 7	0;48	0;32	0;13	*0; 5*	*0;21*	*0;36*
	0	2;39	2;33	2;23	2;10	1;56	1;42	1;27	1;10	0;53	0;35	0;16	0; 0	*0;14*	*0;33*	*0;46*
	6	2;30	2;22	2;11	1;58	1;43	1;28	1;13	0;57	0;40	0;22	0; 3	*0;13*	*0;29*	*0;44*	*0;57*
♒	12	2;20	2;11	1;59	1;46	1;31	1;15	1; 0	0;44	0;27	0; 9	*0;9*	*0;26*	*0;40*	*0;55*	*1; 6*
	18	2; 9	2; 0	1;47	1;34	1;19	1; 3	0;47	0;31	0;14	*0; 3*	*0;20*	*0;36*	*0;50*	*1; 4*	*1;15*
	24	1;58	1;48	1;35	1;22	1; 7	0;51	0;35	0;18	0; 1	*0;15*	*0;31*	*0;46*	*1; 0*	*1;12*	*1;23*
	0	1;46	1;36	1;23	1;10	0;54	0;39	0;23	0; 6	*0;10*	*0;26*	*0;41*	*0;56*	*1; 9*	*1;19*	*1;29*
	6	1;34	1;28	1;11	0;57	0;42	0;27	0;11	*0; 6*	*0;21*	*0;36*	*0;50*	*1; 3*	*1;16*	*1;25*	*1;37*
♓	12	1;21	1;10	0;58	0;44	0;29	0;14	*0; 1*	0;15	0;31	0;45	0;58		*1;22*	*1;31*	*1;38*
	18	1; 9	0;58	0;45	0;32	0;17	0; 2	*0;12*	*0;25*	*0;40*	*0;53*	*1; 6*	*1;16*	*1;28*	*1;36*	*1;41*
	24	0;57	0;46	0;33	0;20	0; 6	*0; 9*	*0;23*	*0;36*	*0;49*	*1; 1*	*1;13*	*1;23*	*1;33*	*1;40*	1;43

♏ 0°, ♐ 24°: **IO** 3;28. ♏ 24°, ♐ 12°: **IO** 2;34 or 2;37. ♐ 18°, ♐ 12°: **IO** 1;27. ♑ 0°, ♎ 18°: **IO** 3;15. ♒ 0°, ♐ 6°: **IO** *0;5*. ♓ 6°, ♐ 6°: all MS's *1;13*. ♓ 12°, ♐ 6°: **IO** *1;48*, **C** *1;18*.

Continuation of the Simplification of the Latitude of Mercury
(fourth quarter, top half)

		♑					♒					♓				
		0	6	12	18	24	0	6	12	18	24	0	6	12	18	24
♈	0	*1;46*	*1;45*	*1;43*	*1;37*	*1;28*	*1;19*	*1; 8*	*0;57*	*0;46*	*0;33*	*0;19*	*0; 5*	0; 9	0;20	0;34
	6	*1;45*	*1;43*	*1;40*	*1;33*	*1;25*	*1;14*	*1; 2*	*0;50*	*0;38*	*0;26*	*0;12*	0; 3	0;18	0;32	0;44
	12	*1;44*	*1;41*	*1;36*	*1;29*	*1;21*	*1; 9*	*0;55*	*0;43*	*0;30*	*0;18*	*0; 4*	0;12	0;27	0;42	0;54
	18	*1;42*	*1;38*	*1;33*	*1;24*	*1;15*	*1; 3*	*0;48*	*0;35*	*0;22*	*0; 9*	0; 5	0;22	0;36	0;52	1; 4
	24	*1;40*	*1;34*	*1;28*	*1;19*	*1; 8*	*0;55*	*0;40*	*0;26*	*0;13*	0; 1	0;16	0;32	0;48	1; 2	1;14
♉	0	*1;36*	*1;29*	*1;22*	*1;12*	*1; 0*	*0;46*	*0;31*	*0;16*	*0; 3*	0;11	0;27	0;43	0;59	1;13	1;24
	6	*1;30*	*1;23*	*1;14*	*1; 4*	*0;51*	*0;36*	*0;20*	*0; 6*	0; 8	0;22	0;38	0;54	1;10	1;24	1;34
	12	*1;23*	*1;15*	*1; 5*	*0;54*	*0;41*	*0;25*	*0; 9*	0; 5	0;19	0;34	0;50	1; 6	1;21	1;34	1;44
	18	*1;16*	*1; 6*	*0;55*	*0;43*	*0;30*	*0;14*	0; 2	0;16	0;31	0;46	1; 2	1;17	1;32	1;44	1;54
	24	*1;8*	*0;57*	*0;45*	*0;32*	*0;18*	*0; 2*	0;13	0;28	0;49	0;58	1;14	1;28	1;42	1;54	2; 3
♊	0	*0;59*	*0;47*	*0;34*	*0;20*	*0; 5*	0;10	0;25	0;40	0;55	1;10	1;25	1;39	1;52	2; 3	2;14
	6	*0;49*	*0;36*	*0;23*	*0; 9*	0; 6	0;22	0;37	0;52	1; 4	1;22	1;36	1;50	2; 2	2;11	2;21
	12	*0;38*	*0;25*	*0;12*	0; 2	0;18	0;34	0;50	1; 5	1;21	1;34	1;47	2; 0	2;11	2;21	2;30
	18	*0;27*	*0;13*	0; 0	0;14	0;30	0;43	1; 3	1;19	1;33	1;46	1;58	2;10	2;21	2;30	2;37
	24	*0;13*	0; 1	0;13	0;27	0;48	1; 1	1;18	1;34	1;46	1;58	2;10	2;21	2;31	2;38	2;43
♋	0	0; 0	0;13	0;27	0;42	0;58	1;15	1;32	1;48	1;59	2;10	2;21	2;31	2;40	2;46	2;49
	6	0;14	0;28	0;42	0;57	1;13	1;30	1;46	2; 1	2;12	2;22	2;31	2;40	2;48	2;53	2;54
	12	0;31	0;44	0;58	1;13	1;29	1;45	2; 0	2;14	2;24	2;33	2;41	2;49	2;55	2;59	2;59
	18	0;43	1; 2	1;16	1;30	1;42	2; 1	2;15	2;27	2;36	2;49	2;51	2;57	3; 2	3; 3	3; 3
	24	1; 4	1;21	1;35	1;48	2; 3	2;18	2;30	2;41	2;49	2;55	3; 0	3; 4	3; 8	3; 8	3; 5
♌	0	1;26	1;38	1;54	2; 2	2;21	2;34	2;45	2;54	3; 1	3; 6	3;9	3;11	3;13	3;11	3; 7
	6	1;46	1;59	2;13	2;27	2;38	2;49	2;59	3;7	3;13	3;16	3;18	3;18	3;17	3;13	3; 6
	12	2; 6	2;19	2;31	2;43	2;54	3; 3	3;12	3;19	3;23	3;24	3;24	3;22	3;19	3;12	3; 3
	18	2;27	2;39	2;49	2;59	3; 9	3;17	3;24	3;25	3;30	3;30	3;27	3;23	3;16	3; 8	2;57
	24	2;46	2;58	3; 7	3;12	3;24	3;29	3;34	3;36	3;35	3;33	3;29	3;22	3;13	3; 1	2;48
♍	0	3; 7	3;18	3;25	3;31	3;34	3;38	3;42	3;41	3;38	3;34	3;28	3;18	3; 6	2;58	2;37
	6	3;26	3;36	3;41	3;44	3;45	3;46	3;48	3;44	3;40	3;32	3;24	3;12	2;58	2;43	2;25
	12	3;42	3;49	3;53	3;54	3;52	3;53	3;51	3;46	3;39	3;29	3;17	3; 3	2;47	2;30	2;10
	18	3;54	3;57	3;59	3;58	3;57	3;54	3;50	3;42	3;34	3;20	3;6	2;50	2;33	2;14	1;52
	24	4; 2	4; 3	4; 4	4; 1	3;55	3;50	3;43	3;33	3;23	3; 8	2;54	2;33	2;14	1;54	1;32

♊ 18°, ♒ 6°: all MS's 1;13. ♋ 6°, ♒ 18°: **IO** 2;42. ♋ 12°, ♓ 0°: **IO** 2;49.

Continuation of the Simplification of the Latitude of Mercury
(fourth quarter, bottom half)

		♑					♒					♓				
		0	6	12	18	24	0	6	12	18	24	0	6	12	18	24
	0	4; 5	4; 3	4; 1	3;57	3;50	3;42	3;34	3;21	3;10	2;52	2;37	2;14	1;53	1;31	1; 9
	6	4; 2	3;59	3;55	3;49	3;40	3;29	3;17	3; 4	2;50	2;34	2;15	1;54	1;30	1; 7	0;45
♎	12	3;54	3;46	3;43	3;35	3;24	3;12	2;58	2;44	2;29	2;13	1;53	1;31	1; 8	0;45	0;25
	18	3;42	3;32	3;29	3;16	3; 4	2;50	2;36	2;21	2; 5	1;48	1;29	1; 9	0;47	0;25	0; 4
	24	3;26	3;17	3; 7	2;55	2;42	2;26	2;13	1;58	1;41	1;24	1; 6	0;46	0;26	0; 5	*0;13*
	0	3; 4	2;56	2;45	2;33	2;19	2; 4	1;49	1;34	1;18	1; 2	0;44	0;26	0; 6	*0;13*	*0;31*
	6	2;44	2;35	2;23	2;10	1;56	1;41	1;25	1;11	0;56	0;40	0;23	0; 6	*0;14*	*0;29*	*0;46*
♏	12	2;27	2;13	1;59	1;47	1;32	1;18	1; 3	0;49	0;35	0;20	0; 5	*0;11*	*0;28*	*0;44*	*1; 0*
	18	2; 6	1;51	1;37	1;24	1;10	0;56	0;42	0;28	0;15	0; 1	*0;18*	*0;27*	*0;42*	*0;56*	*1; 7*
	24	1;46	1;31	1;17	1; 3	0;49	0;35	0;22	0;10	*0; 2*	*0;15*	*0;28*	*0;41*	*0;53*	*1; 3*	*1;15*
	0	1;25	1;12	0;58	0;44	0;30	0;15	0; 2	*0; 7*	*0;17*	*0;29*	*0;40*	*0;52*	*1; 2*	*1;11*	*1;20*
	6	1; 6	0;52	0;39	0;26	0;12	0; 2	*0;12*	*0;22*	*0;31*	*0;41*	*0;50*	*0;59*	*1; 8*	*1;16*	*1;23*
♐	12	0;48	0;34	0;22	0;10	*0; 3*	*0;15*	*0;26*	*0;35*	*0;43*	*0;50*	*0;58*	*1; 5*	*1;12*	*1;18*	*1;25*
	18	0;31	0;18	0; 5	*0; 6*	*0;17*	*0;28*	*0;38*	*0;46*	*0;53*	*0;59*	*1; 5*	*1;10*	*1;15*	*1;20*	*1;27*
	24	0;15	0; 2	*0;10*	*0;21*	*0;31*	*0;41*	*0;49*	*0;06*	*1; 1*	*1; 6*	*1;11*	*1;14*	*1;17*	*1;20*	*1;27*
	0	*0; 0*	*0;12*	*0;23*	*0;34*	*0;44*	*0;52*	*0;59*	*1; 4*	*1; 8*	*1;12*	*1;15*	*1;17*	*1;19*	*1;20*	*1;26*
	6	*0;13*	*0;25*	*0;35*	*0;45*	*0;54*	*1; 2*	*1; 7*	*1;11*	*1;14*	*1;19*	*1;18*	*1;19*	*1;19*	*1;18*	*1;24*
♑	12	*0;25*	*0;38*	*0;46*	*0;55*	*1; 3*	*1; 9*	*1;14*	*1;17*	*1;19*	*1;21*	*1;20*	*1;19*	*1;17*	*1;15*	*1;22*
	18	*0;36*	*0;49*	*0;57*	*1; 3*	*1;10*	*1;15*	*1;19*	*1;21*	*1;22*	*1;22*	*1;19*	*1;17*	*1;14*	*1;11*	*1; 7*
	24	*0;46*	*0;59*	*1; 7*	*1;12*	*1;17*	*1;21*	*1;23*	*1;24*	*1;23*	*1;22*	*1;17*	*1;15*	*1;11*	*1; 7*	*1; 1*
	0	*0;57*	*1; 8*	*1;15*	*1;20*	*1;23*	*1;31*	*1;27*	*1;26*	*1;24*	*1;21*	*1;16*	*1;12*	*1; 7*	*1; 1*	*0;55*
	6	*1; 6*	*1;17*	*1;23*	*1;26*	*1;28*	*1;33*	*1;29*	*1;27*	*1;23*	*1;19*	*1;15*	*1; 8*	*1; 1*	*0;54*	*0;46*
♒	12	*1;15*	*1;24*	*1;29*	*1;31*	*1;32*	*1;34*	*1;30*	*1;28*	*1;23*	*1;17*	*1;11*	*1; 3*	*0;55*	*0;47*	*0;40*
	18	*1;23*	*1;30*	*1;32*	*1;36*	*1;35*	*1;35*	*1;31*	*1;27*	*1;21*	*1;14*	*1; 7*	*0;58*	*0;49*	*0;40*	*0;33*
	24	*1;29*	*1;35*	*1;38*	*1;39*	*1;37*	*1;34*	*1;31*	*1;26*	*1;19*	*1;11*	*1; 2*	*0;53*	*0;43*	*0;33*	*0;25*
	0	*1;34*	*1;39*	*1;41*	*1;41*	*1;38*	*1;32*	*1;30*	*1;24*	*1;16*	*1; 6*	*0;56*	*0;46*	*0;35*	*0;25*	*0;16*
	6	*1;38*	*1;42*	*1;43*	*1;42*	*1;39*	*1;29*	*1;28*	*1;20*	*1;11*	*1; 0*	*0;49*	*0;38*	*0;27*	*0;16*	*0; 5*
♓	12	*1;41*	*1;44*	*1;44*	*1;42*	*1;38*	*1;27*	*1;24*	*1;15*	*1; 5*	*0;54*	*0;42*	*0;30*	*0;18*	*0;6*	0; 5
	18	*1;43*	*1;45*	*1;44*	*1;41*	*1;36*	*1;25*	*1;20*	*1;10*	*0;59*	*0;47*	*0;34*	*0;22*	*0;9*	0;4	0;15
	24	*1;45*	*1;46*	*1;44*	*1;39*	*1;32*	*1;24*	*1;14*	*1; 4*	*0;53*	*0;40*	*0;26*	*0;14*	*0;0*	0;13	0;24

♏ 24°, ♓ 6°: **IO** *0;10.* ♐ 18°, ♒ 0°: *0;25.* ♑ 0°, ♑ 6°: *0;17.* ♑ 6°, ♓ 24°: **IO** *1;16.* ♒ 6°, ♒ 6°: **IO** illegible.

Equation Tables in the *Dṛggaṇita* of Parameśvara

Sho HIROSE

1. Introduction

Parameśvara (*c.* 1360–1460) was an astronomer in Kerala who made significant contributions to astronomy in second millennium India. He was the author of many original works and commentaries on a range of astronomical topics. In 1431–32 CE he composed the *Dṛggaṇita* (literally 'Observation and Computation'; hereafter *DG*) which was intended to improve the parameters underlying the planetary computations of his predecessors. A significant part of the work is made up of planetary equation tables which are presented in versified form. I will examine these tables in detail, comparing them with the rules and parameters for finding such equations presented earlier in Parameśvara's text and use these reconstructions, along with variant readings from the manuscripts, to discuss how this numerical data may have been generated and how this can help us with the critical editing process.

2. Background

2.1. Versified tables

In Sanskrit treatises on mathematics or astronomy, single entry tables, such as Sine[1] tables, can be expressed as verses where numbers are replaced by words or sets of syllables. One system that proved to be popular was the *kaṭapayādi* system which emerged in the south Indian region of Kerala. This system uses the consonant next to the vowel to be taken as a digit (Table 1), thus enabling the composer to describe a number with varieties of meaningful words.[2] Because each digit is associated with several different consonants, the same value could be expressed by different words in this system. The first text that is known to use the *kaṭapayādi* is the *Grahacāranibandhana* (*c.* 690 CE) by Haridatta.[3] This treatise gives various constants for astronomical computation in the *kaṭapayādi* form. It includes a set of versified tables of planetary

[1] Capitalized to distinguish it from the modern notion of the sine where $R = 1$. $\text{Sin}\,\vartheta = R\sin\vartheta$.

[2] See Sarma, 'From My Grandfather's Chest', for details on the *kaṭapayādi* system.

[3] Critical edition in K. V. Sarma, *Grahacāranibandhana*.

Editing and Analysing Numerical Tables: Towards a Digital Information System for the History of Astral Sciences, ed. by Matthieu Husson, Clemency Montelle and Benno van Dalen, PALS 2 (Turnhout, 2021), pp. 331–362
© BREPOLS PUBLISHERS 10.1484/M.PALS-EB.5.127701

Table 1: Correspondence between consonants and numbers in the *kaṭapayādi* system

k = 1	*kh* = 2	*g* = 3	*gh* = 4	*ṅ* = 5
c = 6	*ch* = 7	*j* = 8	*jh* = 9	*ñ* = 0
ṭ = 1	*ṭh* = 2	*ḍ* = 3	*ḍh* = 4	*ṇ* = 5
t = 6	*th* = 7	*d* = 8	*dh* = 9	*n* = 0
p = 1	*ph* = 2	*b* = 3	*bh* = 4	*m* = 5
y = 1	*r* = 2	*l* = 3	*v* = 4	*ś* = 5
ṣ = 6	*s* = 7	*h* = 8	*ḷ* = 9	
vowel without consonant = 0				

equations corresponding to mean anomalies at intervals of 3°45′ (*Grahacāra-nibandhana* 2.1–15 and 3.35–36). Haridatta calls (in 1.34 and 3.34) these verses *vākyas* (literally 'sentence'). Versified tables employing *kaṭapayādi* became very common in Kerala and are frequently referred to as *vākyas*, especially when the text is made up exclusively of versified tables.

An important feature of versified tables are that linguistic or metric structures, instead of spacial alignment, indicate the format of the data. The following is a verse from the *Dṛggaṇita* which we shall look at in the next section. The Sanskrit words in bold are *kaṭapayādi*. I have first translated them literally (whether they make sense for the modern reader or not) and then converted them into numbers.

> ***jāto balāya na naraś***[4] *cacāra* ***nīlāṅgi ruddhaguḥ śramavit*** |
> ***suniśā sumarma punitād dhūliṣu dhūrteṣu dhījitau vinasaḥ*** |
> ***sūnārthī*** *kṣitisūnoḥ ṣaḍaṃśajā māndajā imā jīvāḥ* ||2.26||

> A man is born not for power (**eight-six, three-three-one, zero-zero-two**). A suppressed cow that knows exertion having black limbs has moved (**six-six-two, zero-three-three, two-nine-three, two-five-four**).
> A good night, having good organs, because it has been purified (**seven-zero-five, seven-five-five, one-zero-six**). The noseless is in the dust, in the rust and victorious with knowledge (**nine-three-six, nine-six-six, nine-eight-six, four-zero-seven**). One who desires for a son (**seven-zero-seven**). These are the Sines produced from the 'slow' [anomalies] of Mars, arising for six degree [intervals].

The reader can separate the fifteen entries listed in this verse from one another since an independent word corresponds to a number. The consonants *j* and *t* in the first word *jāto*, literally 'born', indicate eight and six respectively. They are placed from the lower position to the higher, and thus we find sixty-eight. Likewise, *balāya* 'for power' denotes one hundred and thirty-three.

[4] The edition by K. V. Sarma, which uses Devanagari, puts *nanaraś* unseparated to read zero-zero-two or two hundred as a *kaṭapayādi*. But grammatically *na* (negation) and *nara* (man) are separate words.

The verse includes non-number descriptions, but there are cases (such as the *Grahacāranibandhana*) where a whole verse is made of *kaṭapayādi* words. The texts themselves do not always indicate whether a phrase or an entire verse is in *kaṭapayādi*, but it is usually easy for the reader to recognize them, as words in *kaṭapayādi* generally create phrases that are distinctively different from the astronomical context.

Sometimes the words may be carefully chosen so that they form a sentence as a whole. However, it is unclear whether the entries in versified tables were always supposed to be meaningfully connected.[5] Likewise, in the *kaṭapayādi* verses of the *Dṛggaṇita*, I could only sporadically find meaningful passages. Otherwise the stanzas could be somehow construed syntactically but defied semantic analysis, as can be seen in the previous 'translation'. Hereafter I shall focus on the numbers represented by *kaṭapayādi*, and leave the task of analyzing the words semantically for the future.

2.1.1. *The possibility of variants without changing the value*

While our main interest is the numerical values themselves, we must keep in mind that in the *kaṭapayādi* system the same digit could be indicated with different letters (e.g., *g*, *ḍ*, *b* and *l* all mean 3). Furthermore, whenever multiple consonants are conjuncted before a vowel, only the last one followed by the vowel is interpreted as a digit and others are ignored. Thus the same number can be represented with different words. To give an example that occurs in the *Dṛggaṇita*,[6] one thousand nine hundred and fifty-eight is expressed by *durmadhupo* (bad honey-drinker/bee) in one manuscript while others read *himābdhaye* (for the cold lake/ocean).

This type of variant cannot be produced by a simple scribal error; it suggests that numbers were transmitted on their own or computed independently before being compiled into *kaṭapayādi*. And indeed we will see that this seems to be the case in the *Dṛggaṇita*.

2.1.2. *Other ways of representing numbers*

While the versified equation tables that we focus on is entirely in *kaṭapayādi*, other methods existed in Sanskrit traditions, and I would like to refer to two of them which are relevant in our scope.

[5] Plofker, *Mathematics in India*, p. 246, analyzes a versified table attributed to Mādhava and comments: 'The meaning of these phrases as ordinary Sanskrit verse is secondary, and not totally coherent. ... A truly learned commentator could doubtless provide a meaningful literary interpretation for the entire verse collection, but a reader concerned primarily with its mathematical meaning would probably not care.'

[6] This is verse 2.28 that has been used in this study and that appears in p. 19 of K. V. Sarma's edition.

The *bhūtasaṃkhyā* system, or word-numeral system, was popular in both north and south India. In this system, various nouns are used to stand for a number with which it can be associated visually, through myths or in any other way. The number can be a single digit (e.g., 'eye', 'twin', etc., for 2) or two digits (e.g., 'sun', 'zodiacal sign', etc., for 12), and for large numbers the words are listed in compounds, starting from the lowest place (e.g., 'sky-sky-sun' for 'zero-zero-twelve', i.e., 1200). Versified tables can also be composed with this system, although this requires more syllables than *kaṭapayādi*. The readers must know the correspondence between words and numbers, and there is almost no syntax between individual words that would help them. *Bhūtasaṃkhyā* is employed in the first half of the *Dṛgganita*. Thus the parameters for generating the equations, explained in Section 3, are in *bhūtasaṃkhyā*.

One may be curious to know whether the versified tables also circulated in formatted tables using Indo-Arabic numerals. Such tradition of table texts known as *koṣṭhaka*s did exist in north and west India, due to influence by the Islamic tradition, since at least the 12th century.[7] However, table texts found in south India are rare and belong to much later periods.[8] We cannot find any tabulated data in the manuscripts of the *Dṛgganita*. Thus, based on what we have, I assume that the equation number sets were transmitted basically in *kaṭapayādi*.

2.2. The *Dṛgganita*

The *Dṛgganita* (literally 'Observation and computation'; hereafter *DG*) was intended to improve the parameters for planetary computations previously introduced by Haridatta in the *Grahacāranibandhana*. Table 2 lists the contents of the *DG*. The entire treatise is divided into two parts, and the second part is predominantly a restatement of the first part. Parameśvara mentions in the first verse of the second part that he will give a 'clearer version of the *Dṛgganita* in *kaṭapayādi* for the benefit of studies during childhood'.[9] By contrast, numbers in the first part are in *bhūtasaṃkhyā*. It seems that *kaṭapayādi* was recognized as an easier system.

Apart from the number systems, the most significant difference in the two parts is the treatment of planetary equations. The third chapter of the first part focuses on the procedure for correcting the mean planet in order to find the true planet, most of which are rules and parameters for computing

[7] Pingree, *Jyotiḥśāstra*, pp. 41–46.

[8] The *Navagrahapadakāni* compiled around 1798–1833 in Tanjore, Tamil Nadu is one example; see Pingree, 'The Fragments', p. 38.

[9] *spaṣṭīkartuṃ dṛgganitaṃ vakṣye kaṭapayādibhiḥ ... bālābhyāsahitaṃ* (*DG* 2.1).

Part	Chapter	Verses	Topic
1	1	1	Invocation
		2–6	Claims on comparison with previous treatises
		7–9	Finding the time elapsed since the epoch
		10–24	Computing mean planets
	2	1–6	*Dhruvas* (initial position of planets at the epoch)
		7–8	Corrections for observer's location
	3	1–2	'Slow' apogees
		3–4ab	'Fast' apogees
		∗ 4cd–5ab	The 'base', 'upright' and great Sines
		∗ 5cd–9ab	The divisors for the 'Slow' equation
		∗ 9cd–12ab	The divisors for the 'Fast' equation
		∗ 12cd	The 'base' and 'upright' equations
		∗ 13–15ab	Computing the arc of the equations
		∗ 15cd–18ab	Methods for finding the radial distance
		18cd–22	Computing true planets
	4	∗ 1–8	Parameters for computing the divisors
		∗ 9–10	Computing the arc corresponding to a Sine
		11	Reference to further corrections
2		∗ 1	Introduction
		2–12ab	Computing mean planets
		12cd–15	*Dhruvas*
		16–17	'Slow' apogees
		18	'Fast' apogees
		19–24	Computing true planets
		∗ 25–43	Table of equations
		44–45	Interpolation for the Sine
		46	Conclusion
		47–50	Correction to be applied in the treatise *Grahaṇamaṇḍana*

Table 2: Contents of the *Dṛggaṇita* (sections relevant to this article are marked with ∗)

the equation from the mean anomaly (*DG* 1.3.4cd–18ab). The set of parameters, which I shall hereafter refer to as 'divisors (represented by *a*)' and 'corrections (represented by *b*)' in the text,[10] are listed in *bhūtasaṃkhyā*. Interestingly, there is yet another set of parameters in the *DG* that generate the divisors and corrections (*DG* 1.4.1–8).

The second part does not include descriptions of the derivations of equations, but lists the precomputed equations for six degree intervals (*DG* 2.25–43). They are versified tables in *kaṭapayādi*. While there are some studies which compare values in these tables with other texts,[11] their origins, partic-

[10] Further explanation on these parameters will be given in the following passages and especially in Section 3.5.

[11] See, for example, Pingree, 'History', p. 613.

ularly their relations with the procedure in the first part, have not yet been deeply studied. The layer of parameters in the first part should enable us to infer how the tables of equations in the second part were made, and also to evaluate their correctness and analyze the manuscript tradition.

The *DG* was edited by K. V. Sarma using five manuscripts.[12] I have examined four of them (A, C, D and E) but not in full detail.[13] This study is based on the texts in the printed edition. Further inspections on the manuscripts themselves are yet to be done.

It is generally accepted that the '*dṛk* system' of astronomical computations based on the *DG* was very popular in Kerala and replaced, in some domains, the system of Haridatta which was known as the '*Parahita*'. For example, Sarma, *A History*, p. 10 remarks:

> ... a large number of manuals have come to be composed following the *dṛk* system, both in Sanskrit and in Malayalam. The results obtained through this system being more accurate, this system was used for horoscopy (*jātaka*), astrological query (*praśna*) and the computation of eclipses (*grahaṇa*), while the *Parahita* continued to be used for fixing auspicious times for rituals and ceremonies (*muhūrta*).

But the paradox is that manuscripts of the *Dṛggaṇita* itself are scarce. Before the five manuscripts were gathered for the critical edition, the text had even been considered lost.[14] How could a 'system' survive without the text? Our study on tables of equations may provide an insight.

3. Generating the table

3.1. Two types of epicycles

Before analyzing the equation tables, let us first look at the computations behind them. Such procedures can be found in the third chapter of the first part of the *DG*. The explanations are brief, but they are probably based on the planetary theories of the *Āryabhaṭīya* (499 CE, hereafter *Ābh*) by Āryabhaṭa, a treatise that was influential especially in southern India.[15] The *Grahacāranibandhana* follows the *Ābh*, and so do most of the other astronomical texts in Kerala (Sarma, *A History*, pp. 6–9). While the *DG* does not refer to the *Ābh* explicitly, the author Parameśvara frequently cites Āryabhaṭa

[12] Sarma, *Dṛggaṇita*.

[13] Read in the University of Kerala Oriental Research Institute and Manuscripts Library (ORI&MSS) in September 2014. A and E (in the same bundle) were too damaged to scrutinize. Manuscript C was also in a bad condition, and some portions that had been readable in K. V. Sarma's time seemed to have crumbled away. Manuscript D was in a fairly good condition, and apparently there was no significant difference nor lacking information in Sarma's critical edition.

[14] Sarma, *Dṛggaṇita*, p. ix.

[15] Critical edition by Shukla and K. V. Sarma in *Āryabhaṭīya of Āryabhaṭa*.

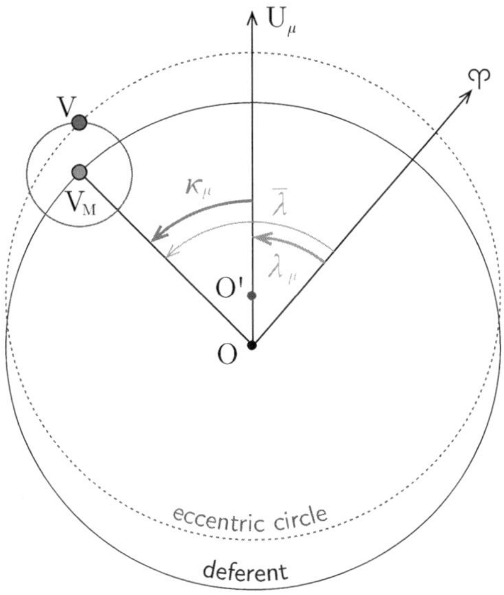

Figure 1: This diagram shows the mean planet V_M on the deferent (centered on O) and the true planet V on the 'slow' epicycle. The true planet does not revolve on this epicycle, and thus the orbit of the true planet is an eccentric circle with a fixed apogee U_μ and centered on O' such that $OO' = VV_M$. When $\bar{\lambda}$ is the longitude of the mean planet and λ_μ that of the 'slow' apogee, $\varkappa_\mu = \bar{\lambda} - \lambda_\mu$ is the mean 'slow' anomaly.

as an authority elsewhere, including his commentary on the *Ābh*.[16] I assume that the basis for computing equations in the *DG* were the methods of the *Ābh*, which I shall explain in the following sections.

In general, Indian astronomical texts compute planetary equations for two different epicycles called the 'slow' (*manda*) and 'fast' (*śīghra*). The true planet on the 'slow' epicycle is either fixed or revolves slowly whereas its rate of revolution on the 'fast' epicycle is faster than that of the mean planet on the deferent, hence their names. The orbit of the true planet on an epicycle can also be represented with an eccentric circle whose center is separated from the Earth towards the apogee at a distance equivalent to the radius of the epicycle. From the modern viewpoint, the 'slow' epicycle (Figure 1) whose apogee does not move accounts for the eccentricity of the orbits while the 'fast' epicycle (Figure 2) transfers the heliocentric motions of planets to a geocentric orbit. The sun and moon only have the 'slow' epicycle while Mars, Mercury, Jupiter, Venus and Saturn also have the 'fast' epicycle. *DG* 1.3.1–4ab gives the rules to find the 'slow' and 'fast' apogees. The 'slow' and 'fast' equations (*phala*), i.e., the difference in longitude between the mean and true planets for the

[16] Critical edition by Kern in *The Âryabhaṭîya*.

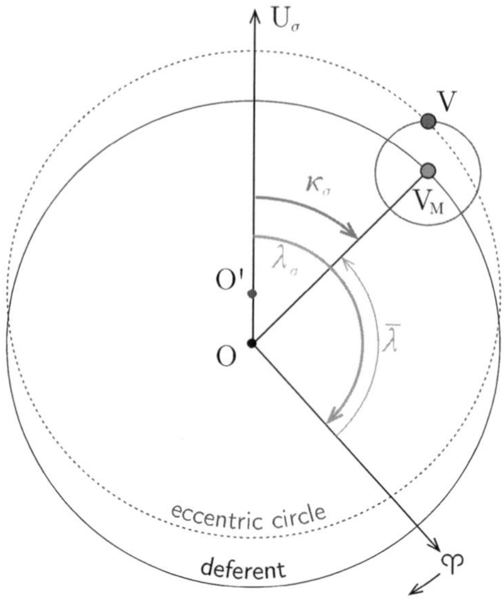

Figure 2: The true planet V does revolve on the 'fast' epicycle (note that we are not taking the 'slow' epicycle into account here). We can draw the eccentric orbit of the true planet if we fix its center O′ and apogee U_σ. $\varkappa_\sigma = \bar{\lambda} - \lambda_\sigma$ is the mean 'fast' anomaly.

corresponding epicycles, are computed independently according to the rules given in *DG* 1.3.4cd–18ab. In the following subsections I shall explain how the equation is derived from the mean anomaly with its associated divisor and correction according to *DG* and *Ābh*.

The two equations are combined in the following manner. In the first two steps, half the equations of the 'slow' and 'fast' epicycles are applied to the mean planet (which comes first depends on the treatise). Then, starting from this half-and-half corrected longitude, the two equations are applied again, this time in whole, to obtain the true planet. Although this scheme might have been founded on a geometrical model of Greek origin,[17] Sanskrit sources do not attempt to ground the procedure geometrically.

3.2. Definition of segments produced by the mean anomaly

DG 1.3.4cd–5ab defines the Sines (*jyā*) of the 'base (*dos* or *bhujā*)' and the 'upright (*koṭi*)'.

> A quadrant is three signs. If [the mean anomaly] is in an odd [quadrant], the 'base' and the 'upright' are the portions elapsed and to come respectively. Opposite when

[17] See Duke, 'The Equant in India'.

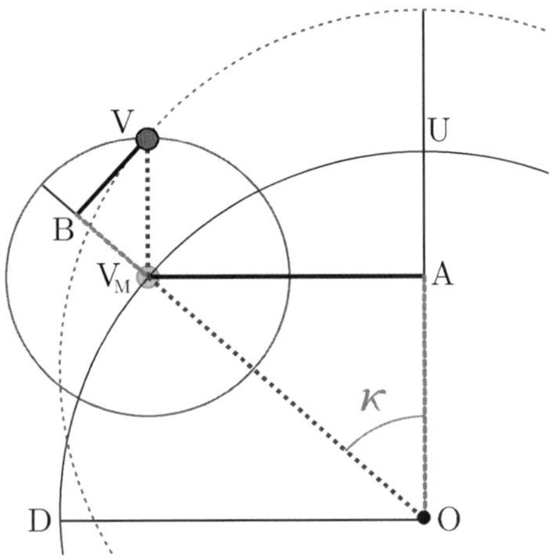

Figure 3: The 'base' Sine AV_M and 'upright' Sine OA and their corresponding 'results' BV and V_MB.

> it is in an even [quadrant]. The [Sine] of a 'base' or an 'upright' produced by the mean anomaly should be understood as a Great Sine. (*DG* 1.3.4cd–5ab)[18]

In Figure 3, U is the direction of the apogee, V_M is the mean planet and therefore $\widehat{UV_M}$ is the arc of the mean anomaly $\varkappa$. Here V_M is in the first quadrant $\widehat{UD}$ counted from the apogee. $\widehat{UV_M}$ and $\widehat{V_MD}$ are the 'base' and 'upright' and $AV_M = \mathrm{Sin}\,\varkappa$ and $OA = \mathrm{Sin}(90^\circ - \varkappa)$ are their Sines. The definitions are different when V_M is in another quadrant, but in any case the two Sines are always $\mathrm{Sin}\,\varkappa$ and $\mathrm{Sin}(90^\circ - \varkappa)$ except for the fact that they are always treated as positive values.

Āryabhaṭa uses $R = 3438$ as the Radius of the great circle when computing Sines, which is an integer approximation when the circumference of the circle is 21600 ($360^\circ \times 60'$). However, Parameśvara refers to a *mahājyā*, 'Great Sine'. This term appears in the Sine table by Mādhava (Table 3) quoted in texts such as Nīlakaṇṭha's commentary on the *Ābh*[19] and is generally interpreted as a reference to its 24 entries.[20] The entries in Mādhava's Sine table are up to the order of the second sexagesimal, and $R = 3437;44,48$. This suggests that Parameśvara might also be using the same table. At least he must be computing 'base' and 'upright' Sines with fractional parts.

DG 1.3.5cd–12ab give a set of parameters called 'divisors (*hara* or *hāra*)', hereafter represented by a, and unnamed parameters which I shall name 'cor-

[18] *rāśitrayaṃ padaṃ syād oje doḥkoṭike gataiṣyāṃśau* ||1.3.4||
yugme 'nyathā mahājyā grāhyā kendrotthayor bhujākoṭyoḥ |

[19] Śāstrī, *The Āryabhaṭīya of Āryabhaṭācārya*, p. 55.

[20] Sarma, *A History*, p. 26.

Table 3: The *Mahājyās* of Mādhava, according to Nīlakaṇṭha's commentary on the *Ābh*

arc	Sine	arc	Sine	arc	Sine
3°45′	224;50,22	33°45′	1909;54,35	63°45′	3083;13,17
7°30′	448;42,58	37°30′	2092;46,03	67°30′	3176;03,50
11°15′	670;40,16	41°15′	2266;39,50	71°15′	3255;18,22
15°	889;45,15	45°	2430;51,15	75°	3320;36,30
18°45′	1105;01,39	48°45′	2584;38,06	78°45′	3371;41,29
22°30′	1315;34,07	52°30′	2727;20,52	82°30′	3408;20,11
26°15′	1520;28,35	56°15′	2858;22,55	86°15′	3430;23,11
30°	1718;52,24	60°	2977;10,34	90°	3437;44,48

rections'[21] and denote as b. They are combined in the form $h = a + \frac{\text{Sin}\,\varkappa}{b}$ which is referred to as the 'corrected divisor (*saṃskṛtahāra*)' or simply 'divisor'. In order to distinguish h from a I shall hereafter call the former 'corrected divisors'.

The corrected divisor and the 'base' and 'upright' Sines are used to compute what are called 'results (*phala*)'. The same Sanskrit term *phala* can also indicate the equation, but I shall distinguish them in the translation.

> The 'upright' and 'base' Sines multiplied by ten and divided by the [corrected] divisor become the 'upright' and 'base' results. (*DG* 1.3.12cd)[22]

The *DG* itself does not mention the geometrical meaning of these results, but we can find explanations in other texts including Parameśvara's commentary on the *Ābh*. In Figure 3, when V is the true planet on the epicycle and B is the foot of a perpendicular dropped on OV_M, $p_\mathcal{B} = \text{BV}$ is the 'base' result and $p_\mathcal{U} = \text{V}_\text{M}\text{B}$ is the 'upright' result. The two results form a right triangle with the radius of the epicycle $\text{V}_\text{M}\text{V}$. Since $\text{VV}_\text{M} \parallel \text{AO}$, corresponding angles $\angle\text{VV}_\text{M}\text{B}$ and $\angle\text{V}_\text{M}\text{OA}$ are equal. Furthermore, $\angle\text{V}_\text{M}\text{BV} = \angle\text{OAV}_\text{M} = 90°$, therefore $\triangle\text{V}_\text{M}\text{BV} \sim \triangle\text{OAV}_\text{M}$. Thus the two results are proportional to the two Sines.

Therefore, if r is the radius of the epicycle, and we define $h = \frac{10R}{r}$, the 'base' and 'upright' results can be denoted as follows:

$$\text{BV} = \frac{\text{VV}_\text{M} \cdot \text{AV}_\text{M}}{\text{V}_\text{M}\text{O}}$$

$$p_\mathcal{B} = \frac{r\,\text{Sin}\,\varkappa}{R} = \frac{10\,\text{Sin}\,\varkappa}{h} \tag{1}$$

[21] Later in the text, as shall be shown in Section 3.6, a set of dividends and divisors that yield this parameter is introduced. There the divisor is called the 'divisor for the correction (*saṃskṛtihara*)'. Thus I have chosen the term 'correction'.

[22] *digghne koṭibhujajye harabhakte koṭidoḥphale bhavataḥ* ||1.3.12||

$$V_M B = \frac{VV_M \cdot OA}{V_M O}$$

$$pu = \frac{r\,\mathrm{Sin}(90^\circ - \varkappa)}{R} = \frac{10\,\mathrm{Sin}(90^\circ - \varkappa)}{h}. \tag{2}$$

By comparing this with the verse, we can see that h is what Parameśvara calls the [corrected] divisor. According to *DG* 1.3.5cd–12ab, the divisors of the sun and moon's 'slow' epicycles are represented by a single value a while those for the epicycles of the other five planets involve the correction b and are given by $h = a + \frac{\mathrm{Sin}\,\varkappa}{b}$. While the *DG* does not tell us why the corrected divisor h changes with the anomaly, it can be explained from the theory of epicycles found in the *Ābh*.

3.3. Size of the epicycle

A unique feature of the *Ābh* is that the epicycles change their size according to the mean anomaly.[23] We can find the circumferences of the epicycles in verses 1.8–9 of the *Ābh*.[24] These circumferences are measured in 'degrees', that is to say, their length when the circumference of the deferent (radius R) is 360 degrees. In other words, when the radius and circumference of the epicycle are r and c respectively, $r : c = R : 360$, and thus

$$c = \frac{360r}{R}. \tag{3}$$

However, Āryabhaṭa condenses his verses by dividing these 'degrees' by $\frac{9}{2}$ (the greatest celestial latitude of the moon). In other words, to find the circumference c of an epicycle from the value c' appearing in *Ābh* 1.8–9, we must multiply c' by $\frac{9}{2}$ ($c = \frac{9}{2}c'$). Its radius r would then be

$$r = \frac{Rc}{360} = \frac{Rc'}{80}. \tag{4}$$

In Table 4 the values c' that actually appear in *Ābh* 1.8–9 are given in parentheses and the circumference in degrees computed by $c = \frac{9}{2}c'$ are written outside the parentheses. Hereafter the term 'circumference' will only refer to c in degrees and not c'.

[23] Apart from direct followers of the *Ābh*, the *Sūryasiddhānta* also adopts this idea but with values considerably different from those of Āryabhaṭa (Burgess and Whitney, 'Translation of the Sûrya-Siddhânta', pp. 205–06)

[24] The numbering of verses in the first chapter differ among commentators. Here I follow the numbering by Parameśvara.

Table 4: Circumference of epicycles in odd (c_o) and even (c_e) quadrants in 'degrees'.

planet	'slow'		'fast'	
	c_o	c_e	c_o	c_e
Sun	13;30 (3)		—	
Moon	31;30 (7)		—	
Mars	63 (14)	81 (18)	238;30 (53)	229;30 (51)
Mercury	31;30 (7)	22;30 (5)	139;30 (31)	130;30 (39)
Jupiter	31;30 (7)	36 (8)	72 (16)	67;30 (15)
Venus	18 (4)	9 (2)	265;30 (59)	256;30 (57)
Saturn	40;30 (9)	58;30 (13)	40;30 (9)	36 (8)

There are two entries for the 'slow' and 'fast' epicycles of the five planets which correspond to their sizes when the mean anomaly is at the beginning of an odd (first or third) or even (second or fourth) quadrant. According to some commentators, the size of the circumference when the mean anomaly is in the middle of a quadrant should be linearly interpolated. Parameśvara also describes a linear interpolation which can be expressed as follows:[25]

$$c = c_o + \frac{c_e - c_o}{R} \operatorname{Sin} \varkappa . \tag{5}$$

Here c is the circumference of an epicycle at a given mean anomaly $\varkappa$, c_o is its circumference when the mean anomaly is at the beginning of an odd quadrant and c_e at the beginning of an even quadrant. The radius would then be

$$r = \frac{Rc_o}{360} + \frac{c_e - c_o}{360} \operatorname{Sin} \varkappa . \tag{6}$$

3.4. Link between the circumference, divisor and correction

In the cases of the sun or the moon, the relation between the divisor $a = h$ and the circumference c is as follows:

$$a = \frac{10R}{r} = \frac{10R \cdot 360}{cR} = \frac{3600}{c} . \tag{7}$$

The values in Tables 4 and 5 satisfy this formula.

Meanwhile, if we assume that the radii of the epicycles of other planets are represented by formula 6, the divisor $h = a + \frac{\operatorname{Sin} \varkappa}{b}$ can be represented as follows:

$$h = a + \frac{\operatorname{Sin} \varkappa}{b} = \frac{3600R}{Rc_o + (c_e - c_o) \operatorname{Sin} \varkappa} . \tag{8}$$

[25] Kern, *The Âryabhaṭîya*, p. 68.

There is no such a and b that satisfy this formula for every $\varkappa$. Thus, I have assumed that the *DG* employs a model where the size of an epicycle changes non-linearly within a quadrant. The following for the circumference would reproduce a divisor with the format $a + \frac{\text{Sin}\,\varkappa}{b}$:

$$c = \frac{c_e c_o}{c_e + \frac{c_o - c_e}{R}\,\text{Sin}\,\varkappa}. \tag{9}$$

Then the radius would be

$$r = \frac{R c_o c_e}{360 c_e + \frac{360(c_o - c_e)}{R}\,\text{Sin}\,\varkappa}, \tag{10}$$

and thus

$$a + \frac{\text{Sin}\,\varkappa}{b} = \frac{3600}{c_o} + \frac{3600(c_o - c_e)}{R c_o c_e}\,\text{Sin}\,\varkappa, \tag{11}$$

where $a = \frac{3600}{c_o}$ and $b = \frac{R c_o c_e}{3600(c_o - c_e)}$.

If we compute a and b using the values of circumferences in the *Ābh* (right-hand side of Table 5), a of the 'slow' epicycles of Jupiter and Venus (in the variant reading) as well as those of the 'fast' epicycles of Mars and Jupiter agree with *DG*. Others are slightly different (within $\pm 10\%$), which suggests that Parameśvara is modifying the parameters. In the case of b, only the 'fast' epicycle of Jupiter matches *DG*, and there are some cases (such as the 'slow' epicycles of Mercury and Saturn) where the values are strikingly different. This is possible because the denominator $c_o - c_e$ in b may change the result greatly. However, it is too far-fetched to assume that Parameśvara replaced the sizes of circumferences in the *Ābh* in order to compute the divisors and their corrections. While it is almost certain that Āryabhaṭa's theory of epicycles with varying sizes is behind the procedure, the parameters themselves could have been chosen with different reasons.[26]

[26] Parameśvara is not the only Indian astronomer who proposed a rule for equations that would necessitate a non-linear change in the epicycle's size. In his *Tantrasaṅgraha*, Nīlakaṇṭha, a student of Parameśvara and his son Dāmodara, explains that the 'base' result for the 'slow' epicycle of Venus is the 'base' Sine divided by $14 + \frac{\text{Sin}\,\varkappa}{240}$ (Ramasubramanian and Sriram, *Tantrasaṅgraha*, pp. 128–29). This is identical with Parameśvara's theory apart from the parameters a and b. However, according to the *Tantrasaṅgraha*, for other epicycles the 'base' Sine is to be *multiplied* by $a + \frac{\text{Sin}\,\varkappa}{b}$ (or by a constant value for the 'slow' epicycles of Mercury and Saturn), which can be explained by linear interpolation for the epicycles' circumference. Nīlakaṇṭha does not explain how he established his rules and whether they had geometrical reasons are yet to be studied. However, considering the scholarly connection between Parameśvara and Nīlakaṇṭha, this exceptional rule for the 'slow' epicycle of Venus could be linked with the methods of *DG*.

3.5. Values of divisors and corrections

For each planet, the divisor a is given up to the first order sexagesimal. Parameśvara calls the integer part 'minutes (*liptā*)' and the fractional part 'seconds (*viliptā*)'.

> The sun's divisor in minutes should be two hundred and sixty-six and forty seconds. (*DG* 1.3.5cd)[27]
>
> Meanwhile, the moon's divisor in minutes is one hundred and fourteen and seventeen seconds. (*DG* 1.3.6ab)[28]
>
> Fifty-five, one hundred and twenty-five, one hundred and fourteen, one hundred and seventy-one (variant: four hundred) and eighty-four are the divisors of the 'slow' [epicycles] in minutes of [the planets] beginning with Mars. There, the seconds are ten, twenty-three, seventeen, zero and thirteen in this order. (*DG* 1.3.6cd–7)[29]
>
> Quotients in minutes of the [planets'] own 'base' Sine divided by five hundred and fifty-six, four hundred and fifty-one, three hundred and eleven, fifteen (variant: sixty) and one thousand five hundred and eighty-six [respectively] are to be added to the divisors of Mercury and Venus, and are to be subtracted from the divisors of Mars, Jupiter and Saturn. (*DG* 1.3.8–9ab)[30]
>
> In the case of 'fast' [epicycles], the divisors in minutes of [the planets] beginning with Mars should be fifteen, twenty-five, fifty, thirteen and ninety. In that order, the seconds are six, forty-nine (variant: zero), zero, forty-four and thirteen. (*DG* 1.3.9cd–10)[31]
>
> The quotient in minutes of the [planet's] own 'fast base' Sine divided by six thousand seven hundred and one, two thousand and sixty, one thousand and thirty-one, three times ten thousand, and three hundred should be added to the divisor.[32]

[27] *bhānor hārakalāḥ syuḥ ṣaḍrasadasrāḥ khavedavikalāś ca* ||1.3.5||

[28] *candrasya tu manucandrā hārakalāḥ saptacandravikalāś ca* |

[29] *iṣubāṇā dviśaraikā manucandrā bhūmisaptacandrās ca* ||1.3.6||
kṛtavasavo bhaumāder māndahārakalā viliptikās tatra |
daśa vikṛtiḥ saptaikā śūnyaṃ viśve ca tāḥ krameṇeti ||1.3.7||
The edition adopts *vyomaśūnyavedāś* (four hundred) in place of *bhūmisaptacandrāś*. My reading is an alteration of the variant in manuscript B, *bhūmicandrasaptāś* (one hundred and seventy-one).

[30] *ṣaḍbāṇeṣubhir ekaprāṇakṛtaiś candrabhūmirāmaiś ca* |
tithibhī rasāhitithibhir labdhā liptāḥ svakīyadorjyātaḥ ||1.3.8||
budhasitaharayoḥ kṣepyāḥ, śodhyā bhaumeḍyaravijahāreṣu |
The edition adopts *ṣaṣṭyā* (sixty) in place of *tithibhī* (fifteen). I have corrected the reading *tithibhiḥ* in manuscript B.

[31] *śaughre hārakalāḥ syur bhaumādeḥ śarabhuvaḥ śarayamāś ca* ||1.3.9||
śūnyeṣavaś ca viśve khāṅkā evaṃ krameṇa vikalās tu |
aṅgāni go'bdhayaṃ khaṃ kṛtavedā rāmabhūmayaś ceti ||1.3.10||
The edition adopts the variant *śūnyam abhraṃ* (zero, zero) instead of the reading *go'bdhayaṃ khaṃ* (forty-nine, zero) in manuscript B.

[32] *kukhanagaṣaḍbhiḥ khāṅkakhayamaiḥ kuguṇakhakubhir ayutanihataguṇaiḥ* |

Table 5: Divisors (a) and corrections (b) given in the *DG* (left) and computed using values in the *Ābh* (right)

	DG				*Ābh* (computed)			
planet	'slow'		'fast'		'slow'		'fast'	
	a	b	a	b	a	b	a	b
Sun	266;40	—	—		266;40	—	—	
Moon	114;17	—	—		114;17	—	—	
Mars	55;10	−556	15; 6	6701	57; 8	−271	15; 6	5808
Mercury	152;23	451	25;49	2060	160; 0	75	25;48	1932
(variant)			(25;0)					
Jupiter	114;17	−311	50; 0	1031	114;17	−241	50; 0	1031
Venus	171; 0	15	13;44	30000	400; 0	17	13;34	7226
(variant)	(400;0)	(60)						
Saturn	84;13	−1586	90;13	300	88;53	−126	88;53	309

The values are listed in Table 5. I have modified five values from K. V. Sarma's edition (hereafter 'edition'), based on the reading in manuscript B, as I have indicated in the footnotes. They are the values a and b for Venus' 'slow' epicycle, a for Mercury's 'fast' epicycle and b for the 'fast' epicycles of Mars and Mercury. With this modification, every value of equations in *DG* 2.26–42 (which follow manuscript B in general) can be accounted for, as we will see later in Section 4.

However, three out of five original readings in the edition (a and b for Venus' 'slow' epicycle and a for Mercury's 'fast' epicycle) turned out to produce variant readings (from manuscript C) in the table of equations. Thus the reading in the edition is also listed in Table 5 with parentheses. The other two proved to be meaningless and thus have been discarded.

3.6. Further parameters for generating the divisors and corrections

There is yet another set of parameters which are also referred to as 'divisors' against constant dividends (either 48000 or $60R$) that give a or b as their

Table 6: Parameters related to equations in *DG*

Parameter	dividend & divisor		divisor a & correction b		equations
Verse no.	*DG* 1.4.1–7	⇒	*DG* 1.3.5cd–11	⇒	*DG* 2.26–42
In this paper	Table 7		Table 5		Tables 8–13

khakharāmair labdhakalā yojyā hāre svaśīghradorjyataḥ ||1.3.11||
The edition adopts the variant *kukha**nakhaṣa**ḍbhiḥ khā**ga**khayamaiḥ* ... (sixty thousand and one, two thousand and seventy). I have chosen the reading in manuscript B.

Table 7: Dividends and divisors for finding *a* and *b*

	'slow' epicycle			
	a		*b*	
planet	dividend	divisor	dividend	divisor
Sun	60*R*	773;33	–	
Moon	60*R*	1805	–	
Mars	48000	870	60*R*	371
Mercury	48000	315	60*R*	457
Jupiter	48000	420	60*R*	663
Venus	48000	281 (120)	60*R*	13752 (3438)
Saturn	48000	570	60*R*	130

	'fast' epicycle			
	a		*b*	
planet	dividend	divisor	dividend	divisor
Mars	48000	3180	60*R*	30;47
Mercury	48000	1859 (1920)	60*R*	100;8
Jupiter	48000	960	60*R*	200;4
Venus	48000	3495	60*R*	6;52,34
Saturn	48000	532	60*R*	687;36

quotient (Table 7). They are given in *DG* 1.4.1–7 as follows:

> For clarification, the computation of the divisors that have been mentioned are stated.[33]
>
> Forty-eight thousand [divided] by eight hundred and seventy, three hundred and fifteen, four hundred and twenty, two hundred and eighty-one (variant: one hundred and twenty) and five hundred and seventy should be the divisors for the 'slow' [equation of the planets in the order] from Mars.[34]
>
> Forty-eight thousand [divided] by three thousand one hundred and eighty, one thousand and fifty-nine (variant: one thousand nine hundred and twenty), nine hundred and sixty, three thousand four hundred and ninety-five and five hundred

[33] *spaṣṭārthaṃ hārāṇām ānayanaṃ cocyate tv ihoktānām* |

[34] *khādryahibhir bāṇavidhutribhir nakhajaladhibhiḥ kuvasudasraiḥ* ||1.4.1||
khādriśarair māndaharā bhaumāt syuḥ khakhakhanāgavedebhyaḥ |
The edition adopts *bāṇavasutribhir* (three hundred and eighty-five) in place of *bāṇavidhutribhir* (three hundred and fifteen). I have chosen the reading that appears in manuscripts A, B and C. It also adopts *khasūryaiś ca* (one hundred and twenty) instead of *kuvasudasraiḥ* (two hundred and eighty-one), which I have put as a variant reading and selected the reading in manuscript B instead.

and thirty-two become the divisors for the 'fast' [equation] of those beginning with Mars. Now for the divisors for their corrections.[35]

Sixty times the Radius [divided] by three hundred and seventy-one, four hundred and fifty-seven, six hundred and sixty-three, thirteen thousand seven hundred and fifty-two (variant: three thousand four hundred and thirty-eight) and one hundred and thirty should be the divisors for the corrections in the case of the 'slow' [equation].[36]

Thirty, one hundred, two hundred, six and six hundred and eighty-seven with seconds of forty-seven, eight, four, fifty-three and thirty-six added to them [respectively;] the Radius multiplied by sixty [and divided by each of these numbers] should be the divisors for the corrections with the 'base'. However, in the case of Venus, the value is decreased by *tatparas* (sixtieths of a second) of twenty-six. Thus are the divisors for the corrections with the 'base' in the case of the 'slow' [equation] of those beginning with Mars.[37]

The sun's divisor is [the Radius multiplied by sixty and divided by] seven hundred and seventy-three plus seconds of thirty-three. The moon's divisor is the Radius multiplied by sixty [and divided by] one thousand eight hundred and five.[38]

The meanings of these dividends and divisors are unclear. The dividend 48000 could be divided into 10, 60 and 80, where 10 is the coefficient as seen in *DG* 1.3.12cd (formulas 1 and 2), 60 is the number of minutes in a degree and 80 is the divisor when we compute the radius of an epicycle from the value c' given in *Ābh* 1.8–9 (formula 4).

Unfortunately, these dividends and divisors add more mystery than clues on how a and b had been found, since we cannot find any convincing mean-

[35] *vyomāhibhūmirāmaiḥ gośaradhṛtibhir nabho'ṅgagobhiś ca* ||1.4.2||
iṣugo'bdhyagnibhir aśvitriśaraiḥ khābhrābhranāgavedebhyaḥ |
bhaumādeḥ śīghraharā bhavanti tatsaṃskṛtiharās tu ||1.4.3||
The edition adopts *khāśvinavaikair* (one thousand nine hundred and twenty) in place of *gośaradhṛtibhir* (one thousand and fifty-nine). I have chosen the reading in manuscript B.

[36] *kvadriguṇaiḥ svarabāṇābdhibhis trirasarasaiḥ dviśaranagaviśvaiḥ* |
khaguṇaikaiḥ ṣaṣṭighnāt triguṇān mānde tu saṃskṛtiharaḥ syuḥ ||1.4.4||
The edition adopts *trirasaṣaḍbhiḥ ahiguṇābdhiguṇaiḥ* (six hundred and sixty-three, three thousand four hundred and thirty-eight) in place of *trirasarasaiḥ dviśaranagaviśvaiḥ* (six hundred and sixty-three, thirteen thousand seven hundred and fifty-two). I have chosen the reading in manuscript B.

[37] *khaguṇaiḥ khābhraśaśāṅkaiḥ khakhayamalaiṣ ṣaḍbhir adrivasuṣaḍbhiḥ* |
vikalānāṃ nagavedair vasubhir vedaiś ca rāmavāṇaiś ca ||1.4.5||
ṣaṭtriṃśātā ca yuktair etaiḥ ṣaṣṭyā hatāt triguṇāt |
doḥsaṃskṛtihārāḥ syuḥ kaves tu rasadasratatparonais taiḥ |
śaighre bhaumādīnāṃ doḥsaṃskṛtihārakā bhavanty evam ||1.4.6||

[38] *guṇasaptādribhir agnitrayavikalāsaṃyutai raver hāraḥ* |
candrasyeṣukhadhṛtibhir hāraḥ ṣaṣṭyā hatāt triguṇāt ||1.4.7||

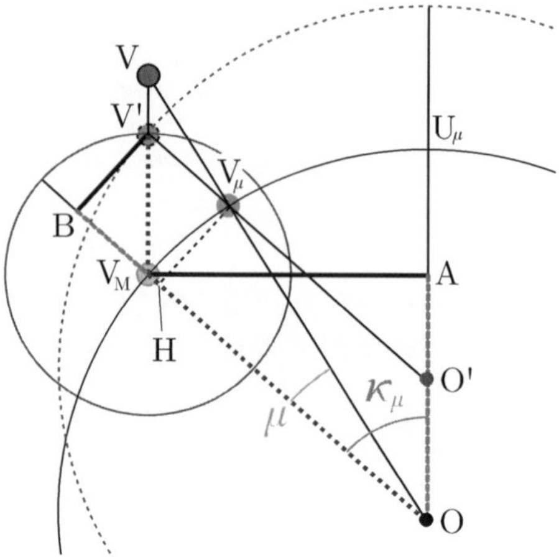

Figure 4: Computing the 'slow' equation Sin μ

ing behind the values. Nonetheless they can help us verify the readings in the manuscripts.

Several variant readings of the divisors were found in manuscripts. We have examined whether divisions with these variant divisors can reproduce the variant readings of a and b and found three such cases. They are the divisors corresponding to a and b of Venus's 'slow' epicycle and the divisor for a of Mercury's 'fast' epicycle. We have indicated them in parentheses in Table 7.

3.7. 'Slow' (*manda*) equation

> In the case of the 'slow' equation, they call it the 'base' result's arc because it is the 'base' result made into an arc.[39]

As is the case with other texts following the *Ābh*, Parameśvara states that the arc of the 'base' result is also the 'slow' equation Sin μ itself. Thus if the mean anomaly from the 'slow' apogee is $\varkappa_\mu$,

$$\operatorname{Sin}\mu = p_B = \frac{10}{a + \frac{\operatorname{Sin}\varkappa_\mu}{b}} \operatorname{Sin}\varkappa_\mu\,. \tag{12}$$

Geometrically, we can interpret that the true planet deviates from the 'slow' epicycle as shown in Figure 4. The true planet is relocated from V′ to V. When V_μ is the intersection of OV with the deferent and O′ is the center of the eccentric circle, V is chosen so that V_μ is on O′V′. When H is the foot of

[39] *mande dohphalacāpas tv iti kathitaṃ cāpitaṃ hi bāhuphalam* ||*1.3.13*||

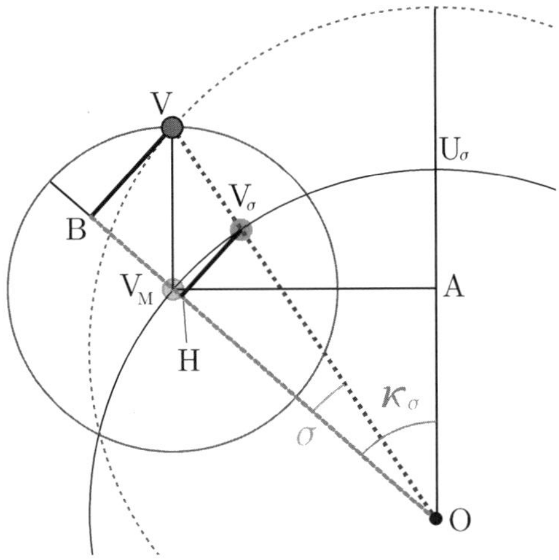

Figure 5: Computing the 'fast' equation J_σ

a perpendicular drawn from V_μ to OV_M, $V_\mu H = \mathrm{Sin}\,\varkappa_\mu$. Since $OO' \parallel V_M V'$ and $OO' = V_M V'$, the quadrilateral $OO'V'V_M$ is a parallelogram and thus $OV_M \parallel O'V'$. $V_\mu H = V'B$ because they are both equivalent to the distance between OV_M and $O'V'$. Therefore $\mathrm{Sin}\,\mu = p_{\mathcal{B}}$. $\widehat{V_\mu V_M} = \mu$ is the arc of the equation, and *DG* 1.3.13 explicitly says that the segment is converted to an arc:

$$\mu = \mathrm{arcSin}\left(\frac{10}{a + \frac{\mathrm{Sin}\,\varkappa_\mu}{b}}\,\mathrm{Sin}\,\varkappa_\mu\right). \tag{13}$$

3.8. 'Fast' (*śīghra*) epicycle

> In the case of the 'fast' [equation, it is] the 'base' result multiplied by the Radius, divided by the radial distance and made into an arc. (*DG* 1.3.14ab)[40]

Unlike the 'slow' epicycle, the true planet stays on the 'fast' epicycle. Therefore we need to take into account its distance from the Earth, or 'radial distance (*karṇa*)' in order to find the 'fast' equation. This is demonstrated in Figure 5.

First of all, we compute the 'base' result $p_{\mathcal{B}} = VB$ from the 'base' Sine of the mean anomaly $\mathrm{Sin}\,\varkappa_\sigma = V_M A$ as previously.

$$VB = \frac{V_M V' \cdot V_M A}{OV_M}$$

$$p_{\mathcal{B}} = \frac{r\,\mathrm{Sin}\,\varkappa_\sigma}{R}. \tag{14}$$

[40] *śaighre trijyāguṇitaṃ karṇahṛtaṃ cāpitaṃ ca bāhuphalam |*

When V_σ is the intersection of OV with the deferent and H the foot of a perpendicular drawn from it on OV_M, $V_\sigma H = \operatorname{Sin}\sigma$ is the Sine of the 'fast' equation. For computing it we use the similar triangles $\triangle VBO \sim \triangle V_\sigma HO$. They are similar because both are right triangles sharing an acute angle. Since $OV = \mathcal{R}_\sigma$ is the radial distance,

$$V_\sigma H = \frac{VB \cdot OV_\sigma}{OV}$$

$$\operatorname{Sin}\sigma = \frac{p_{\mathcal{B}}R}{\mathcal{R}_\sigma}. \tag{15}$$

The next task is to find the radial distance. First Parameśvara describes a procedure using the Pythagorean theorem.

> When in [the quadrants] beginning with Cancer and Capricorn, the 'upright' result is subtracted from or added to the Radius. Its square added with the square of the 'base' result; its square root becomes the radial distance. (*DG* 1.3.15cd–16ab)[41]

This can also be explained from Figure 5. BO in the right triangle $\triangle VBO$ can be found by adding or subtracting the 'upright' result $p_{\mathcal{U}} = BV_M$ to or from the Radius $R = V_M O$. The 'upright' result is subtractive when the mean anomaly is in the 2nd or 3rd quadrant and additive in the 1st or 4th quadrant. Therefore the radial distance is

$$OV = \sqrt{VB^2 + BO^2} = \sqrt{VB^2 (V_M O \mp B_V M)^2}$$

$$\mathcal{R}_\sigma = \sqrt{p_{\mathcal{B}}^2 + (R \mp p_{\mathcal{U}})^2}, \tag{16}$$

where $p_{\mathcal{B}}$ and $p_{\mathcal{U}}$ can be computed using formulas 1 and 2.

Parameśvara also gives an alternative method which does not involve square root computation.

> The 'base' result multiplied by half of itself [is divided] by the Radius diminished by or increased by the 'upright' result. The quotient should be added to the divisor. The quotient, taken separately, multiplied by the half of itself and increased by a thirty-fourth of itself, [is divided] by the corrected divisor. The quotient should be subtracted from this divisor. This, alternatively, is the radial distance. (*DG* 1.3.16cd–18ab)[42]

[41] *koṭiphalenonayutā karkimṛgādyos trirāśijyā* ||1.3.15||
tadvargabāhuphalakṛtiyogasya padaṃ bhavati karṇaḥ |

[42] *koṭiphalenonayutatrijīvayā svārdhanihatabāhuphalāt* ||1.3.16||
labdhaṃ kṣepyaṃ tasminn eva tu hāre pṛthaksthitāl labdhāt |
tasmāt svārdhavinighnān nijacaturagnyaṃśasaṃyuktāt ||1.3.17||
saṃskṛtahāreṇāptaṃ viśodhayet taddharāt sa vā karṇaḥ |

The word divisor in this verse refers not to a, but to the divisor in the first division, i.e., the Radius diminished by or increased by the 'upright' result. Let us denote this as $d = R \mp p_{\mathcal{U}}$. The quotient q_1 in this division is

$$q_1 = \frac{p_{\mathcal{B}} \cdot \frac{p_{\mathcal{B}}}{2}}{R \mp p_{\mathcal{U}}} = \frac{p_{\mathcal{B}}^2}{2d}. \tag{17}$$

By adding the quotient to the divisor we obtain the 'corrected divisor' $d + q_1$. This is used in the next division, which yields another quotient q_2:

$$q_2 = \frac{\frac{q_1^2}{2} + \frac{q_1}{34}}{d + q_1}$$

and finally,

$$\mathcal{R}_\sigma = d + q_1 - q_2. \tag{18}$$

I have not been able to find the origin of this formula, or similar rules in other texts. This is a good approximation when the 'fast' equation σ (not $\mathcal{R}_\sigma$ itself) is small. However, as σ increases the approximate method (formula 18) yields a smaller σ than formula 16, and when σ is larger than around $2000'$ their difference exceeds $1'$. As I shall explain in the next section, I assume that the approximate method is more likely the one adopted for producing the equation values in *DG* 2.26–42.

In any case, the radial distance thus obtained is involved in the computation of the Sine equation as shown in formula 15. But since *DG* 1.3.14ab states that this is 'made into an arc', the final result is the 'fast' equation σ as an arc:

$$\sigma = \text{arcSin}\left(\frac{p_{\mathcal{B}} R}{\mathcal{R}_\sigma}\right). \tag{19}$$

4. Tables of equations

DG 2.26–42 are planetary equations in *kaṭapayādi*. The verses themselves are in the Appendix while their values are listed in Tables 8–13. As previously quoted on page 332, *DG* 2.26 tells us that they are Sines of equations that correspond to anomalies in six-degree intervals. Some of the following verses also refer to the values as Sines (*jyā* or *jīva*), but in fact they are arc lengths, as will be explained in Section 4.1.

Each verse corresponds to a quadrant, and includes fifteen entries ($6° \times 15 = 90°$). There are three verses for each planet: one verse for the 'slow' equations which are symmetrical about both axes (smallest at $\varkappa = 0°$ and $180°$, and largest at $\varkappa = 90°$ and $270°$) and two verses for the 'fast' equations, each corresponding to the semicircle beginning with the sign Capricorn ♑ (270°) and that beginning with Libra ♎ (90°); i.e., the 4th & 1st quadrants and the 2nd & 3rd quadrants. Equations in the two quadrants within each of these semicircles are symmetrical, but the two semicircles are asymmetrical.

Only three manuscripts (C, D and E) out of five contain *DG* 2.26–42 (A does not include the second part and B is broken before these verses). Among the three, readings in C tend to be distinct from D and E. In general, the variant readings still give the same number in *kaṭapayādi*. However, there are three verses where even the numbers are different, namely *DG* 2.30–31 which cover the two hemispheres of Mercury's 'fast' equations and *DG* 2.35 which gives the 'slow' equations of Venus. Interestingly, manuscript D, which follows the reading of E in general, shares the same set of values with C for *DG* 2.35 and probably also for *DG* 2.30–31.[43] The edition adopts the table in E. I follow this and denote the values in C and D as 'variants', but this is not to suggest any difference in their weight.

For each value in the table I also give the difference from values computed using the parameters and procedures as shown in the previous section (+ when the values in the table are larger than computed values, − when smaller). For the 'fast' equations, I have computed the radial distance using a square root (formula 16) as well as with the approximate method (formula 18). I have used Mādhava's Sine table (Table 3) for finding Sines from arcs and vice versa, with linear interpolation. Fractions were kept until the final step, at which I rounded off the values of the equations to integers.

4.1. Sines or arcs?

Despite the statements in the verses, the *kaṭapayādi* values are arcs and not Sines. The Sine can only be approximated by the arc when their value is smaller than approximately 300′. For instance, the 'slow' equation of Mars at 90° is 707′ in integers whereas its Sine is 702. The computed 'fast' equation of Mars is 1959′, not far from the table value 1858, while the Sine is 1854.

DG 1.3.13 and 1.3.14ab clearly state that the Sines of equations should be converted to an arc. Indeed the arc and not the Sine is required for calculating the true planet, and a table of Sine equations would be less practical.

The equation tables in the *Grahacāranibandhana* are also interpreted as Sines by some historians,[44] but there is actually nothing in the text itself that refers to the entries as such. Only the captions in the edition by Sarma say 'Sines (*jyā*)'. On the other hand, there are cases where equations in versified

[43] Manuscript D lacks the folio that includes *DG* 2.30. The next folio begins with the last line of *DG* 2.31, where the reading (and of course the values) mostly coincides with C.

[44] For example, Pingree, 'History', p. 596.

Table 8: Table of equations for Mars (*DG* 2.26–28)

	'slow'		'fast' ♑		'fast' ♋		
$\varkappa$	edition	diff.	edition	diff.	edition	diff.	
6	68[a]	+2	143	0	680	0	
12	133	0	285	0	1256	−1	
18	200	0	426	−1	1697	+2	
24	266	0	566	−1	2001	0	
30	330	−1	706	0	2201	−5	(−4)
36	392	−1	843	0	2328	+1	(+2)
42	452	0	978	−1	2388	−4	(−3)
48	507	0	1113	+1	2414	+1	(+3)
54	557	0	1245	+1	2401	0	(+1)
60	601	0	1372	0	2364	−1	(0)
66	639	+1	1498	+1	2308	0	(+2)
72	669	+2	1622	+3	2236	0	(+1)
78	689	0	1738	0	2154	+1	
84	704	+2	1851[b]	0	2059	0	
90	707	0	1958	−1	1958	−1	

[a] Ms.C: 66 [b] Ms.C: 1351

Table 9: Table of equations for Mercury (*DG* 2.29–31)

	'slow'		'fast' ♑				'fast' ♋			
$\varkappa$	edition	diff.	edition	diff.	variant	diff.	edition	diff.	variant	diff.
6	23	0	101	+1	100[b]	−2	223	0	237	+2
12	46	0	199	0	204	+1	435	+1	456	0
18	69	0	296	0	308	+5	624	0	656	+1
24	90	0	390	−1	401	+1	791	0	830	+1
30	110	0	483	−1	495	0	931	−1	973	−1
36	129	0	573	−1	587	0	1044	−2	1091	0
42	146	0	660	−1	679[c]	+2	1134	−2	1183	+1
48	162	0	744	−1	765	+2	1200	−3	1255	+6
54	175	0	824	−1	847	+1	1245	−3	1291	−3
60	187	0	900[a]	−1	924	0	1274	−2	1321	+1
66	198	+1	971	−2	998	0	1283	−3	1330	+2
72	205	0	1037	−3	1067	+1	1276	−6	1323	+1
78	210	0	1097	−3	1129	0	1262	−3	1327	+24
84	214	0	1151	−2	1184	−1	1233	−4	1273	0
90	215	0	1197	−3	1234	0	1197	−3	1234	0

[a] Ms.E: 890 [b] Ms.C: 130 [c] Ms.C: 749

Table 10: Table of equations for Jupiter (*DG* 2.32–34)

	'slow'		'fast' ♑		'fast' ♋	
ϰ	edition	diff.	edition	diff.	edition	diff.
6	32	0	59	−1	89	0
12	64	0	118	0	175	0
18	96	0	176	+1	256	+1
24	127	0	231	0	330	0
30	158	0	284	0	398	0
36	187	−1	335	0	458	0
42	215[a]	0	384	0	510	0
48	241	0	429	0	553	0
54	265	+1	472	0	589	+1
60	285	0	511	0	615	0
66	303	+1	546	+1	634	0
72	316	0	576	0	646	+1
78	326	0	601	−1	649	0
84	331	−1	621	−1	646	0
90	333	−1	637	0	637	0

[a] Mss.C,D,E: 227

Table 11: Table of equations for Venus (*DG* 2.35–37)

	'slow'				'fast' ♑			'fast' ♋		
ϰ	edition	diff.	variant	diff.	edition	diff.		edition	diff.	
6	18	0	9	0	152	0		927	+6	
12	33	0	17	0	303	0		1657	+2	
18	48	+4	25	0	454	0		2162	+5	
24	53	0	33	0	604	0		2466	−4	(−2)
30	60	0	40	0	753	−1		2645	−12	(−7)
36	66	0	46	−1	901	−1		2753	+4	(+11)
42	71	0	52	0	1051	+2		2783	0	(+8)
48	75	0	57	−1	1197	+2		2788	+16	(+23)
54	78[a]	0	62	0	1341	+2		2734	+4	(+11)
60	81	0	66	0	1483	0		2667	0	(+5)
66	83	0	70	0	1623	+1		2579	−6	(−2)
72	84	0	72	0	1761	+1	(+2)	2483	−6	(−4)
78	85	0	75	+1	1893[b]	−1		2383	−2	(0)
84	85	−1	75	0	2024	+1		2269	−1	(0)
90	86	0	75	0	2150	−1		2150	−1	

[a] Ms.E: 04 [b] Mss.C,D: 1986

Table 12: Table of equations for Saturn (*DG* 2.38–40)

	'slow'		'fast' ♑		'fast' ♋	
$\varkappa$	edition	diff.	edition	diff.	edition	diff.
6	43	0	36	+1	44	0
12	85	0	70	0	87	+1
18	127	0	104	+1	127	+1
24	168	0	135	+1	163	0
30	207	0	164	0	197	0
36	244	0	192	0	227	0
42	278	0	218	0	255	+1
48	309	−1	242	0	277	0
54	337	−1	263	0	297	+1
60	362	0	282	−1	313	+1
66	383	0	299	0	325	+1
72	398	−1	313	0	333	+1
78	410	−1	325[a]	+1	338	+1
84	417	−1	332[b]	0	338	−1
90	420	0	337	0	337	0

[a] Mss.C,D,E: 315 [b] Mss.C,D,E: 312

Table 13: Table of equations for the sun and the moon (*DG* 2.41,42)

	Sun — 'slow'		Moon — 'slow'	
$\varkappa$	edition	diff.	edition	diff.
6	13	0	31	0
12	27	0	63	0
18	40	0	93	0
24	52	0	122	0
30	64	−1	150	−1
36	76	0	177	0
42	86	0	201	0
48	96	0	224	0
54	104	0	243	0
60	112	0	261	0
66	118	0	275	0
72	123	0	286	0
78	126	0	294	−1
84	128	0	299	−1
90	129	0	301	0

tables are undeniably arcs, such as the table of 'fast' equations in *Brāhmasphuṭasiddhānta* 25.47–56ab by Brahmagupta.[45]

4.2. Possible causes for the differences

However, there are two tables where arcs *do* seem to have been approximated by Sines, namely the 'slow' equations of the sun and the moon. All four entries that are smaller than the computed arcs do agree with the Sines when rounded off to integers.

This seems not to be the case elsewhere. Other small differences in the range of ± 2 may be due to rounding off in the middle of the procedure, but I could not locate such step that would thoroughly explain the errors in every table.

The differences between the 'fast' equations of Mercury according to the edition and those computed from $a = 25;49$ and $b = 2060$ are systematic. I have also computed them using $a = 25;48$ and $b = 1932$, which can be derived from the circumferences given in the *Ābh*, and found that they agree better with the table values (Table 14). This is the only case where values from the *Ābh* could account for the differences. Perhaps the equations were computed by someone who did not know the parameters given in the *DG*,[46] and the equations were left unnoticed since the difference is relatively subtle.

Discrepancies are also recognizable in the 'fast' equation of Mercury in group 2. In this case, scribal errors seem to be playing a bigger role. The fact that this table is significantly corrupted in comparison with others suggest that, as was the case with group 1, this table may have a different origin and/or history.

Large deviations can also be found in the 'fast' equations of Mars and Venus, especially when the values are larger than 2000. This is also the point when the difference between results from the approximation method for the radial distance and from square root (non-approximation) computation become perceptible. The latter produces slightly better values, but is far from

[45] Dvivedī, *Brāhmasphuṭasiddhānta*, pp. 436–42.

[46] It is reasonable to think that anyone using the circumference values of the *Ābh* would deploy a method based on linear interpolation for the size of the epicycle (and is thus different from the *DG*). However such a method does not agree any better with the table values. When rounded off to integers, 16 out of 30 values obtained from the *DG* method with $a = 25;48$ and $b = 1932$ (shown in Table 14) agree with the table values, 12 more are within ± 1, 1 is off by $+2$ and another by -2. On the other hand, when the size of the epicycle is linearly interpolated, 14 match the table values, 13 are different by ± 1, 2 of them were 3 minutes larger than the table and 1 was 4 minutes smaller. Thus I assume it more likely that the *DG* method had been used, although it remains a question how the parameters a and b had been derived in that case.

Table 14: 'Fast' equations of Mercury compared with computed values using a and b derived from the *Ābh* and those using a and b in the *DG*

	hemisphere ♑			hemisphere ♋		
$\varkappa$	edition	*Ābh*	*DG*	edition	*Ābh*	*DG*
6	101	100.0	100.0	223	223.5	223.4
12	199	198.6	198.6	435	433.6	433.8
18	296	295.6	295.7	624	623.8	624.4
24	390	390.4	390.7	791	789.8	791.0
30	483	483.2	483.7	931	930.2	932.1
36	573	573.1	573.9	1044	1043.8	1046.4
42	660	660.0	661.0	1134	1132.8	1135.9
48	744	743.9	745.3	1200	1199.4	1203.0
54	824	823.8	825.5	1245	1244.5	1248.4
60	900	899.5	901.6	1274	1271.9	1276.0
66	971	970.7	973.2	1283	1281.6	1285.8
72	1037	1036.7	1039.6	1276	1277.8	1282.0
78	1097	1096.5	1099.6	1262	1261.2	1265.3
84	1151	1149.7	1153.2	1233	1233.3	1237.3
90	1197	1196.4	1200.2	1197	1196.4	1200.2

perfect. It is possible that an approximation different from my interpretation is being used here.

4.3. Variant equation sets and their correspondence with variant parameters

As previously mentioned, there are two different sets of values for the 'fast' equation of Mercury and the 'slow' equation of Venus. The values labeled 'edition' in the previous tables, found in manuscript E and adopted in the edition, agree in general with equations computed from the divisors a and corrections b in manuscript B. As mentioned in Section 4.2, the 'fast' equations of Mercury seem to have been computed from slightly different parameters, and thus we cannot match the values in manuscripts E and B perfectly. Nonetheless, taking into account their distance from the other group, I shall categorize them as group 1.

Meanwhile the 'variants' of equations in manuscripts C and D correspond to a and b in manuscripts A, C and D (and the edition), which I have shown

Table 15: The two groups of values in the manuscripts

	A	B	C	D	E	(edition)
Divisor and correction	2	1	2	2	-	2
Equations	-	-	2	2	1	1

as variants in Section 3.5. I shall call them group 2. The result is given in Table 15. Unfortunately the edition has conflated the two groups.

It is remarkable that manuscripts A and E which are in the same bundle (L.1248-E and L.1248-J of the ORI&MSS) belong to different groups. This suggests that the two parts of the *DG* could be distributed separately at times.

5. Conclusion

While *vākya*s are thought to have been widely used in south India for computing calendars and casting horoscopes, explanations on how these values were computed by contemporary astronomers as well as modern historians have been lacking. This chapter is a first attempt to crack the *vākya*s of the *DG*, one of the most famed astronomical manuals in Kerala.

The verses in the second part of the *DG* refer to the equations, which are actually arc lengths, as Sines. This is such a significant mistake that it makes us doubt whether the entries had really been computed by the composers of the verses themselves. A possible scenario is that these arc values were first stored in a non-versified form, after which someone who misunderstood them as Sines, versified the values and incorporated them into the *DG*. However, this is countered by the fact that some variants in the table of equations can be explained by variants in *a* and *b*. Computations for generating the tables have been done more than twice, and it is difficult to imagine that every historical mathematician overlooked the difference between Sines and arcs.

The variants in *a* and *b* could have been the results of either the correction by Parameśvara himself or alterations (either accidental or intentional) in the transmission. Concerning the first scenario, it is worth considering the claim by Sarma that manuscript C might have been 'derived from a preliminary draft', on the basis of its readings.[47] Improvements from this draft did not involve modifications in the parameters, because the same values of divisors and equations appear in other manuscripts. This opens the possibility that the table of equations as found in the edition was the result of changes made by someone else. If so, we must investigate other texts that are assumed to have adopted the '*dṛk* system' in order to reassess the influence of this text.

Our study has shown that parameters underlying the table play an important role when working on equations. The next step is to investigate other texts that are assumed to have adopted the '*dṛk* system'. We should consider which of the tables or parameters were essential in the 'system', and also reassess the impact of the *DG* itself.

[47] Sarma, *Dṛgganita*, p. xviii.

Bibliography

Burgess, Ebenezer, and William Dwight Whitney, 'Translation of the Sûrya-Siddhânta, A Text-Book of Hindu Astronomy; With Notes, and an Appendix', *Journal of the American Oriental Society* 6 (1858), pp. 141–498.

Duke, Dennis, 'The Equant in India: The Mathematical Basis of Ancient Indian Planetary Models', *Archive for History of Exact Sciences* 59 (2005), pp. 563–76.

Dvivedī, Sudhākara, *Brāhmasphuṭasiddhānta and Dhyānagrahopadeṣādhyāya by Brahmagupta Edited with His Own Commentary*, Benares: Medical Hall Press, 1902.

Kern, Hendrick, *The Âryabhaṭîya, a Manual of Astronomy, with the Commentary Bhaṭadîpikâ of Paramâdîçvara*, Leiden: E.J. Brill, 1874.

Pingree, David, 'The Fragments of the Works of Al-Fazārī', *Journal of Near Eastern Studies* 29 (1970), pp. 103–23.

Pingree, David, 'History of Mathematical Astronomy in India', in Charles C. Gillespie (ed.), *Dictionary of Scientific Biography*, vol. XV (Supplement I), New York: Scribner's Sons, 1978, pp. 533–633.

Pingree, David, *Jyotiḥśāstra: Astral and Mathematical Literature*, Wiesbaden: Harrassowitz, 1981.

Plofker, Kim, *Mathematics in India*, Princeton: Princeton University Press, 2009.

Ramasubramanian, K., and M. S. Sriram, *Tantrasaṅgraha of Nīlakaṇṭha Somayājī*, London: Springer / New Delhi: Hindustan Book Agency, 2011.

Sarma, K. V., *Grahacāranibandhana: A Parahitagaṇita Manual by Haridatta*, Madras: Kuppuswami Sastri Research Institude, 1954.

Sarma, K. V., *Dṛggaṇita of Parameśvara*, Hoshiarpur: Vishveshvaranand Vedic Research Institute, 1963.

Sarma, K. V., *A History of the Kerala School of Hindu Astronomy (in Perspective)*, Hoshiarpur: Vishveshvaranand Vedic Research Institute, 1972.

Sarma, Sreeramula Rajeswara, 'From My Grandfather's Chest of Palmleaf Books', in: Saraju Rath (ed.), *Aspects of Manuscript Culture in South India*, Leiden / Boston: Brill, 2012, pp. 207–33.

Śāstrī, K. Sāmbaśiva, *The Āryabhaṭīya of Āryabhaṭācārya with the Bhāṣya of Nīlakaṇṭhasomasutvan - Part 1. Gaṇitapāda*, Trivandrum: Government Press, 1930.

Shukla, Kripa Shankar, and K. V. Sarma, *Āryabhaṭīya of Āryabhaṭa, Critically Edited with Translation*, New Delhi: Indian National Science Academy, 1976.

Appendix: Versified tables of equations

For the sake of brevity, I shall only list the Sanskrit verses with variant readings that give different numbers. Note that many variant readings, notably in manuscript C, that are different words but give the same number, are not indicated. For more details, the reader is advised to consult Sarma, *Dṛggaṇita*, pp. 18–25. The *kaṭapayādi* words are in ***bold***.

Mars: 'slow' epicycle

jāto balāya nanaraś cacāra nīlāṅgi ruddhaguḥ śramavit |
sunisā sumarma punitād dhūliṣu dhūrteṣu dhījitau vinasaḥ |
sūnārthī *kṣitisūnoḥ ṣaḍaṃśajā māndajā imā jīvāḥ* ||2.26||

Mars: 'fast' epicycle (4th and 1st quadrants)

gavayo madirā taruvit kṣitimāṃ stenārthī lavaja dāsārdhāḥ |
gokaṭakaṃ śivarūpaṃ rathālaye dugdhavṛṣṭi khuratāpam |
jalasevyaṃ karmajyaṃ himābdhaye *bhaumajā mṛgādibhavāḥ* ||2.27||

Mars: 'fast' epicycle (2nd and 3rd quadrants)

nādārta stṛṇarūpe sindhutaṭe kānanendra yānakharaḥ |
h***arigātre dehagurau vṛkabhadre yānavairi vittaguruḥ*** ||
hīnagara stailakharo vaṃśakarī dharmanetri hemadhiyaḥ |
karkyādibhavā evaṃ bhūmisutasya prakīrtitā jīvāḥ ||2.28||

Mercury: 'slow' epicycle

gātre kṣobhaṃ dhūrte nāḷaṃ nayakṛd dharāḍhya tadvaśyau |
raktapa māṃsapa sūdapa dugdhapa mānāri nṛkari bhūputraḥ |
śaṅkara *etā jīvā māndabhavā śaśisutasya pañcadaśa* ||2.29||

Mercury: 'fast' epicycle (4th and 1st quadrants)

yānapa dhūḷipa taddhari nidhigaṃ gajavad bisāśi nakṣāntaḥ |
vivasu rvaradan [nanaḷaḥ][48] ***kusudhīḥ saṅgānakṛt sudhājñatya*** |
niśipuṇyaṃ sādhupuṭaṃ *śaśijasya mṛgādijā imā jīvāḥ* ||2.30||

(Variant in Manuscript C)

[ānava][49] ***vānara henila kunavā mudhuvit sudeśa [dhīsākṣī]***[50] |
mokṣārthī saṃvādaṃ bhadrābdhir gandhadhīḥ sutajño'yam |
dhātrīpuṭaṃ bhujapuṭaṃ bhogaripur *jñasya caivam eṇādau* ||2.30||

Mercury: 'fast' epicycle (2nd and 3rd quadrants)

gururāḍ mṛgavad bharato yuddhārthī yogadhī bhavānnāyam |
bhṛgupuṇyaṃ nanurūpaṃ śivarūpaṃ vatsarasya gajarūpam |
tithirapi ratirapi gaṅgārūpaṃ sudhiyo'pi *śaśijakarkyādau* ||2.31||

[48] Suggested emendation by Sarma. Manuscripts read *naḷadaḥ* (zero-nine-eight).

[49] Suggested by Sarma. Manuscript reads *ālaya* (zero-three-one).

[50] Suggested by Sarma. Manuscript reads *dhāsvārthī* (nine-four-seven).

(Variant in Manuscript C)

sagara stṛṇavit sitamitaṃ nīlābjaṃ gosudhā kaḷānīkaḥ |
gajapo'yaṃ śiśurūpaḥ kuḷīrapa pauraloka nagalokau |
gurulokaḥ suraloko gathāpriya bhṛguripū *jñakarkyādau* ||2.31||

Jupiter: 'slow' epicycle

rāgī vitte stabdhaḥ surapo homasya sañjayaḥ [śukarād][51] |
kumbhāriḥ śītāri rmadirā līnāṅgi cadula tārāṅgī |
kubalaṃ balāṅgi *jīvā imā guror māndajāḥ kramaśaḥ* ||2.32||

Jupiter: 'fast' epicycle (4th and 1st quadrants)

dhīmān jayāya tithipaḥ kugirau bhūdātri śabala bhojāṅgam |
dhīreva strīsevā karkaśi tadvaṃśi tatsamaṃ punātu |
puruṣe saṅgatir *etā jīvasya mṛgādijā bhavanti jyāḥ* ||2.33||

Jupiter: 'fast' epicycle (2nd and 3rd quadrants)

dhījin māṃsapa timiraṃ nṛbalaṃ dugdhāṅgi homavin niyamaḥ |
lomaśa dhīdṛśi śakṛtā vigataṃ tadvṛtti dhībhṛtaṃ kṣubhitaḥ |
saṅgatir *etā jīvāḥ karkyādibhavāḥ surendrapūjyasya* ||2.34||

Venus: 'slow' epicycle

dīpa balaṃ java bāṇā netā kṣāntiḥ kathaṃ mitho [dāsāḥ][52] |
pāde guhā bhuja madaṃ mudā tadā *śukramandajīvāḥ syuḥ* ||2.35||

(Variant in Manuscripts C and D)

dhenuḥ sevyā mitrair bāla navau stambha rāma soma ratāḥ |
kṣiti nāthau rasa māsau māse māse *sitasya mandajyā* ||2.35||

Venus: 'fast' epicycle (4th and 1st quadrants)

śrīśuki līnagu vaśabhṛd vinatā gomāṃsi kunidhi narmanaṭī |
sindhupuṭe navaloke guhavṛki gorakṣayā pitṛsthasya |
gandhodaki[53] ***vīranare narmapure*** *iti bhṛgor mṛgādijyāḥ* ||2.36||

Venus: 'fast' epicycle (2nd and 3rd quadrants)

sakrodhi sāmatuṣṭā prītikarī cūtavakri śivacārī |
bāṇasukhaṃ guhasatre dehasukhaṃ valgusūtri sutacārī ||
dhīsamare gajabhadre lohagurau dhūrtaraudri narmapure |
bhṛgusūnoḥ karkyādau jīvā etāḥ krameṇoktāḥ ||2.37||

Saturn: 'slow' epicycle

garbhaṃ mahat sukhāḍhyaṃ dūtāyāḥ sunakhi vivari dāsāreḥ |
dhanagaḥ sthalago ratigo gajago dugdhāṅgi nākavit sukaviḥ |
naravid *[iti] sūryasutasya*[54] *jīvā mandodbhavāḥ kramādetāḥ* ||2.38||

[51] Suggested by Sarma. Manuscripts read *surarād* (seven-two-two).

[52] Suggested by Sarma. Manuscripts read *bhānoḥ* (four-zero).

[53] Manuscript C reads *tundāḷika* and D reads *tundāḷaki* (both six-eight-nine-one)

[54] Manuscripts read *naravit sūryasutasya* which lacks two syllables. Sarma inserts *hi* after *sūryasutasya* which is still insufficient. Thus I suggest inserting *iti* instead.

Saturn: 'fast' epicycle (4th and 1st quadrants)

> ***tailaṃ nārthe vinayaḥ śailāḍhye varṣake praḷaye*** |
> ***hayarāṭ prabhurāḍ gatirāḍ rājaśrī rdhīdharo lipigaḥ*** |
> ***[śubhrāṅgī rāgāṅgī]***[55] ***subalo*** *jīvāḥ*[56] *śaner mṛgādibhavāḥ* ||2.39||

Saturn: 'fast' epicycle (2nd and 3rd quadrants)

> ***bhāvaṃ sehe surapo gatikṛt sandhārya satkharaḥ śiśiram*** |
> ***satsūtraṃ siddhāgre lipigaḥ śarago balago helāṅgī*** |
> ***jālāṅgī sābālā*** *jīvāḥ karkyādijāḥ śaner etāḥ* ||2.40||

Sun: 'slow' epicycle

> ***kole sukhī nava ramā varṣe tīrthe tadā kṣudhā vinayaḥ*** |
> ***priyakṛd jayakṛd gurukṛt cirāya haraye dharāḍhya*** *ravijīvāḥ* ||2.41||

Moon: 'slow' epicycle

> ***kula gati gandhaṃ kharapo nṛmānya satsevya kunakhi vīraśrīḥ*** |
> ***gobhadre kīrtiśrī rmatsūtrī tajjarā vidhurā*** |
> ***dhīdhāri yānagaur*** *iti ṣaḍaṃśajā jyāḥ krameṇa śītaruceḥ* ||2.42||

[55] Suggested by Sarma. Manuscripts read *śukāṅgī rākāṅgī* (five-one-three, two-one-three).

[56] Sarma suggests *jīvā[ś ca]* but this does not fit the meter.

Cracking the *Tabulae permanentes* of John of Murs and Firmin of Beauval with Exploratory Data Analysis

Richard L. Kremer

Introduction

Historians of mathematics have developed, over the past generation, very effective statistical methods to 'squeeze' unknown parameters from astronomical tables whose underlying mathematical algorithms are known. Other techniques have been invented to determine the dependency of one table on another, for example, the dependency of a table of solar declinations on a particular table of sines. The challenge, in both instances, is to assure that the assumptions required for modern probability theory are met when sampling historical tables with relatively few entries, unknown computational errors, and many layers of scribal errors or physical deterioration of successive manuscript witnesses. Nonetheless, by using such techniques historians of early and medieval astronomy have been able to map through space and time, in considerable detail, the spread of parameters, tabular algorithms and tabular content, both within and among various cultural traditions.[1]

Some astronomical tables, however, exhibit irregularities that might reveal interesting computational practices that modern statistical methods would erase or not detect. Consider, for example, a table whose differing sections were computed with different parameters; or a table in which every n^{th} entry was computed and the intervening entries found by interpolation or smoothing by eye; or a table whose basic algorithm is unknown. In such cases, simple application of least squares or Monte Carlo methods might not elucidate how the tables were composed by historical actors. In such cases it may often prove more effective to apply 'exploratory data analysis' (EDA), a set of techniques developed in the 1960s at Bell Telephone Laboratories and formalized by the Princeton mathematician John W. Tukey in a well-known textbook that first appeared in the late 1960s. EDA looks for patterns in 'messy' data and precedes the application of confirmatory inference and testing procedures rooted in probability theory. EDA methods seek to resist 'wild' values of localized 'misbehavior' in data; to look for patterns in residuals between data and a potential fit (i.e., a

[1] Cf. Van Brummelen and Butler, 'Determining the Interdependence'; van Dalen, *Islamic Astronomical Tables*. For a much earlier discussion, see Kennedy, 'The Digital Computer'.

Editing and Analysing Numerical Tables: Towards a Digital Information System for the History of Astral Sciences, ed. by Matthieu Husson, Clemency Montelle and Benno van Dalen, PALS 2 (Turnhout, 2021), pp. 363–422

10.1484/M.PALS-EB.5.127703

known algorithm); to re-express data in scales or axes that uncover symmetries or other structures; and to reveal patterns by visual displays of each step of an analysis. These four tools – resistance, residuals, re-expression and revelation – can often contribute much to data analysis before one turns to more robust statistical tools to test hypotheses.[2]

In this paper, I will employ techniques from EDA, especially visual patterns in residuals, to seek the algorithm that was used by two astronomers of the early Alfonsine era, John of Murs and Firmin of Beauval, to compute a new double-entry table for finding the time correction between mean and true syzygy of the luminaries. Although they wrote a brief canon that explains how to use the table, the authors provided no hints about how they had constructed their table (most canons to medieval astronomical tables do not describe how said tables were constructed). Earlier true syzygy tables had been developed; yet none had reduced the problem to a function of solar and lunar anomaly, i.e., one enters the table with these two variables. Indeed, the structure of John's and Firmin's table is unlike anything discussed in the *Almagest* or any known Arabic zij. 'Cracking' the *Tabulae permanentes* (henceforth TP) thus may require tools other than least squares.

The TP (see Plate 12) are a large table, filling between 6 and 24 pages in the manuscript witnesses I have examined.[3] All 15 known witnesses identically lay out the table with 60 columns for the solar anomaly (angular distance of the mean Sun from its apogee or aux) at 6° intervals from 0° to 354° and 31 rows for the lunar anomaly (angular distance of the Moon from its mean apogee of the epicycle) at 6° intervals from 0° to 180° (the second half of the lunar anomalies, from 180° to 360°, are symmetrical to the first half and thus were never copied in the manuscripts). The double-entry table has a total of 1860 entries. Each sexagesimal entry specifies the time correction in hours and minutes, listed in adjacent columns. The column and row headers are generally specified by the number of physical signs of 30° and the number of degrees (only one manuscript uses natural signs of 60°, the basic format of the Parisian Alfonsine Tables, henceforth PAT). Hence, the top row and left column headers describe entries, respectively, for solar anomaly from $0^s\ 0°$ to $11^s\ 24°$ and lunar anomaly from $0^s\ 0°$ to $6^s\ 0°$; the bottom row and right column headers, respectively, for solar anomaly from $12^s\ 0°$ to $0^s\ 6°$ and lunar anomaly from $12^s\ 0°$ to $6^s\ 0°$ (where the sign of the entries is reversed). Signs of the entries are marked with 'm' for values less than zero, with 'a' for values greater than zero.

[2] Cf. Tukey, *Exploratory Data Analysis*; Hoaglin et al., *Understanding Robust and Exploratory Data Analysis*, pp. 2–4.

[3] In the manuscript of the Servitenkloster depicted in Plate 12, the table is copied in red and black ink with tabulated differences for both rows and columns. It has headings for *Argumentum solis* at top and bottom, and for *Argumentum lune* at left and right sides.

Mastering these layers of content in the TP must not have been obvious; most of the manuscripts include a set of canons or instructions that describe the formatting features of the tables. Most of the manuscripts use red and brown (or black) ink, successively by column, presumably to enhance legibility. Some, but not all, of the manuscripts also include rows and columns of tabulated differences (in minutes of time) between the successive entries. Tabulated differences might enable scribes to control for errors as they copied the tables; and they surely reduce computation required for double-interpolation. Three of the manuscripts tabulate differences only for the first column; presumably their scribes decided the benefit of that information was not worth the labor of adding it to their manuscript.

Once a user understood the format, handling the TP required only three steps: i) by means of a separate table of mean syzygies for the luminaries, find the date and time of mean syzygy and the values of the solar and lunar anomalies at that time (all quantities presumably given to minutes, the usual precision of Ptolemaic and Alfonsine astronomy); ii) with these latter two values enter the TP and extract the four contiguous entries that border the 'exact' values of the anomalies; iii) double-interpolate to extract the 'exact' value (to minutes of time) for the time correction.[4]

The TP and their canons have previously been edited.[5] In 2001, Porres and Chabás published an edition of the Latin canons, based on 8 manuscripts, with an English translation. As noted above, the canon provides no hints about the construction of the TP so we need not here consider that text. Porres's 2003 unpublished dissertation presents an edition of the TP, collating 5 manuscripts and including tabulated differences as well as the entries. Lacking knowledge of the algorithm, however, Porres could not control her edition for errors; it contains at least 59 deviations that do not appear in her 5 manuscripts nor the additional 10 examined here. Thus I have prepared a new edition of the TP, collating the 15 known manuscripts (see the Appendix).

Our investigation will proceed in several steps. Our first section situates the TP in early Alfonsine astronomy and tentatively suggests a date for its composition. The second section reviews slightly earlier methods for finding true syzygy, as offered in the Toledan Tables and by other Alfonsine astronomers, and will quantitatively compare the results of those methods with the entries found in the TP. By employing techniques of exploratory data analysis, we will move toward John's and Firmin's algorithm and will identify the sub-tables they incorporated into that algorithm. A third section (and the Appendix) presents

[4] For an epistemological analysis of the procedures needed to use the TP, see Husson, *Les domaines d'application*, pp. 203–20.

[5] Porres and Chabás, 'John of Murs's *Tabulae permanentes*'; Porres, *Les tables astronomiques*, pp. 395–406.

a new edition of the TP and comments on some of the scribal practices exhibited in the manuscript witnesses.

We will conclude that the TP represent yet another example, within the Alfonsine astronomical tradition, of seeking 'user-friendly' tabular formats to reduce computational labor for astronomers using the PAT to calculate eclipses or cast weather-predicting horoscopes for times of true syzygy.[6] Yet unlike many other Alfonsine user-friendly innovations, the TP introduce approximations into the computation that, as we will see, slightly degrade the precision of the results.

Finally, let me clarify several conventions used in this paper. First, I refer to individual entries in the TP by the notation 'solar anomaly:lunar anomaly'. For example, '294:24' refers to the entry for a solar anomaly of 294° and a lunar anomaly of 24°. Second, in counting what I shall call scribal errors, I consider each sexagesimal digit recorded in the table. Hence, if the vulgate entry 8h 45m is written 7h 45m, I record one scribal error; if it is written 7h 38m, I record two scribal errors. Errors for row or column slippage are similarly counted by sexagesimal digit. Third, all computation for this study has been realized via Microsoft Excel spreadsheets and their internal trigonometric functions. I have not tried to implement computational techniques that fourteenth-century Alfonsine astronomers would have employed. To a precision of minutes, my computations and theirs will generally agree. I have not found, for example, any sine table in medieval Latin manuscripts whose values differ, in minutes of arc, from my sine table. We may differ in how we round internal computations but their procedures remain opaque to us and any variations would rarely exceed ±1 minute of arc or ±1 minute of time.

Situating the TP in Early Alfonsine Astronomy

Given the task of this essay, we need not join the historiographical quest to fit together ever more biographical tessera for our two authors.[7] A native of

[6] The concept of 'user-friendliness' in Alfonsine astronomy has been proposed and extensively explored by Chabás and Goldstein. Cf. Chabás and Goldstein, 'Computing Planetary Positions'.

[7] Scholarship on John of Murs is vast and not always consistent in its claims. The initial surveys by Duhem, *Le Système Du Monde*, vol. IV, pp. 30–38, 54–60, and Thorndike, *A History of Magic*, vol. VII, pp. 294–324, must be used with caution. More recently, cf. Gushee, 'New Sources'; Michels, *Die Musiktraktate*; Busard, 'Die "Arithmetica speculativa"'; Poulle, 'John of Murs'; Beaujouan, 'Observations et calculs'; Poulle, 'Jean de Murs et les tables alphonsines'; L'Huillier, *Le Quadripartitum numerorum*; Saby, 'Mathématique et métrologie'; Gack-Scheiding, *Johannes de Muris Epistola*; Schabel, 'John of Murs and Firmin of Beauval's Letter'; Gushee, 'Jehan des Murs and His Milieu'; Hentschel, 'Johannes de Muris'; Chabás and Goldstein, *The Alfonsine Tables of Toledo*, pp. 277–81; Lejbowicz, 'Présentation de Jean de Murs'; Poulle, 'Les astronomes parisiens', pp. 5–35. For the latest surveys, cf. Desmond, *Music*

Normandy, John of Murs wrote works dating from about 1321 to 1347, during which he spent much time in Paris at the Sorbonne where he may have earned a *magister atrium* as early as 1321.[8] Yet in 1319, he measured solar altitudes during a stay in Évreux (Normandy). In 1326–27, he made astronomical observations at the royal abbey of Fontevraud at the behest of its abbess Alienor de Bretagne. From 1338 to 1342, he was a clerk of Philippe III d'Évreux, king of Navarre. In the 1340s, he was called to Avignon by Pope Clement VI to work on calendar reform; from 1342 to 1344, he served as canon of Mézières-en-Brenne in the diocese of Bourges. Clearly, the milieu of his career extended beyond Paris.

In addition to widely copied, elementary texts on music and mathematics, John wrote ten known astronomical works:[9]

1321	*Expositio tabularum Alfonsi regis Castelle*[10]
1321	*Kalendarium solis et lune*[11]
late 1320s	*Tabule principales* ('Tables of 1321')[12]
1329–32	*Kalendarium et patefit* ('Patefit Tables')[13]
1332	*Sermo de regulis computistarum*[14]
1339	*Canones tabularum Alfonsii* (i.e., for the PAT)[15]
1345	*Epistola super reformatione antique kalendarii, Tractatus de reformatione kalendarii* (with Firmin of Beauval)[16]
1345	*Prognosticatio super coniunctione Saturni* (triple conjunction of 1345)[17]

and the moderni, pp. 70–114, 246–47, 259–61; Nothaft, *Scandalous Error*, pp. 205–34; and the dedicated issue of *Erudition and the Republic of Letters* 4 (2019), edited by Nothaft, Desmond and Husson.

[8] Laure Miolo recently has hypothesized that John later became a *hospes* (a *magister* who rooms at the Sorbonne) rather than a *socius* (a fellow who teaches). See Miolo, 'In Quest of Jean de Murs's Library', pp. 17–18.

[9] I follow Nothaft in rejecting the long-held attribution of a 1317 computistic text with a sixteenth-century explicit, finis *kalendarii Ioannis de Muris de observantia termini pascalis*, to John of Murs. Cf. Nothaft, 'The Chronological Treatise'; Nothaft, 'John of Murs and the Treatise'.

[10] Edited by Poulle, 'Jean de Murs et les tables alphonsines'. Cf. Husson, 'L'astronomie alphonsine'.

[11] Excerpted by Chabás and Goldstein, 'John of Murs Revisited'.

[12] Excerpted by Chabás and Goldstein, 'John of Murs's Tables of 1321'. I accept Lejbowicz's and Desmond's dating (see note 7).

[13] Canons edited by Plassard, *Projets de réforme*, text 3, a work I have not seen. Cf. Kremer, 'John of Murs, Wenzel Faber'; Chabás and Goldstein, 'John of Murs's Tables of 1321', pp. 313–17.

[14] Edited by Plassard, *Projets de réforme*, text 4.

[15] Extant in two versions, neither edited. See Nothaft, 'Jean des Murs's Canones'.

[16] Edited by Schabel, 'John of Murs and Firmin of Beauval's Letter'.

[17] Edited by Pruckner, *Studien zu den astrologischen Schriften*, pp. 222–26. English translation in Goldstein and Pingree, 'Levi Ben Gerson's Prognostication', pp. 35–39.

1347?	*Epistola magistri Iohannis de Muris ad Clementen sextum* (conjunctions of 1357 and 1365)[18]
late 1340s	*Tabulae permanentes* (with Firmin de Beauval)[19]

Less is known about Firmin of Beauval, born in the diocese of Amiens and later canon at its cathedral.[20] In 1343, Pope Clement VI granted him a benefice in the diocese of Cambrai. In addition to collaborating with John of Murs on calendar reform and the TP, he independently authored a prognostication for the 1345 triple conjunction. He was widely known for his 1338 compendium of meteorological astrology, *De mutatione aeris* (Th/K 1220), a work that would be printed (without attribution) in 1485 by Erhard Ratdolt in Venice and in 1529 (with attribution) by J. Kerver in Paris.[21] Borrowing from at least 23 ancient Arabic and Latin authorities, with al-Kindī (28 references), Ptolemy (25), Albumasar (20), and Haly (17) most frequently cited, *De mutatione aeris* offers a comprehensive, clearly-written overview of its topic. Since the work draws on so many specialized sources, historian Gustav Hellmann has suggested that Firmin, while drafting the text, must have spent time in Paris with its large libraries. John of Murs also loaned Firmin astrological works, on the lunar mansions (Th/K 1095?) and others by al-Kindī on weather (Th/K 1364, 1383, 1385, 1515), topics treated extensively in *De mutatione aeris*. Presumably, the two men had worked together since at least the mid 1330s.

In any case, knowledge of the TP emerged rather late in the historiography of John of Murs. The early studies by Duhem and Thorndike (cf. note 7) do not mention the TP. Ernst Zinner's 1925 catalog of astronomical manuscripts lists 3 codices containing unnamed 'tables' with the incipit *Omnis utriusque sexus armonium* (my sigla BMeM) but does not attribute them to John and Firmin even though both men are named in the explicits of all three manuscripts. By 1937 Thorndike/Kibre listed Zinner's manuscripts and added V_1 but with no attribution; in 1948 Thorndike quoted from the explicit (*Canones pemanentium*) but wrote nothing about the structure of the tables and did not attribute the work to anyone. As far as I know, the TP were first attributed to John and Firmin (albeit with '?') in the 1963 revised edition of Th/K.[22]

[18] Edited in Boudet, 'La papauté d'Avignon', pp. 281–84. French translation in Duhem, *Le système du monde*, vol. IV, pp. 35–37.

[19] Canons edited in Porres and Chabás, 'John of Murs's *Tabulae permanentes*'; tables edited in Porres, *Les tables astronomiques*, pp. 395–406.

[20] Hellmann, 'Die Wettervorhersage'; Thorndike, *A History of Magic*, vol. III, pp. 268–80, 304–05; Boudet, *Le recueil des plus célèbres*, vol. I, pp. 511–12; Miolo, 'In Quest of Jean de Murs's Library', pp. 27, 37.

[21] Thorndike and Kibre, *Catalogue of Incipits*, rev. and augmented ed. (henceforth Th/K), col. 1220.

[22] Zinner, *Verzeichnis*, p. 343, nos 11196–98; Thorndike and Kibre, *Catalogue of Incipits*, 1st ed., col. 469; Thorndike, 'Some Little Known', p. 43; Th/K, col. 1004.

The first detailed biographical study of John of Murs, by Gushee in 1969, does not refer to the TP. Not until 1970 would the musicologist Ulrich Michels publish a chronological list of John's works including the TP that he dated to 1347–48. Michels justified this date by reference to John's collaboration with Firmin on the calendar, which is firmly dated to 1344–45. To strengthen his dating, Michels cited a short treatise *Contra tabulatores tabularum Alphonsi*, dated 1348 and attributed to John of Murs in one manuscript, that, he claimed, was a 'Traktat einer Gegengruppe', a vague reference I cannot understand. However, the content of this text is not related to the TP and the *Contra tabulatores* is no longer attributed to John.[23] Nonetheless, Michels was the first scholar to propose a date for the TP.

Subsequently, historians of astronomy have struggled to date the TP. As we will see below, the tables are indeed 'permanent' and contain no radices or quantitative information that could internally suggest a date. And none of the 15 currently known manuscripts presents a date. Poulle in 1973 briefly described the work but suggested no date. In 1977, John D. North, without justification, dated the TP to 'a. 1320/1'. Twenty years later, Chabás and Goldstein found our tables in several manuscripts (VC), but attributed them to John of Gmunden, dated to 1440 in the explicit for Gmunden's canon in V. In 2001 Porres and Chabás attributed the TP to John and Firmin, guessing the work was completed in 1321 along with John's (dated) *Tabula tabularum* (the text cited by Michels) and that both were intended to be part of his (dated) Tables of 1321.[24] Equally vaguely, Poulle in 2005 suggested that the TP 'perhaps' were composed between 1321 and 1327. But in this same article, Poulle parsed the explicit in the TP canons and wondered whether the TP might better be dated toward the end of John's career, i.e., in the 1340s.[25]

As can be seen above, I propose to date the TP to the late 1340s, perhaps after the 1345 collaborative treatise on calendar reform. This speculation relies solely on contextual evidence. First, as I will describe below, John of Murs from the beginning expressed interest in formulating astronomical tables around the

[23] Gushee, 'New Sources'; Michels, *Die Musiktraktate*, p. 15; Poulle, 'John of Murs', p. 133; Nothaft, 'Critical Analysis'.

[24] As Desmond, *Music and the* moderni, pp. 86–87 noticed, this early dating apparently derives from several misreadings. In his transcription of the canons to the *Tabula tabularum*, Poulle, 'Jean de Murs et les tables alphonsines', p. 144, mistakenly read 'calogia (*sic*)' for 'genealogia' in the list of four works John says he had composed in 1321 (on musical notation, squaring the circle, an *expositio tabularum Alfonsi* and a *genealogia astronomie*). And Porres and Chabás, 'John of Murs's *Tabulae permanentes*', pp. 65–66, mistakenly assigned the TP to one of these four works.

[25] Poulle, 'John of Murs', p. 130; North, 'The Alfonsine Tables in England', p. 298; Chabás and Goldstein, 'Computational Astronomy', pp. 99–100; Poulle, 'Les astronomes parisiens', pp. 22, 28–30.

phenomena of syzygies. His Tables of 1321 offer a novel method for computing true planetary and lunar longitudes from the times of their mean conjunctions with the Sun (as far as I know, no other astronomer ever employed this 'contratabula' approach, and John's tables are known from only two manuscripts).[26] Second, his so-called Patefit Tables explore two methods for finding times of true syzygy of the luminaries, one requiring the computation of separate solar and lunar corrections, the other employing lunar velocities (that he borrows from al-Battānī or the Toledan Tables). Both procedures start with a list of 251 dated consecutive mean syzygy times, beginning in 1321; both procedures are cumbersome, computationally, when compared to the TP.[27] Third, in his computation of the solar eclipses of 1333 and 1337 (presumably performed in those two years), John found the times of true syzygy using computational procedures outlined in John of Saxony's 1327 canons to the PAT.[28] Fourth, Firmin's 1338 *De mutatione aeris* contains two sections on the astrology of true syzygies of the luminaries (nearly one-quarter of the text). Fifth, John's 1339 very short canons to the PAT describe an iterative method for finding true syzygy, much more complex than that presented in the TP. Finally, for their proposed calendar reform, John and Firmin computed mean syzygy times for a full 19-year cycle, recently demonstrated by Nothaft to have been calculated for the years 1349–1367.[29] Given these circumstances, it seems highly unlikely that the TP, had they been constructed in 1321, would not have been mentioned in the Patefit Tables or John's canons of 1339 or the calendar treatise. It seems much more likely that John and Firmin composed the TP after (or during) their time together in Avignon from 1344–45. My supposition agrees, that is, with Michels's 1970 dating of the TP.

Finally, the explicit to the TP canons, found in six manuscripts, sheds further light on the collaborative projects of John and Firmin. The explicit appears as an elegiac distich, a literary form not displayed in other explicits in their works, or for that matter, in any other text I know of Alfonsine astronomy:

Expliciunt canones tabularum permanencium.

Ista Johannes equat, cepit Firminus et implet
Lux gaudet, reprobat livor, amicus habet.

John calculates them, Firmin began and finishes them
Light [knowledge] rejoices, envy rejects them, a friend cherishes them.

[26] North, 'The Alfonsine Tables in England', pp. 284–85; Poulle, 'The Alfonsine Tables and Alfonso X', pp. 3–4; Chabás and Goldstein, 'John of Murs's Tables of 1321'.

[27] Kremer, 'John of Murs, Wenzel Faber', pp. 148–55.

[28] Husson, 'Exploring the Temporality'.

[29] Nothaft, 'The Astronomical Data', pp. 118–21.

John and Firmin undoubtedly based this distich on two lines found in the twelfth-century epic poem, *Tobias*, by the French author Matthew of Vendôme, composed in hexameter couplets:

Transfert Hieronymus, exponit Beda, Matthaeus
Metrificat, reprobat livor, amicus habet.

Jerome translates, Bede explains, Matthew
versifies, envy rejects them, a friend cherishes them.

Much of Matthew's heavily didactic verse was intended for the schools and would have been widely known to schoolboys across France. The opening sentences of John's and Firmin's canon describe the 'science of predicting conjunctions and oppositions' as something important for everyone, common people (*vulgares*), men and women (*omnis utrisque sexus*). John and Firmin apparently saw in the poetry of Matthew of Vendôme a rhetorical frame to help situate the new TP in this public space.[30]

The Toledan Tables and True Syzygy

Translated from Hispano-Arabic materials into Latin during the twelfth century, the Toledan Tables circulated widely in medieval Europe. With canons extant in three versions and tables for most tasks faced by mathematical astronomers, the Toledan Tables strongly shaped the context in which Alfonsine astronomy would emerge, first in late thirteenth-century Castile and then in 1320s Paris. To explore the construction of the *Tabulae permanentes*, therefore, we begin with the Toledan Tables and their procedures for finding times of true syzygy.[31] We know that John of Murs and other Parisian astronomers of the 1320s were well aware of the Toledan Tables. Interestingly, their modern editor denigrated the true syzygy methods found in the Toledan Tables as 'triv-

[30] Müldener, *Matthaei Vindocinensis Tobias*, p. 22, lines 53–54; Porres and Chabás, 'John of Murs's *Tabulae permanentes*', pp. 67, 71. Among John's 24 known works in music, mathematics and astronomy, five include what musicologist Karen Desmond has called 'puzzle-like explicits in verse format'. His earliest work, the *Notitia artis musicae*, *c.* 1320, offers clever Leonine hexameters: *Nomen factoris signat deca signa doloris munda / Nec est mirum quia de cognomina firmum* ('The author's name signifies fittingly the ten signs of sorrow, it is not surprising for the name is firm'). As explicated by Michels, the tenth letter in the Latin alphabet is 'j'; the exclamation of pain is 'o', i.e., the abbreviation for 'Johannes'. The 'firm name' is de Muris, i.e., strengthened by walls. See Desmond, *Music and the* moderni, pp. 32–33; Michels, *Johannes de Muris, Notitia*, pp. 107, 110; Gushee, 'New Sources', p. 22.

[31] The methods of Ptolemy, 'slightly more unlike the truth but easier', and al-Battānī, 'more laborious indeed but more similar to the truth', are also discussed in the *Almagesti minor*, an early thirteenth-century commentary on Ptolemy's treatise. See Zepeda, *The First Latin Treatise*, pp. 446–55.

ial', a judgment that may or may not be true; nonetheless, those methods would be considered carefully by Alfonsine astronomers of the fourteenth century.[32]

The canons to the Toledan Tables describe three different procedures for finding true syzygy, albeit without identifying the historical roots of the methods. The canon identified by Pedersen as version Ca presents what was undoubtedly the best-known medieval method, borrowed from the *Almagest* VI.4.[33] Assuming that the ratio of the lunar to solar velocity, on average, is 13 to 1 and that these velocities remain constant over the time interval between mean and true syzygy, Ptolemy offered an iterative solution to find the interval between mean and true syzygy (Δt):

$$\Delta t(t) = \frac{-13\eta(t)}{12v_m(t)}, \qquad (1)$$

where t is the time of mean syzygy, η is the elongation between the luminaries at t or the difference of the corrections $c_m - c_s$ at t, and $v_m(t)$ is the lunar velocity at that time.[34] One enters this algorithm with the time of mean syzygy, computes Δt and then, for $t+\Delta t$, again computes η. If not zero (to whatever desired degree of precision, i.e., arcminutes or arcseconds), one then computes Δt at time $t+\Delta t$ and iterates the computation until the true distance between the luminaries (η) is reduced to zero.

As Chabas and Goldstein noted, Ptolemy's method converges quickly; to achieve a precision of one arcminute of longitude, one rarely needs more than two or three iterations. In Table 1, I compute for the 12 new moons of 1336 the time from mean to true syzygy, using the solar and lunar equations and velocities of the Toledan Tables (very similar to al-Battānī's equations and velocities).[35] After one iteration, the Ptolemaic method gets to within ±0;19h of the time of true syzygy; after two iterations, it gets close to the desired value. The computed true elongations after the second iteration are within 1 arcminute of longitude (one case of 2 arcminutes). But Ptolemy's algorithm is laborious; each iteration requires computation of the solar and lunar equations followed by sexagesimal multiplication and division.

[32] John of Murs annotated a copy of the Toledan Tables now in Paris, BnF, lat. 16211, fols 22r–98r, and in annotating other works referred to those tables. See Hentschel, *Sinnlichkeit und Vernunft*, pp. 271–78; Desmond, *Music and the* moderni, p. 54; Miolo, 'In Quest of Jean de Murs's Library', pp. 20–21, 26–27, 35–36; and Pedersen, *The Toledan Tables*, vol. I, p. 76.

[33] Pedersen, *The Toledan Tables*, vol. I, pp. 274–75; vol. II, pp. 450–51.

[34] See Chabás and Goldstein, 'Nicholaus de Heybech', pp. 93–94.

[35] See Suter, *Die astronomischen Tafeln*, pp. 175–80; Pedersen, *The Toledan Tables*, vol. IV, p. 1412; Goldstein, 'Lunar Velocity in the Middle Ages', p. 90. Velocity tables of Alkhwarizmi also circulated with the Toledan Tables, but al-Battānī's appear much more frequently in extant manuscripts. Cf. Pedersen, *The Toledan Tables*, vol. IV, pp. 1410, 1417, 1419.

	TT	TT	Ptol		Ptol–TT	Kam–TT	Batt – TT
Date	t (h)	Δt (h)	Δt (h)		Δt (m)	Δt (m)	Δt (m)
13.01	21;30	10;19	10;10	+0;09	0	−7	−6
12.02	10;14	13;24	13;19	+0;04	1	−4	−5
12.03	22;58	13;20	13;29	−0;08	−1	6	4
11.04	11;42	10;26	10;45	−0;18	−1	14	13
11.05	0;26	5;39	5;50	−0;10	−1	8	8
9.06	13;10	−0;12	−0;11	+0;00	−1	1	1
9.07	1;54	−6;00	−6;11	+0;13	−2	−8	−6
7.08	14;38	−10;43	−11;01	+0;18	0	−13	−9
6.09	3;22	−13;30	−13;38	+0;08	0	−6	2
5.10	16;06	−13;26	−13;18	−0;07	−1	7	14
4.11	4;50	−10;14	−10;01	−0;11	−2	9	12
3.12	17;34	−4;37	−4;29	−0;08	0	6	6

Table 1 : Times and time corrections for mean and true conjunctions of the luminaries for 1336, computed for the meridian of Toledo to a precision of minutes, computed with equations and velocities of the Toledan Tables (TT). Col. 3 gives 'exact' time corrections that reduce η to zero, rounded to arcmins. Col. 4 shows time corrections from the first and second iteration of Ptolemy's method. Col. 5 compares Ptolemy's and the 'exact' corrections; col. 6 compares the second method in the TT canons (Ibn al-Kammād) and the 'exact' corrections; col. 7 compares the third method in the TT canons (al-Battānī) and the 'exact' corrections.

A slightly less laborious method was proposed early in the twelfth century by the Andalusian astronomer Ibn al-Kammād, which uses the actual rather than averaged solar and lunar velocities and could be deployed in a tabular format to reduce computational labor.[36] As with Ptolemy's method, the velocities are assumed to remain constant over Δt. This method is described in canons Ca and Cc of the Toledan Tables.[37] As rendered by Chabas and Goldstein:

$$\Delta t(t) = \frac{-\eta(t)}{v_m(t) - v_s(t)}. \tag{2}$$

Double-entry tables, based on Ibn al-Kammād's method, perform the division, for arguments of η from 0;30 to 12;00 in half-degree steps and differences in velocity in 30 arcsec/hour steps from 0;27,30 to 0;33,30. But users of these tables still must compute, for the time of mean syzygy, solar and lunar corrections, look up the solar and lunar velocities, and then interpolate in the double-entry table to achieve a precision of minutes. As can be seen in Table 1,

[36] Chabás and Goldstein, 'Computational Astronomy', pp. 94–96.
[37] Pedersen, *The Toledan Tables*, vol. I, pp. 274–75, vol. II, pp. 642–43.

Ibn al-Kammād's method can misestimate the times of true syzygy by up to one-quarter of an hour.

A third method, that I shall call al-Battānī's, appears in canons Cb and Cc of the Toledan Tables.[38] Described, respectively, as 'a more refined and a surer manner' and 'more precise than any other', this method computes true elongation at the time of mean syzygy and then adds 13/12 of half of this amount to the lunar argument at the time of mean syzygy ($\bar{\alpha}$). This corrected lunar argument is then used to find the corrected lunar velocity, to be employed in an Ibn al-Kammād division to get the time correction. In other words, rather than setting the (ever-changing) lunar velocity at time t, the rate of lunar velocity is set at the midpoint of the interval between mean and true syzygy, presumably a rate closer to the average rate of the velocity over that entire interval. Although not explicated as such in the canons, we can easily demonstrate the assumptions behind this algorithm.

As per Ptolemy's method:

$$\Delta t(t) = \frac{-13\eta(t)}{12v_m(t)}, \text{ where } \eta(t) = c_m(t) - c_s(t). \tag{3}$$

Half of this time interval is:

$$\frac{\Delta t(t)}{2} = \frac{-13\eta(t)}{24v_m(t)}. \tag{4}$$

Now, the Toledan Tables' lunar argument moves in the epicycle at a fixed rate of 0;32,39,45°/hour; the lunar longitude, viewed from the Earth, moves at an average rate of 0;32,56,27°/hour. To find the amount moved by the lunar argument in $\frac{1}{2}\Delta t$, we compute:

$$\begin{aligned}\bar{\alpha}_{\text{corr}}(t) &= \bar{\alpha}(t) + 0;32,39,45 \cdot \frac{-13\eta}{24v_m} \\ &= \bar{\alpha}(t) + \frac{0;32,39,45}{0;32,56,27} \cdot \frac{-13\eta}{24} \approx \bar{\alpha}(t) - \frac{13\eta}{24}\end{aligned} \tag{5}$$

The Toledan Tables canons do not attribute this method to anyone. But as Pedersen noted, its basic outlines can be found in Chapter 42 of al-Battānī's zij.[39]

[38] Ibid., pp. 452–53, 708–09.

[39] Ibid., p. 76; Nallino, *Al-Battānī sive Albatenii*, vol. I, p. 96, lines 10–17, where al-Battānī says the corrected lunar anomaly is selected for a time 'inter' (but not the midpoint) mean and true syzygy. Interestingly, al-Battānī (vol. I, p. 93, line 38 – p. 94, line 3 and p. 95, line 33 – p. 96, line 2) also proposed that the lunar anomaly be corrected by a factor of 7/24 rather than 13/24. He did not clearly justify this value, but Schiaparelli (in Nallino, *Al-Battānī sive Albatenii*, vol. I, pp. 273–74) showed how it can be derived from Ptolemy's final lunar model and parameters and application of the small-angle approximation, tan(x) = x. Al-Battānī also briefly described Ptolemy's method (see Nallino, *Al-Battānī sive Albatenii*, vol. I, p. 94, lines 26–34). The authors of the Toledan Tables canon thus selected from among various methods in al-Battānī's zij.

One further correction, also not justified in the canons, is described. After the correction of Eq. 5, one enters a small auxiliary table for the 'seconds of difference', provided in many but not all Toledan Tables manuscripts (also in al-Battānī's zij), that corrects the lunar velocity, set as a function of the corrected lunar argument, by one arcsec of longitude/hour less than the degrees (from 1 to 7) of absolute value of the true elongation at t.[40] Hence:

$$v_{m\,\mathrm{corr}}(\bar{\alpha}_{\mathrm{corr}}(t)) = v_m(\bar{\alpha}_{\mathrm{corr}}(t)) \pm 0;00,01(|\eta(t)| - 1^\circ), \tag{6}$$

where the final value in Eq. 6 is positive for the lower half of the epicycle $90^\circ < \bar{\alpha} < 270^\circ$) and negative for the upper half. If $|\eta(t)| < 1^\circ$, no correction is made. Neither the canons to the Toledan Tables nor al-Battānī's zij justify this correction; both simply describe the procedure for using the auxiliary table. The al-Battānī method, in the Toledan Tables, thus becomes:

$$\Delta t(t) = \frac{-\eta(t)}{v_{m\,\mathrm{corr}}(\bar{\alpha}_{\mathrm{corr}}(t)) - v_s(t)}. \tag{7}$$

As suggested by Giovanni Schiaparelli in his notes to Nallino's edition of al-Battānī, the auxiliary table compensates for the second lunar anomaly, i.e., for the fact that as time passes between mean and true syzygy, the lunar epicycle does not remain at the aux (solar apogee), where the lunar equation of center is null. As the epicycle leaves the aux, its distance from Earth decreases and the apparent lunar velocity, viewed from the Earth, proportionally increases. This idea of correcting the apparent lunar velocity is very similar to Ptolemy's general treatment of epicycles changing their distance from the Earth by tabulating the maximal or minimal distances and then using the method of proportional parts to find the intermediate distances.[41]

In Ptolemy's (and al-Battānī's) lunar model, the eccentricity (e) is 10;19 parts, the radius of the deferent (R) is 49;11 parts, and the radius of the epicycle (r) is 5;15 parts. The maximal Earth-epicycle distance is $R + e$, the minimal distance is $R - e$. As the center of the epicycle moves 180° from apogee to perigee, the Earth-epicycle distance thus varies by 20;38 parts. The angle being swept out, from the Earth, in this motion is double the mean elongation or $2\bar{\eta}$.

[40] Al-Battānī's version, as edited in Nallino, *Al-Battānī sive Albatenii*, vol. II, p. 88., presents entries one arcsec larger than those in the Toledan Tables, as edited by Pedersen, *The Toledan Tables*, vol. IV, p. 1414. The auxiliary table is frequently recorded in the Toledan corpus.

[41] For a useful analysis of Ptolemy's approach to varying epicycle distances, see North, *Richard of Wallingford*, vol. III, pp. 185–92. As North shows, the method of proportional parts assumes that the parameter in question varies linearly between its exactly determined minimal and maximal values and, as such, generally introduces errors (deviations from the geometry of the Ptolemaic models) of several arc minutes into the final longitudes being computed. For other analyses of such approximations, see Petersen, 'The Three Lunar Models'; Van Brummelen, 'Lunar and Planetary Interpolation'; Husson, *Les domaines d'application*, pp. 267–70.

The change in distance, starting from the apogee, as $2\bar{\eta}$ increases from 0 to 180°, is thus:

$$\Delta d(2\bar{\eta}) = 20;38\left(\frac{2\bar{\eta}}{180}\right). \tag{8}$$

Viewed from the center of the epicycle, the lunar anomaly (radius vector in the epicycle) moves at a rate of about 33 arcmins/hour or 1980 arcsecs/hour [$v(\alpha)$]. With the epicycle fixed at the apogee, the apparent lunar velocity viewed from the Earth [$v(2\bar{\eta})$] is reduced because of the greater distance between the center of the epicycle and the Earth ($R + e$). The ratio of the velocities is inversely proportional to the distances from the centers:

$$\frac{v_1}{v_2} = \frac{d_2}{d_1}. \tag{9}$$

For the situation at apogee in Ptolemy's lunar model:

$$\frac{v(2\bar{\eta})}{v(\alpha)} = \frac{5;15}{60}, \quad \text{or} \quad v(2\bar{\eta}) \approx \frac{v(\alpha)}{12}. \tag{10}$$

As the epicycle moves toward perigee, its distance from the Earth decreases and the apparent lunar velocity, $v(2\bar{\eta})$, proportionally increases. The change in apparent lunar velocity, as a function of $2\bar{\eta}$, is thus given by the velocity at maximal distance multiplied by the change in distance, as a function of $2\bar{\eta}$, divided by the minimal distance:

$$\begin{aligned}\Delta v(2\bar{\eta}) &= \frac{v(\alpha)}{12}\cdot\frac{2e}{R-e}\cdot\frac{2\bar{\eta}}{180}\\ &= \frac{1980\ \text{arcsecs/hr}}{12}\cdot\frac{20;38}{39;22}\cdot\frac{2\bar{\eta}}{180}\\ &= 0.96\bar{\eta} \approx \bar{\eta}\ \text{arcsecs/hr}.\end{aligned} \tag{11}$$

This derivation is based on Ptolemy's final lunar model, in which the distance of the epicycle from the Earth is a function of $2\bar{\eta}$. Yet our canon Cb instructs us to enter Eq. 11 with the true elongation, 2η, at the time of mean syzygy. The auxiliary table thus selects a velocity correction for the *time of true syzygy*, when $2\bar{\eta}$ will have moved from the aux roughly the same amount as 2η had at the time of mean syzygy, although the method selects the average lunar velocity for the *midpoint* in the time interval between mean and true syzygy. Whoever formulated this method fully understood Ptolemy's terse discussion (*Almagest* V.7) of the method of proportional minutes and its simplifying approximations and was willing to combine what look to us as inconsistent techniques in an effort to estimate the changing lunar velocity over the time between mean and true syzygy. This method clearly reflects a more sophisticated understanding of the problem of approximating a changing lunar velocity than those revealed in Ptolemy's and Ibn al-Kammād's methods.

On the other hand, a comparison of time corrections computed with Ibn al-Kammād's uncorrected and al-Battānī's corrected lunar velocities (Table 1, columns 6 and 7) shows that the effects of the latter are small, shifting the Δt values by no more than 8 minutes of time and not necessarily closer to the 'correct' values. Two iterations of Ptolemy's method yield Δt corrections closer to the 'correct' values than do either of the other two methods presented in the canons of the Toledan Tables. We might guess that whoever formulated the third method thought carefully about the geometry of Ptolemy's lunar model but did not seek to evaluate the efficacy of the approximations by direct computation of true syzygy time corrections, such as we have offered in Table 1.

Early Alfonsine Astronomers and True Syzygy

In addition to the Toledan Tables, John and Firmin surely had access to early Alfonsine materials as they formulated the *Tabulae permanentes.* As is well known, only the canons of the Castilian Alfonsine Tables have survived. Chapter 30 of this text discusses conjunctions and oppositions of the luminaries.[42] It mentions tables that provide, for collected years and months, the dates and hours of mean syzygies and for those times the mean longitudes of the luminaries, lunar arguments of anomaly and lunar arguments of latitude.[43] And it describes Ptolemy's method for computing true syzygy, using solar and lunar equation tables to find the true elongation at the time of mean syzygy and then dividing 13/12 of this amount by the 'hourly lunar velocity' not otherwise specified. The canons do not attribute the method to anyone and do not explain its rationale. The Castilian canons offer nothing not already found in the Toledan Tables on the true syzygy question.

The Tables of 1322, compiled in Paris probably near that date by John of Lignères, are devoted primarily to eclipses. John borrowed most of this material (28 of the 32 tables) directly from the Toledan Tables. Several chapters of the canons treat true syzygies, describing the methods of Ptolemy and Ibn al-Kammād much as they had been presented in the Toledan Tables. Interestingly, John includes the small table for correcting the lunar velocity as a function of elongation, implying that it should be used for both methods; but he did not describe al-Battānī's method, where the lunar correction table had originated, as found in the Toledan Tables.[44]

John of Lignère's *Tabule magne*, compiled in Paris *c.* 1325, include two tables and a canon for finding true syzygy, presenting now only the method of Ibn al-Kammād, i.e., dividing the true elongation at mean syzygy by the *superatio* (the term had originated in the Toledan Tables) or difference between the

[42] Chabás and Goldstein, *The Alfonsine Tables of Toledo*, pp. 56–57, 188–89.

[43] For an example of such tables, long found in zijes, see Pedersen, *The Toledan Tables*, vol. IV, pp. 1327–40.

[44] Erfurt, UFB, CA Q377, fol. 44v; Saby, *Les canons de Jean de Lignères*, pp. 221–25, 426. An edition of these tables and canons is currently being prepared by M.-M. Saby and J. Chabás.

lunar and solar velocities at that time. A single-entry table gives the solar and lunar equations and velocities as functions of the solar and lunar anomalies at mean syzygy. A double-entry table gives the time correction (to seconds!) as a function of the *superatio* (from 0;27, 0;28 ... to 0;34°/hr) and the true elongation. To reduce the labor of interpolation, John broke the latter table into two parts, one with entries for elongations of 1, 2, ..., 8 degrees, the other with entries for elongations of 1, 2, ..., 36 minutes (for arguments greater than 36 minutes, one must enter this table twice; perhaps he employed this format so that the minutes table could be written into a single folio?). Ibn al-Kammād had included a similar table in his twelfth-century zij, which was translated into Latin in 1260. But such a table had not circulated with the Toledan Tables and it is unknown whether John of Lignères knew about Ibn al-Kammād's table.[45] The *Tabule magne* thus reveal a Parisian astronomer crafting (or reformatting) a double-entry table for true syzygy.

Not surprisingly, the earliest and most widely copied canons to the PAT, John of Saxony's 1327 *Tempus est mensura motus primi mobilis* (*editio princeps*, 1483), also include a chapter on true syzygy.[46] John outlined a three-step procedure that does not require any dedicated tables. First, using the corrected lunar argument and correcting the lunar velocity with an algorithm he describes in words rather than introducing the small table (Eq. 5 and 6), one computes the time correction (τ), following al-Battānī's method. Second, one computes the elongation at time $t + \tau$ and again at time $t + \tau + 0;01$ day. He then computes a final time correction:

$$\tau^* = \frac{-\eta^*}{d\eta},$$

where η^* is the true elongation at $t + \tau$, and $d\eta$ is the change in true elongation over 1/60th of a day or 24 minutes. Rather than using approximated hourly lunar velocities as in the first step, now the exact elongations are found for a 24-minute interval. The time of true syzygy is given by $t + \tau + \tau^*$. Previously I have shown that John's procedure invariably yields time corrections that agree, to the nearest minute, with times generated by brute-force computation of true

[45] See the article by Husson in this volume; Chabás and Goldstein, 'Andalusian Astronomy', p. 3.

[46] Two earlier studies examined John of Saxony's method, but neither fully understood its procedures or its relation to al-Battānī's method as presented in the Toledan Tables. See Chabás and Goldstein, 'Nicholaus de Heybech', pp. 269–71; Kremer, 'Thoughts on John of Saxony's Method'. Both of these studies were hampered by accepting the text of the *editio princeps* as authoritative, although its printer, Erhard Ratdolt, presumably used only one, late manuscript that at times had been inconsistently reworded by scribes. In a later publication, I will offer a critical edition of John's true syzygy chapter, collating the early and more reliable manuscripts to construct a text closer to John's original.

elongations at one-minute intervals with the PAT.[47] Thus, by 1327, Alfonsine astronomers had available a computation technique that yielded 'exact' results to minutes and required no dedicated tables. Yet John's method is very laborious; one computes the true elongation and divides by the velocities not once but three times. The TP would reduce this labor considerably.

Finally, we must mention John of Murs's 1339 *Canones tabularum Alfonsii*. They tersely describe the true syzygy problem, in only 134 words. As explicated by Nothaft, this canon merely summarizes John of Saxony's iterative method without fully describing all the steps. Only an 'expert-level reader', well practiced in John of Saxony's method, could have followed the 1339 canon, concluded Nothaft.[48] Evidently, by 1339 John of Murs had not yet invented the TP.

Cracking the *Tabulae permanentes*

Rather than simply announcing the algorithm behind the *Tabulae permanentes*, I will instead summarize the steps I followed diachronically in seeking that algorithm. Tracing this 'exploratory data analysis' will enable us more fully to compare the various methods available, in the early fourteenth century, to find true syzygy. Lacking sources, we have no way of knowing, of course, what 'exploratory' steps John of Murs himself followed as he worked on the problem.

To visualize the performance of the various algorithms, I will present only the residuals, in minutes of time, between their results. I will not deploy statistical tools but rather show the differences at 30° intervals in the two arguments of the TP, viz., the solar and lunar anomalies ($\bar{\kappa}$ and $\bar{\alpha}$). Because the time interval is symmetrical around 180° of lunar anomaly, we shall consider the performance of the algorithms only for $0° \leq \bar{\kappa} \leq 360°$ and $0° \leq \bar{\alpha} \leq 180°$. This limited display of the data adequately illustrates the performance of the various algorithms for our purposes. Given our dating of the TP, we will initially assume that John of Murs took his solar and lunar equations from the PAT and velocity tables of John of Genoa (formulated at least by 1332).[49]

We begin with the *Tabulae permanentes* themselves, the tables whose algorithm we are seeking (see the Appendix for our edition of the full set of tables).

[47] Kremer, 'Wenzel Faber's Table', p. 14; Kremer, 'Thoughts on John of Saxony's Method', pp. 271–76.

[48] Nothaft, 'Jean de Murs's Canones', p. 116.

[49] The PAT equations are given at one-degree intervals, John of Genoa velocities at six-degree intervals. In both cases, I linearly interpolate as required and do not round intermediate results. I began with John of Genoa velocities, as edited from Paris, BnF, lat. 7282, fols 129r-v by Goldstein, 'Lunar Velocity in the Ptolemaic Tradition', pp. 12–13, since Nothaft recently has shown how John of Murs in his canons of 1339 may have been influenced by John of Genoa's *Canones eclipsium* of 1332. Cf. Nothaft, 'Jean de Murs's Canones', pp. 117–22, and the forthcoming study of John of Genoa's oeuvre by Laure Miolo.

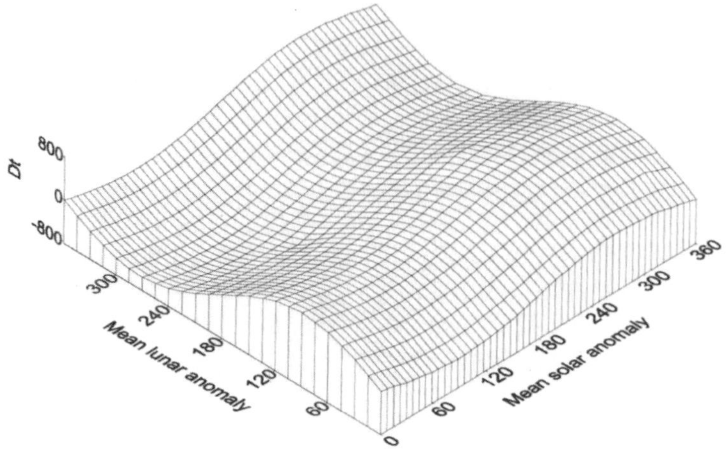

Fig. 1: TP in minutes of time. From Porres and Chabás, 'John of Murs's *Tabulae*', p. 64.

	Argumentum solis											
Arg lune	**0**	**30**	**60**	**90**	**120**	**150**	**180**	**210**	**240**	**270**	**300**	**330**
0	0;00	-2;19	-4;04	-4;47	-4;14	-2;29	0;00	2;29	4;14	4;47	4;04	2;19
30	4;59	2;42	0;58	0;17	0;49	2;34	5;01	7;26	9;10	9;4”	8;59	7;15
60	8;31	6;21	4;41	4;01	4;33	6;14	8;34	10;54	12;33	13;03	12;22	10;42
90	9;40	7;38	6;40	5;27	5;57	7;32	9;44	11;55	13;28	13;56	13;18	11;44
120	8;15	6;19	4;50	4;16	4;44	6;13	8;18	10;21	11;50	12;1”	11;40	10;11
150	4;42	2;51	1;26	0;53	1;19	2;45	4;44	6;42	8;07	8;33	7;59	6;34
180	0;00	-1;49	-3;13	-3;46	-3;20	-1;57	0;00	1;57	3;20	3;46	3;13	1;49

Table 2 : Abridged TP in hours, for $0 < \bar{\alpha} < 180°$. Signs are reversed for $180 > \bar{\alpha} > 360°$.

Table 2 abridges my edition to 30° intervals. As can be seen, the time corrections range from 0 to ±14 hours of time. Fig. 1 offers a graphical representation of the tables, in minutes of time.

Seeking John of Murs's algorithm, I began with Ptolemy's method, widely known by both Arabic and Latin astronomers and well described in the Toledan Tables and John of Lignères's Tables of 1322. As can be seen in Table 3, Ptolemy's results, in their first iteration, differ systematically from the TP, with deviations reaching a maximum of 10 minutes of time. Given that the maximal time interval is 14 hours, we might assume that a match to 10 minutes could capture the underlying algorithm of the TP; i.e., we do not know at what level of precision John of Murs was working. However, the distribution of the residuals is systematic, not random, and we must suspect that Ptolemy's algorithm misses some of John's algorithm. And iterating Ptolemy's algorithm a second time nicely reduces the true elongations to less than one arcminute of longitude but increases the deviations from the TP to more than 12 minutes of time.

	Argumentum solis											
Arg lune	**0**	**30**	**60**	**90**	**120**	**150**	**180**	**210**	**240**	**270**	**300**	**330**
0	0	1	1	2	2	2	0	-2	-2	-2	-1	-1
30	-1	0	0	-1	0	-1	-3	-3	-3	-2	-1	-1
60	2	0	0	0	-1	-1	-1	0	1	2	3	3
90	7	4	3	1	1	1	3	5	8	9	9	8
120	7	5	4	2	2	3	4	7	8	10	10	9
150	4	2	2	0	1	1	2	5	6	6	6	5
180	0	-2	-2	-2	-2	-1	0	1	2	2	2	2

Table 3: Residuals, Ptolemy's method minus TP, in minutes of time.

	Argumentum solis											
Arg lune	**0**	**30**	**60**	**90**	**120**	**150**	**180**	**210**	**240**	**270**	**300**	**330**
0	0	0	0	0	0	0	0	0	0	0	0	0
30	0	0	0	-1	0	0	0	1	1	1	1	1
60	2	0	1	1	1	1	2	3	4	4	4	3
90	4	2	1	1	1	2	4	6	8	8	7	5
120	2	1	1	0	0	1	2	5	4	5	4	4
150	1	0	1	0	0	0	0	2	2	2	1	0
180	0	0	1	0	0	0	0	0	0	0	-1	0

Table 4: Residuals, Ibn al-Kammād's method minus TP, in minutes of time.

Ibn al-Kammād's method requires slightly more work as it adds the solar velocity to the computation (Eq. 2). As can be seen in Table 4, this method also differs systematically from the TP, although by several minutes of time less than Ptolemy's deviations. Incorporating the solar velocity, rather than estimating this value as did Ptolemy, improves, but only slightly, our fit to the TP. A comparison of Tables 3 and 4 shows how well Ptolemy estimated the relative velocities of the luminaries with his 13/12 factor.

The third method described in the Toledan Tables sets the lunar anomaly at the midpoint rather than the beginning of the time interval between mean and true syzygy and corrects the lunar velocity for the second lunar anomaly (Eq. 5). We begin by considering the size of the adjustment of the lunar anomaly ($\bar{\alpha}$). As can be seen in Table 5, correction to the middle of the time interval can shift the lunar anomaly by up to 4 degrees, with the magnitude of the shift directly proportional to true elongation between the luminaries.

If we use this corrected lunar anomaly in the Ibn al-Kammād algorithm (i.e., the third method of the Toledan Tables but without correcting the lunar velocity in Eq. 6) we find a much improved fit to the values of the TP (see Table 6).

	Argumentum solis											
Arg lune	**0**	**30**	**60**	**90**	**120**	**150**	**180**	**210**	**240**	**270**	**300**	**330**
0	0	-1	-1	-1	-1	-1	0	1	1	1	1	1
30	1	1	0	0	0	1	1	2	2	2	2	2
60	2	2	1	1	1	2	2	3	3	3	3	3
90	3	2	2	1	2	2	3	3	4	4	4	3
120	2	2	1	1	1	2	2	3	3	4	3	3
150	1	1	0	0	0	1	1	2	2	3	2	2
180	0	-1	-1	-1	-1	-1	0	1	1	1	1	1

Table 5: $\bar{\alpha}_{\text{corr}}$ minus $\bar{\alpha}$, in degrees.

	Argumentum solis											
Arg lune	**0**	**30**	**60**	**90**	**120**	**150**	**180**	**210**	**240**	**270**	**300**	**330**
0	0	0	0	0	0	0	0	0	0	0	0	0
30	-1	0	0	-1	0	0	-1	1	0	0	0	0
60	0	-1	0	0	0	0	0	0	0	0	0	0
90	1	0	0	0	0	0	0	1	1	0	0	0
120	0	0	1	-1	0	0	0	1	0	0	0	0
150	0	0	1	0	0	0	0	1	0	0	-1	0
180	0	0	1	0	0	0	0	0	0	0	-1	0

Table 6: Residuals, Ibn al-Kammād's method with $\bar{\alpha}_{\text{corr}}$ minus TP, in minutes of time.

Indeed, this procedure matches, to the nearest minute of time, eighty percent of the entries in our abridged table; residuals for the remaining 16 entries amount to only ±1 minute of time. These residuals show no obvious symmetries around 180° of the solar anomaly or 90° of the lunar anomaly. Clearly, we have moved much closer to John of Murs's algorithm for the TP.

We next complete the third method by including the corrected lunar velocity in our computation, a correction presented in John of Lignères's Tables of 1322 and John of Saxony's canons of 1327 (Eq. 6). Table 7 indicates the magnitude of these corrections, rounded to arcsecs/hour of lunar velocity, as spread across the arguments of the TP. As a function of the true elongation, this correction will never exceed 6 arcsecs/hour for Ptolemy's geometrical models and the parameters of the PAT.

Table 8 presents the residuals between the TP and the fully implemented third method of the Toledan Tables, i.e., adding the small corrections to the lunar velocity. Comparison of Tables 6 and 8 suggests that John of Murs did not correct the lunar velocities in his algorithm; as the size of the lunar velocity corrections grows, so do the residuals between our algorithm and the TP. We conclude that John of Murs simplified the third method by ignoring the small

	Argumentum solis											
Arg lune	**0**	**30**	**60**	**90**	**120**	**150**	**180**	**210**	**240**	**270**	**300**	**330**
0	0	0	1	1	1	0	0	0	1	1	1	0
30	1	0	0	0	0	0	1	2	3	3	3	2
60	3	2	1	1	1	2	3	4	5	5	5	4
90	4	3	2	2	2	3	4	5	6	6	6	5
120	3	2	2	1	2	2	3	5	5	6	5	4
150	2	1	0	0	0	1	2	3	4	4	4	3
180	0	-2	-3	-3	-3	-2	0	0	1	1	1	0

Table 7: Corrections to lunar velocity, in arcsecs/hour.

	Argumentum solis											
Arg lune	**0**	**30**	**60**	**90**	**120**	**150**	**180**	**210**	**240**	**270**	**300**	**330**
0	0	0	0	0	-1	0	0	0	1	0	0	0
30	0	0	0	-1	0	0	0	1	1	1	1	0
60	1	0	0	0	0	0	1	2	2	2	2	1
90	-1	-1	0	-1	-1	-1	-1	-1	-2	-2	-3	-2
120	-1	-1	0	-1	-1	0	-1	-1	-2	-2	-2	-1
150	0	0	1	0	0	0	0	0	-1	-1	-1	-1
180	0	0	0	0	0	0	0	0	0	0	-1	0

Table 8: Residuals, third method minus TP, in minutes of time.

table presented in al-Battānī's zij and the Toledan Tables and described by both John of Lignères and John of Saxony.

To test this finding, I next examined the performance of my algorithm, replacing the equations of the PAT with those of the Toledan Tables (maximum solar equation of 1;59,10 rather than 2;10,00 and maximum lunar equation of argument of 5;01,00 rather than 4;56,00) and with other velocity tables for the luminaries that appear in fourteenth-century manuscripts. We need not rehearse the particulars of these explorations. As we might expect, using the Toledan equations introduces systematic shifts of more than ±30 minutes of time in the results of the algorithm and destroys its close match to the values of the TP. Inserting the lunar velocities of the Toledan Tables (essentially those of al-Battānī), which range from 0;30,18 to 0;36,04 °/h, increases the differences in Table 8 to ±13 minutes of time. Inserting lunar velocities attributed in a single manuscript to John of Lignères,[50] that essentially truncate to two

[50] Apparently, the solar and lunar velocity tables in Oxford, BL, Canon. Misc. 499, fols 41v–42r, 154v–155r, were first attributed to John of Lignères by Rosińska, *Scientific Writings*, p. 408, based on two marginal annotations by a later hand in this manuscript. Goldstein,

significant digits the velocities of John of Genoa and correct significant errors in the two final values of the latter (see Table 9), slightly worsens the fit to the TP, especially for the lunar anomaly of 180°. I have found no other lunar velocity table, circulating in Paris the first third of the fourteenth century, that improves the fit in Table 6.

To complete this exploratory analysis, I considered finally the performance of my proposed algorithm against the full TP with 1860 entries. Using my initial version of John of Genoa's velocities, my algorithm matched, to the nearest minute of time, 1400 of the entries. The residuals, mostly ±1 minute of time, were randomly scattered across the table; however, for lunar arguments of 96° and 102°, deviations of 1–2 minutes of time appear for nearly every entry (a pattern not seen in my earlier analysis using the abbreviated 30-degree intervals of the table). Clearly one component of my algorithm, a function of the lunar anomaly, was not behaving properly.

Two components of the algorithm are functions of the lunar anomaly, viz., the lunar equations of PAT and the lunar velocities of John of Genoa. Presumably the former are well known. Some years ago, Goldstein showed that the latter include some discrepant values. Goldstein proposed an algorithm for John of Genoa's lunar velocities, based on Ptolemy's final lunar model, i.e., taking into account both anomalies:

$$v_m(\alpha) = 0;32,56 - 0;41,49 \cdot \left[c_m(\alpha + 1) - c_m(\alpha)\right] \qquad (12)$$

where c_m are the lunar equations of PAT and the velocity is a function of the true rather than mean lunar anomaly (however, at mean syzygy, the mean and true lunar anomalies are equal).[51] Using Eq. 12, I found four significantly discrepant values in John of Genoa's lunar velocities at 6° intervals. And I began to collate additional copies of these lunar velocities, expanding beyond Goldstein's single witness and confirming the four discrepant values he had identified (see Table 9).[52]

'Lunar Velocity in the Ptolemaic Tradition', pp. 11–14, accepted this attribution and noted that the same velocity table is found in an early printed edition of the *Tabulae resolutae*. This velocity table, however, does not appear in the two sets of tables firmly attributed to John of Lignères (the *Tabule magne* and the Tables of 1322). In a talk at the Paris Observatoire in September 2018, Alena Hadravova argued that Oxford, BL, Canon. Misc. 499, is associated with several other manuscripts (Toruń, University Library, 74; Prague, National Library, X.B.3 and Cracow, Biblioteka Jagiellońska, 610), all originating in Prague around 1450. I would guess that the abridged (only to seconds) and corrected version of John of Genoa's lunar velocities, found in Oxford, BL, Canon. Misc. 499, originated in Prague around 1450 and was copied into some manuscripts containing the *Tabulae resolutae*. At some point well after 1450, an anonymous annotator attributed the tables to John of Lignères, a claim I find dubious. We have yet to find another scribe who repeated the attribution in Oxford, BL, Canon. Misc. 499.

[51] Goldstein, 'Lunar Velocity in the Ptolemaic Tradition', pp. 8–9.

[52] Using PAT equations and factors of 0;32,56,28 and 0;41,48,00 in Eq. 12, my computed lunar velocities match, to the nearest third, the values in my working edition collating 14 manuscripts, except for one case with a deviation of 1 third. Laure Miolo, currently editing John of Genoa's astronomical works, kindly provided me with a list of relevant manuscripts. I have

Lunar argument ($\bar{\alpha}$)	Mss value	Computed value
96	0;33,08,31	0;33,05,31
102	0;33,30,36	0;33,23,36
174	0;36,53,15	0;36,51,15
180	0;36,58,54	0;36,53,20

Table 9: Deviations in John of Genoa's table of lunar velocities.

Interestingly, if I insert the recomputed 'correct' values into John of Murs's algorithm for lunar arguments of 96° and 102°, I increase the number of exact fits in the TP for those rows from 6 to 84 (of the 120 entries). But the 'correct' velocities for lunar arguments of 174° and 180° decreases the number of exact fits from 106 to 89 (of the 120 entries). It appears as if John of Murs used a copy of John of Genoa's lunar velocities with 'correct' values for lunar arguments 96° and 102° and 'incorrect' values for lunar arguments 174° and 180°. Of my 14 witnesses for John of Genoa's velocities, the two Oxford manuscripts give the 'correct' values for lunar arguments 102° and 174°. I have not found a copy of John of Genoa's velocity tables that lists 'correct' values for lunar arguments 96° and 102° and 'incorrect' values for lunar arguments 174° and 180°. But apparently John of Murs had such a table as he computed entries for the TP.[53]

In any case, if we 'correct' John of Genoa's lunar velocities only for lunar arguments 96° and 102°, my algorithm matches the values of the TP to the nearest minute of time in 1504 of the 1860 cases (81 percent) cases (Table 10). The deviations of ±1 minute of time are scattered randomly across the table, a claim from exploratory data analysis that I ground not in statistics but simply from observing that the sums of the non-zero cases by row and by column are fairly uniform and the distribution of the black and shaded cells across the table reveals no obvious patterns. The non-zero residuals in Table 10 reflect 'noise' in John's computational procedures (rounding, interpolation, truncation, etc.) and not any systematic variation arising from differences between his and my algorithm. Only in two cases do John's results differ by 2 minutes of time from mine.

collated Paris, BnF, lat. 7282, fol. 129r; BnF, lat. 7286C, fol. 56v; BnF, lat. 7295A, fol. 137r; BnF, lat. 7284, fol. 55r; Vatican, BAV, Reg. lat. 1241, fol. 152v; Vatican, BAV, Pal. lat. 446, fol. 93r; Vatican, BAV, Pal. lat. 1374, fol. 47r; Vatican, BAV, Ott. lat. 1826, fol. 148v; Cracow, Biblioteka Jagiellońska, 459, fol. 30v; Cracow, Biblioteka Jagiellońska, 563, 93r; BJ 613, fol. 60r; Prague, National Library, XIII-C-17, fol. 169r (all fifteenth century) and Oxford, BL, Hertford 4, fol. 148r; Oxford, BL, Digby 97, fol. 130v (both fourteenth century).

[53] Interestingly, John of Gmunden computed a table of lunar velocities at 1° intervals to thirds, with factors very close to those used by John of Genoa. His values vary by no more than several thirds from John's. Edited by Porres, *Les tables astronomiques*, pp. 329–33. John of Gmunden's velocities frequently appear in codices also bearing the TP (BPSVUV$_2$). If I use John of Gmunden's velocities in my algorithm, I match to the nearest minute 1471 entries (75 percent), only slightly fewer than when I use John of Genoa's velocities with two 'incorrect' values. We might wonder whether any early users of the 6 manuscripts holding both the TP and John of Gmunden's velocities ever used the latter to explore the former.

Table 10: Residuals, best-fit algorithm minus TP, in minutes of time. Residuals in black cells are 1 minute, in shaded cells are –1 minute. The bottom row indicates the number of non-zero residuals (out of 31) in each column; the right-most column indicates the number of non-zero residuals (out of 60) in each row.

	Argumentum solis																													
Arg lune	**0**	**6**	**12**	**18**	**24**	**30**	**36**	**42**	**48**	**54**	**60**	**66**	**72**	**78**	**84**	**90**	**96**	**102**	**108**	**114**	**120**	**126**	**132**	**138**	**144**	**150**	**156**	**162**	**168**	**174**	**180**
0	0	0	0	0	-1	0	0	-1	0	0	0		0	0	0	0	0	0	0	0	0	0	0		0	0	0	0	0	0	0
6	0		0	0	0	0	0	0	0	0	0		0	0	0	0	0	0	0	0	0	0	0	0	0		0	0	0	0	0
12			0	0	0	0	0	-1	0	0	0	0	0	0	0	0	0	0	0				0		0	0	0		0	0	0
18	0	0	0	0	0	0	0	-1	0	-1	0	0	0	0	0	0	0	0	0	0	0	0	0	0	0	0	0		0	0	0
24	0	0	0	0	0	0	0	-1	0	-1	-1		0	0	0		0	0	0		0	0	0	0	0		0		0		0
30			0	0	0	0	0	0	0	0	0		0		0		0		0		0	0	0	0	0	0	-1	0	0		
36	0		0	0	0	0	0	0	0	-1	-1	0	0	0	0	0	0	0	0	0	0	0	0	0	0	0	0	0	0	0	0
42	0		0	0	0		0	0	0	0	0		0	0	0	0	0	0	0	0	-1	0	0	0	-1	0	0	0	0	0	0
48	0	0	0	0	0	0	0	-1	0	0	0	0	0	0	0	0	0	0	0		0	0	0	0	0		0				0
54	0	0	0	0		0	0	-1	0	-1	0	0	-1	-1	0	0	0	0	0	0	0	0	0	0	0	0	0	0	0		0
60	0	0	-1	0	0			0	0	0	0	0	0	0	0	0	-1	0	0	0	0		0		0	0	0	0	0	0	0
66	0		0	0	0		0	-1	-1	0	0	0	0	0	0	0	0	0	0	0	0	0	0	0	-1	0	0	0	-1	0	0
72	0	0	0	0	0	0	0	0	0	0	0	0	0	0	0	0	-1	0	0	0			0	-1	-1	0	0		0		0
78	0		0	0	0	0	0	0	0	0	0	0	0	-1	0	-1	-1	0	0	0			0	0	0	0	0	0	0	0	0
84	0		0	0	0	0	0	0	0	0	0	0	0	-1	0		0	0	0	0	0	0	0	0	0		0		0	0	0
90	-1	0	0	0	0	0	0	0	0	0	0		0	-1	0	0	0	0	0	0	0		0			0	0			0	0
96	0	0	0	0	0	0	0	-1	0		0	0	0	0	0		0	0	0	0			0		0	0	0	0	0	-1	-2
102	-1	0	0	0	0	0	0	-1	0	0	0	0	0	0	0		0	0	0		0	0	0		0	0	0	0	0	0	0
108	0	0			0			0	0	0	0	0	0	0	0	0	0	-1	0	0	0	0	0	0	0	0	0		0		
114	0	0	0	0	0	0	0	0	0	0	0	0	0	0	0	0	0	0	0	0	0	0	0		0	0	0	0	0	0	0
120	0	0	0	0	0	0		0	0	0	-1	0	0	0	0		0	0	0	0	0		0		0	0	0		0	0	0
126	0		0	0	0	0	0	0	0	0	-1	0	0	0	0	0	0	0	0	0	0	0	0		0	0	0	0	0	0	0
132	0		0	0	0	0	0	0	-1	0	0	0	0	0	0	0	0	0	0	0	0	-1	0	0	0	0	0	0	0	0	0
138			0	0	0	0	0	0	0	0	0		0	0	0		0	0	0	0	0	0	0	0	0	0	0	0		0	0
144	0	0	0	0	0	0	0	0	0	-1	-1	0	0	-1	0		0	0	0	0	0	0	0	0	0	0	0	0	0	0	
150	0	0	0	0	0	0	0	0	0	-1	-1		0	0	0	0	0	0	0	0	0	0	0	0	0	0	0	0	0		0
156	0	0	0	0	0	0	0	0	0	0	-1		0	0		0	0	0	0	0	0	0	0		0	0	0	0	0		0
162	0	0	0	0	0	0	0	0	0	0	0	0	0	-1	0		0	0	0		0	0	0		0	0	0	0	0	0	0
168	0	0	0	0	0	0	0	0	0		0		0	0	0	0	0	0	0	0	0	0	0	0	0	0	0	0	0	0	0
174	0		0	0	0	0	0	0	0	0	0	0	0	-1	0	0	0	0	0	0	0	0	0		0	0	0		0	0	0
180	0	0	0	0	0	0	0	0	0	0	-1	0	0	0	0	0	0	0	0	0	0	0	-1	0	-1	0	0	0	0	0	0
Non-0	**5**	**12**	**2**	**1**	**2**	**4**	**3**	**9**	**2**	**8**	**8**	**10**	**1**	**8**	**1**	**10**	**3**	**2**	**0**	**6**	**5**	**8**	**1**	**13**	**5**	**4**	**1**	**10**	**4**	**9**	**4**

	Argumentum solis																												
Arg lune	**186**	**192**	**198**	**204**	**210**	**216**	**222**	**228**	**234**	**240**	**246**	**252**	**258**	**264**	**270**	**276**	**282**	**288**	**294**	**300**	**306**	**312**	**318**	**324**	**330**	**336**	**342**	**348**	**354**	**Non-0**
0	0	0	0	0	0	0	-1	0	0	0	0	0	0	0	0	0	0	0	-1	0	0	0		0	0		0	0	0	**8**
6	0	0	0	0	-1	0	0	-1	0	0	-1	0	-1	0	0	0	0	0	0	0	0	0	0	0	0	0	0	0	0	**7**
12	0	0	0	0	0	0	0		0	0	0	0	0	0	0	0	0	0	0	0	0	0		0	0	0	0	0	0	**10**
18	0		0	0	0	0	0	-1	0		0	0	0	-1	-1	0	0	0	0	0		0		0	0	0	0	0	0	**10**
24	0	0	0	0	-1	0	-1	0	0	0	-1	0	0	-1	-1	0	0	0	0	0	0	0	0			0	0	0	0	**16**
30	0	0	0	0	-1	0	0	0	0	0	-1	0	0	0	0	0	0	0	0	0	0	0	0	0	0	0	0	0	0	**12**
36	0	0	-1	0	-1	0	0	0	-1		0	0	0	0	0	0	0	0	0			0	0	0	0	0	0	0	0	**9**
42	0	0	-1	0	-1	-1	-1	0	-1		0	0	0			0	0	0	0			0	0	0	0	0	0	0	0	**15**
48	0	0	0	-1	-1	0	0	0	0		0	0	0	0	0	-1	-1	-1	-1	0	0	0	0	0					0	**17**
54	0	0		0	0	0	0	0	-1	0	0	0	0	0	0	0	0	0	-1	0	0	0	-1	0	0	0		0	0	**11**
60	0	0		0	0	0	0	0	-1	0	0	0	0	0	0	0	0	0	-1	0		0	0	0	0	0	0	0	0	**10**
66	0	0	0	0	-1	0	-1	-1	-2	0	0	0		0	0		0	0	-1	0	0	0	0	0	0	0	0	-1	0	**14**
72	0	0	0	0	0	0	-1	-1	-1	0	-1	0	0	0	0	0	0	0	0	0	0	0		0	0	0	0	0	0	**12**
78	0	0	0	0	0	-1	-1	-1	-1	0	0	0	0	0	0	0	0	0	0	0	0	-1	0	0	0	0	0	0	0	**11**
84	0	0	0	0	0	0	-1	0	0	0	-1	0		0	0	0	0	0	0		0	0		0	0			0	-1	**13**
90	0	0	0	0	-1	-1	-1	-1	-1	-1	-1	0	0	-1	0	0	0	0	-1	0		0			0		0	0	-1	**22**
96	-1	0	0	0	0	0	0	0	0	0	-1	0	0	0	0	0	0	0	0	0		0	0	0	0					**15**
102	0	0	-1	0	-1	-1	0	0	0	0	0	0	0	-1	0	0	0	0	0	0	0		0	0	0	0	0	0	0	**10**
108	0	0	0	0	-1	-1	-1	0	-1	0	0	0	0	-1	0	0	0	0	0	0	0	0		0	0	0	0	0	-1	**15**
114	0	0	0	0	-1	0	-1	0	0	0	0	0	0	0	0	0	0	0				0		0	-1	0	0	0	0	**8**
120	0	0	-1	0	-1	0	0	0	0	0	0	0	0	0	0	0	0	0	0	0	0	-1		0	0	0	0		0	**11**
126	0		0			0	0	0	0		0	0	0	0	0	0	0	-1	-1	0		0	0	0	0	0	0	0	0	**10**
132	0			0	0	0	0	0	0	0	0	0	0	0	0	0	0	0	0			0		0		0	0	0	0	**9**
138		0			0	0	-1	0	0	0	0	0	0	0	0	0	0	0	0			0		0	0	0		0	0	**13**
144			0	0	0	0	0	0	0	0	0	0	0	0	0	0	0	0	-1	0	0	0		0	0		0	0	-1	**11**
150	0	0	-1	0	-1	-1	-1	0	0	0	0	0	0	0	0	0	0	-1	0			0		0	0	0		0	0	**13**
156	0	0	0	0	0	0		0	0	0	0	0	0	-1	-1	-1	0	0	-1	0	0	-1	0	0	0	0	0	0	0	**11**
162	0	0	0	0	0	0	0	0	0	0	0	0	0	0	0	0	0	0	0			0		0	0	0	0	0	-1	**8**
168	0	0	0	0	0	0		0	0	0	0	0	0	0	0	0	0	0	0		0	0	0	0	0	0	0	0	0	**4**
174	0	-1	-1	0	0	0		0	0		0	0	0	0	0		0	0	0		0	0	0	0	0		0		0	**12**
180	0	0	0	0	0	0	0	0	0	0	0	0	0	0	0	0	0	0	0		0	0	0	0	0	0	0	0	0	**4**
	3	**5**	**10**	**3**	**14**	**6**	**15**	**7**	**9**	**7**	**7**	**0**	**3**	**7**	**4**	**4**	**1**	**3**	**10**	**11**	**12**	**4**	**15**	**2**	**4**	**7**	**6**	**5**	**6**	

	Argumentum solis											
Arg lune	**0**	**30**	**60**	**90**	**120**	**150**	**180**	**210**	**240**	**270**	**300**	**330**
0	0	0	0	0	0	0	0	0	0	0	0	0
30	-1	0	0	-1	0	0	-1	0	0	-1	0	0
60	0	-1	0	0	0	0	0	0	4	3	4	-1
90	1	0	0	0	0	0	0	6	5	5	5	5
120	0	0	1	-1	0	0	0	1	4	4	4	0
150	0	0	1	0	0	0	0	1	0	0	-1	0
180	0	0	0	-1	-1	0	0	0	1	1	0	0

Table 11: Time correction residuals, 'computed directly' minus TP, in minutes of time. Cf. Table 5.

Finally, we might ask how closely the approximative algorithm of the TP reproduces the time correction for true syzygies computed directly (i.e., no velocity tables required) with the PAT, i.e., manually varying the time correction by one-minute intervals until the true elongation is less than one minute of longitude (Table 11). These directly computed time corrections consistently match those of the abridged TP to ±1 minute of time for corrections less than about 11 hours. For corrections exceeding 11 hours, the TP corrections differ from the directly computed values by up to 6 minutes of time. Hence, if Alfonsine astronomers had sought to compute to a precision of minutes, the approximations in the algorithm of the TP would have thwarted their goal for about two or three of the 24 or 25 syzygies in any given year. Only the more laborious methods of Ptolemy or John of Saxony could have achieved a precision of minutes for all PAT syzygy computations.

Peurbach and the *Tabulae permanentes*

Georg Peurbach in the 1450s prepared an expanded version of the TP, decreasing the intervals of solar argument from 6 to 2 degrees and the intervals of lunar arguments from 6 to 1 degrees. Peurbach's version thus increases the size of the table from John of Murs's 1860 to 32,400 entries. With smaller distances between respective entries, Peurbach's version makes it easier for users to interpolate in the double-entry table. However, the differences between successive entries can still exceed 10 minutes of time and this user, at least, cannot perform the double interpolations simply with mental arithmetic.[54]

Many manuscript copies of Peurbach's expanded version are known, including:

[54] For a pioneering study of interpolation procedures for double-entry tables of the early fourteenth century, see Husson, 'Ways to Read a Table'.

Nuremberg, SB, Cent. V 57, fols 112v–136r. Includes Peurbach's eclipse canon (Th/K 1562) and tables. Regiomontanus autograph ('Tabula eclipsium M. Georgii Peurbach preceptoris mei', fol. 108r), Vienna, dated 1460–1461. Also contains Regiomontanus's autograph of Giovanni Bianchini's astronomical tables.[55]

Paris, BnF, lat. 7288, fols 17r–40r. Includes Peurbach's eclipse canon (Th/K 1562) and tables.

Munich, BSB, Clm 19550, fols 163v–187r. Includes Peurbach's eclipse canon (Th/K 1562) and tables, John of Saxony's canon and parts of the PAT. Mid 15th c., Tegernsee.[56]

Vienna, ÖNB, lat. 5412, fols 186v–211r. Includes Peurbach's eclipse canon (Th/K 1562), explicit dated 1501, and tables, in same hand.[57]

Venice, Biblioteca Nazionale Marciana, lat. 342. Peurbach's autograph eclipse canon (Th/K 1562) and tables for the meridian of Grosswardein (Tabulae Waradienses), dated 1460.[58]

Vienna, ÖNB, lat. 5291, fols 100–63. Includes Peurbach's eclipse canon and tables for Grosswardein. Regiomontanus autograph.

Editio princeps, 1514, sig. a3v-d3r.[59] Part of Peurbach's eclipse canon and tables.

Did Peurbach know John of Murs's algorithm and compute the additional entries in the expanded version? Or did he simply create the additional values by linear interpolation from the entries at six-degree intervals in John's table? Comparison of the two versions indicates that Peurbach carefully copied all the values John had provided; even if Peurbach knew John's algorithm, he certainly did not independently recompute the values John had provided.[60]

[55] Neske, *Die Handschriften der Stadtbibliothek*, pp. 90–91.

[56] Porres, *Les tables astronomiques*, p. 79.

[57] Porres and Chabás, 'John of Murs's *Tabulae permanentes*, p. 65, date this copy of the true syzygy table to 1444. That date appears in this codex, in the explicit of a copy of John of Gmunden's treatise on the albion (fol. 154v), but the quires containing Peurbach's canon and expanded version of the TP clearly were copied in 1501.

[58] I have not seen this manuscript that was, according to Valentinelli, *Bibliotheca Manuscripta*, vol. II, pp. 265–66, owned by Bessarion, who wrote on the flyleaf: *Tabulae eclypsium solis et lunae, noviter compositae per non minus philosophum quam doctum, doctissimum tamen virum Georgium de Peuerbach.*

[59] Tanstetter, *Tabulae eclypsium*.

[60] I have found only three instances where Peurbach's times differ from those of the TP. For entry 126:138, three manuscripts and the printed version read 3;17h for the correct 3;07h (Vienna, ÖNB, lat. 5412 gives the correct value); for 300:156, four manuscripts and the printed version read 7;05h for the correct 7;04h; and all four manuscripts and the printed version make a two-place column slip at 354:60. None of these differences match those I find in the

To explore whether Peurbach interpolated or recomputed the additional entries we can compare his values against those computed, by linear interpolation, from the TP and against those computed according to the algorithm we have found for John of Murs's tables. To avoid complications of trying to determine which scheme for double-entry interpolation may have been used, I limit this analysis to interpolations within a single column in the table, i.e., to single-entry interpolation.[61] Since gaps between entries are largest at the ends rather than middle of a column, I consider three columns from Peurbach's table, for solar arguments of 0°, 6° and 108° and lunar arguments in one-degree intervals from 0° to 42°. If I linearly interpolate between John's values at 6° intervals, I match Peurbach's values (to the nearest minute of time) for our three test columns in 67, 77, and 72 percent of the 43 cases. If I recompute the columns with John's algorithm, I match Peurbach's values in 54, 42 and 86 percent of the same cases. These results do not conclusively prove that Peurbach interpolated or recomputed. But since I consider it quite improbable for Peurbach to have known about John of Genoa's lunar velocities, it seems likely that Peurbach interpolated rather than recomputed his expanded table. Perhaps he smoothed the interpolations with mental rather than pencil-and-paper operations, which might explain why about one-third of his cases vary from my exact interpolation by 1 minute of time.[62]

The Manuscript Witnesses

At present, the *Tabulae permanentes* are found in fifteen manuscripts, all dating from the middle third of the fifteenth century (see the Appendix for full descriptions and details).[63] No other astronomical work by John of Murs has so many surviving witnesses.[64] As can be seen from Table 12, the manuscripts fall into several groups. The earliest witness (B), from the 1430s and offering the most mathematically consistent version of the tables (i.e., fewest scribal errors), is an autograph by the early fifteenth-century Viennese astronomer, John of Gmunden. Another five manuscripts (SPUVG) embed the TP within copies of Gmunden's tables, date from the 1430–50s, have relatively few scribal errors, and in all cases but one tabulate differences between successive entries in both

15 manuscript witnesses for the TP, so I cannot add the Peurbach manuscripts to my stemma in Fig. 2.

[61] Cf. Husson, 'Ways to Read a Table'.

[62] I will offer further analysis of Peurbach's interpolation techniques in a later publication.

[63] Porres and Chabás, 'John of Murs's *Tabulae permanentes*', pp. 64–65. E was first identified by Husson, *Les domaines d'application*, p. 242. I thank José Chabás for generously helping to identify additional manuscripts not listed in these earlier studies.

[64] John's *Musica speculativa*, written at the Sorbonne in 1323 and revised in 1325, survives in 58 copies according to Desmond, *Music and the* moderni, p. 100. For critical editions of this text, see Falkenroth, *Die 'Musica speculativa'*; Fast, *Johannes de Muris Musica*.

MS sigla	Canon	Placement in codex	Date	Scribal errors	Tabulated differences
B	Omnis (explicit)	within Gmunden's tables	1433–1437	1	yes
S	no	within Gmunden's tables	1437	2	yes
P	Omnis (explicit)	within Gmunden's tables	1436–1439	14	yes
V	Sciendum	within Gmunden's tables	1440–1443	9	no
U	Omnis abbreviated	within Gmunden's tables	1444	11	yes
G	Omnis	within Gmunden's tables	1458	8	yes
Pa	no	with JL's Tables of 1322	1442	9	yes
E	no	with JL's *Tabule magne*	1446	32	no
Me	Omnis (explicit)	with Hermann of Saxony's tables	1450	54	yes
N	Omnis	with Oxford Tables	1452	31	yes
C	no	with Bianchini's tables	1452	21	yes
V_3	no	with JL's Tables of 1322	1450	41	first col.
M	Omnis (explicit)	within Oxford Tables	1450	62	first col.
V_2	Omnis (explicit)	within Oxford Tables	1458	25	first col.
V_1	Omnis (explicit)	with mean syzygies	1463–1464	25	no

Table 12: Overview of the fifteen manuscripts containing the TP.

rows and columns. With the explicit referring to John of Murs and Firmin appearing only in B and P, scribes copying the 'Gmunden manuscripts' may well have assumed that Gmunden had authored the TP (as did some twentieth-century historians).

A seventh manuscript (N), copied by Regiomontanus in 1452, is not bound with Gmunden's tables. But Regiomontanus, who had matriculated at the university of Vienna in 1450 and in 1452 copied Gmunden's tables in a separate codex, surely was well acquainted with Gmunden's works and must have encountered a copy of the TP in Vienna. Interestingly, Regiomontanus, known to be a very fastidious computer of ephemerides, introduced relatively many scribal errors into his autograph; only one of his deviations, however, corresponds with those in other manuscripts, so presumably his original was close to B or S.

A second group of manuscripts (PaMeE), written in the 1440s, uniquely have two scribal deviations in common; PaMe have another 3 deviations in common. These 5 deviations differ by only ±1 minute from my vulgate edition based on B in the Gmunden family. Pa alone contains another two ±1 minute deviations from my edition, which prompts us to ask whether this group of manuscripts might be closer to John of Murs's original version than is the Gmunden family with its Viennese roots. Of the seven ±1 minute deviations in this group, only four match my computed values; I hesitate to conclude, therefore, that the second group of witnesses is 'closer' to the original than is the Gmunden group.

The provenance of the manuscripts in this group is also uncertain. E was copied in Erfurt in 1446. The large collection of astronomical manuscripts, assembled by Amplonius Rating de Berka and donated to Erfurt's university library in 1410, includes nothing by Gmunden.[65] The source of E is a mystery. The quire bearing the TP in Me is written on paper dating to *c.* 1450 from Metz (current location of Me) and to *c.* 1440 from Vienna; Me also has the canon with explicit. Perhaps a student from Metz, studying in Vienna, made this copy during his sojourn in the latter city?

A third group of manuscripts ($MV_1V_2V_3$) was produced in the 1450s by scribes at the St Emmeram cloister in Regensburg, where, as is well known, interest in astronomy, mathematics and geography had surged in that decade due to the efforts of two monks, Friedrich Amann and Hermann Pötzlinger. The latter in 1439 had earned a baccalaureate degree in Vienna and was an avid bibliophile, bequeathing at his death in 1469 more than 100 manuscripts to St Emmeram. He surely could have conveyed Gmunden material, including a copy of the TP, from Vienna to Regensburg.[66] The 'Regensburg manuscripts' have more scribal errors in the table (generally not tabulating differences), yet consistently include the canon with explicit.

Despite the flourishing of Alfonsine astronomy in fifteenth-century Italy, only one witness (C), now bound with the astronomical tables of Giovanni Bianchini, is known today in Italian libraries. C apparently was copied around 1452 in Italy. Its rubric, alone among all the witnesses, states that the TP were 'composita' (compiled or arranged) in Erfurt. I have found no evidence that corroborates such a claim. Of our 15 witnesses, only C presents the arguments in natural signs of 60 degrees; the others all use physical signs of 30°. As is well known, copies of the PAT by the end of the fourteenth century increasingly revise the mean motions from sexagesimal days to collected years and from physical to natural signs. Apparently the Erfurt source for C used physical signs; the Vienna source for B and the Gmunden manuscripts used physical signs.

An analysis of variant entries in the 15 witnesses broadly corroborates the groupings we have suggested on codicological grounds. In these copies, I have identified 253 variant entries (14 percent of the total 1860 entries), of which

[65] Cf. Schum, *Beschreibendes Verzeichniss.*

[66] For example, Munich, BSB, Clm 14583, largely an autograph by Amann written from 1447–1454, is filled with Gmunden's astronomical texts and tables, some reworked for the meridian of Regensburg. Munich, BSB, Clm 14111 also contains much astronomical material in the hands of Amann and Pöltzinger. Cf. Folkerts, 'Fridericus Amann'; Rumbold, 'The Library of Hermann Pötzlinger'; Rumbold, 'Lehren und Lernen'.

Manuscripts	Number
$BPVUMV_3GV_1V_2$, writing '10' for '19'	1
$PMEV_3V_1V_2$, writing '16' for '56'	1
$PMV_3V_1V_2$	6
$MV_3V_1V_2$, writing '59' for '49'	1
PMMeE	2
UCN, writing '32' for '22'	1
UGC, writing '49' for '41'	1
V_1V_2	6
MV_3, writing '9' for '8'	1
MePa	5
MeE	3
MeV_3, writing '2' for '3'	1
MeV_1, writing '57' for '59'	1
VM, writing '48' for '44'	1

Table 13: Numbers of shared variant entries among the fifteen manuscripts. Sets in bold font cannot be directly explained by my proposed stemma.

208 are variations that occur in only one manuscript.[67] The remaining 49 variants occur in 2 or more witnesses (see Table 13). Given the relatively large changes in value between successive entries in either rows or columns of the TP, it seems unlikely that different scribes would enter identical variant values unless they were working from a common source. A conscientious scribe, on the other hand, might refer to the tabulated differences listed in his source and correct errors in his copy.[68] The 49 common variants thus can at best suggest only a tentative genealogy for the manuscripts (see Fig. 2).[69]

[67] Statistical analyses of scribal errors in astronomical tables remains a desideratum for scholarly research. But cf. the chapter by Husson in this volume and Chabás, 'The Astronomical Tables of Jacob Ben David Bonjorn', p. 281, who found in the late fifteenth-century tables of Jacob ben David Bonjorn a scribal error rate of 1.15 percent (1227 errors in 100,000 entries). Error rates in our witnesses range from less than 0.05 percent (B) to 3.2 percent (M).

[68] Our proposed stemma requires that MeE corrected one, V_1V_2 corrected one, and V_3 two values. Another five shared sets of variants cannot easily be explained by our stemma.

[69] Note (Table 1) that the differences some scribes placed between successive entries in both rows and columns did not necessarily reduce variants; yet most of the low-variant manuscripts do display differences and most of the high-variant manuscripts do not, with several exceptions.

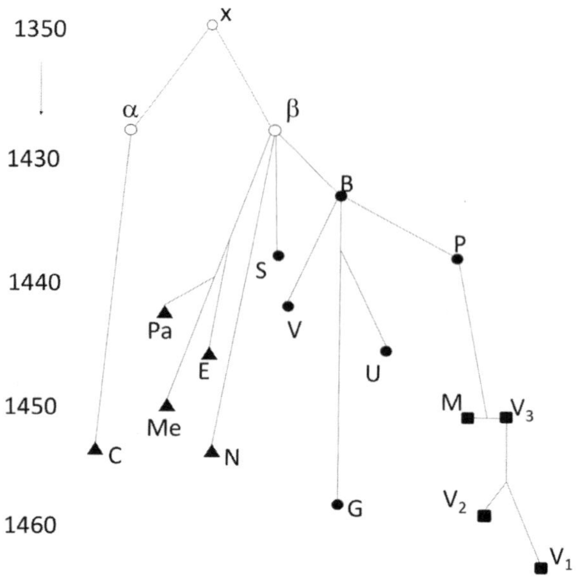

Fig. 2: Tentative stemma of the TP manuscripts. 'x' refers to the original copy by John and Firmin. Transmission to unknown copies in Erfurt (*α*, with natural signs) and Vienna (β, with physical signs) cannot be traced. Filled circles signify the 'Gmunden group' of manuscripts, squares the 'Regensburg group'. Triangles share relatively few variants with other manuscripts.

Readers might wonder about differentiating between scribal and computational 'errors'. However, the algorithm I propose for the TP can reproduce, to ±1 minute of time, all 1860 values recorded in witness B, except for two cases where the residuals reach −2 minutes of time and one case (294:24) where B joins eight other manuscripts in writing 8;10 hours for the 'correct' 8;19 hours. If my algorithm is correct, we can assume that B records the computationally 'correct' values with a single exception (the correct 8;10 hours appears in the remaining six manuscripts). Since it seems highly unlikely that another medieval author recomputed the TP, we can thus consider all variants from B to be scribal in nature.

The six 'Gmunden manuscripts' contain relatively few scribal errors and these rarely overlap among the set. We might surmise that the Gmunden set were copied, independently, from a 'good' original in Vienna (B shows 1 scribal error, S has 2). The four 'Regensburg manuscripts', on the other hand, have considerably more deviations, with 8 in common and another in three (V_1, V_2, V_3) of the four witnesses. V_1 and V_2 further share 6 common errors; surely one was copied from the other. Only one error (294:24) occurs in all the 'Regensburg manuscripts' and five of the 'Gmunden manuscripts' (not in S); this suggests that both sets shared a common progenitor with that error. It seems unlikely that scribes writing the other four manuscripts (Me, E, C, and N) caught and corrected that error in their source; their progenitors may have been α and β, one generation earlier than B. These deviations suggest a stemma for our 15 manuscripts (see Fig. 2). It remains mysterious how a copy as good as B could have reached Vienna nearly a century after John and Firmin authored the text in northern France. Perhaps Gmunden understood the TP well enough to have expunged scribal errors from his source (e.g., by smoothing entries 'by eye')?

As indicated in Table 12, two canons circulated with the TP. Those attributed to John of Murs and Firmin in the explicit appear in nine of the manuscripts, with the incipit: *Omnis utrisque sexus armoniam* ... (Th/K 1004). In 1440, John of Gmunden abbreviated this canon, dropping some sentences and expanding instructions for using the double-entry table, with the incipit: *Sciendum quod in hiis tabulis per supponintur tempus medie* ... This version appears, however, only in one of the 'Gmunden manuscripts'. Five of the manuscripts lack canons; but given the vagaries of binding, we cannot conclude that these copies of the TP originally circulated without instructions.

All fifteen of our manuscripts are astronomical miscellany; some include more or less complete sets of eclipse tables. Evidently, the TP by the middle of the fifteenth century had become a useful tool for astronomers computing true syzygies (required for monthly astrological weather prediction) and eclipses. Peurbach by 1459 would complete a new set of eclipse tables which feature an expanded version of the TP, reducing the intervals between successive entries and thereby making double interpolation easier. Since we have no copies of the TP written after the 1450s, we might guess that by this date, Peurbach's version, to be printed in 1514, had quickly replaced the TP for most working astronomers.

Finally, the 253 variant entries in the manuscripts reveal insights into scribal practices of copying astronomical tables in the final decades before the advent of European printing with moveable type. As can be seen in Table 12, 198 of these variants (78 percent) occur in manuscripts that do not tabulate differences between successive entries. Apparently the extra labor required to record differences reduced the number of scribal errors; whether the scribes themselves realized this, however, is another question.

About 88 percent of all the scribal errors (Table 14) involve miscopying a single digit in an entry. The most common errors appear in the first digit of the minutes of an entry, generating errors of 10, 20, etc. minutes in total value (85 cases). Roughly half this number of errors (51 cases) appear in the hours of an entry, creating errors of 1, 2, etc. hours (surely a more recognizable error for scribes or users of the tables). Another 106 cases involve miscopying the final digit of the minutes, generating errors of 1 to 9 minutes in the total absolute value. Among these single digit errors, the most frequent replacements are writing a '3' for a '2' (39 cases), a '2' for a '3' (23 cases), a '5' for a '4' (13 cases), a '4' for a '5' (9 cases), a '1' for a '2' (10 cases) and a '2' for a '1' (7 cases). Summing

Absolute value	Number		Absolute value	Number
10′	66		1′	43
20′	8		2′	18
30′	5		3′	10
40′	5		4′	9
1h	47		5′	7
3h	2		6′	6
4h	2		8′	8
			9′	5

Table 14: Frequency of scribal errors in the 15 manuscripts, sorted by absolute value of the scribal error.

the total number of cases in which a given digit appears in a single-digit scribal error yields similar results. The digit '2' is involved in 96 cases, '3' in 81 cases, '4' in 59 cases, '5' in 38 cases, and '1' in 35 cases. Scribes of mid fifteenth-century astronomical tables appear to have shared difficulties in recognizing given numerical digits.

Scribal errors involving two or three incorrect sexagesimal digits in a given entry appear far less frequently than the single-digit errors. Scribes occasionally transposed digits (e.g., writing 11;04 for 11;40). More generally, multi-digit errors result from what we might call 'column slippage', i.e., the successive values in a column have been correctly copied, relative to each other, but shifted up or down by several rows or columns in the table. These vertical slippages strongly suggest that scribes copied successive entries moving down the folio, writing the numbers column-by-column and not row-by-row.

N, an autograph of Regiomontanus, contains an example of column slippage at solar argument 126°, starting at lunar argument 0° and continuing until lunar argument 72° (13 entries in the minutes column). The first or top minutes entry in the column should be '58'; but Regiomontanus placed the next entry '55' in the top position, thereby shifting each minutes entry up one row until writing the entry 5;58h at lunar argument 72°. He then returned to the correct place in his original and wrote 5:58h again at lunar argument 78° and wrote correctly the final two entries in the column. Regiomontanus also correctly wrote all the differences in this column, obviously not checking his entries against the differences.

Similarly, the scribe of V_3 erroneously repeated a minutes entry and then returned to his original, thereby shifting the minutes down one row for the next 13 entries. He then skipped an entry and returned to placing entries in their proper rows for the remainder of the column.[70]

[70] V fol. 57r, 102:48 to 102:126.

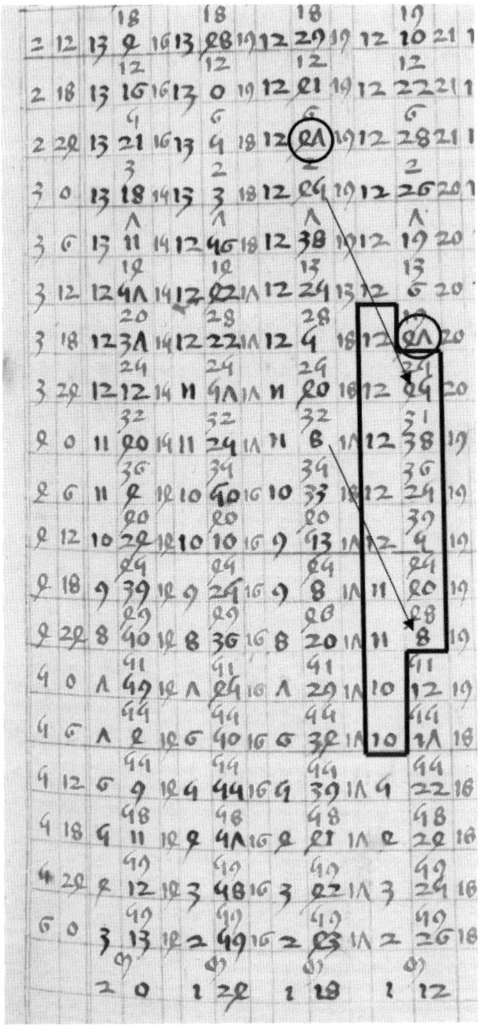

Fig. 3: Me, fol. 82r detail, showing column slippage. All entries within the black outline are erroneous. The erroneous minutes entries from 45 to 8 are shifted downward from the previously written column as are the erroneous hours entries 11, 11, 10.

Our third example combines both column and row slippage in Me, a manuscript showing 52 total errors, the second highest number among our witnesses. This scribe correctly copied all entries in the columns for solar arguments 306° and 312° down to lunar argument 102° (312:102 = 12;06h). As can be seen in Fig. 3, he had just written the digit '12' six times sequentially; he then erroneously inserted the digit '12' for five additional rows before returning to his original and copying correctly the sequence '11 11 10 10'. He then skipped 5 entries for the hours and correctly completed the column by copying '5 4 3 2 1'. The column for hours thus contains five extra '12's and lacks the sequence '9 8 8 7 6'. The minute entries for this section show a different slippage, now from the previous row. After correctly copying 12;06, he then recorded the next 7 minutes entries from the previous column of his original, shifted up 4 rows, writing successively '47 45 38 25 5 40 8' rather than the correct '47 22 51 15 36 51 3'. He records correct minutes for the remaining 6 minutes entries of the column, including the first two that have erroneous hours. Our scribe also recorded the 'correct' differences between successive entries in this column, i.e., presumably correctly following his original manuscript. The only way to explain this pattern of slippage, I think, is to assume that the scribe of Me copied digits sequentially by column, the hours (brown ink), minutes (brown), and differences (red), and that he did not use the differences to control for errors in successive entries down the column.

Given our identification of scribal errors by using B and our proposed algorithm for the TP, we are thus able to i) suggest a stemma for the surviving 15 manuscripts and sort them into groups by provenance; ii) gain new insights into the types of scribal errors that populate mid-fifteenth-century astronomical tables; and iii) argue that at least some scribes wrote successive entries by moving down the columns and not across the rows of each folio. And it appears that the tabulated differences sometimes, but not always, reduced scribal errors; we might hypothesize that the primary function of the differences was not to help scribes but rather to aid users in double-entry interpolation.[71]

Conclusion

Based on this exploratory data analysis, I conclude that the TP were computed with the following algorithm:

$$\Delta t(t) = \frac{-\eta(t)}{v_m(\bar{\alpha}_{\text{corr}}(t)) - v_s(t)}, \quad \text{where} \tag{13}$$

$$\bar{\alpha}_{\text{corr}}(t) = \bar{\alpha}(t) - \tfrac{13\eta(t)}{24}, \tag{14}$$

and t = the time of mean syzygy, $\eta(t) = c_m(t) - c_s(t)$ with the PAT equations, and the solar and lunar velocities are John of Genoa's, with 'correct' values for lunar arguments 96° and 102° and 'incorrect' values for lunar arguments 174° and 180°. With this algorithm, I match to ±1 minute of time 1858 of the 1860 entries of the TP; in 1535 cases (83 percent), the algorithm exactly matches the TP entries of my edition.

[71] Occasionally errors appear that demonstrate that scribes did attend to the tabulated differences. Me 48:66 reads 5;05h, an erroneous entry because the scribe (incorrectly) subtracted the listed difference of 28 minutes rather than adding it to the previous entry of 5;36h. The correct value should be 5;36h plus 28 minutes or 6;04h.

As we have argued, this algorithm is a compromise. It implements al-Battānī's correction of the lunar anomaly to the midpoint of the time interval between mean and true syzygy, as described in the Toledan Tables and in John of Saxony's 1327 canon. But it does not include al-Battānī's next step of correcting the apparent lunar velocity as the distance between the Earth and the lunar epicycle shifts. Unlike many cases in which Alfonsine astronomers constructed user-friendly formats that obtain identical quantitative results with 'easier' computations, the TP represent a case where an increase in user-friendliness is accompanied by approximations that decrease the quantitative precision of the results.[72]

In earlier publications, I showed that the TP, in the expanded form of Peurbach, generally match the results of John of Saxony's iterative method to ±1 minute of time; however, when the time correction is large, more than about 12 hours, the TP (by neglecting to correct for shifting Earth-lunar epicycle distance) can differ from the iterative method by 4 to 6 minutes of time.[73] Had John of Murs realized this behavior in his algorithm and concluded that a precision of ±6 minutes was adequate for the practice of Alfonsine computation? Or had he concluded that the small correction (reaching a maximum of only 6 arcsecs/hour in lunar velocity) could be ignored? No sources allow us to answer this question. However, in the Escorial manuscript filled with John of Murs's annotations and computations, we find in John's hand a copy of the small al-Battānī/Toledan table for correcting the lunar velocity, written beside a set of solar and lunar velocities that are known only from two earlier Islamic zijes but not in any Latin manuscripts.[74] This suggests that John of Murs knew the al-Battānī lunar velocity correction table but decided not to insert it into the TP algorithm. To save computational labor, John of Murs apparently was willing to introduce approximation into the syzygy problem.

In any case, John and Firmin were quite proud of the TP. As their literary-minded readers might have recognized, the explicit to the TP canons elevates their work to the plane of Jerome's Vulgate, Bede's scriptural commentaries, and Matthew of Vendôme's elegiac poetry. Not bad company for two astronomer-astrologer-computists of the 1340s.

[72] For another example of an Alfonsine astronomer increasing user-friendliness by introducing approximations that reduce quantitative precision, see Kremer, 'Abbreviating the Alfonsine Tables'.

[73] Kremer, 'John of Murs, Wenzel Faber', pp. 155–57.

[74] Escorial, RBMSL, O.II.10, fol. 204v. The lunar velocities range from 0;30,20–0;36,01, very close to the 0;30,21–0;36,01 found in the zijes of Ibn Isḥāq (fl. early thirteenth century) and Ibn al-Raqqām (d. 1315), both active in Tunis. Interestingly, both these zijes give the maximal lunar equation as 4;55,59°, very close to that found in the PAT of 4;56°. Cf. Goldstein, 'Lunar Velocity in the Middle Ages', pp. 184, 190. See also Husson and Miolo, 'Tables in the Margin'. Another early Parisian manuscript, BnF, lat. 7286C, fol. 49r, also includes the small table for adjusting the lunar velocity.

Acknowledgements

Long in gestation, this paper has greatly benefited from generous assistance and critical advice of many colleagues. José Chabás, Matthieu Husson, Laure Miolo, and Alberto Bardi helped locate manuscripts. Alexandre Tur and Marie-Madeline Saby assisted my codicological analyses. Karen Desmond and Monika Otter enriched my understanding, respectively, of John of Murs's texts on music theory and of thirteenth-century Latin poetry. Falk Eisermann, Nick Jacobson and Michael Shank improved my translations. C.P.E. Nothaft brought other fifteenth-century true syzygy tables to my attention. Glen van Brummelen elucidated the geometrical reasoning implicit in al-Battānī's methods for finding true syzygy. I learned much from responses to earlier versions of this chapter presented at the Paris Observatoire's CNRS-Syrte Tables Workshop in June 2015, the TAMAS Workshop in Granada in February 2017, and the Rio de Janeiro International Congress of History of Science and Technology in July 2017. I thank the relevant libraries for providing digital scans of their manuscripts. This research was supported by Dartmouth College and by the European Research Council (ERC)-funded ALFA Project (Consolidator grant 2016 agreement n. 723085), 'Shaping a European Scientific Scene: Alfonsine Astronomy' (PI Matthieu Husson).

Bibliography

Ábel, Jenö, 'Eperjesi János', *Egyetemes philologiai közlöny* 5 (1881), p. 259.

Academia Caesarea Vindobonensis, *Tabulae codicum manu scriptorum praeter graecos et orientales in Bibliotheca Palatina Vindobonensi Asservatorum*, 10 vols, Vienna: Gerold, 1864–1899.

Beaujouan, Guy, 'Observations et calculs astronomiques de Jean de Murs (1321–1344)', *Actes du 14. Congrès International d'Histoire des Sciences (Tokyo-Kyoto 1974)*, Tokyo: Science Council of Japan, 1975, pp. 27–30.

Bell, Philip, and Richard L. Kremer, 'An Early Sixteenth-Century Drawing of a Spring Clock in a Fifteenth-Century Astronomical Manuscript', *Antiquarian Horology* 42 (2021), pp. 467–76.

Bond, Edward A., *Catalogue of Additions to the Manuscripts in the British Museum in the Years 1854–1875*, 2 vols, London: British Museum, 1875–1877.

Boudet, Jean-Patrice, 'La papauté d'Avignon et l'astrologie', in Edith Pasztor (ed.), *Fin du monde et signes des temps: Visionnaires et prophètes en France méridionale, fin XIIIe-début XVe siècle*, Toulouse: Privat, 1992, pp. 257–93.

Boudet, Jean-Patrice, *Le 'Recueil des plus célèbres astrologues' de Simon de Phares*, 2 vols, Paris: Librairie honoré Champion, 1997–1999.

Briquet, Charles-Moïse, *Les filigranes: Dictionnaire historique des marques du papier dès leur apparition vers 1282 jusq'en 1600*, 4 vols, Paris: A. Picard & fils, 1907.

Burnett, Charles, 'Arabic into Latin in Twelfth Century Spain: The Works of Hermann of Carinthia', *Mittellateinisches Jahrbuch* 13 (1978), pp. 100–34.

Busard, Hubertus L. L., *Der Traktat De Sinibus, chordis et arcubus von Johannes von Gmunden*, Wien: Springer, 1971. [Österreichische Akademie der Wissenschaften, math.-naturwiss. Kl., Denkschriften 116, no. 3].

Busard, Hubertus L. L., 'Die "Arithmetica speculativa" des Johannes de Muris', *Scientiarum Historia* 13 (1971), pp. 103–32.

Busonero, Paola, et al., *I manoscritti datati delle Biblioteche Casanatense e Vallicelliana di Roma*, Florence: SISMEL / Edizioni del Galluzzo, 2016. [Manoscritti datati d'Iitalia, 25].

Chabás, José, 'The Astronomical Tables of Jacob Ben David Bonjorn', *Archive for History of Exact Sciences* 42 (1991), pp. 279–314.

Chabás, José, and Bernard R. Goldstein, 'Nicholaus de Heybech and His Table for Finding True Syzygy', *Historia Mathematica* 19 (1992), pp. 265–89.

Chabás, José, and Bernard R. Goldstein, 'Andalusian Astronomy: Al-Zīj Al-Muqtabis of Ibn Al-Kammād', *Archive for History of Exact Sciences* 48 (1994), pp. 1–41.

Chabás, José, and Bernard R. Goldstein, 'Computational Astronomy: Five Centuries of Finding True Syzygy', *Journal for the History of Astronomy* 28 (1997), pp. 93–105.

Chabás, José, and Bernard R. Goldstein, *The Alfonsine Tables of Toledo*, Dordrecht: Kluwer, 2003.

Chabás, José, and Bernard R. Goldstein, 'John of Murs's Tables of 1321', *Journal for the History of Astronomy* 40 (2009), pp. 297–320.

Chabás, José, and Bernard R. Goldstein, *The Astronomical Tables of Giovanni Bianchini*, Leiden: Brill, 2009.

Chabás, José, and Bernard R. Goldstein, 'John of Murs Revisited: The *Kalendarium Solis et Lune* for 1321', *Journal for the History of Astronomy* 43 (2012), pp. 411–37.

Chabás, José, and Bernard R. Goldstein, 'Computing Planetary Positions: User-Friendliness and the Alfonsine Corpus', *Journal for the History of Astronomy* 44 (2013), pp. 257–76.

D'Ancona, Paolo, and Erardo Aeschlimann, *Dictionnaire des miniaturistes du Moyen Âge et de la Renaissance, dans les différentes contrées de l'Europe*, 2nd enlarged edition, Milan: Hoepli, 1949.

Dalen, Benno van, *Islamic Astronomical Tables. Mathematical Analysis and Historical Investigation*, Farnham: Ashgate, 2013.

Desmond, Karen, *Music and the* moderni, *1300–1500. The* ars nova *in Theory and Practice*, Cambridge: Cambridge University Press, 2018.

Duhem, Pierre, *Le système du monde. Histoire des doctrines cosmologiques de Platon à Copernic*, 10 vols, Paris: A. Hermann, 1913–1958.

Falkenroth, Christoph, *Die 'Musica speculativa' des Johannes de Muris: Kommentar zur Überlieferung und kritische Edition*. Stuttgart: Steiner, 1992. [Beihefte zum Archiv für Musikwissenschaft, 34].

Fast, Susan, *Johannes de Muris Musica (speculativa)*, Ottawa: Institute of Medieval Music, 1994.

Folkerts, Menso, 'Fridericus Amann und seine Bedeutung für die mathematischen Wissenschaften im 15. Jahrhundert', in Peter Schmid and Rainer Scharf (eds), *Gelehrtes Leben im Kloster: Sankt Emmeram als Bildungszentrum im Spätmittelalter*, Munich: Martin Meidenbauer, 2012, pp. 187–211.

Gack-Scheiding, Christine, *Johannes de Muris 'Epistola super reformatione antiqui kalendarii'. Ein Beitrag zur Kalenderreform im 14. Jahrhundert*, Hannover: Hahn, 1995. [Monumenta Germaniae Historica, Studien und Texte, 11].

Gerl, Armin, 'Fridericus Amann', in Rainer Gebhardt (ed.), *Rechenbücher und mathematische Texte der frühen Neuzeit*, Annaberg-Buchholz: Adam-Ries-Bund, 1999, pp. 1–12.

Goldstein, Bernard R., 'Lunar Velocity in the Ptolemaic Tradition', in Peter M. Harman and Alan E. Shapiro (eds), *The Investigation of Difficult Things: Essays on Newton and the History of the Exact Sciences*, Cambridge: Cambridge University Press, 1992, pp. 3–17.

Goldstein, Bernard R., 'Lunar Velocity in the Middle Ages: A Comparative Study', in Josep Casulleras and Julio Samsó (eds), *From Baghdad to Barcelona: Studies in the Islamic Exact Sciences in Honour of Prof. Juan Vernet*, Barcelona: Instituto Millas Vallicrosa de Historia de la Ciencia Arabe, 1996, pp. 181–94.

Goldstein, Bernard R., and David Pingree, 'Levi Ben Gerson's Prognostication for the Conjunction of 1345', *Transactions of the American Philosophical Society* 80 (1990), pp. 1–60.

Gushee, Lawrence, 'Jehan des Murs and His Milieu', in Frank Hentschel (ed.), *Musik und die Geschichte der Philosophie und Naturwissenschaften im Mittelalter: Fragen zur Wechselwirkung von* Musica *und* Philosophia *im Mittelalter*, Leiden: Brill, 1998, pp. 339–71.

Gushee, Lawrence, 'New Sources for the Biography of Johannes de Muris', *Journal of the American Musicological Society* 22 (1969), pp. 3–26.

Hellmann, Gustav, 'Die Wettervorhersage im ausgehenden Mittelalter', *Beiträge zur Geschichte der Meteorologie* 2 (1917), pp. 169–229.

Hentschel, Frank, 'Johannes de Muris', in Ludwig Finscher (ed.), *Die Musik in Geschichte und Gegenwart*, Kassel: Bärenreiter, 2003, vol. IX, pp. 1102–07.

Hentschel, Frank, *Sinnlichkeit und Vernunft in der mittelalterlichen Musiktheorie: Strategien der Konsonanzwertung und der Gegenstand der Musica sonora um 1300*, Stuttgart: Steiner, 2000. [Beihefte zum Archiv für Musikwissenschaft, 47].

Hoaglin, David C., Frederick Mosteller, and John W. Tukey, *Understanding Robust and Exploratory Data Analysis*, New York: Wiley, 1983.

Husson, Matthieu, *Les domaines d'application des mathèmatiques dans la première moitié du quatorzième siècle*, Thèse de doctorat, Paris: École pratique des Hautes Études, 2007.

Husson, Matthieu, 'L'astronomie alphonsine dans *L'expositio intentionis regis alfonsii circa tabulas ejus* de Jean de Murs', *Archives d'histoire doctrinale et littéraire du moyen âge* 78 (2011), pp. 229–45.

Husson, Matthieu, 'Ways to Read a Table: Reading and Interpolation Techniques in Canons of Early Fourteenth-Century Double-Argument Tables', *Journal for the History of Astronomy* 43 (2012), pp. 299–319.

Husson, Matthieu, 'Exploring the Temporality of Complex Computational Practice: Two Eclipse Notes by John of Murs in the Ms Escorial O.II.10', *Centaurus* 58 (2016), pp. 46–65.

Husson, Matthieu, and Laure Miolo, 'Tables in the Margin: John of Murs's Astronomical Tables in the Margin of Escorial O.II.10' (forthcoming).

Juste, David, 'MS Vatican, Biblioteca Apostolica Vaticana, Pal. lat. 1376' (update: 27.10.2021), Ptolemaeus Arabus et Latinus, http://ptolemaeus.badw.de/ms/719.

Keller, Karl Heinz, *Katalog der lateinischen Handschriften der Staatlichen Bibliothek (Schloßbibliothek) Ansbach*, vol. I, Wiesbaden: Harrassowitz, 1994.

Kennedy, Edward S., 'The Digital Computer and the History of the Exact Sciences', *Centaurus* 12 (1968), pp. 107–13.

Kremer, Richard L., 'Thoughts on John of Saxony's Method for Finding True Syzygy', *Historia Mathematica* 30 (2003), pp. 263–77.

Kremer, Richard L., 'Wenzel Faber's Table for Finding True Syzygy', *Centaurus* 45 (2003), pp. 305–29.

Kremer, Richard L., '"Abbreviating" the Alfonsine Tables in Cracow: The *Tabulae Aureae* of Petrus Gaszowiec (1448)', *Journal for the History of Astronomy* 38 (2007), pp. 283–304.

Kremer, Richard L., 'John of Murs, Wenzel Faber and the Computation of True Syzygy in the Fourteenth and Fifteenth Centuries', in Joseph W. Dauben, Stefan Kirschner, Andreas Kühne, Paul Kunitzsch and Richard P. Lorch (eds), *Mathematics Celestial and Terrestrial: Festschrift für Menso Folkerts zum 65. Geburtstag*, Stuttgart: Wissenschaftliche Verlagsgesellschaft GmbH, 2008, pp. 147–60.

Kremer, Richard L., 'How Did the Turketum (or Torquetum) Get Its Name?', in Neil Brown, Silke Ackermann and Feza Günergun (eds), *Scientific Instruments Between East and West*, Leiden: Brill, 2019, pp. 80–107.

Kremer, Richard L., and Jerzy Dobrzycki, 'Alfonsine Meridians: Tradition Versus Experience in Astronomical Practice *c.* 1500', *Journal for the History of Astronomy* 29 (1998), pp. 187–99.

L'Huillier, Ghislaine, *Le Quadripartitum numerorum de Jean de Murs. Introduction et édition critique*, Geneva: Droz, 1990. [Mémoires et documents publiés par la Société de l'École des Chartes, 32].

Lejbowicz, Max, 'Présentation de Jean de Murs, observateur et calculateur sagace et laborieux', in Christophe Grellard (ed.), *Méthodes et statut des sciences à la fin du moyen âge*, Villeneuve d'Ascq: Presses universitaires du Septentrion, 2004, pp. 159–80.

Michels, Ulrich, *Die Musiktraktate des Johannes de Muris*, Wiesbaden: Steiner, 1970. [Beihefte zum Archiv für Musikwissenschaft, 8].

Michels, Ulrich, *Johannes de Muris, Notitia artis musicae et Compendium musicae practicae. Petrus de Sancto Dionysio, Tractatus de musica*, Dallas: American Institute of Musicology, 1972.

Miolo, Laure, 'In Quest of Jean de Murs's Library. An Overview of His Readings and Uses of Manuscripts', *Erudition and the Republic of Letters* 4 (2019), pp. 13–39.

Müldener, Friedrich Wilhelm August, *Matthaei Vindocinensis Tobias*, Göttingen: Dieterich, 1859.

Nallino, Carlo Alfonso, *Al-Battānī sive Albatenii opus astronomicum*, 3 vols, Milan: Hoepli, 1899–1907. Reprinted Hildesheim: Georg Olms Verlag, 1977.

Neske, Ingeborg, *Die Handschriften der Stadtbibliothek Nürnberg*, vol. V: *Die lateinischen mittelalterlichen Handschriften, Varia, 13.-15. und 16–18. Jahrhundert*, Wiesbaden: Harrossowitz Verlag, 1997.

North, John D., 'The Alfonsine Tables in England', in Yasukatsu Maeyama and Walter G. Saltzer (eds), *Prismata. Naturwissenschaftsgeschichtliche Studien. Festschrift für Willy Hartner*, Wiesbaden: Steiner, 1977, pp. 269–301.

North, John D., *Richard of Wallingford. An Edition of His Writings, with Introductions, English Translation and Commentary*, 3 vols, Oxford: Clarendon Press, 1976.

Nothaft, C. Philipp E., 'The Astronomical Data in the *Très Riches Heures* and Their Fourteenth-Century Source', *Journal for the History of Astronomy* 46 (2015), pp. 113–29.

Nothaft, C. Philipp E., 'The Chronological Treatise *Autores Kalendarii* of 1317, Attributed to John of Murs', *Cahiers de l'Institut du moyen âge grec et latin* 82 (2013), pp. 1–89.

Nothaft, C. Philipp E., 'Critical Analysis of the Alfonsine Tables in the Fourteenth Century. The Parisian *Expositio Tabularum Alfonsii* of 1347', *Journal for the History of Astronomy* 46 (2015), pp. 76–98.

Nothaft, C. Philipp E., 'Jean des Murs's *Canones Tabularum Alfonsii* of 1339', *Erudition and the Republic of Letters* 4 (2019), pp. 98–122.

Nothaft, C. Philipp E., 'John of Murs and the Treatise *Autores Kalendarii* (1317). A Problem of Authorship', *Sudhoffs Archiv* 99 (2015), pp. 209–29.

Nothaft, C. Philipp E., *Scandalous Error. Calendar Reform and Calendrical Astronomy in Medieval Europe*, Oxford: Oxford University Press, 2018.

Pedersen, Fritz S., *The Toledan Tables. A Review of the Manuscripts and the Textual Versions with an Edition*, 4 vols, Copenhagen: Reitzels, 2002. [Det Kongelige Danske Videnskabernes Selskab, Historisk-filosofiske skrifter, 24].

Petersen, Viggo M., 'The Three Lunar Models of Ptolemy', *Centaurus* 14 (1969), pp. 142–71.

Plassard, Joël, *Projets de réforme du calendrier à Paris au début du XIVe siècle. Textes édités et commentés*, Thèses diplôme d'archiviste-paléographe, Paris: École Nationale des Chartes, 1975.

Porres de Mateo, Beatriz, *Les tables astronomiques de Jean de Gmunden: Édition et étude comparative*, Thèse de doctorat, Paris: École pratique des Hautes Études, 2003.

Porres de Mateo, Beatriz, and José Chabás, 'John of Murs's *Tabulae permanentes* for Finding True Syzygies', *Journal for the History of Astronomy* 32 (2001), pp. 63–72.

Poulle, Emmanuel, 'John of Murs', in Charles C. Gillispie (ed.), *Dictionary of Scientific Biography*, vol. IX, New York: Scribners, 1974, pp. 128–33.

Poulle, Emmanuel, 'Jean de Murs et les tables alphonsines', *Archives d'histoire doctrinale et littéraire du moyen âge* 47 (1980), pp. 241–71.

Poulle, Emmanuel, 'The Alfonsine Tables and Alfonso X of Castille', *Journal for the History of Astronomy* 19 (1988), pp. 97–113.

Poulle, Emmanuel, 'Les astronomes parisiens au XIVe siècle et l'astronomie alphonsine', *Histoire littéraire de la France* 43 (2005), pp. 1–54.

Pruckner, Herbert, *Studien zu den astrologischen Schriften des Heinrich von Langenstein*, Leipzig: Teubner, 1933. [Studien der Bibliothek Warburg, 14].

Reuter, Marianne, *Die lateinischen mittelalterlichen Handschriften der Universitätsbibliothek München*, vol. V: *Die Handschriften aus der Quartreihe*, Wiesbaden: Harrassowitz Verlag, 2000.

Riezler, Sigmund, *Geschichte Baierns*, 6 vols, Gotha: Perthes, 1878–1903.

Rosińska, Grazyna, *Scientific Writings and Astronomical Tables in Cracow. A Census of Manuscript Sources (XIVth-XVIth Centuries)*, Wrocław: Ossolineum, 1984. [Studia Copernicana, 22].

Rumbold, Ian, 'Lehren und Lernen in der St Emmeramer Klosterschule im Spätmittelalter: Hermann Pötzlinger als Rector Scolarum in der Zeit der Melker Reform', in Peter Schmid and Rainer Scharf (eds), *Gelehrtes Leben im Kloster: Sankt Emmeram als Bildungszentrum im Spätmittelalter*, Munich: Meidenbauer, 2012, pp. 243–59.

Rumbold, Ian, 'The Library of Hermann Pötzlinger (ca. 1415–1469), Rector Scolarium at the Monastery of St Emmeram, Regensburg', *Gutenberg Jahrbuch* 60 (1985), pp. 329–40.

Rupprich, Hans, *Der Briefwechsel des Konrad Celtis*, Munich: Beck, 1934.

Saby, Marie-Madeleine, *Les canons de Jean de Lignères sur les tables astronomiques de 1321. Édition critique, traduction et étude*, Diplome d'archiviste paleographe, Paris: École nationale des Chartes, 1987.

Saby, Marie-Madeleine, 'Mathématique et métrologie Parisiennes au début du XIVe siècle. Le calcul du volume de la mer d'airain de Jean de Murs', *Archives d'histoire doctrinale et littéraire du moyen âge* 66 (1991), pp. 197–213.

Schabel, Chris, 'John of Murs and Firmin of Beauval's Letter and Treatise on Calendar Reform for Clement VI', *Cahiers de l'Institut du moyen-âge grec et latin* 66 (1996), pp. 187–215.

Schneider, Karin, *Die deutschen Handschriften der Bayerischen Staatsbibliothek München, Cgm 691–867*, Wiesbaden: Otto Harrassowitz, 1984.

Schuba, Ludwig, *Die Quadriviums-Handschriften der Codices Palatini latini in der Vatikanischen Bibliothek (Cod. Pal. lat. 1340–1458)*, Wiesbaden: Reichert, 1992.

Schum, Wilhelm, *Beschreibendes Verzeichniss der Amplonianischen Handschriftensammlung zu Erfurt*, Berlin: Weidmannsche Buchhandlung, 1887.

Suter, Heinrich, *Die astronomischen Tafeln des Muḥammed ibn Mūsā al-Khwārizmī in der Bearbeitung des Maslama ibn Aḥmed Al-Madjrīṭī und der lateinischen Übersetzung des Athelhard von Bath*, Copenhagen: Høst, 1914.

Tanstetter Collimitius, Georg, *Tabulae eclypsium magistri Georgij Peurbachij. Tabula primi mobilis Joannis de Monte Regio*, Vienna: Johann Winterburger, 1514.

Thorndike, Lynn, *A History of Magic and Experimental Science*, 8 vols, New York: Macmillan, 1923–1958.

Thorndike, Lynn, 'Astronomy at Paris Around 1485 and 1360 (as Reflected in MS Berne 454)', *Humanisme et Renaissance* 3 (1936), pp. 165–68.

Thorndike, Lynn, and Pearl Kibre, *Catalogue of Incipits of Mediaeval Scientific Writings in Latin*, 1st ed., Cambridge: Mediaeval Academy of America, 1937.

Thorndike, Lynn, 'Some Little Known Astronomical and Mathematical Manuscripts', *Osiris* 8 (1948), pp. 41–72.

Th/K = Lynn Thorndike and Pearl Kibre, *Catalogue of Incipits of Mediaeval Scientific Writings*, rev. and augmented ed., Cambridge: Mediaeval Academy of America, 1963.

Tukey, John W., *Exploratory Data Analysis*, Reading, Mass.: Addison-Wesley, 1977.

Unterkircher, Franz, *Katalog der datierten Handschriften in lateinischer Schrift in Österreich*, 8 vols, Vienna: Verlag der österreichischen Akademie der Wissenschaften, 1969–1988.

Valentinelli, Joseph, *Bibliotheca manuscripta ad S. Marci Venetiarum*, 6 vols in 2, Venice: Typographia commercii, 1868–1875.

Van Brummelen, Glen, 'Lunar and Planetary Interpolation Tables in Ptolemy's *Almagest*', *Journal for the History of Astronomy* 25 (1994), pp. 297–311.

Van Brummelen, Glen, and Kenneth Butler, 'Determining the Interdependence of Historical Astronomical Tables', *Journal of the American Statistical Association* 92 (1997), pp. 41–48.

Vogel, Kurt, *Die Practica des Algorismus Ratisbonensis. Ein Rechenbuch des Benediktinerklosters St Emmeram aus der Mitte des 15. Jahrhunderts nach den Handschriften der Münchner Staatsbibliothek und der Stiftsbibliothek St Florian*, Munich: Beck, 1954. [Schriftenreihe zur Bayerischen Landesgeschichte, 50].

Zepeda, Henry, *The First Latin Treatise on Ptolemy's Astronomy: The Almagesti Minor (c. 1200)*, Turnhout: Brepols, 2018. [Ptolemaeus Arabus et Latinus – Texts, 1]

Zinner, Ernst, *Verzeichnis der astronomischen Handschriften des deutschen Kulturgebietes*, Munich: Beck, 1925.

Appendix: A new edition of the TP

This edition, collating all currently known witnesses of the TP, reproduces the 1860 entries of the TP. It does not include the tabulated differences copied into some of the manuscripts and does not describe differing formats of labeling the solar arguments (horizontal axis) and lunar arguments (vertical axis) found in the manuscripts. Porres de Mateo's 2003 edition does incorporate the differences and the axes.[75] But she collated only five of the 'Gmunden manuscripts' (BPSVU) and introduced at least 59 deviations which appear in none of those manuscripts nor in the additional nine I have examined. Hence, a new edition of the TP, controlled by B and my recomputation of the table, is required. But my edition seeks to capture the numerical content of the table, not all the details of its format.[76]

Manuscript sigla and description

(B) Vienna, ÖNB, 5268, fols 45v–48r (table), 48v (canon, *Omnis utriusque sexus*, Th/K 1004). Bound (139 fols) with John of Gmunden's *Tabulae maiores* (second version, 3r–34r, 41r–45r, 49r-v, including Gmunden's velocities, 27r) and canons (39r–40v, 50r–83v), incomplete and reordered in an early restoration; Gmunden's treatises on sines and chords (84r–97v), the albion (99r–130r), and the equatoria of Campanus (131r–139r); Gmunden's table of mean syzygies for 1433, 1473, …, 2433, for the meridian of Vienna, i.e., 80 minutes of time east of Toledo (6v–7r); and John of Murs's table of proportions, the *Tabula tabularum* (35r–36v) with canon (Th/K 1461, 37r–39r): *Explicit canon tabule tabularum edite a magistro Johanne de Muris anno domini 1321 et scriptus et finitus per magistrum Johannem de Gmunden in die cineris anno Domini 1433* (39r). A Gmunden autograph, dated 1433–1437; annotated by the Hungarian mathematician, Johannes de Epperies, who in 1520 updated a list of astrolabe stars in the manuscript (122r).[77] Paper (fols 45, 47, 48) Dreiburg im Kreis,

[75] Porres, *Les tables astronomiques*, pp. 395–406.

[76] Digital scans of the entire manuscripts are openly accessible for PaMMe$V_1V_2V_3$.

[77] Academia Caesarea Vindobonensis, *Tabulae codicum*, vol. IV, p. 80; Porres, *Les tables astronomiques*, pp. 118–23, 133, 137, reproduces fol. 48v as Fig. 18. For an edition of John of Gmunden's table of mean syzygies, see ibid., pp. 157–60. Interestingly, a later annotator, Johannes de Epperies, who in 1520 converted some of Gmunden's star positions to equatorial coordinates (fol. 122r), added a note on the front flyleaf (fol. 1r) describing a laborious, iterative procedure for finding the time of true syzygy. Did he not know about (or understand) the *Tabulae permanentes* copied in the codex? See ibid., pp. 547–49, Fig. 18; Busard, *Der Traktat* De sinibus; Ábel, 'Eperjesi János'; Rupprich, *Der Briefwechsel*, pp. 93–95, 387; Kremer, 'How Did the Turketum'; Bell and Kremer, 'An Early Sixteenth-Century Drawing'.

DE2520-PO-153644, dated 1436, Heilsbronn Abbey, near Nuremberg.[78]

(S) Innsbruck, ULB Tirol, Servitenkloster I.b. 62, tables (pp. 74–79). Bound (175 pp.) with Gmunden's *Tabulae maiores* (1–107, including Gmunden's velocities, 61) and canons (113–65); Gmunden's treatise on sines and chords; Gmunden's table of mean syzygies for 1433, 1473, …, 2433 (7–8). Dated Vienna, 1437 (165).[79]

(P) London, BL, Add. 24070, fols 52r–54v (tables), fols 55r, 57v (canon, *Omnis utriusque sexus*). Bound (77 fols) with Gmunden's calendar (1–7) and *Tabulae maiores* (8–51, 70–75, including Gmunden's velocities, 44r); John of Murs's *Tabula tabularum* (63r) with canon dated 1321 (64r–67r) and Gmunden's canon to the same table (67v–69v); Gmunden's table of mean syzygies for 1433, 1473, …, 2433 (11r-v). Mostly in the hand of Georg Prunner of Lower Ruspach, a student of Gmunden's and scribe at the Klosterneuburg, dated 1436–39 (7v, 67r).[80]

(U) Munich, UB, 4° 737, fols 121v–133r (table), fol. 136r-v (abbreviated canon, *Omnis utriusque sexus*[81]). Bound (149 fols) in an astronomical miscellany that includes Gmunden's *Tabulae breviores* (third version, 1–73, Gmunden's velocities incomplete, 37r) with radices for Vienna, dated 1440 complete, and canon (75–116, Th/K 46); Gmunden's table of mean syzygies for 1433, 1473, …, 2433 (133v–135r); Johannes Swab de Wutzbeich, *Practica eclipsium solis et luna, a. 1412* (139r–148r, Th/K 1225). Copied 1444 in Vienna, mostly by one hand, annotated by several hands including the first owner's.[82]

[78] I identify watermarks using the Wasserzeichen-Informationssystem, available at www.wasserzeichen-online.de/wzis/index.php.

[79] Porres, *Les tables astronomiques*, pp. 80–84.

[80] Prunner, a student of Gmunden's in Vienna, is known to have copied several other astronomical manuscripts now at Klosterneuburg. See Bond, *Catalogue of Additions*, vol. II, pp. 6–7; Porres, *Les tables astronomiques*, pp. 22, 80, 87–91.

[81] Includes sentences 7–33 of the *Omnis* canon, as edited by Porres and Chabás, 'John of Murs's *Tabulae permanentes*', pp. 67–72, with slightly different opening and closing phrases.

[82] Keller, *Katalog der lateinischen Handschriften*, pp. 217–18; Reuter, *Die lateinischen mittelalterlichen Handschriften*, pp. 186–89; Porres, *Les tables astronomiques*, pp. 80, 94–97. Ex libris (fol. 1r) magister Matthias Rem de Weinsberg (d. 1495), 1433 matriculated Vienna (Bacc 1436, Mag 1444), 1444 professor of arts Heidelberg, 1454 doctor theology Vienna, from 1454 preacher at the Chorherrenstift Gumbertus in Ansbach, near Nuremberg. *Repertorium Academicum Germanicum*, s.v. 'Matthias Rem', www.rag-online.org/gelehrter/id/1413222073 (accessed July 2017).

(V) Vienna, ÖNB, 5151, fols 119v–122r (table), 117v–119r (Gmunden's expanded version of John of Murs's canon,[83] *Tempus distancie inter coniunccionem aut opposicionem veram et mediam solis et lune per tabulas ad hoc factas invenire. Sciendum quod in hiis tabulis ... Iste canon editus et scriptus est wienne per magistum Johannem de Gmunden die 20 mensis maii anno Domini 1440 currente*, autograph). Bound (168 fols) with the first version of Gmunden's tables with 60° signs (1–62, 104–17, Gmunden's velocities, 13v–14v) and canons (63–103); Gmunden's tables of mean syzygies for 1433, 1473, ..., 2433 (111r–113v); and excerpts from the Oxford Tables (131v–146r) with Gmunden's canon (130r-v). Colophons frequently dated Vienna, 1440–1443.[84]

(G) Munich, BSB, Cgm 739, fols 105v–117r (table), fol. 118r-v (*Omnis utriusque sexus*, lacks sentences 33–34). Bound (167 fols) in an astronomical-computus-mathematical miscellany with Gmunden's calendar and three cycles of mean syzygies for 1439–1496 for Vienna (5–22); Gmunden's text on the astrolabe (80–100, Th/K 1294); miscellaneous tables (21 of the 47 are from Gmunden's *Tabulae breviores* with radices for Vienna for 1440, some updated to 1456, in several hands (102–45); table of mean syzygies *ad meridiem Ertfordensem*, computed for 62 minutes east of Toledo (142v); Gmunden's table of mean syzygies for 1433, 1473, ..., 2433 (103v–105r, same hand as 105v–117r); Nicholaus de Heybech's tables for true syzygy (145v–147v); an anonymous text on an astronomical equatorium (148r–155r, *Hirnach volgt der allersubtilist weg dy tholicen zw machen auff ainen newen sin*). No eclipse tables. Schneider dates the manuscript mid-fifteenth century, the paper of quire fols 103–18 (all same hand) is Waage, dated to Nuremberg 1458.[85]

(N) Nuremberg, SB, Cent. VI 23, fols 99v–111r (table), fols 114r–116r (*Omnis utriusque sexus*, lacks sentences 33–34). Bound (122 fols)

[83] Porres, *Les tables astronomiques*, p. 112, calls this text a 'version agrégée' of John of Murs's canon. Gmunden deleted John's introduction (sentences 1–8), copied about half of John's remaining text verbatim and considerably expanded the instructions for using tables of proportion to solve the double-entry interpolations.

[84] Academia Caesarea Vindobonensis, *Tabulae codicum*, vol. IV, p. 41; Porres, *Les tables astronomiques*, pp. 80, 111–15, 130. Since several colophons in this manuscript are dated to within the same five days in 1440, Unterkircher, *Katalog der datierten Handschriften*, vol. II, p. 176, has suggested that these dates are 'doubtful'.

[85] Schneider, *Die deutschen Handschriften*, pp. 203–09; Riezler, *Geschichte Baierns*, vol. III, p. 458; Porres, *Les tables astronomiques*, pp. 73, 666, 669. Most of these quires were collected by Ulrich Greimolt (1413–1495), a Master of Arts from Weilheim who in 1452 became tutor

with a partial set of the Oxford Tables (1v–97v), no mean motion or eclipse tables. Regiomontanus autograph, copied *c.* 1452 in Vienna.[86]

(C) Rome, Biblioteca Casanatense, 1673, fols 89v–92r (table with 60° signs, *Composita Erfordie Duringie*). Bound (120 fols) with Giovanni Bianchini's canons (1r–10v) and a partial set of his tables (20r–89r); a planetary ephemerides beginning in 1456; a table of mean syzygies with a radix of 1452 complete for Vienna (93r); Jacob ben David Bonjorn's syzygy tables and canons for 1361 (97r–102r); and a partial set of the Oxford Tables (fols 109r–120r). No eclipse tables. Recent binding.[87]

(V_2) Vatican, BAV, Pal. lat. 1376, fols 389r–391v (table), fol. 392r (canon, *Omnis utriusque sexus*, dated *frater fridericus 1458*). Bound (410 fols) in an astronomical-astrological-mathematical miscellany, including the PAT, dated 1406 (1–18); a fragment of John of Saxony's canon (221r–223v); various eclipse tables (34–40, 51–56); Gmunden's solar and lunar velocity tables (57v–60r); mean syzygies for 1321, 1345, ..., 1609 (45v–46r, computed for Paris, 48 minutes east of Toledo); John of Lignères's *Tabule magne* (102r–134v), canon for *Tabulae Erfordiensis* (135r–136v) with lunar radices (128v, 129v) 62 minutes east of Toledo;[88] the posthumous version of Gmunden's tables (138r–170r, Th/K 1164); John of Lignères's various canons (170v–177v); Johannes Schindel, *Tractatus de quantitate trium solidorum*, dated 1420 (181r–184v, Th/K 1232); astronomical and astrological notes from Leopold of Austria's *De astrorum scientia* and other sources (191r–193r, Th/K 68, 1409); star catalog for 1444 (194r–207v); *Theorica planetarum* (212r–218v, Th/K 223); John of Saxony's canon to the PAT (221r–223v); Sacrobosco's sphere (224r–236r, Th/K 1524); Alfraganus, *Liber de aggregationibus scientiae stellarum* (238r–253v, Th/K 960); Alcabitius, *Liber introductorius ad iudicia astrorum* (256r–286r,

---

to the sons of Albrecht III the Pious, Duke of Bavaria. Upon his death, Greimolt's books went to the monastery in Tegernsee where they were rebound in 1504.

[86] Neske, *Die Handschriften der Stadtbibliothek*, p. 181; Porres, *Les tables astronomiques*, p. 671.

[87] Chabás and Goldstein, 'Computational Astronomy', pp. 99–100; Chabás and Goldstein, *The Astronomical Tables of Giovanni Bianchini*, pp. 14, 24–26; Chabás, 'The Astronomical Tables of Jacob Ben David Bonjorn'; Busonero et al., *I manoscritti datati*, pp. 63–64; Paolo d'Ancona and Aeschlimann, *Dictionnaire des miniaturistes*, p. 159. Bianchini's tables were copied by Nicolaus Germanus (fols 10v, 83v, 84v), known as a cartographer and miniaturist due to the handsome manuscript of Ptolemy's *Geography* that he made in 1466 for Ludovico Casella in Ferrara.

[88] Most glossators of the PAT located Erfurt 64 time minutes east of Toledo. See Kremer and Dobrzycki, 'Alfonsine Meridians', p. 96.

Th/K 1078); John of Lignères's *Algorismus minutiarum* (300r–308r, Th/K 878); Messahalla, *Compositio et usus astrolabii* (335r–342v, Th/K 1409); Prophatius Judaeus, *De quadrante novo* (343r–345v, Th/K 827); Nicholaus de Heybech's syzygies tables and canon (350v–352v; Th/K 1478); Oxford Tables and canon[89] (355–388v, Th/K 1686); tables and canon (*Si verum locum lune volueris invenire*, Th/K 1468) for finding lunar true longitude 1–30 days after mean syzygy (393v–408r), copied 1458 by 'Fridericus';[90] Gmunden's mean syzygies for 1433, 1473, ..., 2423 (408v–409v, but times computed for Regensburg, 69 minutes of time east of Toledo). In the hand of Friedrich Amann (236r), a monk at St Emmeram, Regensburg. Various colophons dated 1447–58.[91]

(M) Munich, BSB, Clm 14783, fols 189v–195r (table), fols 198v–200v (canon, *Omnis utriusque sexus*, followed by several unidentified canons on solar and lunar motion with explicit: *Iste Johannes equat, cepit Firminus... amicus habet 1450 in die luca evangelistice. Frater Fridericus professus monasterii Emmerami Ratispoenensis diocis*, 203v). Bound (568 fols) in a masssive astronomical-mathematical-astrological miscellany containing, among other things, Gmunden's calendar with mean syzygies for two cycles 1450–1488 (2r–15r), computed for Regensburg 69 minutes east of Toledo; scattered tables and canons of Gmunden, including his table of mean syzygies for 1433, 1473, ..., 2433 (26v); mean motions and equations from the PAT (62v–91r); Oxford Tables (126v–198r); various eclipse tables including parallax tables *ad Nurembergensis* (215v–217v); 'posthumous version' of Gmunden's tables (249r–409v, Th/K 1164); *Algorismus Ratisbonensis* (411r–441v); several geometrical texts (455r–505v); astrological tables (523r–538v); Leopold of Austria's *De mutacione aeris* (539r–547r, Th/K 381); tables and canon (*Si verum locum lune volueris invenire*, Th/K 1468) for finding lunar true longitude 1–30 days after mean

[89] No colophon follows the canon, but the same hand that copied the Oxford Tables added at the end of the text (fol. 383r) *Deinde Auctor vel compositor / Iste Johannes equat cepit firmicus / et complet Lux gaudet reprobat / lux amicus habet*, thereby linking John of Murs to the canon of the Oxford Tables.

[90] These tables closely follow the 'contratabula' procedures of John of Murs's Tables of 1321. See Chabás and Goldstein, 'John of Murs's Tables of 1321', p. 303.

[91] Schuba, *Die Quadriviums-Handschriften*, pp. 94–102; Chabás and Goldstein, 'Nicholaus de Heybech'; Gerl, 'Fridericus Amann'; Porres, *Les tables astronomiques*, pp. 78–79; Folkerts, 'Fridericus Amann'; Juste, 'MS Vatican, Biblioteca Apostolica Vaticana, Pal. lat. 1376'.

syzygy (548v–562v; similar to V_2, 393v–408r). Primarily in the hand of Friedrich Amann, colophons dated 1449–1456.[92]

(Pa) Paris, BnF, lat. 7285, fols 110v–112v (table). Bound (118 fols) in an astronomical miscellany, including daily calendars of mean motions for 1448 and 1451 for Paris (3v–6v); the PAT (9r–13v, 27r–29v, 46r–60v) with radices for Paris, 48 minutes east of Toledo; John of Saxony's pedagogical examples for John of Lignères's canons on the *primum mobile* (30r–36r, Th/K 1228); John of Lignères's Tables of 1322 (39r, 62r–83v, lacks the eclipse tables); John of Saxony's canon to the PAT (84r–90r); *Theorica planetarum* (90r–93r); Nicholas de Heybach's canon and tables for true syzygy (93r–94r, Th/K 1478); Oxford Tables for planetary latitudes (94v–107r); table for the duration of pregnancy (116v–117r); sundial text (117v–118r, Th/K 753). Paper (fols 37, 46) unicorn, Briquet 10013, dated 1443–1445, northern France; (fols 110, 112) NL0360-PO-118477, Culemborg (Netherlands), dated 1442.[93]

(Me) Metz, Bibliothéque municipale, 287, fols 80v–83r (table), fols 79v–80r (canon, *Omnis utriusque sexus*). Bound (428 fols) in an astronomical-astrological miscellany that includes John of Saxony's canons to the PAT (20r–27r); horoscope dated Erfurt, 1361 (27r); table of mean syzygies, 1369–1609 (27v), computed for Paris, about 48 minutes east of Toledo; Hermann of Saxony's Alfonsine computation of planetary positions for 1361 (49r–51v, Th/K 102) and his *Tabulae de motibus stellarum*, that incorporate much content from the PAT (52r–70v, Th/K 688);[94] Haly Abenragel, *De judiciis astrologiae* (88r–277v); Bartolomeo da Parma, *Breviloquium de fructu artis tocius astronomiae* (280r–317v); Sahl ibn Bishr's *Fatidica = Liber sextus astronomie*, transl. Hermann of Carinthia (334r–351v);[95] Leopold of Austria, *De astrorum scientia* (354r–363v), blank leaves (fols 364–428). Paper Waage (77) type FR5460-PO-116623, dated 1450, Metz; Dreiburg (79, 80, 82) type DE1935-Mscr_Dresd_P_33_124, dated *c.* 1440, Vienna.

(V_1) Vatican, BAV, Pal. lat. 1354, fols 50v–53r (table), fol. 60r-v (canon, *Omnis utriusque sexus*). Bound (252 fols) in an astronomical-astrological-medical miscellany copied by an unknown scribe at St Emmeram, Regensburg, *c.* 1463–64. Includes Nicholas de Heybech's syzygy

[92] Vogel, *Die Practica des Algorismus Ratisbonensis*, pp. 10–11; Porres, *Les tables astronomiques*, pp. 79, 670.

[93] Briquet, *Les filigranes*.

[94] Thorndike, 'Astronomy at Paris'.

[95] Burnett, 'Arabic into Latin', pp. 13–14.

tables (46v–47v), John of Saxony's canons to the PAT (109r–19v), tables of mean syzygies for 1433, 1473, ..., 1673 (46r), computed for Regensburg, 68 minutes of time east of Toledo, calendar of mean syzygies for 1463–1547 (37r–42v), computed for Prague, 74 minutes east of Toledo; tables and canon (*Si verum locum lune volueris invenire*, Th/K 1468) for finding lunar true longitude 1–30 days after mean syzygy (61v–78v, similar to V_2, 393v–408r); Leopold of Austria, *De astrorum scientia* (169r–233r).[96] Bound 18th century.

(V_3) Vatican, BAV, Pal. lat. 1367, fols 56v–59r (table). Bound (179 fols) in an astronomical-medical miscellany that includes the PAT (1r–26v), with radices added for the date of the Council of Basel, 1440 complete, and meridians of Konstanz, Basel, Freiburg im Breisgau, as well as Nuremberg, Wrocław, Prague and Paris; John of Lignères's Tables of 1322 (27v–40v); planetary latitudes from the Oxford Tables (49r–56r); several tables of mean syzygies (in the same hand as 56v–59r) for 1433, 1473, ..., 2433 (59v–60v), computed for Vienna, 80 minutes east of Toledo; for 1321, 1345, ..., 1609, computed *meridiani Parisiensis* (61r–62r) 48 minutes east of Toledo; and for 1393, 1417, ..., 1609, computed for *orisontem pragenseni* (62r-v) 74 minutes east of Toledo; tables and canon (*Si verum locum lune volueris invenire*, Th/K 1468) for finding lunar true longitude 1–16 days after mean syzygy (64r–71v; similar to V_2, 393v–408r).[97] Also contains many short texts or excerpts on astrological and medical topics. Southwest Germany, *c.* 1450.[98] Bound 18th century.

(E) Erfurt, UFB, Amplon. F. 388, fols 39v–42r (table). Bound (42 fols) with John of Lignères, *Tabule magne* (1r–35r), Erfurt meridian (25r, 38r). Dated Erfurt, 1446 complete (35r).[99] Paper Dreiburg mit Kreuz (24), DE4860-Rep_V_5_6, dated 1446, Leipzig.

[96] Schuba, *Die Quadriviums-Handschriften*, pp. 27–33; Thorndike, 'Some Little Known', pp. 43–44; Goldstein, 'Lunar Velocity in the Middle Ages', p. 190; Chabás and Goldstein, 'John of Murs's Tables of 1321', p. 306.

[97] These values for the three meridians are canonical within the Alfonsine corpus; see Kremer and Dobrzycki, 'Alfonsine Meridians', p. 196.

[98] Schuba, *Die Quadriviums-Handschriften*, pp. 58–62, dates to mid-fifteenth century. The astronomical materials of fols 1–84 are all copied on parchment by the same hand. Ex libris (1r) Nikolaus Pruckner (1488–1557), an Augustinian monk who in the 1520s introduced the Reformation to Mulhausen, authored calendars in the 1530–40s printed in Strasbourg, and in 1553 was named professor of astronomy at the university in Tübingen.

[99] Schum, *Beschreibendes Verzeichniss*, pp. 273–74. This codex is not included in the 1410 catalog of the Amploniana Collection. Donated, *c.* 1450, to library of the Collegium Porta Coeli by Peter of Cassel, vicar at the Saint Severi in Erfurt. Cf. Schum, *Beschreibendes Verzeichniss*, pp. 798–808.

Usually found in codices of astronomical/astrological miscellany, the TP are bound with some of the key works of Alfonsine astronomy: the PAT ($PaMV_1V_2V_3$), the tables of 1322 of John of Lignères (PaV_2V_3), tables of Nicholaus de Heybech, which also are used to compute time corrections to true syzygy ($PaGV_1V_2$), or tables of John of Gmunden ($BSPUVGV_2M$). In other cases, the TP are bound with copies of Bianchini's astronomical tables (C) or the Oxford Tables ($PaNVCMV_2V_3$). Only rarely are they found with major astrological (Me) or medical treatises (V_1V_3). From their placement in codices, it appears as if some mid fifteenth-century astronomers considered the TP to be part of Gmunden's tables, which explains Porres's decision to include them in her edition of Gmunden's work.

In our edition, all entries with manuscript variants are underlined. The variants are listed at the end of the edition, in order of solar argument, then lunar argument. I refer to individual entries by the notation 'solar argument:lunar argument'. Gray-shaded values are negative, unshaded values are positive for lunar arguments from 0 to 180°; the signs are reversed for lunar arguments from 180 to 360°.

Tabula ostendens distanciam vere coniunctionis vel oppositionis a media

	Arg Solis (degrees)																			
Arg Lune (degrees)	**0**		**6**		**12**		**18**		**24**		**30**		**36**		**42**		**48**		**54**	
	h	m	h	m	h	m	h	m	h	m	h	m	h	m	h	m	h	m	h	m
0	0	0	0	28	0	57	1	25	1	53	2	19	2	43	3	6	3	27	3	47
6	1	3	0	35	0	6	0	22	0	50	1	16	1	40	2	3	2	24	2	44
12	2	5	1	37	1	8	0	40	0	12	0	14	0	38	1	1	1	22	1	42
18	3	5	2	37	2	9	1	41	1	13	0	47	0	23	0	0	0	21	0	41
24	4	3	3	35	3	7	2	39	2	12	1	46	1	22	0	59	0	38	0	18
30	4	59	4	31	4	3	3	35	3	8	2	42	2	18	1	56	1	35	1	15
36	5	50	5	23	4	55	4	27	4	0	3	35	3	11	2	49	2	28	2	8
42	6	38	6	11	5	43	5	16	4	49	4	24	4	0	3	38	3	17	2	58
48	7	21	6	54	6	26	5	59	5	33	5	8	4	44	4	22	4	2	3	43
54	7	59	7	32	7	5	6	38	6	12	5	47	5	23	5	1	4	41	4	22
60	8	31	8	4	7	37	7	11	6	45	6	21	5	58	5	36	5	16	4	57
66	8	58	8	32	8	5	7	39	7	13	6	49	6	26	6	4	5	44	5	26
72	9	18	8	52	8	26	8	0	7	34	7	10	6	48	6	27	6	7	5	49
78	9	32	9	7	8	41	8	15	7	50	7	26	7	4	6	43	6	24	6	6
84	9	40	9	15	8	49	8	24	7	59	7	35	7	13	6	53	6	34	6	16
90	9	40	9	15	8	50	8	25	8	1	7	38	7	16	6	56	6	37	6	19
96	9	36	9	11	8	46	8	21	7	57	7	34	7	12	6	52	6	34	6	17
102	9	24	9	0	8	35	8	11	7	47	7	24	7	3	6	43	6	25	6	8
108	9	7	8	43	8	19	7	55	7	31	7	9	6	48	6	28	6	10	5	53
114	8	44	8	20	7	56	7	32	7	9	6	47	6	26	6	7	5	49	5	32
120	8	15	7	51	7	27	7	4	6	41	6	19	5	59	5	40	5	22	5	5
126	7	41	7	18	6	54	6	31	6	8	5	46	5	26	5	7	4	49	4	33
132	7	3	6	40	6	16	5	53	5	30	5	9	4	49	4	30	4	12	3	56
138	6	20	5	57	5	33	5	10	4	48	4	27	4	7	3	49	3	31	3	15
144	5	33	5	10	4	47	4	24	4	2	3	41	3	21	3	3	2	45	2	29
150	4	42	4	20	3	57	3	34	3	12	2	51	2	32	2	14	1	56	1	40
156	3	49	3	27	3	4	2	42	2	20	1	59	1	39	1	21	1	4	0	48
162	2	54	2	32	2	9	1	47	1	25	1	4	0	45	0	27	0	10	0	6
168	1	57	1	35	1	12	0	50	0	28	0	7	0	12	0	30	0	47	1	2
174	0	59	0	37	0	14	0	8	0	30	0	51	1	10	1	28	1	45	2	0
180	0	0	0	22	0	45	1	7	1	29	1	49	2	8	2	26	2	43	2	59

	Arg Solis (degrees)																			
Arg Lune (degrees)	**60**		**66**		**72**		**78**		**84**		**90**		**96**		**102**		**108**		**114**	
	h	m	h	m	h	m	h	m	h	m	h	m	h	m	h	m	h	m	h	m
0	4	4	4	18	4	29	4	38	4	44	4	47	4	47	4	43	4	37	4	27
6	3	1	3	15	3	26	3	35	3	41	3	44	3	44	3	40	3	34	3	24
12	1	59	2	13	2	24	2	32	2	38	2	41	2	41	2	37	2	31	2	21
18	0	58	1	12	1	23	1	31	1	37	1	40	1	40	1	36	1	30	1	20
24	0	1	0	12	0	23	0	31	0	37	0	40	0	40	0	36	0	30	0	20
30	0	58	0	45	0	34	0	26	0	20	0	17	0	17	0	21	0	27	0	37
36	1	51	1	38	1	27	1	19	1	13	1	10	1	10	1	14	1	20	1	30
42	2	41	2	28	2	17	2	9	2	3	2	0	2	0	2	4	2	10	2	20
48	3	26	3	13	3	2	2	54	2	49	2	46	2	46	2	50	2	56	3	6
54	4	6	3	53	3	42	3	34	3	29	3	26	3	26	3	30	3	36	3	46
60	4	41	4	28	4	18	4	10	4	4	4	1	4	1	4	5	4	11	4	21
66	5	10	4	57	4	47	4	39	4	34	4	31	4	31	4	35	4	41	4	50
72	5	33	5	20	5	10	5	2	4	57	4	54	4	54	4	58	5	4	5	13
78	5	50	5	37	5	27	5	19	5	14	5	11	5	11	5	15	5	21	5	30
84	6	0	5	48	5	38	5	30	5	25	5	23	5	23	5	26	5	32	5	41
90	6	4	5	52	5	42	5	34	5	29	5	27	5	27	5	30	5	36	5	45
96	6	1	5	49	5	39	5	32	5	27	5	25	5	25	5	28	5	34	5	43
102	5	53	5	41	5	31	5	24	5	19	5	17	5	17	5	20	5	26	5	35
108	5	38	5	26	5	17	5	10	5	5	5	2	5	2	5	5	5	11	5	20
114	5	17	5	5	4	56	4	49	4	44	4	42	4	42	4	45	4	51	4	59
120	4	50	4	39	4	30	4	23	4	18	4	16	4	16	4	19	4	24	4	33
126	4	18	4	7	3	58	3	51	3	46	3	44	3	44	3	47	3	52	4	1
132	3	42	3	31	3	22	3	15	3	10	3	8	3	8	3	11	3	16	3	24
138	3	1	2	50	2	41	2	34	2	29	2	27	2	27	2	30	2	35	2	43
144	2	15	2	4	1	55	1	48	1	44	1	42	1	42	1	45	1	50	1	58
150	1	26	1	16	1	7	1	0	0	55	0	53	0	53	0	56	1	1	1	9
156	0	34	0	24	0	15	0	8	0	4	0	1	0	1	0	4	0	9	0	17
162	0	20	0	31	0	40	0	47	0	51	0	53	0	53	0	50	0	45	0	37
168	1	16	1	27	1	36	1	43	1	47	1	50	1	50	1	47	1	42	1	34
174	2	14	2	25	2	34	2	41	2	45	2	48	2	48	2	45	2	40	2	32
180	3	13	3	24	3	32	3	39	3	43	3	46	3	46	3	43	3	38	3	30

Arg Lune (degrees)	Arg Solis (degrees)																			
	120		126		132		138		144		150		156		162		168		174	
	h	m	h	m	h	m	h	m	h	m	h	m	h	m	h	m	h	m	h	m
0	4	14	3	58	3	39	3	18	2	55	2	29	2	1	1	32	1	3	0	32
6	3	11	2	55	2	36	2	15	1	51	1	25	0	57	0	29	0	0	0	31
12	2	8	1	52	1	33	1	12	0	49	0	23	0	5	0	34	1	3	1	34
18	1	7	0	51	0	32	0	11	0	12	0	38	1	6	1	35	2	4	2	35
24	0	8	0	8	0	27	0	48	1	12	1	38	2	5	2	34	3	3	3	34
30	0	49	1	5	1	24	1	45	2	8	2	34	3	1	3	30	3	59	4	30
36	1	43	1	59	2	18	2	38	3	1	3	27	3	54	4	22	4	51	5	22
42	2	32	2	48	3	7	3	27	3	50	4	16	4	43	5	11	5	40	6	10
48	3	18	3	34	3	52	4	12	4	35	5	1	5	28	5	56	6	24	6	54
54	3	58	4	14	4	32	4	52	5	15	5	40	6	7	6	34	7	2	7	32
60	4	33	4	49	5	7	5	27	5	49	6	14	6	40	7	7	7	35	8	4
66	5	2	5	18	5	36	5	55	6	17	6	42	7	8	7	35	8	2	8	31
72	5	26	5	41	5	58	6	17	6	39	7	4	7	30	7	57	8	24	8	53
78	5	43	5	58	6	15	6	34	6	56	7	20	7	46	8	12	8	39	9	7
84	5	53	6	8	6	25	6	44	7	6	7	30	7	55	8	21	8	47	9	15
90	5	57	6	12	6	29	6	48	7	9	7	32	7	57	8	23	8	49	9	16
96	5	55	6	9	6	26	6	45	7	5	7	28	7	53	8	18	8	44	9	10
102	5	46	6	0	6	17	6	36	6	56	7	19	7	43	8	8	8	33	9	0
108	5	31	5	45	6	2	6	20	6	40	7	3	7	27	7	52	8	17	8	44
114	5	10	5	24	5	41	5	59	6	19	6	41	7	5	7	29	7	54	8	20
120	4	44	4	58	5	14	5	32	5	51	6	13	6	37	7	1	7	25	7	51
126	4	12	4	25	4	41	4	59	5	18	5	40	6	3	6	27	6	51	7	17
132	3	35	3	48	4	4	4	22	4	41	5	3	5	26	5	49	6	13	6	39
138	2	54	3	7	3	23	3	40	3	59	4	20	4	43	5	7	5	31	5	56
144	2	8	2	21	2	37	2	54	3	13	3	34	3	57	4	20	4	44	5	9
150	1	19	1	32	1	48	2	5	2	24	2	45	3	7	3	30	3	54	4	19
156	0	27	0	40	0	56	1	13	1	31	1	52	2	14	2	37	3	1	3	26
162	0	27	0	14	0	1	0	18	0	36	0	57	1	19	1	42	2	5	2	30
168	1	24	1	11	0	56	0	39	0	21	0	0	0	22	0	45	1	8	1	33
174	2	22	2	9	1	54	1	37	1	19	0	58	0	36	0	13	0	10	0	34
180	3	20	3	8	2	53	2	36	2	18	1	57	1	35	1	12	0	49	0	25

	Arg Solis (degrees)																			
Arg Lune	**180**		**186**		**192**		**198**		**204**		**210**		**216**		**222**		**228**		**234**	
(degrees)	h	m	h	m	h	m	h	m	h	m	h	m	h	m	h	m	h	m	h	m
0	0	0	0	32	1	3	1	32	2	1	2	29	2	55	3	18	3	39	3	58
6	1	3	1	35	2	6	2	35	3	3	3	31	3	57	4	21	4	41	5	1
12	2	5	2	37	3	8	3	37	4	5	4	33	4	59	5	23	5	44	6	3
18	3	6	3	38	4	9	4	38	5	6	5	34	6	0	6	23	6	43	7	3
24	4	5	4	36	5	7	5	36	6	4	6	31	6	57	7	20	7	41	8	0
30	5	1	5	32	6	2	6	31	6	59	7	26	7	52	8	15	8	36	8	54
36	5	53	6	24	6	54	7	22	7	50	8	17	8	43	9	6	9	26	9	44
42	6	41	7	11	7	41	8	9	8	37	9	4	9	29	9	52	10	12	10	30
48	7	24	7	54	8	24	8	52	9	19	9	46	10	11	10	34	10	54	11	12
54	8	2	8	32	9	2	9	30	9	56	10	23	10	48	11	10	11	30	11	47
60	8	34	9	4	9	34	10	2	10	28	10	54	11	19	11	41	12	0	12	17
66	9	1	9	31	10	0	10	27	10	53	11	19	11	44	12	5	12	24	12	41
72	9	22	9	51	10	19	10	46	11	13	11	38	12	2	12	23	12	42	12	59
78	9	36	10	5	10	33	11	0	11	26	11	51	12	14	12	35	12	54	13	11
84	9	44	10	12	10	40	11	6	11	32	11	57	12	20	12	41	13	0	13	17
90	9	44	10	12	10	40	11	6	11	31	11	55	12	18	12	39	12	57	13	14
96	9	37	10	6	10	34	11	0	11	25	11	49	12	12	12	33	12	51	13	7
102	9	28	9	55	10	22	10	47	11	12	11	36	11	58	12	19	12	37	12	53
108	9	11	9	38	10	4	10	29	10	54	11	17	11	39	11	59	12	17	12	33
114	8	47	9	14	9	40	10	5	10	29	10	52	11	14	11	34	11	52	12	8
120	8	18	8	44	9	10	9	34	9	58	10	21	10	43	11	3	11	20	11	36
126	7	44	8	10	8	36	9	0	9	24	9	47	10	8	10	27	10	44	11	0
132	7	5	7	31	7	57	8	21	8	44	9	7	9	28	9	47	10	4	10	20
138	6	22	6	48	7	13	7	37	8	0	8	22	8	43	9	2	9	19	9	35
144	5	35	6	1	6	26	6	49	7	12	7	34	7	55	8	14	8	31	8	46
150	4	44	5	9	5	34	5	57	6	20	6	42	7	3	7	22	7	39	7	54
156	3	51	4	16	4	41	5	4	5	27	5	49	6	10	6	29	6	45	7	0
162	2	55	3	20	3	45	4	8	4	31	4	53	5	13	5	32	5	49	6	4
168	1	58	2	23	2	47	3	10	3	33	3	55	4	16	4	35	4	51	5	6
174	0	59	1	24	1	48	2	11	2	34	2	56	3	17	3	36	3	52	4	7
180	0	0	0	25	0	49	1	12	1	35	1	57	2	17	2	36	2	52	3	8

	Arg Solis (degrees)																			
Arg Lune	**240**		**246**		**252**		**258**		**264**		**270**		**276**		**282**		**288**		**294**	
(degrees)	h	m	h	m	h	m	h	m	h	m	h	m	h	m	h	m	h	m	h	m
0	4	14	4	27	4	37	4	43	4	47	4	47	4	44	4	38	4	29	4	18
6	5	17	5	29	5	39	5	45	5	49	5	49	5	46	5	40	5	32	5	21
12	6	19	6	31	6	41	6	47	6	51	6	51	6	47	6	41	6	33	6	22
18	7	19	7	31	7	41	7	47	7	50	7	50	7	47	7	41	7	33	7	22
24	8	16	8	28	8	38	8	44	8	47	8	47	8	44	8	38	8	30	8	19
30	9	10	9	22	9	32	9	38	9	42	9	42	9	38	9	32	9	24	9	13
36	10	1	10	13	10	22	10	28	10	32	10	32	10	28	10	22	10	14	10	3
42	10	47	10	59	11	8	11	14	11	18	11	18	11	14	11	8	11	0	10	49
48	11	28	11	40	11	49	11	55	11	58	11	58	11	54	11	48	11	40	11	29
54	12	3	12	15	12	24	12	30	12	33	12	33	12	30	12	24	12	16	12	5
60	12	33	12	45	12	54	13	0	13	3	13	3	13	0	12	54	12	46	12	35
66	12	57	13	9	13	18	13	24	13	27	13	27	13	24	13	18	13	10	12	59
72	13	15	13	26	13	35	13	41	13	44	13	44	13	41	13	35	13	27	13	17
78	13	27	13	38	13	47	13	52	13	55	13	55	13	52	13	47	13	39	13	29
84	13	31	13	42	13	51	13	57	14	0	14	0	13	56	13	51	13	43	13	33
90	13	28	13	39	13	48	13	53	13	56	13	56	13	53	13	48	13	40	13	30
96	13	21	13	32	13	41	13	46	13	49	13	49	13	46	13	41	13	33	13	23
102	13	7	13	18	13	26	13	31	13	34	13	34	13	31	13	26	13	19	13	9
108	12	47	12	58	13	6	13	11	13	14	13	14	13	11	13	6	12	59	12	49
114	12	21	12	32	12	40	12	45	12	48	12	48	12	45	12	40	12	33	12	24
120	11	50	12	0	12	8	12	13	12	16	12	16	12	13	12	8	12	1	11	52
126	11	14	11	24	11	32	11	37	11	40	11	40	11	37	11	32	11	24	11	15
132	10	33	10	43	10	51	10	56	10	59	10	59	10	56	10	51	10	44	10	35
138	9	48	9	58	10	6	10	11	10	14	10	14	10	11	10	6	9	59	9	50
144	8	59	9	9	9	17	9	22	9	25	9	25	9	22	9	17	9	10	9	1
150	8	7	8	17	8	25	8	30	8	33	8	33	8	30	8	25	8	18	8	10
156	7	13	7	23	7	31	7	36	7	38	7	38	7	35	7	31	7	24	7	15
162	6	17	6	27	6	34	6	39	6	42	6	42	6	39	6	35	6	28	6	20
168	5	19	5	29	5	36	5	41	5	44	5	44	5	42	5	37	5	30	5	22
174	4	20	4	30	4	37	4	42	4	45	4	45	4	43	4	38	4	31	4	23
180	3	20	3	30	3	38	3	43	3	46	3	46	3	43	3	39	3	32	3	24

	Arg Solis (degrees)																			
Arg Lune (degrees)	300		306		312		318		324		330		336		342		348		354	
	h	m	h	m	h	m	h	m	h	m	h	m	h	m	h	m	h	m	h	m
0	4	4	3	47	3	27	3	6	2	43	2	19	1	53	1	25	0	57	0	28
6	5	7	4	49	4	29	4	8	3	45	3	21	2	55	2	27	1	59	1	31
12	6	8	5	51	5	31	5	10	4	47	4	23	3	57	3	29	3	1	2	33
18	7	8	6	51	6	31	6	10	5	47	5	23	4	57	4	29	4	1	3	33
24	8	5	7	48	7	28	7	7	6	45	6	21	5	54	5	27	4	59	4	31
30	8	59	8	42	8	22	8	1	7	39	7	15	6	49	6	22	5	54	5	26
36	9	50	9	33	9	13	8	52	8	30	8	6	7	40	7	13	6	46	6	18
42	10	36	10	19	9	59	9	38	9	16	8	52	8	27	8	0	7	33	7	5
48	11	16	10	59	10	40	10	19	9	58	9	35	9	10	8	43	8	16	7	48
54	11	52	11	35	11	16	10	55	10	34	10	11	9	46	9	20	8	53	8	26
60	12	22	12	6	11	47	11	26	11	5	10	42	10	17	9	51	9	25	8	58
66	12	46	12	30	12	11	11	51	11	30	11	7	10	43	10	17	9	50	9	24
72	13	4	12	48	12	29	12	10	11	49	11	26	11	2	10	36	10	10	9	44
78	13	16	13	0	12	41	12	22	12	1	11	39	11	15	10	50	10	24	9	58
84	13	21	13	5	12	47	12	28	12	7	11	45	11	22	10	57	10	31	10	5
90	13	18	13	3	12	45	12	26	12	6	11	44	11	21	10	56	10	31	10	5
96	13	11	12	56	12	38	12	19	11	59	11	38	11	15	10	51	10	26	10	1
102	12	57	12	42	12	25	12	6	11	46	11	25	11	2	10	38	10	14	9	49
108	12	37	12	22	12	5	11	47	11	27	11	6	10	44	10	20	9	56	9	31
114	12	12	11	57	11	40	11	22	11	2	10	41	10	19	9	56	9	32	9	8
120	11	40	11	25	11	8	10	51	10	32	10	11	9	49	9	26	9	3	8	39
126	11	4	10	50	10	33	10	15	9	56	9	36	9	14	8	51	8	28	8	4
132	10	24	10	10	9	53	9	36	9	17	8	57	8	35	8	12	7	49	7	26
138	9	39	9	25	9	8	8	51	8	32	8	12	7	51	7	29	7	6	6	43
144	8	50	8	36	8	20	8	3	7	44	7	25	7	4	6	41	6	18	5	55
150	7	59	7	45	7	29	7	12	6	53	6	34	6	13	5	51	5	28	5	5
156	7	4	6	50	6	34	6	17	5	59	5	40	5	19	4	57	4	35	4	12
162	6	9	5	55	5	39	5	22	5	4	4	44	4	23	4	1	3	39	3	16
168	5	11	4	57	4	41	4	24	4	6	3	47	3	26	3	4	2	42	2	19
174	4	12	3	58	3	42	3	25	3	7	2	48	2	28	2	6	1	44	1	21
180	3	13		59	2	43	2	26	2	8	1	49	1	29	1	7	0	45	0	22

Apparatus

The tabular arguments are given in the form solar arg:lunar arg, the deviant entries in hrs;mins.

0:102 V_2 9;20, **0:120** E 8;25, **0:156** E 2;49, **0:162** N 2;44

6:42 C 8;53, **6:48** V_1V_2 6;57, **6:72** N 8;04, **6:78** N 9;32, **6:120** PM$V_3V_1V_2$ 7;41

12:114 PME$V_3V_1V_2$ 7;16

18:48 E 5;50 C 5;58, **18:72** E 8;08

24:24 V_1 1;12

36:18 U 0;13, **36:126** M 5;20, **36:138** M 3;47, **36:144** M 3;51

42:66 MePa 5;05, **42:90** V_3 6;55

48:0 N 0;27, **48:6** N 3;24, **48:180** M 1;43

54:12 Me 1;43, **54:42** E 3;58, **54:48** E 4;43, **54:54** E 4;12, **54:72** E 6;49, **54:90** MV_3V_1 2;19, **54:126** M 4;34, **54:144** V_1 2;19, **54:150** V_2 1;10, **54:180** Me 3;59

60:156 E 0;24, **60:180** MeE 3;12

66:180 E 3;34 C 3;23

72:18 G 1;33, **72:54** E 3;04, **72:126** U 3;48 V_2 3;56, **72:132** UCN 3;32, **72:144** M 1;52, **72:174** V_3 2;24, **72:180** V_2 3;22

78:48 M 2;52, **78:174** M 2;51 V_3 2;42

84:12 E 2;36, **84:144** M 1;49, **84:180** MeV_3 2;43

90:30 M 0;57, **90:126** M 3;24, **90:168** Me 0;50, **90:174** C 2;40

96:174 C 2;40

102:6 V_2 3;30, **102:48** V_3 2;04, **102:54** V_3 2;05, **102:60** V_3 3;30, **102:66** V_3 4;05, **102:72** V_3 4;35, **102:78** V_3 4;58, **102:84** V_3 5;15, **102:90** V_3 5;26, **102:96** V_3 5;30, **102:102** V_3 5;28, **102:108** V_3 5;20, **102:114** V_3 4;05, **102:120** V_3 4;45

108:138 V_3 2;36, **108:150** E 0;01, **108:180** M$V_3V_1V_2$ 3;08 E 3;36

114:108 Me 5;30 V_1 20;20

120:54 V_1V_2 3;56, **120:132** V_3 3;54, **120:138** M3 3;14, **120:162** G 0;00, **120:168** C 1;27, **120:174** C 2;24

126:0 N 3;55, **126:6** N 2;52, **126:12** N 1;51, **126:18** N 0;08, **126:24** N 0;05, **126:30** N 1;59, **126:36** N 1;48 V 1;58, **126:42** N 2;34, **126:48** N 3;14, **126:54** N 4;49, **126:60** N 4;18, **126:66** N 5;41, **126:72** N 5;58

132:48 M 3;53, **132:72** V_1 5;56, **132:180** MePa 2;52

138:60 PMMeE 5;17, **138:102** V_2 6;35

144:18 C 0;13, **144:24** G 0;12, **144:66** V_2 6;57, **144:180** MePa 2;17

150:36 M 3;37, **150:96** M 7;38

156:78 V_3 7;47, **156:108** M 7;47, **156:120** M 6;33, **156:180** V_1 1;25

162:24 U 2;54, **162:66** E 7;34, **162:114** Me 7;22, **162:156** V_1 2;27, **162:162** V_3 1;41

168:42 U 5;44, **168:54** U 7;35, **168:78** V_3 8;34, **168:114** Me 7;55, **168:144** Me 4;54, **168:150** Me 3;44, **168:162** C 2;11

174:66 C 8;21, **174:174** C 0;33, **174:180** Me 0;35

180:6 V 2;03, **180:42** N 6;51, **180:102** Me 9;18, **180:162** E 2;?? (the scribe did not enter minutes)

186:78 M 10;02, **186:84** P 10;14, **186:138** M 7;48, **186:168** V_3 2;33, **186:180** V_2 0;31

192:12 M 3;09, **192:18** M 4;08, **192:30** E 6;03, **192:60** M 9;38 Me 10;34, **192:144** V 6;36 N 6;24, **192:162** P 3;55, **192:180** P 0;59

198:0 C 1;12, **198:36** E 7;23

204:30 MeV_1 6;57, **204:60** E 10;26

210:54 M 9;23, **210:120** M 10;51, **210:126** M 9;57, **210:132** M 9;02, **210:162** M 3;57

216:30 V 7;55, Me 6;52, **216:36** Me 7;43, N 8;53, **216:42** Me 8;29 C 9;59, **216:48** Me 9;11, **216:60** Me 10;19, **216:108** E 11;34

222:6 V_3 5;23, **222:12** V_3 6;23, **222:18** $V_3$7;23, **222:102** Pa 12;10

228:6 MeEPa 4;42, **228:84** M 12;00

234:36 Pa 9;45, **234:42** Pa 10;31, **234:78** M 12;11, **234:84** M 12;17, **234:90** M 12;14, **234:96** M 12;07, **234:132** E 9;20

240:108 M 12;46, **240:132** Me 10;22, E 10;32, **240:180** E 3;30, **240:96** G 13;27

246:126 M 11;34, **246:144** N 8;09, **246:150** N 7;17, **246:156** N 6;23, **246:162** N 5;27, **246:168** N 4;29 Me 5;20

252:0 N 4;34, **252:36** M 10;32, **252:78** C 13;57, **252:96** M 13;51, **252:102** M 13;36

258:0 M 4;42, **258:36** M 10;38, **258:72** Me 13;01, **258:84** S 14;57, **258:114** M 12;25, **258:144** C 9;23, **258:156** Me 7;26

264:54 M 12;32, **264:66** V_3 13;37, **264:156** Me 7;36, **264:162** V_2 6;32, **264:168** VM 5;48

270:114 P 12;38, **270:156** PM$V_3V_1V_2$ 7;37, **270:168** M 5;34

276:30 M 9;48, **276:48** Me 11;56, **276:84** V_3 13;52

282:36 E 10;32, **282:48** C 11;44

288:30 G 9;34, **288:138** M 10;59, **288:156** Me 7;34, **288:162** Me 6;20

294:24 BPVUM$V_3V_1V_2$ 8;10, **294:60** E 12;25, **294:72** S 12;17, **294:96** PM$V_3V_1V_2$ 13;33 U 13;13, **294:138** M 10;50, **294:180** U 3;23 V_1V_2 **3:34**

300:42 V_1V_2 10;38, **300:126** V_2 11;40, **300:138** V_3 9;34

306:12 E 6;51, **306:36** Me 9;38, **306:54** Me 12;35, **306:72** Me 13;48

312:54 Me 11;03, **312:138** V 9;20, **312:144** V 8;39, **312:156** C 6;36, **312:162** V_1 5;29, **312:180** U 2;42

318:36 MV_3 9;52, **318:54** PMMeE 11;55, **318:108** Me 12;47, **318:114** M 11;42 Me 12;45, **318:120** Me 12;38 V_3 11;51, **318:126** Me 12;25, **318:132** Me 12;05, **318:138** Me 11;40, M 8;56, **318:144** Me 11;08, **318:150** Me 10;12, **318:156** Me 10;17 V 6;19

324:0 V_1V_2 3;43, **324:30** PM $V_3V_1V_2$ 7;49, **324:60** N 11;50, **324:72** G 11;41, **324:156** C 5;49

330:90 Me 11;34, **330:126** Me 9;35 N 10;36, **330:144** M 7;35

336:24 N 5;44, **336:30** PM $V_3V_1V_2$ 6;59, **336:120** Me 9;41, **336:174** P 2;38

342:42 Pa 8;04, **342:54** M$V_3V_1V_2$ 9;22, **342:144** UGC 6;49

348:78 Me 10;34, **348:84** MeEPa 10;32, **348:162** Me 3;37

354:66 V_1 9;34, **354:84** M 10;56

Part 4

Pushing Approaches to Table Analysis Further

Computing with Manuscripts: Time between Mean and True Syzygies in John of Lignères' *Tabule magne**

Matthieu HUSSON

Introduction

General questions

When performing computations, ancient and medieval astronomers worked with astronomical tables accessible to them in manuscripts. Not only did they perform their computations 'by hand', but also the numbers they manipulated were 'read' from handwritten documents, with all the potential complexity of this act. Manuscript transmission produces variability in many ways, but, for the purposes of our research, we will treat it here primarily on two different levels. On one level, scribal variants can occur when numbers are copied or when the rows and columns of tables are inadvertently shifted. On a second, more structural, level, variations arise, especially in Latin sources, in the ways tables and related canons are assembled in manuscripts. This produces not only variants in different manuscript witnesses of a given table, but also deeper variants in the ways tables can be combined in astronomical procedures. Computation, on the other hand, is sensitive to differences in numbers and procedures. Even small procedural variations in rounding and truncation for elementary arithmetical operations might yield different results. How does the variability inherent to manuscript transmission affect the process of computing and the results produced by computation? Can we build tools to isolate these effects? What kind of critical edition and analysis of the tabular material can be constructed in order to grasp these effects? How can a digital information system on astronomical tables be helpful in this respect? How can such a set of tools then inform us on what 'reading' a table, 'precision', and 'errors' may have meant for historical actors? Is it possible, for instance, to collect supporting evidence indicating that some particularly skilled 'readers' were able to adjust scribal variants in the manuscript they were using in a computation to avoid results they would

* I am grateful to the many colleagues who participated in the TAMAS project. Their insights and our stimulating discussions shaped this work in significant ways. Clemency Montelle, Benno van Dalen and Glen van Brummelen, as well as Nick Jacobson, Samuel Gessner, Eleonora Andriani, José Chabas and Richard Kremer, made helpful comments on drafts of the text. The research presented in this chapter was supported by the TAMAS project ('Jeunnes chercheurs-nouvelles équipes' PSL, 2017–2018, PI Matthieu Husson).

Editing and Analysing Numerical Tables: Towards a Digital Information System for the History of Astral Sciences, ed. by Matthieu Husson, Clemency Montelle and Benno van Dalen, PALS 2 (Turnhout, 2021), pp. 425–468
© BREPOLS PUBLISHERS 10.1484/M.PALS-EB.5.127704

consider inappropriate? How does the layout and presentation of a table set influence its possible use in computations or the frequency of specific types of scribal variants? What level of accuracy in computation could historical actors produce with their computation tools? How different can these results be when faced with two different manuscript versions of the 'same' table set? What are the relations between procedures described in canons and the actual astronomical tables found in manuscripts? These are the kinds of questions this chapter asks in the context of a specific case study: determination of the time between mean and true syzygies[1] with the computational tools presented by the known manuscript tradition of John of Lignères' *Tabule magne*.[2]

Brief introduction to syzygy computations

'Syzygy' is a general term pointing to the conjunction or opposition of the sun and the moon in ecliptic longitude. Computing syzygy is thus, from a mathematical perspective, a pursuit problem. Two points are moving on a given circle and the moment they coincide or are diametrically opposed is to be determined. Two aspects make this computation complex in the context of ancient astronomy. First, the velocities of the sun and moon are not constant. Second, the mathematical tools of ancient astronomy were generally designed to compute heavenly positions at a fixed time. Syzygy computation requires the opposite: find the time at which a given configuration of heavenly objects is satisfied. Ancient astronomers usually address these issues by decomposing the computation into two steps. In the first step, they compute what they call 'mean syzygy', that is the point and moment of conjunction or opposition of the sun and moon if they are considered to move at a constant mean pace along the ecliptic. The second step starts from the result of the first. It takes into account the 'true' position and changing velocities of the sun and moon at the time of 'mean syzygy' by considering their respective 'equations' (correction terms for the varying paces of the luminaries) and determines from this the time of 'true syzygy'. Many different procedures were proposed by ancient actors to address this second step: some use tables of the solar and lunar motions directly, others rely on different types of tables specifically designed for this purpose. Most

[1] The literature on this topic is large. Cf. Chabás and Goldstein, 'Computational Astronomy: Five Centuries'; Kremer, 'Thoughts on John of Saxony's' Method'; Chabás and Goldstein, *A Survey of European Astronomical Tables*, pp. 139–54. My selection of syzygy computations to explore these general questions is inspired by this scholarship, which also analyses the mathematical properties and astronomical foundations of different historical procedures.

[2] Chabás and Goldstein, *The Alfonsine Tables*; Chabás and Goldstein, *A Survey of European Astronomical Tables*; North, 'The Alfonsine Tables in England'; Poulle, 'John of Lignères'; Saby, *Les canons de Jean de Lignères*; Husson, 'Ways to Read a Table'. For an overview of the *Tabule magne* and the identification of the manuscripts containing them, see Chabás, *Computational Astronomy in the Middle Ages*, pp. 199–206.

procedures are iterative and propose successive evaluations of the time of true syzygy. This second step is by far the more complex of the two. The present chapter is concerned with the way in which the manuscript tradition of John of Lignères' *Tabule magne* framed the computation of the second step of syzygy computation: finding the time between mean and true syzygy.

The *Tabule magne* and true syzygy computations

The *Tabule magne* is a set of astronomical tables with their canons. It was created by John of Lignères and dedicated in 1325 to Robert the Lombard, Dean of Glasgow, with two other texts on astronomical instruments (a *saphea* and an *equatorium*).[3] The *Tabule magne* are concerned with computations of syzygies (mean and true) and planetary positions (mean and true). A particular feature of the *Tabule magne* is related to planetary positions, for which double-argument equation tables are used for the first time in the tradition of Latin sources. The *Tabule magne* also have a strong link to the various versions of the Oxford Tables and were important in the transmission of Parisian Alfonsine astronomy to England.

In the four known manuscript witnesses of the canons, a chapter with the title *Tempus vere coniunctionis et oppossitionis solis et lune invenire* ('To find the time of the true conjunction or opposition of the sun and the moon') is found.[4] This canon describes a method to compute the time from mean to true syzygy. The procedure in this canon instructs the reader to compute two astronomical quantities at the time of mean syzygy: first the distance between the true sun and the true moon; and second the difference between the velocities of the sun and the moon (the *superatio*). In order to compute these quantities, one needs to rely on tables that are not specified and only implied in the context of this particular chapter. Then these two quantities are used as the arguments of a specific table. This table is described in the canons with the name *tabula longitudinis horarum* ('table of the longitudes of hours'). It simply tabulates the quotient of the first argument by the second.[5] The result of the reading of this table is then taken as an approximation of the time between mean and true syzygy. The process is to be iterated until one estimates that the true sun and true moon are equal.

[3] Husson, *Les domaines d'application.*

[4] Cambridge, Gonville and Caius College, MS 110, pp. 1–5; Erfurt, UFB, Amplon. Q 366, 28r–32v (see for a digital copy: https://dhb.thulb.uni-jena.de/receive/ufb_cbu_00022114); Paris, BnF, MS 7281, 201v–205v (see for a digital copy: https://gallica.bnf.fr/ark:/12148/btv1b52 5030045); Paris, BnF, lat. 10263, 70r–78r (see for a digital copy: https://gallica.bnf.fr/ark:/ 12148/btv1b9072582f).

[5] Tables of this type for the computation of true syzygies are related to a tradition that can be traced back to the twelfth-century Andalusian astronomer Ibn al Kammād (Chabás and Goldstein, *A Survey of European Astronomical Tables*, pp. 145–46).

Scholarship has identified fifteen manuscript witnesses of the tables of the *Tabule magne*.[6] Six witnesses include specific tables for the computation of true syzygy.[7] Among these, five contain the same group of four tables presenting respectively the solar and lunar equations and velocities. This group is always presented in a single grid[8] and allows one to carry out the first part of the procedure described by the canons.[9] In this paper I will refer to it as the 'equations and velocities grid'. Two among the six witnesses present the division table described in the canons.[10] However, the format of the table in the manuscripts does not correspond exactly to the description of the table in the canons. The canon's description implies that the table is presented in a single grid, where the table's arguments and entries have no specific order of magnitude. Only the procedure in the canons fixes the rule according to which these orders of magnitude are determined. In the manuscripts, the table is divided into two distinct grids: one in which the first argument (i.e., the distance between the sun and the moon) is in degrees, the other where the first argument is in minutes. In both cases the units of the entries are also specified directly in the grid

[6] Paris, BnF, Latin 7300A, 94v–112r (https://gallica.bnf.fr/ark:/12148/btv1b100271684); Bernkastel-Kues, Cusanusstift, MS 210, 89v, 103r–117r; Bernkastel-Kues, Cusanusstift, MS 212, 91v–92r and 93r (tables for radices and syzygies); Cambridge, Gonville and Caius College, MS 110, pp. 7–18; Erfurt, UFB, Amplon. F 376, 30v–53v; Erfurt, UFB, Amplon. F 388, 1r–42v; Lisbon, Biblioteca da Ajuda, MS 52-XII-35, 67r–92v; London, British Library, Add. 24070, 24v–42v; Paris, BnF, lat. 7286C, 10v–11r, 23v–24r (see for a digital copy: https://gallica.bnf.fr/ark:/12148/btv1b10035226k); Paris, BnF, lat. 10264, 1r–28v (see for a digital copy: https://gallica.bnf.fr/ark:/12148/btv1b10036926k); Segovia, Biblioteca de la Catedral, MS 84, pp. 680–91; Vatican, BAV, Pal. lat. 1367, 60v–62r (see for a digital copy: https://digi.ub.uni-heidelberg.de/diglit/bav_pal_lat_1367); Vatican, BAV, Pal. lat. 1374, 26r–27v, 51v (see for a digital copy: https://digi.ub.uni-heidelberg.de/diglit/bav_pal_lat_1374); Vatican, BAV, Pal. lat. 1376, 46r, 102r–130r (see for a digital copy: https://digi.ub.uni-heidelberg.de/diglit/bav_pal_lat_1376); Vatican, BAV, Pal. lat. 1412, 102r–116v (see for a digital copy: https://digi.ub.uni-heidelberg.de/diglit/bav_pal_lat_1412). Note that table sets are often very mixed in manuscripts from the Latin tradition. I here give extensive folio ranges in which tabular material from the *Tabule magne* are found along with tabular material from different origins.

[7] Cambridge, Gonville and Caius College, MS 110; Erfurt, UFB, Amplon. F 388; Paris, BnF, lat. 10264; Vatican, BAV, Pal. lat. 1367; Vatican, BAV, Pal. lat. 1374; Vatican, BAV, Pal. lat. 1376, 46r, 102r–130r; Vatican, BAV, Pal. lat. 1412.

[8] In this paper I use the word 'table' in order to point to a set of arguments and entries (mathematically) related to each other, and I use the word 'grid' to point to the particular layout in which tables are written. A single table can be displayed in several grids. Several tables can be grouped into a single grid.

[9] This arrangement is probably linked directly to John of Lignères and a very similar type of tables arrangement is found in his tables of 1321 (Saby, *Les canons de Jean de Ligneres*), except with argument every 6° instead of every 1°.

[10] Cambridge, Gonville and Caius College, MS 110; Paris, BnF, lat. 10264.

displaying the table. Finally, three manuscript witnesses among the six present a double-argument table by John of Murs relying on a completely different logic for the computation of true syzygies.[11] While the canons of the *Tabule magne* describe a unique and quite simple procedure to compute true syzygy, the manuscript tradition of the table set shows that users of the *Tabule magne* did not feel compelled by the canons, and made the computations along diverse and different lines.[12]

This situation, which was quite typical for the respective traditions of canons and table sets, demonstrates that the relation between the tables described or implied by the canons and those actually found in table sets was not straightforward. It also shows that the relation between the procedures described in the canons and those that can actually be performed with the tables found in table sets is likewise not straightforward. This complexity of course has a documentary aspect: it depends heavily on what is usually described as the 'accident of transmission'. This points to the history of each individual manuscript, which can, to some extent, be analysed and partly recovered from a careful examination of the document's material, graphical and intellectual dimensions. Like scribal variants, some of these 'accidents' are completely unintended or the result of events that occurred long after the manuscripts were actually used by medieval astronomers. However, some aspects of these 'accidents' are consequences of the status of these writings and the way they were used, among other things in computations, by historical actors. In any case, the complexity also reflects, at least partially, the conditions under which these actors could perform computations.

In light of this complexity I will focus here on the equation and velocity grid and on the *Tabula longitudinis horarum* as they are found in the manuscripts. I will study their mathematical properties in the context of the algorithm described by the related canon in the *Tabule magne*. In particular, I will not explore the full range of possibilities the tables offer for syzygy computations, which could rely also on other tables or on other procedures. This focused approach has two main steps. First, I will offer a critical edition of the true syzygy tables in John of Lignères' *Tabule magne*. I will transcribe and study each manuscript witness and then conduct an astronomical and mathematical analysis of the tables. Second, I will use this critical edition to propose tools that can help address general issues of computing with manuscripts by

[11] Erfurt, UFB, Amplon. F 388, Vatican, BAV, Pal. lat. 1367, and Vatican, BAV, Pal. lat. 1376, 46r, 102r–130r contain a copy of John of Murs' *Tabule permanentes*. The Erfurt manuscript includes this in the same set of folios bearing the *Tabule magne*. The *Tabule permanentes* are another set of tables designed for the same purpose and produced also in Paris by John of Murs; see Richard Kremer's contribution to this volume.

[12] It will be shown that numerical performance may also be an important aspect of the limited success of the *Tabula longitudinis horarum* among Latin astronomers.

hand by reconstructing the computations of medieval astronomers with these tables and analysing their properties of convergence and robustness within the procedure described by the canons.[13]

Critical edition of the tables

Figure 1 lists the five manuscripts that contain the equation and velocity grid and the *tabula longitudinis horarum*. Their details and *sigla* are as follows:

- **C**: Cambridge, Gonville and Caius College, MS 110
- **P**: Paris, BnF, Latin 10264
- **V1**: Vatican, BAV, Pal. lat. 1367
- **V2**: Vatican, BAV, Pal. lat. 1374
- **V3**: Vatican, BAV, Pal. lat. 1412

There are many ways to build critical editions from this material. A common aim of a critical edition is to propose a version of the table as close as possible to what could have been the intention of its original compiler. This usually requires a careful study of variants in order to build a stemma on which the critical edition will rely. In this study, my aim is different. I need to build a critical edition that supports an analysis of the practice of computing with manuscripts. In particular, I want to understand how scribal agency is related to the uses of tables for computation. The way to achieve this begins with a careful description of the presentation of the tables in the different manuscripts. I also want to identify scribal variants in witnesses that have different kinds of numerical effects on the computational procedure or its result. The main tool for this will be a mathematical and astronomical analysis of the tables. Manuscript descriptions, mathematical and astronomical analysis of tables, and the critical edition are the three main parts of this section.

Description of the manuscripts

Scribal variants in a copy and, more generally, the way a given manuscript can be used in a computation depend on different diplomatic features of the copy. By diplomatic features, I mean those aspects of the copy that depend mainly on scribal agency, intentional or not, constrained by the context of scribal work. Exactly which diplomatic features may have an impact on the table copy as a computational tool remains a new and mostly open set of questions in research.[14] I will take the opportunity of this edition to explore these issues and hope to contribute the beginnings of answers as well as more accurate formulations of the questions involved.

[13] Husson, 'Astronomers' Elementary Computations'.

[14] See the articles by Montelle and van Dalen in this volume.

Tables \ Sigla	C	P	V1	V2	V3
Solar equation	X	X	X	X	X
Lunar equation	X	X	X	X	X
Solar velocity	X	X	X	X	X
Lunar velocity	X	X	X	X	X
Tabula longitudinis horarum	X	X			

Figure 1: The different versions of the tables to be critically edited.

For these purposes I am relying on a general distinction between 'digit', 'number' and 'quantity' that can be most easily understood in a specific instance. In the equation and velocity grid the maximum value of the solar equation can be read as 2;10 arc degrees for arguments 92 to 95. This is a specific astronomical quantity. Among other things, it is expressed using the units 'degree' and 'minute' and the number '2;10'. This number in turn is expressed with the digits '2' and '10'. The ways these different elements are related and presented on the page are essential features of the general layout of tables with respect to their use in computations. This will be a first point of attention in my descriptions. A second important point of attention will be to describe those aspects that point to the use of a group of tables in a single procedure. These aspects include, for instance, the positioning of the tables on the pages and their display in different types of grid, their positioning in the manuscript quires, their possible titles, headings and accompanying paratexts, etc.

The manuscript tradition of the *Tabule magne*, at least for the tables here under consideration, is not very original. The scribes responsible for the copies I will describe below have not created new designs for grids or new ways to express astronomical quantities in tabular format. The purpose of my description is not to identify originality. It is rather to understand how, in particular situations, tables are presented as a tool for computation. The fact that most of the diplomatic features I describe here are common to many tables may just attest to the fact that they are part of a common body of tacit knowledge related to how astronomical tables are to be presented and read. It is interesting to point out that on many occasions aspects of this tacit knowledge are shared across linguistic domains and are the object of a specific transmission process.[15] I also hope that these descriptions will be useful in discussions about the diplomatic features of table copies that need to be taken into account in the diplomatic transcription of tables.

[15] Husson, 'Remarks on Two Dimensional Array Tables'; Li Liang, 'Tables with "European" Layout'.

Cambridge, Gonville and Caius College, MS 110

Cambridge, Gonville and Caius College, MS 110 (C) is a vellum composite manuscript containing five codicological units of the fourteenth and fifteenth centuries. Although composite, the manuscript is intellectually coherent as it contains only material related to astral sciences and especially to mathematical astronomy.[16] The initial and final leaves of the manuscript are remnants of a fourteenth-century document. They contain astronomical material produced in the 1320s around the Parisian faculty of arts, viz., John of Lignères's *Tabule magne* from pp. 1 to 18 and a fragment of John of Saxony's almanac from pp. 363 to 368.[17] This witness of Lignères' *Tabule magne* is the only known one where canons and tables are found together in a set of quires that results from a single production act.

Several features of C's copy of the equation and velocity tables point to their common use in a single computation. First, these tables are presented on two pages facing each other. The four tables only occupy the top three quarters of the page and two other sets fill the remaining space. The equation and velocity table is presented under a unified title *Tabula equationis solis et lune et ad inveniendum motus solis et lune in una hora* ('Table of the equation of the sun and moon and for finding the motion of the sun and moon in one hour'). This title is repeated on pp. 16 and 17. The use of the singular for table is interesting for a grid that gathers two equation and two velocity tables. A further feature unifying the four tables into a single computational tool in C is the arrangement of the arguments and entries. Similar to what is usually done for planetary equation tables, all four tables share the same argument headed *linee numeri communis*. Under this heading, two columns are found; the first runs from 1 to 30 with a step of one, the second starts symmetrically at 29 and runs down to 0 with a step of one. After this we read three times the same set of four headings: *equatio solis, equatio lune, motus solis in una hora, motus lune in una hora*. With the symmetry of the argument this covers six zodiacal signs on each page, thus the full zodiac on pp. 16 and 17. These column headings identify the astronomical quantities that are tabulated. The *tabula longitudinis horarum* is split into two grids. The first is at the bottom of page 16, the other on page 18 presented in a landscape format. The title of the table appears only at the top of the first grid on page 16. Neither on page 16 nor on page 18 do headings indicate the type of astronomical quantities to read for the first argument in the first row or for the second argument in the first column. The type of astronomical quantity that one reads in the entries is not mentioned either. The two grids display difference columns that are distinguished, among other

[16] James, *A Descriptive Catalogue*, vol. I, pp. 114–15. I am in debt also to Sebastian Falk for providing me with pictures of the relevant manuscript folios for my study.

[17] Chabás and Goldstein, *A Survey of European Astronomical Tables*, pp. 91–92.

things, by the heading *differentia*. Thus, the only heading of this table points to a mathematical property of the number presented. The use of the table and its interpretation can only rely on a good comprehension of the procedure of syzygy computation in general and in particular of the one described in the canons to the *Tabule magne*.

A second set of diplomatic features of C is related to the ways astronomical quantities are expressed by sexagesimal numbers and used in computations. In the equation and velocity grid, the first two columns under the heading *linee numeri communis* represent different kinds of astronomical quantities depending on the specific table for which they will be used as an argument. For instance, if read in relation to the *equatio solis* the number in the *linee numeri communis* refers to the mean argument of the sun; if read in relation to the *motus solis in una hora* the same number refers to the true solar longitude. In each case, however, the value is an arc of the zodiac. The zodiacal signs are implicitly expressed by the layout of the table on the facing pages 16 and 17. Only the numbers of degrees are written, but no unit for them is explicitly mentioned. The situation with respect to the astronomical quantities displayed in the entries of the tables is different. Under each table heading that marks the quantities, you have either three columns (for the *equatio solis* and the *equatio lune*) or two columns (for the *motus solis* and the *motus lune*). Each of these columns stands for a particular unit of the sexagesimal numbers expressing the astronomical quantity. For instance, the *equatio lune* is expressed in degrees, minutes and seconds marked by their abbreviations ('g' for *gradus*, 'm' for *minuta*, and '2' for *secunda*) as sub-headings in the first row of the table. This shows that what each column means in such a table and the way columns work together to constitute astronomical quantities is very different for columns used as arguments and those used as entries.

The unit sub-headings are nicely and systematically written on p. 17 but with less rigor on p. 16. This variation shows that these features are truly 'diplomatic', i.e., they depend on scribal agency. There is no use of colours in this manuscript to mark the numbers (numbers are marked in black, the table ruling is in red). However, the scribe used another feature to reduce his labour and produce a readable page. For all the tabulated astronomical quantities the first sub-column (in degrees for the two equations, in minutes for the two velocities) is highly repetitive. For instance, on p. 17 the first column of the *equatio solis* should have a sequence of twenty-four '2's for the digit of the degree of the equation. The scribe decided to write only every second, third or fourth digit in the sequence. He used this strategy almost exclusively for the first column of each astronomical quantity. This produces a visual effect that underlines the astronomical quantities to the reader. On the other hand, this way to omit 'repetitive' numerical information in specific columns of the tables amplifies the risk of introducing shift variants into the copy. Because these potential shift variants will occur in the column that has the greatest order of magnitude, their impact on the computation will be significant. Some instances of this will be identified later.

At the bottom of the columns for the *equatio solis* and *equatio lune* one finds the word *adde* to indicate how the entries are to be managed in the larger procedure to find the true longitude of the sun or moon. It is also related to the symmetry of the *linee numeri communis* columns. If the table is read from top to bottom the equations will be subtracted, if the table is read from bottom to top the equations will be added.

The two grids displaying the *tabula longitudinis horarum* treat the arguments and entries differently. The single digit arguments are not marked with a heading or unit. The first argument ranges from 27 to 34 in both grids. The second argument ranges from 1 to 8 in the first grid and from 1 to 34 in the second grid. Differences and actual entries are also treated in distinct ways. Units of the entries are expressed using abbreviations and top column headings similar to the grid for the equations and velocities. In the first grid on p. 16, these units are hours, minutes and seconds. In the second grid on p. 18, these units are minutes and seconds. In terms of layout, the first argument is treated as a kind of table heading in both grids. The differences have no explicit units. In the first grid on p. 16, each cell contains a two-position number, e.g., 4 46, which the context allows to be identified as 4 minutes 46 seconds. Thus while quantities are separated in different cells for the entry, they are gathered in one cell for the differences. Because of the very small differences between two successive entries, this particular diplomatic feature does not appear in the second grid on p. 18.

Paris, BnF, Latin 10264

P is a fifteenth-century paper manuscript of 286 folios. It is closely related to Paris, BnF, Latin 10263 as both are linked to the fifteenth-century printer and humanist Arnaud of Brussels.[18] This coupling of manuscripts is interesting because Latin 10263 is one of the four witnesses of the canons to the *Tabule magne*. The close relationship between the witnesses of the table set and of the canons might explain why P has, like C, a table set coherent with the canons concerning syzygy computations. P has an interesting intellectual profile: along with the *Tabule magne*, it contains cosmological, cosmographical and geographical texts of Alfargani and Albertus Magnus. The way the five tables are arranged in this manuscript is almost identical to that found in C (cf. Plate 13). Only three relevant variants will be mentioned here. First, the two grids of the *tabula longitudinis horarum* are copied together on fol. 30v. Thus, in this

[18] Poulle, *La bibliothèque scientifique*.

Parisian witness, the two steps of the procedure correspond to two different openings of the manuscript. Second, in the equation and velocity grid the zodiacal signs are marked with numbers from 0 to 5 at the top of the grid and from 6 to 11 in reverse order at the bottom of the grid. Third, the scribe of this copy did not use the technique of omitting repetitive numbers in columns, as was done by the scribe of C.

Vatican, BAV, Pal. Lat. 1367

The three Vatican witnesses only have the equation and velocity grid. V1 is a fifteenth-century composite manuscript with the first 84 folios of parchment and the last 79 of paper. Despite the composite nature of the manuscript, its intellectual profile is coherent with texts of astronomy, astrology and medicine. The parchment section contains only astronomical tables. It is opened by a version of the Parisian Alfonsine Tables as described in John of Saxony's canons of 1327[19] and is completed with material from other sets, including parts of the *Tabule magne*. The equation and velocity grid is spread over three pages (70v–71v). The layout of the grid is similar to that of C but some relevant variants can be noted. The first striking diplomatic aspect is the use of colours. Black and red are linked to numerical quantity; all digits that need to be read together in order to form a number and signify an astronomical quantity are of the same colour. Colours separate the columns of the tables. The argument columns common to all four tables are repeated for each zodiacal sign on 70v.

Vatican, BAV, Pal. Lat. 1374

V2 is a paper manuscript of 126 folios copied in Prague in 1407. It contains exclusively astronomical tables. Like in V1, the equation and velocity grid is spread over three pages. However, the use of colours is different and not systematic in this manuscript. The integrity of quantities is not respected. For instance, the *motus solis* is written in red for the minutes and in black for the seconds. This use of colour enhances the possibility of shifts when copying the table column by column.

Vatican, BAV, Pal. Lat. 1412

The last witness, V3, is a 138-folio paper manuscript copied in Paris in 1453–54. It is entirely concerned with mathematical astronomy and includes texts like the *Theorica planetarum gerardi*, various canons of John of Lignères and different tables from the Alfonsine traditions. The equation and velocity grid is spread over three pages. The diplomatic features that are of interest in this

[19] Poulle, *Les tables alphonsines*.

study are almost identical to those of V1. One may, however, note that on fol. 111r the scribe has chosen to copy the *linee numeri communis* only once in the middle of the page.

Scribal agency affects the ways a given manuscript can be used in a computation. The association of tables in grids and the distribution of these grids on pages is one important aspect. The organisation of these grids is also relevant especially with respect to the relation between arguments and entries and with respect to the symmetry of the tables. Scribal agency also affects the way digits, numbers and units are related to the astronomical quantities they express. The practice of avoiding the copy of repetitive numbers and use of colours that does not respect the integrity of astronomical quantities are two interesting instances. These different diplomatic features also may induce specific types of column shift variants in the copy. Finally, the above survey has confirmed the manuscript grouping that was already apparent from the contents of the manuscripts with respect to syzygy computation. The Cambridge and Paris manuscripts are distinct from the three Vatican witnesses especially with respect to the use of colours.

Mathematical and astronomical analysis of the tables

Astronomical and mathematical understanding of the tables is enhanced by 'recomputing' the table according to historically pertinent methods. 'Understanding' here means identifying the astronomical models, parameters and mathematical methods on which the tables rely. Different contributions in this volume illustrate this type of inquiry on original tables. Recomputations can be done at different levels of accuracy depending on the evidence of the manuscript, the precision of the table, and the aim of the study. In the context of this chapter, I need to recompute values that will give me a point of comparison from which the manuscript variants can be analysed and a critical edition established. In particular, the recomputed values need to help me identify those scribal variants that could have been identified also by especially 'skilled' table users.

Note that the tables analysed here are either already fairly well known or mathematically simple, so that it will not be necessary for me to develop this part of the analysis very far. The *Tabula longitudinis horarum* is a division table. The top row argument, i.e., the *superatio* or velocity difference between the moon and the sun, is divided by the left column argument, i.e., the elongation. The first grid of this table has the elongation running up to 8° (somewhat more than the maximum elongation in a half-day). The second grid of this table has the elongation running up to 34 arcminutes (enough to let the result of the division between the *superatio* and the elongation reach 60 minutes). At the level of precision required for this analysis the recomputation of this table is trivial.[20] The solar and lunar equations are those of the Parisian Alfonsine Tables as described for instance in John of Saxony's canons of 1327 and printed in the 1483 *editio princeps*. It is not necessary, for my purpose, to elaborate further on the existing literature on these equations.[21] The solar and lunar velocities are also well documented.[22] However, their recomputation presents interesting methodological aspects in a simple situation. The cases of the sun and moon are fairly similar. Thus, I will here present only the recomputation of the lunar velocity. The minimum value of the lunar velocity in the table is 0;30,18 and the maximum value 0;36,04. This pair of values indicates that the table is related to the corpus of the Toledan Tables.[23] Thus the first logical step in exploring how this table could have been computed is to recompute the lunar velocity starting from the Toledan Tables. One may apply the following formula, where $\bar{\alpha}$ is the mean anomaly of the moon, v is the lunar velocity, m is the mean lunar motion expressed in degrees per hour, m_a is the mean lunar motion in anomaly in degrees per hour and c is the lunar equation of anomaly:[24]

$$v(\bar{\alpha}) = m + m_a\big(c(\bar{\alpha} + 1) - \big(c(\bar{\alpha})\big).$$

Figure 2 displays the differences in seconds (the precision of the velocity tabulated in the manuscripts) between the results obtained using this formula[25] and the values in manuscript C. The results obtained with the other manuscript witnesses are qualitatively identical. A pattern is apparent in this diagram that shows that probably an interpolation grid was used in computing the velocity table.

[20] In the recomputation I have performed an exact division and have rounded the result to the precision of the recomputed table. I do not study specifically the practice of division among astronomers of the late medieval period in the Latin tradition.

[21] For a short state of the art and excellent starting point for the scholarship on the solar and lunar equations in European traditions, see Chabás and Goldstein, *A Survey of European Astronomical Tables*, pp. 63–73.

[22] See Chabás and Goldstein, *A Survey of European Astronomical Tables*, pp. 95–99.

[23] Pedersen, *The Toledan Tables*. From the point of view of astronomical theory the table set is not homogenous. In particular, it is not possible to derive the velocity tables from the equation tables as could be expected. The former rely on Alfonsine parameters while the latter rely on different Toledan parameters. This situation is not uncommon in table sets that circulated in Latin sources during the latter part of the Middle Ages.

[24] Different formulas are possible especially with respect to the use of c but it will not be necessary to explore them here. See Goldstein, 'Lunar Velocity in the Ptolemaic tradition'.

[25] I used the lunar equation of anomaly from Pedersen, *The Toledan Tables*, vol. IV, pp. 1453–58: 0;32,56,0 °/h for the lunar mean motion in longitude and 0;32,40,0 °/h for the mean motion in anomaly.

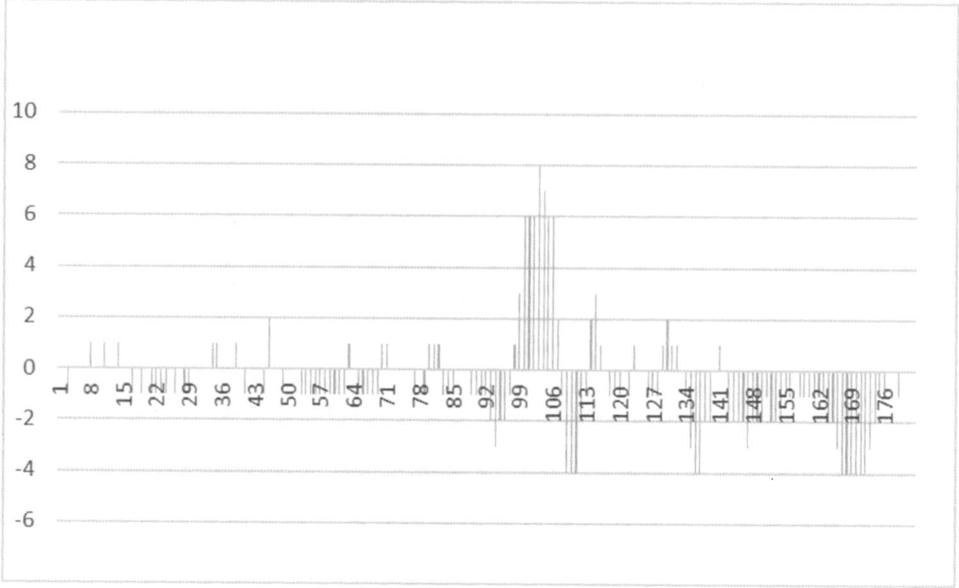

Figure 2: Differences in seconds between C and results obtained with the first recomputation scenario.

The lunar velocity table usually presented in the Toledan Tables is given with a step of 6 degrees for the argument. A close comparison of the values of the tables edited by Pedersen[26] and those for multiples of 6 degrees in the manuscripts here under consideration shows that both sets of values coincide exactly except in three cases:

- For argument 30, Pedersen's edition has 30,55, while John of Lignères' version reads 30,36.
- For argument 102, Pedersen's edition has 33,17, while John of Lignères' version reads 30,27.
- For argument 168, Pedersen's edition has 35,58 while John of Lignères' version reads 35,54.

Thus, the possibility that the table in our manuscripts was computed using interpolation in between nodes at every 6 degrees taken from the Toledan Tables (with the adjustments listed above) is worth exploring. Figure 3 shows the differences in seconds between the lunar velocity table in C and the results of such a recomputation. The agreement is obviously much better. And the results obtained with the other manuscripts are qualitatively similar. I have chosen to stop my recomputation effort at this point.[27]

---

[26] Pedersen, *The Toledan Tables*, vol. IV, p. 1412.

[27] Recomputed values for all tables are provided in the critical apparatus of the edition of the tables in Appendix A.

Figure 3: Differences in seconds between C and the results obtained with a second recomputation scenario.

Critical edition

I have two types of material for each table: first, manuscript transcriptions from each witness of the table; second, recomputed or expected values for each table. In the context of this study, the goal of the critical edition is not to restore a version of the table intended by John of Lignères, thus I do not need to study the interrelations of the manuscripts and provide a stemma. Rather, the goal of the edition is to provide a reference point for the analysis of manuscript variants and a tool to analyse the effect of these variants on the computations and their results. As a consequence of the first goal, the edition must be neutral with respect to the different manuscripts, because if the edition is by construction closer to one manuscript than to the others, the specificities of computing with this particular manuscript will not be in sufficient contrast with those of computing with the critical edition. The second goal of the edition implies that the edition must be close enough to the expected values so that computations made from the edition do not potentially lead to the computational effects I want to isolate in each particular manuscript. Finally, the computational effects we need to isolate are small in most cases, thus the critical edition must also remain close to the manuscripts. In the end, the edition needs to be a middle term of some sort between the different manuscript versions and the expected or recomputed version.

In light of these requirements, I have adopted a simple algorithmic rule to construct the critical edition. Each value is determined according to a majority rule in which the expected values are weighted with a coefficient two and the manuscript witnesses are weighted with a coefficient one. This ensures that the expected values have more weight than any manuscript, that all the manuscripts have the same weight, and, because the weights given to each source are close to each other, the critical edition will remain close enough to the manuscript witnesses. Following such a procedure has the supplementary advantage that the resulting critical edition is entirely transparent as no *ad hoc* emendations are made. Since the computation of the *tabule longitudinis horarum* is trivial I have chosen to give a weight of three to the recomputed values and kept the two manuscript versions with a weight of one. This implies that the edited version is identical with the recomputed table.

In the next section, this particular choice will allow the comparison of various means of division: modern division, exact division using the algorithmic rules of ancient actors, division using an exact version of the table (i.e. with values rounded to seconds), division using the quotient tables as they appear in the manuscripts with their variant. More generally, the choice of the above algorithmic rule to construct the critical edition will prove efficient with respect to our goal: it will be possible, for instance, to distinguish between 'obvious' variants that any 'skilled' user would have 'corrected' during or before the computation and 'non-obvious' variants that would silently go into the computation. It will also be possible to isolate various kinds of variants and their effects on the computation by comparing the manuscript version to the critical edition.

As, to my knowledge, this is the first time such a procedure has been used to generate a critical edition of a table with respect to a given research objective, it is not clear if the set of weights I have selected is the only or even the best to achieve these goals and I make no claim about this. I do note that this kind of weighted procedures could easily be implemented as a tool to generate a critical edition directly from queries to a database of astronomical tables. In this context such issues could be investigated systematically. These weighted procedures are in principle flexible enough to fit many different kinds of research goals for critical editions, including the classical stemma-oriented type of critical edition. Naturally these types of 'computer-assisted' editions are not to replace the expertise of the historian. Even if he decides to use such a tool, the choice of the scholarly goal of the edition, the weight to be given to the various sources, and whether or not to follow the result of the weighted procedure in individual cases will remain up to the researcher.

A typology of scribal variants in tables

My edition of the five tables is given in Appendices A to E and the conventions used to mark the variants are specified there. In order to explore how manuscript variants affect results of computations and practices with tables I distinguish two directions along which they can be analysed.[28] First, some vari-

[28] An interesting analysis of manuscript variants in a tradition of a table set involving Arabic and Latin sources is found in van Dalen and Pedersen, 'Re-editing the Tables'.

ants affect one number at a time while others affect in correlated ways a group of numbers. These phenomena are connected in different ways to diplomatic choices when copying the table and may point in some cases to specific mathematical practices. These two categories of variants will affect the computation with tables differently since variants where sets of numbers are changed in correlated ways are likely to have a greater impact. A second manner of analysing variants relies on the specific way quantities are shaped from numbers in astronomical tables. Quantities are expressed as a set of numbers (i.e., digits) each having a different order of magnitude in sexagesimal arithmetic. The numbers at each of these orders of magnitude will usually have different diplomatic properties. The number of degrees will change very slowly and thus parts of the column will be repetitive, the number of seconds will usually behave more randomly except for some arithmetical tables where cycles are likely to appear that will help reading and copying, etc. A variant affecting the number of seconds in a quantity will presumably have a smaller impact on the computation than a variant affecting the number of degrees in the same quantity. Thus one might consider that from a computational perspective a table does not usually present one type of variant but has as many as there are positions in its entries.

Among variants affecting one number at a time, some remain inscrutable. For instance, in the table for the solar equation, V3 gives the value 2;7,18 for argument 104. However, all the other witnesses and the recomputation give a '14' instead of the final '18'. There is no obvious palaeographical or mathematical explanation for this variant. On the other hand, some variants clearly have a palaeographical cause. For instance, in the table for the solar equation the '29' in the number of minutes for argument 138 varies as '20' in V1 and V2 and as '39' in P. These types of variants are linked to the script used to denote the number. They are more and more frequent as the order of magnitude of the digit is smaller and control of the value of the number from local parsing becomes more difficult. There are practically no errors of this type affecting the number of degrees in the table set here under analysis. Some variants affect the last position of a number by a value of plus or minus one. In some of those cases, no palaeographical explanation is available. For instance, in the table for the lunar velocity the value for argument 65 was most probably interpolated in between the values for arguments 60 and 66. All versions except V3 give the value '36' in the last position, V3 gives '37' instead. There is no simple Latin palaeographical explanation for a '7' instead of a '6', so that a 'rounding effect' is much more likely to account for this variant. This suggests that some actors along the transmission of the table did recompute the interpolated values and thus produced small variant traditions of the tables because of variations in their rounding practices.

There are different types of variants that affect sets of numbers in correlated ways. The most common type shifts a block of values vertically. For instance, the table for the solar equation in V2 has a shift affecting the numbers of minutes and the numbers of seconds between arguments 45 and 48.[29] Overall, this type of shift is more frequent in the three Vatican manuscripts than in the two others. This might be related to the specific use of colours made by the Vatican group of manuscripts. Some other types of variants similar to this block shift also occur, and they especially affect the numbers of degrees. They are linked to the repetitive character of the table for this order of magnitude. For instance, in the Cambridge manuscript, the degree column in the lunar equation table shows a block of '3's instead of '2's for arguments 146 to 150. Similar effects on the numbers of degrees are attested in V2 for the solar equation table for arguments 115 to 120 and arguments 151 to 153. Another curious effect of repetitive character affecting a block of values is seen in the solar velocity table in V3. For arguments 151 to 159 a '23' replaces '32' in the second column. Finally, some mathematical effects also produce variants that create block shifts. A clear instance is given in the table for the solar velocity in V3. For arguments 160 to 170, V3 gives '32' as the number of seconds while all the other manuscripts give '33'. One might remember that for arguments 174 and 180, which are interpolation nodes, all manuscripts read 2′32″ instead of the expected 2′33″. Thus in all manuscripts but V3 the solar velocity shows an unexpected decrease from 2′33″ to 2′32″ in the seconds' column. One of the actors along the chain of transmission that produced V3 felt this was to be corrected. However, instead of changing the value of the two last interpolation nodes to 2′33″, he changed the value of the two preceding interpolation nodes to 2′32″ instead of 2′33″. All the values dependent on these nodes are then affected.

Computing ΔT with manuscript tables of the *Tabule magne*

Preparing the computations

In order to explore the practice of computing the time between mean and true syzygy (hereafter ΔT) with these manuscripts we need to consider at least three different types of variability that must be addressed: manuscript variability, procedural variability, and arithmetical variability. The first sections of this paper were devoted to building the tools needed to manage manuscript variability especially for the critical edition of the tables.

As far as procedural variability is concerned, as noted above, the table set of the *Tabule magne* could be used to compute the position between mean and true syzygy in many ways. In this section I want to focus on the variability stemming from manuscripts' variants. Thus, procedural variability is to be con-

[29] It is not clear in this specific case how the entry for argument 49, namely 1;34,34, was generated: simple linear interpolation would produce 1;34,36 or 1;34,37.

trolled rather than explored. The canons of the *Tabule magne* are here helpful because they describe only one method for computing the time between mean and true syzygy, which I will thus follow closely:

1. Starting from the mean anomaly of the sun and moon compute the difference in longitude of the true sun and true moon.
2. Compute the difference of the velocities of the sun and moon.
3. Divide the first by the second result and obtain an estimation of ΔT.
4. If the true sun and moon are at syzygy after that estimation of ΔT (i.e. their true positions are equal or 180° removed at the desired level of accuracy), stop; if not, iterate.

Following this procedure closely allows me to avoid dependence on tables that are not in the set I have critically edited above, and thus to keep a good control over manuscript variability. In particular, starting the procedure directly from the mean anomaly of the sun and moon allows me to avoid depending on a set of mean motion tables from which to derive a set of mean conjunctions and oppositions for the sun and moon for a time period that would also need to be justified. There is a second consequence of the choice to closely follow the canons and the table set. After the first iteration the sun and moon are not at mean syzygy any more. In principle the first and second lunar model are then not equivalent, and the second lunar model should be used to compute the moon's true position and its velocity. However, there is no lunar equation of centre in the table set here under consideration, as would be necessary to use the second lunar model, and we have seen that the lunar velocity table is also dependent on the first lunar model drawn from the Toledan Tables. Thus computing the time between mean and true syzygy according to the table proposed in the *Tabule magne* set implies that one should follow the first lunar model. In concrete terms, I will compute ΔT according to the above procedure for every pair of mean anomaly of the sun and moon in the range 0 to 360 with a step of 10 degrees.

I am fully conscious that it would be important and interesting to explore other possible procedural scenarios with the same accuracy, and to compute ΔT or even true syzygy times for other table sets. These other ways to use the tables (and to associate them with other velocity, equation and mean motion tables) will likely lead to different values of ΔT.[30] However, I am not trying to compare the results found in that way to 'exact Alfonsine results' or to explore this procedural variability. My aim is only to understand what kind of variability in the computational procedure and in its results derive from the manu-

[30] The impact of this kind of alternatives on the computed time between mean and true syzygy has been investigated for the case of John of Saxony in Kremer, 'Thoughts on John of Saxony's Method'.

script variants in this corpus. In this respect, the results obtained in this study are likely to be qualitatively valid also for other table sets and procedures as the manuscript variability will be of the same kind for other procedures and so too will be the arithmetical practices on which the computations rely.

The last type of variability to consider is arithmetical variability. Arithmetical practices and their variability are an important factor here because their order of magnitude is likely to be of the same order as that produced by manuscript variants. In order to control this variability I have made two different choices.

First, a competent table user presumably corrects some manuscript variants, especially those that affect the number of degrees in an entry. It is thus important to see how these skilled corrections affect the computation and its results. For this reason, I have repeated my computations for four different situations:

1. The table set as found in the manuscripts.
2. The table set corrected for variants that only affect degrees.[31]
3. The table set corrected for variants that affect degrees and minutes.
4. The critically edited table set.

A comparison of the results produced in these situations will help me understand how a competent table user is able to improve the accuracy of results by amending some easily spotted variants in his tables.

Second, the arithmetic of historical actors is not that of modern computational software. I have thus also explored different arithmetical algorithms in my computations:

1. Computation with floating numbers and modern arithmetic.
2. Simulated computations of historical actors, in which all numbers are converted to integer multiples of the smallest sexagesimal unit used (here the second) and regular integer arithmetic along with a specified type of rounding is used.[32]
3. The same as situation 2) but with the use of the *tabula longitudinis horarum* for the final division.

Thus, multiple manuscript versions and their corrected versions are combined with different possible arithmetical algorithms in an iterative procedure. The space of computations explored is huge and in some cases, different scenarios

[31] Here 'corrected' means that the number is set to be identical to the one chosen in the critical edition: I eliminate the variant. Consequently, 'corrected' does not mean that the number is set to be equal to the one I would expect from a recomputation.

[32] Husson, 'Astronomers' Elementary Computations'.

lead to the same results. I will not fully describe this space here.[33] Rather I will explore two aspects of the computation of the time between mean and true syzygy with this set of manuscripts. First, it is important to see if the computations actually produce a result. In other words, I will consider how the convergence of the iterative process is affected by variants in the manuscripts and by arithmetical practices. For this, I will consider manuscript versions one by one and, after each iteration, I will inspect the difference in ecliptic longitude between the true sun and true moon. If the procedure converges, this distance should approach zero. Once the convergence issue is clarified, I will be in a position to analyse how manuscript variants propagate iteration after iteration, and compare the results produced by the different manuscript versions. In other words, I will ask the question: are the ΔT values produced by different manuscripts the same? And if different, by how much and why? When asking this second question, I will compare the actual values of ΔT produced by different manuscripts after different numbers of iterations. For each of these two questions, the properties and results of computation with the edited version of the table will be the paradigm against which the phenomena linked to the different manuscript versions are identified. All effects that I will point out are local: they are caused by individual variants in the manuscripts and influence only the ΔT found for specific solar and lunar positions.

Computing with one manuscript: convergence issues

In this section I will present three different results in order to grasp how convergence occurs when computing with the tables attested in the corpus of the *Tabule magne*. The first thing is to measure the effect of arithmetic variability on the convergence. The second is to measure the effect of manuscript variability on the speed of the convergence. The third and last case will be a curious situation of non-convergence.

In order to isolate the particular effect of arithmetic variability on the convergence, three iterations of the process were made using the critically edited table set with the three different types of arithmetic here considered. As can be seen in Figure 4, the distance between the true sun and true moon after three iterations is zero everywhere when floating number arithmetic is used.

[33] I have used spreadsheets to explore the space of computations because I needed to follow each step of the computations with accuracy. These spreadsheets are available on demand. Soon the DISHAS platform (see the introduction of this volume) will allow researchers to manipulate tables and different related procedures directly in a Python environment. This will allow a much more efficient exploration of the space of computations created by historical tables and procedures. It will also foster collaborative research on these topics as the research data and procedures will be publicly available.

	0	30	60	90	120	150	180	210	240	270	300	330	360
0	0	0	0	0	0	0	0	0	0	0	0	0	0
30	0	0	0	0	0	0	0	0	0	0	0	0	0
60	0	0	0	0	0	0	0	0	0	0	0	0	0
90	0	0	0	0	0	0	0	0	0	0	0	0	0
120	0	0	0	0	0	0	0	0	0	0	0	0	0
150	0	0	0	0	0	0	0	0	0	0	0	0	0
180	0	0	0	0	0	0	0	0	0	0	0	0	0
210	0	0	0	0	0	0	0	0	0	0	0	0	0
240	0	0	0	0	0	0	0	0	0	0	0	0	0
270	0	0	0	0	0	0	0	0	0	0	0	0	0
300	0	0	0	0	0	0	0	0	0	0	0	0	0
330	0	0	0	0	0	0	0	0	0	0	0	0	0
360	0	0	0	0	0	0	0	0	0	0	0	0	0

Figure 4: Distance between the true sun and true moon in arcsecs after three iterations when using floating number arithmetic and the critically edited tables.

If the arithmetic of an historical actor is used, this same convergence occurs after three iterations with residual arithmetical noise. Figure 5 shows that when using integer arithmetic with rounding to seconds this noise has a maximum magnitude of ±2 seconds. This noise persists also after six iterations.

	0	30	60	90	120	150	180	210	240	270	300	330	360
0	0	-1	-2	-1	-1	0	0	0	0	0	0	1	0
30	1	0	0	0	0	0	1	0	1	0	0	0	1
60	-1	-1	-2	1	0	0	0	-1	0	1	0	0	-1
90	1	1	1	1	0	1	1	0	-1	0	-1	2	1
120	1	0	-1	0	0	0	1	0	0	0	2	0	1
150	0	0	0	0	0	0	1	0	0	0	1	1	0
180	0	0	0	0	0	-1	0	1	0	0	0	0	0
210	1	-1	-1	0	0	0	-1	0	0	0	0	0	0
240	0	0	-2	0	0	0	-1	0	0	0	1	0	-1
270	-1	-2	1	0	1	0	-1	0	0	-1	-1	-1	-1
300	0	0	0	-1	0	1	0	0	0	-1	2	1	1
330	-1	0	0	0	-1	0	-1	0	0	0	0	0	-1
360	0	-1	0	0	0	0	0	0	0	0	0	1	0

Figure 5: Distance between the true sun and true moon in arcsecs after three iterations when using integer arithmetic and the critically edited tables.

	0	30	60	90	120	150	180	210	240	270	300	330	360
0	0	0	14	6	-5	-10	0	-5	-12	-22	31	-16	0
30	9	28	-17	-28	-28	26	21	8	-5	-27	29	-13	9
60	-1	27	-4	-12	-8	30	15	-18	17	-16	-12	18	-1
90	28	2	-23	29	-25	9	-8	8	29	-10	-1	-22	28
120	8	-23	1	-7	-18	-20	26	-7	29	-4	1	28	8
150	26	-21	-9	-22	-20	-24	-22	29	18	-1	-7	8	26
180	0	3	7	-4	-10	-6	0	6	10	4	-7	-3	0
210	8	-8	7	1	-18	-29	22	24	20	22	9	21	-26
240	6	-28	-1	4	-28	7	-26	20	18	7	-1	23	-8
270	-20	22	1	10	-29	-8	8	-9	25	-29	23	-2	-28
300	17	-18	12	16	-17	18	-15	-30	8	12	4	-27	1
330	25	13	-29	27	5	-8	-21	-26	28	28	17	-28	-9
360	0	16	-31	22	12	5	0	-5	-12	-22	31	-16	0

Figure 6: Distance between the true sun and true moon in arcsecs after three iterations when using integer arithmetic, the *tabula longitudinis horarum* and the critically edited tables.

However, if the *tabula longitudinis horarum* is used for the final division, the magnitude of the arithmetical noise is around fifteen times larger with a maximum of ±28 seconds (Figure 6). This noise remains robust even after six iterations. This is already a significant result because the effects of the various arithmetical choices available to the actors show that even on arithmetical grounds they must have had a nuanced understanding of convergence.

These arithmetical effects are robust enough to be preserved across the manuscript variability. Arithmetical variability and manuscript variability do not compound and appear to be, in the situation analysed here, independent. Whatever manuscript version is used to compute ΔT, it will not be possible to go below the arithmetical noise that was isolated above. When using integer arithmetic there will always be a residual arithmetical noise of up to two arcsecs and when using the *Tabula longitudinis horarum* the residual noise can be as large as half an arcmin. This arithmetical noise, especially as shown in the last case, may seem large. However, most ephemerides were computed to the nearest minute during the late medieval period in Europe, thus these computations met a standard that was state of the art at the time.[34]

Now that this arithmetical effect is known, I will give results only using integer arithmetic provided that it is easy to conceive what the result would be if other arithmetics were used. The second effect I want to consider is that of manuscript variability on convergence speed. In this respect, it is important

[34] Personal discussion with Richard Kremer (January 2019).

	0	30	60	90	120	150	180	210	240	270	300	330	360
0	0	-1	-2	-1	-1	0	0	0	0	0	0	1	0
30	1	0	1	0	0	0	1	0	0	2	0	1	1
60	-1	-1	-2	1	0	0	0	-1	0	1	-1	0	-1
90	1	1	1	1	0	1	1	0	-1	0	-1	2	1
120	1	0	-1	0	0	0	1	0	0	0	2	0	1
150	0	0	1	0	0	0	1	0	0	0	1	1	0
180	0	0	0	0	0	-1	0	1	0	0	0	0	0
210	1	-1	-1	0	0	0	-1	0	0	0	-1	0	0
240	0	0	-2	0	0	0	-1	0	0	0	1	0	-1
270	-1	-2	1	0	1	0	-1	-1	0	-1	-1	-1	-1
300	0	0	1	-1	0	1	0	0	0	-1	2	1	1
330	-1	-1	0	-2	0	0	-1	0	0	0	-1	0	-1
360	0	-1	0	0	0	0	0	0	0	0	0	1	0

Figure 7: Distance between the true sun and true moon in arcsecs after three iterations when using integer arithmetic and manuscript P.

to understand that the first two steps of the iterative procedure proposed by the canons to the *Tabule magne* are complex processes requiring in most cases several readings in tables, linear interpolations and multiple elementary arithmetical operations on the numbers composing the astronomical quantities. Depending on the situation and the skills of the computer, adding one iteration to the computation will require between five to fifteen minutes of arithmetical work on paper. In order to isolate and illustrate this effect I compare manuscripts P and V3. Figure 7 shows that after three iterations with the tables in manuscript P the distance between the true sun and true moon are everywhere within the arithmetical noise proper to the use of integer arithmetic. Thus with this manuscript, just as with the critical edition, we obtain a final value for ΔT after three iterations.

The situation is different when we use V3 instead of P. Figure 8 shows that after three iterations the distance between the true sun and true moon can rise in some cases up to almost 10 arcmins. It is only after two more iterations that these worst-case scenarios are finally settled, as is shown in Figure 9. In other words, an historical actor's computing with V3 rather than P would in most cases find a stable value of ΔT in three iterations but could in some cases need as many as five. This certainly makes a significant difference. In this respect the computation of ΔT with V2 is qualitatively similar to that of V3.

It is interesting to note that the effects of manuscript variability disappear when the errors in the degree columns of V3 are corrected, as is shown in Figure 10. A user able to make these 'skilled corrections' to the tables would avoid the inconvenience of having to do the additional iterations.

	0	30	60	90	120	150	180	210	240	270	300	330	360
0	0	-1	-2	-1	-1	0	0	0	0	0	0	1	0
30	1	0	0	0	0	33	1	0	1	0	0	0	1
60	9	-7	-6	-9	-3	355	13	-18	-10	-8	-16	-16	9
90	-3	2	1	1	0	70	-3	0	-1	0	-1	2	-3
120	1	0	-1	-1	0	71	1	0	0	0	2	0	1
150	0	0	0	0	0	32	1	0	0	0	1	1	0
180	0	0	0	0	0	-1	0	1	0	0	0	0	0
210	1	-1	-1	0	0	0	-1	-32	0	0	0	0	0
240	0	0	-2	0	0	0	-1	-71	0	1	1	0	-1
270	2	-2	1	0	1	0	3	-70	0	-1	-1	-2	3
300	-20	16	16	8	10	18	-13	-355	3	9	6	7	-9
330	-1	0	0	0	-1	0	-1	-33	0	0	0	0	-1
360	0	-1	0	0	0	0	0	0	0	0	0	1	0

Figure 8: Distance between the true sun and true moon in arcsecs after three iterations when using integer arithmetic and manuscript V3.

	0	30	60	90	120	150	180	210	240	270	300	330	360
0	0	0	0	1	0	0	0	0	0	0	0	0	0
30	0	0	0	0	0	2	1	0	0	0	0	0	0
60	1	0	0	0	0	1	0	0	0	0	0	0	1
90	0	0	0	0	0	0	0	0	0	0	0	0	0
120	0	0	0	0	0	-1	0	0	0	0	0	0	0
150	0	0	0	0	0	2	0	0	0	0	1	0	0
180	0	0	0	0	0	0	0	0	0	0	0	0	0
210	0	0	-1	0	0	0	0	-2	0	0	0	0	0
240	0	0	0	0	0	0	0	1	0	0	0	0	0
270	0	0	0	0	0	0	0	0	0	0	0	0	0
300	0	0	0	0	0	0	0	-1	0	0	0	0	-1
330	-1	0	0	0	0	0	-1	-2	0	0	0	0	0
360	0	0	0	0	0	0	0	0	0	0	0	0	0

Figure 9: Distance between the true sun and true moon in arcsecs after five iterations when using integer arithmetic and manuscript V3.

With respect to convergence speed, the case of V1 also shows an interesting mathematical phenomenon. The distance between the true sun and true moon in arc-seconds after three iterations using integer arithmetic is presented in Figure 11. It shows a set of results within the expected arithmetical noise.

	0	30	60	90	120	150	180	210	240	270	300	330	360
0	0	-1	-2	-1	-1	0	0	0	0	0	0	1	0
30	1	0	0	0	0	0	1	0	1	0	0	0	1
60	0	-1	-1	1	0	0	0	0	-1	1	-1	1	0
90	-3	2	1	1	0	2	-3	0	-1	0	-1	2	-3
120	1	0	-1	-1	0	0	1	0	0	0	2	0	1
150	0	0	0	0	0	0	1	0	0	0	1	1	0
180	0	0	0	0	0	-1	0	1	0	0	0	0	0
210	1	-1	-1	0	0	0	-1	0	0	0	0	0	0
240	0	0	-2	0	0	0	-1	0	0	1	1	0	-1
270	2	-2	1	0	1	0	3	-2	0	-1	-1	-2	3
300	1	-1	1	-1	1	0	0	0	0	-1	1	1	0
330	-1	0	0	0	-1	0	-1	0	0	0	0	0	-1
360	0	-1	0	0	0	0	0	0	0	0	0	1	0

Figure 10: Distance between the true sun and true moon in arcsecs after three iterations when using integer arithmetic and a version of manuscript V3 in which discrepancies in the degrees are corrected.

	0	30	60	90	120	150	180	210	240	270	300	330	360
0	0	-1	-2	-1	-1	0	0	0	0	0	0	1	0
30	1	0	0	0	0	0	1	0	1	0	0	0	1
60	-1	-1	-1	0	-1	0	0	-1	0	1	0	0	-1
90	-3	1	1	1	0	2	-3	0	-1	0	-1	2	-3
120	1	0	-1	0	0	0	1	0	0	0	2	0	1
150	0	0	0	0	0	0	1	0	0	0	1	1	0
180	0	0	0	0	0	-1	0	1	0	0	0	0	0
210	1	-1	-1	0	0	0	-1	0	0	0	0	0	0
240	0	0	-2	0	0	0	-1	0	0	0	1	0	-1
270	2	-2	1	0	1	0	3	-2	0	-1	-1	-1	3
300	0	0	0	-1	0	1	0	0	1	0	1	1	1
330	-1	0	0	0	-1	0	-1	0	0	0	0	0	-1
360	0	-1	0	0	0	0	0	0	0	0	0	1	0

Figure 11: Distance between the true sun and true moon in arcsecs after three iterations when using integer arithmetic and manuscript V1.

However, the situation at the fourth iteration becomes even better, as shown in Figure 12. It remains stable at least until the sixth iteration. Thus, in the grid of the distance between the true sun and true moon after the third iteration, some of the non-zero values are caused by the manuscript variants, but

	0	30	60	90	120	150	180	210	240	270	300	330	360
0	0	0	0	-1	0	0	0	0	0	0	0	0	0
30	0	0	0	0	0	0	-1	0	0	0	0	0	0
60	0	1	0	0	0	0	0	-1	0	0	0	0	0
90	0	1	0	0	0	0	0	0	0	0	0	0	0
120	0	0	0	0	0	0	0	0	0	0	0	0	0
150	0	0	0	0	0	0	0	0	0	0	-1	0	0
180	0	0	0	0	0	0	0	0	0	0	0	0	0
210	0	0	1	0	0	0	0	0	0	0	0	0	0
240	0	0	0	0	0	0	0	0	0	0	0	0	0
270	0	0	0	0	0	0	0	0	0	0	0	-1	0
300	0	0	0	0	0	1	0	0	0	0	0	-1	0
330	1	0	0	0	0	0	1	0	0	0	0	0	0
360	0	0	0	0	0	0	0	0	0	0	0	0	0

Figure 12: Distance between the true sun and true moon in arcsecs after four iterations when using integer arithmetic and manuscript V1.

	0	30	60	90	120	150	180	210	240	270	300	330	360
0	0	0	0	1	0	0	0	0	0	0	0	0	0
30	0	0	1	0	0	0	1	0	0	0	0	0	0
60	0	1	0	0	0	0	0	1	0	0	0	0	0
90	0	0	0	0	0	0	0	0	0	0	0	0	0
120	0	0	0	0	0	0	0	0	0	0	0	0	0
150	0	0	675	1860	929	0	0	0	0	0	-1	0	0
180	0	0	0	0	0	0	0	0	0	0	0	0	0
210	0	0	1	0	0	0	0	0	-929	-1860	-675	0	0
240	0	0	0	0	0	0	0	0	0	0	0	0	0
270	-1	0	0	0	0	0	0	0	0	0	0	0	0
300	0	0	0	0	0	-1	0	0	0	0	0	-1	0
330	-1	0	0	0	0	0	-1	0	0	0	-1	0	0
360	0	0	0	0	0	0	0	0	0	0	0	0	0

Figure 13: Distance between the true sun and true moon in arcsecs after five iterations when using integer arithmetic and manuscript C.

these values are at the level of the arithmetical noise. Most of these are eliminated by a further iteration. It should be noted that no table in V1 shows discrepancies from the critical edition in the numbers of degrees. Thus, this effect is produced only by variants in minutes and seconds and probably difficult to avoid even for a 'skilled' user.

	0	30	60	90	120	150	180	210	240	270	300	330	360
0	0	0	0	-1	0	0	0	0	0	0	0	0	0
30	0	0	-1	0	0	0	-1	0	0	0	0	0	0
60	0	-1	0	0	0	0	0	-1	0	0	0	0	0
90	0	0	0	0	0	0	0	0	0	0	0	0	0
120	0	0	0	0	0	0	0	0	0	0	0	0	0
150	0	0	-674	-1861	-929	0	0	0	0	0	1	0	0
180	0	0	0	0	0	0	0	0	0	0	0	0	0
210	0	0	-1	0	0	0	0	0	929	1861	674	0	0
240	0	0	0	0	0	0	0	0	0	0	0	0	0
270	1	0	0	0	0	0	0	0	0	0	0	0	0
300	0	0	0	0	0	1	0	0	0	0	0	1	0
330	1	0	0	0	0	0	1	0	0	0	1	0	0
360	0	0	0	0	0	0	0	0	0	0	0	0	0

Figure 14: Distance between the true sun and true moon in arcsecs after six iterations when using integer arithmetic and manuscript C.

In the different situations presented so far, the procedure converges and the manuscripts produce a value for ΔT. However, this is not always the case. Figure 13 shows the distance between the true sun and moon after five iterations using C and integer arithmetic. Most of the values are within the arithmetical noise (and this was already the case at the third iteration, just as it was for P and V1), but six values show a distance of up to 30 arcmins.

The next iterations do not improve this situation, as shown in Figure 14. We get the same six outliers with the same order of magnitude, except that their sign is reversed. This oscillation between two sets of outliers continues endlessly as further iterations are performed.

Probably the most skilled users would have tried to correct the values in the table. This attitude would have been rewarded because the effect disappears with a version of C corrected for the variants in the degrees. This in turn shows that this oscillation is truly an effect of manuscript variants.

This first set of results clearly shows that not all manuscript versions have the same arithmetical features. Some versions, because of their variants, require additional iterations to obtain convergence in certain cases. There are even cases where no result is obtained. The largest effects of manuscript variants, such as oscillation instead of convergence, are corrected when variants in the largest order of magnitude are amended, but some effects remain even when variants only in minutes and seconds are kept. Finally, the amplitude of the arithmetical noise is related to specific computational methods. This noise usually swamps the effects of manuscript variants somewhere between the third and fourth iteration and implies that the practical notion of convergence and

	0	30	60	90	120	150	180	210	240	270	300	330	360
0	0,00	0,00	0,00	0,00	129,42	0,00	0,00	0,00	-129,42	0,00	0,00	0,00	0,00
30	0,00	0,00	0,00	0,00	128,04	0,00	0,00	0,00	-128,04	0,00	0,00	0,00	0,00
60	0,00	0,00	0,00	0,00	124,50	0,00	0,00	0,00	-124,50	0,00	0,00	0,00	0,00
90	0,00	0,00	0,00	0,00	119,14	0,00	0,00	0,00	-119,14	0,00	0,00	0,00	0,00
120	0,00	0,00	0,00	0,00	113,39	0,00	0,00	0,00	-113,39	0,00	0,00	0,00	0,00
150	0,00	0,00	0,00	0,00	108,98	0,00	0,00	0,00	-108,98	0,00	0,00	0,00	0,00
180	0,00	0,00	0,00	0,00	107,20	0,00	0,00	0,00	-107,20	0,00	0,00	0,00	0,00
210	0,00	0,00	0,00	0,00	108,98	0,00	0,00	0,00	-108,98	0,00	0,00	0,00	0,00
240	0,00	0,00	0,00	0,00	113,39	0,00	0,00	0,00	-113,39	0,00	0,00	0,00	0,00
270	0,00	0,00	0,00	0,00	119,14	0,00	0,00	0,00	-119,14	0,00	0,00	0,00	0,00
300	0,00	0,00	0,00	0,00	124,50	0,00	0,00	0,00	-124,50	0,00	0,00	0,00	0,00
330	0,00	0,00	0,00	0,00	128,04	0,00	0,00	0,00	-128,04	0,00	0,00	0,00	0,00
360	0,00	0,00	0,00	0,00	129,42	0,00	0,00	0,00	-129,42	0,00	0,00	0,00	0,00

Figure 15: Differences in minutes between ΔT produced by V2 and by the critical edition after one iteration using integer arithmetic.

zero distance between the sun and moon must have been nuanced for the historical actors if only on arithmetical grounds.

This study also allows to propose manuscript groupings, based on computation performance. In this respect V2 and V3 are put together as less efficient versions requiring in specific cases five or six iterations while C, P, and V1 all have similar convergence speeds reasonably close to that of the edited version.[35]

Comparing computational features of manuscripts: variant propagation and coherency issues

The preceding section portrayed an individual historical actor computing with one manuscript at a time. In this final section, I shall consider a collective of historical actors and compare results given by the different manuscript versions. More precisely, the values of ΔT given by each manuscript version are compared with those given by the edited version.

The case of V2 is representative of the different phenomena I was able to identify with respect to this question. Figure 15 shows the differences in minutes between the values of ΔT produced with V2 and those produced with the critical edition after one iteration. In most cases the two values of ΔT agree, except for two large sets of discrepancies of around 2 hours.

If the degrees of the values in V2 are 'corrected' to match those of the critical edition these two large sets of discrepancies disappear, thus showing that they are an effect of manuscript variability at this level.

[35] If we exclude the very specific oscillation case of C.

	0	30	60	90	120	150	180	210	240	270	300	330	360
0	0,00	0,00	0,00	0,00	129,24	0,00	0,00	0,00	-129,28	0,00	0,00	0,00	0,00
30	0,00	0,00	0,00	0,00	128,00	-13,21	0,00	0,00	-127,61	0,00	0,00	0,00	0,00
60	0,00	0,00	0,00	0,00	112,00	-31,28	0,00	0,00	-123,27	0,00	0,00	0,00	0,00
90	1,37	0,33	0,00	0,00	99,28	-36,57	1,35	0,00	-117,59	0,00	0,00	0,00	1,37
120	0,00	0,00	0,00	0,00	99,78	-29,28	0,00	0,01	-112,36	0,00	0,01	0,01	0,00
150	0,00	0,00	0,00	0,00	109,13	-12,56	0,00	0,00	-108,62	0,00	0,00	0,00	0,00
180	0,00	0,00	0,00	0,00	107,46	0,00	0,00	0,00	-107,46	0,00	0,00	0,00	0,00
210	0,00	0,00	0,00	0,00	108,62	0,00	0,00	12,56	-109,13	0,00	0,00	0,00	0,00
240	0,00	-0,01	-0,01	0,00	112,36	-0,01	0,00	29,28	-99,78	0,00	0,00	0,00	0,00
270	-1,37	0,00	0,00	0,00	117,59	0,00	-1,35	36,57	-99,28	0,00	0,00	-0,33	-1,37
300	0,00	0,00	0,00	0,00	123,27	0,00	0,00	31,28	-112,00	0,00	0,00	0,00	0,00
330	0,00	0,00	0,00	0,00	127,61	0,00	0,00	13,21	-128,00	0,00	0,00	0,00	0,00
360	0,00	0,00	0,00	0,00	129,24	0,00	0,00	0,00	-129,28	0,00	0,00	0,00	0,00

Figure 16: Differences in minutes between ΔT produced by V2 and by the critical edition after two iterations using integer arithmetic.

	0	30	60	90	120	150	180	210	240	270	300	330	360
0	0,00	0,00	0,00	0,00	0,00	0,00	0,00	0,00	0,00	0,00	0,00	0,00	0,00
30	0,00	0,00	0,00	0,00	0,00	0,00	0,00	0,00	0,00	0,00	0,00	0,00	0,00
60	0,00	0,00	0,00	0,00	0,00	0,00	0,00	0,00	0,00	0,00	0,00	0,00	0,00
90	1,37	0,33	0,00	0,00	0,00	0,23	1,35	0,00	0,00	0,00	0,00	0,00	1,37
120	0,00	0,00	0,00	0,00	0,00	0,00	0,00	0,01	0,01	0,00	0,01	0,01	0,00
150	0,00	0,00	0,00	0,00	0,00	0,00	0,00	0,00	0,00	0,00	0,00	0,00	0,00
180	0,00	0,00	0,00	0,00	0,00	0,00	0,00	0,00	0,00	0,00	0,00	0,00	0,00
210	0,00	0,00	0,00	0,00	0,00	0,00	0,00	0,00	0,00	0,00	0,00	0,00	0,00
240	0,00	-0,01	-0,01	0,00	-0,01	-0,01	0,00	0,00	0,00	0,00	0,00	0,00	0,00
270	-1,37	0,00	0,00	0,00	0,00	0,00	-1,35	-0,23	0,00	0,00	0,00	-0,33	-1,37
300	0,00	0,00	0,00	0,00	0,00	0,00	0,00	0,00	0,00	0,00	0,00	0,00	0,00
330	0,00	0,00	0,00	0,00	0,00	0,00	0,00	0,00	0,00	0,00	0,00	0,00	0,00
360	0,00	0,00	0,00	0,00	0,00	0,00	0,00	0,00	0,00	0,00	0,00	0,00	0,00

Figure 17: Differences in minutes between ΔT produced by V2 with values corrected to match the critical edition in degrees and minutes and the critical edition after two iterations using integer arithmetic.

On the other hand, adding one iteration to the computation does not smoothen the differences between the critical edition and V2. As can be seen in Figure 16, not only are the two columns of large differences maintained at the same order of magnitude, but new discrepancies appear of a smaller but yet significant magnitude of more than 30 minutes. For the second iteration new values from the tables are used. This explains the appearance of new discrepancies. Discrepancies appearing at one stage of the iteration are usually maintained in the following iterations. This is linked to the fact that this iterative process is generally convergent. If the convergence target of a manuscript version is slightly distinct from that of an edited text, successive iterations will only confirm the tendency and thus discrepancies will be conserved.

Finally, it is interesting to see that even when V2 is 'corrected' so that its degree values match those of the critical edition, these minor discrepancies between values of ΔT remain although their magnitude is much smaller and in most cases less than a minute, as can be seen in Figure 17.

Conclusion

In this article my aim was to explore the practice of computing with manuscripts that contain variant entries. For this I have selected a specific astronomical issue and a single set of astronomical tables.

From a methodological perspective, this goal required a complete astronomical understanding of the table set and I thus relied on standard recomputation approaches. I have associated in specific ways these recomputation approaches with that of diplomatic description of tables and variants typology. I also designed my critical edition not as research of a hypothetical genuine version of the tables but as a tool to explore variants and potential 'skilled corrections' of these variants by table users. Finally, I developed new tools to 'restore' astronomical tables as computational devices by attending to their accompanying canons.

The results obtained are encouraging. My approach allows us to isolate computational effects proper from the effects of manuscript variants and to compare differing arithmetical procedures. Analysis suggests that historical actors engaged with these types of tables and computations could probably have detected some of these effects. For instance, some manuscript versions generate results that converge more rapidly than others do; and some tables, like the *tabula longitudinis horarum*, have arithmetical performances significantly different from that of other ways to perform the computation.

I hope that some of the approaches and tools developed and used here manually and on a small scale could be further refined and integrated into tool boxes for digital humanities that are being developed for the history of the astral sciences. For instance, the various alternative arithmetical algorithms may be useful in other contexts as well. The use of weighted majority rules to produce a specific type of critical edition could also be usefully implemented as a general digital tool. Finally, research on those diplomatic features of astronomical tables that might have an impact on the way manuscripts can be used in computations, if only by being linked to certain kinds of manuscript variants, ought also to be pursued. In the end, these tools might enable the design of digital diplomatic transcriptions of astronomical tables that could help us explore the use of tables in computations and as such provide more insight into actors' practices than does a simple digital *facsimile* of a manuscript.

Bibliography

Chabás, José, *Computational Astronomy in the Middle Ages. Sets of Astronomical Tables in Latin*, Madrid: Editorial CSIC, 2019.

Chabás, José, and Bernard R. Goldstein, 'Computational Astronomy: Five Centuries of Finding True Syzygy', *Journal for the History of Astronomy* 28 (1997), pp. 93–105. [Reprinted in Id., *Essays on Medieval Computational Astronomy*, Leiden: Brill, 2015, pp. 40–56 (newly typeset)].

Chabás, José, and Bernard R. Goldstein, *The Alfonsine Tables of Toledo*, Dordrecht: Kluwer, 2003.

Chabás, José, and Bernard R. Goldstein, *A Survey of European Astronomical Tables in the Late Middle Ages*, Leiden: Brill, 2012.

van Dalen, Benno, and Fritz S. Pedersen, 'Re-editing the Tables in the *Ṣābi' Zīj* by al-Battānī', in Joseph W. Dauben, Stefan Kirschner, Andreas Kühne, Paul Kunitzsch and Richard P. Lorch (eds), *Mathematics Celestial and Terrestrial. Festschrift für Menso Folkerts zum 65. Geburtstag*, Stuttgart: Wissenschaftliche Verlagsgesellschaft, 2008, pp. 405–28. [Reprinted in Id., *Islamic Astronomical Tables. Mathematical Analysis and Historical Investigation*, Farnham/Burlington (Ashgate Variorum), 2013, ch. VIII].

Goldstein, Bernard R., 'Lunar Velocity in the Ptolemaic Tradition', in Peter M. Harman and Alan E. Shapiro (eds), *The Investigation of Difficult Things: Essays on Newton and the History of the Exact Sciences in Honour of T. D. Whiteside*, Cambridge: Cambridge University Press, 1992, pp. 3–17.

Husson, Matthieu, *Les domaines d'application des mathèmatiques dans la première moitié du quatorzième siècle*, Thèse de doctorat, École pratique des Hautes Études, 2007.

Husson, Matthieu, 'Ways to read a Table: Reading and Interpolation Techniques in Canons of Early Fourteenth-Century Double-Argument Tables', *Journal for the History of Astronomy* 43 (2012), pp. 299–319.

Husson, Matthieu, 'Remarks on Two Dimensional Array Tables in Latin Astronomy: A Case Study in Layout Transmission', *Suhayl* 13 (2014), pp. 103–17.

Husson, Matthieu, 'Astronomers' Elementary Computations in Paris between 1321 and 1327: the Circle around John of Lignères and John of Saxony', in Karine Chemla, Agathe Keller and Matthieu Husson (eds), *Mathematical Practices in Relation to the Astral Sciences*, Springer, forthcoming.

James, Montague Rhodes, *A Descriptive Catalogue of the Manuscripts in the Library of Gonville and Caius College*, 2 vols, Cambridge: Cambridge University Press, 1907–1908.

Kremer, Richard L., 'Thoughts on John of Saxony's Method for Finding Times of True Syzygy', *Historia Mathematica* 30 (2003), pp. 263–77.

Li Liang, 'Tables with "European" Layout in China: A Case Study in Tabular Layout Transmission', *Suhayl* 13 (2014), pp. 83–101.

North, John D., 'The Alfonsine Tables in England', in Yasukatsu Maeyama and Walter G. Saltzer (eds), *Prismata. Naturwissenschaftsgeschichtliche Studien. Festschrift für Willy Hartner*, Wiesbaden: Steiner, 1977, pp. 269–301.

Pedersen, Fritz S., *The Toledan Tables. A Review of the Manuscripts and the Textual Versions with an Edition*, 4 vols, Copenhagen: Reitzels, for Det Kongelige Danske Videnskabernes Selskab, 2002.

Poulle, Emmanuel, *La bibliothèque scientifique d'un imprimeur humaniste au XVe siècle. Catalogue des manuscrits d'Arnaud de Bruxelles à la Bibliothèque nationale de Paris*, Genève: Droz, 1963.

Poulle, Emmanuel, 'John of Lignères', in Charles C. Gillispie (ed.), *Dictionary of Scientific Biography*, New York: Scribner's Sons, 1973, vol. VII, pp. 122–28.

Poulle, Emmanuel, *Les Tables Alphonsines avec les canons de Jean de Saxe. Édition, traduction et commentaire*, Paris: Éditions du Centre National de la Recherche Scientifique, 1984.

Saby, Marie-Madeleine, *Les canons de Jean de Ligneres sur les tables astronomiques de 1321. Edition critique, traduction et étude [I]*, PhD dissertation, 3 vols, Ecole Nationale des Chartes, 1987.

Appendix A: Critical edition of the tables

A1. Conventions

Digits with variants are bold in the table. In the critical apparatus, variant positions are identified in parentheses using the line and column marks written around the grid. Variant digits are given after the sigla of the witnesses that include them. Variants are listed column by column, from top to bottom.

In all tables, arguments and headings are marked in grey shaded cells.

A2. Edition of the solar equation

	A	*B*	*C*	*D*	*E*	*F*	*G*	*H*	*I*	*J*	*K*	*L*	*M*	*N*	*O*
	argument		Solar Equation			argument		Solar Equation			argument		Solar Equation		
	d	d	d	m	s	d	d	d	m	s	d	d	d	m	s
1	1	179	0	2	10	31	149	1	4	46	61	119	1	51	51
2	2	178	0	4	19	32	148	1	6	37	62	118	1	52	56
3	3	177	0	6	27	33	147	1	8	28	63	117	1	54	**9**
4	4	176	0	8	**36**	34	146	1	10	19	64	116	1	55	6
5	5	175	0	10	44	35	145	1	12	9	65	115	1	56	9
6	6	174	0	12	53	36	144	1	13	58	66	114	1	57	**11**
7	7	173	0	15	**2**	37	143	1	15	41	67	113	1	58	2
8	8	172	0	17	10	38	142	1	17	24	68	112	1	58	52
9	9	171	0	19	19	39	141	1	19	6	69	111	1	59	41
10	10	170	0	21	28	40	140	1	20	48	70	110	2	0	26
11	11	169	0	23	36	41	139	1	**22**	**29**	71	109	2	1	**16**
12	12	168	0	25	45	42	138	1	**24**	10	72	108	2	2	**2**
13	13	167	0	27	53	43	137	1	25	50	73	107	2	2	41
14	14	166	0	30	1	44	136	1	27	29	74	106	2	3	21
15	15	165	0	32	8	45	135	1	**29**	**8**	75	105	2	3	59
16	16	164	0	34	16	46	134	1	**30**	**46**	76	104	2	4	36
17	17	163	0	36	23	47	133	1	**32**	**23**	77	103	2	5	16
18	18	162	0	38	30	48	132	1	**33**	**59**	78	102	2	5	48
19	19	161	0	40	37	49	131	1	**35**	**30**	79	101	2	6	17
20	20	160	0	42	43	50	130	1	37	0	80	100	2	6	45
21	21	159	0	44	49	51	129	1	38	30	81	99	2	7	12
22	22	158	0	46	55	52	128	1	39	58	82	98	2	7	**37**
23	23	157	0	48	59	53	127	1	41	**57**	83	97	2	8	2
24	24	156	0	51	4	54	126	1	**42**	54	84	96	2	8	27
25	25	155	0	53	4	55	125	1	44	14	85	95	2	8	45
26	26	154	0	55	2	56	124	1	45	34	86	94	2	9	1
27	27	153	0	57	1	57	123	1	46	**53**	87	93	2	9	17
28	28	152	0	**58**	59	58	122	1	**48**	**10**	88	92	2	9	32
29	29	151	1	0	57	59	121	1	49	28	89	91	2	9	45
30	30	150	1	2	54	60	120	1	50	45	90	90	2	9	57

(D28) V3:59 (E4) V1:38 (E7) P:3 (I11) V1:24 (I12) P:23 (I15) V2:30 (I16) V2:32 (I17) V1:30; V2:33 (I18) V1, V2:32 (I19) V2:34 (I24) P,V2,V3:43 (I28) V3:46 (J11) V1,V2,V3:10 (J15) V2,V3:46 (J16) V2:23 (J17) V1:46; V2:59 (J18) V1,V2:23 (J19) V2:34 (J23) E:27 (J27) C,P:52 (J28) V2:18 (O3) V3:0 (O6) P:12 (O11) V1,V2,V3:13 (O12) P:3 (O22) V2,V3:36

	A	*B*	*C*	*D*	*E*	*F*	*G*	*H*	*I*	*J*	*K*	*L*	*M*	*N*	*O*
	argument		Solar Equation			argument		Solar Equation			argument		Solar Equation		
	d	d	d	m	s	d	d	d	m	s	d	d	d	m	s
1	91	89	2	9	59	121	59	1	53	**46**	151	29	**1**	5	1
2	92	88	2	10	0	122	58	1	**52**	35	152	28	**1**	2	54
3	93	87	2	10	0	123	57	1	**52**	24	153	27	**1**	0	57
4	94	86	2	10	0	124	56	1	50	**15**	154	26	0	58	40
5	95	85	2	9	57	125	55	1	48	59	155	25	0	56	33
6	96	84	2	9	51	126	54	1	47	**46**	156	24	0	54	25
7	97	83	2	9	**36**	127	53	1	46	**20**	157	23	0	52	17
8	98	82	2	9	20	128	52	1	44	**53**	158	22	0	50	**9**
9	99	81	2	9	2	129	51	1	43	26	159	21	0	48	**11**
10	100	80	2	8	45	130	50	1	41	57	160	20	0	45	**54**
11	101	79	2	8	**25**	131	49	1	40	27	161	19	0	43	44
12	102	78	2	8	6	132	48	1	38	**57**	162	18	0	**41**	**35**
13	103	77	2	7	41	133	47	1	37	25	163	17	0	39	26
14	104	76	2	7	**14**	134	46	1	35	53	164	16	0	37	16
15	105	75	2	6	46	135	45	1	34	20	165	15	0	35	6
16	106	74	2	6	18	136	44	1	**32**	**46**	166	14	0	32	51
17	107	73	2	5	48	137	43	1	31	**12**	167	13	0	30	35
18	108	72	2	5	**18**	138	42	1	**29**	**37**	168	12	0	28	19
19	109	71	2	4	42	139	41	1	27	50	169	11	0	26	1
20	110	70	2	4	5	140	40	1	26	3	170	10	0	23	42
21	111	69	2	3	37	141	39	1	24	16	171	9	0	21	22
22	112	68	2	**2**	37	142	38	1	22	28	172	8	0	19	1
23	113	67	2	1	**45**	143	37	1	20	40	173	7	0	16	40
24	114	66	2	0	51	144	36	1	18	**51**	174	6	0	14	19
25	115	65	**1**	59	53	145	35	1	17	0	175	5	0	11	**58**
26	116	64	**1**	58	55	146	34	1	15	8	176	4	0	9	36
27	117	63	**1**	57	57	147	33	1	13	16	177	3	0	7	12
28	118	62	**1**	56	57	148	32	1	11	13	178	2	0	4	**48**
29	119	61	**1**	55	57	149	31	1	9	10	179	1	0	2	24
30	120	60	**1**	54	57	150	30	1	7	**0**	180	0	0	0	0

(C25) V2:2 (C26) V2:2 (C27) V2:2 (C28) V2:2 (C29) V2:2 (C30) V2:2 (D22) V3:1 (E7) V1:56 (E11) C,V1:26 (E14) V3:18 (E18) V2,V3:16 (E23) V3:47 (I2) P,V1:53 (I3) P,E:51 (I16) C,P:33 (I18) V1,V2:20 (J1) V1:48 (J4) E:12 (J6) E:40 (J7) P:30 (J8) V1:52 (J12) C:53; V2:53 (J16) V2:26 (J17) V2:13 (J18) P:39; V2:40 (J24) V2:11 (J30) E:7 (M1) V2,V3:0 (M2) V1,V3:0 (M3) V1,V2:0 (N12) p:42 (O8) C:11; P:2 (O9) C,P:9; E:1 (O10) E:53 (O12) P:38 (O16) V3:50 (O28) V1,V2,V3:28

A3. Edition of the lunar equation

	A	*B*	*C*	*D*	*E*	*F*	*G*	*H*	*I*	*J*	*K*	*L*	*M*	*N*	*O*
	argument		Lunar equation			argument		Lunar equation			argument		Lunar equation		
	d	d	d	m	s	d	d	d	m	s	d	d	d	m	s
1	1	179	0	4	46	31	149	2	21	**46**	61	119	4	7	47
2	2	178	0	9	31	32	148	2	25	55	62	118	4	10	**27**
3	3	177	0	14	15	33	147	2	30	5	63	117	4	13	3
4	4	176	0	19	0	34	146	2	34	12	64	116	4	15	35
5	5	175	0	23	44	35	145	2	38	17	65	115	4	18	**3**
6	6	174	0	28	28	36	144	2	42	21	66	114	4	20	27
7	7	173	0	**33**	11	37	143	2	46	22	67	113	4	22	**47**
8	8	172	0	37	54	38	142	2	**50**	19	68	112	4	25	**2**
9	9	171	0	42	**27**	39	141	2	54	14	69	111	4	**27**	12
10	10	170	0	47	19	40	140	2	58	7	70	110	4	29	18
11	11	169	0	52	0	41	139	3	**1**	**58**	71	109	4	31	20
12	12	168	0	56	**41**	42	138	3	5	46	72	108	4	33	18
13	13	167	1	1	20	43	137	3	9	31	73	107	4	35	11
14	14	166	1	**5**	59	44	136	3	13	13	74	106	4	36	**59**
15	15	165	1	10	38	45	135	3	**16**	**51**	75	105	4	38	43
16	16	164	1	**15**	15	46	134	3	**19**	26	76	104	4	40	23
17	17	163	1	19	51	47	133	3	**23**	**59**	77	103	4	41	58
18	18	162	1	**25**	**24**	48	132	3	**27**	**30**	78	102	4	43	**28**
19	19	161	1	29	**0**	49	131	3	**30**	**57**	79	101	4	44	**53**
20	20	160	1	33	32	50	130	3	**34**	**20**	80	100	4	46	13
21	21	159	1	38	3	51	129	3	**37**	**40**	81	99	4	47	26
22	22	158	1	42	33	52	128	3	**40**	**57**	82	98	4	48	35
23	23	157	1	**46**	1	53	127	3	44	19	83	97	4	**49**	**38**
24	24	156	1	51	27	54	126	3	47	20	84	96	4	50	41
25	25	155	1	55	**52**	55	125	3	50	**26**	85	95	4	51	38
26	26	154	2	0	15	56	124	3	53	29	86	94	4	52	28
27	27	153	2	4	37	57	123	3	56	30	87	93	4	53	11
28	28	152	**2**	8	**57**	58	122	3	59	26	88	92	4	**53**	50
29	29	151	2	13	14	59	121	**4**	2	17	89	91	4	54	**25**
30	30	150	2	17	**29**	60	120	**4**	**4**	**5**	90	90	4	54	**58**

(C28) C,P:1 (D7) C:32 (D14) P:15 (D16) V3:19 (D18) C:22; P:27; E:24 (D23) E:47 (E9) E:37 (E12) V2,V3:14 (E18) C:34; E:27 (E19) V1:1 (E25) C,P:51 (E28) C,P:59 (E30) C,P:39 (H29) V3:3 (H30) V3:3 (I8) V2:30 (I11) V1,V2:5 V3:2 (I15 V2,V3:19 (I16) E:20 (I17) V1,V2:19 (I18) V1,V2:23 (I19) V1,V2:27 (I20) V1,V2:30 (I21) V1,V2:34 (I22) V1,V2:47 (I30) V3,E:5 (J1) E:43 (J11) V1,V2,V3:46 (J15) V2,V3:26 (J1è) C,P:29; V1,V2:26 (J18) V1,V2: 59 (J19) V1,V2:30 (J20) V1,V2:57 (J21) V1,V2:20 (J22) V1,V2:40 (J25) V2:27 (J30) C:2; E:4 (N9) V1:57 (N23) V3:45 (N28) V2:54 (O2) V1:37 (O5) V3:30 (O7) C,P:57 (O8) C,P:20 (O14) V2,V3:50 (O18) C,P,V1:48 (019):V2:51 (O23) p:28 (O26) V2:38 (O29) P:35 (O30) E:54

	A	*B*	*C*	*D*	*E*	*F*	*G*	*H*	*I*	*J*	*K*	*L*	*M*	*N*	*O*
	argument		Lunar equation			argument		Lunar equation			argument		Lunar equation		
	d	d	d	m	s	d	d	d	m	s	d	d	d	m	s
1	91	89	4	55	18	121	59	4	24	38	151	29	2	34	52
2	92	88	4	55	37	122	58	4	22	**7**	152	28	2	30	6
3	93	87	4	55	49	123	57	4	19	38	153	27	2	25	16
4	94	86	4	55	**55**	124	56	4	16	58	154	26	2	20	23
5	95	85	4	**56**	0	125	55	4	14	13	155	25	2	15	26
6	96	84	4	55	56	126	54	4	11	23	156	24	2	10	26
7	97	83	4	55	43	127	53	4	8	28	157	23	2	5	**22**
8	98	82	4	55	25	128	52	4	5	31	158	22	2	**0**	17
9	99	81	4	55	**4**	129	51	4	2	30	159	21	1	55	9
10	100	80	4	54	41	130	50	3	59	**29**	160	20	1	49	**58**
11	101	79	4	54	12	131	49	3	56	5	161	19	1	44	44
12	102	78	4	53	38	132	48	3	**52**	47	162	18	1	39	27
13	103	77	4	52	59	133	47	3	49	23	163	17	1	34	9
14	104	76	4	52	14	134	46	3	45	**52**	164	16	1	28	49
15	105	75	4	51	22	135	45	3	42	17	165	15	1	23	26
16	106	74	4	50	22	136	44	3	38	37	166	14	1	18	1
17	107	73	4	49	17	137	43	3	34	53	167	13	1	12	34
18	108	72	4	48	10	138	42	3	31	3	168	12	1	7	**6**
19	109	71	4	46	**54**	139	41	3	27	10	169	11	1	1	36
20	110	70	4	**45**	33	140	40	3	23	12	170	10	0	56	5
21	111	69	4	44	7	141	39	3	**19**	9	171	9	0	50	32
22	112	68	4	42	34	142	38	3	15	**2**	172	8	0	44	58
23	113	67	4	40	56	143	37	3	10	**50**	173	7	0	39	23
24	114	66	4	39	15	144	36	3	6	**35**	174	6	0	33	47
25	115	65	4	37	29	145	35	3	2	**15**	175	5	0	28	**10**
26	116	64	4	35	37	146	34	2	57	**51**	176	4	0	22	33
27	117	63	4	33	41	147	33	2	53	**23**	177	3	0	16	56
28	118	62	4	31	34	148	32	**2**	48	**51**	178	2	0	11	18
29	119	61	4	29	20	149	31	**2**	44	15	179	1	0	5	40
30	120	60	4	27	0	150	30	**2**	39	35	180	0	0	0	0

(D5) V1,V2,V3:55 (D20) P:55 (E4) C,P:56; V1:51 (E9) C:41 (E19) C:59 (H28) C:3 (H:29) C:3 (H30) C:3 (I12) V1,V2,V3:53 (I21) V1:10 (I21) V1:10 (J2) V3,E:11 (J10) C,E:20 (J14) V2:53 (J22) V3:8 (J23) C,P:35 (J24) C,P:15; V1:25 (J25) C,P:51 (J26) C,P:23 (J27) C,P:51 (J28) C,P:14 (N8) VA,V2:9 (O1) C,P:0 (O7) V3:25 (O10) V2,V3:59 (O11) V3:24 (O18) V1:16 (O25) P:50

A4. Edition of the solar velocity

	A	*B*	*C*	*D*	*E*	*F*	*G*	*H*	*I*	*J*	*K*	*L*
	argument		solar velocity		argument		solar velocity		argument		solar velocity	
	d	d	m	s	d	d	m	s	d	d	m	s
1	1	179	2	23	31	149	2	24	61	119	2	25
2	2	178	2	23	32	148	2	24	62	118	2	**25**
3	3	177	2	23	33	147	2	24	63	117	2	26
4	4	176	2	23	34	146	2	24	64	116	2	26
5	5	175	2	23	35	145	2	24	65	115	2	26
6	6	174	2	23	36	144	2	24	66	114	2	26
7	7	173	2	23	37	143	2	24	67	113	2	26
8	8	172	2	23	38	142	2	24	68	112	2	26
9	9	171	2	23	39	141	2	24	69	111	2	26
10	10	170	2	23	40	140	2	24	70	110	2	26
11	11	169	2	23	41	139	2	24	71	109	2	26
12	12	168	2	23	42	138	2	24	72	108	2	26
13	13	167	2	23	43	137	2	24	73	107	2	26
14	14	166	2	23	44	136	2	24	74	106	2	26
15	15	165	2	23	45	135	2	**25**	75	105	2	27
16	16	164	2	23	46	134	2	25	76	104	2	27
17	17	163	2	23	47	133	2	25	77	103	2	27
18	18	162	2	23	48	132	2	25	78	102	2	27
19	19	161	2	23	49	131	2	25	79	101	2	27
20	20	160	2	23	50	130	2	25	80	100	2	27
21	21	159	2	23	51	129	2	25	81	99	2	27
22	22	158	2	23	52	128	2	25	82	98	2	27
23	23	157	2	23	53	127	2	25	83	97	2	27
24	24	156	2	23	54	126	2	25	84	96	2	27
25	25	155	2	23	55	125	2	25	85	95	2	27
26	26	154	2	23	56	124	2	25	86	94	2	27
27	27	153	2	24	57	123	2	25	87	93	2	28
28	28	152	2	24	58	122	2	25	88	92	2	28
29	29	151	2	24	59	121	2	25	89	91	2	28
30	30	150	2	24	60	120	2	25	90	90	2	28

(H15) V2:24 (L2) C,P:26 (L14) V2:27

	A	*B*	*C*	*D*	*E*	*F*	*G*	*H*	*I*	*J*	*K*	*L*
	argument		solar velocity		argument		solar velocity		argument		solar velocity	
	d	d	m	s	d	d	m	s	d	d	m	s
1	91	89	2	28	121	59	2	**30**	151	29	2	**32**
2	92	88	2	28	122	58	2	**30**	152	28	2	**32**
3	93	87	2	28	123	57	2	30	153	27	2	**32**
4	94	86	2	28	124	56	2	30	154	26	2	**32**
5	95	85	2	28	125	55	2	30	155	25	2	**32**
6	96	84	2	28	126	54	2	30	156	24	2	**32**
7	97	83	2	28	127	53	2	30	157	23	2	**32**
8	98	82	2	28	128	52	2	30	158	22	2	**32**
9	99	81	2	29	129	51	2	30	159	21	2	**33**
10	100	80	2	29	130	50	2	30	160	20	2	**33**
11	101	79	2	29	131	49	2	30	161	19	2	**33**
12	102	78	2	29	132	48	2	30	162	18	2	**33**
13	103	77	2	29	133	47	2	30	163	17	2	**33**
14	104	76	2	29	134	46	2	**30**	164	16	2	33
15	105	75	2	29	135	45	2	**30**	165	15	2	**33**
16	106	74	2	29	136	44	2	**30**	166	14	2	**33**
17	107	73	2	29	137	43	2	**30**	167	13	2	**33**
18	108	72	2	29	138	42	2	**32**	168	12	2	**33**
19	109	71	2	29	139	41	2	**32**	169	11	2	**33**
20	110	70	2	29	140	40	2	**32**	170	10	2	**33**
21	111	69	2	29	141	39	2	**32**	171	9	2	**32**
22	112	68	2	29	142	38	2	**32**	172	8	2	32
23	113	67	2	29	143	37	2	**32**	173	7	2	32
24	114	66	2	29	144	36	2	**32**	174	6	2	32
25	115	65	2	29	145	35	2	32	175	5	2	32
26	116	64	2	29	146	34	2	32	176	4	2	32
27	117	63	2	29	147	33	2	32	177	3	2	32
28	118	62	2	29	148	32	2	32	178	2	2	32
29	119	61	2	29	149	31	2	32	179	1	2	32
30	120	60	2	29	150	30	2	32	180	0	2	32

(H1) E:29 (H2) E:29 (H14) E:31 (H15) E:31 (H16 E:31 (H17) E:32 (H18) C,P:30 (H19) C,P:30 (H20 C,P:30 (H21) C,P:30 (H22) C,P:30 (H23) C,P:30 (H24) C,P:30 (L1) V3:23 (L2) V3:23 (L4) V3:23 (L5) V3:23 (L6) V3:23 (L7) V3:23 (L8) V3:23 (L9) V3:23 (L10) V1,V2:33; V3:23 (L11) V2:32; V3:23 (L12) V3:32 (L13) V3:32 (L14) V3:32 (L15) V3:32 (L16) V3:32 (L17) V3:32 (L18) V3:32 (L19) V3:32 (L20) V3:32 (L21) E:33

A5. Edition of the lunar velocity

	A	*B*	*C*	*D*	*E*	*F*	*G*	*H*	*I*	*J*	*K*	*L*
	argument		lunar velocity		argument		lunar velocity		argument		lunar velocity	
	d	d	m	s	d	d	m	s	d	d	m	s
1	1	179	30	18	31	149	30	37	61	119	31	26
2	2	178	30	18	32	148	30	38	62	118	31	29
3	3	177	30	18	33	147	30	40	63	117	31	31
4	4	176	30	**19**	34	146	30	**41**	64	116	31	33
5	5	175	30	19	35	145	30	**42**	65	115	31	**36**
6	6	174	30	19	36	144	30	**43**	66	114	31	38
7	7	173	30	**20**	37	143	30	**44**	67	113	31	41
8	8	172	30	20	38	142	30	46	68	112	31	**43**
9	9	171	30	20	39	141	30	47	69	111	31	46
10	10	170	30	**21**	40	140	30	48	70	110	31	48
11	11	169	30	21	41	139	30	50	71	109	31	51
12	12	168	30	21	42	138	30	51	72	108	31	53
13	13	167	30	22	43	137	30	53	73	107	31	56
14	14	166	30	22	44	136	30	54	74	106	31	58
15	15	165	30	23	45	135	30	**58**	75	105	32	1
16	16	164	30	23	46	134	**30**	**58**	76	104	32	**3**
17	17	163	30	24	47	133	**31**	**0**	77	103	32	6
18	18	162	30	24	48	132	31	1	78	102	32	8
19	19	161	30	25	49	131	31	3	79	101	32	**11**
20	20	160	30	25	50	130	31	5	80	100	32	**14**
21	21	159	30	26	51	129	31	7	81	99	32	17
22	22	158	30	27	52	128	31	8	82	98	32	19
23	23	157	30	27	53	127	31	10	83	97	32	22
24	24	156	30	28	54	126	31	12	84	96	32	25
25	25	155	30	29	55	125	31	14	85	95	32	**28**
26	26	154	30	**31**	56	124	31	16	86	94	32	31
27	27	153	30	32	57	123	31	18	87	93	32	34
28	28	152	30	33	58	122	31	20	88	92	32	36
29	29	151	30	35	59	121	31	22	89	91	32	**38**
30	30	150	30	36	60	120	31	24	90	90	32	42

(D4) C:19 (D7) E:19 (D10) E:20 (D26) E:30 (G16) V1,V2,V3:31 (G17) E:30 (H4) V2,V3:42 (H5) V2,V3:43 (H6) V2,V3:44 (H7) V2,V3:45 (H15) E:56 (H16) V1,V2,V3:0 (H17) E:59 (L5) V3:37 (L8) V3:42 (L16) V2:2 (L19) C,P:15 (L25) V2:26 (L29) E:39

	A	*B*	*C*	*D*	*E*	*F*	*G*	*H*	*I*	*J*	*K*	*L*
	argument		lunar velocity		argument		lunar velocity		argument		lunar velocity	
	d	d	m	s	d	d	m	s	d	d	m	s
1	91	89	32	45	121	59	34	17	151	29	35	**33**
2	92	88	32	48	122	58	34	20	152	28	35	35
3	93	87	32	51	123	57	34	23	153	27	35	37
4	94	86	32	53	124	56	34	26	154	26	35	39
5	95	85	32	56	125	55	34	**29**	155	25	35	41
6	96	84	32	59	126	54	34	**32**	156	24	35	43
7	97	83	33	4	127	53	34	35	157	23	35	45
8	98	82	33	8	128	52	34	38	158	22	35	**46**
9	99	81	33	13	129	51	34	41	159	21	35	**48**
10	100	80	33	**18**	130	50	34	43	160	20	35	**49**
11	101	79	33	22	131	49	34	46	161	19	35	51
12	102	78	33	27	132	48	34	49	162	18	35	52
13	103	77	33	29	133	47	34	52	163	17	35	52
14	104	76	33	30	134	46	34	54	164	16	35	53
15	105	75	33	32	135	45	34	57	165	15	35	53
16	106	74	33	**33**	136	44	34	59	166	14	35	53
17	107	73	33	35	137	43	35	2	167	13	35	54
18	108	72	33	36	138	42	35	4	168	12	35	54
19	109	71	33	39	139	41	35	**7**	169	11	35	**55**
20	110	70	33	42	140	40	35	9	170	10	35	57
21	111	69	33	46	141	39	35	11	171	9	35	58
22	112	68	33	49	142	38	35	13	172	8	35	59
23	113	67	33	**53**	143	37	35	16	173	7	36	1
24	114	66	33	55	144	36	35	18	174	6	36	2
25	115	65	33	58	145	35	35	20	175	5	36	2
26	116	64	34	1	146	34	35	22	176	4	36	3
27	117	63	34	5	147	33	35	**25**	177	3	36	3
28	118	62	34	8	148	32	35	27	178	2	36	3
29	119	61	34	11	149	31	35	29	179	1	36	4
30	120	60	34	14	150	30	35	31	180	0	36	4

(D10) V2:16; V3:19 (D16) V3:32 (D23) V1,vé:54; E:52 (H5) V2:20 (H6) V2:22 (H19) E:6 (H27) V2:22 (L1) V1:34 (L8) V2,V3:47 (L9) V1:47 (L10) V1:48 (L19) V3:57

A6. Edition of the Tabula longitudinis horarum

	A	*B*	*C*	*D*	*E*	*F*	*G*	*H*	*I*	*J*	*K*	*L*	*M*	*N*	*O*	*P*	*Q*	*R*	*S*	*T*	*U*	*V*	*W*	*X*	*Y*
1		27			28			29			30			31			32			33			34		
2	d	h	m	s	h	m	s	h	m	s	h	m	s	h	m	s	h	m	s	h	m	s	h	m	s
3	1	2	***13***	20	2	8	34	2	4	8	2	0	0	1	56	8	1	52	30	1	49	5	1	45	**53**
4	2	4	26	40	4	17	9	4	8	17	4	0	0	3	52	15	3	45	0	3	38	11	3	31	46
5	3	6	40	0	6	25	43	6	12	25	6	0	0	5	48	23	5	37	30	5	27	16	5	17	39
6	4	8	53	20	8	34	17	8	16	33	8	0	0	7	44	31	7	30	0	7	16	22	7	3	32
7	5	11	6	40	10	42	51	10	20	41	10	0	0	9	40	39	9	22	30	9	5	27	8	49	25
8	6	13	20	0	12	51	26	12	24	50	12	0	0	11	36	46	11	15	0	10	54	33	10	35	18
9	7	15	33	20	15	0	0	14	28	58	14	0	0	13	32	54	13	7	30	12	43	38	12	21	11
10	8	17	46	40	17	8	34	16	33	6	16	0	0	15	29	2	15	0	0	14	**32**	44	14	***7***	4

(C3) C, P: 12; (U10) C:37; (V8) C: 32; (X10) C, P:3; (Y3) C:54;

	A	*B*	*C*	*D*	*E*	*F*	*G*	*H*	*I*	*J*	*K*	*L*	*M*	*N*	*O*	*P*	*Q*
1		27		28		29		30		31		32		33		34	
2	m	m	s	m	s	m	s	m	s	m	s	m	s	m	s	m	s
3	1	2	13	2	9	2	4	2	0	1	56	1	52	1	49	1	46
4	2	4	27	4	17	4	8	4	0	3	52	3	45	3	38	3	32
5	3	6	40	6	26	6	12	6	0	5	48	5	***38***	5	27	5	18
6	4	8	***53***	8	34	8	17	8	0	7	45	7	30	7	16	7	4
7	5	11	7	10	43	10	21	10	0	9	41	9	***23***	9	5	8	49
8	6	13	20	12	51	12	25	12	0	11	37	11	15	10	55	10	35
9	7	15	33	15	0	14	29	14	0	13	33	13	***8***	12	44	12	21
10	8	17	47	17	9	16	33	16	0	15	29	15	0	14	33	14	7
11	9	20	0	19	17	18	37	18	0	17	25	16	***53***	16	22	15	53
12	10	22	13	21	26	20	41	20	0	19	21	18	54	18	11	17	39
13	11	24	27	23	34	22	46	22	0	21	17	20	***38***	20	0	19	25
14	12	26	40	25	***43***	24	50	24	0	23	14	22	30	21	49	21	11
15	13	28	53	27	51	26	54	26	0	25	10	24	***23***	23	38	22	56
16	14	31	7	30	0	28	58	28	0	27	6	26	15	25	27	24	42
17	15	33	20	32	9	31	2	30	0	29	2	28	***8***	27	16	26	***28***
18	16	35	33	34	17	33	6	32	0	30	58	30	0	29	5	28	14
19	17	37	47	36	26	35	10	34	0	32	24	31	***53***	30	55	30	0
20	18	40	0	38	34	37	14	36	0	34	***50***	33	***45***	32	44	31	46
21	19	42	13	40	43	39	19	38	0	36	***46***	35	***38***	34	33	33	32
22	20	44	27	42	51	41	23	40	0	38	43	37	30	36	22	35	18
23	21	46	40	45	0	43	27	42	0	40	39	39	***23***	38	11	37	4
24	22	48	53	47	9	45	31	44	0	42	35	41	15	40	0	***38***	44
25	23	51	7	49	17	47	35	46	0	44	31	43	***8***	41	49	40	35
26	24	53	20	51	26	49	39	48	0	46	27	45	0	46	38	42	21
27	25	55	***33***	53	34	51	43	50	0	48	***23***	46	***53***	45	27	44	7
28	26	57	47	55	43	53	48	52	0	50	***19***	48	45	47	16	45	53
29	27	60	0	57	51	55	52	54	0	52	***15***	50	***38***	49	5	47	39
30	28			60	0	57	56	56	0	54	12	52	30	50	55	49	25
31	29					60	0	58	0	56	8	54	***23***	52	44	51	11
32	30							60	0	58	4	56	15	54	33	52	56
33	31									60	0	58	***8***	56	22	54	42
34	32											60	0	58	11	56	***28***
35	33													60	0	58	14
36	34															60	0

(C6) C:52; (C27) P:32; (E14) P:42; (K20) p:20; (K21) p:48; (K27) C, P:24; (K28) C,P:20; (K29) C,P:16; (M5) C, P:37; (M7) C,P:22; (M9) C, P:52; (M11) C, P:37; (M13) C, P:22; (M15) C, P:7; (M17) C, P:52; (M18) p:44; (M19) C, P:37; (M21) C, P:22; (M23) C,P:7; (M25) C:52; (M27) C, P:37; (M29) C,P:22; (M31) C, P:7; (P24) C, P:39; (Q17) C, P:26; (Q34) P:26;

Reverse Engineering Applied to Ephemerides

Analysis and Edition of the Arabic Ephemeris of 1326/7 CE (MS Cairo, Dār al-Kutub, *mīqāt* 817)

Johannes THOMANN

1. Introduction

Most contributions in this volume deal with astronomical handbooks and primary tables (i.e., tables that serve for calculating solar, lunar and planetary positions, and also other astronomical phenomena such as the length of the day and the time and magnitude of eclipses). These are the main sources for the teaching of astronomy, the development of astronomical theory and historical observations. However, handbooks and tables often do not contain much information about their practical use. New astronomical tables might not have been used immediately everywhere. Old tables might have survived in some places and among some groups of astronomers. For addressing such historical issues, other sources need to be consulted. Horoscopes are the final products of astronomical practice and bear important information on the use of astronomical tools. Birth horoscopes and historical horoscopes were most probably calculated by means of primary astronomical tables, but for daily tasks such as elections, interrogational astrology or year horoscopes, yearbooks were the more convenient tools. These are tables that contain pre-calculated solar, lunar and planetary positions either for each day of the year or at longer time intervals, such as for each month or for the days on which a heavenly body entered a new zodiacal sign.

Astronomical handbooks and primary tables are preserved in great quantity, but in most cases our knowledge of these texts is based on manuscripts that were copied centuries later than their original production. Horoscopes and yearbooks, in contrast, lose their usefulness relatively quickly and were not copied by later generations. Consequently, their rate of survival is relatively small, and most of the existing documents are only fragments. However, the exclusive information they provide justifies any effort to analyse and reconstruct them.

Ancient Greek horoscopes and ephemerides have received much attention in research over the last 60 years, whereas the study of Arabic horoscopes and ephemerides is only at its beginning.

The focus of this chapter is on astronomical yearbooks. Their advantage is that, even if they are fragmentary, they usually contain astronomical informa-

Editing and Analysing Numerical Tables: Towards a Digital Information System for the History of Astral Sciences, ed. by Matthieu Husson, Clemency Montelle and Benno van Dalen, PALS 2 (Turnhout, 2021), pp. 469–510
 10.1484/M.PALS-EB.5.127705

tion for a series of days. This allows for a deeper analysis than information for just a single day and hour, as is normally the case with horoscopes. In fact, if sufficient data exist, the tables used for the calculation of yearbooks can be fully reconstructed.[1]

The present contribution is mainly concerned with an Arabic ephemeris but, in general, the demonstrated methods can also be applied to Greek or Latin ephemerides as long as they are based on Ptolemaic models. For the sake of brevity and simplicity, the complete yearbook for the year 1326/7 CE that is extant in a manuscript in Cairo has been chosen as the main example.

At this point, it seems appropriate to say something about the terms 'ephemeris' and 'almanac'. Often these are used almost indistinctly. Here the distinction made by Alexander Jones in his edition of the Greek astronomical papyri from Oxyrhynchus will be adopted.[2] 'Ephemeris' is only used to denote a yearbook that provides a complete set of solar, lunar and planetary positions for every day, since it is derived from the Greek word for 'daily'. 'Almanac' or 'almanac-ephemeris' is used for a yearbook in a more compact format in which the intervals between consecutive entries consist of more than one day. This distinction made between Greek documents is even more appropriate for Arabic documents. In the tenth century, two clearly distinct types of yearbooks existed. One type very much resembled the Greek ephemerides in form and content, and therefore can justly be called 'ephemeris', since no precise Arabic term was in use. Another type did not contain solar and planetary positions, but only the zodiacal signs in which the moon was located on a certain day. Nevertheless, solar and planetary positions are implicitly present, since the lunar transits across the sun and the planets and their aspects are indicated. Traditionally, this type of yearbook is called an 'almanac'.

2. Ptolemy's planetary models

For those readers who are not familiar with the details of Ptolemy's models for the motions of the heavenly bodies, a brief summary may be welcome. The experts can safely skip it without missing any argument beyond the basic concepts of Ptolemaic astronomy. The models looked at in this chapter are two-dimensional; the lunar and planetary orbits in space are projected onto the plane of the ecliptic (i.e., the plane of the solar orbit). Common to all models is the ancient Greek premise that the irregular motions of the planets must be generated by uniform circular motions. In the case of the lunar nodes, the two points where the lunar orbit crosses the ecliptic, the motion is simply a uniform circular motion and the only underlying parameters are their daily velocity

[1] For a list of Greek and Arabic ephemerides, see Appendix D.

[2] Jones, *Astronomical Papyri from Oxyrhynchus*, vol. I, pp. 175–77.

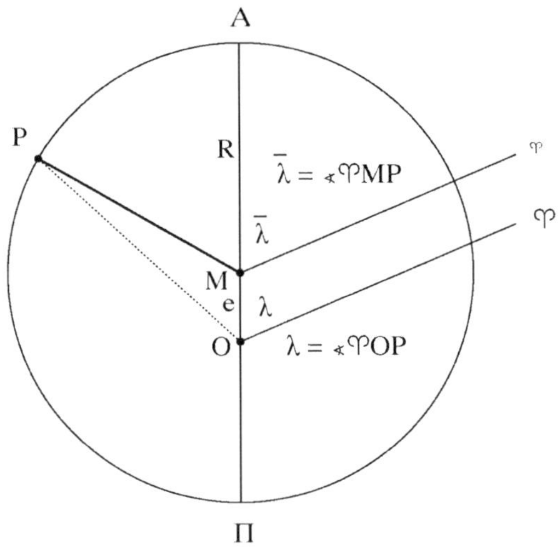

Figure 1: Ptolemy's solar model.

and positions at epoch. Of the nonlinear motions, the solar motion can be modelled in a simpler way than the motion of the moon and the planets. The solar velocity oscillates between a minimum and a maximum value once every year. Ptolemy accounted for this by making the orbit of the sun an eccentric circle (see Figure 1).

Point *O* marks the place of the observer on the earth. The sun moves with a constant velocity on a circle with centre *M*, which has a distance *e* from the observer at *O*; *e* is called the solar eccentricity. Point *A* marks the apogee, the point where the sun has its maximal distance from the earth (*O*); point *Π* is the perigee, the point where it has its minimal distance. On its way from *A* to *Π*, the velocity of the sun *P* increases, then it decreases on the way back from *Π* to *A*. When the sun is situated at *P*, it is seen by the observer at *O* in the direction of the dotted line *OP*. The true ecliptic longitude λ of the sun corresponds to the angle ♈*OP*. The direction to ♈ points to the starting point of the ecliptic, the position of the sun at the vernal equinox. The mean solar longitude $\bar{\lambda}$ corresponds to the angle ♈*MP* and is a linear function of time. This is the starting value for the computation of the true longitude. This calculation consists of three steps. From the table of mean solar motion, the motions corresponding to the given years, months, days, hours and fractions of hours are simply added together. Next, the longitude of the apogee (which is obtained in a similar way from the table of apogee motion) is subtracted from the mean longitude. The result is the mean solar anomaly, which corresponds to the angle *AMP* in Figure 1. Next, in order to obtain the true longitude, the angle *MPO*, called the solar equation, must be found. This is done by entering the table of the solar equation with the solar anomaly as the argument. The result is subtracted from the mean longitude if the anomaly is

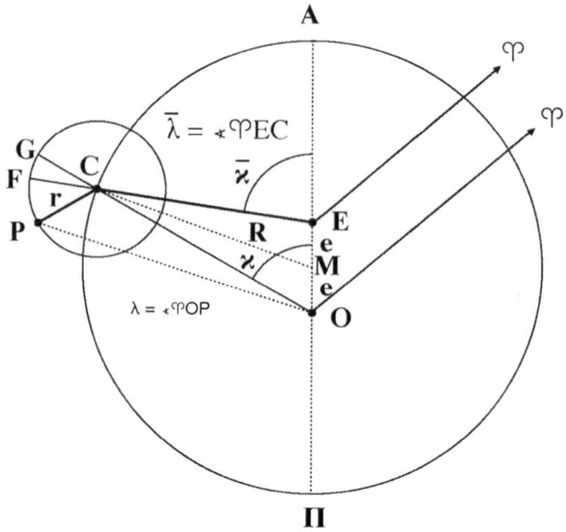

Figure 2: Ptolemy's planetary model.

in the interval from 0° to 180°, and it is added to the mean longitude in the interval from 180° to 360°.

The topic of this article is how to carry out this procedure backwards. Ephemerides contain only true longitudes. From these the corresponding mean longitudes, and in the end the value of the solar eccentricity *e*, should be extracted. This is possible because the equation is a nonlinear function with a characteristic periodical pattern. Therefore, it can be separated from the linear component of the mean longitude. Before explaining how this can be done efficiently for all planets, a quick summary of the planetary and lunar models is given.

The motion of each of the five planets as seen from the earth shows two irregularities. According to modern theory, the first irregularity is produced by the eccentricity of the orbit of the earth, and the second by the motion of the earth around the sun, which is mirrored in the geocentric motions of the planets. In the case of the upper planets – Saturn, Jupiter and Mars – Ptolemy accounted for the first irregularity with an eccentric arrangement, as in the case of the sun, but with a significant difference, which will be explained in a moment. For the second irregularity, he introduced an epicycle. The centre of the epicycle moves on the deferent ('carrier', i.e., the circle that accounts for the eccentricity), and the planets perform a uniform circular motion on the periphery of the epicycle. In the case of Venus and Mercury, the epicycle accounts for the first irregularity and the eccentricity for the second, since – from the modern perspective – the earth is inside the orbits of the upper planets but outside the orbits of the lower planets. In Figure 2, the point *M* is the centre of the deferent and *C* is the epicycle centre. The point *O* is the location of the terrestrial observer. The planet *P* moves with a constant angular velocity on the

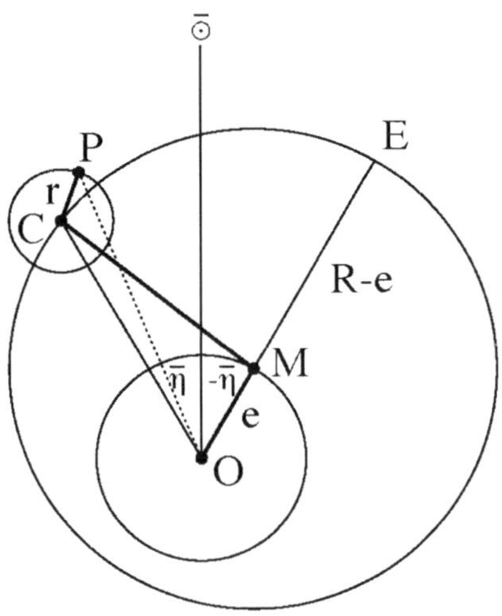

Figure 3: Ptolemy's lunar model.

epicycle. However, the epicycle centre *C* does not move uniformly with respect to the centre *M* of the deferent, as in the case of the sun, but with respect to a point *E*, which has the same distance from *M* as the point *O*. This means that the bold line *EC* rotates with a constant velocity. The point *E* is called in Latin the *punctum aequans*, the equant point, i.e., 'the point which makes [the motion] equal [to itself]'. Both the centre of the epicycle on the deferent and the planet on the epicycle move counter-clockwise. In the case of the upper planets (Saturn, Jupiter and Mars), the line *CP* from the centre of the epicycle to the planet always points to the point of mean solar longitude. This is because, from the modern perspective, the motion of the epicycle, mirrors the motion of the earth around the sun. In the case of Venus and Mercury, the line *EC* from the equant point to the centre of the epicycle points to the mean solar longitude, since the motion on the deferent mirrors the motion of the earth around the sun. The model for Mercury shows a further complication, which will be explained below. The true longitude of a planet is a function of four quantities: the mean longitude of the planet (angle ♈*EC* in Figure 2), the mean solar longitude (angle ♈*MP*), the eccentricity of the planet (the length of ME and MO), and the radius of the epicycle (the length *r* of *CP*).

The lunar model shows further complications as well. The irregularity of the lunar motion produced by the eccentricity of its orbit around the earth was noticed early on. In the Ptolemaic model, it is generated by the moon's motion on the epicycle. However, Ptolemy discovered two more irregularities by his analysis of observations. According to modern theory, evection, which is the second irregularity, is produced by the gravitation of the sun. It reaches its extreme values approximately at the first quarter and the third quarter of the synodic month, and it vanishes at the syzygies. Its period is half a synodic month. Ptolemy accounted for this effect with a kind of crank mechanism. In the case of the planets (except for Mercury), the equant point was fixed on the line connecting the apogee and perigee. However, in the lunar model the equant point M moves in a clockwise direction on a small circle around the terrestrial observer at O (see Figure 3). The centre of the epicycle C again rotates uniformly with respect to M and is simultaneously drawn back and forth towards the eye of the observer by the crank mechanism. This has the effect that the first irregularity produced on the epicycle is increased at certain times and diminished at others. Ptolemy did not change the geometrical design of the model for the third irregularity, since it has the same period as the second; therefore, he used the same device for its generation. On the circle around O, a point B is constructed opposite to M. From there, a line is drawn through C, the centre of the epicycle meeting the periphery at Point $\bar{A}$. The crank mechanism moves the point $\bar{A}$ back and forth relative to point A. The moon at point P moves with a constant velocity relative to point $\bar{A}$. Despite these additional complications, the parameters of the lunar model are similar to those of the planetary model, but rather than feeding in the mean longitude of the moon and the mean longitude of the sun separately, the true longitude of the moon can be calculated as a function of the difference of the two, namely, the elongation. This is the arc $\overline{\odot OC}$ in Figure 3.

Of all the planets, the orbit of Mercury posed the greatest problems to astronomers. Because of its small maximal elongation from the sun, observing it was difficult, and because the eccentricity of its orbit was very large, it could not be ignored in the model, as Ptolemy did in the case of Venus. Based on partly erroneous observations, he came to the conclusion that Mercury had one apogee but two perigees 120° apart. For Mercury he used a similar model to that of the moon, but with the inner circle having M as its centre instead of O. The point E midway between M and O has the same function as point B in the case of the lunar model, and the point F on the epicycle has the same function as the point $\bar{A}$ in the lunar model. The extension of the line EC (cf. Figure 2) meets the periphery of the epicycle at the point F, in relation to which Mercury at P moves with constant velocity. Despite these differences from the other planetary models, the set of quantities is the same as for the other planets but the formula for calculating the true longitude is different.

3. Extraction of the Implicit Parameters of Solar, Lunar and Planetary Motions

Fragments of ephemerides usually contain a series of daily solar, lunar and planetary positions. The task of extracting implicit parameters from such data can be done by hand, but this is not recommended, since it is cumbersome to do and barely executable for the moon and the planets.[3] A more fundamental

[3] An example of such a manual approach can be found in Thomann, 'An Arabic Ephemeris for the year 931–932 CE'.

approach does not require any pre-assumptions concerning the eccentricity and epicycle radius. For primary astronomical tables, Benno van Dalen proposed using nonlinear least squares estimation.[4] The advantage of this method is that only the function through which the data were produced needs to be defined, and the closest fit is searched for in the multidimensional parameter space. In the optimal case, this leads to perfect results. However, in its traditional form, using the Gauss–Newton method, the algorithm often does not terminate, since it is very sensitive to the choice of initial values, which must be defined for the parameters to be estimated.

A method based on the Levenberg–Marquardt algorithm is more stable and less sensitive to the choice of initial values.[5] In the following, this method will be used exclusively for estimating parameters. All calculations and graphics are produced in R.[6] Other implementations are available, for instance, in Scilab.[7]

The Gauss–Newton method of nonlinear least squares estimation uses the gradient with the first derivatives of the function for approximation. It is called the method of steepest descent. In each step, it chooses the direction in which the gradient is most inclined towards the desired minimum. Since the model function is known, the second derivatives are also available. In cases where the method of steepest descent fails, the matrix of second derivatives (Hessian matrix) can be used instead. In this way, the curvature of the function is taken into consideration. The Levenberg–Marquart method switches, if necessary, between the method of steepest descent and use of the curvature matrix (one half of the Hessian matrix). Thus the Levenberg–Marquart method is more robust than the Gauss–Newton method, even if it does not guarantee convergence.

The method presupposes that the underlying function is differentiable twice at all data points. This is not the case with all historical ephemerides. For instance, P. Oxy. 4179, an ephemeris for the year 349 CE, is not based on Ptolemaic models but on Babylonian zig-zag functions, which, at some points, are not differentiable. The lunar positions in this ephemeris are calculated according to the Standard Lunar Scheme.[8] The daily lunar velocities exhibit the pattern shown in Figure 5.

[4] van Dalen, *Ancient and Mediaeval Astronomical Tables*, pp. 55–60.

[5] Press et al., *Numerical Recipes*, pp. 799–806.

[6] https://www.r-project.org.

[7] https://www.scilab.org.

[8] Jones, 'Studies in the Astronomy of the Roman Period I'; Jones, *Astronomical Papyri from Oxyrhynchus*, vol. I, p. 187.

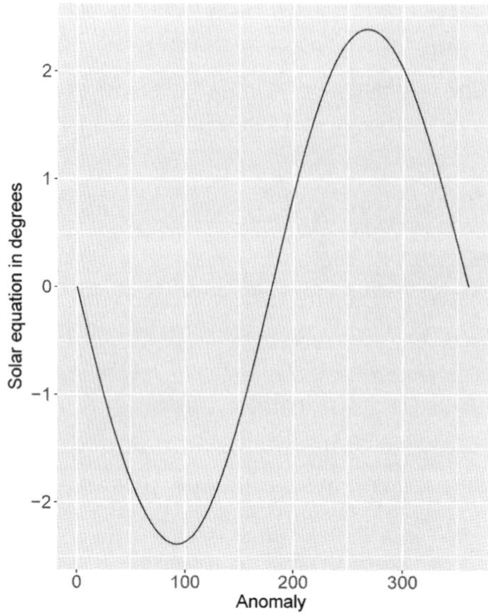

Figure 4: The solar equation as a function of the solar anomaly in Ptolemy's model.

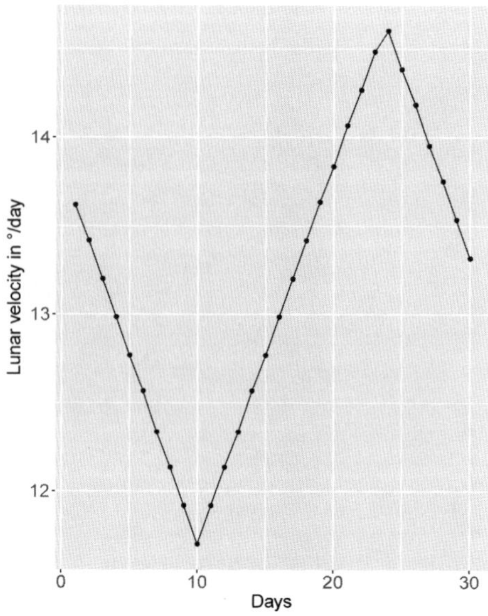

Figure 5: Daily lunar velocities based on Babylonian zig-zag functions.

In contrast, velocities according to the Ptolemaic model show an entirely smooth pattern, as is clear from Figure 4, which displays the solar equation.

Besides the positions of the sun, moon and planets, medieval ephemerides also contain the positions of the ascending lunar node. In Ptolemaic astronomy, the velocity of the lunar node is assumed to be constant. The acceleration of the lunar node is indeed very small: the quadratic component in its calculation is +0.0020754°, if time is measured in Julian centuries.[9] This means that it would take about 300 years to deviate from the linear model by one arc-minute. Thus a linear least squares method can be used for estimating the parameters of the motion of the lunar node, and two values suffice for an estimate. The positions calculated by different historical tables differ from each other by at least several arc-minutes. Generally, this allows for an identification if a known astronomical table was used.

4. Tests of Parameter Estimations with Synthetic Data

Up to now, no parameter estimations of ephemeris data by means of the nonlinear least squares method have been carried out. Before applying this method

[9] Meeus, *Astronomical Algorithms*, p. 342.

to historical data, it is advisable to investigate its efficiency in controlled tests with synthetic data. These data are obtained by using the Ptolemaic formulae and by rounding the results to arc-minutes. This includes a simplification, since the results may differ from the values obtained by calculations based on tables rounded to arc-minutes. In the next section, intermediate rounding errors will be simulated by adding random noise. Our first test concerns solar longitudes. The parameters to be estimated are the value of the solar mean longitude at the beginning of the time interval covered by the ephemeris (referred to as 'the value at epoch'), the longitude of the apogee at epoch and the eccentricity. In the following, the days of the ephemerides will be counted as 1, 2, 3, ..., and the epoch values are defined as being for Day 0. The mean velocity is assumed to be a constant, since for short periods of time, the differences found from historical tables are far below the accuracy of ephemerides, which generally give positions in signs, degrees and minutes. Thus the differences can safely be neglected for time spans of a few months. From antiquity to the end of the thirteenth century, only fragments of ephemerides exist that contain less than two months of data; the majority of the fragments contain less than 15 days. Therefore, the tests with synthetic data are made with samples of different length.

In the case of the epoch values of the mean solar longitude and the solar apogee, the accuracy of the estimates is very different. One may assume that it matters where in the solar orbit the positions covered by the ephemeris are situated, since the change in solar velocity is smallest when the sun is in the vicinity of the apogee and largest close to the perigee. Therefore, all simulations are carried out for positions around the apogee, quadrature and perigee. In each case, the simulations are carried out for 20 consecutive years at approximately the same positions in the solar orbit. The boxplots display the range of estimates found for each particular situation. The bold horizontal line indicates the median of the estimates. The box contains the lower and upper quartile of the estimates, which, between them, cover half of the estimates. The vertical lines, called the 'whiskers', either mark the extreme values or they end at 1.5 times the interquartile distance from the end of the box, if there are values farther away. In the latter case, the extreme values are represented as single dots. These are considered to be outliers.

Indeed, the errors in the estimates of the mean solar longitude at epoch are significantly greater around the apogee than around the quadrature and perigee (see Figure 6). For a reliable identification of the mean motion tables used for the calculation, at least 30 values appear to be necessary.

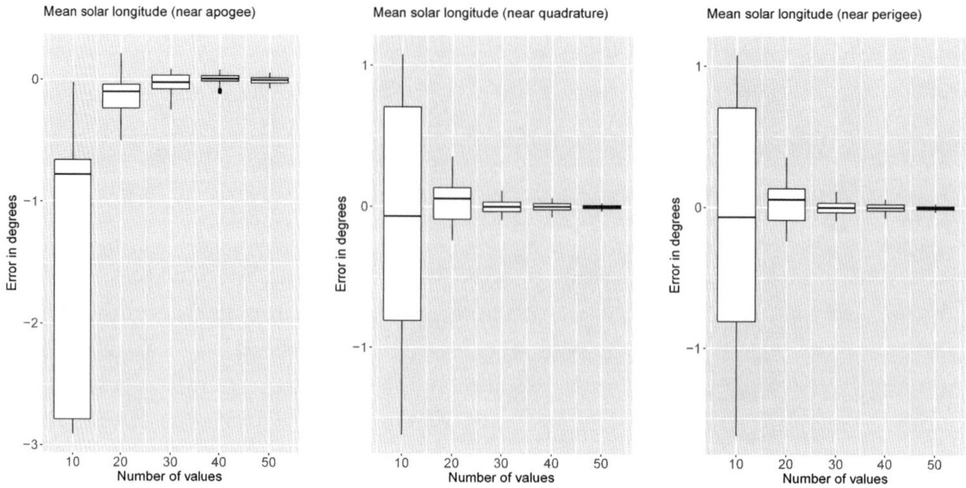

(a) (b) (c)

Figure 6: Errors in estimates of the mean solar longitude at epoch (simulated data) when the given solar positions lie (a) around the apogee, (b) around quadrature, and (c) around the perigee.

The errors in the longitude of the apogee are about 10 times as large as the errors in the mean longitudes at epoch (Figure 7). This reflects the fact that the longitude of the apogee contributes much less to the value of the true longitude than the mean longitude. Note that the size of the errors in the apogee longitude does not depend on the part of the solar orbit for which the ephemeris positions are given.

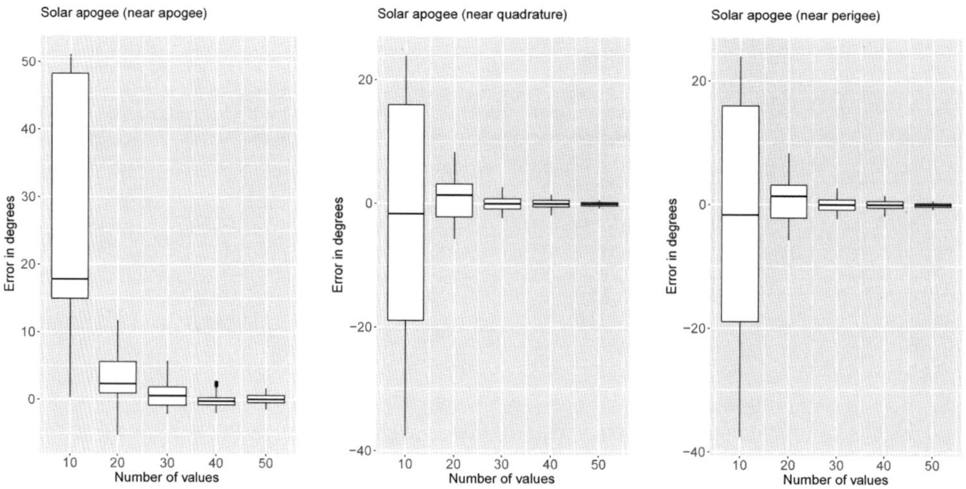

(a) (b) (c)

Figure 7: Errors in estimates of the solar apogee (simulated data) when the given solar positions lie (a) around the apogee, (b) around quadrature, and (c) around the perigee.

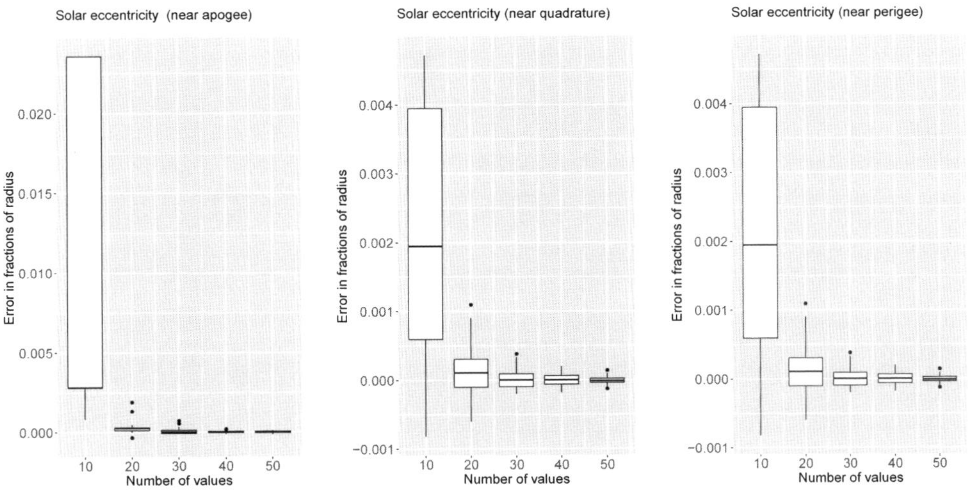

(a) (b) (c)

Figure 8: Errors in estimates of the solar eccentricity (simulated data) when the given solar positions lie (a) around the apogee, (b) around quadrature, and (c) around the perigee.

For the estimates of the solar eccentricity, the increase in accuracy from 10 data points to 20 is particularly significant (Figure 8).

The lunar and planetary motions depend on the mean solar motion. Therefore, if estimates for the mean solar longitudes based on the true solar positions given in the ephemeris are available, they can be used to estimate the parameters underlying the lunar and planetary positions. The differences in the daily mean motions in longitude and in anomaly among different astronomical tables are too small for being relevant for the true longitudes in the timespan of a month or even a year. Thus only four parameters need to be estimated: the epoch values of the mean longitude and the mean anomaly, the eccentricity, and the epicycle radius. If no solar positions are available, the mean solar longitudes must be estimated from the lunar and planetary positions themselves. In the following simulations, true solar longitudes are supposed to be available, and the mean solar longitudes are estimated from them as explained above.

Figures 9 to 12 display the results of the estimation of the lunar parameters from synthetic ephemeris data. For all parameters, significantly fewer positions are necessary for a useful estimate than in the case of the sun. This reflects the larger and quicker change in the lunar velocities in longitude and anomaly. Ten values suffice to estimate the epoch values of the mean lunar longitude and anomaly with an accuracy of, respectively, 2 and 20 arc-minutes. With 15 values, the errors are below an arc-minute.

The errors in the estimates of the mean solar longitudes influence the estimates of the motion of the inferior planets directly, since their mean longitude is equal to the mean solar longitude. Therefore, no higher accuracy is to be

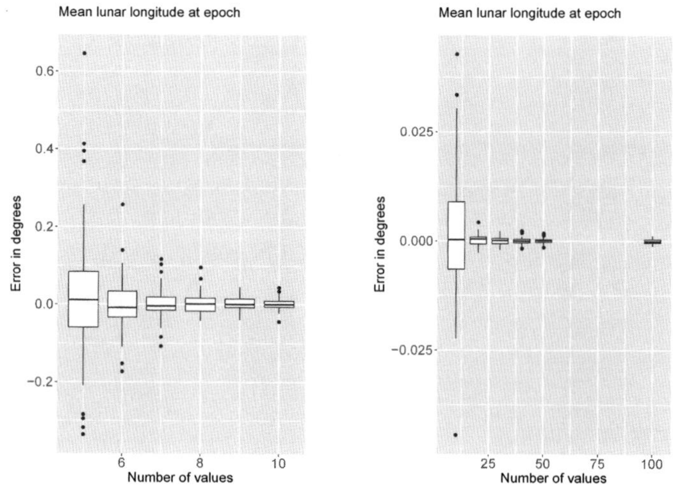

(a) (b)

Figure 9: Errors in estimates of the mean lunar longitude at epoch (simulated data): (a) with 5 to 10 values, and (b) with 20 to 100 values.

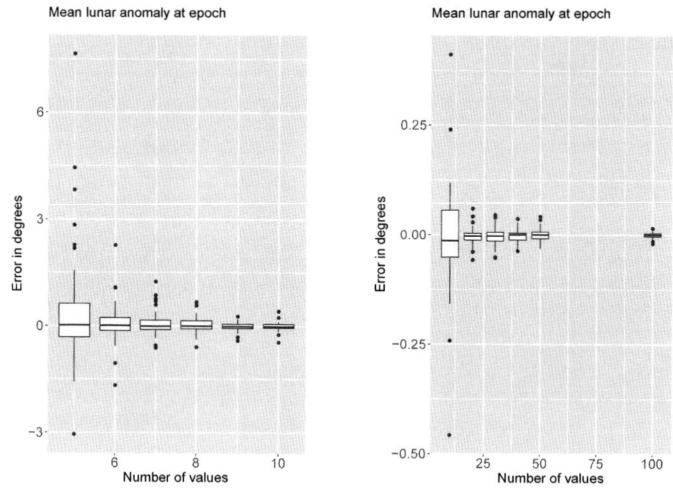

(a) (b)

Figure 10: Errors in estimates of the mean lunar anomaly at epoch (simulated data), (a) with 5 to 10 values, and (b) with 20 to 100 values.

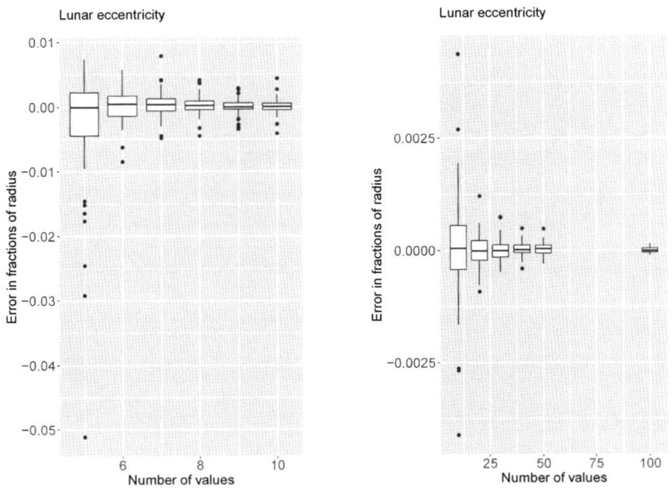

(a) (b)

Figure 11: Errors in estimates of the lunar eccentricity (simulated data), (a) with 5 to 10 values, and (b) with 20 to 100 values.

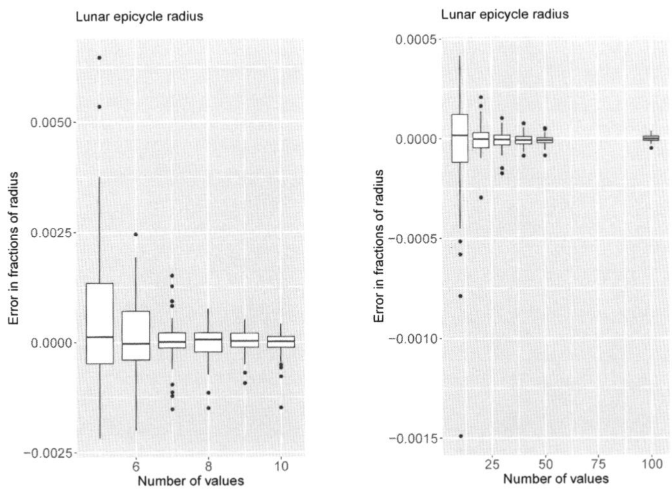

(a) (b)

Figure 12: Errors in estimates of the lunar epicycle radius (simulated data), (a) with 5 to 10 values, and (b) with 20 to 100 values.

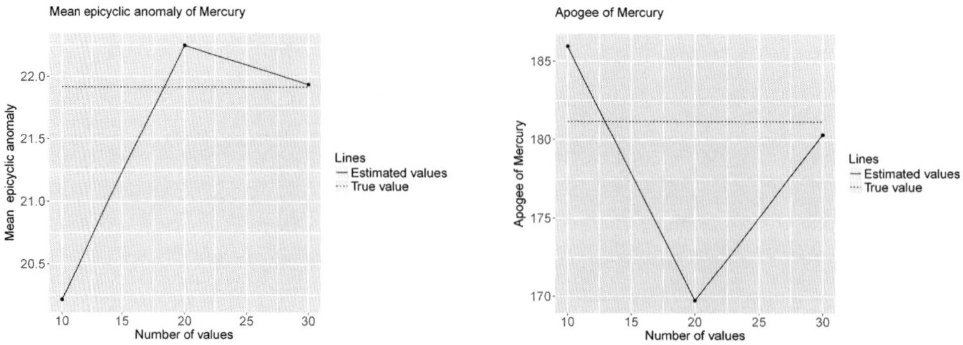

(a) (b)

Figure 13: Estimates of the parameters of the Mercury model (simulated data): (a) the mean epicyclic anomaly at epoch, and (b) the apogee.

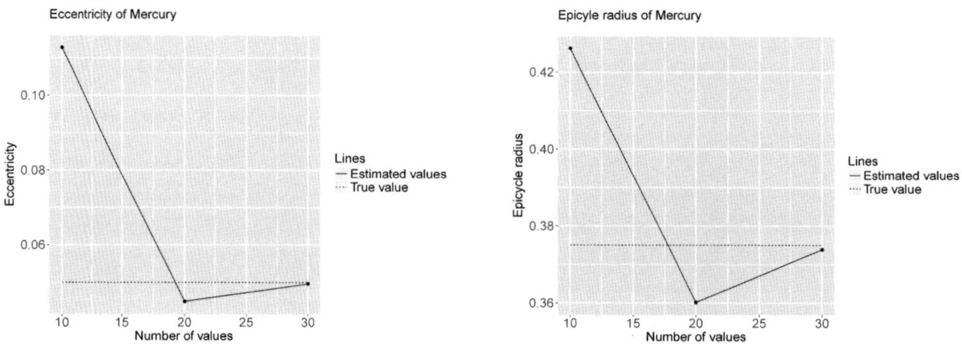

(a) (b)

Figure 14: Estimates of the parameters of the Mercury model (simulated data): (a) the eccentricity of Mercury, and (b) the epicycle radius.

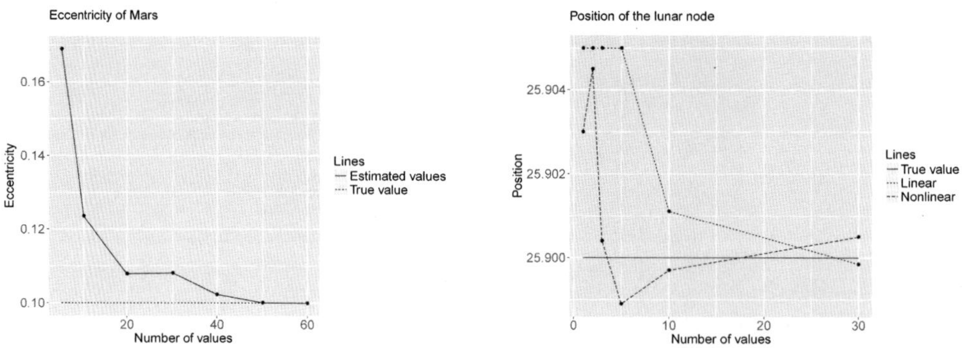

(a) (b)

Figure 15: (a) Estimates of the eccentricity of Mars (simulated data); (b) Estimates of the position of the lunar node at epoch based on linear and nonlinear models (simulated data).

expected. The slightly simplified diagrams for Mercury in Figures 13 and 14 confirm this.

For the upper planets, larger errors are to be expected, since their changes in velocity are slower. The single example for Mars in Figure 15a indeed shows that the errors are about 50% greater than in the case of Mercury.

The tests show that the performance of the method varies significantly for the sun, moon and planets. The number of values necessary to reach an accuracy of one arc-minute in the case of angular parameters, and minutes of parts ($R = 60$) in the case of eccentricities and epicycle radii, are very different. The results of the tests with synthetic data make clear that each case needs to be examined separately.

As we have seen above, the longitudes of the moon and the planets all depend on the mean solar longitude, even if their sensitivities are different. The only positions independent of the sun are those of the lunar node. In the Ptolemaic lunar model, the motion of the lunar node was assumed to be constant; therefore, its approximation from very few data points is particularly efficient. Actually, to use a nonlinear least squares method for fitting data to a linear function is not standard practice, because linear models are generally used instead. A comparison of the tests with both methods shows a remarkable difference (Figure 15b).

If 30 values are available, the linear method produces a slightly better result. If fewer values are available, the results of the nonlinear method are clearly better. Even if only one value is available, the estimate is accurate to within arc-minutes. Different primary tables in general produce much larger differences and therefore values of the node turn out to be the best first criterium for their identification, especially in the case of smaller fragments.

Nevertheless, it is worth the effort to also estimate the parameters of lunar and planetary motions. No positions of the lunar node may be available, as is the case with Greek ephemerides from antiquity, or with smaller fragments from later periods. Alternatively, an ephemeris may have been based on an otherwise unknown set of tables.

5. Preparatory simulations for the analysis of the 1326/7 ephemeris

Before analysing the Arabic ephemeris of 1326/7 CE, it seems advisable to execute some further simulations tailored to this particular historical case in order to see if, under the given conditions, an identification of the underlying primary tables may be assumed to be reliable. For reasons that will become clear in the next section, positions of the sun in the neighbourhood of 0° are to be avoided. The positions are calculated for every day from 16 March 1327 to 13 March 1328. In order to make the simulations more realistic, intermediate rounding errors are simulated by adding random values generated according

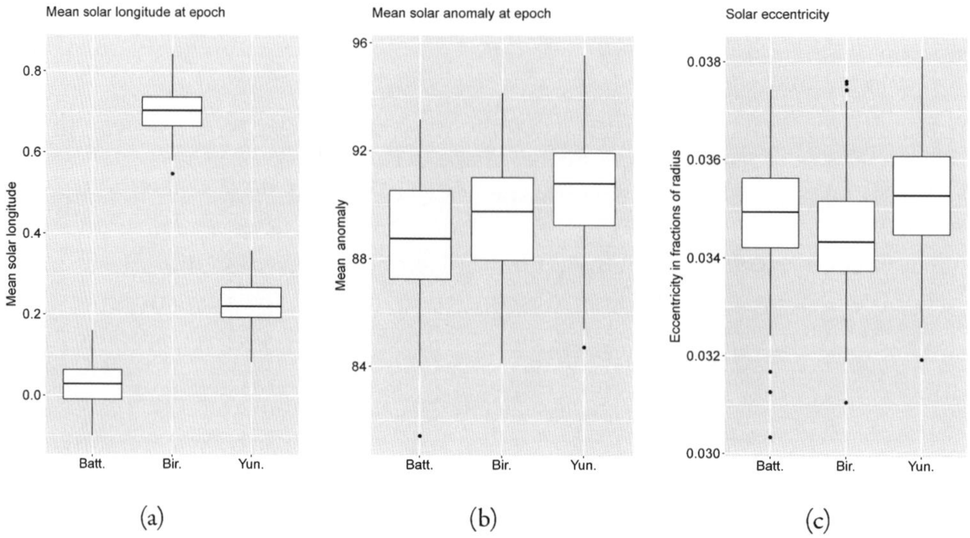

Figure 16: Estimation of the solar parameters according to data of al-Battānī, al-Bīrūnī and Ibn Yūnus: (a) the mean solar longitude at epoch, (b) the mean solar anomaly at epoch, and (c) the solar eccentricity.

to a normal distribution with a mean of 0 and a standard deviation of 1′. Next the values are rounded to the nearest arc-minute. The historical sets of tables against which the results of the estimations will be checked are the *zīj* of al-Battānī, *al-Qānūn al-Masʿūdī* of al-Bīrūnī and *al-Zīj al-Ḥākimī* of Ibn Yūnus. Here this limitation is possible because we already know that the tables of Ibn Yūnus were used (cf. p. 487). In a case where no such information is available, a larger number of tables may need to be checked. The parameters from these *zīj*es are taken from Raymond Mercier's program 'Devplo'. For each parameter set, 100 sets of positions for the entire solar year are calculated, which differ in the random noise added before they are rounded to the nearest arc-minute. A nonlinear least squares fit of the parameters is executed for each of the 100 sets of positions. In the boxplots in Figures 16 to 20, the distribution of the parameter estimates obtained in this way are shown.

In the case of the sun (Figure 16), it turns out that only the estimates of the mean longitudes at epoch are sufficiently different to allow a distinction between the sources. In the cases of the mean anomaly at epoch and the eccentricity, the distributions overlap too much for a reliable identification.

In the case of the moon (Figures 17 and 18), all four parameters, namely the mean longitude and the mean anomaly at epoch, the eccentricity and the radius of the epicycle, are clearly separated among the three sources.

In the case of Mercury (Figures 19 and 20), all four parameters show well separated distributions, except in the case of the epicycle radius, for which al-Battānī and al-Bīrūnī use the same value (cf. Figure 20b).

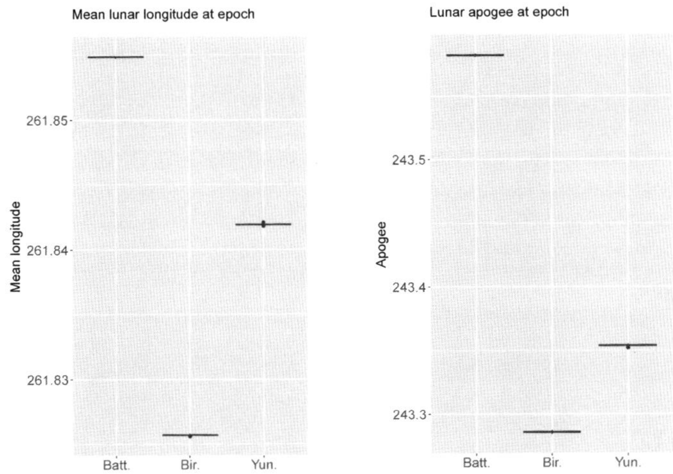

(a) (b)

Figure 17: Estimation of the lunar parameters according to data of al-Battānī, al-Bīrūnī and Ibn Yūnus: (a) the mean lunar longitude at epoch, and (b) the lunar apogee at epoch.

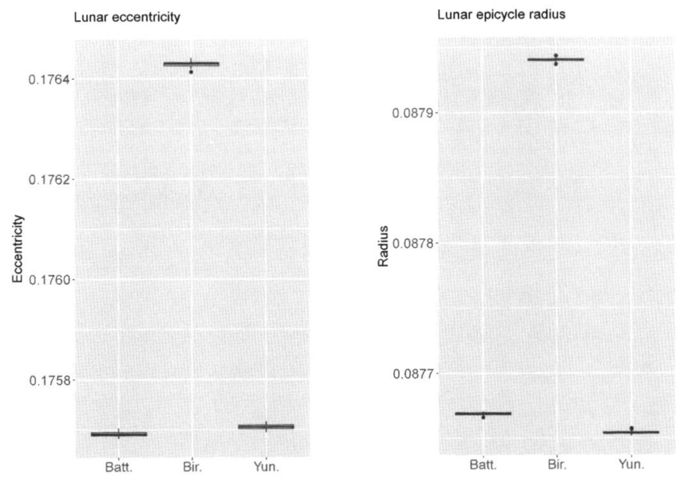

(a) (b)

Figure 18: Estimation of the lunar parameters according to data of al-Battānī, al-Bīrūnī and Ibn Yūnus: (a) the lunar eccentricity, and (b) the lunar epicycle radius.

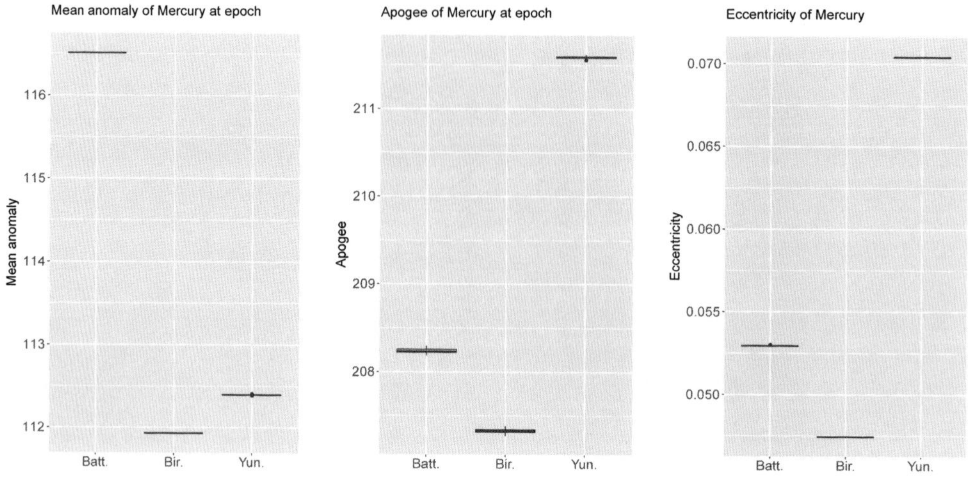

(a) (b) (c)

Figure 19: Estimation of the parameters of the Mercury model according to data of al-Battānī, al-Bīrūnī and Ibn Yūnus: (a) the mean anomaly at epoch, (b) the apogee, and (c) the eccentricity.

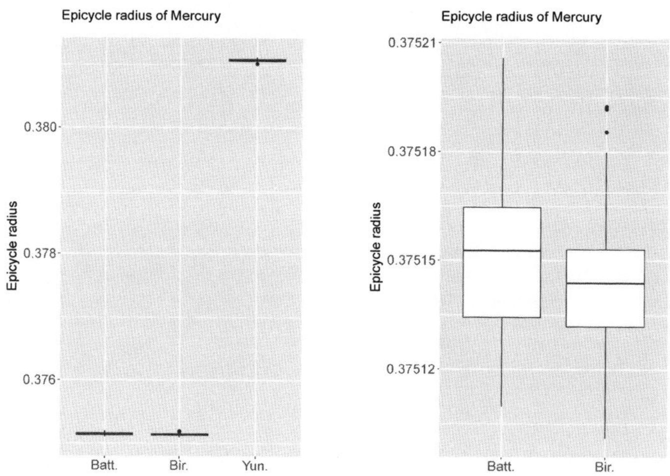

(a) (b)

Figure 20: Estimation of the epicycle radius of Mercury: (a) according to data of al-Battānī, al-Bīrūnī and Ibn Yūnus, and (b) according to data of al-Battānī and al-Bīrūnī only.

The tests discussed in this section have shown, that (i) even relatively small sets of data allow for useful parameter estimates (Figures 6 to 15), and that (ii) for identifying the use of a particular table, the epoch values of the mean solar longitude, the lunar mean anomaly, the lunar node and the planetary anomaly, as well as the longitude of the planetary apogee and the planetary eccentricity, are suitable (Figures 16 to 20).

6. The 1326/7 Ephemeris as a test case for parameter estimation

The ephemeris for 727 Hijra (1326/7 CE) is the earliest complete Arabic ephemeris that has survived (see Appendix A for a description of the Cairo manuscript in which it is included, and Plate 14 for the first page of the ephemeris with planetary positions for the month Ramaḍān). It seems particularly suited to be used as a test case for parameter estimation because the text accompanying one of the four seasonal horoscopes immediately preceding the ephemeris (see Plate 15) states that the horoscopes were established by means of the 'Ḥakimite tables', i.e., the tables of Ibn Yūnus. Therefore one may assume that this also holds for the following ephemeris.

In order to start with the most stable estimation, the values for the ascending lunar node will be analysed first. The linear least squares method does not require initial values. In calculating with R, a linear model is fitted to a data frame containing values for the variable t (which runs from 1 to 354 days in our case) and the positions of the node are converted from sexagesimal to decimal notation.

In the following table, the position estimated by the linear model is compared with the positions based on a selection of historically plausible astronomical tables. Since the tables of al-Battanī and al-Bīrūnī produce almost identical results, al-Battanī has been replaced here by al-Ṭūsī:

Source	*Position*	*Difference from estimation*
Estimation ephemeris 1326/7	183;16,56	–
Ibn Yūnus	183;16,28	0; 0,28
Al-Bīrūnī	183;53,32	0;36,36
Al-Ṭūsī	183;27,45	0;10,49

The agreement of the value of Ibn Yūnus with the value of the ephemeris for 727 Hijra is perfect: the calculated difference (28″) is below the precision of the ephemeris (±0.5′). Furthermore, the values based on the other tables are significantly different (37′ and 11′, respectively). Thus the use of the Ḥakimite tables by Ibn Yūnus, as explicitly indicated in the text, is corroborated by the analysis of the positions of the lunar node in the ephemeris. In other words, the test shows that, in this particular historical case, the tables used for calculating the ephemeris are identified correctly by an analysis of the positions of the lunar node alone.

Next, the positions of the sun will be used as a second test. In the following report, all problems encountered will be documented for giving a realistic impression of the process that leads to the final results, even if a different approach would have led more directly to the result, as will become clear afterwards. The slightly rounded values obtained from the tables of Ibn Yūnus are used as initial values for the nonlinear least squares estimation. An initial attempt with all 354 solar positions leads to some historically impossible results:

Source	*Mean Longitude*	*Apogee*	*Eccentricity*
Ephemeris (*t*=1 to 354)	249;36,27	91;46, 5	0;0,56
Ibn Yūnus	251;45,21.74	90;45,48.10	0;2, 6.16
Al-Bīrūnī	252;17, 2.41	89;27,36.22	0;1,59
Al-Ṭūsī	251;48,10.43	89;46,12.49	0;2, 6.16

The estimated values for the mean longitude and apogee at epoch differ significantly from the values found from Ibn Yūnus, al-Bīrūnī and al-Ṭūsī, but could easily correspond to another historical set of parameters. However, the estimated value for the eccentricity (0;0,56) is historically impossible and points to a problem either in the estimation process or in the data themselves. One possibility to localise this problem is to split the 354 values into bins of 50 values (54 in the last bin). The mean of the estimates derived from each of the bins can then be calculated:

Range of t	*Mean Longitude*	*Apogee*	*Eccentricity*
1–50	251;50,35"	92;32,19	0;2, 4
51–100	251;42,27	89;35,19	0;2, 9
101–150	249;36,27	97;53,18	0;2, 1
151–200	251;51,23	88;43,37	0;2, 2
201–250	251;46,40	90;54,56	0;2, 7
251–300	251;45,26	91; 6,16	0;2, 5
301–354	251;51,35	91;58,20	0;2,11
mean	251;29,13	91;25,14	0;2, 5

The estimates of the eccentricity in all seven intervals now vary within historically possible limits – in contrast to the overall estimation. Only the estimates calculated from the solar positions in the interval 101–150, especially that for the mean longitude, seem to indicate a problem in the data. Therefore these are split into bins of 25 values and the estimations are repeated:

Range of t	*Mean Longitude*	*Apogee*	*Eccentricity*
101–125	249;41,50	108;32,31	0;2, 5
126–150	251;28, 7	93;29,54	0;2,24

The results are now somewhat contradictory. The mean longitude in the second bin is close to that of the other bins and the value of the apogee is only slightly too high, but the eccentricity is much too high. None of the values of the first bin is within the limits of the other bins. For this reason, it seems sensible to exclude the values of the bin 101–150 entirely and to use the means of the other bins:

Range of t	*Mean Longitude*	*Apogee*	*Eccentricity*
Mean of bins 1–50, 51–100, 151–200, ...	251;48,1	90;48,28	0;2,6
Difference from Ibn Yūnus	+0;2,39	+0;2,40	0;0,0

Instead of taking the means of the binned data, an overall estimate can be made from all data without those from the interval 101–150:

Range of t	*Mean Longitude*	*Apogee*	*Eccentricity*
All except 101–150	251;46,31	90;41,47	0;2,6
Difference from Ibn Yūnus	+0;1,9	−0;4,1	0;0,0

The mean longitude is now closer to Ibn Yūnus, but the apogee differs slightly more. The eccentricity again fits perfectly. This corroborates the suspicion that there is a problem with the data for the interval 101–150. Inspection of the differences between these problematic data and the solar positions calculated with the parameters estimated from the other bins may give a clue to the solution of the problem. These differences are plotted in Figure 21. The 101–150 bin is the only interval with a difference of more than one arc-minute.

However, this two-minute difference occurs in the second half of the interval and thus cannot account for the problems diagnosed in the first half. Furthermore, two-minute differences could hardly be the cause of the significant deviation of the estimates. In order to verify this, the estimation for this bin has been repeated with simulated data based on the parameters estimated from the other bins:

Range of t	*Mean Longitude*	*Apogee*	*Eccentricity*
Ephemeris (101–150)	249;36,27	95;5,53	0;1,56
Simulated data	249;40,20	97;53,1	0;2,1

The mean longitudes at epoch are almost identical but the apogee in the simulated data is even worse. Only the eccentricity is slightly better. Consequently, the problem cannot be caused by errors in the data, but its cause must lie in the process of estimation. After some trials, the problematic interval could be reduced to 108–109:

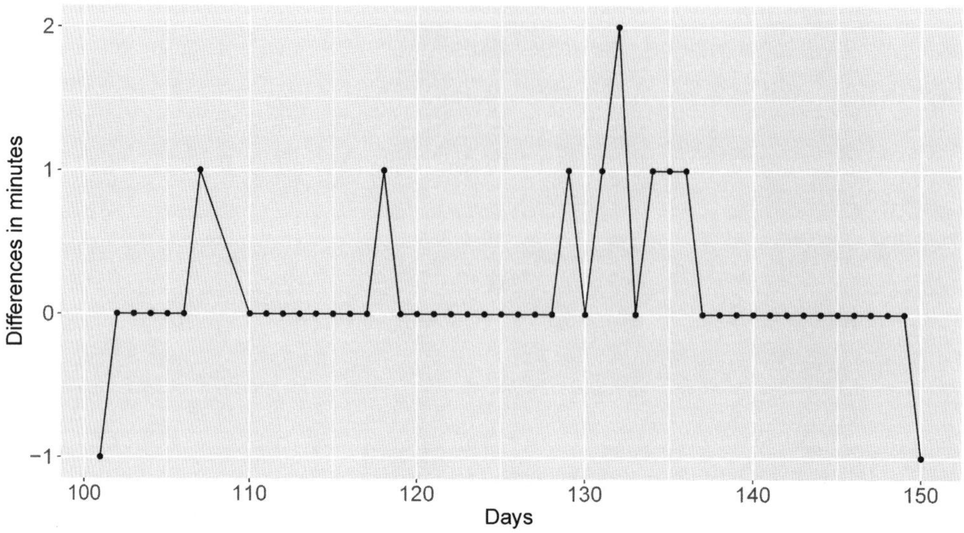

Figure 21: Differences between the problematic mean solar longitudes in the ephemeris and the calculated positions.

Range of t	*Mean Longitude*	*Apogee*	*Eccentricity*
Simulated (60–107)	251;47,47	90;18,29	0;2,5
Simulated (110–150)	251;48, 2	90;15,20	0;2,4

If the values of the true solar longitudes for the interval 107–110 are considered, the reason for the problem becomes clear:

t=107	*t=108*	*t=109*	*t=110*
359;14,0	0;14,0	1;12,60	2;12,0

The discontinuity from 359;59 to 0;0 seems to disturb the estimation process. Therefore, it is advisable to improve the results in cases where the data include this discontinuity by excluding some data points in its neighbourhood.[10] In our case, the omission of two values (108 and 109) suffices:

Range of t	*Mean Longitude*	*Apogee*	*Eccentricity*
Ephemeris (1–107, 110–354)	251;46,32	90;42, 6	0;2,6
Ibn Yūnus	251;45,22	90;45,48	0;2,6

The differences between the estimated values and the values found from Ibn Yūnus are now very small (+0;1, −0;4 and 0;0,0, respectively), and lie well within the range of intermediate rounding errors that are customary in such calculations. The use of the other historical tables taken into consideration here can be excluded.

[10] An alternative would be to operate with angles greater than 360° in order to avoid the discontinuity.

The test has shown that the method applied is effective and leads to unambiguous results. However, the example has shown that the method is not foolproof. It is advisable to check the data by looking at the pattern of differences in order to detect scribal errors or mistakes in the calculation. If enough data are available, as in the present case, it is advisable to make estimates of subsets of the data in order to test for their consistency.

Bibliography

Bosworth, Clifford Edmund, *The New Islamic Dynasties. A Chronological and Genealogical Manual*, New York: Colombia University Press, 1996.

Curtis, Heber D., and Frank E. Robbins, 'An Ephemeris of 467 A.D.', *Publications of the Observatory of the University of Michigan* 6/9 (1935), pp. 77–100.

van Dalen, Benno, *Ancient and Mediaeval Astronomical Tables: Mathematical Structure and Parameter Values*, PhD dissertation, University of Utrecht, Mathematical Institute, 1993 (available from http://www.bennovandalen.de/).

Gerstinger, Hans, and Otto Neugebauer, 'Eine Ephemeride für das Jahr 348 oder 424 n. Chr. in den PER, Pap. Graec. Vindob. 29370', *Sitzungsberichte der Österreichischen Akademie der Wissenschaften, philosophisch-historische Klasse* 240 (1962), pp. 3–25.

Heilen, Stephan, *Hadriani Genitura – Die astrologischen Fragmente des Antigonos von Nikaia*, 2 vols, Berlin/Boston: De Gruyter, 2015.

Jones, Alexander, 'An Astronomical Ephemeris for A.D. 140: P. Harris I.60', *Zeitschrift für Papyrologie und Epigraphik* 100 (1994), pp. 59–63.

Jones, Alexander, 'Two Astronomical Papyri Revisited', *Analecta Papyrologica* 6 (1994), pp. 111–26.

Jones, Alexander, 'On the Planetary Table, Dublin TCD Pap. F. 7', *Zeitschrift für Papyrologie und Epigraphik* 107 (1995), pp. 255–58.

Jones, Alexander, 'Studies in the Astronomy of the Roman Period I. The Standard Lunar Scheme', *Centaurus* 39 (1997), pp. 1–36.

Jones, Alexander, *Astronomical Papyri from Oxyrhynchus (P. Oxy. 4133–4300a). Edited with Translations and Commentaries*, 2 vols, Philadelphia: American Philosophical Society, 1999.

King, David A., *Fihris al-makhṭūṭāt al-ʿilmiyya al-maḥfūẓa bi-Dār al-Kutub al-Miṣriyya* (in Arabic), 2 vols, Cairo: General Egyptian Book Organisation, 1981–1986.

King, David A., *Mathematical Astronomy in Medieval Yemen. A Biobibliographical Survey*, Malibu: Undena Publications, 1983.

King, David A., *In Synchrony with the Heavens. Studies in Astronomical Timekeeping and Instrumentation in Medieval Islamic Civilization*, vol. I: *The Call of the Muezzin (Studies I–IX)*, Leiden: Brill, 2004.

Meeus, Jean, *Astronomical Algorithms*, 2nd ed., Richmond: Willman-Bell, 1998.

Mercier, Raymond P., *An Almanac for Trebizond for the Year 1336*, Louvain-la-Neuve: Academia, 1994.

Press, William H., Saul A. Teukolsky, William T. Vetterling and Brian P. Flannery, *Numerical Recipes. The Art of Scientific Computing*, 3rd ed., Cambridge: Cambridge University Press, 2007.

Rosenfeld, Boris A., and Ekmeleddin İhsanoğlu, *Mathematicians, Astronomers, and other Scholars of Islamic Civilization and their Works (7th–19th c.)*, Istanbul: Research Centre for Islamic History, Art and Culture (IRCICA), 2003.

Schmidl, Petra G., *Volkstümliche Astronomie im islamischen Mittelalter. Zur Bestimmung der Gebetszeiten und der Qibla bei al-Aṣbaḥī, Ibn Raḥīq und al-Fārisī*, 2 vols, Leiden: Brill, 2007.

Thomann, Johannes, 'An Arabic Ephemeris for the Year 954/955 CE and the Geographical Latitude of al-Bahnasā/Oxyrhyncus (P.Stras. Inv. Ar. 446)', *Chronique d'Égypte* 88 (2013), pp. 385–96.

Thomann, Johannes, 'An Arabic Ephemeris for the Year 1026/1027 CE. in the Vienna Papyrus Collection', in Gianna Katsiampoura (ed.), *5th International Conference of the European Society for the History of Science. Scientific Cosmopolitanism and Local Cultures: Religions, Ideologies, Societies. Proceedings. Athens, 1–3 November 2012*, Athens: National Hellenic Research Foundation, 2014, pp. 54–60.

Thomann, Johannes, 'An Arabic Ephemeris for the Year 931–932 CE', in Andreas Kaplony, Daniel Potthast and Cornelia Römer (eds), *From Bāwīṭ to Marw. Documents from the Medieval Muslim World*, Leiden: Brill, 2015, ch. 6, pp. 115–53.

Thomann, Johannes, 'Kat.-Nr. 62: Ephemeride für das persische Jahr 363 (994/995 n. Chr.)', in Angelika Zdiarsky (ed.), *Orakelsprüche, Magie und Horoskope: Wie Ägypten die Zukunft sah*, Vienna: Phoibos, 2015, pp. 136–37.

Thomann, Johannes, 'Kat.-Nr. 63: Ephemeride für das persische Jahr 371 (1002/1003 n. Chr.)', in Angelika Zdiarsky (ed.), *Orakelsprüche, Magie und Horoskope: Wie Ägypten die Zukunft sah*, Vienna: Phoibos, 2015, p. 137.

Thomann, Johannes, 'Kat.-Nr. 65: Ephemeride für das persische Jahr 413 (1044/1045 n. Chr.): P.Vindob. ACh 1252 + Ach 14324', in Angelika Zdiarsky (ed.), *Orakelsprüche, Magie und Horoskope: Wie Ägypten die Zukunft sah*, Vienna: Phoibos, 2015, pp. 138–40.

Thomann, Johannes, 'The Arabic Ephemeris for the Year 1149/1150 CE (P.Cambridge UL Inv. Michael. Chartae D 58) and the Arabic *Baḥnīṭas*, Greek Παχνίτης and Coptic ΠαϣοΝϹ', *Chronique d'Égypte* 90 (2015), pp. 207–24.

Thomann, Johannes, 'A Fragment of an Unusual Arabic Almanac for 297 AH/910 CE (P.Berl. inv. 12793)', in Sobhi Bouderbala, Sylvie Denoix and Matt Malczycki (eds), *New Frontiers of Arabic Papyrology. Arabic and Multilingual Texts from Early Islam*, Leiden/Boston: Brill, 2017, ch. 10, pp. 179–96.

Thomann, Johannes, 'From *katarchai* to *ikhtiyārāt*: The Emergence of a New Arabic Document Type Combining Ephemerides and Almanacs', in Alberto Nodar and Sofía Torallas Tovar (eds), *Proceedings of the 28th Congress of Papyrology (Barcelona 1–6 August 2016)*, Barcelona: Publicacions de l'Abadia de Montserrat, 2019, pp. 342–54.

Appendix A: Description of MS Cairo, Dār al-Kutub, *mīqāt* 817

The manuscript in the Egyptian National Library that contains the ephemeris for the year 727 Hijra was discovered and described by David King.[11] It consists of 85 folios and measures 20 × 27 cm. The manuscript contains five parts:

(1) fols 1v–54r: *Kitāb maʿārij al-fikr al-wahīj fī ḥall mushkilāt al-zīj* ('Ascension of Flaming Thoughts on the Resolution of the Difficulties of a *Zīj*') by Muḥammad ibn Abī Bakr al-Fārisī (d. 1278/9 CE).[12] He was in the service of the Rasulid Sultan al-Malik al-Muẓaffar Shams al-Dīn Yūsuf I (1249–1295 CE). The *Maʿārij* are extant in a number of copies but have not yet been edited. In the present manuscript, it was copied approximately in 725/1325.[13]

(2) fols 55v–68v: Tables on astronomical-astrological subjects by an anonymous author. The section starts with a richly decorated dedication to the caliph written in a mandorla: *li-l-sulṭān ibn al-sulṭān ibn al-sulṭān al-malik al-Muʾayyad Hazīr al-Dunyā wa-l-Dīn Dāʾūd ibn khalīfa amīr al-Muʾminīn Yūsuf ibn ʿUmar ibn ʿAlī ibn Rasūl khullad Allāh mulkahu* ('For the Sultan, son of the Sultan, son of the Sultan, the King al-Muʾayyad Hazīr al-Dunyā wa-l-Dīn Dāʾūd, son of the caliph, the commander of the believers Yūsuf ibn ʿUmar ibn Rasūl, may God make his kingdom eternal'). This is the Rasulid Sultan al-Malik al-Muʾayyad Dāwūd ibn Yūsuf I, who ruled from 1296 until 1321 CE.[14] After the dedication come tables on astronomical–astrological subjects such as the lunar mansions, etc.

(3) fols 69r–80v: An astronomical yearbook with an ephemeris for the Hijra year 727 (26 November 1326–14 November 1327). Since this year is five years later than the death of al-Muʾayyad Dāʾūd ibn Yūsuf it seems to be independent from the previous section. Folio 69r contains four horoscope diagrams with introductory texts. The text for the top left horoscope for Sunday, 24 Rajab 727 AH/14 June 1327 CE, the day of the summer solstice, closes with the sentence *al-falak al-aʿẓam ʿalā mā huwa muthbat fī hādhihi l-zāʾirja maʿmūl dhālika bi-l-zīj al-ḥākimī wa-bi-llāhi l-tawfīq* ('The largest sphere concerning what is registered in this horoscope diagram, this is done by means of the Ḥākimite Tables, may God be reconciled'). The ephemeris begins on fol. 59v and ends on

[11] King, *Fihris al-makhṭūṭāt*, vol. I, p. 145; vol. II, p. 148. I thank Aymon Kreil for conveying digital images of a black-and-white microfilm of the manuscript, and Flora Vafea for obtaining the colour images here reproduced as Plates 14 and 15 with kind permission of the directorate of the Egyptian National Library.

[12] For the life and works of al-Fārisī, see Schmidl, *Volkstümliche Astronomie im islamischen Mittelalter*, vol. I, pp. 18–23; Rosenfeld and İhsanoğlu, *Mathematicians, Astronomers, and other Scholars*, pp. 219–20 (no. 608).

[13] King, *Fihris al-makhṭūṭāt*, vol. II, p. 112.

[14] Bostworth, *The New Islamic Dynasties*, p. 108.

fol. 80v. Every double page contains the data for one month. The calendarium with the days of the weeks and the dates in four calendars, together with the solar, lunar and planetary positions, are found on the right-hand pages. The right margin contains further calendrical data and data concerning the lunar mansions. The top header contains the positions of the ascendant and mid-heaven at the time of the opposition (*istiqbāl*) of sun and moon. The footer contains the corresponding data for the conjunction (*ijtimāʿ*). The prognoses based on the aspects of the moon, the *ikhtiyārāt*, are found on the left-side pages. For the last month, the page with the *ikhtiyārāt* is missing. The central fields contain information on the entry of the moon into the zodiacal signs and its aspects with the sun and the planets. To the left, aspects of the planets and the sun with themselves (*muzājāt*) are found. To the right is a reduced calendarium with the number of days in the Arabic month and the days of the week written in words.

(4) fols 81r–84v: Drawings and horoscopes.

(5) fol 85v: *Manẓūma fī bayān al-dukhūl bi-l-zawja ʿalā ḥasab al-manzila fīhā l-qamar* ('Poem on the explanation of having intercourse with the spouse, depending on the mansion the moon is in').[15] The text was added much later (*c.* 1100 AH) on the originally empty page.

A critical edition of the entire yearbook in part (3) would go far beyond the frame of the present study. If we consider the difficulties caused by the many worn parts of the manuscript and the often cursive and careless script without dots, the effort of producing such an edition would be considerable and cannot be promised in the near future. However, in Appendix B, the numerical material of the planetary positions is presented in a standardised format in the conventional notation of sexagesimal numbers.

[15] King, *Fihris al-makhṭūṭāt*, vol. II, p. 802.

Appendix B: Edition of the ephemeris for 727 Hijra

The following table contains only the numerical material of the ephemeris parts. The astrological parts, which follow on the pages to the left of the ephemeris parts, are not included, nor is the calendrical information in the right margins of the ephemeris parts. The following abbreviations are used in the header of the tables:

wd Day of the week

ar Date in the Arabic calendar

sy Date in the Syriac calendar

co Date in the Coptic calendar

ya Date in the Persian Yazdgird calendar

so Solar position

lu Lunar position

sa Position of Saturn

ju Position of Jupiter

ma Position of Mars

ve Position of Venus

me Position of Mercury

no Position of the ascending lunar node

dl Length of the day from sunrise to sunset

mh Maximal height of the sun at noon

In one case an entry has been corrected: The solar position for Jumādā 15 is given in the MS as $0^s24;42$, but it should be $0^s24;43$ in order to fit in smoothly between the adjacent values.

Muḥarram 727 AH

wd	*ar*	*sy*	*co*	*ya*	*so*	*lu*	*sa*	*ju*	*ma*	*ve*	*me*	*no*	*dl*	*mh*
4	1	26	30	2	8^s12; 8	8^s21;45	3s11;48	4s10;49	0^s20;48	7^s 2;33	8^s28; 7	6^s 3;14	11;10	52;52
5	2	27	1	3	8^s13; 9	9^s 3;43	3^s11;44	4^s10;49	0^s21; 0	7^s 3;45	8^s28;17	6^s 3;11	11; 9	52;45
6	3	28	2	4	8^s14;10	9s15;40	3^s11;40	4^s10;48	0^s21;12	7^s 4;57	8^s28;21	6^s 3; 8	11; 9	52;36
7	4	29	3	5	8^s15;12	9^s27;43	3^s11;36	4^s10;47	0^s21;24	7^s 6; 9	8^s28;18	6^s 3; 4	11; 9	52;28
1	5	30	4	6	8^s16;13	10^s10; 0	3^s11;32	4^s10;47	0^s21;35	7^s 7;21	8^s28;10	6^s 3; 1	11; 9	52;21
2	6	1	5	7	8^s17;14	10^s22;40	3^s11;28	4^s10;46	0^s21;47	7^s 8;34	8^s28; 0	6^s 2;58	11; 8	52;15
3	7	2	6	8	8^s18;16	11^s 5;44	3^s11;24	4^s10;45	0^s21;59	7^s 9;46	8^s27;46	6^s 2;54	11; 8	52; 8
4	8	3	7	9	8^s19;17	11^s19; 5	3^s11;19	4^s10;43	0^s22;12	7^s10;58	8^s27;28	6^s 2;51	11; 8	52; 2
5	9	4	8	10	8^s20;18	0^s 2;53	3^s11;15	4^s10;41	0^s22;26	7^s12;11	8^s27; 8	6^s 2;48	11; 8	51;57
6	10	5	9	11	8^s21;19	0^s17;15	3^s11;10	4^s10;38	0^s22;40	7^s13;24	8^s26;48	6^s 2;45	11; 8	51;53
7	11	6	10	12	8^s22;20	1^s 2; 6	3^s11; 5	4^s10;35	0^s22;55	7^s14;36	8^s26;12	6^s 2;42	11; 7	51;49
1	12	7	11	13	8^s23;22	1^s17; 6	3^s11; 0	4^s10;32	0^s23;10	7^s15;48	8^s25;37	6^s 2;39	11; 7	51;45
2	13	8	12	14	8^s24;23	2^s 2; 2	3^s10;55	4^s10;28	0^s23;26	7^s17; 1	8^s24;59	6^s 2;35	11; 7	51;42
3	14	9	13	15	8^s25;24	2^s16;44	3^s10;50	4^s10;24	0^s23;42	7^s18;14	8^s24; 7	6^s 2;32	11; 7	51;39
4	15	10	14	16	8^s26;25	3^s 1;15	3^s10;45	4^s10;20	0^s23;59	7^s19;27	8^s23;15	6^s 2;29	11; 7	51;36
5	16	11	15	17	8^s27;27	3^s15;43	3^s10;40	4^s10;16	0^s24;16	7^s20;40	8^s22;25	6^s 2;26	11; 7	51;34
6	17	12	16	18	8^s28;28	3^s29;50	3^s10;35	4^s10;11	0^s24;34	7^s21;54	8^s21;35	6^s 2;23	11; 7	51;33
7	18	13	17	19	8^s29;29	4^s13;50	3^s10;30	4^s10; 6	0^s24;53	7^s23; 7	8^s20;46	6^s 2;20	11; 7	51;32
1	19	14	18	20	9^s 0;30	4^s27;27	3^s10;24	4^s10; 1	0^s25;13	7^s24;20	8^s19;52	6^s 2;16	11; 7	51;31
2	20	15	19	21	9^s 1;32	5^s10;43	3^s10;19	4^s 9;56	0^s25;33	7^s25;33	8^s19; 2	6^s 2;13	11; 7	51;32
3	21	16	20	22	9^s 2;33	5^s23;43	3^s10;14	4^s 9;51	0^s25;53	7^s26;47	8^s18;25	6^s 2;10	11; 7	51;33
4	22	17	21	23	9^s 3;34	6^s 6;41	3^s10; 9	4^s 9;46	0^s26;14	7^s28; 1	8^s17;53	6^s 2; 7	11; 7	51;35
5	23	18	22	24	9^s 4;35	6^s19;23	3^s10; 3	4^s 9;41	0^s26;34	7^s29;14	8^s17;27	6^s 2; 4	11; 7	51;38
6	24	19	23	25	9^s 5;36	7^s 1;42	3^s 9;58	4^s 9;35	0^s26;55	8^s 0;28	8^s17; 3	6^s 2; 1	11; 7	51;41
7	25	20	24	26	9^s 6;37	7^s13;32	3^s 9;53	4^s 9;31	0^s27;16	8^s 1;42	8^s16;46	6^s 1;57	11; 7	51;45
1	26	21	25	27	9^s 7;39	7^s 25;7	3^s 9;47	4^s 9;26	0^s27;38	8^s 2;56	8^s16;30	6^s 1;54	11; 7	51;49
2	27	22	26	28	9^s 8;40	8^s 6;49	3^s 9;42	4^s 9;21	0^s28; 1	8^s 4;10	8^s16;22	6^s 1;51	11; 8	51;54
3	28	23	27	29	9^s 9;41	8^s18;41	3^s 9;36	4^s 9;16	0^s28;23	8^s 5;24	8^s16;26	6^s 1;48	11; 8	51;58
4	29	24	28	30	9^s10;42	9^s 0;41	3^s 9;30	4^s 9;11	0^s28;45	8^s 6;38	8^s16;36	6^s 1;45	11; 8	52; 3
5	30	25	29	1	9^s11;43	9^s13; 1	3^s 9;25	4^s 9; 5	0^s29; 7	8^s 7;52	8^s16;51	6^s 1;42	11; 8	52; 8

Ṣafar 727 AH

wd	*ar*	*sy*	*co*	*ya*	*so*	*lu*	*sa*	*ju*	*ma*	*ve*	*me*	*no*	*dl*	*mh*
6	1	26	30	2	9^{s}12;44	9^{s}25; 4	3^s 9;19	4^s 9; 0	0^{s}29;29	8^s 9; 7	8^{s}17;14	6^s 1;38	11; 8	52;14
7	2	27	1	3	9^{s}13;46	10^s 7;28	3^s 9;14	4^s 8;54	0^{s}29;52	8^{s}10;21	8^{s}17;42	6^s 1;35	11; 8	52;20
1	3	28	2	4	9^{s}14;47	10^{s}19;52	3^s 9; 9	4^s 8;48	1^s 0;15	8^{s}11;36	8^{s}18;15	6^s 1;32	11; 9	52;28
2	4	29	3	5	9^{s}15;48	11^s 2;40	3^s 9; 4	4^s 8;42	1^s 0;39	8^{s}12;50	8^{s}18;54	6^s 1;29	11; 9	52;36
3	5	30	4	6	9^{s}16;49	11^{s}15;52	3^s 8;59	4^s 8;35	1^s 1; 3	8^{s}14; 5	8^{s}19;40	6^s 1;26	11; 9	52;44
4	6	31	5	7	9^{s}17;50	11^{s}29;16	3^s 8;54	4^s 8;29	1^s 1;27	8^{s}15;19	8^{s}20;30	6^s 1;22	11;10	52;52
5	7	1	6	8	9^{s}18;52	0^{s}12;48	3^s 8;49	4^s 8;22	1^s 1;52	8^{s}16;33	8^{s}21;22	6^s 1;19	11;10	53; 1
6	8	2	7	9	9^{s}19;53	0^{s}26;38	3^s 8;44	4^s 8;15	1^s 2;16	8^{s}17;48	8^{s}22;18	6^s 1;16	11;10	53;11
7	9	3	8	10	9^{s}20;54	1^{s}11; 5	3^s 8;39	4^s 8; 8	1^s 2;41	8^{s}19; 2	8^{s}23;20	6^s 1;13	11;11	53;20
1	10	4	9	11	9^{s}21;55	1^{s}25;59	3^s 8;35	4^s 8; 1	1^s 3; 6	8^{s}20;16	8^{s}24;26	6^s 1;10	11;11	53;30
2	11	5	10	12	9^{s}22;56	2^{s}11; 3	3^s 8;30	4^s 7;54	1^s 3;32	8^{s}21;30	8^{s}25;38	6^s 1; 7	11;11	53;41
3	12	6	11	13	9^{s}23;58	2^{s}25;51	3^s 8;25	4^s 7;47	1^s 3;59	8^{s}22;44	8^{s}26;53	6^s 1; 3	11;12	53;53
4	13	7	12	14	9^{s}24;59	3^{s}10;20	3^s 8;20	4^s 7;40	1^s 4;26	8^{s}23;58	8^{s}28;10	6^s 1; 0	11;12	54; 5
5	14	8	13	15	9^{s}26; 0	3^{s}24;28	3^s 8;15	4^s 7;33	1^s 4;52	8^{s}25;13	8^{s}29;30	6^s 0;57	11;13	54;16
6	15	9	14	16	9^{s}27; 1	4^s 8;24	3^s 8;10	4^s 7;25	1^s 5;19	8^{s}26;27	9^s 0;52	6^s 0;54	11;13	54;28
7	16	10	15	17	9^{s}28; 2	4^{s}22; 4	3^s 8; 5	4^s 7;18	1^s 5;47	8^{s}27;41	9^s 2;17	6^s 0;51	11;14	54;41
1	17	11	16	18	9^{s}29; 3	5^s 5;30	3^s 8; 1	4^s 7;10	1^s 6;14	8^{s}28;56	9^s 3;45	6^s 0;48	11;14	54;53
2	18	12	17	19	10^s 0; 4	5^{s}18;37	3^s 7;56	4^s 7; 3	1^s 6;42	9^s 0;10	9^s 5;16	6^s 0;44	11;15	55; 1
3	19	13	18	20	10^s 1; 5	6^s 1;26	3^s 7;52	4^s 6;55	1^s 7; 9	9^s 1;25	9^s 6;49	6^s 0;41	11;15	55;10
4	20	14	19	21	10^s 2; 6	6^{s}14; 3	3^s 7;48	4^s 6;48	1^s 7;37	9^s 2;39	9^s 8;25	6^s 0;38	11;16	55;22
5	21	15	20	22	10^s 3; 7	6^{s}26;40	3^s 7;44	4^s 6;40	1^s 8; 5	9^s 3;54	9^{s}10; 2	6^s 0;35	11;17	55;35
6	22	16	21	23	10^s 4; 8	7^s 9;21	3^s 7;40	4^s 6;32	1^s 8;34	9^s 5; 9	9^{s}11;40	6^s 0;32	11;17	55;46
7	23	17	22	24	10^s 5; 9	7^{s}21;48	3^s 7;36	4^s 6;25	1^s 9; 3	9^s 6;23	9^{s}13;21	6^s 0;28	11;18	56; 6
1	24	18	23	25	10^s 6;10	8^s 3;48	3^s 7;33	4^s 6;17	1^s 9;32	9^s 7;38	9^{s}15; 3	6^s 0;25	11;18	56;21
2	25	19	24	26	10^s 7;11	8^{s}15;34	3^s 7;29	4^s 6; 9	1^{s}10; 1	9^s 8;53	9^{s}16;45	6^s 0;22	11;19	56;38
3	26	20	25	27	10^s 8;11	8^{s}27;23	3^s 7;25	4^s 6; 1	1^{s}10;30	9^{s}10; 7	9^{s}18;31	6^s 0;19	11;19	56;54
4	27	21	26	28	10^s 9;12	9^s 9;25	3^s 7;21	4^s 5;54	1^{s}10;59	9^{s}11;20	9^{s}20;19	6^s 0;16	11;20	57; 9
5	28	22	27	29	10^{s}10;13	9^{s}21;41	3^s 7;17	4^s 5;46	1^{s}11;28	9^{s}12;37	9^{s}22; 7	6^s 0;13	11;21	57;26
6	29	23	28	30	10^{s}11;13	10^s 4; 9	3^s 7;13	4^s 5;39	1^{s}11;57	9^{s}13;52	9^{s}23;57	6^s 0; 9	11;22	57;42

Rabīʿ I 727 AH

wd	*ar*	*sy*	*co*	*ya*	*so*	*lu*	*sa*	*ju*	*ma*	*ve*	*me*	*no*	*dl*	*mh*
7	1	24	29	1	10^{s}12;14	10^{s}16;45	3^s 7;10	4^s 5;32	1^{s}12;27	9^{s}15; 7	9^{s}25;46	6^s 0; 6	11;22	58; 0
1	2	25	30	2	10^{s}13;15	10^{s}29;34	3^s 7; 6	4^s 5;24	1^{s}12;56	9^{s}16;22	9^{s}27;36	6^s 0; 3	11;23	58;17
2	3	26	1	3	10^{s}14;16	11^{s}12;37	3^s 7; 3	4^s 5;17	1^{s}13;25	9^{s}17;36	9^{s}29;29	6^s 0; 0	11;23	58;25
3	4	27	2	4	10^{s}15;16	11^{s}25;50	3^s 7; 0	4^s 5; 9	1^{s}13;55	9^{s}18;51	10^s 1;25	5^{s}29;57	11;24	58;52
4	5	28	3	5	10^{s}16;17	0^s 9;49	3^s 6;57	4^s 5; 1	1^{s}14;24	9^{s}20; 6	10^s 3;21	5^{s}29;54	11;25	59;12
5	6	29	4	6	10^{s}17;17	0^{s}23;52	3^s 6;54	4^s 4;53	1^{s}14;54	9^{s}21;21	10^s 5;17	5^{s}29;51	11;25	59;31
6	7	30	5	7	10^{s}18;18	1^s 7;47	3^s 6;51	4^s 4;45	1^{s}15;24	9^{s}22;36	10^s 7;13	5^{s}29;47	11;26	59;49
7	8	31	6	8	10^{s}19;19	1^{s}21;37	3^s 6;48	4^s 4;37	1^{s}15;55	9^{s}23;50	10^s 9; 9	5^{s}29;44	11;27	60; 8
1	9	1	7	9	10^{s}20;19	2^s 5;48	3^s 6;45	4^s 4;30	1^{s}16;26	9^{s}25; 5	10^{s}11; 5	5^{s}29;41	11;28	60;28
2	10	2	8	10	10^{s}21;20	2^{s}20;33	3^s 6;42	4^s 4;22	1^{s}16;57	9^{s}26;20	10^{s}13; 0	5^{s}29;38	11;28	60;48
3	11	3	9	11	10^{s}22;20	3^s 5;20	3^s 6;39	4^s 4;15	1^{s}17;28	9^{s}27;35	10^{s}14;54	5^{s}29;35	11;29	61; 8
4	12	4	10	12	10^{s}23;21	3^{s}19;43	3^s 6;36	4^s 4; 8	1^{s}18; 0	9^{s}28;50	10^{s}16;48	5^{s}29;32	11;30	61;28
5	13	5	11	13	10^{s}24;21	4^s 3;43	3^s 6;33	4^s 4; 1	1^{s}18;31	10^s 0; 5	10^{s}18;41	5^{s}29;28	11;30	61;48
6	14	6	12	14	10^{s}25;22	4^{s}17;33	3^s 6;31	4^s 3;53	1^{s}19; 2	10^s 1;19	10^{s}20;35	5^{s}29;25	11;31	62; 8
7	15	7	13	15	10^{s}26;22	5^s 0;46	3^s 6;28	4^s 3;46	1^{s}19;33	10^s 2;34	10^{s}22;30	5^{s}29;22	11;32	62;28
1	16	8	14	16	10^{s}27;22	5^{s}13;57	3^s 6;26	4^s 3;38	1^{s}20; 4	10^s 3;49	10^{s}24;22	5^{s}29;19	11;33	62;49
2	17	9	15	17	10^{s}28;23	5^{s}26;50	3^s 6;24	4^s 3;31	1^{s}20;36	10^s 5; 4	10^{s}26;12	5^{s}29;16	11;33	63;10
3	18	10	16	18	10^{s}29;23	6^s 9;31	3^s 6;22	4^s 3;24	1^{s}21; 7	10^s 6;18	10^{s}27;59	5^{s}29;13	11;34	63;31
4	19	11	17	19	11^s 0;23	6^{s}21;52	3^s 6;20	4^s 3;18	1^{s}21;39	10^s 7;33	10^{s}29;40	5^{s}29; 9	11;35	63;52
5	20	12	18	20	11^s 1;23	7^s 4; 0	3^s 6;18	4^s 3;12	1^{s}22;11	10^s 8;48	11^s 1;25	5^{s}29; 6	11;36	64;13
6	21	13	19	21	11^s 2;24	7^{s}16;13	3^s 6;16	4^s 3; 5	1^{s}22;44	10^{s}10; 3	11^s 3; 7	5^{s}29; 3	11;36	64;31
7	22	14	20	22	11^s 3;24	7^{s}28;43	3^s 6;14	4^s 2;59	1^{s}23;17	10^{s}11;17	11^s 4;48	5^{s}29; 0	11;37	64;50
1	23	15	21	23	11^s 4;24	8^{s}11;21	3^s 6;12	4^s 2;53	1^{s}23;49	10^{s}12;31	11^s 6;27	5^{s}28;57	11;38	65;12
2	24	16	22	24	11^s 5;24	8^{s}23;42	3^s 6;11	4^s 2;47	1^{s}24;22	10^{s}13;46	11^s 8; 9	5^{s}28;54	11;39	65;34
3	25	17	23	25	11^s 6;24	9^s 5;46	3^s 6; 9	4^s 2;41	1^{s}24;55	10^{s}15; 2	11^s 9;50	5^{s}28;50	11;40	65;56
4	26	18	24	26	11^s 7;24	9^{s}17;51	3^s 6; 7	4^s 2;35	1^{s}25;27	10^{s}16;17	11^{s}11;30	5^{s}28;47	11;40	66;20
5	27	19	25	27	11^s 8;24	10^s 0;10	3^s 6; 6	4^s 2;30	1^{s}25;59	10^{s}17;31	11^{s}13;10	5^{s}28;44	11;41	66;44
6	28	20	26	28	11^s 9;24	10^{s}12;51	3^s 6; 5	4^s 2;25	1^{s}26;31	10^{s}18;46	11^{s}14;49	5^{s}28;41	11;42	67; 9
7	29	21	27	29	11^{s}10;23	10^{s}25;44	3^s 6; 4	4^s 2;20	1^{s}27; 3	10^{s}20; 1	11^{s}16;28	5^{s}28;38	11;43	67;32
1	30	22	28	30	11^{s}11;23	11^s 8;50	3^s 6; 4	4^s 2;15	1^{s}27;35	10^{s}21;16	11^{s}18; 4	5^{s}28;34	11;43	67;55

Rabīʿ II 727 AH

wd	*ar*	*sy*	*co*	*ya*	*so*	*lu*	*sa*	*ju*	*ma*	*ve*	*me*	*no*	*dl*	*mh*
2	1	23	29	1	11^{s}12;23	11^{s}22; 9	3^s 6; 3	4^s 2;10	1^{s}28; 8	10^{s}22;31	11^{s}19;44	5^{s}28;31	11;44	68;27
3	2	24	30	2	11^{s}13;23	0^s 5;43	3^s 6; 3	4^s 2; 5	1^{s}28;40	10^{s}23;45	11^{s}21;20	5^{s}28;28	11;45	68;51
4	3	25	1	3	11^{s}14;22	0^{s}19;45	3^s 6; 3	4^s 2; 1	1^{s}29;13	10^{s}25; 0	11^{s}22;59	5^{s}28;25	11;46	69;14
5	4	26	2	4	11^{s}15;22	1^s 4;10	3^s 6; 3	4^s 1;56	1^{s}29;46	10^{s}26;15	11^{s}24;35	5^{s}28;21	11;47	69;37
6	5	27	3	5	11^{s}16;21	1^{s}18;50	3^s 6; 3	4^s 1;52	2^s 0;19	10^{s}27;30	11^{s}26;12	5^{s}28;18	11;48	70; 1
7	6	28	4	6	11^{s}17;21	2^s 3;20	3^s 6; 3	4^s 1;48	2^s 0;52	10^{s}28;44	11^{s}27;47	5^{s}28;15	11;49	70;24
1	7	1	5	7	11^{s}18;21	2^{s}17;20	3^s 6; 4	4^s 1;43	2^s 1;25	10^{s}29;59	11^{s}29;20	5^{s}28;12	11;49	70;47
2	8	2	6	8	11^{s}19;20	3^s 1;12	3^s 6; 4	4^s 1;39	2^s 1;58	11^s 1;13	0^s 0;52	5^{s}28; 9	11;50	71;11
3	9	3	7	9	11^{s}20;20	3^{s}15;20	3^s 6; 4	4^s 1;35	2^s 2;32	11^s 2;28	0^s 2;22	5^{s}28; 6	11;51	71;35
4	10	4	8	10	11^{s}21;19	3^{s}29;33	3^s 6; 5	4^s 1;32	2^s 3; 6	11^s 3;42	0^s 3;50	5^{s}28; 2	11;52	71;59
5	11	5	9	11	11^{s}22;19	4^{s}13;19	3^s 6; 6	4^s 1;29	2^s 3;40	11^s 4;56	0^s 5;17	5^{s}27;59	11;53	72;23
6	12	6	10	12	11^{s}23;18	4^{s}26;46	3^s 6; 6	4^s 1;26	2^s 4;14	11^s 6;11	0^s 6;40	5^{s}27;56	11;54	72;47
7	13	7	11	13	11^{s}24;18	5^s 9;54	3^s 6; 7	4^s 1;24	2^s 4;48	11^s 7;25	0^s 8; 0	5^{s}27;53	11;55	73;11
1	14	8	12	14	11^{s}25;17	5^{s}22;46	3^s 6; 8	4^s 1;21	2^s 5;22	11^s 8;40	0^s 9;21	5^{s}27;50	11;56	73;36
2	15	9	13	15	11^{s}26;17	6^s 5;27	3^s 6; 9	4^s 1;19	2^s 5;55	11^s 9;54	0^{s}10;43	5^{s}27;46	11;56	73;59
3	16	10	14	16	11^{s}27;16	6^{s}17;57	3^s 6;11	4^s 1;16	2^s 6;29	11^{s}11; 8	0^{s}12; 0	5^{s}27;43	11;57	74;23
4	17	11	15	17	11^{s}28;15	7^s 0;14	3^s 6;12	4^s 1;14	2^s 7; 3	11^{s}12;23	0^{s}13;12	5^{s}27;40	11;58	74;47
5	18	12	16	18	11^{s}29;15	7^{s}12;15	3^s 6;13	4^s 1;12	2^s 7;37	11^{s}13;38	0^{s}14;22	5^{s}27;37	11;59	75;11
6	19	13	17	19	0^s 0;14	7^{s}24; 5	3^s 6;15	4^s 1;10	2^s 8;10	11^{s}14;52	0^{s}15;29	5^{s}27;34	12; 0	75;35
7	20	14	18	20	0^s 1;13	8^s 5;58	3^s 6;16	4^s 1; 8	2^s 8;44	11^{s}16; 6	0^{s}16;28	5^{s}27;31	12; 1	75;59
1	21	15	19	21	0^s 2;12	8^{s}18;10	3^s 6;18	4^s 1; 6	2^s 9;18	11^{s}17;21	0^{s}17;24	5^{s}27;27	12; 2	76;24
2	22	16	20	22	0^s 3;11	9^s 0;45	3^s 6;20	4^s 1; 4	2^s 9;52	11^{s}18;35	0^{s}18;19	5^{s}27;24	12; 3	76;48
3	23	17	21	23	0^s 4;10	9^{s}13;24	3^s 6;22	4^s 1; 3	2^{s}10;26	11^{s}19;50	0^{s}19;12	5^{s}27;21	12; 4	77;13
4	24	18	22	24	0^s 5; 9	9^{s}25;52	3^s 6;24	4^s 1; 2	2^{s}11; 0	11^{s}21; 4	0^{s}20; 6	5^{s}27;18	12; 5	77;37
5	25	19	23	25	0^s 6; 8	10^s 8;23	3^s 6;27	4^s 1; 1	2^{s}11;34	11^{s}22;18	0^{s}21; 1	5^{s}27;15	12; 5	78; 0
6	26	20	24	26	0^s 7; 7	10^{s}21; 9	3^s 6;29	4^s 1; 1	2^{s}12; 8	11^{s}23;33	0^{s}21;40	5^{s}27;11	12; 6	78;24
7	27	21	25	27	0^s 8; 6	11^s 4;19	3^s 6;31	4^s 1; 2	2^{s}12;43	11^{s}24;48	0^{s}22;16	5^{s}27; 8	12; 7	78;47
1	28	22	26	28	0^s 9; 5	11^{s}17;39	3^s 6;34	4^s 1; 2	2^{s}13;17	11^{s}26; 2	0^{s}22;50	5^{s}27; 5	12; 8	79;11
2	29	23	27	29	0^{s}10; 4	0^s 1;16	3^s 6;36	4^s 1; 3	2^{s}13;51	11^{s}27;16	0^{s}23;20	5^{s}27; 2	12; 9	79;35

Jumādā I 727 AH

wd	*ar*	*sy*	*co*	*ya*	*so*	*lu*	*sa*	*ju*	*ma*	*ve*	*me*	*no*	*dl*	*mh*
3	1	24	28	30	0^{s}11; 2	0^s 15;5	3^s 6;39	4^s 1; 3	2^{s}14;26	11^{s}28;31	0^{s}23;38	5^{s}26;59	12; 9	79;56
4	2	25	29	1	0^{s}12; 1	0^{s}29;15	3^s 6;41	4^s 1; 4	2^{s}15; 1	11^{s}29;45	0^{s}24;16	5^{s}26;56	12;10	80;20
5	3	26	30	2	0^{s}13; 0	1^{s}13;42	3^s 6;44	4^s 1; 4	2^{s}15;35	0^s 1; 0	0^{s}24;40	5^{s}26;53	12;10	80;44
6	4	27	1	3	0^{s}13;58	1^{s}28;34	3^s 6;47	4^s 1; 5	2^{s}16;10	0^s 2;14	0^{s}24;30	5^{s}26;49	12;11	81; 8
7	5	28	2	4	0^{s}14;57	2^{s}13;36	3^s 6;50	4^s 1; 6	2^{s}16;44	0^s 3;28	0^{s}24;18	5^{s}26;46	12;12	81;32
1	6	29	3	5	0^{s}15;56	2^{s}28;22	3^s 6;53	4^s 1; 7	2^{s}17;19	0^s 4;43	0^{s}24; 6	5^{s}26;43	12;13	81;55
2	7	30	4	6	0^{s}16;54	3^{s}12;33	3^s 6;56	4^s 1; 8	2^{s}17;53	0^s 5;57	0^{s}23;52	5^{s}26;40	12;13	82;17
3	8	31	5	7	0^{s}17;53	3^{s}26;44	3^s 6;59	4^s 1;10	2^{s}18;28	0^s 7;11	0^{s}23;36	5^{s}26;36	12;14	82;40
4	9	1	6	8	0^{s}18;51	4^s 9;53	3^s 7; 3	4^s 1;11	2^{s}19; 3	0^s 8;25	0^{s}23;10	5^{s}26;33	12;15	83; 2
5	10	2	7	9	0^{s}19;50	4^{s}23;25	3^s 7; 6	4^s 1;13	2^{s}19;37	0^s 9;39	0^{s}22;46	5^{s}26;30	12;16	83;25
6	11	3	8	10	0^{s}20;49	5^s 6;43	3^s 7; 9	4^s 1;15	2^{s}20;12	0^{s}10;53	0^{s}22;20	5^{s}26;27	12;17	83;48
7	12	4	9	11	0^{s}21;47	5^{s}19;44	3^s 7;13	4^s 1;18	2^{s}20;47	0^{s}12; 7	0^{s}21;52	5^{s}26;24	12;18	84;10
1	13	5	10	12	0^{s}22;46	6^s 2; 0	3^s 7;17	4^s 1;20	2^{s}21;23	0^{s}13;21	0^{s}21;19	5^{s}26;21	12;19	84;32
2	14	6	11	13	0^{s}23;45	6^{s}14;24	3^s 7;20	4^s 1;23	2^{s}21;58	0^{s}14;36	0^{s}20;49	5^{s}26;17	12;20	84;54
3	15	7	12	14	0^{s}24;43	6^{s}26;42	3^s 7;24	4^s 1;25	2^{s}22;33	0^{s}15;50	0^{s}20;17	5^{s}26;14	12;21	85;16
4	16	8	13	15	0^{s}25;41	7^s 8;44	3^s 7;28	4^s 1;28	2^{s}23; 7	0^{s}17; 4	0^{s}19;45	5^{s}26;11	12;22	85;38
5	17	9	14	16	0^{s}26;39	7^{s}20;54	3^s 7;32	4^s 1;30	2^{s}23;42	0^{s}18;18	0^{s}19;19	5^{s}26; 8	12;22	85;59
6	18	10	15	17	0^{s}27;37	8^s 2;44	3^s 7;37	4^s 1;33	2^{s}24;17	0^{s}19;32	0^{s}18;54	5^{s}26; 5	12;23	86;21
7	19	11	16	18	0^{s}28;35	8^{s}14;24	3^s 7;42	4^s 1;35	2^{s}24;52	0^{s}20;46	0^{s}18;34	5^{s}26; 2	12;23	86;42
1	20	12	17	19	0^{s}29;33	8^s 26;4	3^s 7;47	4^s 1;38	2^{s}25;28	0^{s}22; 0	0^{s}18;20	5^{s}25;58	12;24	87; 2
2	21	13	18	20	1^s 0;31	9^s 7;58	3^s 7;52	4^s 1;41	2^{s}26; 3	0^{s}23;13	0^{s}18;18	5^{s}25;55	12;25	87;23
3	22	14	19	21	1^s 1;29	9^{s}20;27	3^s 7;57	4^s 1;44	2^{s}26;39	0^{s}24;27	0^{s}18;12	5^{s}25;52	12;26	87;45
4	23	15	20	22	1^s 2;27	10^s 3;20	3^s 8; 2	4^s 1;47	2^{s}27;14	0^{s}25;41	0^{s}18; 6	5^{s}25;49	12;26	88; 6
5	24	16	21	23	1^s 3;25	10^{s}16;10	3^s 8; 7	4^s 1;51	2^{s}27;50	0^{s}26;54	0^{s}18; 0	5^{s}25;46	12;27	88;26
6	25	17	22	24	1^s 4;23	10^{s}29;10	3^s 8;12	4^s 1;55	2^{s}28;25	0^{s}28; 8	0^{s}17;55	5^{s}25;43	12;28	88;46
7	26	18	23	25	1^s 5;21	11^{s}12;24	3^s 8;17	4^s 1;59	2^{s}29; 1	0^{s}29;22	0^{s}17;50	5^{s}25;39	12;28	89; 7
1	27	19	24	26	1^s 6;19	11^{s}26; 1	3^s 8;22	4^s 2; 4	2^{s}29;36	1^s 0;36	0^{s}17;58	5^{s}25;36	12;29	89;27
2	28	20	25	27	1^s 7;17	0^s 9;57	3^s 8;27	4^s 2; 9	3^s 0;13	1^s 1;50	0^{s}18;23	5^{s}25;33	12;29	89;46
3	29	21	26	28	1^s 8;15	0^{s}24; 5	3^s 8;32	4^s 2;15	3^s 0;49	1^s 3; 4	0^{s}19; 0	5^{s}25;30	12;30	90;14
4	30	22	27	29	1^s 9;13	1^s 8;25	3^s 8;37	4^s 2;20	3^s 1;26	1^s 4;18	0^{s}19;41	5^{s}25;27	12;31	90;38

Jumādā II 727 AH

wd	*ar*	*sy*	*co*	*ya*	*so*	*lu*	*sa*	*ju*	*ma*	*ve*	*me*	*no*	*dl*	*mh*
5	1	23	28	30	1^{s}10;11	1^{s}22;54	3^s 8;42	4^s 2;26	3^s 2; 3	1^s 5;31	0^{s}20;29	5^{s}25;24	12;31	89;18
6	2	24	29	1	1^{s}11; 8	2^s 7;38	3^s 8;48	4^s 2;32	3^s 2;39	1^s 6;45	0^{s}21;14	5^{s}25;20	12;32	88;59
7	3	25	30	2	1^{s}12; 6	2^{s}22;37	3^s 8;54	4^s 2;39	3^s 3;16	1^s 7;59	0^{s}22; 4	5^{s}25;17	12;33	88;40
1	4	26	1	3	1^{s}13; 4	3^s 7;40	3^s 9; 0	4^s 2;45	3^s 3;52	1^s 9;12	0^{s}22;56	5^{s}25;14	12;34	88;22
2	5	27	2	4	1^{s}14; 2	3^{s}22;27	3^s 9; 6	4^s 2;52	3^s 4;28	1^{s}10;26	0^{s}23;50	5^{s}25;11	12;34	88; 4
3	6	28	3	5	1^{s}15; 0	4^s 6;16	3^s 9;13	4^s 2;58	3^s 5; 4	1^{s}11;40	0^{s}24;49	5^{s}25; 8	12;35	87;40
4	7	29	4	6	1^{s}15;57	4^{s}20;22	3^s 9;19	4^s 3; 5	3^s 5;40	1^{s}12;54	0^{s}25;53	5^{s}25; 4	12;35	87;31
5	8	30	5	7	1^{s}16;55	5^s 3;37	3^s 9;25	4^s 3;12	3^s 6;16	1^{s}14; 8	0^{s}27; 2	5^{s}25; 1	12;36	87;14
6	9	1	6	8	1^{s}17;53	5^{s}16;39	3^s 9;30	4^s 3;19	3^s 6;52	1^{s}15;22	0^{s}28;13	5^{s}24;58	12;36	86;57
7	10	2	7	9	1^{s}18;50	5^{s}29;22	3^s 9;36	4^s 3;26	3^s 7;28	1^{s}16;36	0^{s}29;26	5^{s}24;55	12;37	86;41
1	11	3	8	10	1^{s}19;48	6^{s}11;38	3^s 9;42	4^s 3;33	3^s 8; 4	1^{s}17;50	1^s 0;42	5^{s}24;52	12;38	86;26
2	12	4	9	11	1^{s}20;46	6^{s}23;46	3^s 9;48	4^s 3;40	3^s 8;40	1^{s}19; 4	1^s 2; 0	5^{s}24;49	12;38	86;10
3	13	5	10	12	1^{s}21;43	7^s 5;46	3^s 9;53	4^s 3;47	3^s 9;16	1^{s}20;18	1^s 3;21	5^{s}24;45	12;39	85;55
4	14	6	11	13	1^{s}22;41	7^{s}17;45	3^s 9;59	4^s 3;54	3^s 9;52	1^{s}21;32	1^s 4;46	5^{s}24;42	12;40	85;40
5	15	7	12	14	1^{s}23;38	7^{s}29;43	3^{s}10; 5	4^s 4; 1	3^{s}10;28	1^{s}22;46	1^s 6;13	5^{s}24;39	12;40	85;26
6	16	8	13	15	1^{s}24;36	8^{s}11;38	3^{s}10;11	4^s 4; 9	3^{s}11; 4	1^{s}24; 0	1^s 7;42	5^{s}24;36	12;41	85;11
7	17	9	14	16	1^{s}25;33	8^{s}23;24	3^{s}10;18	4^s 4;16	3^{s}11;40	1^{s}25;14	1^s 9;10	5^{s}24;33	12;42	84;57
1	18	10	15	17	1^{s}26;31	9^s 5; 2	3^{s}10;25	4^s 4;24	3^{s}12;16	1^{s}26;28	1^{s}10;40	5^{s}24;29	12;42	84;42
2	19	11	16	18	1^{s}27;28	9^{s}16;43	3^{s}10;31	4^s 4;32	3^{s}12;53	1^{s}27;41	1^{s}12;12	5^{s}24;26	12;43	84;28
3	20	12	17	19	1^{s}28;26	9^{s}28;40	3^{s}10;38	4^s 4;40	3^{s}13;29	1^{s}28;55	1^{s}13;46	5^{s}24;23	12;43	84;15
4	21	13	18	20	1^{s}29;23	10^{s}11; 4	3^{s}10;45	4^s 4;49	3^{s}14; 6	2^s 0; 8	1^{s}15;20	5^{s}24;20	12;44	84; 1
5	22	14	19	21	2^s 0;21	10^{s}23;55	3^{s}10;51	4^s 4;57	3^{s}14;42	2^s 1;22	1^{s}16;55	5^{s}24;17	12;44	83;48
6	23	15	20	22	2^s 1;18	11^s 7; 1	3^{s}10;58	4^s 5; 6	3^{s}15;18	2^s 2;35	1^{s}18;32	5^{s}24;14	12;45	83;36
7	24	16	21	23	2^s 2;15	11^{s}20;23	3^{s}11; 6	4^s 5;14	3^{s}15;55	2^s 3;49	1^{s}20;12	5^{s}24;10	12;45	83;25
1	25	17	22	24	2^s 3;13	0^s 4; 2	3^{s}11;13	4^s 5;23	3^{s}16;31	2^s 5; 2	1^{s}21;54	5^{s}24; 7	12;46	83;14
2	26	18	23	25	2^s 4;10	0^{s}18;13	3^{s}11;20	4^s 5;31	3^{s}17; 7	2^s 6;15	1^{s}23;35	5^{s}24; 4	12;46	83; 3
3	27	19	24	26	2^s 5; 8	1^s 2;43	3^{s}11;28	4^s 5;40	3^{s}17;43	2^s 7;28	1^{s}25;17	5^{s}24; 1	12;47	82;54
4	28	20	25	27	2^s 6; 5	1^{s}17;17	3^{s}11;35	4^s 5;49	3^{s}18;20	2^s 8;41	1^{s}27; 0	5^{s}23;58	12;47	82;44
5	29	21	26	28	2^s 7; 2	2^s 1;56	3^{s}11;43	4^s 5;57	3^{s}18;56	2^s 9;54	1^{s}28;44	5^{s}23;54	12;47	82;34

Rajab 727 AH

wd	*ar*	*sy*	*co*	*ya*	*so*	*lu*	*sa*	*ju*	*ma*	*ve*	*me*	*no*	*dl*	*mh*
6	1	22	27	29	2^s 7;59	2^{s}16;35	3^{s}11;51	4^s 6; 6	3^{s}19;33	2^{s}11; 6	2^s 0;30	5^{s}23;51	12;48	82;20
7	2	23	28	30	2^s 8;57	3^s 1;21	3^{s}11;58	4^s 6;16	3^{s}20;10	2^{s}12;19	2^s 2;15	5^{s}23;48	12;48	82;12
1	3	24	29	1	2^s 9;54	3^{s}16;11	3^{s}12; 6	4^s 6;25	3^{s}20;46	2^{s}13;33	2^s 4; 0	5^{s}23;45	12;48	82; 3
2	4	25	30	2	2^{s}10;51	4^s 0;55	3^{s}12;13	4^s 6;34	3^{s}21;23	2^{s}14;46	2^s 5;45	5^{s}23;42	12;49	81;55
3	5	26	1	3	2^{s}11;49	4^{s}15;28	3^{s}12;20	4^s 6;44	3^{s}22; 0	2^{s}16; 0	2^s 7;30	5^{s}23;39	12;49	81;48
4	6	27	2	4	2^{s}12;46	4^{s}29;45	3^{s}12;27	4^s 6;53	3^{s}22;36	2^{s}17;13	2^s 9;15	5^{s}23;35	12;49	81;41
5	7	28	3	5	2^{s}13;43	5^{s}12;55	3^{s}12;35	4^s 7; 3	3^{s}23;13	2^{s}18;26	2^{s}11; 0	5^{s}23;32	12;50	81;33
6	8	29	4	6	2^{s}14;40	5^{s}26; 8	3^{s}12;42	4^s 7;13	3^{s}23;50	2^{s}19;40	2^{s}12;48	5^{s}23;29	12;50	81;25
7	9	30	5	7	2^{s}15;38	6^s 8;54	3^{s}12;49	4^s 7;24	3^{s}24;27	2^{s}20;53	2^{s}14;36	5^{s}23;26	12;50	81;18
1	10	31	6	8	2^{s}16;35	6^{s}21;16	3^{s}12;57	4^s 7;35	3^{s}25; 4	2^{s}22; 6	2^{s}16;26	5^{s}23;23	12;51	81;12
2	11	1	7	9	2^{s}17;32	7^s 3;17	3^{s}13; 4	4^s 7;45	3^{s}25;41	2^{s}23;20	2^{s}18;17	5^{s}23;20	12;51	81; 6
3	12	2	8	10	2^{s}18;29	7^{s}15; 6	3^{s}13;11	4^s 7;56	3^{s}26;18	2^{s}24;33	2^{s}20; 8	5^{s}23;16	12;51	81; 1
4	13	3	9	11	2^{s}19;27	7^{s}26;49	3^{s}13;18	4^s 8; 7	3^{s}26;56	2^{s}25;46	2^{s}22; 0	5^{s}23;13	12;51	80;57
5	14	4	10	12	2^{s}20;24	8^s 8;39	3^{s}13;25	4^s 8;17	3^{s}27;33	2^{s}26;59	2^{s}23;50	5^{s}23;10	12;51	80;52
6	15	5	11	13	2^{s}21;21	8^{s}20;33	3^{s}13;32	4^s 8;28	3^{s}28;10	2^{s}28;12	2^{s}25;40	5^{s}23; 7	12;51	80;48
7	16	6	12	14	2^{s}22;18	9^s 2;26	3^{s}13;39	4^s 8;38	3^{s}28;47	2^{s}29;25	2^{s}27;31	5^{s}23; 4	12;52	80;45
1	17	7	13	15	2^{s}23;16	9^{s}14;20	3^{s}13;47	4^s 8;48	3^{s}29;24	3^s 0;38	2^{s}29;25	5^{s}23; 1	12;52	80;41
2	18	8	14	16	2^{s}24;13	9^{s}26; 6	3^{s}13;54	4^s 8;59	4^s 0; 1	3^s 1;52	3^s 1;20	5^{s}22;57	12;52	80;38
3	19	9	15	17	2^{s}25;10	10^s 8; 0	3^{s}14; 1	4^s 9;10	4^s 0;38	3^s 3; 5	3^s 3;14	5^{s}22;54	12;52	80;36
4	20	10	16	18	2^{s}26; 7	10^{s}20; 8	3^{s}14; 9	4^s 9;21	4^s 1;15	3^s 4;18	3^s 5; 9	5^{s}22;51	12;52	80;35
5	21	11	17	19	2^{s}27; 4	11^s 2;21	3^{s}14;17	4^s 9;32	4^s 1;52	3^s 5;31	3^s 7; 4	5^{s}22;48	12;52	80;33
6	22	12	18	20	2^{s}28; 1	11^{s}15;27	3^{s}14;25	4^s 9;44	4^s 2;29	3^s 6;45	3^s 8;58	5^{s}22;45	12;52	80;32
7	23	13	19	21	2^{s}28;59	11^{s}28;39	3^{s}14;33	4^s 9;55	4^s 3; 6	3^s 7;58	3^{s}10;52	5^{s}22;41	12;52	80;31
1	24	14	20	22	2^{s}29;56	0^{s}12;12	3^{s}14;41	4^{s}10; 7	4^s 3;43	3^s 9;11	3^{s}12;47	5^{s}22;38	12;52	80;30
2	25	15	21	23	3^s 0;53	0^{s}26;45	3^{s}14;49	4^{s}10;18	4^s 4;21	3^{s}10;24	3^{s}14;41	5^{s}22;35	12;52	80;31
3	26	16	22	24	3^s 1;50	1^{s}11; 2	3^{s}14;57	4^{s}10;30	4^s 4;58	3^{s}11;38	3^{s}16;33	5^{s}22;32	12;52	80;32
4	27	17	23	25	3^s 2;47	1^{s}25;54	3^{s}15; 6	4^{s}10;42	4^s 5;35	3^{s}12;51	3^{s}18;25	5^{s}22;29	12;52	80;33
5	28	18	24	26	3^s 3;44	2^{s}10;48	3^{s}15;14	4^{s}10;54	4^s 6;13	3^{s}14; 4	3^{s}20;15	5^{s}22;25	12;52	80;34
6	29	19	25	27	3^s 4;41	2^{s}25;30	3^{s}15;22	4^{s}11;6	4^s 6;50	3^{s}15;17	3^{s}22; 3	5^{s}22;22	12;52	80;36
7	30	20	26	28	3^s 5;38	3^{s}10;12	3^{s}15;31	4^{s}11;17	4^s 7;28	3^{s}16;30	3^{s}23;51	5^{s}22;19	12;52	80;39

Shaʿbān 727 AH

wd	*ar*	*sy*	*co*	*ya*	*so*	*lu*	*sa*	*ju*	*ma*	*ve*	*me*	*no*	*dl*	*mh*
1	1	21	27	29	3^s 6;35	3^{s}24;47	3^{s}15;39	4^{s}11;28	4^s 8; 5	3^{s}17;43	3^{s}25;40	5^{s}22;16	12;52	80;42
2	2	22	28	30	3^s 7;33	4^s 9;12	3^{s}15;47	4^{s}11;40	4^s 8;43	3^{s}18;57	3^{s}27;27	5^{s}22;13	12;52	80;45
3	3	23	29	1	3^s 8;30	4^{s}23;33	3^{s}15;55	4^{s}11;52	4^s 9;20	3^{s}20;10	3^{s}29;11	5^{s}22;10	12;52	80;49
4	4	24	30	2	3^s 9;27	5^s 7;32	3^{s}16; 3	4^{s}12; 4	4^s 9;58	3^{s}21;23	4^s 0;53	5^{s}22; 6	12;51	80;52
5	5	25	1	3	3^{s}10;24	5^{s}21; 3	3^{s}16;12	4^{s}12;16	4^{s}10;35	3^{s}22;37	4^s 2;34	5^{s}22; 3	12;51	80;56
6	6	26	2	4	3^{s}11;21	6^s 4; 4	3^{s}16;20	4^{s}12;28	4^{s}11;13	3^{s}23;50	4^s 4;14	5^{s}22; 0	12;51	81; 0
7	7	27	3	5	3^{s}12;19	6^{s}17;17	3^{s}16;28	4^{s}12;40	4^{s}11;50	3^{s}25; 3	4^s 5;55	5^{s}21;57	12;51	81; 5
1	8	28	4	6	3^{s}13;16	7^s 0;10	3^{s}16;37	4^{s}12;53	4^{s}12;28	3^{s}26;16	4^s 7;32	5^{s}21;54	12;51	81;11
2	9	29	5	7	3^{s}14;13	7^{s}12;28	3^{s}16;45	4^{s}13; 5	4^{s}13; 5	3^{s}27;30	4^s 9; 6	5^{s}21;51	12;50	81;17
3	10	30	6	8	3^{s}15;10	7^{s}24;25	3^{s}16;54	4^{s}13;17	4^{s}13;43	3^{s}28;43	4^{s}10;37	5^{s}21;47	12;50	81;23
4	11	1	7	9	3^{s}16; 7	8^s 6; 3	3^{s}17; 2	4^{s}13;29	4^{s}14;21	3^{s}29;56	4^{s}12; 5	5^{s}21;44	12;50	81;31
5	12	2	8	10	3^{s}17; 4	8^{s}17;41	3^{s}17;10	4^{s}13;42	4^{s}14;59	4^s 1; 9	4^{s}13;33	5^{s}21;41	12;50	81;38
6	13	3	9	11	3^{s}18; 1	8^{s}29;30	3^{s}17;18	4^{s}13;54	4^{s}15;37	4^s 2;22	4^{s}15; 0	5^{s}21;38	12;49	81;45
7	14	4	10	12	3^{s}18;58	9^{s}11;24	3^{s}17;26	4^{s}14; 7	4^{s}16;14	4^s 3;35	4^{s}16;22	5^{s}21;35	12;49	81;53
1	15	5	11	13	3^{s}19;56	9^{s}23;25	3^{s}17;34	4^{s}14;19	4^{s}16;52	4^s 4;48	4^{s}17;40	5^{s}21;32	12;48	82; 3
2	16	6	12	14	3^{s}20;53	10^s 5;29	3^{s}17;42	4^{s}14;32	4^{s}17;30	4^s 6; 1	4^{s}18;55	5^{s}21;28	12;48	82;11
3	17	7	13	15	3^{s}21;50	10^{s}17;36	3^{s}17;51	4^{s}14;44	4^{s}18; 8	4^s 7;14	4^{s}20; 8	5^{s}21;25	12;48	82;20
4	18	8	14	16	3^{s}22;47	10^{s}29;52	3^{s}17;59	4^{s}14;56	4^{s}18;46	4^s 8;27	4^{s}21;19	5^{s}21;22	12;47	82;30
5	19	9	15	17	3^{s}23;45	11^{s}12;16	3^{s}18; 7	4^{s}15; 9	4^{s}19;24	4^s 9;40	4^{s}22;27	5^{s}21;19	12;47	82;40
6	20	10	16	18	3^{s}24;42	11^{s}25; 8	3^{s}18;14	4^{s}15;21	4^{s}20; 2	4^{s}10;53	4^{s}23;19	5^{s}21;16	12;47	82;50
7	21	11	17	19	3^{s}25;39	0^s 8;19	3^{s}18;22	4^{s}15;34	4^{s}20;40	4^{s}12; 7	4^{s}24; 7	5^{s}21;12	12;46	83; 1
1	22	12	18	20	3^{s}26;37	0^{s}21;35	3^{s}18;29	4^{s}15;45	4^{s}21;18	4^{s}13;20	4^{s}25; 0	5^{s}21; 9	12;46	83;12
2	23	13	19	21	3^{s}27;34	1^s 5;24	3^{s}18;38	4^{s}15;57	4^{s}21;56	4^{s}14;33	4^{s}25;50	5^{s}21; 6	12;46	83;24
3	24	14	20	22	3^{s}28;31	1^{s}19;43	3^{s}18;46	4^{s}16;10	4^{s}22;34	4^{s}15;46	4^{s}26;35	5^{s}21; 3	12;45	83;36
4	25	15	21	23	3^{s}29;29	2^s 4;36	3^{s}18;54	4^{s}16;24	4^{s}23;12	4^{s}16;59	4^{s}27; 8	5^{s}21; 0	12;45	83;47
5	26	16	22	24	4^s 0;26	2^{s}19;38	3^{s}19; 2	4^{s}16;37	4^{s}23;50	4^{s}18;12	4^{s}27;33	5^{s}20;57	12;44	83;59
6	27	17	23	25	4^s 1;24	3^s 4;35	3^{s}19;10	4^{s}16;50	4^{s}24;28	4^{s}19;25	4^{s}27;53	5^{s}20;53	12;44	84;12
7	28	18	24	26	4^s 2;21	3^{s}19;15	3^{s}19;18	4^{s}17; 2	4^{s}25; 6	4^{s}20;38	4^{s}28; 5	5^{s}20;50	12;43	84;26
1	29	19	25	27	4^s 3;19	4^s 3;34	3^{s}19;27	4^{s}17;15	4^{s}25;44	4^{s}21;51	4^{s}28;12	5^{s}20;47	12;42	84;39

Ramaḍān 727 AH

wd	*ar*	*sy*	*co*	*ya*	*so*	*lu*	*sa*	*ju*	*ma*	*ve*	*me*	*no*	*dl*	*mh*
2	1	20	26	28	4^s 4;16	4^{s}17;50	3^{s}19;35	4^{s}17;28	4^{s}26;22	4^{s}23; 3	4^{s}28;16	5^{s}20;44	12;42	84;53
3	2	21	27	29	4^s 5;13	5^s 1;50	3^{s}19;43	4^{s}17;41	4^{s}27; 0	4^{s}24;16	4^{s}28;10	5^{s}20;41	12;41	85; 7
4	3	22	28	30	4^s 6;11	5^{s}15;34	3^{s}19;50	4^{s}17;54	4^{s}27;38	4^{s}25;29	4^{s}27;58	5^{s}20;37	12;41	85;22
5	4	23	29	1	4^s 7; 8	5^{s}29; 0	3^{s}19;58	4^{s}18; 7	4^{s}28;17	4^{s}26;42	4^{s}27;38	5^{s}20;34	12;40	85;37
6	5	24	30	2	4^s 8; 6	6^{s}12; 4	3^{s}20; 6	4^{s}18;20	4^{s}28;55	4^{s}27;56	4^{s}27; 3	5^{s}20;31	12;40	85;53
7	6	25	1	3	4^s 9; 3	6^{s}24;55	3^{s}20;13	4^{s}18;33	4^{s}29;33	4^{s}29; 9	4^{s}26;25	5^{s}20;28	12;39	86;6
1	7	26	2	4	4^{s}10; 1	7^s 7;40	3^{s}20;21	4^{s}18;46	5^s 0;12	5^s 0;22	4^{s}26;19	5^{s}20;25	12;38	86;23
2	8	27	3	5	4^{s}10;58	7^{s}20;29	3^{s}20;28	4^{s}19; 0	5^s 0;50	5^s 1;35	4^{s}26; 8	5^{s}20;22	12;38	86;37
3	9	28	4	6	4^{s}11;56	8^s 2;55	3^{s}20;36	4^{s}19;13	5^s 1;28	5^s 2;48	4^{s}24;23	5^{s}20;18	12;37	86;54
4	10	29	5	7	4^{s}12;53	8^{s}14;51	3^{s}20;43	4^{s}19;26	5^s 2; 6	5^s 4; 1	4^{s}23;30	5^{s}20;15	12;36	87;11
5	11	30	6	8	4^{s}13;51	8^{s}26;33	3^{s}20;50	4^{s}19;39	5^s 2;44	5^s 5;14	4^{s}22;28	5^{s}20;12	12;36	87;29
6	12	31	7	9	4^{s}14;49	9^s 8;17	3^{s}20;58	4^{s}19;52	5^s 3;22	5^s 6;27	4^{s}21;25	5^{s}20; 9	12;35	87;46
7	13	1	8	10	4^{s}15;46	9^{s}20;10	3^{s}21; 5	4^{s}20; 6	5^s 4; 1	5^s 7;40	4^{s}20;22	5^{s}20; 6	12;34	88; 4
1	14	2	9	11	4^{s}16;44	10^s 2;14	3^{s}21;12	4^{s}20;19	5^s 4;39	5^s 8;53	4^{s}19;22	5^{s}20; 3	12;34	88;21
2	15	3	10	12	4^{s}17;42	10^{s}14;31	3^{s}21;20	4^{s}20;33	5^s 5;17	5^{s}10; 6	4^{s}18;27	5^{s}19;59	12;33	88;39
3	16	4	11	13	4^{s}18;40	10^{s}26;54	3^{s}21;27	4^{s}20;46	5^s 5;56	5^{s}11;19	4^{s}17;37	5^{s}19;56	12;32	88;58
4	17	5	12	14	4^{s}19;38	11^s 9;29	3^{s}21;34	4^{s}21; 0	5^s 6;34	5^{s}12;32	4^{s}16;52	5^{s}19;53	12;31	89;15
5	18	6	13	15	4^{s}20;36	11^{s}22;10	3^{s}21;41	4^{s}21;13	5^s 7;12	5^{s}13;45	4^{s}16; 9	5^{s}19;50	12;31	89;33
6	19	7	14	16	4^{s}21;34	0^s 5;10	3^{s}21;48	4^{s}21;26	5^s 7;50	5^{s}14;58	4^{s}15;39	5^{s}19;47	12;30	89;50
7	20	8	15	17	4^{s}22;32	0^{s}18;37	3^{s}21;56	4^{s}21;39	5^s 8;29	5^{s}16;11	4^{s}15;14	5^{s}19;44	12;29	89;48
1	21	9	16	18	4^{s}23;30	1^s 2;20	3^{s}22; 3	4^{s}21;53	5^s 9; 7	5^{s}17;24	4^{s}14;54	5^{s}19;40	12;29	89;28
2	22	10	17	19	4^{s}24;28	1^s 16;8	3^{s}22;10	4^{s}22; 6	5^s 9;46	5^{s}18;37	4^{s}14;39	5^{s}19;37	12;28	89; 8
3	23	11	18	20	4^{s}25;26	1^{s}29;52	3^{s}22;17	4^{s}22;19	5^{s}10;25	5^{s}19;50	4^{s}14;29	5^{s}19;34	12;28	88;48
4	24	12	19	21	4^{s}26;24	2^s 14;4	3^{s}22;24	4^{s}22;32	5^{s}11; 4	5^{s}21; 3	4^{s}14;25	5^{s}19;31	12;27	88;27
5	25	13	20	22	4^{s}27;22	2^{s}28;50	3^{s}22;31	4^{s}22;45	5^{s}11;43	5^{s}22;16	4^{s}14;30	5^{s}19;28	12;26	88; 7
6	26	14	21	23	4^{s}28;20	3^{s}13;48	3^{s}22;37	4^{s}22;58	5^{s}12;22	5^{s}23;29	4^{s}14;43	5^{s}19;25	12;25	87;50
7	27	15	22	24	4^{s}29;18	3^{s}28;30	3^{s}22;44	4^{s}23;11	5^{s}13; 1	5^{s}24;42	4^{s}15; 3	5^{s}19;21	12;24	87;30
1	28	16	23	25	5^s 0;16	4^{s}12;47	3^{s}22;50	4^{s}23;24	5^{s}13;40	5^{s}25;55	4^{s}15;30	5^{s}19;18	12;23	87; 9
2	29	17	24	26	5^s 1;14	4^{s}26;49	3^{s}22;57	4^{s}23;37	5^{s}14;19	5^{s}27; 8	4^{s}16; 5	5^{s}19;15	12;22	86;46
3	30	18	25	27	5^s 2;12	5^{s}10;53	3^{s}23; 4	4^{s}23;50	5^{s}14;59	5^{s}28;21	4^{s}16;40	5^{s}19;12	12;21	86;21

Shawwāl 727 AH

wd	*ar*	*sy*	*co*	*ya*	*so*	*lu*	*sa*	*ju*	*ma*	*ve*	*me*	*no*	*dl*	*mh*
4	1	19	26	28	5^s 3;10	5^{s}23;57	3^{s}23;11	4^{s}24;3	5^{s}15;38	5^{s}29;34	4^{s}17;19	5^{s}19; 9	12;21	86; 4
5	2	20	27	29	5^s 4; 9	6^s 7; 4	3^{s}23;18	4^{s}24;16	5^{s}16;17	6^s 0;47	4^{s}18; 4	5^{s}19; 5	12;20	85;42
6	3	21	28	30	5^s 5; 7	6^{s}20; 4	3^{s}23;24	4^{s}24;29	5^{s}16;57	6^s 2; 0	4^{s}18;52	5^{s}19; 2	12;19	85;20
7	4	22	29	31	5^s 6; 5	7^s 2;43	3^{s}23;31	4^{s}24;42	5^{s}17;36	6^s 3;13	4^{s}19;47	5^{s}18;59	12;19	84;58
1	5	23	30	32	5^s 7; 3	7^s 15;5	3^{s}23;37	4^{s}24;55	5^{s}18;15	6^s 4;26	4^{s}20;46	5^{s}18;56	12;18	84;36
2	6	24	31	33	5^s 8; 1	7^{s}27;23	3^{s}23;43	4^{s}25; 8	5^{s}18;54	6^s 5;39	4^{s}21;50	5^{s}18;53	12;17	84;14
3	7	25	32	34	5^s 9; 0	8^{s}10; 0	3^{s}23;50	4^{s}25;21	5^{s}19;33	6^s 6;52	4^{s}22;56	5^{s}18;49	12;16	83;52
4	8	26	33	35	5^s 9;58	8^{s}22;36	3^{s}23;56	4^{s}25;34	5^{s}20;13	6^s 8; 5	4^{s}24; 6	5^{s}18;46	12;16	83;30
5	9	27	34	1	5^{s}10;57	9^s 4;53	3^{s}24; 2	4^{s}25;47	5^{s}20;52	6^s 9;18	4^{s}25;19	5^{s}18;43	12;15	83; 7
6	10	28	35	2	5^{s}11;55	9^{s}16;50	3^{s}24; 8	4^{s}26; 0	5^{s}21;31	6^{s}10;31	4^{s}26;33	5^{s}18;40	12;14	82;45
7	11	29	36	3	5^{s}12;54	9^{s}28;46	3^{s}24;14	4^{s}26;13	5^{s}22;10	6^{s}11;44	4^{s}27;50	5^{s}18;37	12;13	82;23
1	12	30	1	4	5^{s}13;52	10^{s}10;52	3^{s}24;19	4^{s}26;26	5^{s}22;49	6^{s}12;57	4^{s}29;11	5^{s}18;34	12;12	82; 1
2	13	31	2	5	5^{s}14;51	10^{s}23;17	3^{s}24;25	4^{s}26;40	5^{s}23;28	6^{s}14;10	5^s 0;37	5^{s}18;30	12;11	81;38
3	14	1	3	6	5^{s}15;49	11^s 5;53	3^{s}24;31	4^{s}26;53	5^{s}24; 7	6^{s}15;23	5^s 2; 2	5^{s}18;27	12;10	81;15
4	15	2	4	7	5^{s}16;48	11^{s}18;44	3^{s}24;37	4^{s}27; 6	5^{s}24;47	6^{s}16;36	5^s 3;30	5^{s}18;24	12;10	80;51
5	16	3	5	8	5^{s}17;47	0^s 1;45	3^{s}24;42	4^{s}27;18	5^{s}25;26	6^{s}17;49	5^s 4;58	5^{s}18;21	12; 9	80;28
6	17	4	6	9	5^{s}18;45	0^s 15;5	3^{s}24;48	4^{s}27;30	5^{s}26; 5	6^{s}19; 2	5^s 6;29	5^{s}18;18	12; 8	80; 4
7	18	5	7	10	5^{s}19;44	0^{s}28;39	3^{s}24;54	4^{s}27;42	5^{s}26;45	6^{s}20;15	5^s 8; 0	5^{s}18;15	12; 7	79;40
1	19	6	8	11	5^{s}20;43	1^{s}12;45	3^{s}24;59	4^{s}27;55	5^{s}27;24	6^{s}21;28	5^s 9;33	5^{s}18;11	12; 6	79;17
2	20	7	9	12	5^{s}21;42	1^s 27;6	3^{s}25; 4	4^{s}28; 7	5^{s}28; 4	6^{s}22;41	5^{s}11; 5	5^{s}18; 8	12; 6	78;55
3	21	8	10	13	5^{s}22;40	2^{s}11;21	3^{s}25; 9	4^{s}28;20	5^{s}28;44	6^{s}23;54	5^{s}12;38	5^{s}18; 5	12; 5	78;31
4	22	9	11	14	5^{s}23;39	2^{s}25;20	3^{s}25;14	4^{s}28;32	5^{s}29;24	6^{s}25; 7	5^{s}14;12	5^{s}18; 2	12; 5	78; 7
5	23	10	12	15	5^{s}24;38	3^s 9;17	3^{s}25;19	4^{s}28;44	6^s 0; 4	6^{s}26;20	5^{s}15;46	5^{s}17;59	12; 4	77;43
6	24	11	13	16	5^{s}25;37	3^{s}23;40	3^{s}25;24	4^{s}28;57	6^s 0;44	6^{s}27;33	5^{s}17;21	5^{s}17;56	12; 3	77;20
7	25	12	14	17	5^{s}26;36	4^s 8; 9	3^{s}25;29	4^{s}29; 9	6^s 1;24	6^{s}28;46	5^{s}18;57	5^{s}17;53	12; 2	76;56
1	26	13	15	18	5^{s}27;35	4^{s}22;22	3^{s}25;34	4^{s}29;22	6^s 2; 4	6^{s}29;59	5^{s}20;34	5^{s}17;49	12; 1	76;32
2	27	14	16	19	5^{s}28;34	5^s 6; 8	3^{s}25;39	4^{s}29;35	6^s 2;44	7^s 1;12	5^{s}22;12	5^{s}17;46	12; 1	76; 8
3	28	15	17	20	5^{s}29;33	5^{s}19;33	3^{s}25;44	4^{s}29;47	6^s 3;24	7^s 2;25	5^{s}23;52	5^{s}17;43	12; 0	75;41
4	29	16	18	21	6^s 0;32	6^s 2;42	3^{s}25;49	4^{s}29;59	6^s 4; 4	7^s 3;38	5^{s}25;36	5^{s}17;40	12; 0	75;18

Dhū l-Qaʿda 727 AH

wd	*ar*	*sy*	*co*	*ya*	*so*	*lu*	*sa*	*ju*	*ma*	*ve*	*me*	*no*	*dl*	*mh*
5	1	17	19	22	6^s 1;31	6^s15;42	3^s25;54	5^s 0;11	6^s 4;45	7^s 4;51	5^s27;15	5^s17;36	11;59	74;54
6	2	18	20	23	6^s 2;31	6^s28;26	3^s25;58	5^s 0;23	6^s 5;25	7^s 6; 4	5^s28;56	5^s17;33	11;58	74;30
7	3	19	21	24	6^s 3;30	7^s10;57	3^s26; 3	5^s 0;35	6^s 6; 5	7^s 7;17	6^s 0;38	5^s17;30	11;57	74; 5
1	4	20	22	25	6^s 4;29	7^s23; 9	3^s26; 7	5^s 0;46	6^s 6;45	7^s 8;30	6^s 2;20	5^s17;27	11;56	73;41
2	5	21	23	26	6^s 5;29	8^s 5; 7	3^s26;11	5^s 0;58	6^s 7;26	7^s 9;42	6^s 4; 3	5^s17;24	11;55	73;17
3	6	22	24	27	6^s 6;28	8^s17; 8	3^s26;15	5^s 1; 9	6^s 8; 6	7^s10;55	6^s 5;45	5^s17;20	11;54	72;53
4	7	23	25	28	6^s 7;27	8^s29;26	3^s26;18	5^s 1;20	6^s 8;47	7^s12; 8	6^s 7;28	5^s17;17	11;54	72;29
5	8	24	26	29	6^s 8;27	9^s12; 2	3^s26;22	5^s 1;32	6^s 9;27	7^s13;20	6^s 9;10	5^s17;14	11;53	72; 5
6	9	25	27	30	6^s 9;26	9^s24;36	3^s26;26	5^s 1;43	6^s10; 8	7^s14;33	6^s10;53	5^s17;11	11;53	71;41
7	10	26	28	1	6^s10;26	10^s 6;54	3^s26;29	5^s 1;54	6^s10;48	7^s15;45	6^s12;35	5^s17; 8	11;52	71;17
1	11	27	29	2	6^s11;25	10^s19; 7	3^s26;32	5^s 2; 6	6^s11;29	7^s16;58	6^s14;18	5^s17; 5	11;51	70;53
2	12	28	30	3	6^s12;25	11^s 1;40	3^s26;35	5^s 2;17	6^s12; 9	7^s18;10	6^s16; 0	5^s17; 1	11;50	70;30
3	13	29	1	4	6^s13;24	11^s14;31	3^s26;39	5^s 2;29	6^s12;49	7^s19;23	6^s17;43	5^s16;58	11;49	70; 7
4	14	30	2	5	6^s14;24	11^s27;33	3^s26;42	5^s 2;40	6^s13;30	7^s20;35	6^s19;25	5^s16;55	11;48	69;44
5	15	1	3	6	6^s15;23	0^s10;52	3^s26;45	5^s 2;51	6^s14;10	7^s21;48	6^s21; 6	5^s16;52	11;47	69;19
6	16	2	4	7	6^s16;23	0^s24;25	3^s26;48	5^s 3; 3	6^s14;51	7^s23; 1	6^s22;46	5^s16;49	11;46	68;55
7	17	3	5	8	6^s17;22	1^s 8;17	3^s26;52	5^s 3;15	6^s15;31	7^s24;14	6^s24;27	5^s16;46	11;45	68;31
1	18	4	6	9	6^s18;22	1^s22;28	3^s26;55	5^s 3;26	6^s16;12	7^s25;27	6^s26; 9	5^s16;43	11;44	68; 8
2	19	5	7	10	6^s19;22	2^s 7; 3	3^s26;58	5^s 3;37	6^s16;52	7^s26;40	6^s27;51	5^s16;39	11;43	67;46
3	20	6	8	11	6^s20;21	2^s21;57	3^s27; 1	5^s 3;48	6^s17;33	7^s27;52	6^s29;31	5^s16;36	11;43	67;23
4	21	7	9	12	6^s21;21	3^s 6;42	3^s27; 3	5^s 3;58	6^s18;13	7^s29; 4	7^s 1;10	5^s16;33	11;42	67; 1
5	22	8	10	13	6^s22;21	3^s20;54	3^s27; 6	5^s 4; 9	6^s18;54	8^s 0;16	7^s 2;47	5^s16;30	11;41	66;38
6	23	9	11	14	6^s23;21	4^s 4;41	3^s27; 9	5^s 4;19	6^s19;34	8^s 1;28	7^s 4;25	5^s16;26	11;40	66;16
7	24	10	12	15	6^s24;22	4^s18;32	3^s27;12	5^s 4;30	6^s20;15	8^s 2;40	7^s 6; 3	5^s16;23	11;40	65;53
1	25	11	13	16	6^s25;22	5^s 2;26	3^s27;14	5^s 4;40	6^s20;55	8^s 3;52	7^s 7;39	5^s16;20	11;39	65;31
2	26	12	14	17	6^s26;22	5^s16; 1	3^s27;16	5^s 4;50	6^s21;36	8^s 5; 4	7^s 9;14	5^s16;17	11;38	65; 9
3	27	13	15	18	6^s27;22	5^s29;12	3^s27;17	5^s 5; 0	6^s22;17	8^s 6;16	7^s10;50	5^s16;14	11;38	64;47
4	28	14	16	19	6^s28;22	6^s12; 1	3^s27;19	5^s 5;10	6^s22;57	8^s 7;28	7^s12;25	5^s16;11	11;37	64;25
5	29	15	17	20	6^s29;22	6^s24;41	3^s27;20	5^s 5;20	6^s23;38	8^s 8;39	7^s13;57	5^s16; 7	11;36	64; 3
6	30	16	18	21	7^s 0;22	7^s 7;10	3^s27;21	5^s 5;30	6^s24;19	8^s 9;51	7^s15;30	5^s16; 4	11;35	63;41

Dhū l-Ḥijja 727 AH

wd	*ar*	*sy*	*co*	*ya*	*so*	*lu*	*sa*	*ju*	*ma*	*ve*	*me*	*no*	*dl*	*mh*
7	1	17	19	22	7^s 1;22	7^{s}19;30	3^{s}27;22	5^s 5;40	6^{s}25; 0	8^{s}11; 2	7^{s}17; 2	5^{s}16; 1	11;34	63;18
1	2	18	20	23	7^s 2;22	8^s 1;40	3^{s}27;22	5^s 5;50	6^{s}25;40	8^{s}12;14	7^{s}18;32	5^{s}15;58	11;33	62;58
2	3	19	21	24	7^s 3;23	8^{s}13;37	3^{s}27;23	5^s 5;59	6^{s}26;21	8^{s}13;26	7^{s}20; 0	5^{s}15;55	11;32	62;37
3	4	20	22	25	7^s 4;23	8^{s}25;22	3^{s}27;23	5^s 6; 8	6^{s}27; 2	8^{s}14;37	7^{s}21;25	5^{s}15;52	11;32	62;16
4	5	21	23	26	7^s 5;23	9^s 7; 6	3^{s}27;23	5^s 6;17	6^{s}27;42	8^{s}15;49	7^{s}22;50	5^{s}15;48	11;31	61;56
5	6	22	24	27	7^s 6;23	9^{s}19; 3	3^{s}27;23	5^s 6;26	6^{s}28;23	8^{s}17; 0	7^{s}24;12	5^{s}15;45	11;30	61;37
6	7	23	25	28	7^s 7;24	10^s 1;10	3^{s}27;24	5^s 6;36	6^{s}29; 3	8^{s}18;11	7^{s}25;32	5^{s}15;42	11;29	61;16
7	8	24	26	29	7^s 8;24	10^{s}13;22	3^{s}27;25	5^s 6;45	6^{s}29;44	8^{s}19;23	7^{s}26;50	5^{s}15;39	11;29	60;56
1	9	25	27	30	7^s 9;25	10^{s}25;55	3^{s}27;26	5^s 6;54	7^s 0;25	8^{s}20;34	7^{s}28; 6	5^{s}15;36	11;28	60;34
2	10	26	28	1	7^{s}10;25	11^s 8;40	3^{s}27;26	5^s 7; 3	7^s 1; 6	8^{s}21;46	7^{s}29;20	5^{s}15;32	11;27	60;14
3	11	27	29	2	7^{s}11;26	11^{s}22;37	3^{s}27;26	5^s 7;12	7^s 1;48	8^{s}22;57	8^s 0;32	5^{s}15;29	11;27	59;54
4	12	28	30	3	7^{s}12;26	0^s 5;56	3^{s}27;26	5^s 7;20	7^s 2;29	8^{s}24; 8	8^s 1;43	5^{s}15;26	11;26	59;33
5	13	29	1	4	7^{s}13;27	0^{s}19;32	3^{s}27;25	5^s 7;29	7^s 3;11	8^{s}25;20	8^s 2;53	5^{s}15;23	11;25	59;13
6	14	30	2	5	7^{s}14;28	1^s 3;24	3^{s}27;25	5^s 7;37	7^s 3;53	8^{s}26;31	8^s 4; 0	5^{s}15;20	11;24	58;53
7	15	31	3	6	7^{s}15;29	1^{s}17;29	3^{s}27;24	5^s 7;46	7^s 4;34	8^{s}27;42	8^s 5; 3	5^{s}15;17	11;24	58;34
1	16	1	4	7	7^{s}16;29	2^s 1;44	3^{s}27;24	5^s 7;54	7^s 5;16	8^{s}28;53	8^s 6; 0	5^{s}15;13	11;23	58;17
2	17	2	5	8	7^{s}17;30	2^{s}16;29	3^{s}27;23	5^s 8; 2	7^s 5;58	8^s 0; 4	8^s 6;52	5^{s}15;10	11;22	58; 0
3	18	3	6	9	7^{s}18;31	3^s 1;15	3^{s}27;23	5^s 8;10	7^s 6;39	8^s 1;14	8^s 7;47	5^{s}15; 7	11;22	57;43
4	19	4	7	10	7^{s}19;31	3^{s}16;19	3^{s}27;22	5^s 8;18	7^s 7;21	8^s 2;25	8^s 8;24	5^{s}15; 4	11;21	57;28
5	20	5	8	11	7^{s}20;32	4^s 1;13	3^{s}27;21	5^s 8;25	7^s 8; 3	8^s 3;35	8^s 9; 4	5^{s}15; 1	11;21	57;14
6	21	6	9	12	7^{s}21;33	4^{s}15;35	3^{s}27;21	5^s 8;34	7^s 8;45	8^s 4;45	8^s 9;40	5^{s}14;58	11;20	56;58
7	22	7	10	13	7^{s}22;33	4^{s}29;24	3^{s}27;20	5^s 8;42	7^s 9;27	8^s 5;55	8^{s}10;12	5^{s}14;54	11;19	56;42
1	23	8	11	14	7^{s}23;34	5^{s}12;48	3^{s}27;19	5^s 8;49	7^{s}10; 9	8^s 7; 5	8^{s}10;40	5^{s}14;51	11;19	56;26
2	24	9	12	15	7^{s}24;35	5^{s}26; 7	3^{s}27;18	5^s 8;56	7^{s}10;51	8^s 8;15	8^{s}11;1	5^{s}14;48	11;18	56;12
3	25	10	13	16	7^{s}25;36	6^s 9; 9	3^{s}27;18	5^s 9; 3	7^{s}11;34	8^s 9;25	8^{s}11;18	5^{s}14;45	11;18	55;58
4	26	11	14	17	7^{s}26;37	6^{s}21;40	3^{s}27;17	5^s 9;10	7^{s}12;17	8^{s}10;35	8^{s}11;32	5^{s}14;42	11;17	55;44
5	27	12	15	18	7^{s}27;38	7^s 4; 2	3^{s}27;16	5^s 9;18	7^{s}13; 0	8^{s}11;45	8^{s}11;41	5^{s}14;39	11;16	55;30
6	28	13	16	19	7^{s}28;39	7^{s}16;14	3^{s}27;15	5^s 9;25	7^{s}13;44	8^{s}12;54	8^{s}11;44	5^{s}14;35	11;16	55;16
7	29	14	17	20	7^{s}29;40	7^{s}28;22	3^{s}27;14	5^s 9;32	7^{s}14;27	8^{s}14; 4	8^{s}11;40	5^{s}14;32	11;15	55; 2

Appendix C: A brief timeline of astronomical yearbooks in the Mediterranean and the Middle East

- Babylon, 700 BCE. Yearbooks with daily information on the motions of the sun, moon and planets were already produced in Mesopotamia from the seventh century BCE onwards. In modern research, they are called 'astronomical diaries' or 'almanacs'. These contain instances of the rising phenomena of the moon and the planets, such as first and last appearance above the eastern and western horizons. For the planets, the zodiacal sign in which these phenomena occur is noted.
- Egypt, 100 BCE. Similar yearbooks exist among Greek papyri from the first century BCE until the second century CE, called 'almanacs', or 'almanac-ephemerides'.
- Egypt, 100 CE. At the beginning of the second century CE, a new type of yearbook arose, which was called 'ephemerides'. These contain, for each day, complete sets of ecliptic longitudes for the sun, moon and planets in degrees and minutes. The latest known ephemeris on papyrus was produced for the year 489 CE.
- After that, for a period of more than 400 years, no traces of astronomical yearbooks exist among Egyptian documents.
- Egypt, 910 CE. The first extant Arabic yearbook was produced for the year 910 CE.[1] It is of an unknown type, probably based on a model from the East. It contains only the daily positions of the moon; therefore, it must be called an almanac.
- Egypt, 931 CE. An Arabic yearbook for the year 931/2 CE is of another type and shows a layout similar to that of ancient Greek ephemerides.[2] I have dubbed this an ephemeris of Type I.
- Egypt, 1044 CE. In the eleventh century CE, a new type of ephemeris appeared, called Type II. The first known example was made for the year 1044/5 CE.[3] In Type II ephemerides, the content of the former Type I ephemerides was placed on the right side of a double page. On the left side, the content of the almanacs was presented. Although the Type I ephemerides disappeared – the last known example was made for 1026 CE[4] – , the almanacs survived. They seem to have been the more popular form of yearbook in the following centuries.

[1] Thomann, 'A Fragment of an Unusual Arabic Almanac'.
[2] Thomann, 'An Arabic Ephemeris for the year 931–932 CE'.
[3] Thomann, 'Kat.-Nr. 65: Ephemeride für das persische Jahr 413'.
[4] Thomann, 'An Arabic Ephemeris for the Year 1026/1027 CE'.

- Several examples from the twelfth century have been found among the documents from the Cairo Geniza, but no Type II ephemerides are included among them.
- Ottoman Empire, 1500 CE. Later on, Type II ephemerides were produced as ambitious deluxe manuscripts. The first example of an ephemeris outside Egypt is a document found in a book binding. It was made for the year 1182/3 CE, probably in Northern Syria, possibly in Aleppo.[5]
- Trebizond, 1336 CE. Type II ephemerides found their way to the Byzantine world. A Greek ephemeris for the year 1336 CE was made in Trebizond. Mercier's analysis has shown that it was calculated according to methods found in Islamic *zījes*.[6] The form and content of this ephemeris correspond closely to those of the Arabic Type II ephemerides with the characteristic double page layout. Of course, the writing direction was changed to left-to-right.
- Paris, 1294 CE. Type II ephemerides also found their way to Western Europe. The stages of transmission are entirely unknown. A Latin ephemeris for the years 1293–1312 was produced in Paris. I thank David Juste, who pointed the manuscript of this ephemeris out to me.[7] It shows a high similarity in form and content to its Arabic Type II predecessors. Again, naturally, the writing direction was changed to left-to-right.
- The Paris ephemeris became the model for the early modern printed ephemerides by Regiomontanus, Stöffler and Kepler. Later, after the decline of astrology as a scientific discipline, European astronomers returned to the Type I format in their ephemerides.

Appendix D: Greek and Arabic Ephemerides[8]

This list contains all known and surviving ephemerides in Greek and Arabic up to the ephemeris for 1326/7 CE, which is the main topic of this article. Abbreviation for the references: *Oxy* = Jones, *Astronomical Papyri from Oxyrhynchus*.

24 BCE P. Oxy. 4175: *Oxy*, vol. I, pp. 177–79; vol. II, pp. 170–73.

100 CE P. Dublin TCD F.7: Jones, 'On the Planetary Table'.

[5] Thomann, 'From *katarchai* to *ikhtiyārāt*', p. 349.

[6] Mercier, *An Almanac for Trebizond*.

[7] MS Paris, BnF, lat. 16210; see https://archivesetmanuscrits.bnf.fr/ark:/12148/cc766833.

[8] Information pertaining to Greek ephemerides is found in Jones, *Astronomical Papyri from Oxyrhynchus*, vol. I, p. 175; almanacs that do not contain the daily planetary positions are not included in the present list. For a list of ancient and some medieval horoscopes, see Heilen, *Hadriani genitura*, vol. I, pp. 204–333.

111 CE P. Oxy. 4176: *Oxy*, vol. I, pp. 179–80; vol. II, pp. 174–75.

121 CE P. Oxy. 4177: *Oxy*, vol. I, pp. 180–85; vol. II, pp. 176–83.

140 CE P. Harris I 60: Jones, 'An Astronomical Ephemeris for A.D. 140'.

161 CE P. Oxy. 4181: *Oxy*, vol. I, pp. 191–93; vol. II, pp. 200–03.

245 CE P. Oxy. 4177a: *Oxy*, vol. I, p. 185; vol. II, pp. 184–85.

261 CE P. Oxy. 4178: *Oxy*, vol. I, p. 186; vol. II, pp. 186–87.

348 CE P. Oxy. 4179: *Oxy*, vol. I, pp. 186–90; vol. II, pp. 188–91.

465 CE P. Oxy. 4180: *Oxy*, vol. I, pp. 190–91; vol. II, pp. 192–99.

467 CE P. Mich. Inv. 1454: Curtis and Robbins, 'An Ephemeris'.

471 CE P. Vind. G. 29370b: Gerstinger and Neugebauer; Jones, 'Two Astronomical Papyri Revisited'.

489 CE P. Vind. G. 29370: Gerstinger/Neugebauer 1962; Jones, 'Two Astronomical Papyri Revisited'.

– P. Oxy. 4182: *Oxy*, p. 193.

– P. Oxy. 4183: *Oxy*, p. 194.

– P. Oxy. 4184: *Oxy*, p. 194.

– P. Oxy. 4184a: *Oxy*, pp. 194–95.

– P. Dublin TCD F.7: Jones, 'On the Planetary Table'.

931 CE P. Vind. A.Ch. 12868: Thomann, 'An Arabic Ephemeris for the year 931–932 CE'.

954 CE P. Stras. Inv. Ar. 446: Thomann, 'An Arabic Ephemeris for the Year 954/955 CE'.

994 CE P. Vind. A.Ch. 13577: Thomann, 'Kat.-Nr. 62'.

1002 CE P. Vind. A.Ch. 32363: Thomann, 'Kat.-Nr. 63'.

1026 CE P. Vind. A.Ch. 25613g: Thomann, 'An Arabic Ephemeris for the Year 1026/1027 CE'.

1044 CE P. Vind. A.Ch. 1252 + P. Vind. A.Ch. 14324: Thomann, 'Kat.-Nr. 65'.

1149 CE P. Cambridge UL Inv. Michael. Chartae D 58: Thomann, 'The Arabic Ephemeris for the Year 1149/1150 CE'.

1182 CE Berlin, Islamisches Museum (no inventory number): Thomann, 'From *katarchai* to *ikhtiyārāt*', p. 324.

1326 CE Cairo, Dār al-Kutub, *mīqāt* 817: King, *Mathematical Astronomy in Medieval Yemen*, p. 33, Plate 2; King, *In Synchrony with the Heavens I*, p. 420, note 13.

The Geographical Table in the *Shāmil Zīj*

Tackling a Thirteenth-Century Arabic Source with the Aid of a Computer Database

Benno van Dalen

In fond memory of Ted and Mary Helen Kennedy

1. The *Shāmil Zīj*

The *Shāmil Zīj* (i.e., 'Comprehensive *Zīj*') is an Arabic astronomical handbook with tables from the first half of the thirteenth century.[1] It was primarily intended for the practising astronomer or astrologer, since in most of its surviving manuscripts it consists of, on average, 12 folios of very compact instructions for solving the common problems in spherical and planetary astronomy and astrology as well as a standard set of tables covering 60 folios. The title of the work is explicitly mentioned in the introduction of only three of the twelve more or less complete manuscripts that I have consulted, whereas one other, the earliest surviving manuscript, has ornamented title pages stating the title explicitly for both the explanatory text and the tables.[2] The author of the *Shāmil Zīj* is unknown, but judging from similarities between the *Shāmil Zīj* and the two contemporary *zīj*es by the well-known philosopher Athīr al-Dīn al-Mufaḍḍal ibn ʿUmar al-Abharī (fl. in Mosul from *c.* 1230 onwards, d. 1263–1265),[3] it is possible that al-Abharī was also the author of the *Shāmil Zīj*. The fact that the geographical table in the *Shāmil Zīj*, as one of only relatively few Islamic geographical sources, also presents coordinates for the small town Abhar in northwestern Iran, and that the mean motion tables are set up for its meridian, may be further evidence for this attribution.[4]

[1] For Islamic *zīj*es and their contents, and for overviews of the most important *zīj*es that are extant or known from references, see Kennedy, 'A Survey' and King & Samsó, 'Astronomical Handbooks'.

[2] Furthermore, the title page of Tehran, Majlis Library, MS 6445 gives a distorted form of the title. For extensive information on all manuscripts, see Section 2.

[3] For al-Abharī, see the *BEA* article by Hüseyin Sarıoğlu; the *EI*³ article by Heidrun Eichner; *MAOSIC*, no. 595, pp. 209–10, and Hasse, 'Mosul and Frederick II'.

[4] *GAS*, vol. XIII, p. 381 presents al-Abharī as the author of the *Shāmil Zīj* on the basis of a second entry for the work (with the same incipit) in the Yaltkaya edition of Ḥājjī Khalīfa's *Kashf al-ẓunūn* (see Yaltkaya and Bilge, *Keşf-el-zunun*, vol. II, cols 968–69 and the *EI*² article

Editing and Analysing Numerical Tables: Towards a Digital Information System for the History of Astral Sciences, ed. by Matthieu Husson, Clemency Montelle and Benno van Dalen, PALS 2 (Turnhout, 2021), pp. 511–566
 10.1484/M.PALS-EB.5.127706

The author of the *Shāmil Zīj* states both in the preface and in a later section that he used the planetary mean motion parameters of the important tenth-century mathematician and astronomer Abū l-Wafāʾ al-Būzjānī (Baghdad, 940–997/8).[5] He criticizes the author of the *ʿAlāʾī Zīj* (i.e., al-Fahhād, Shirwan in northwestern Iran, *c.* 1176) for presenting his planetary parameters as having been observed by himself, but instead actually having taken them from Abū l-Wafāʾ.[6] Since the chapters on planetary theory and all tables of Abū l-Wafāʾ's major *zīj*, entitled *al-Majisṭī* after Ptolemy's *Almagest*, are missing from the unique manuscript Paris, Bibliothèque nationale de France, arabe 2494, and planetary mean motion parameters attributed to Abū l-Wafāʾ in the margins of the thirteenth-century Berlin manuscript of a revision of the *zīj* of

---

'Kātib Čelebi' by Orhan Şaik Gökyay, esp. vol. IV, p. 761b, no. 12). This second entry is not found in the edition by Flügel (*Kashf al-ẓunūn*, vol. III, p. 565) and in the online Bologna manuscript of the work, which both contain only the attribution to Abū l-Wafāʾ. However, Flora Vafea kindly confirmed for me that the second entry is found in one of the two manuscripts used by Yaltkaya and Bilge, namely Istanbul, Süleymaniye, Carullah 1619, fol. 70v. This is an 'autograph draft' dated AH 1051 (AD 1641) and consists of a main text with numerous marginal additions. The second entry for the *Shāmil Zīj* is written upside down and diagonally in the margin. It does not repeat the title of the work, but is linked to the main entry by a line drawn with the same black ink.

GAS, ibid., also states that the *Shāmil Zīj* was the earliest work to integrate corrections of coordinates in the eastern part of the Islamic world with those from the Maghrib and al-Andalus relative to the meridian of water (cf. footnote 10). However, since the geographical table in the *Shāmil Zīj* associated with al-Abharī does not include any localities west of Constantinople, I suspect that this merit should rather be assigned to the *Shāmil Zīj* by Ibn al-Raqqām, as indicated in Comes, 'The "Meridian of Water"', p. 47.

[5] For Abū l-Wafāʾ, see the *DSB* article by A. P. Youschkevitch, the *BEA* article 'Būzjānī: Abū al-Wafāʾ' by Behnaz Hashemipour, the *EI*[3] article by Ulrich Rebstock, and *MAOSIC*, no. 256, pp. 96–98. On Abū l-Wafāʾ's astronomical work *al-Majisṭī*, see, among others, Carra de Vaux, 'L'Almageste'; van Dalen, *Ancient and Mediaeval Astronomical Tables*, Chapter 4; Moussa, 'Mathematical Methods', and several sections and footnotes in van Dalen, *Ptolemaic Tradition*. The explicit mention of Abū l-Wafāʾ in the preface of the *Shāmil Zīj* misled several cataloguers, and thence also authors of modern biobibliographical works, into naming him as the author of the *zīj*.

[6] This paragraph is transcribed from the manuscript Paris, Bibliothèque nationale de France, arabe 2528 in de Slane, *Catalogue des manuscrits arabes*, p. 451 (with a French translation) and in King, *Fihris al-makhṭūṭāt*, Part II, pp. 107–08. See also Suter, 'Nachträge', no. 167, pp. 166–67. For the *ʿAlāʾī Zīj*, see Pingree, *The Astronomical Works* (containing an edition of the Byzantine version) and van Dalen, 'The *Zīj-i Nāṣirī*'. An incomplete Persian manuscript of the *ʿAlāʾī Zīj* was noted by Sonja Brentjes to be present in the library of the Salar Jung Museum in Hyderabad. Mohammad Mozaffari informs me that the introduction of the *ʿAlāʾī Zīj* makes clear that al-Fahhād did make regular observations and that he compared the results with calculations based on the *Mumtaḥan Zīj* by Yaḥyā ibn Abī Manṣūr (Baghdad/Damascus, *c.* 830), the lost *ʿAḍūdī Zīj* by Ibn al-Aʿlam (Baghdad, *c.* 960) and the *Sanjarī Zīj* by al-Khāzinī (Marw, *c.* 1120).

Ḥabash al-Ḥāsib (Damascus/Samarra, *c.* 870) appear unreliable, it has not yet been possible to verify this claim. In any case, it can be seen that the author of the *Shāmil Zīj* makes use of the same parameters as the *ʿAlāʾī Zīj*, which he lists in a table together with epoch values for the beginning of the year 600 Yazdigird (AD 1231/2).[7] The *Shāmil Zīj* is therefore of interest both for recovering the values of Abū l-Wafāʾ's mean motion parameters and for studying the transmission of the *ʿAlāʾī Zīj* to Byzantium, where it was translated into Greek by Gregory Chioniades around the year 1300.

All extant manuscripts of the *Shāmil Zīj* contain basically the same explanatory text and the same set of accompanying tables.[8] The explanatory text, headed *al-qawl fī muʿāmarat al-aʿmāl* 'Statement on the Restoration of Astronomical Operations', consists of ten numbered chapters (*abwāb*), further divided into unnumbered sections (*fuṣūl*), and an epilogue (*khātima*). The topics of the chapters are as follows:

1. Calendars: year and month lengths in the Arabic, Persian, Byzantine and Maliki calendars; numbers of days since epoch, date conversion, days of the week of month beginnings.
2. Sines and versed sines: finding the sine and versed sine from an arc and vice versa, by means of the sine table.
3. Tangents: finding the tangent and cotangent of an arc and vice versa, by means of the tables of the first and second tangents.
4. Fundamental arcs on the heavenly sphere: first and second declinations, right ascensions, geographical longitudes and latitudes, equation of daylight, and oblique ascensions.
5. True planetary positions: mean motions and epoch positions, apogee longitudes, true longitudes of the sun, the moon, the lunar nodes, and the five planets.
6. Progressive and retrograde motion and planetary latitudes: first and second stations, lunar latitude, latitudes of the superior and the inferior planets, positions of the fixed stars.
7. Preliminaries for the operations on ascendants: distance of a planet or star from the equator, equation of daylight, maximum altitude, half arc

[7] See van Dalen, 'The *Zīj-i Nāṣirī*', pp. 830–36.

[8] Unlike early European sets of astronomical tables, most extant Arabic and Persian *zīj*es appear in a fixed form with a consistent set of instructions and tables. Only occasionally tables were replaced for use at a different geographical latitude; more often additional tables were provided at the end of the manuscript of a *zīj*. The major exceptions among extant manuscripts are the two recensions of the *Mumtaḥan Zīj* by Yaḥyā ibn Abī Manṣūr (cf. footnote 6) and those of the *zīj* by Ḥabash al-Ḥāsib (see above), which were all copied four centuries after the original works were written and also include materials from later centuries.

of daylight, hours of daylight, degree of transit, degree of rising and setting.

8. Ascendants: calculating the ascendant from the altitude of a star and from the time of day or night, ascendant of a year transfer.
9. Conjunctions and eclipses: time of a conjunction or opposition, solar eclipses, lunar eclipses, lunar crescent visibility, lunar conjunctions with planets or stars.
10. Other astronomical operations: equalisation of the houses, projection of the rays, rising amplitude, altitude of the ecliptic pole, proportion and equation of the azimuth, azimuth and its direction, latitude of the incident horizon, prorogations (*tasyīrs*).
11. Epilogue. On the azimuth of the *qibla* (i.e., the direction of prayer towards Mecca).

The standard set of tables provided in the *Shāmil Zīj* includes all tables that are necessary to carry out the basic operations of astronomy and astrology, but nothing further. Various tables have a non-standard arrangement, as indicated in the following overview. Some tables (especially the first and second declination and solar equation) are tabulated not only for integer degrees but for every 6′ of arc, several others (especially the lunar and planetary equations) for every 12′.

1. Sines and versed sines (these are combined into a single table, in which shared additive values for 2, 4, 6, ..., 60′ must be added to the sine and versed sine values for integer degrees given at the top of each column).
2. First tangent, second tangent.
3. First declination, second declination.
4. Right ascension.
5. Longitudes and latitudes of localities.
6. Universal table of the equation of daylight (with values for every 10° of the ecliptic and the range of geographical latitudes 28, 29, 30, ..., 45°).[9]
7. Universal table of oblique ascensions, with tabular differences for carrying out linear interpolation (with the same general set-up as the equation of daylight).
8. Table of planetary mean motion parameters said to be taken from al-Būzjānī.

[9] Tables and instruments are called 'universal' if they cannot only be used for specific geographical latitudes, but for all or at least a wide range of latitudes. Cf. King, 'Universal Solutions'.

9. Table of planetary apogee motion.
10. Mean motion tables: solar centrum; lunar centrum, anomaly and longitude and lunar node; centrum and anomaly for each of the five planets. These display the planetary positions for the beginnings of the Yazdigird years 600, 601, 602, ..., 699 together with the motions in 100, 200, ..., 1400 Persian years and in 1, 2, 3, ..., 365 days (arranged in 13 columns for the months Farwardīn to Isfandār(mudh) plus the five 'stolen days' *al-mustaraqa*). The mean motions are given for a longitude of 84° measured from the Fortunate Isles, i.e., the same as that used by al-Fahhād in the *'Alā'ī Zīj*.[10]
11. Equation tables: sun; lunar first and second equation, interpolation minutes and variation (*ikhtilāf*); planetary first and second equation, interpolation minutes and variation (*ikhtilāf*).
12. Equation of time (for an apogee longitude of $2^s\,28°$, with the true solar longitude as the argument).
13. Planetary stations.
14. Lunar latitude and planetary latitudes.
15. Fixed stars.
16. Parallax for the 4th and 5th climates.
17. Eclipse tables: radius of the luminaries and of the shadow, 'general digits' and 'amount of darkness' (for calculating eclipse magnitudes), times of solar eclipses and of lunar eclipses (for calculating eclipse durations).
18. The velocity of the moon and its distance (i.e., the distance traveled by the moon in multiples of half an hour as a function of its true velocity).
19. *Ittiṣālāt* (i.e., conjunctions of the moon and planets).

As in many surviving manuscripts of early *zīj*es, in most of the manuscripts of the *Shāmil Zīj* we find additional tables, in particular of types that are not included in the work itself, for example chronological tables, the equation of

[10] Longitudes in medieval Islamic sources were traditionally measured from the Fortunate Isles (*al-jazā'ir al-khālidāt*), i.e., the Canaries, or from the 'Western Shore (of the encompassing sea)' (*sāḥil al-baḥr al-muḥīṭ al-gharbī*), i.e., a point on the Atlantic coast of Africa that was assumed to be 10° east of the Fortunate Isles. (The actual situation was more complicated and stood in relation to a rescaling of Ptolemy's longitudes, especially by the geographers of the caliph al-Ma'mūn around AD 830, which made the Mediterranean 10° shorter. See *GAS*, vol. X, Chapter 2 for an extensive description of this process, and Robles Macías, 'The Longitude of the Mediterranean' for an analysis based on geographical tables, maps and instruments. In western-Islamic sources the so-called 'meridian of water' became in use due to similar reasons; cf. Comes, 'The "Meridian of Water"', which also explains that in many geographical tables longitudes measured from different meridians would occur together).

daylight and oblique ascensions for specific latitudes between 36 and 39°, and astrological tables (for example, duration of gestation, lots, and lunar mansions).

Only scattered materials from the *Shāmil Zīj* have been treated in the literature.[11] In this article I will turn my attention to one of the only very few non-mathematical tables in the *Shāmil Zīj*, namely the geographical table with longitudes and latitudes of 79 localities, concentrated in and around northwestern Iran. Geographical tables lack the standard features that allow us to analyse and restore many types of mathematically computed tables. Foremost, they cannot be recomputed, and since the individual coordinates tended to be copied uncountable times during many centuries, an unusually large number of scribal errors would creep in, including of types that are almost impossible to correct without multiple copies of a table or the original source at hand. Although many geographical tables were arranged by climate,[12] and within each climate by increasing longitude, in practice this only very rarely allows one to correct coordinates or the order of entries. As a result, we will see that even an exhaustive critical edition incorporating eleven manuscript copies of the table from the *Shāmil Zīj* leaves several cases in which we cannot decide on the correct or original entries on the basis of the evidence in the table alone. Therefore it will be necessary to look at a database of a much wider range of Islamic geographical sources, which will not only allow us to make well-founded decisions for the edition, but will also show us on which earlier sources the author of the geographical table in the *Shāmil Zīj* based himself, and that he included several coordinates that appear to be entirely original.

Such a database of Islamic geographical data can be found in an indispensable secondary source for the study of Islamic geographical tables, namely *Geographical Coordinates of Localities from Islamic Sources* by Edward S. Kennedy

[11] Individual methods and tables were discussed in Kennedy, 'Comets', pp. 47–48; Berggren, 'The Origins', pp. 5–7; and King, 'Some Early Islamic Tables', pp. 217–218. In van Dalen, 'A Statistical Method', pp. 106–13, I used the solar equation table from the *Shāmil Zīj* as an example for the application of various statistical estimators that I had developed. (The result of my estimation of the solar eccentricity was confirmed in Bellhouse, 'An Analysis of Errors', pp. 287–92, by means of a different statistical method making use of integer-valued residuals). In van Dalen, 'The *Zīj-i Nāṣirī*' (with a summary of results on p. 857), I compared the planetary tables in the *Shāmil Zīj* and their underlying parameters with those in several earlier *zīj*es and found that the planetary equations were mostly taken from Kūshyār ibn Labban's *Jāmiʿ Zīj* (Iran, *c.* 1025) and in some cases from al-Khāzinī's *Sanjarī Zīj* (Marw, *c.* 1120), possibly through the intermediary of the *ʿAlāʾī Zīj*.

[12] The climates (or climes) were seven bands of latitude values determined from round numbers for the maximum lengths of daylight. See Honigmann, *Die sieben Klimata*; the *EI*[2] article 'Iḳlīm' by A. Miquel; Dallal, *Al-Bīrūnī on Climates*; and King, *Bringing Astronomical Instruments*, pp. 6–9. For a more general introduction to Islamic mathematical geography, see Kennedy, 'Mathematical Geography' or King, *World-Maps*, Section 1.6, pp. 23–28.

and his wife Mary Helen Kennedy (here further abbreviated as K&K). This book contains lists of place names, longitudes and latitudes as found in about 80 Islamic works, including *zīj*es, instruments and several others. As will be explained in more detail in Section 5, the sources are indicated by a three- or six-letter alphabetical code as well as a chronologically ordered numerical one, with the data arranged in four different ways. The geographical table from the *Shāmil Zīj* is included in K&K as source SML, on the basis of a manuscript that my edition will show to be defective in various respects.[13]

2. The Geographical Table in the *Shāmil Zīj*

The geographical table in the *Shāmil Zīj* is found among the trigonometric and spherical-astronomical tables that immediately follow the explanatory text and precede the planetary tables. It thus facilitates the use of the *zīj* for the computation of arcs on the celestial sphere and planetary positions for any locality covered by the table and, as described in Chapter 4 of the explanatory text, in other localities by means of interpolation between the longitudes of two localities with the same latitude and known distance (the text prescribes to find the latitude of other localities by direct observation of the solar altitude at noon). The table covers a single page in the manuscripts and includes 26 localities in the first and second columns, and originally 27 in the third (some minor deviations are specified in the apparatus to my edition in Section 3). With its total of 79 localities it is one of the smaller geographical tables found in Islamic *zīj*es.

For critically editing the table, I will make use of nearly every single copy that I have been able to get hold of. The *Shāmil Zīj* is extent in more or less complete form in the following twelve manuscripts, eight of which include the geographical table:

- Cairo, Dār al-kutub, *mīqāt* Ṭalʿat 138 (58 fols, only the tables, copied *c.* 900 Hijra; see King, *Fihris al-makhṭūṭāt*, Part I, p. 476 and Part II, pp. 107–09, as well as King, *A Survey*, B100, p. 52 and plate XVIII, p. 238). A fine copy in clear *naskh* with a variety of additional tables. An inlaid sheet misled some scholars in attributing the work in this manuscript to Ibn Yūnus. — This manuscript omits the geographical table and the universal tables for the equation of daylight and oblique ascensions and replaces them by specific tables for latitude 38°.
- **C** = Cairo, Dār al-Kutub, *riyāḍī* Taymūr 296/1 (pp. 1–160 [220 pp. in total], copied in 1123 Hijra = AD 1711/2; see King, *Fihris al-makhṭūṭāt*, Part I, p. 609 and Part II, pp. 107–09, as well as King, *A Survey*, B100,

[13] The description of source SML is found in Kennedy and Kennedy, *Geographical Coordinates*, p. xxxii, and the coordinates from the table, arranged alphabetically by location, on pp. 471–73.

p. 52). Written in a clear *naskh*, with several additional tables. — The geographical table is on fol. 21v. The place names are frequently written with diacritical dots, but occasionally the shapes of the letters are completely wrong. All text is in black; only the frame of the table, somewhat carelessly drawn, is in red. The first column with place names is preceded by a column containing only the word *madīna* 'city' for every entry. **C** has an unusually large number of scribal errors, nearly all of which are shared by manuscript **J** of a *zīj* written in Mosul, likewise in the 18th century (see below). Different from the common form ے for *abjad* 10 found in all other witnesses, **C** and **J** both use the standard form ى used for *yā'* in ordinary text. Since the tables in **C** and **J** can be seen to be descendants of the already rather faulty manuscript $\mathbf{T}_2$, to which they add further scribal errors of their own, variants from **C** and **J** are only included in the apparatus of my edition if they differ from $\mathbf{T}_2$ or show very typical or informative deviations.

- $\mathbf{F}_1$ = Florence, Biblioteca Medicea Laurenziana, Or. 95b (previously Palatina 289, 116 fols separately numbered from Or. 95a in the same volume, copied in 747 Hijra = AD 1347; see Assemanus, *Bibliothecae Medicae Laurentianae*, no. 289, p. 394). Written in a clear *naskh* on uneven lines. The explanatory text (up to fol. 18r) is followed by a similar but unrelated text in Persian and some other smaller ones. This manuscript has an unusually large number of additional tables beyond the basic set. — The geographical table is found on fol. 41v. It omits the second column of the original table and distributes the localities from the first and third columns over three shorter columns, adding some further localities at the end. The place names are mostly written without diacritical dots and the definite article *al-* is omitted even from place names for which most other manuscripts include it. As the only witness, this manuscript includes a label *iqlīm awwal* 'first climate' before the first entry and Hindu-Arabic numbers 2 to 7 roughly at the places where the other climates start. These undoubtedly do not stem from the original work, but their exact placement will be mentioned in the general comments to my edition.
- Florence, Biblioteca Medicea Laurenziana, Or. 106/1 (fols 1–71r [170 fols in total], 8th c. Hijra; see Assemanus, *Bibliothecae Medicae Laurentianae*, no. 120, pp. 197–98). Written in a clear *naskh* on uneven lines. The last four pages of the explanatory text were replaced by a copy written on ruled lines within an outer frame in a different hand, apparently that of Or. 106/2. The second half of this manuscript contains one of the two *zījes* explicitly attributed to al-Abharī with a copy of the geographical table from the *Shāmil Zīj*; see $\mathbf{F}_2$ below. —

The geographical table is missing together with the second half of the right ascension table that would have immediately preceded it, possibly because the sheet concerned was lost or torn out. The two universal tables are included in this manuscript, but the oblique ascensions were never filled in.

- Istanbul, Süleymaniye Kütüphanesi, Carullah 1479 (65 fols, undated; very brief notice in Krause, *Stambuler Handschriften*, p. 466). Written in a clear, slightly cursive *naskh*. The title 'Shāmil Zīj' is explicitly mentioned in the preface. — The geographical table is missing from this manuscript, together with all spherical astronomical tables.
- Meshhed, Holy Shrine Library, MS 12086 (96 fols, copied in 781 Hijra = AD 1379/80; see ʿIrfānian, *Fihrist-i kutub-i khaṭṭī*, no. 536, pp. 60–63). A fine copy in a very clear *naskh*, generally with diacritical dots. The title 'Shāmil Zīj' is explicitly mentioned in the preface. Both text and tables have a large number of well-informed glosses in the margins. The explanatory text is followed by sections on lunar crescent visibility from the *(Il)khānī* and *Shāhī Zīj*es copied by a different hand. The tables are supplemented by an unusually large set of additional tables, including several for latitudes 38° and 39°, a table for the equalisation of the houses and an uncommon double-argument table for the lunar equation covering 35 pages. — The geographical table is missing from this manuscript together with all trigonometrical and spherical-astronomical tables.
- **P**$_8$ = Paris, Bibliothèque nationale de France, arabe 2528 (73 fols, manuscript from the 15th/16th c.; see de Slane, *Catalogue des manuscrits arabes*, pp. 451–52 and http://archivesetmanuscrits.bnf.fr/ark:/12148/cc30444x). Written in a clear *naskh* on carefully drawn lines. Numbers that would have been in red are systematically missing from the headings of the tables, probably because the manuscript from which this copy was made already lacked those. — The geographical table appears on fol. 19v. The diacritical dots on the place names are frequently missing. Each of the three columns with place names is preceded by a column containing only the word *madīna* 'city' for every entry. The coordinates from this copy of the geographical table from the *Shāmil Zīj* were included in K&K as source SML (cf. p. 517).
- **P**$_9$ = Paris, Bibliothèque nationale de France, arabe 2529 (76 fols, manuscript from the 16th c.; see de Slane, *Catalogue des manuscrits arabes*, p. 452 and http://archivesetmanuscrits.bnf.fr/ark:/12148/cc304455). Written in a clear and careful *naskh*. The title 'Shāmil Zīj' is explicitly mentioned in the preface. — The geographical table is included on fol. 27v (see Plate 16). It omits the entry for Siwas, but gives four

additional localities in the same hand as the 78 original ones. It provides diacritical dots for most place names with some exceptions near the end of the first column, where also the original latitudes of three localities appear to have been corrected to alternative ones. The definite article *al-* is omitted from all place names except al-Raqqa and al-Rayy.

- **P**$_0$ = Paris, Bibliothèque nationale de France, arabe 2540 (fols 7v–15 and 29v–99 [99 fols in total], manuscript from the 15th c.; see de Slane, *Catalogue des manuscrits arabes*, p. 454 and https://archivesetmanuscrits.bnf.fr/ark:/12148/cc30455c). This manuscript was obtained in Aleppo in 1673. The remaining folios contain a rather sloppy copy of the explanatory text and tables of *al-Dustūr al-ʿajīb* by Naṣir al-Dīn ibn ʿĪsā ibn al-Ḥiṣkafī, pointing to a strong relationship to the Vatican manuscript (**V** below), which likewise contains this work together with the *Shāmil Zīj*. The text is written in a somewhat untidy but readable *naskh*. The tables of the *Shāmil Zīj* are nicely laid out as in most of the other manuscripts, but those of *al-Dustūr* are in a small and often unclear script. Numbers in red are systematically missing from the headings of the tables belonging to the *Shāmil Zīj*. In general the calligraphy of the headings of these tables is very similar to **P**$_8$. *Abjad* 3 (ح) and 4 (د) look very much alike in this manuscript when written separately. — The geographical table is on fol. 38v. It generally includes the diacritical dots on the place names. Each of the three columns with place names is preceded by a column containing only the word *madīna* 'city' for every entry.
- **T**$_1$ = Tehran, Majlis Library, MS 6422 (76 fols, copied in 672 Hijra = AD 1273/4; see Ḥusaynī Ashkawarī, *Fihrist-i nuskhahā-yi khaṭṭi*, pp. 15–16 and https://dlib.ical.ir/site/catalogue/835022). This is the oldest surviving copy of the *Shāmil Zīj*. Fol. 1r gives the title with ornaments in gold and blue, fol. 11r the title of the second *maqāla* 'On the tables of the *Shāmil Zīj*' in gold. Folios 2–9, and therewith almost the entire explanatory text, are missing from the manuscript. Fol. 10 contains the last part of the epilogue and several additional texts in different hands. The original set of tables is included in its entirety in the correct order. The tables were copied in an elegant *naskh*, possibly different from the explanatory text. Some additional tables and texts are found from fol. 71v onwards, including oblique ascensions for latitudes 38° and 39° (Konya). — The geographical table is on fol. 20v. It generally includes the diacritical dots on place names. Different from all other manuscripts, the place names are here alternately written in black and in red. In the last column, the minutes of four longitude and six latitude values and the degrees for one latitude value have been omitted.

My edition of the table will make clear that these must in each case be taken to be equal to the last written digit above them (all minutes concerned are zero, and the degrees for Aqsaray are 38). These omissions have not been noted in the apparatus.

- **T_2** = Tehran, Majlis Library, MS 6445 (91 fols, copied in AD 1880 according to the catalogue, but this is probably incorrect because the earlier manuscripts **C** and **J** depend on it (cf. below); see Ḥusaynī Ashkawarī, *Fihrist-i nuskhahā-yi khaṭṭi*, p. 29 and https://dlib.ical.ir/site/catalogue/836523). I only became aware of this manuscript after an earlier version of this article had been submitted. The title is given on the title page (fol. 2r) in the distorted form *al-Zāʾirja al-shāmila li-l-maḥāsin al-mukāmila* 'The Comprehensive *zāʾirja* for the Perfect Merits'.[14] A different hand on fol. 1r describes the contents as 'The *zīj* of Abū l-Wafāʾ Būzjānī and the introduction of the *Athīrī Zīj*'. The *Shāmil Zīj* is here mixed with some texts, tables and diagrams from very different sources written on different paper by different hands. Text and tables of the *zīj* are in a *naskh* hand. The introduction starts at fol. 7r and is indeed followed by the opening lines of the *Athīrī Zīj* on fol. 15v; the opening section of the *Shāmil Zīj* is repeated on fol. 16v on different paper. — The geographical table is on fol. 27r. The place names are generally written with diacritical dots. The first two columns with place names (but not the third) are preceded by a column containing only the word *madīna* 'city' for every entry. Since this table is clearly an ancestor of the even faultier copies **C** and **J**, I have included all variants from **T_2** in my edition but those from **C** and **J** only when they differ from **T_2** or provide typical or informative further deviations.
- **V** = Vatican, Biblioteca Apostolica Vaticana, Vat. ar. 1499 (fols 3v–10v and 14r–101v [102 fols in total], 984–87 Hijra = AD 1576–1580; see Levi della Vida, *Secondo elenco*, pp. 2–3). Like in **P_0**, the *Shāmil Zīj* appears here mixed up with a copy of al-Ḥiṣkafī's *al-Dustūr al-ʿajīb*. The *Shāmil* tables are neatly laid out, those from the *Dustūr* in a very small script. — The geographical table is found on fol. 48v. The scribe omitted most of the diacritical dots on place names and used a form for *abjad* zero (ﻩ) that is very close to the letter *ʿayn* ع. Instances in which the manuscript indeed has an actual *ʿayn* rather than a zero are not separately included in the apparatus. Each of the three columns with place names is preceded by a column containing only the word *madīna* 'city' for every entry. The first two longitude values (for Haba-

[14] The *zāʾirja* is a divinatory device involving letter magic, geomancy and astrology; cf. the *EI*² article 'Zāʾirdja' by T. Fahd.

sha and Sanaa, in black) were written over further copies of this word that were mistakenly inserted in red.

Some further fragments of the *Shāmil Zīj* are contained in London, British Library, Add. 7492/3 (fols 51v–67r with only the explanatory text, copied in 912 Hijra = AD 1506/7), and possibly in Mumbai, Cama Oriental Institute, R I.86. The commentary on the *Shāmil Zīj* by al-Qumnāṭī is extant in Paris, Bibliothèque nationale de France, arabe 2530 and Istanbul, Süleymaniye Kütüphanesi, Laleli 2137, and the commentary by Ḥasan Muḥammad Ṭūsī in Isparta, Halil Hamit Paşa İl Halk Kütüphanesi, MS 2252.[15]

The geographical table from the *Shāmil Zīj* is also included in several works that are strongly related to the *zīj*. Of these works I have used the following manuscripts:

- **F_2** = Florence, Biblioteca Medicea Laurenziana, Or. 106/2 (fols 72v–170r, copied in the 8th c. Hijra; see Assemanus, *Bibliothecae Medicae Laurentianae*, no. 120, pp. 197–98) of the *Athīrī Zīj* (described as a shortened version of *al-Zīj al-mulakhkhaṣ ʿalā arṣād al-ʿAlāʾī* in the opening sentences) by Athīr al-Dīn al-Abharī. This work appears to be contemporary and to share several tables and other characteristics with the *Shāmil Zīj*, but it also shows a large number of significant differences. The explanatory text of the *Athīrī Zīj* consists of 15 sections (*fuṣūl*). The planetary mean motions are for geographical longitude 70°, most likely Damascus, and the planetary positions are given for the year 600 Yazdigird. A detailed investigation is necessary in order to establish the exact relationships between the *Shāmil Zīj*, its slight reworking in the manuscript Dublin, Chester Beatty, Arabic 4076, and the two *zīj*es of al-Abharī. This part of the manuscript is written in a clear *naskh*; the first half contains a copy of the *Shāmil Zīj* that is listed above but lacks the geographical table. — The geographical table from the *Shāmil Zīj* is included by al-Abharī on fol. 141v just after a set of spherical astronomical tables for latitude 36°. Different from the *Shāmil Zīj*, the geographical table here comes after the planetary tables. Most of the diacritical dots on place names are omitted and some names are completely miswritten. Every single occurrence of a digit 50 in the coordinates is written as an unambiguous 55, making it plausible that in one of the ancestors of this manuscript the shapes of 50 and 55 were extremely close.

- **J** = Mosul, Pāshā Mosque (*Jāmiʿ al-Bāshā*), MS 323 (90 fols, copied in 1141 Hijra = AD 1728/9; see al-Shantī, *Fihris al-makhṭūṭāt al-muṣaw-*

[15] See *GAS*, vol. V, pp. 324–25 and vol. VI, pp. 223–24 (under Abu l-Wafāʾ). For the commentaries see also *OALT*, vol. I, no. 6, p. 22 and vol. II, pp. 809–10, as well as *MAOSIC*, no. 766, p. 259 and no. 859, p. 290.

wara, no. 319, p. 121) of the *zīj* by ʿAbd al-Qādir ibn Ṣafāʾī al-Mawṣilī. This manuscript might now very well have been lost for eternity if the Cairo Institute of Arabic Manuscripts had not prepared a microfilm of it many decades ago. Professor Edward S. Kennedy was kind enough to lend me a copy of this film. A more extensive description of the *zīj* of al-Mawṣilī will appear in my *A New Survey of Islamic Astronomical Handbooks*. Written in a very clear *naskh*. — The geographical table is on fol. 8v. The place names generally include the diacritical dots, *shadda*s and *hamza*s. However, the tabular frame was drawn in an amateurish way without a ruler and with one column too many and two rows too few. The first column of place names (but not the second and third) is preceded by a column containing only the word *madīna* in every row. The shape of *abjad* zero (ؠ) is somewhat similar to the Hindu-Arabic numeral 5, but is also repeatedly confused with *abjad* 8 (ح). This table can be seen to be a descendant from manuscript $\mathbf{T}_2$ through the intermediary of manuscript **C**. The three manuscripts share a large number of peculiar errors that do not appear in any of the other eight witnesses. These include grave mistakes in the spelling of place names, the confusion of *abjad* numbers 0 and 8 (especially in **C** and **J**), and the slide of 14 place names towards the end of the third column which made it impossible for Kennedy to recognize that **J** is in fact a copy of the table from the *Shāmil Zīj*. In my critical edition of the geographical table I have only included variants from **C** and **J** if they are different from those in $\mathbf{T}_2$ or show very typical or informative deviations.

- **O** = Oxford, Bodleian Library, Laud Or. 253 (88 fols, autograph; see Nicoll, *Bibliothecae Bodleianae*, no. 274, pp. 242–46) of *al-Durr al-muntakhab* by the Priest Cyriacus (in Arabic: al-Qiss Qiryāqus). The author presumably worked in Mardin, now in southeastern Turkey, toward the end of the 15th century. The introduction of his *zīj* states that it was based on the *Athīrī Zīj* by al-Abharī (cf. $\mathbf{F}_2$ above) and the planetary mean motion parameters of Abū l-Wafāʾ. The highly original lunar and planetary equations with double arguments, displacements and further adjustments, as well as some other individual tables, were studied by Saliba and Kennedy.[16] The author's hand is a clear *naskh*. — The geographical table, basically identical with that in the *Shāmil Zīj*, appears on fol. 84v. As in most of the manuscripts described above, the title is written in suprascript; most of the diacritical dots on the place names are provided.

[16] See Saliba, 'The Double-Argument Lunar Tables'; Saliba, 'The Planetary Tables'; Kennedy, 'Comets'; Kennedy and Agha, 'Planetary Visibility Tables', and Saliba, 'Easter Computation'.

As has already been mentioned in Section 1, K&K includes the geographical table from the *Shāmil Zīj* as source SML on the basis of manuscript $\mathbf{P}_8$. Furthermore, since they appear in different works, K&K separately presents the table from manuscript **J** as source ABD and the table from manuscript **O** as source QIR. Besides these, it includes one further copy of the geographical table from the *Shāmil Zīj*, namely the source that it abbreviates as ULE because it is included on the inside of the back cover of the manuscript Oxford, Bodleian Library, Greaves 5 of the *Zīj* of Ulugh Beg (Samarqand, *c.* 1440).[17] The title and place names in this table were copied in Arabic by a European hand, but most of the numbers are in European numerals rather than in the standard Arabic alphabetical notation. This is most likely a copy by the Savilian professor of astronomy John Greaves himself of a table from another manuscript from his own collection or from the Bodleian Library, or one that he inspected during his travels in the Levant from 1638 to 1640.[18] A direct comparison seems to exclude that the table was copied from the Bodleian manuscript **O** of the *zīj* of the Priest Cyriacus. Whereas seven place names are incorrectly spelled, the tabular values agree almost entirely with those in my edition, with the exception of a possible scribal error for Qum and glitches for Malatiya and Qaysariyya. For Egypt (*Miṣr*, i.e., Cairo), Greaves gives the correct longitude 64;40°, which is further only found as an obvious later correction in manuscript $\mathbf{P}_8$.[19] So in some respects, ULE (or GRV as I will further call it) can be considered the earliest edition of the geographical table from the *Shāmil Zīj*. Because of the uncertainty of its exact sources I will nevertheless omit it from my own edition.

[17] Kennedy and Kennedy, *Geographical Coordinates*, p. xxxv mistakenly states that this table appears after the colophon of the manuscript used for source ULG (the *Zīj* of Ulugh Beg), namely Oxford, Bodleian Library, Marsh 396. The coordinates are listed in ibidem, pp. 564–65.

[18] See Maddison, 'Greaves'; R. Mercier, 'English Orientalists', pp. 261–77, and the further literature mentioned by Maddison. Cf. also Greaves, *Binae tabulae geographicae*. The astronomical sources used by Greaves and the astronomical marginalia in his manuscripts are currently being investigated in detail by Taha Yasin Arslan as part of the pilot project 'The Arabic Books and Astronomy in Seventeenth Century Oxford' led by Julia Bray.

[19] Although most digits in Greaves' copy are written with European numerals, their order remains as in the Arabic. Thus '30 73' stands for the longitude 73;30° of Sanaa. The mistakes in the place names include curious ones such as مكيه for Mecca, اسكنديّه for Alexandria and بادته for Qadisiyya. The longitude of Qum is given as 82;55°, different from all three values found in the eleven witnesses that I have used for my edition; the latitude of Malatiya is listed as 38;30°, possibly miscopied from the entry for Qaysariyya; and the latitude of Qaysariyya is written in a mixed form in incorrect order as '39 ى'. Seven incorrect digits of coordinates are underlined and were corrected in standard Arabic alphabetical notation above the numbers.

3. Editing the Geographical Table

I present my edition of the entries from the geographical table in the *Shāmil Zīj* in Table 1 on pp. 529–31 and that of the further textual elements such as headings and marginal notes below. For easier reference I have indicated all entries in the original table with a letter (A, B and C for the first to third columns in the manuscripts) and a running number within each column (up to 26 for the first two columns and up to 27 for the third). For the modern equivalents of the place names in the second column of the edition I have used for easier comparison the exact forms as found in Kennedy and Kennedy, *Geographical Coordinates* (K&K) with only very few exceptions.[20] The apparatus to the table is given in the form of notes in the last column of the edition. Here any Arabic form gives a variant to the place names, 'long.' refers to the longitude and 'lat.' to the latitude. In a pair of coordinates separated by a slash the longitude precedes the latitude. The symbols ° and ′ indicate variants in respectively the degrees and the minutes of longitude or latitude.[21] In addition to the sigla for the manuscript sources introduced in Section 2, I use **K** for cases where K&K deviates from my reading of $\mathbf{P_8}$ (in all other cases K&K gives the value that I present for $\mathbf{P_8}$). A question mark indicates a reading that is uncertain or ambiguous.

I have applied the following general editing policies:

- Any variants in the apparatus for place names, longitudes and latitudes are given in the order $\mathbf{T_1P_9F_1F_2OP_8P_0VT_2CJ}$, i.e., as we will see, in the order of general correctness (or smallest number of errors) of the manuscripts. Note that variants from **CJ** are only given explicitly when they differ from $\mathbf{T_2}$ or are otherwise of interest.
- If an entry is unclear but can be read as what I consider to be correct on the basis of the entire manuscript evidence or the correct spelling of place names, I will assume that the correct entry was intended and not mark the unclarity in the apparatus.
- If an entry was corrected in the main hand, I will not include the incorrect original entry in the apparatus unless it appears relevant, for example because other witnesses have the same incorrect entry.

[20] Specifically, I write Sanaa for Sana, Madā'in for Ctesiphon, and Qum instead of Qumm.

[21] As most numbers in Islamic astronomical sources, the coordinates in geographical tables are written in the Arabic alphabetical (*abjad*) notation, in which letters *alif* to *ṭā'* denote the numbers 1 to 9, letters *yā'* to *ṣād* the numbers 10 to 90, and letters *qāf* to *ghayn* the numbers 100 to 1000 (with small variations in a system that was mostly used in the western Arabic world). By combining the letters for one thousand, hundreds, tens and units, any number up to 1999 can be written. For example, *ghayn-shīn-nūn-wāw* غشنو denotes 1356. See Irani, 'Arabic Numeral Forms' and Thomann, 'Scientific and Archaic Arabic Numerals'. For the types of scribal errors that may result from the similarities between certain letters, see Section 4.

- If an entry was corrected in a different hand, I will indicate this whenever the correction is clear.

In editing the place names I have applied the following specific policies:

- In the edition I generally write the place names with correct diacritical dots and *shadda*s, unless all witnesses concerned write them differently (e.g., several manuscripts write زيحان Zayhān with a dotted *yā'* for زنجان Zanjān, B17). I write dots on *tā' marbūṭa* even though they are almost never written in the manuscripts.
- In the apparatus I will write the variant names exactly as they appear in the manuscripts, i.e., often without diacritical dots. If two or more manuscripts have the same letter shapes for a place name, I will add any diacritical dots that are found in at least one of the witnesses.
- I have not been able to recognize any systematic patterns in the addition or omission of the definite article *al-* before certain place names and have therefore decided generally not to indicate such additions or omissions in the apparatus of the table.

In editing the coordinates, I have used the following general rules:

- All variants caused by the omission or inclusion of diacritical dots are indicated in the apparatus.
- A *fā'* or *qāf* without dot is read as 80; thus a reading as 100 requires both dots. Only for the longitudes of Kirman and Khwarizm is the presence of the dots explicitly specified in the apparatus, because the writing in most manuscripts is incorrect.
- Variants in place names or coordinates that are part of a slide in some of the manuscripts (cf. Section 4) are given for the locality for which they were originally intended, i.e., after the errors resulting from the slide itself have been corrected. Whenever a variant is the direct result of a slide, this is explicitly indicated.

Title of the table

جدول أطوال البلدان من الجزائر الخالدات وعروضها عن خطّ الاستواء

jadwal aṭwāl al-buldān min al-jazā'ir al-khālidāt wa-'urūḍihā 'an khaṭṭ al-istiwā'

Table of the Longitudes of Cities from the Fortunate Isles and their Latitudes from the Equator

Apparatus. جدول] $\mathbf{P_8P_0VT_2C}$ om., **J** هذا جدول أطوال] $\mathbf{P_0V}$ لطول الجزائر] $\mathbf{T_1F_1OP_8T_2CJ}$ جزائر, $\mathbf{P_0V}$ خزاين الخالدات] $\mathbf{F_2}$ الخالده وعروضها] $\mathbf{F_2}$ وعرضها عن] $\mathbf{T_2CJ}$ من عن خط الاستواء] $\mathbf{F_1}$ om. $\mathbf{P_8T_2}$ add عرض لز 'latitude 37°' above البلدان.

Column headers

al-buldān / al-aṭwāl / al-ʿurūḍ البلدان / الأطوال / العروض

Cities / Longitudes / Latitudes

Apparatus. الأطوال] $\mathbf{P_9F_1}$ الطول العروض] $\mathbf{P_9F_1T_2}$ العرض.

Other general characteristics

$\mathbf{P_8P_0V}$ add a column containing only the word *madīna* for every entry before each of the three columns with place names. $\mathbf{T_2}$ adds such a column before the first and second columns of place names, **CJ** only before the first column.

$\mathbf{F_1}$ adds indications of the climates as follows: *iqlīm awwal* above the first column, and Hindu-Arabic numerals '2' before Madina (A5), '3' in the cell of Egypt and Alexandria (A7–A8), '4' before Tarsus (A23), '5' before Khwarizm (C6), '6' before Konya (C26) and '7' before the unidentified *al-ʿzh* (see below) at the end of the table.

Marginal notes

$\mathbf{T_1}$ (right margin in two different hands):

عرض الموصل على ما ذكر العمّالون بها ووجد في حسابهم وتحريرهم له نه مح

ʿarḍ al-Mawṣil ʿalā mā dhakara al-ʿammālūn bihā wa-wujida fī ḥisābihim wa-taḥrīrihim ***lh nh mḥ***

The latitude of Mosul according to what those who work at ⟨the city⟩ stated and to what was found in their calculation and their redaction, is 35;55,48°.[22]

انحراف دمشق شرق وجنوب لا ى / حلب يو م / ترابلس كح ى / حماه كح ل / انطاكية كا مه

inḥirāf dimashq sharq wa-janūb lā y / ḥalab yw m / tarābulus kḥ y / ḥamāh kḥ l / anṭakiya kā mh

The inclination ⟨of the *qibla*⟩ is ⟨for⟩ Damascus south-east 31;10, Aleppo 16;40, Tripoli 28;10, Hama 28;30, Antioch 21;45.[23]

A note in the margin of the table in $\mathbf{P_0}$ computes the longitude and latitude differences between Mecca (rounded coordinates 78° / 21°) and possibly Mardin (74° / 33°, correct latitude 37°), apparently with the purpose of calculating the *qibla*. The numbers are here written with Hindu-Arabic numerals.

[22] This latitude is found in a section on solar eclipses (fols 7v–8r) and in an oblique ascension table for Mosul (fol. 100r–v) in the manuscript Escorial, RBMSL, árabe 927 of a thirteenth-century recension of the *Mumtaḥan Zīj* by Yaḥyā ibn Abī Manṣūr (cf. footnote 6).

[23] These values are in full agreement with the *qibla* values found in the geographical table from al-Khāzinī's *Sanjarī Zīj* (see King, *World-Maps*, pp. 71–75 and Appendix D).

Additional localities

Several of the manuscripts give additional localities and coordinates in the table itself or in the bottom margin. In some cases these were copied into some of the other extant manuscripts as well, in others they appear to be incidental additions by the scribe or a user.[24]

P_0V insert entries Mardin ماردين with longitude 74;30° (**V** 37′) and al-Ḥiṣn الحصن with longitude 75;35° after the first and third entries of the second column. **O** adds the same entries in the opposite order with coordinates 75;30° / 38;15° for al-Ḥiṣn and 75;0° / 37;30° for Mardin. In **P_0** these additions are clearly in a slightly lighter red and a different hand, in **VO** they are in the main hand. Because of its relative position with respect to Mardin and Harran, al-Ḥiṣn can be assumed to stand for Ḥiṣn Kayfā (now Hasankeyf), which is found in only very few Islamic sources with coordinates 74;35° / 37;35°.

P_9 adds an entry لادبق و سركى with coordinates 62;30° / 41;30° at the end of the first column and an entry اياثلوغ with coordinates 61;0° / 41;0° at the end of the second column. These can be recognized as two localities in western Anatolia, namely respectively Laodicea (*Lādhiq*) of Lycos (or Phrygia), whose ruins are just north of present-day Denizli, and Ayathulūg or Ayasulūk, now Selçuk (cf. the *EI*[2] articles 'Lādhiḳ' and 'Aya Solūk'). I have not been able to interpret the word following *Lādhiq*, but the excellent relative coordinates of the two cities leave little doubt about their identification. Neither locality occurs in K&K.

Both **P_9** and **F_1** add entries Bulghar بلغار with coordinates 68;0° / 49;30° and Saray with coordinates 72;20° / 46;10° (**F_1** 20′) at the end of the third column. The former stands for the Turkic people that founded a state on the Volga in the early Middle Ages, and the latter most probably for one of the two capitals of the Mongol Golden Horde, likewise in modern southern Russia. The coordinates for Bulghar stem from al-Khāzinī, those for Saray are not attested. **F_1** furthermore adds العزه *al-ʿzh* with longitude 43;11° and latitude 38;22°, which I have not been able to identify.[25]

F_2 adds under the table مدينه باجوج *madīna Bājūj* with longitude 172;30 (for ياجوج *Yājūj* (Gog) or perhaps also ماجوج *Mājūj* (Magog), since 172;30° is the latter's commonly used longitude measured from the Western Shore). **F_1** adds مدنه ياجوح *madīna Yājūḥ* (for Gog) with coordinates 65;0° / 21;5° which are unattested and nonsensical.

O writes under the last entry بغير تغيير 'without any change'.

[24] Marginal notes with geographical coordinates that allow the use of, for example, planetary tables at a different locality, may be found in many manuscripts of Islamic astronomical works.

[25] Judging from its coordinates this locality should be near Ammuriya (Amorion) in western Anatolia. Note that Ghazza غزّة is an unlikely candidate both because of the bad agreement of the coordinates (it appears in most sources with longitude 64;50° and latitude 32;0°) and because it is generally written without definite article.

Table 1: Edition of the geographical table from the *Shāmil Zīj*

Column 1

	locality		long.	lat.	apparatus
A1	Habasha	حبشة	51;40	19;30	long. $\mathbf{F}_1$ 11°
A2	Sanaa	صنعا	73;30	14;30	$\mathbf{P}_8\mathbf{P}_0\mathbf{V}$ صنعه
A3	Aden	عدن	75; 0	13; 0	$\mathbf{P}_0\mathbf{V}$ عند, lat. $\mathbf{T}_2$ 8′
A4	Oman	عمان	94;30	19;45	long. $\mathbf{P}_8\mathbf{P}_0\mathbf{V}\mathbf{T}_2$ 74°, lat. $\mathbf{F}_1$ 15′
A5	Madina	(ال)مدينة	75;20	25; 0	–
A6	Mecca	مكّة	77;10	21;40	long. **K** 74°, lat. $\mathbf{P}_9$ 30′
A7	Egypt	مصر	54;40	29;45	long. $\mathbf{T}_1$ 14° $\mathbf{P}_8$ 64° (corrected) $\mathbf{P}_9$ 55′, lat. $\mathbf{P}_9$ 55′ (?) **O** 42′ $\mathbf{P}_8\mathbf{P}_0\mathbf{V}\mathbf{T}_2$ 47′ (**CJ** 40′)
A8	Alexandria	اسكندريّة	60;30	30;20	long. $\mathbf{P}_8\mathbf{P}_0\mathbf{V}\mathbf{T}_2$ 65°
A9	Jerusalem	بيت المقدس	66;30	32; 0	–
A10	Damascus	دمشق	70; 0	33; 0	lat. $\mathbf{P}_9$ 30′
A11	Kufa	(ال)كوفة	79;30	31;50	long. $\mathbf{T}_1\mathbf{F}_2$ 0′, lat. $\mathbf{F}_2$ 55′
A12	Baghdad	بغداد	80; 0	33;25	long. $\mathbf{P}_8\mathbf{P}_0\mathbf{V}\mathbf{T}_2$ 10′, lat. $\mathbf{P}_0\mathbf{V}$ 46′ (beginning of slide[+3])
A13	Wasit	واسط	81;30	32;20	lat. $\mathbf{P}_0\mathbf{V}$ 10′ (due to slide)
A14	Basra	(ال)بصرة	84; 0	31; 0	lat. $\mathbf{P}_0\mathbf{V}$ 10′ (due to slide)
A15	Qadisiyya	قادسيّة	79;25	31;46	$\mathbf{P}_0\mathbf{V}$ فادسه, lat. $\mathbf{P}_0\mathbf{V}$ 0′ (due to slide) $\mathbf{T}_2$ 47′
A16	Hilla	الحلّة	79;10	32;10	lat. $\mathbf{P}_0\mathbf{V}$ 0′ (due to slide)
A17	Madā᾽in	مدائن	80;20	33;10	lat. $\mathbf{P}_0\mathbf{V}$ 0′ (end of slide[+3], here or at one of the next two entries)
A18	Ahwaz	الاهواز	85; 0	30; 0	$\mathbf{T}_1\mathbf{F}_2$ entry inserted after Shiraz
A19	Shiraz	شيراز	88; 0	32; 0	$\mathbf{P}_0\mathbf{V}$ سبران, lat. $\mathbf{T}_2$ 30°
A20	Sabur	سابور	88;40	30; 0	$\mathbf{T}_1\mathbf{F}_2$ entry inserted before Shiraz, $\mathbf{T}_1\mathbf{F}_1\mathbf{P}_9$ شابور, $\mathbf{P}_0\mathbf{V}$ سانوز, long. $\mathbf{T}_2$ 0′
A21	Kirman	كرمان	100; 0	30; 0	long. $\mathbf{T}_1$ 105°, $\mathbf{F}_1\mathbf{P}_8$ 80° (undotted) $\mathbf{O}\mathbf{T}_2$ 80° (dotted)
A22	Kabul	كابل	110; 0	28; 0	$\mathbf{P}_8\mathbf{P}_0\mathbf{V}$ بحابد $\mathbf{T}_2$ عاوند, long. $\mathbf{T}_2$ 61°, lat. $\mathbf{P}_8\mathbf{P}_0\mathbf{V}\mathbf{T}_2$ 37;0
A23	Tarsus	طرسوس	67;40	37;15	$\mathbf{P}_9$ طرطوس, long. $\mathbf{F}_2\mathbf{O}$ 60°, lat. $\mathbf{P}_9$ 36° (corrected) $\mathbf{T}_2$ 45′
A24	Aleppo	حلب	71;00	35;50	lat. $\mathbf{F}_2$ 55′, $\mathbf{P}_9$ 36;0 (corrected)
A25	Manbij	منبج	63;45	35;30	$\mathbf{F}_1$ Homs حمص (but the coordinates are correct for Manbij), long. $\mathbf{P}_8\mathbf{P}_0\mathbf{V}\mathbf{T}_2$ 15′, lat. $\mathbf{P}_9$ 36;10 (corrected) **O** 34;0
A26	Homs	حمص	71; 0	33;40	$\mathbf{F}_1$ Sarakhs سرحر (but the coordinates are correct for Homs), lat. $\mathbf{F}_2$ 0′

Column 2

	locality		long.	lat.	apparatus
B1	Raqqa	الرقّه	73;15	36; 0	$\mathbf{P_8}$ لرقه $\mathbf{T_2}$ مرقه
B2	Amid	آمد	75;15	38; 0	$\mathbf{P_9}$ امل
B3	Harran	حرّان	77; 0	37; 0	long. **OV** 76;0
B4	Hama	حماه	72;40	36; 0	$\mathbf{P_9P_8P_0VT_2}$ حما
B5	Mosul	(ال)موصل	78; 0	36;30	–
B6	Antioch	انطاكيه	79; 0	35;30	$\mathbf{P_9}$ انطاله
B7	Hulwan	حلوان	81;45	34; 0	$\mathbf{P_8P_0VT_2}$ Shahrazur (beginning of slide[+1] of place names); **K** 87;0 / 35;0 (due to the slide in $\mathbf{P_8}$)
B8	Shahrazur	شهرزور	80;20	37;15	$\mathbf{P_8P_0VT_2}$ Nahawand (due to slide), $\mathbf{F_2T_2C}$ شهرور $\mathbf{OP_8P_0V}$ شهروز **J** شهرروز, long. $\mathbf{F_2J}$ 100°, lat. $\mathbf{T_2}$ 45′; **K** Shahrud, 81;45 / 34;0 (due to the slide in $\mathbf{P_8}$)
B9	Nahawand	نهاوند	82; 0	36;10	$\mathbf{P_8P_0VT_2}$ Hamadan (due to slide); **K** 80;20 / 37;15 (due to the slide in $\mathbf{P_8}$)
B10	Hamadan	همذان	83; 0	36;10	$\mathbf{P_8P_0VT_2}$ Qum (due to slide), lat. $\mathbf{P_8T_2}$ 37;0 $\mathbf{P_0V}$ 0′; **K** long. 82;0 (due to the slide in $\mathbf{P_8}$, lat. correct by coincidence)
B11	Qum	قم	80;15	34; 0	$\mathbf{P_8P_0VT_2}$ Hulwan (end of slide[+1] of place names), long. $\mathbf{P_8T_2}$ 87;0 $\mathbf{P_0V}$ 84;15, lat. $\mathbf{P_8P_0VT_2}$ 35°; **K** 83;0 / 37;0 (due to the slide in $\mathbf{P_8}$)
B12	Sawa	ساوه	87; 0	35; 5	**V** ساده **CJ** ساده, lat. $\mathbf{F_2}$ 34;5, $\mathbf{P_8P_0VT_2}$ 34;0
B13	Isfahan	اصفهان	84;40	32; 0	$\mathbf{OP_8P_0VT_2C}$ اسفهان
B14	Rayy	الريّ	85; 0	35;30	lat. $\mathbf{T_2}$ 0′
B15	Qazwin	قزوين	85; 0	37; 0	–
B16	Abhar	ابهر	85; 0	36; 0	$\mathbf{F_2}$ الهد **O** الهري, long. **K** 84°
B17	Zanjan	زنجان	84; 0	36; 0	$\mathbf{P_9OP_8P_0VT_2C}$ زيحان $\mathbf{F_2}$ ابحان, long. $\mathbf{OP_8T_2}$ 85°
B18	Daylam	الديلم	85; 0	38; 0	$\mathbf{P_8}$ الرسله $\mathbf{P_0V}$ الرسبله $\mathbf{T_2}$ الرمله
B19	Ruyan	رويان	86;35	37;10	$\mathbf{P_9}$ رونان **O** روسانه $\mathbf{P_8P_0VT_2}$ روسان, long. **O** 30′, lat. $\mathbf{P_9}$ 0′
B20	Sariya	سارية	87;50	38; 0	long. $\mathbf{F_2}$ 55′
B21	Damghan	دمغان	88;37	37; 0	$\mathbf{T_1P_9}$ دامغان, $\mathbf{F_2}$ دابعان
B22	Bistam	بسطام	89;15	35;40	$\mathbf{F_2}$ نطام $\mathbf{P_8P_0VT_2}$ اسطام, long. **V** 55′
B23	Astarabad	استراباد	89;50	38;45	**O** استارياب $\mathbf{P_8}$ سيراز $\mathbf{P_0V}$ شيراز $\mathbf{T_2}$ سيزاز **CJ** سيرات, long. $\mathbf{F_2}$ 55′, lat. $\mathbf{P_8P_0VT_2}$ 15′
B24	Jurjan	جرجان	90; 0	36;50	$\mathbf{F_2}$ (؟) خرطان $\mathbf{P_8P_0VT_2}$ اباد, long. $\mathbf{P_8T_2}$ 95°, lat. $\mathbf{F_2}$ 55′; **K** Anar/Abad?, 92;0 / 37;0 (coordinates of Tus, due to distortion of entry B23)
B25	Tus	طوس	92; 0	37; 0	$\mathbf{P_8P_0VT_2}$ Jurjan **J** جرحرن; **K** om. entry
B26	Nishabur	نيسابور	92;30	36;21	$\mathbf{T_2}$ نيابور, long. $\mathbf{OP_0V}$ 0′, lat. **O** 37°

Column 3

	locality		long.	lat.	apparatus
C1	Sarakhs	سرخس	93;20	36; 0	$\mathbf{F}_1\mathbf{F}_2$ سرحر ($\mathbf{F}_1$ with coordinates for Homs, see above) $\mathbf{P}_8$ حرحسر (?) $\mathbf{P}_0\mathbf{V}$ جرجس $\mathbf{T}_2$ حرحر
C2	Marv	مرو	94;20	37;30	**O** مرف $\mathbf{P}_8$ Balkh crossed out $\mathbf{P}_0\mathbf{V}$ مرد $\mathbf{T}_2$ بكح
C3	Bukhara	بخارا	97;20	36;50	$\mathbf{OT}_2$ بخاره, long. **O** 96°, lat. $\mathbf{F}_2$ 55′
C4	Balkh	بلخ	98;30	38;40	lat. $\mathbf{P}_9\mathbf{F}_1$ 33°; **K** om. entry
C5	Samarqand	سمرقند	99;30	36;30	lat. $\mathbf{T}_2$ 7′ **CJ** 4′
C6	Khwarizm	خوارزم	101;50	42;10	**O** حوار $\mathbf{T}_2$ خورزم, long. $\mathbf{F}_1$ 81° (undotted), $\mathbf{P}_9\mathbf{OP}_8\mathbf{P}_0\mathbf{VT}_2$ 81° (dotted) $\mathbf{F}_2$ 55′ $\mathbf{P}_8\mathbf{P}_0\mathbf{VT}_2$ 30′
C7	Darband	دربند	83; 0	42;05	–
C8	Tiflis	تفليس	83; 0	43; 0	lat. $\mathbf{T}_1\mathbf{F}_2$ 33° **O** 38° $\mathbf{T}_2$ 83° **CJ** 88°
C9	Kanja	كنجه	81;30	41; 0	$\mathbf{P}_8\mathbf{T}_2$ طنجه
C10	Baylaqan	بيلقان	82;30	38; 0	**O** سلقان **V** صلقان $\mathbf{T}_2$ تىلقان, long. $\mathbf{T}_1\mathbf{O}$ 81°, lat. $\mathbf{P}_0\mathbf{V}$ 33°
C11	Bardhaah	بردع	84; 0	43; 0	$\mathbf{P}_0\mathbf{V}$ بروح, long. $\mathbf{T}_1\mathbf{O}$ 81°, lat. $\mathbf{F}_1\mathbf{CJ}$ 48°
C12	Shirwan	شروان	84; 0	35; 0	$\mathbf{F}_1$ شروار $\mathbf{F}_2$ سرفان $\mathbf{P}_8\mathbf{P}_0\mathbf{V}$ شيروان, long. $\mathbf{P}_0\mathbf{V}$ 85°
C13	Tabriz	تبريز	83;10	37;30	$\mathbf{T}_2$ Maragha (beginning of slide[+1] of place names); $\mathbf{P}_0$ place name partially erased (…سرا ?) **V** سرليف, long. $\mathbf{F}_1\mathbf{CJ}$ 88°, lat. **K** 10′
C14	Maragha	مراغة	83;20	37;25	$\mathbf{T}_2$ Ardabil (due to slide); lat. $\mathbf{P}_9$ 20′ modified to 24 or 25?
C15	Ardabil	اردبيل	83; 0	38; 0	$\mathbf{T}_2$ Marand (due to slide); $\mathbf{P}_0\mathbf{V}$ اردنيل $\mathbf{T}_2$ اردبفد
C16	Marand	مرند	83;10	37;15	$\mathbf{T}_2$ Salmas (due to slide); $\mathbf{T}_2$ مراند, lat. **OJ** 45′
C17	Salmas	سلماس	83;10	38;30	$\mathbf{T}_2$ Khuway (due to slide); $\mathbf{P}_8\mathbf{P}_0\mathbf{VT}_2$ سلاماس, lat. $\mathbf{P}_8\mathbf{P}_0\mathbf{V}$ 33° $\mathbf{T}_2$ 7′
C18	Khuway	خوي	82; 0	38;30	$\mathbf{T}_2$ Akhlat (due to slide)
C19	Akhlat	اخلاط	74;50	39;50	$\mathbf{T}_2$ Arzan Rum (due to slide); $\mathbf{T}_1\mathbf{F}_2$ خداك **O** حدال $\mathbf{P}_8\mathbf{P}_0\mathbf{VT}_2$ خلاط, long. $\mathbf{F}_2$ 55′, lat. $\mathbf{P}_9$ 19° $\mathbf{F}_2$ 55′
C20	Arzan Rum	ارزان الروم	76; 0	39; 0	$\mathbf{T}_2$ Arzingan (due to slide); $\mathbf{F}_2\mathbf{OP}_8\mathbf{P}_0\mathbf{VT}_2$ ارز الروم, lat. $\mathbf{P}_8\mathbf{P}_0\mathbf{VT}_2$ 44′ **K** 45′
C21	Arzingan	ارزنجان	74; 0	38; 0	$\mathbf{T}_2$ Siwas (due to slide); **V** ارزنجانه $\mathbf{T}_2$ ارنجان **CJ** om., long. **V** 75°, lat. $\mathbf{P}_8\mathbf{P}_0\mathbf{VT}_2$ 37°
C22	Siwas	سيواس	71; 0	39; 0	$\mathbf{T}_2$ Malatiya (due to slide); $\mathbf{P}_9$ om. entry, lat. $\mathbf{P}_8\mathbf{P}_0\mathbf{V}$ 30° $\mathbf{T}_2$ 37°
C23	Malatiya	ملطية	71; 0	39; 0	$\mathbf{T}_2$ Qaysariyya Rum (due to slide); $\mathbf{P}_8\mathbf{T}_2$ ملاطيه
C24	Qaysariyya Rum	قيصريّة	69; 0	38;30	$\mathbf{T}_2$ Konya (end of slide[+1] of place names); **O** قيساريّه $\mathbf{F}_1\mathbf{T}_2$ قيسريه, lat. $\mathbf{P}_8\mathbf{P}_0\mathbf{VT}_2\mathbf{C}$ 2′
C25	Aqsaray	اقسرا	68; 0	38; 0	$\mathbf{P}_8\mathbf{P}_0\mathbf{VT}_2$ om. entry (or mixed up with Konya), $\mathbf{F}_2$ اقصرل (?)
C26	Konya	قونيه	65; 0	38; 0	$\mathbf{P}_8\mathbf{T}_2$ قومه $\mathbf{P}_0\mathbf{V}$ قوم, $\mathbf{P}_8\mathbf{P}_0\mathbf{V}$ coordinates for Aqsaray, $\mathbf{T}_2$ coordinates for Qaysariyya (due to slide); **K** reads *Qūma* as *Aqsarā*, presumably due to the correct coordinates.
C27	Constantinople	قسطنطنيّة	59;50	45; 0	$\mathbf{P}_0\mathbf{V}$ قسطنطنه **CJ** قسطنطينيّه, long. $\mathbf{F}_1$ 65° (?) (from Konya?) $\mathbf{P}_8$ degrees (؟) فك $\mathbf{P}_0\mathbf{V}$ degrees فها $\mathbf{F}_2$ 55′ **O** 0′, lat. $\mathbf{P}_9$ 41;0 (later modification) $\mathbf{F}_1$ 59;50 (from longitude)

4. Classifying the Errors in the Table

Variants in the place names in the geographical table from the *Shāmil Zīj* are rarer than those in the coordinates. It is often quite obvious from the omission of diacritical dots and ambiguous shapes of certain letters when a scribe did not actually know the localities. In some cases we see clear mistakes such as عند for عدن Aden (A3), the various forms for Kabul (A22) given in Table 1, Ṭanja طنجة for Kanja كنجة (C9), سلقان and صلقان for بيلقان Baylaqan (C10), etc. In some other cases we find slightly different but acceptable spellings of well-known place names, such as صنعا and صنعه for Sanaa (A2), حما and حماه for Hama (B4), or اسفهان and اصفهان for Isfahan (B13). These could have been adjusted independently by any scribe, possibly depending on his or her philological background, and therefore coincidence of the way of writing does not need to point to a close relationship of the manuscripts concerned. As mentioned in Section 3, I have not generally indicated additions or omissions of the definite article *al-* in certain place names in the apparatus of the table.

For most of the localities in the table it is relatively easy to decide on the correct values of the coordinates. For these, a clear majority of our ten witnesses are in agreement, and most of the deviating digits are obvious scribal errors. A list of common scribal errors can be found in Table 2. In this table *t* stands for a number of tens (possibly also none) and *u* for a number of units not equal to zero. For example, the scribal error denoted by 1u–5u indicates all confusions 11–51, 12–52, ..., 19–59 and the scribal error t2–t4 includes the confusions 2–4, 12–14, 22–24, ... The somewhat less common and more specific scribal errors such as the confusion of 0 and 8 (especially in sources **CJ**) and 0 and the letter *ʿayn* (in source **V**) are not included in the table. The same holds for the confusion of digits such as 0 with 30 and 30 with 40 and several others that occur again and again in geographical tables. It is less likely that these were the result of a misreading of the correct digit; instead they may be the result of miscopying from a different entry in the table, or, whenever a digit is given as zero or is entirely omitted, from truncation or rounding of values given to a higher precision. Since the possibilities for such mistakes are almost infinite, I have not attempted to explain these unless other sources provide evidence for their likeliness.[26]

[26] The eleven witnesses for the geographical table from the *Shāmil Zīj* that I have used show seven instances in which an original digit 0 was written incorrectly as 4, 5, 8, 10, 30 or 44 (of these errors only the confusion with 5 is a common scribal one). In eleven cases were original digits 10, 15, 30 and 40 minutes written as 0, which in no case can be considered a common scribal error.

scribal errors in units (t ≥ 0)			
t 0 – t 2, mainly:			
ك	20 – 22	كب	
ل	30 – 32	لب	ǀ
	t 0 – t 5 (t≠1)		ǀǀǀǀǀ
ح	0 – 5	ه	ǀ
ن	50 – 55	نه	ǀǀǀǀǀ ǀǀǀǀǀ
t 0– t 7, mainly:			
م	40 – 47	مر	
س	60 – 67	سر	ǀ
ف	80 – 87	فر	ǀ
ٮ	t 2 – t 4	ٮد	
ٮ	t 2 – t 7	ٮر	ǀǀ
ٮح	t 3 – t 4	ٮد	
ٮح	t 3 – t 8	ٮح	ǀǀǀǀǀ ǀǀǀǀǀ ǀǀǀǀǀ ǀǀ
ٮد	t 4 – t 5	ٮه	ǀǀǀǀǀ ǀ
ٮد	t 4 – t 6	ٮو	ǀ
ٮد	t 4 – t 7	ٮر	
ٮو	t 6 – t 7	ٮر	ǀǀǀǀǀ ǀǀǀǀ

scribal errors in tens (u > 0)			
ٮد	1u – 3u	لد	ǀǀ
ٮد	1u – 4u	مد	
ٮه	15 – 45	مه	ǀǀǀǀǀ ǀ
ٮد	1u – 5u	ند	ǀǀǀ
كد	2u – 3u	لد	
لد	3u – 5u	ند	
مد	4u – 5u	ند	ǀ
فد	8u – 5u	ند	ǀǀ
صد	9u – 7u	عد	ǀ
فد	8u – 10u	قد	ǀǀǀǀ
other common scribal errors			
ح	0 – 31	لا	
د	4 – 20	ك	
د	4 – 30	ل	ǀ
ز	7 – 30	ل	ǀ
ز	7 – 40	م	
ز	7 – 50	ن	
ط	9 – 20	ك	ǀǀ
ط	9 – 21	كا	
ل	30 – 50	ن	ǀ

Table 2: Common scribal errors in Arabic and Persian numerical tables in *abjad* notation. Note that the forms of the letters as printed here in some cases deviate from those written in the manuscripts, and that the probability of certain errors further depends on the particular type of script (cf. Irani, 'Arabic Numeral Forms'). Diacritical dots have been omitted from letters that often do not carry them in the manuscripts, especially: final ٮ = 2, ح = 3, ر = 7, initial or medial ٮ = 10. The examples for errors of the form t2–t4, etc. are given with an initial undotted *yā'*, the examples for errors of the form 1u–3u etc. with a final *dāl*. The ticks after each error indicate the number of occurrences in the geographical table in the Shāmil Zīj. For some general forms particular occurrences that are especially common have been listed and counted separately (for example, 0–5 and 50–55 for t0–t5).

In Section 5, I will briefly discuss the usefulness of a frequency distribution of scribal errors in astronomical tables for judging the likeliness that one table is dependent on another. As an example I have indicated by tally marks after each possible error in Table 2 its number of occurrences in the geographical table from the *Shāmil Zīj*. Because of the clear interdependence of the manuscripts (see further below), the occurrence of the same error in multiple manuscripts is counted only once. The largest frequencies are found for scribal errors that are also known from general experience with numerical tables to be among the most common ones, namely t3–t8 and t6–t7. However, some other errors known to be common do not show up here so clearly due to the special characteristics of the table. In particular, the vast majority of all numbers of minutes in the coordinates are multiples of 10 and/or 15,[27] while the degrees of the longitudes lie overwhelmingly between 60 and 99 (leaving only three values between 50 and 60 and three values above 100) and those of the latitudes exclusively between 13 and 45. As a result, such common confusions as 1u–5u, t2–t4, t3–t4, t4–t7, 40–47 and 7–50 are found only rarely in the geographical table from the *Shāmil Zīj*, whereas 1u–4u only appears in the form 15–45. Also peculiarities of the handwriting of the manuscripts or their (often unknown) precursors will influence the probability of certain scribal errors. Thus all ten errors 50–55 are due to source $\mathbf{F}_2$, whereas the confusion of digits 0 and 3, which is not included in the table, appears very frequently in, for example, Abū l-Fidā'’s *Taqwīm al-buldān*.[28]

For a significant number of localities in the geographical table from the *Shāmil Zīj*, the errors in some of the manuscripts cannot so easily be recognised as scribal errors. In these cases there are multiple plausible scribal variants or even on first sight inexplicable variants and no clear majority of the sources in favour of any one of them. Examples are the wide variety of longitude and latitude values for Qum (B11), even after correction of the slide that will be discussed below, the differing latitudes for Tiflis (C8), and a range of significant differences on which we will see that the eleven witnesses are divided into two groups of five and six manuscripts.

A type of scribal error that may cause great problems for a reliable transmission of coordinates is what the late Fritz S. Pedersen dubbed a ‘slide’:[29] while copying a row or column of tabular values or digits, the scribe skipped or repeated one or more items. As a result, all following values or digits in the row or column concerned would ‘slide’ by a number of columns or rows until the scribe discovered the mistake and continued with the correct values. In most manuscripts of mathematical tables slides allow us to see that scribes generally

[27] The minutes of only five coordinates in the geographical table in the *Shāmil Zīj* are a multiple of 5 which is *not* a multiple of 10 and/or 15. Furthermore, the table contains three ‘irregular’ numbers of minutes that can nevertheless be confirmed from most of the witnesses, and even on the basis of other sources, namely for Qadisiyya (A15), Damghan (B21) and Nishabur (B26). 74 of the 168 coordinates (i.e., nearly half of the total) have a number of minutes equal to zero.

[28] For the historian, geographer and gouvernor-prince Abū l-Fidā' (1273–1331), see the *DSB* article by Juan Vernet, the *EI*[2] article by H. A. R. Gibb, or the *EI*[3] article by Daniella Talmon-Heller. His *Taqwīm al-buldān* is particularly important for the history of geography because it systematically compares the information about many hundreds of localities on the basis of four or five earlier sources, of which some very important ones are now lost (see further Section 5). The *Taqwīm al-buldān* was edited in Reinaud and MacGuckin de Slane, *Géographie d'Aboulféda. Texte Arabe* and translated in Reinaud and Guyard, *Géographie d'Aboulféda. Traduite.*

[29] See for example, Pedersen, *The Toledan Tables*, vol. I, pp. 30–32.

	place names	*correct*		$\mathbf{P_0V}$		
A12	Baghdad	80;00	33;25	80;10	33;46	
A13	Wasit	81;30	32;20	81;30	32;10	
A14	Basra	84;00	31; 0	84; 0	31;10	
A15	Qadisiyya	79;25	31;46*	79;25	31; 0	* 47′ in **CJO**
A16	Hilla	79;10	32;10	79;10	32; 0	
A17	Madā'in	80;20	33;10	80;20	33; 0	
A18	Ahwaz	85;00	30; 0	85; 0	30; 0	
A19	Shiraz	88;00	32; 0	88; 0	32; 0	
A20	Sabur	88;40	30; 0	88;40	30; 0	

Table 3: Slide of the minutes of latitude in sources $\mathbf{P_0V}$ (slid values underlined).

	place names		coordinates	
	$\mathbf{T_1P_9F_2O}$	$\mathbf{P_8P_0VT_2CJ}$		
B7	Hulwan	Shahrazur	81;45	34; 0
B8	Shahrazur	Nahawand	80;20	37;15
B9	Nahawand	Hamadan	82; 0	36;10
B10	Hamadan	Qum	83; 0	36;10
B11	Qum	Hulwan	80;15	34; 0

Table 4: Slide of five place names in sources $\mathbf{P_8P_0VT_2CJ}$ (differences underlined).

copied tables column by column. Especially in tables with slowly increasing tabular values the slides of values in a column would only be discovered close to the end of the column.

Among the manuscripts of the geographical table in the *Shāmil Zīj* we find three examples of slides, in one case of digits, in the two other cases of place names (or, theoretically but much less plausibly, of the longitudes *and* latitudes corresponding to these place names). Table 3 illustrates the upward slide over three rows of the minutes (but not the degrees) of the latitudes of the localities Baghdad (A12) to Madā'in (A17, Ctesiphon) in sources $\mathbf{P_0V}$. This slide is relatively easy to recognize because in these witnesses the highly uncommon number of minutes '46' appears for Baghdad instead of for Qadisiyya. Table 4 illustrates an apparent slide over one row of the place names for Hulwan (B7) to Qum (B11). In this case the sources are more or less equally divided over the two variants and a decision on the correct form cannot be made without further information. We will later see that the scribe of a common ancestor of sources $\mathbf{P_8P_0VT_2CJ}$ must first have skipped Hulwan and when he noticed this mistake four rows further down, apparently 'corrected' it by inserting Hulwan with the coordinates of Qum.[30]

[30] In mathematical tables the 'lost' entries at the end of an upward slide were usually filled up with the correct ones, which would thus be repeated from the rows above. Apparently this was not considered an appropriate strategy for a slide in the place names in a geographical

	$T_1P_9F_2O$				$P_8P_0VT_2CJ$			
	place names		coordinates		place names		coordinates	
B22	Bistam	بسطام	89;15	35;40	Astam	اسطام	89;15	35;40
B23	Astarabad	أستراباد	89;50	38;45	Shiraz*	شيراز	89;50	38;15
B24	Jurjan	جرجان	90;00	36;50	Anar	اباد	95;00**	36;50
B25	Tus	طوس	92;00	37;00	Jurjan	جرجان	92;00	37;00

* سيزاز in T_2, سيزات in **CJ** ** 90;0 in P_0V

Table 5: Two significantly different versions of four rows near the bottom of the 2nd column (deviations in the second version underlined).

A set of remarkable differences extending over four consecutive lines can be found near the bottom of the second column in six of the ten witnesses (see Table 5). Although the coordinates are basically in agreement and contain only two scribal mistakes, witnesses $T_1P_9F_2O$ on the one hand and $P_8P_0VT_2CJ$ on the other here partially give entirely different localities (note that the second column of the table is missing from F_1). A plausible explanation for this confusion that requires some imagination, is that in one early manuscript Astarabad (B23) was written with a rather large vertical descent, for example:

ا سر
اباد

In the common ancestor manuscript of $P_8P_0VT_2CJ$, this might have led to the following mistakes: the initial *alif* was prepended to the preceding entry Bistam (B22) in order to produce اسطام Astam; the middle part سر was restored to a well-known locality, namely Shīrāz, although this city already appears in the first column; the final اباد became a separate locality with the coordinates of the next one, Jurjan (B24); Jurjan received the coordinates of Tus (B25), and Tus was discarded. Having only the manuscript P_8 at his disposal, Kennedy could not do any better than identifying Abād with Anār, formerly Abān, in the province of Kirman (mentioned in Le Strange, *The Lands of the Eastern Caliphate*, p. 286), which does not appear in any other of the sources in K&K.

Tables 4 and 5 illustrate only two of a rather large number of cases in which witnesses $T_1P_9F_1F_2O$ (from here on to be referred to as Group A) differ significantly from manuscripts $P_8P_0VT_2CJ$ (Group B). Table 6 lists all of these cases. It turns out that only incidentally, especially where the differences concern place names, is it possible to decide which of the two groups provides the better variant on the basis of the geographical table in the *Shāmil Zīj* alone. As for the other non-trivial cases discussed above, also here a comparison with further geographical data from Islamic sources is necessary in order to make

table, although the insertion of the omitted place name four rows further down can hardly be considered a better one.

localities		*Group A*	*Group B*
entire table		—	columns filled only with *madīna*
third column		27 localities	26 localities (Aqsaray omitted)
A4	Oman	long. 94;30	long. 74;30
A7	Egypt	lat. 29;45	lat. 29;47 ($\mathbf{P_8P_0VT_2}$) or 29;40 (**CJ**)
A8	Alexandria	long. 60;30	long. 65;30
A12	Baghdad	long. 80;00	long. 80;10
A24	Kabul	كابل	غاوند or بحابد
		lat. 28;00	lat. 37;00
A27	Manbij	long. 63;45	lat. 63;15
B7	Hulwan to B11 Qum	—	upward slide of place names
B12	Sawa	lat. 35;5*	lat. 34;0
B18	Daylam	الديلم	الرسله ($\mathbf{P_8P_0V}$) or الرمله ($\mathbf{T_2CJ}$)
B22	Bistam to B25 Tus	—	name Astarabad mutilated
B23	Astarabad / Shiraz	lat. 38;45	lat. 38;15
C1	Sarakhs	سرخس*	حرحر or حرحس
C6	Khwarizm	long. 101;50*	long. 81;30
C17	Salmas	سلماس	سلاماس
C20	Arzan Rum	lat. 39;0	lat. 39;44 (**K&K** 39;45)
C21	Arzingan	lat. 38;0	lat. 37;0
C22	Siwas	lat. 39;0	lat. 37;0 or 30;0
C23	Qaysariyya Rum	lat. 38;30	lat. 38;2
C25	Aqsaray	اقسرا	قومه (presumably for Konya)
C26	Konya	65;0 / 38;0	om. (or combined with previous line?)

Table 6: Significant differences in place names and coordinates between Group A (manuscripts $\mathbf{T_1P_9F_1F_2O}$) and Group B (manuscripts $\mathbf{P_8P_0VT_2CJ}$). The presence of further scribal errors that are not essential for the differences between the two groups is indicated in this table by an asterisk, but these errors are further ignored. They are, of course, included in the apparatus to the edition of the table in Table 1.

plausible judgements about the correctness of the variants. In the next section I will therefore introduce in more detail the Kennedys' *Geographical Coordinates of Localities from Islamic Sources* as well as the computer program that I have written on the basis of its raw data.

5. The Kennedys' Database of Islamic Geographical Coordinates

In 1987, Edward S. Kennedy and his wife Mary Helen Kennedy published their *Geographical Coordinates of Localities from Islamic Sources* (Frankfurt: Institute for the History of Arabic-Islamic Sciences, abbreviated as K&K). This book includes a total of more than 13,000 entries from 74 geographical tables and several other types of Islamic sources covering more than 2500 different localities. Every entry consists of:

- the place name in a standardised modern form;

- the Arabic form of the place name as found in the original Arabic sources (as opposed to translated and Latin sources);
- a numerical code and a three- or six-letter abbreviation for the source;
- a reference to the source usually consisting of the page or folio, column and line number at which the locality is found;
- the longitude and the latitude of the locality in degrees and minutes; and
- a field with brief comments.

The book contains four different listings of the entries, namely:

1. an alphabetical one by place name (pp. 1–386, with entries for each locality ordered by source code, i.e., roughly chronologically);
2. a listing by source (pp. 387–594, with the sources ordered chronologically and the place names for each source alphabetically);
3. a listing by increasing longitude (pp. 595–655, with abbreviated entries for each longitude ordered by increasing latitude); and
4. a listing by increasing latitude (pp. 657–709, with abbreviated entries for each latitude ordered by increasing longitude).

The Kennedys presented a measure for the dependence of a source A on a source B defined as the percentage of latitude values in A that is also found in B for the same localities. They did not include longitudes in this measure because of the different base meridians that these may refer to. They also did not consider the possibility of scribal errors. For example, the latitude value 38;25° for Baghdad in our witness **J** of the geographical table in the *Shāmil Zīj* (source ABD in K&K) is undoubtedly a scribal error for the common 33;25°, but would not contribute to the Kennedys' measure of dependence of **J** on any source with latitude 33;25°.[31]

In the early 1990s I received from Professor Kennedy a complete dump of the DBASE3 database in which the geographical data were stored at the time K&K was published. Since the collection of data was started long before computers became as versatile as they are now, the only characters used in the original data sets and in the book are capital letters, digits, periods and commas. Thus doubtful readings are indicated by a Q, standing for a question mark. Arabic place names (in the book reproduced in Arabic script) are in the database encoded by means of the letters of the alphabet and the 10 digits. For

[31] In fact, K&K gives the latitude of Baghdad in source ABD as 33;25° although it is unambiguously written as 38;25°. However, the value 38;25° is assigned to Baghdad for source GT2 (a manuscript from Gotha). Some further exploratory statistical analysis on the data of K&K were carried out in Regier, 'Kennedy's Geographical Tables'.

example, the letter C stands for *ṣād*, X for *khā'*, the letter O or the digit 0 for undotted medial *bā'*, *tā'*, *thā'*, *nūn* or *yā'*, 1 for *ghayn*, 4 for *ḍād*, etc.[32]

In 1995, I programmed in Turbo Pascal a first version of an application KaK that displays the data from K&K and allows sorting in the same four ways as in the listings in the book (and in addition by decreasing longitudes and latitudes). In order not to make standard operations such as sorting and searching too time consuming, I made it possible to load into memory only the data from particular groups of sources, e.g., early Islamic ones, late Persian ones, western Islamic ones, or instruments. Consecutive selections of the data in memory could be made by specifying the sources, place names and ranges of longitudes and latitudes to be included or excluded. In this way it became possible, for example, to select all localities from the *zīj*es of al-Battānī, Ibn Yūnus and al-Bīrūnī with longitudes between 75 and 85° and latitudes between 30 and 35°, sorted first by longitude and then by latitude.

At around the same time, Mercè Comes took over the geographical coordinates project from the Kennedys with as main purposes to convert the DBASE3 database into a more modern format, to replace the limited set of characters by a more extensive one and add Arabic transcriptions for the place names, to correct the existing entries for errors that had crept in during the process of reading the sources, and to expand the database with further sources, especially also those from the various western Islamic *zīj*es that her colleagues in Barcelona had explored. Unfortunately, Comes's untimely death in 2010 did not allow her to finish this huge project. The partially corrected data, with a number of additional sources, are now available in Microsoft Access format, but are not yet ready for publication.

Since 2012, I have continued to work on my program KaK on and off. Most importantly, I made it possible to display the longitudes from all sources with respect to the Fortunate Isles by adjusting them for their particular base meridian. Converted longitudes are indicated in a slightly different colour, and comparison with modern longitudes measured from Greenwich is made easier by making it possible to adjust these in such a way that Baghdad receives a longitude of exactly 80;0° (i.e., by adding 35;34°). Making use of these adjustments, I implemented a function for comparing any two sources, displaying the differences either in alphabetical order of place name or in the order in which the localities appear in one of the sources. I have also experimented with a measure for the dependence of geographical sources that compares the degrees and minutes of both latitude and longitude values and furthermore also attaches a weight to corresponding coordinates that differ only by a plausible scribal error. As an example, the above-mentioned pair of latitude values

[32] These conventions, together with the developments in the early stage of the project, are described in Haddad and Kennedy, 'Geographical Tables'.

38;25° and 33;25° for Baghdad might be attached a weight of 0.8 or 0.9 (as compared with 1.0 if the values were equal). I have not yet attempted to apply such weights consistently; they could be based on frequency counts of scribal errors in a large number of sources (cf. Table 2), possibly to be distinguished by certain categories such as western and eastern Islamic manuscripts. Needless to say, a database containing multiple witnesses for a large number of Islamic mathematical, astronomical and geographical tables would enormously facilitate carrying out such frequency counts.

I also added to the program KaK the entries from the original database for the headers and the cross-references from K&K, so that an export to a text file of the whole database, sorted first by numerical source code and then alphabetically by place name, produces exactly the alphabetical listing from the book. The headers and cross-references also make it possible to select, for example, all entries for Constantinople by searching for 'Istanbul', or to add localities to the current selection that appear in displayed headers with the indication 'cf.' or 'see also' by invoking the command 'Follow'. I systematized and cleaned up the references for each source, so that the entries from a particular source cannot only be displayed in alphabetical order of place name (as in the book), but also in the order in which they appear in the source. Finally, I started to make corrections to the coordinates as published in the book, in several cases by using multiple manuscripts of the sources concerned, and added several new geographical tables, such as those from the Leipzig manuscript of the *Mumtaḥan Zīj* (based on al-Battānī), the Salar Jung Museum (Hyderabad) manuscript of the *ʿAlāʾī Zīj*, and the huge Timurid table (TMR) discovered and edited by David A. King.[33] KaK is still a DOS program in a beta version, but thanks to FreePascal can also be compiled to an executable that runs on 32bit versions of Windows 7 and later. This I am happy to make available on request. Obviously, inclusion of the geographical data in a larger database of astronomical data and re-programming of the functionality of KaK in a platform-independent, mouse-driven application that accesses such a database is a desideratum.

Although my more comprehensive measure of dependence between geographical sources explained above sharpens the Kennedys' results (for example, the measure for the basically identical sources SML (*Shāmil Zīj*) and QIR (Qyriacus) increases from K&K's 68% and 71% to around 95%), the basic conclusions about dependences and the identification of the three main families of geographical sources presented on pp. xl–xliii of K&K remain valid. The Kennedys' conclusions that the likelihood of scribal errors in geographical data that were copied over and over again is particularly large, is also confirmed by my edition of the geographical table in the *Shāmil Zīj* (as well as by my edition of the table from the *Jāmiʿ Zīj* by Kūshyār ibn Labbān, recently published in the

[33] King, *World-Maps*, Section 3.3, pp. 149–61 and Appendix A, pp. 456–77.

book series *Ptolemaeus Arabus et Latinus*,[34] and by the experiences described by David A. King in his *World-Maps*). For this reason it is not possible to take the (presumably) earliest source from each of the three main traditions that Kennedy established (to which one may add Ptolemy's *Handy Tables* and *Geography* because of the influence they had on certain traditions of Islamic coordinates for particular regions and localities) and use those as reference points for all other Islamic sources. Rather, it is necessary to look at the entire evidence of coordinates for each locality in order to determine whether the surviving witnesses of the founding source for each tradition do not already include scribal errors, and which scribal errors that occurred in the transmission of the coordinates became accepted and were included in later dependent sources as well. By thus analysing the data for all important localities, one may obtain a reference table of geographical coordinates in the three main Islamic traditions to which the coordinates in all other sources may be compared, and which may hence be used to make reasonable decisions on the correct coordinates in cases where the sources yield multiple possibilities. In the following section, I will attempt the compilation of such a reference table for all localities included in the geographical table in the *Shāmil Zīj*.

6. Creating a Reference Table for the Main Islamic Geographical Traditions

The main purpose of this section is to establish a geographical reference table, presented as Table 8, showing for all localities appearing in the *Shāmil Zīj* the coordinates in the three main Islamic traditions as well as in some smaller ones. For the purpose of convenient comparison, all longitudes are given with respect to the base meridian of the Fortunate Isles. This means that 10° has been added to the longitudes from all sources that have the Western Shore as their base meridian.[35] Most of the abbreviations for the sources used in the table are briefly introduced below in the description of the traditions.[36] Since only less than half of the localities from the *Shāmil Zīj* is found in Ptolemy's *Geography* (PTO) or *Handy Tables* (HTP), and Ptolemy's coordinates were in

[34] See van Dalen, *Ptolemaic Tradition*, Section IV.13, esp. pp. 508–09.

[35] Cf. footnote 10. The possibility of different underlying meridians needs to be borne in mind in particular when it comes to judging the possibility of scribal errors in the coordinates. For example, the frequent confusion of degrees of longitude of the forms 1u and 5u (cf. Section 4) may be visible in the reference table as a confusion of longitudes 2u and 6u if the underlying sources have the Western Shore as their base meridian.

[36] Further information on these sources may be found in Kennedy and Kennedy, *Geographical Coordinates*, pp. xv–xxxvii. Note that, besides seven new sources, I use the following abbreviations different from those in K&K: ATW for ATH FID, BIRF for BIR FID, and SNJ for entries confirmed by SNB, SNH and SNS (all three conventions as in King, *World-Maps*), MAM for entries confirmed by KHU, KHZ, RES and SUH, DIM for QBL, WAB for MUN and MOS for ABD (i.e., witness **J** for the table from the *Shāmil Zīj*).

only very few cases taken over by Islamic geographers, I have here omitted the data from Ptolemy's works. Whenever coordinates in Islamic sources are equal, or close, to Ptolemy's, this will be mentioned in the comments to the reference table.

The three main Islamic traditions of geographical coordinates as they were already established by the Kennedys are the following:

1. The generic abbreviation MAM stands for the results of the geographical survey carried out under the Abbasid caliph al-Maʾmūn (r. 813–833) with the purpose of creating a world map. The largest and presumably earliest set of coordinates from this survey is contained in KHU, the *Kitāb Ṣūrat al-arḍ* by al-Khwārizmī. KHZ is a table found together with treatises attributed to al-Khwārizmī in MS Istanbul, Süleymaniye Kütüphanesi, Ayasofia 4830. RES, the *Kitāb Rasm al-rubʿ al-maʿmūr*, is said by Abū l-Fidāʾ (1273–1331) to have been translated from Greek into Arabic under al-Maʾmūn, but turns out to be another name for KHU or the corresponding world-map. SUH, the *Kitāb ʿAjāʾib al-aqālim al-sabʿa* by Suhrāb or Serapion (*c.* 930) is a reworking of KHU with some significant differences in the coordinates. K&K also includes several dozens of variant readings for MAM as found in Nallino's edition of al-Battānī's *zīj* and his article on al-Khwārizmī's geography, as well as in Honigmann's *Die sieben Klimata*.[37] For many localities the four sources that constitute the tradition of MAM in the reference table agree with each other. But in many other cases they show differences that cannot all be explained as common scribal errors. In deciding on a value for MAM in such cases, I have given preference to KHU, but may have chosen a value from one of the other three sources (or even a mixed one) if the values in later sources that usually borrow from MAM (especially YUN and KUS, for which see below) give reason to do so. In such cases relevant deviating values from KHU, KHZ, RES and SUH are given in the comments. All Maʾmūnic sources measure the longitudes from the Western Shore.
2. BIR indicates the tradition of the geographical table in al-Bīrūnī's major astronomical work, *al-Qānūn al-Masʿūdī* (Ghazna, *c.* 1030). In his *Kitāb Taḥdīd nihāyāt al-amākin*,[38] al-Bīrūnī explained the methods for establishing the longitudes and latitudes of localities by astronomical observation (solar altitude for latitudes, lunar eclipses for longitude differences) and surveying (triangulation of distances between localities).

[37] See Nallino, *al-Battānī sive Albatenii*; Nallino, 'Al-Khuwārizmī e il suo rifacimento', and Honigmann, *Die sieben Klimata*.

[38] See the edition in Bulgakov and Aḥmad, *Kitāb Nihāyāt al-amākin*; the translation in Ali, *The Determination*, and the elucidations in Kennedy, *A Commentary*.

With 604 localities, *al-Qānūn* contains one of the largest Islamic geographical tables, and it provides mostly new coordinates as compared to MAM.[39] Many of the coordinates are cited in Abū l-Fidā'’s *Taqwīm al-buldān*, which hence provides a second source for al-Bīrūnī; in K&K this source is referred to as BIR FID, in King's *World-Maps* and here as BIRF. BIR, with longitudes measured from the Western Shore, was the basis for the tradition of the geographical table from the *Sanjarī Zīj* by al-Khāzinī (Marw, *c.* 1125), which contains a selection of localities from BIR in a different arrangement by region, to which rather inaccurate *qibla* values were added. Al-Khāzinī's main table is contained in SNB, the manuscript London, British Library, Or. 6669 of the *Sanjarī Zīj*, which lacks one folio of the table. Two somewhat different extracts of the main table are found in SNH (Istanbul, Süleymaniye Library, Hamidiye 859) and SNS (Tehran, Madrasa-yi ʿĀlī-i Shahīd Muṭahharī (previously Sipāhsalār), MS 682), two abridged versions of al-Khāzinī's *zīj*. These contain some, but by far not all localities missing from SNB. Therefore the copy of al-Khāzinī's table in SHA, the *Jadīd Zīj* of Ibn al-Shāṭir (Damascus, *c.* 1365), which in several of the surviving manuscripts even maintains the exact page layout of al-Khāzinī's table, is our main source for the missing part of the table. I will follow King, *World-Maps* in using the abbreviation SNJ for al-Khāzinī's coordinates whenever the three copies of the *Sanjarī Zīj* agree or the original values can be plausibly deduced from the complete set of available sources. Two somewhat different extracts of SHA can be found in NUZ, the *Nuzhat al-nāẓir* by Shihāb al-Dīn al-Ḥalabī (Damascus, *c.* 1435), and in the slightly later HLB, *al-ʿIqd al-yamānī* by the same author. NUZ is generally more faithful to SHA. HLB, already used by King, was obviously based on the table in SHA since it maintains the same arrangement of the localities within regions and includes the same *qibla* values, but while all other sources in this tradition are based on the meridian of the Western Shore, the coordinates in HLB were adjusted to the meridian of the Fortunate Isles and corrected on the basis of al-Bīrūnī's original geographical table. Thus whenever SNJ deviates from BIR, HLB will generally follow BIR rather than SHA.

SNJ was also copied into ASH, the *Ashrafī Zīj* by Sayf-i munajjim Yazdī (Shiraz, *c.* 1303), which, however, has several distortions and in some cases appears to have chosen the coordinates from the tradition of ATW (see below). Occasionally, when the coordinate tradition for

[39] K&K does not make use of the useful edition of this table in Togan, *Bīrūnī's Picture.* For a recent study of al-Bīrūnī's coordinates of localities in Pakistan and India, see Weber, 'Neue Analysen und Identifikationen'.

a given locality is unclear, I may also resort to the evidence from two further sources that appear to have relied upon BIR to some extent, namely TAJ, the *Tāj al-azyāj* by Muḥyi 'l-Dīn al-Maghribī (Damascus, *c.* 1258), and TUQ, a treatise on the astrolabe by the otherwise unknown al-Ṭūqānī (*GAS*, vol. XIII, pp. 415–16 names him al-Ṭūqātī ('from Tokat' in present-day Turkey) and tentatively dates him to the 14th century).

3. By ATW (in K&K: ATH FID) I refer to the tradition of the anonymous *Kitāb al-Aṭwāl wa l-ʿurūḍ li-l-Furs*, which is itself lost but from which Abū l-Fidāʾ quotes the coordinates of 452 localities. These coordinates are usually different from those in MAM or BIR and are in many cases of a remarkable accuracy. ATW has been dated to the 12th or early 13th century,[40] but this leaves unexplained that the coordinates in the much smaller table (although incomplete in the unique Paris manuscript) in DST, the Ismāʿīlī astronomical handbook *Dustūr al-munajjimīn* (Alamut (?), *c.* 1110) that is almost entirely derivative from earlier works, almost fully coincide with ATW. Besides DST, a whole range of Persian *zīj*es depended heavily on ATW, especially TUS (al-Ṭūsī's *Īlkhānī Zīj*, Maragha, 1271/2), WAB (Shams al-Dīn al-munajjim al-Wābkanawī's *Muḥaqqaq Zīj*, Tabriz, *c.* 1320; MUN in K&K and King, *World-Maps*), KAS (al-Kāshī's *Khāqānī Zīj*, Kashan/Shiraz, 1413/4), ULG (Ulugh Beg's *Sulṭānī Zīj*, Samarqand, *c.* 1440), TMR (a Timurid table from the second half of the 15th century discovered, edited and analysed by David A. King; cf. footnote 33), THF (the *Tuḥfat-i sulaymānī* by Muḥammad Zamān, Meshhed, 1667/8), ZAH (an anonymous collection in a Ẓāhiriyya manuscript, now in the al-Assad National Library in Damascus), and AIN (the Mughal administrative manual *Āʾīn-i Akbarī* by Abū l-Faḍl ʿAllāmī, India, *c.* 1580).[41] All sources in this tradition, except for ATW itself, present the longitudes with respect to the Fortunate Isles. Of course, it is possible that Abu'l-Fidāʾ adjusted the longitudes of the original *Kitāb al-Aṭwāl*, since all longitudes he quotes in the *Taqwīm al-buldān* are with respect to the Western Shore.

Besides the three main traditions, several smaller coordinate traditions can be recognized as well. Some of these, and especially those that are relevant for

[40] A survey of the literature on this source may be found in *GAS*, vol. XIII, pp. 369–75. See also King, *World-Maps*, pp. 42–43.

[41] The geographical tables of al-Ṭūsī and Ulugh Beg were edited in Greaves, *Binae tabulae geographicae*, that of al-Kāshī in Kennedy and Kennedy, *Al-Kāshī's Geographical Table*.

understanding the sources of the *Shāmil Zīj*, are included in the comments for each locality in the reference table. It concerns the following works:

4. BAT⁺ is the small tradition of the *Ṣābiʾ Zīj* by al-Battānī (Raqqa, *c.* 900).[42] This further consists of MUM, the recension of al-Battānī's table in the second known manuscript copy of Yaḥyā ibn Abī Manṣūr's *Mumtaḥan Zīj* (cf. footnote 6) extant in Leipzig, and DIM, al-Dimyāṭī's (Egypt, 12th century) treatise on the determination of the *qibla* (in K&K and King, *World-Maps*: QBL). This is the only tradition in which some coordinates from Ptolemy's *Geography* (PTO) are preserved, especially for the regions around the eastern part of the Mediterranean. In some cases BAG, the *zīj* by Jamāl al-Dīn al-Baghdādī (Baghdad or Wasit, 1286), is helpful in confirming readings from BAT. Note that in his editions of al-Battānī's table Nallino applied corrections based on KHU and FID which produced geographically better coordinates that, however, are in most cases not historically attested or justified. K&K gives these corrected coordinates without further comment, but I have used those from the unique Escorial manuscript of the *Ṣābiʾ Zīj*, which are indicated in the footnotes to Nallino's editions. The longitudes in this tradition are given with respect to the meridian of the Fortunate Isles, as in Ptolemy.
5. YUN, the *Ḥākimī Zīj* by Ibn Yūnus (Cairo, *c.* 1000) stood at the basis of a tradition of several centuries of Egyptian and Yemeni *zīj*es that also included extracts of its geographical table. YUN generally follows MAM, but there are frequent exceptions that cannot all be explained as scribal errors. In almost every single case the derivative works, namely SHR, BNA=FAR, MUH=ZAD and SAN,[43] follow YUN to the letter with incidental scribal errors. Only in very few cases has a subset of these *zīj*es another particular error in common, which is then separately mentioned in the comments to the reference table. All sources in this tradition measure the longitudes from the meridian of the Western Shore.
6. KUS is the *Jāmiʿ Zīj* by Kūshyār ibn Labbān (Iran, *c.* 1025). As mentioned above (cf. footnote 34), I have edited the tables from this work, including the geographical table, from the eight extant manuscripts that contain them. KUS was obviously heavily influenced by MAM, but in

[42] After Lelewel had been the first to investigate al-Battānī's table in *Géographie du moyen âge*, Tome V (Épilogue), pp. 60–108, Nallino edited it in 'Le tabelle geografiche' and *al-Battānī sive Albatenii*, vol. II, pp. 33–54, whereas Honigmann attempted to identify some additional localities in 'Bemerkungen'.

[43] For details of these sources, see Kennedy and Kennedy, *Geographical Coordinates*, pp. xix, xxvii, xxxi–xxxii and xxxvi.

numerous cases shows small differences, often but not always common scribal errors, that were copied into later sources. The table in MUF, the *Mufrad Zīj* by al-Ṭabarī (Amul in northern Iran, *c.* 1100), is clearly based on KUS but has enough modifications to be called a separate source. As we will see, the author of the *Shāmil Zīj* took a large part of his geographical coordinates from KUS. All three tables measure their longitudes from the Fortunate Isles.

7. ALA is the *ʿAlāʾī Zīj* by al-Fahhād (Shirwan, *c.* 1176; not in K&K), which is extant in a Persian manuscript at the Salar Jung Museum in Hyderabad as well as in a Byzantine Greek translation (the latter does not contain the geographical table). The smaller table in ABH/UTT, the *Mulakhkhaṣ Zīj* by al-Abharī (northern Iraq, *c.* 1240), is clearly dependent on ALA.[44] Since the *Shāmil Zīj* shares with Ibn al-Fahhād the mean motion parameters as well as some of the planetary equations, and because the Florence manuscript of the other *zīj* by al-Abharī includes the geographical table from the *Shāmil Zīj*, it is worth checking possible dependences of the geographical table in the *Shāmil Zīj* on ALA as well. While a separate study is necessary of the rather large number of unique coordinates in ALA and ABH/UTT, occasionally the coordinates from this small tradition will be mentioned in the comments as confirmation of coordinates in other traditions. Whereas ALA has the Fortunate Isles as its base meridian, ABH/UTT is unique in measuring the longitudes with respect to a meridian 84° east of the Fortunate Isles. Of the two longitude differences of 0° in the table, Kennedy chose Basra as the most likely candidate for the base meridian of UTT. Since we now know that the author of this work is Athīr al-Dīn al-Abharī, Abhar may appear to be the more plausible candidate. However, all coordinates concerned were taken from ALA, in which al-Fahhād also assigned the longitude of 84° to the city of Bardhaah and the region of Azerbaijan in which he was active. Since both surviving manuscripts of ABH/UTT omit the indications in red ink of additive values, doubt about the intended longitudes exists in particular for some localities close to the meridian of 84;0°.

In order to establish the coordinates used in each of the above traditions, I omitted twelve small sources from K&K that seemed to fit less into the overall traditions, in particular the two Latin sources and all instruments. On the other hand, I added the seven new sources that have been mentioned above, two of which were already used in King, *World-Maps*. Of the remaining 69

[44] However, as indicated above, the Yemenī *zīj* by al-Fārisī (FAR), which uses al-Fahhād's planetary mean motions and equations, and the reworking of this *zīj* extant in the same Cambridge manuscript (SHR), borrowed their geographical data from Ibn Yūnus.

sources, several more (especially ZAY, SAA, LYD and MAR) turned out not to be particularly suited for systematic inclusion in the comparisons because they often deviate from the most common coordinates, apparently at least to some extent due to serious defects in the transmission of the sources. These works should be investigated independently in order to discover the origin of the deviations and to place their geographical data into their proper historical context. Of course, for no geographical table can it be excluded that the coordinates were more or less arbitrarily assembled from a variety of sources (possibly with different base meridians) available to the author.

Underlying my decisions on the representative coordinates in each tradition is the basic assumption, also applied by Kennedy and King, that the vast majority of coordinates were simply copied from another source of the types included in K&K rather than newly observed or taken from an entirely different kind of source. It follows that most of the deviations between coordinates within the same tradition may be expected to be the result of scribal confusions. In many cases these will be common scribal errors as listed in Table 2, but in others they may be of more complex types as we have seen in Section 4, for example a slide of place names, coordinates or individual digits, or a miscopy from a different column. Using this basic assumption, it is inevitable that some newly introduced coordinates will be overlooked if they are not attested in sources known to us or clearly stand out in a different way. For the time being, I accept this risk and use the following specific criteria for deciding on the coordinates in each tradition.

- If a majority of the sources within a tradition agree on the same coordinates for a given locality, these coordinates are chosen as the representative ones. In the Maʾmūnic tradition I attach a larger weight to the coordinates from KHU and RES and a smaller weight to those from SUH. If the Maʾmūnic sources differ among them in a significant way, I may use YUN (and incidentally some of the other sources generally dependent on MAM such as KUS and ALA) to decide in favour of one of the candidates. In the traditions of al-Bīrūnī and the *Sanjarī Zīj*, I distinguish between the original coordinates as found in BIR/BIRF/HLB (with occasional confirmations by TAJ and TUQ) on the one hand and those from SNJ/ASH/SHA/NUZ on the other. Al-Bīrūnī's coordinates are given in the main table, those from the *Sanjarī* tradition in the comments ('SNJ=BIR' indicating that they are the same). In the tradition of the *Kitāb al-Aṭwāl wa l-ʿurūḍ li-l-Furs* I will consider coordinates not quoted by Abū l-Fidāʾ but included in DST as deriving from ATW.[45] If TUS includes the earliest coordinates for a locality, or

[45] It is almost certain that Abū l-Fidāʾ did not cite all coordinates from ATW. Examples are the important cities Harran and Homs, which are missing from ATW, but for which DST

ones different from ATW/DST, these will be separately mentioned in the comments.

- Incidental deviating coordinates within each of the three main traditions are generally ignored, especially when they seem inexplicable and may, for example, result from a slide or a miscopy that we have no means of identifying. If the deviations can be explained as common scribal errors of the most likely candidates for the coordinates in a tradition, they will be taken to support these candidates.
- If two or more independent sources from one of the three main traditions and that of Ibn Yūnus share coordinates that deviate from the representative ones, these will be mentioned as a sub-tradition, also in the cases that the difference can be explained as a scribal error. The same holds if the deviating sources differ among themselves by a common scribal error but all differ from the representative coordinates of the tradition in a non-trivial way.
- For the tradition of al-Battānī I select the coordinates (possibly the longitude and latitude separately) that are found in a majority of the three sources BAT, MUM and DIM, or are confirmed by BAG. For Kushyār I always use the reliable values that I have established in my edition of KUS, which are often confirmed by MUF and frequently adopted by SML. For the *ʿAlāʾī Zīj* I take the coordinates from ALA if they are confirmed by ABH/UTT, or those from ABH/UTT if they agree among themselves and have more plausible values than ALA.
- In the reference table I use the following notations: A superscript + after a source code indicates this source together with the sources usually dependent on it. Thus SNJ$^+$ (or also SNB$^+$ if the other sources for al-Khāzinī's table do not contain coordinates for the locality concerned) stands for SNJ, ASH, SHA and NUZ, and SHA$^+$ for SHA, NUZ and HLB (whenever HLB differs from SHA, SHA/NUZ will be written out and HLB mentioned separately). ATW$^+$ includes DST, TUS, WAB, KAS, ULG, TMR, AIN, THF and ZAH. The notation TUS$^+$ is used for coordinates that are not yet present in ATW or DST but appear in TUS and the later sources from the tradition of ATW. KAS$^+$ stands for KAS and AIN, since the latter appears to have followed specific variants in the former relatively often. ULG$^+$ indicates the sources that appear to have followed Ulugh Beg's table specifically, namely TMR, THF and ZAH. In the smaller tradition of al-Battānī's table, BAT indicates that the coordinates are found in BAT or MUM,

already presents the coordinates that were also used by TUS and later sources from the ATW tradition.

BAT^{+} that they are also included in DIM. 'Not in BAT' (as opposed to 'not in BAT^{+}') therefore means that the locality concerned is only contained in DIM. KUS^{+} stands for KUS and MUF, 'not in KUS' implies that the locality is found in MUF. ALA^{+} stands for ALA together with ABH/UTT, ABH includes UTT since they are two copies of the same table. To indicate the sources generally dependent on a source XYZ without the source itself I write $\langle$XYZ$\rangle^{+}$.

- In the comment column, BIR, SNJ and ATW and the four sources incorporated in MAM are written in bold face to indicate that there are deviations in these sources from the coordinates that I established for their tradition. A notation such as SML=BIR means that the *Shāmil Zīj* uses the coordinates from the *tradition* of al-Bīrūnī, rather than the possibly different specific values from BIR mentioned in the comment column. Individual coordinates may be referred to by λ for longitude and ϕ for latitude. Deviations or elucidations of main variants given in the comment column are placed between parentheses and may be given in the form x° or y′ if they involve only individual digits. For example, the entry '**TUS**$^{+}$ 86;55/30;0 (WAB/ULG λ 15′)' for Sabur indicates that WAB and ULG have for the longitude of this city the common scribal error 86;15° instead of 86;55°.

Please bear in mind that the purpose of the reference table is to establish the particular pairs of coordinates that were most likely generally used in each of the three main and four smaller traditions, not to provide an overview of all variants found in all sources belonging to each tradition. Also note that in most cases I have only considered the coordinates of the localities that the sources have in common, but not the evidence that the layout of the tables, the order in which the entries appear, the absence or presence of certain localities, and other characteristics of the tables may provide concerning the relationships between them. It is possible that the relations that I have noticed are valid only for localities in the Middle East as covered by the *Shāmil Zīj*, whereas further differences within the traditions might be present, for instance, for localities in the Maghrib and al-Andalus.[46]

It turns out that the localities appearing in the geographical table in the *Shāmil Zīj* (and in Islamic geographical tables in general) are of a widely varying character. Certain groups of localities show only very little variation in their coordinates over the entire set of available sources and some appear in only one or two of the three major traditions. These particularly include cities (or regions) that are far away from the central Islamic lands and where no

[46] Further research in this direction and the compilation of reference tables for a much wider range of localities will be necessary to clarify such issues. See already Robles Macías, 'The Longitude of the Mediterranean' and E. Mercier, 'Mathematical Geography'.

Samarqand			Homs		
HTP	112;30	39;45	PTO	69;40	34;00
KHU	99;30	37;30	KHU	71;00	34;00
RES	99;30	37;30	KHUB	71;10	34;00
KHZ	99;30	36;30	KHZ	71;00	33;10
SUH	99;30	36;30	SUH	71;00	34;00
ATW	99;00	40;00	BAT	69;05	34;00
YUN	99;30	36;30	MUM	69;00	34;00
BNY	99;30	37;30	YUN	71;35	33;10
DIM	99;00	36;00	BNY	71;35	33;10
KUS	99;30	36;30	DIM	69;05	34;00
BIR	98;20	40;00	KUS	71;00	33;40
BIRF	98;20	40;00	BIR	71;00	33;40
MUF	99;30	37;30	ZAY	71;00	33;45
SNB	98;20	40;00	MUF	71;00	33;40
SNH	98;20	40;00	DST	70;45	34;00
SNS	98;20	40;00	SNB	71;00	33;40
ALA	99;30	36;30	SML	71;00	33;40
ABH	99;30	36;30	SAA	71;31	34;00
UTT	99;30	36;30	TUQ	71;00	33;40
SML	99;30	36;30	TAJ	71;00	34;00
YAQ	99;30	36;30	FAR	69;00	34;50
SAA	101;52	36;30	TUS	70;45	34;00
TUQ	98;20	40;00	BNA	69;00	34;50
TAJ	98;20	40;00	LYD	71;00	34;00
SHR	99;30	36;30	MAG	70;40	34;30
FAR	99;30	36;30	MAR	69;05	33;40
TUS	99;00	40;00	BAG	71;00	33;00
BNA	99;30	36;30	MUH	71;35	33;10
MAG	98;20	39;00	SAN	69;00	34;50
BAG	97;00	37;00	QYSF	71;00	34;20
MUH	99;30	36;30	ASH	71;00	33;40
SAN	99;30	36;30	WAB	70;40	34;40
ASH	98;20	40;00	MSR	71;13	33;40
WAB	98;20	40;00	SHA	71;00	33;40
SHA	98;20	40;00	MIZ	71;00	34;20
GT1	49;00	37;30	KHL	71;00	34;20
GT2	99;30	39;30	GT1	71;00	33;40
KAS	99;00	40;00	KAS	70;45	34;00
ULG	99;16	39;37	ULG	70;45	34;00
TMR	99;16	39;37	TMR	70;45	34;00
NUZ	98;20	40;00	NUZ	71;00	34;20
HLB	98;20	40;00	HLB	71;00	34;40
QIR	99;30	36;30	QIR	71;00	33;40
ZAD	99;30	36;30	ZAD	71;35	33;10
AIN	99;00	40;00	AIN	70;15	34;20
THF	99;17	39;37	MOS	71;00	33;40
MOS	99;30	36;30			
ZAH	99;16	39;37			

Table 7: Examples of a rather clean (Samarqand, on the left) and a very convoluted coordinate tradition (Homs, on the right). The coordinates were taken from K&K with incidental corrections and additions on the basis of the actual sources.

astronomical activity may be assumed to have taken place (e.g., Ethiopia and Kabul), famous cities from antiquity that were of lesser importance or were even in ruins through much of the Islamic period or not continuously in Muslim hands (e.g., Tarsus or Qaysariyya), and localities of a mythical nature such as Kang (the cupola of the world) or Yājūj wa-Mājūj (Gog and Magog from the Bible). For many of the large and important cities in the Islamic lands the variation of the coordinates is much larger, partially due to a larger number of scribal errors that occurred during a more extensive copying history, but certainly also because of borrowing from additional sources or adoption of newly observed coordinates, especially latitudes. Obviously, localities of the first category are most suitable for establishing the basic relationships between sources and the outline of the larger traditions of coordinates. Once these have been established, localities from the second category can be tackled with a little more prescience. Table 7 shows examples of a very clean coordinate tradition (Samarqand, on the left) and a much more convoluted one (Homs, on the right).[47] For many other localities one would consider such deviating coordinates as 99;16° / 39;37° for Samarqand in ULB and sources dependent on it to be the result of a non-obvious copying error, but in this case these are of course the results of the extensive, highly accurate observations at the observatory of Ulugh Beg himself.

7. Results and Conclusions

By now comparing the uncertain cases in my edition of the *Shāmil Zīj* with the reference table on pp. 552–56, it is possible to resolve most of them.

For Oman (A4), MAM and a clear majority of all other sources have λ = 94;30°, although 74;30° appears not only in half of the witnesses for the *Shāmil Zīj* but also in the *ʿAlāʾī Zīj*. We conclude that 94;30° was the original value and 74;30° a scribal mistake.

For Egypt (A7), the longitude 54;40° that persists through the entire tradition of the *Shāmil Zīj* (and was apparently later corrected only in manuscript $\mathbf{P}_8$) is obviously a mistake for the value 64;40° from MAM/BIR, most likely due to a forgotten correction for the different meridian that these sources use. Interestingly enough, with longitude 73;0° for Fustat, which in Islamic sources usually receives coordinates very close to the ones assigned to Egypt, KUS errs consistently by around 10° in the other direction. For Alexandria (A8), the longitude 60;30° in the *Shāmil Zīj* is that used by PTO, BAT and KUS and is obviously the one intended, although it places Alexandria way too far west with respect to Egypt/Fustat.

For Baghdad (A12), the longitude value 80;10° does not appear in any other traditions and is hence most likely a copying mistake (possibly also from a nearby

[47] As in the reference table, all longitudes are here given with respect to the Fortunate Isles.

Table 8: Reference table for the localities from the *Shāmil Zīj* in the main coordinate traditions.
The column headed # gives the total number of K&K database entries that were considered for each locality

place name	#	al-Maʾmūn = MAM		al-Bīrūnī = BIR		*Kitāb al-aṭwāl* ATW		*Shāmil Zīj* SML	Other traditions / Comments
Habasha/ Jarmi	17 +26	51;40	19;40	= MAM		65;00	09;30	= KUS	Not in BAT$^+$. **YUN**$^+$,KUS$^+$ 51;40/19;30. **BIR**,SNJ$^+$ Jarmi.
Sanaa	50	73;30	14;30	77;00	14;30	= BIR		= MAM/KUS	**KHU/RES** 78;30/14;30. **BAT**$^+$,BAG 73;0/14;30. **YUN**$^+$,KUS$^+$=MAM. **BIR** λ also 77;20, **SNJ**$^+$ 77;30/14;30.
Aden	49	75;00	13;00	76;00	11;00	= BIR		= MAM/KUS	**BAT**$^+$ 74;0/13;38 (DIM φ 13;0). **YUN**$^+$,BAG 75;30/13;00. KUS$^+$=MAM. MAG=BIR. **SNJ**$^+$ 76;30/11;00. **ATW** 77;0/11;0.
Oman	31	94;30	19;45	–		–		= MAM/KUS	Not in SNJ$^+$. **BAT**,**YUN**,KUS$^+$=MAM. ⟨YUN⟩$^+$ also 95;30/19;20.
Madina	56	75;20	25;00	77;30	24;00	= MAM		= MAM/KUS	**BAT**,BAG 75;0/25;0. **YUN**$^+$,TAJ,MAG 75;20/24;0. KUS$^+$=MAM. **ATW** 75;20/25;3. **SNJ**$^+$ 77;30/24;45 (**SHA**/NUZ λ 76;30).
Mecca	57	77;00	21;00	77;00	21;20	77;13	21;40	= KUS	**BAT**$^+$ 71;0/21;40 (DIM λ 71;13). **YUN**$^+$=MAM. KUS$^+$,ASH,KAS/ULG$^+$ 77;10/21;40. **SNJ**$^+$=BIR.
Egypt / Fustat	34 +24	64;40	29;55	64;40	29;55	63;00	30;10	54;40 (!) / 29;45	**KHU** Fustat 64;50[55′]/30;0. **BAT** Fustat 63;0/31;0. **YUN**$^+$ 65;0/30;0. KUS Fustat 73;0 (sic!) / 31;0. **BIR** Fustat, **SNJ**$^+$ Fustat 64;50/29;30. **ATW** Fustat, AIN=ATW. **TUS**$^+$ 63;20/30;20 (KAS/AIN λ 63;0).
Alexandria	53	61;20	31;00	62;00	30;58	61;54	30;58	= KUS	**BAT**$^+$ 60;30/30;18 (DIM φ 30;58). **YUN**$^+$=MAM. KUS$^+$ 60;30/30;20. **BIR**$^+$ φ also 30;18, **SNJ**$^+$ 61;50/30;58.
Jerusalem	58	66;00	32;00	66;00	33;00	66;30	31;50	= KUS	**BAT**$^+$=ATW. **YUN**$^+$ 67;50/32;0. KUS 66;30/32;0. MAG/WAB 68;0/31;0. **SNJ**$^+$=BIR (**SHA**$^+$=MAM).
Damascus	53	70;00	33;00	70;00	33;30	–		= MAM/KUS	**BAT**$^+$ 69;0/33;0 (=PTO). **YUN**$^+$,KUS$^+$,**DST**=MAM. **SNS**/ASH 70;0/32;30, **SHA**/NUZ 70;0/33;25. TAJ/MAG/ WAB/KAS 70;0/33;20, **TUS**/AIN=BIR, ULG$^+$ 70;0/33;15.
Kufa	54	79;30	31;50	= MAM		79;30	31;30	= MAM/KUS/BIR	**BAT**$^+$=ATW. **YUN**$^+$,KUS=MAM.
Baghdad	54	80;00	33;09	80;00	33;25	= BIR		= BIR/ATW	**BAT**$^+$=MAM. **YUN**$^+$ 80;0/33;10 (YUN also φ 33;25). KUS 75;0 (sic! corrected to 80;0 in only 2 mss) /33;0. TAJ/MAG 80;0/33;21.
Wasit	53	81;30	32;20	= MAM		= MAM		= MAM/KUS/BIR	**BAT**$^+$,BAG 81;30/32;30 (?). **YUN**$^+$ 81;30/31;30 (φ also 32;0). KUS=MAM. **BIRF**/NUZ 81;30/32;25.
Basra	53	84;00	31;00	= MAM		84;00	30;00	= MAM/KUS/BIR	**BAT**$^+$ 80;10/31;0. **YUN**$^+$,KUS$^+$,TAJ/MAG=MAM/BIR. **ATW** 84;0/30;3. **SNJ**$^+$ 85;0/31;0.
Qadisiyya	16	–		79;25	31;45	79;25	31;10	79;25 / 31;46	Not in BAT$^+$, KUS$^+$. **YUN** om./32;0, not in ⟨YUN⟩$^+$. AIN=BIR. **ATW** om. λ. Not in TUS$^+$.
Hilla[a]	7	–		–		–		79;10 / 32;10	Not in BAT$^+$, YUN$^+$,KUS$^+$. KAS 79;15/32;15. AIN 79;0/32;0.

[a] According to the *EI*^{2} this is a town on the Euphrate between al-Kufa and Baghdad, founded in 496 Hijra = AD 1102. Of the only very few sources including this locality, the *Shāmil Zīj* is the earliest.

place name	#	al-Maʾmūn = MAM		al-Bīrūnī = BIR		*Kitāb al-aṭwāl* ATW		*Shāmil Zīj* SML	Other traditions / Comments
Madāʾin	23	80;00	33;00	80;20	33;10	80;00	33;10	= BIR	**BAT**$^+$ 80;0/35;55. Not in YUN$^+$,KUS$^+$. Not in SNJ$^+$. WAB/ULG 79;0/33;10, KAS/AIN 80;20/33;0.
Ahwaz	36	85;00	32;00	= MAM		85;00	31;00	= KUS	Not in BAT. **YUN**$^+$=MAM. KUS,DIM 85;0/30;0. Not in SNJ$^+$. **ATW** 85;0/31;8.
Shiraz[b]	50	88;00	32;00	88;35	29;36	–		= MAM/KUS	**YUN**$^+$,KUS=MAM (KUS (Fatih ms)/MUF 88;0/30;0). TAJ=BIR. **SNJ**$^+$ 85;30/30;0 (HLB φ 29;30). **TUS**$^+$ 88;0/29;36. BAG/ASH/KAS 88;0/29;30.
Sabur	22	88;15	31;00	–		–		= KUS	**YUN**$^+$=MAM. KUS$^+$ 88;40/30;0. **TUS**$^+$ 86;55/30;0 (WAB/ULG λ 86;15).
Kirman	32	100;00	30;00	–		–		= MAM/KUS	Not in BAT$^+$. **YUN**$^+$,KUS$^+$=MAM. KAS/AIN 91;30/30;5.
Kabul	34	110;00	28;00	105;20	33;45	104;40	34;30	= MAM/KUS	**BAT**$^+$,**YUN**$^+$,KUS$^+$=MAM. Not in SNJ$^+$. Not in TUS/ULG.
Tarsus	35	68;00	36;15	= MAM		–		= KUS	**KHU** 68;0/36;55, **KHZ** 68;0/38;0, **SUH** 68;0/37;35. **BAT**$^+$ 67;40/36;55 (≈PTO). **YUN**=MAM. KUS 67;40/37;15. Not in SNJ, ASH/SHA 68;0/36;0. **TUS**$^+$ 68;40/36;42 (KAS/AIN 68;40/36;50).
Aleppo	52	73;00	34;30	= MAM		72;10	35;50	= KUS/ALA	**BAT**$^+$ 71;0/34;50. **YUN**/BNY 73;0/35;30. **KHZ**/⟨YUN⟩$^+$ 73;0/33;30. KUS$^+$/ALA$^+$ 71;0/35;50. **SNH** 73;0/34;50, ASH=BIR, **SHA**/NUZ 73;0/35;50. TUQ/TAJ 72;0/35;50.
Manbij	35	73;45	35;30	= MAM		–		63;45 (!) / 35;30	**BAT**$^+$ 71;15/36;15 (=HTP). **YUN**=MAM, ⟨YUN⟩$^+$ also 71;20/35;40. Not in KUS$^+$. **TUS**$^+$ 72;15/36;15 (KAS/AIN 72;50/36;30). Not in SNJ, ASH 73;0/34;50, **SHA** 73;0/35;50 (NUZ φ 36;0, HLB φ 36;15).
Homs	45	71;00	34;00	71;00	33;40	70;45	34;00	= KUS/BIR	**KHZ** 71;0/33;10. **BAT**$^+$ 69;5/34;0. KUS$^+$=BIR. **YUN**$^+$ 71;35/33;10 (⟨YUN⟩$^+$ also 69;0/34;50). Not in SNJ, ASH/**SHA**$^+$=BIR. Not in **ATW** (but in **DST**/**TUS**$^+$).
Raqqa	50	76;00	36;00	73;15	36;01	73;15	36;00	= BAT/KUS/ATW	**BAT**, KUS$^+$,ALA, TUQ,TAJ/MAG 73;15/36;0 (=ATW). **YUN**=MAM, ⟨YUN⟩$^+$ also 83;15 (sic!) / 36;0.[c] **SNJ**$^+$=BIR.
Amid	39	75;50	37;52	67;30	37;45	77;20	37;00	= BAT/KUS	**BAT**$^+$, KUS 75;15/38;0. **YUN**$^+$ 87;20 (sic!) / 37;45. **SNJ**$^+$=BIR (SNH/SNS φ 45;30). Not in TUS,WAB,ULG. KAS/AIN 77;20/37;52[42′]. TMR/ZAH 73;40/38;0.
Harran	46	77;00	37;00	67;00	37;00	–		= MAM/KUS	**KHU** 75;0/36;40, **SUH** 77;30/36;30. **BAT**$^+$,**DST**/**TUS**$^+$ 73;0/36;40. **YUN**$^+$ 77;0/39;0 (sic!). KUS$^+$=MAM. **SNJ**$^+$ 83;0/37;0 (SNH/SNS φ 32;0).
Hama	23	72;15	36;00	72;40	36;00	–		= BIR	Not in YUN$^+$,KUS$^+$,ALA$^+$. **BAT**$^+$,BAG 69;30/35;20. Not in SNJ$^+$, TUQ/HLB 71;0/35;0.
Mosul / Nineveh	44 + 13	79;00	35;30	–		77;00	36;30	= KUS	**BAT**$^+$,BAG 78;10/36;30. **YUN**,**SNH**/**SNS**=MAM. [YUN]$^+$ also 79;35/35;30. KUS 78;0/36;30. KAS/AIN 77;0/36;50, TMR/THF/ZAH 77;0/34;30. Niniveh: **BIR** 79;0/36;0, **SNB**$^+$ 79;0/35;15.

[b] Note that the modern latitude of Shiraz is 29;36°.

[c] The table in FAR does not specify from which meridian the longitudes are measured, but for nearly all localities this is obviously from the Western Shore. There are, however, some more localities (including such major ones as Madina and Mosul) for which apparently a longitude measured from the Fortunates Isles was adopted without adjustment, in this case most likely from al-Battānī or Kūshyār.

place name	#	al-Ma'mūn = MAM		al-Bīrūnī = BIR		*Kitāb al-aṭwāl* ATW		*Shāmil Zīj* SML	Other traditions / Comments
Antioch	44	71;35	34;10	71;30	34;10	–		= KUS	**KHZ/YUN** 71;35/33;10. ⟨YUN⟩$^+$ 71;35/35;30. **BAT**$^+$ 69;0/35;30 (=PTO). KUS$^+$,BAG 79;0 (sic!) / 35;30. **BIR** 71;35/34;10 (=MAM), **SNJ**$^+$=BIR. **TUS**$^+$ 71;26/35;30 (TUS φ 35;35, KAS/AIN φ 35;40).
Hulwan	45	81;45	34;00	82;15	34;00	= BIR		= MAM/KUS	**BAT**$^+$ 81;0/38;0 (MUM φ 35;0). **YUN**$^+$,KUS$^+$=MAM. **SNJ**$^+$=BIR. **ATW** 82;55/34;0.
Shahrazur	24	80;20	37;45	–		80;20	35;30	= KUS	Not in BAT. **YUN**$^+$=MAM. KUS 80;20/37;15. Not in SNJ$^+$. ULG/TMR 82;20/32;30.
Nahawand	37	82;00	36;00	86;20	35;00	83;15	34;20	= KUS	Not in BAT$^+$. **KHU/SUH** 84;0/36;0. **YUN**$^+$=MAM. KUS$^+$ 82;0/36;10. Not in SNJ$^+$. **ATW**/MAG/KAS 83;45/34;20.
Hamadan	51	83;00	36;00	85;20	34;40	83;00	35;10	= KUS	**BAT** 60;20 (sic!) / 36;0.[d] **YUN**$^+$/AIN=MAM. KUS 83;0/36;10. **SNJ**$^+$=BIR (SNH/SNS 85;20/37;30). **ATW**/MAG 84;0/35;0.
Qum	45	84;15	35;40	87;00	34;10	85;40	34;45	= KUS	**BAT**$^+$ 84;0/36;0 (DIM λ 85;0). **YUN**$^+$ 85;15/35;40. KUS 80;15/34;0. **SNB**$^+$ 87;0/35;10, ASH=BIR.
Sawa	23	–		87;00	35;00	85;00	35;00	= BIR	Not in BAT$^+$, YUN$^+$,KUS$^+$. ALA$^+$ 83;0 or 85;0/32;0.[e] **BIR** φ 35;5, **SNB**$^+$=BIR. **TUS**/KAS 85;0/36;0.
Isfahan	45	84;40	34;30	87;20	33;30	86;40	32;25	= KUS	Not in BAT. **YUN**$^+$=MAM. KUS 84;40/32;0. **SNJ**$^+$ 87;20/32;30 (ASH=ATW). **ATW** 86;40/32;40.
Rayy	51	85;00	35;45	88;00	35;35	86;20	35;35	= KUS	**BAT** 66;0 (sic!) / 36;30 (MUM λ 73;0, DIM λ 96;0). **YUN**=MAM, ⟨YUN⟩$^+$ also 93;53 (sic!) / 35;35. KUS 85;0/35;30 (λ also 81;0, 83;0, 83;15, 84;0). **SNB**$^+$=BIR (SNH/SNS φ 35;40). KAS/ULG/TMR/THF 86;20/35;0.
Qazwin	44	85;00	37;00	= MAM		85;00	36;00	= MAM/KUS	**BAT**$^+$, **YUN**$^+$,KUS=MAM. **SNB**$^+$=BIR/MAM.
Abhar	17	–		84;00	38;00	84;30	36;45	85;0 / 36;0	Not in BAT$^+$, YUN$^+$,KUS$^+$. Not in SNJ$^+$. **ATW**/AIN 84;30/36;55.
Zanjan	21	–		83;00	38;00	83;40	36;30	84;0 / 36;0	Not in BAT$^+$, YUN$^+$,KUS$^+$. **SNB**$^+$ 84;0/38;0. SML λ also 85;0. KAS/AIN 83;0/36;30.
Daylam	19	85;00	38;10	–		–		= KUS	Region. Not in BAT, YUN$^+$. DIM/KUS 85;0/38;0. **SNH/SNS** 87;0/35;0.
Ruyan	21	86;35	36;15	86;00	36;10	85;50	37;03	= KHZ/KUS	**KHZ**, BAT$^+$, KUS 86;35/37;10. **YUN**=MAM, not in ⟨YUN⟩$^+$. **BIRF** λ 87;0. Not in SNJ$^+$. Not in ⟨ATW⟩$^+$.
Sariya	38	87;50	38;00	88;00	36;15	88;00	37;00	= MAM/KUS	**BAT**$^+$, **YUN**,KUS=MAM, not in ⟨YUN⟩$^+$. **BIR**/HLB φ also 36;55, **SNJ**$^+$ 83;0 (sic!) / 36;15.
Damghan	26	–		89;30	36;20	88;55	36;20	= ABH	Not in BAT$^+$, YUN$^+$,KUS. **SNJ**$^+$=BIR. ALA 88;16/37;0, ABH 88;37/37;0.

[d] Because of the Maghribian numerals in the Escorial manuscript of BAT, 60 is denoted by *ṣād* ص, 90 by *ḍād* ض. Nallino corrected this to the more correct, but historically unjustified, 83;20 by comparing with KHU and ATW. The geographical table in the Leipzig manuscript of the *Mumtaḥan Zīj*, heavily based on al-Battānī, has longitude 79;20, al-Dimyāṭī does not give coordinates for Hamadan.

[e] Also for Darband ALA seems to have taken the longitude difference from 84° in the wrong direction, suggesting that the original geographical table from the *ʿAlāʾī Zīj* may have had 84° as its base meridian just as the *Mulakhkhaṣ Zīj* by al-Abharī (ABH/UTT), which is obviously based on it.

place name	#	al-Maʾmūn = MAM		al-Bīrūnī = BIR		*Kitāb al-aṭwāl* ATW		*Shāmil Zīj* SML	Other traditions / Comments
Bistam	25	–		89;15	36;40	89;30	36;10	89;15 / 35;40	Not in BAT^{+}, YUN^{+},KUS^{+}. ALA^{+}=BIR. **SNB**$^{+}$=BIR (**BIR**/SNB λ 89;55).
Astarabad	31	89;50	38;45	89;20	37;00	89;35	36;50	= MAM/KUS	Not in BAT. YUN,KUS=MAM. **SNJ**$^{+}$=BIR (**BIR**/**BIRF**/SNS 89;20/37;5).
Tus	38	92;50	37;00	94;30	36;20	92;30	37;00	= KUS	**KHZ**/**SUH**, **BAT**$^{+}$, KUS^{+}, WAB 92;0/37;0. **YUN**$^{+}$ 92;45/37;0. **BIR** 'Ṭābarān, capital of Tus', **SNJ**$^{+}$=BIR.
Jurjan	50	90;45	38;50	90;10	38;10	90;00	36;50	= KUS/ATW	**BAT** 95;0/40;0. **YUN**$^{+}$ 90;45/37;45. KUS^{+}=ATW. **SNJ**$^{+}$ 92;10/38;10, TAJ/HLB=BIR.
Nishabur	42	90;45	37;00	94;00	36;10	92;30	36;21	= ATW	Not in BAT, KUS^{+}. **YUN**$^{+}$ 90;45/36;0. **BIR** 'Abarshahr, citadel of Nīshāpūr', in K&K under Iranshahr, TAJ=BIR, **SNJ**$^{+}$ 92;30/36;20 (≈ATW), ASH=ATW.
Sarakhs	40	93;20	37;00	95;00	36;40	94;30	37;00	= KUS	**KHU** 93;20/38;0. **BAT**$^{+}$, **YUN**=MAM. KUS 93;20/36;0. **SNJ**$^{+}$=BIR.
Marv	48	94;20	37;35	96;30	37;40	94;00	37;40	= KUS/ALA	Not in BAT. **KHZ**/**YUN**$^{+}$ 95;27/37;35. KUS^{+},ALA^{+} 94;20/37;30. **SNJ**$^{+}$=BIR. **ATW**/TMR/THF 97;0/37;40, MAG=ATW.
Bukhara	46	97;20	36;50	97;30	39;20	= BIR		= MAM/KUS/ ALA	**KHU**/**RES** 97;20/37;50. Not in BAT. KUS^{+},ALA^{+}=MAM. **YUN**$^{+}$ 97;9/38;50. TAJ/MAG/HLB=BIR, **SNJ**$^{+}$ 96;50/39;0.[f] **ATW** 97;50/39;20, not in **DST**, ULG^{+} 97;50/39;50.
Balkh	46	98;35	38;40	101;00	36;41	= BIR		= KUS	Not in BAT. **YUN**$^{+}$ 98;35/37;40. KUS^{+},ALA^{+} 98;30/38;40. TAJ/MAG=BIR, **SNJ**$^{+}$/HLB 101;0/36;40.
Samarqand	48	99;30	36;30	98;20	40;00	99;00	40;00	= MAM/KUS/ ALA	**KHU**/**RES**,MUF 99;30/37;30. Not in BAT. **YUN**$^{+}$,KUS,ALA^{+}=MAM. **SNJ**$^{+}$,TAJ,WAB=BIR. ULG^{+} 99;16/39;37.
Khwarizm/ Gurganj	41 +15	101;50	42;10	94;01	42;17	94;05	42;45	= MAM/KUS	Not in BAT. **YUN**$^{+}$,KUS^{+}=MAM (⟨YUN⟩$^{+}$ also 106;0/40;15). TAJ,MAG,HLB=BIR, not in SNJ^{+}. Gurganj: **SNJ**$^{+}$ 94;0/42;16. **TUS**$^{+}$ 93;45/42;35 (KAS/AIN 94;0/42;45[15′] ≈ATW).
Darband	6	–		–		–		83;0 / 42;5	Only in ALA^{+},SML^{+} (not in BAT^{+}, YUN^{+},KUS^{+}). ALA 85;0/42;0, ABH 83;0 or 85;0/42;0.[g]
Tiflis	37	–		72;00	42;00	83;00	43;00	= ATW/ALA	Not in YUN,KUS^{+}. **BAT**$^{+}$ 82;0/43;0. ALA^{+},MAG=ATW. ⟨YUN⟩$^{+}$ 83;0 from African shore / 43;0. **SNJ**$^{+}$=BIR.
Kanja	19	–		–		–		= ABH	Not in MAM, BAT^{+}, YUN,KUS^{+}. ABH 81;30/41;0. ⟨YUN⟩$^{+}$ 81;30 from African shore (!) / 41;0. **BIRF** 84;0/43;10, not in SNJ^{+}. **TUS**$^{+}$ 83;0/41;20, not in KAS/AIN.
Baylaqan	16	–		74;00	39;50	–		82;30 / 38;0	Not in BAT^{+}, YUN^{+},KUS^{+},ALA. **SNJ**$^{+}$=BIR. **TUS**$^{+}$ 83;30/39;50.

[f] Apparently al-Khāzinī (or the scribe of his copy of the *Qānūn*) miscopied the coordinates of Bukhara from those for Baykand, found one line higher up in the table.
[g] cf. note e to this table.

place name	#	al-Maʾmūn = MAM		al-Bīrūnī = BIR		*Kitāb al-aṭwāl* ATW		*Shāmil Zīj* SML	Other traditions / Comments
Bardhaah	44	83;00	43;00	73;00	43;00	83;00	40;30	= KUS	**SUH**/KUS 84;0/43;0. **BAT**, ALA 84;0/42;0. **BIRF**/YUN$^+$ 83;0/48;0. **SNJ**$^+$=BIR.
Shirwan	14	–		77;30	40;50	–		84;0 / 35;0[h]	Not in BAT$^+$, YUN$^+$,KUS$^+$. **SNJ**$^+$=BIR.
Tabriz	23	–		83;10	37;30	–		= BIR/ALA	Not in BAT$^+$, YUN$^+$,KUS$^+$. ALA$^+$=BIR. **SNB**$^+$ 83;10/37;40, ASH=ATW. **TUS**$^+$ 82;0/38;0.
Maragha	25	–		83;20	37;25	81;20	37;40	= BIR	Not in BAT$^+$, YUN$^+$,KUS$^+$. ALA 82;10/37;25 (ABH φ 84;10). Not in DST, **TUS**$^+$ 82;0/37;20. **BIRF** 83;10/37;20, **SNB**$^+$ 83;10/36;25 (ASH=TUS).
Ardabil	24	–		83;00	38;00	82;30	38;00	= BIR	Not in BAT$^+$, YUN$^+$,KUS. **SNJ**$^+$ 83;0/37;50.
Marand	16	–		83;00	37;50	80;45	37;50	83;10 / 37;15	Not in BAT$^+$, YUN$^+$,KUS$^+$. **SNB**$^+$ 83;0/37;15 (SNB φ 37;55). **ATW** 82;45/37;50.[i]
Salmas	16	–		83;10	38;30	79;15	37;40	= BIR	Not in BAT$^+$, YUN$^+$,KUS$^+$. **BIRF** 83;0/38;25, not in SNJ$^+$.
Khuway	20	–		–		79;40	37;40	82;0 / 38;30	**BAT**$^+$ 82;0/41;40. Not in YUN$^+$,KUS$^+$. Not in BIR,SNJ$^+$. **BIR**$^+$/**SNB**$^+$ (Khūnaj *wa-huwa*) Khūna (included under Khuway in K&K) 83;20/38;20 (BIR φ 33;20 and 37;20).
Akhlat	47	74;50	39;50	74;50	39;40	75;50	39;20	= MAM/KUS	**BAT**$^+$ 78;0/39;20. **YUN**$^+$,KUS$^+$=MAM. **SNJ**$^+$=BIR.
Arzan Rum	16	–		–		79;00	41;00	76;0 / 39;0	**RES** 76;0/39;15. Not in BAT$^+$, YUN$^+$,KUS$^+$. Not in SNJ$^+$. **TUS**$^+$ 77;0/39;40 (KAS 79;0/41;15≈ATW).
Arzingan	11	–		–		73;00	39;50	74;0 / 38;0	Not in BAT, YUN,KUS. Not in SNJ$^+$. **TUS**$^+$ 74;0/38;0 (=SML) (KAS/AIN 78;0/39;50≈ATW).
Siwas	18	–		–		71;30	40;10	71;0 / 39;0	Not in BAT$^+$, YUN$^+$,KUS$^+$. Not in SNJ$^+$. **TUS** 71;0/39;0 (=SML). MAG/WAB 71;30/39;30, ULG/TMR 71;40/39;0. KAS/AIN=ATW.
Malatiya	39	71;00	39;00	61;00	39;00	71;00	37;00	= MAM/KUS	**BAT**$^+$, **YUN**,KUS=MAM. **BIRF**/HLB,MAG=MAM. Not in **SNJ**. **SHA**/NUZ,TAJ=BIR.
Qaysariyya Rum	17	–		–		70;00	40;00	69;0 / 38;30	Not in BAT$^+$, YUN$^+$,KUS$^+$. Not in SNJ$^+$. Not in TUS/ULG/TMR. MAG/WAB 69;0/39;30. KAS 67;15/40;40, AIN 60;15/40;0.
Aqsaray	9	–		–		67;08	40;00	68;0 / 38;0	Not in BAT$^+$, YUN$^+$,KUS$^+$. Not in SNJ$^+$. Not in TUS/ULG/TMR. MAG/WAB/HLB=SML. KAS/AIN 67;45/40;15[0′].
Konya	16	–		–		66;30	41;00	65;0 / 38;0	Not in BAT$^+$, YUN$^+$,KUS$^+$. Not in SNJ$^+$. **TUS**/ULG=SML, KAS/AIN 66;30/41;40.
Constan-tinople	53	59;50	45;00	= MAM		= MAM		= MAM/KUS	**BAT**$^+$ 56;40/43;10. **YUN**$^+$,KUS$^+$=MAM. **SNJ**$^+$=MAM. Significantly better values of φ are found in: TUQ 41;15, LYD/NUZ 41;0, BAG 41;20, ZAH 40;58.

[h] The latitude in SML is so far away from a realistic one that we may assume that some kind of mistake has occurred (note that the modern city Shirvan, in the southern part of the historical region, has longitude 48;55° (4;32° east of Baghdad, i.e. virtually at 84;32° from the Fortunate Isles) and latitude 39;55°).

[i] Since the longitude quoted by Abū l-Fidāʾ from ATW places Marand just east of Ardabil rather than several hundreds of kilometers east of it, it is unlikely that ATW had this value originally.

entry in the table from which it was copied) for the common value 80;0°. The reference table also confirms that the apparent slide of the minutes of latitude in $\mathbf{P_0V}$ from Baghdad onwards is indeed in these two sources and not in the other eight.

For Kirman (A21), 100;0° is the longitude already found in PTO, MAM and KUS. It places the city (or region) around 8° too far east with respect to Baghdad, but the variants 80° and 105° are neither elsewhere attested nor geographically better. For Kabul (A22), the latitude 28° was the traditional value in MAM, but PTO, ATW and BIR all had values much closer to the actual latitude of 34;30°. So here we can not entirely exclude that a subtradition of SML introduced the improvement 37;0° (possibly a common scribal error for the even better value 34;0°).

For Tarsus (A23), the longitude 67;40° is already found in PTO, BAT and KUS, and the values in ATW and BIR (68;0°) are very near. So 60;40° and 66;40° are undoubtedly scribal confusions. For Manbij (A25), the author of the *Shāmil Zīj* appears to have taken the longitude 63;45° from MAM and omitted to carry out the adjustment for the different meridian. Like for Egypt, he could here not directly rely on KUS. Note that the confusion of 15′ and 45′ found in five of the manuscripts is a very common one.

The coordinates that are found in $\mathbf{T_1P_9F_2O}$ for Hulwan, Shahrazur, Nahawand and Hamadan (B7–B10) are in each case those from Kūshyār's *Jāmiʿ Zīj*, which for Hulwan are the same as MAM and for the other three localities differ minimally from it by what may have been small scribal mistakes. It thus seems certain that the slide of these place names took place in Group B (i.e., sources $\mathbf{P_8P_0VT_2CJ}$). For Qum (B11) the situation is more complicated. Again $\mathbf{T_1P_9F_2O}$ present the coordinates also found in KUS, whose longitude was certainly distorted. Thus the longitudes 84;15° (=MAM) and 87;0° (=BIR) and the latitude 35;0° found in Group B may in fact have been improvements. Note that the manuscripts $\mathbf{P_8P_0VT_2C}$ also have an incorrect rendering of the name Shahrazur. Since Kennedy made use of $\mathbf{P_8}$, the combination of the incorrect spelling and the slid coordinates for Hulwan led him to identify the locality with Shahrud, which in medieval times was a tiny village just south of Bistam and hence does not appear in any other Islamic geographical tables.

For Sawa (B12) the decision is difficult because all latitude values found in the sources for the *Shāmil Zīj* are common scribal errors of each other. I decided for al-Bīrūnī's value 35;5°, also because most other cases show that $\mathbf{T_1P_9F_2O}$ are more reliable than the other witnesses. The value 35;0° does appear in some manuscripts of *al-Qānūn al-Masʿūdī*, in BIRF and in the entire tradition of al-Khāzinī. For Zanjan (B17) the coordinates in the *Shāmil Zīj* are otherwise unattested. I preferred the longitude value of 84;0° because $\mathbf{T_1P_9F_2O}$

are generally the more reliable witnesses and because this value, which is very accurate with respect to Baghdad, was also used in the tradition of al-Khāzinī.

The confusion between the entries for Bistam, Astarabad, Jurjan and Tus has already been explained in Section 4. Again the correct, undistorted entries appear in the sources from Group A (witnesses $\mathbf{T_1P_9F_2O}$). Note that the latitude 38;45° for Astarabad (as compared to 38;15° in Group B) is the value from MAM and the longitude 90;0° for Jurjan (95;0° in three sources from Group B) the value from KUS and ATW. For Khwarizm the intended longitude may be assumed to be the value 101;50° from MAM, from which also the latitude was taken. The 30′ found in Group B can be explained as a scribal error.

For Tiflis it is now clear that the intended latitude is the 43;0° from the tradition of ATW. Only very few Islamic sources give coordinates for Arzan Rum (Erzurum). The more reliable sources from Group A all have latitude 39;0°, but the value 39;44° in Group B is very close to the 39;45° quoted by Abū l-Fidāʾ for the *Kitāb Rasm al-rubʿ al-maʿmūr* (RES). Also for Qaysariyya Rum the *Shāmil Zīj* presents new coordinates. I chose the latitude 38;30° from Group A because of the lesser reliability of Group B in general and the unlikeliness of 2′ in a geographical coordinate in particular (in fact, four of the five manuscripts here have a plain horizontal bar rather than a correctly shaped *bāʾ*). Finally, the manuscripts from Group B omit the entry for Aqsaray but write its coordinates after Konya. For both cities the *Shāmil Zīj* has coordinates not found in the main tradition but in some incidental later sources.

The final version of my edition of the geographical table in the *Shāmil Zīj* as presented in Table 1 (pp. 529–31) allows us to establish a stemma for the eleven witnesses of the table, displayed in Figure 1. It is obvious that Group A ($\mathbf{T_1P_9F_1F_2O}$) and Group B ($\mathbf{P_8P_0VT_2CJ}$) form two independent branches, with Group A being generally more correct and hence closer to the original table.

Group B stands out especially because of the slide of five place names in the middle of the second column and the distorted entries, starting with Astarabad (B23), at the end of the second column. But the group also shares a large number of other peculiarities and errors not found in Group A, ranging from additional columns with only the word *madīna* 'city' and deviating spellings of place names to the mix-up of the entries Aqsaray and Konya and numerous mistakes in the coordinates (cf. Table 6). Within Group B some subgroups can be recognized. As was already to be expected on the basis of the description of the entire manuscripts, $\mathbf{P_0}$ and $\mathbf{V}$ are obviously strongly related. They share a number of deviations that do not appear in other witnesses, namely لطول instead of اطول in the title, the incorrect spelling of Aden (A3) as عند, the slide of the minutes of latitude from Baghdad (A12) to Madāʾin (A17), the addition of interlinear entries for Mardin and al-Ḥiṣn in the first two cells of the second column, the coordinates for Qum (B11), and various other devi-

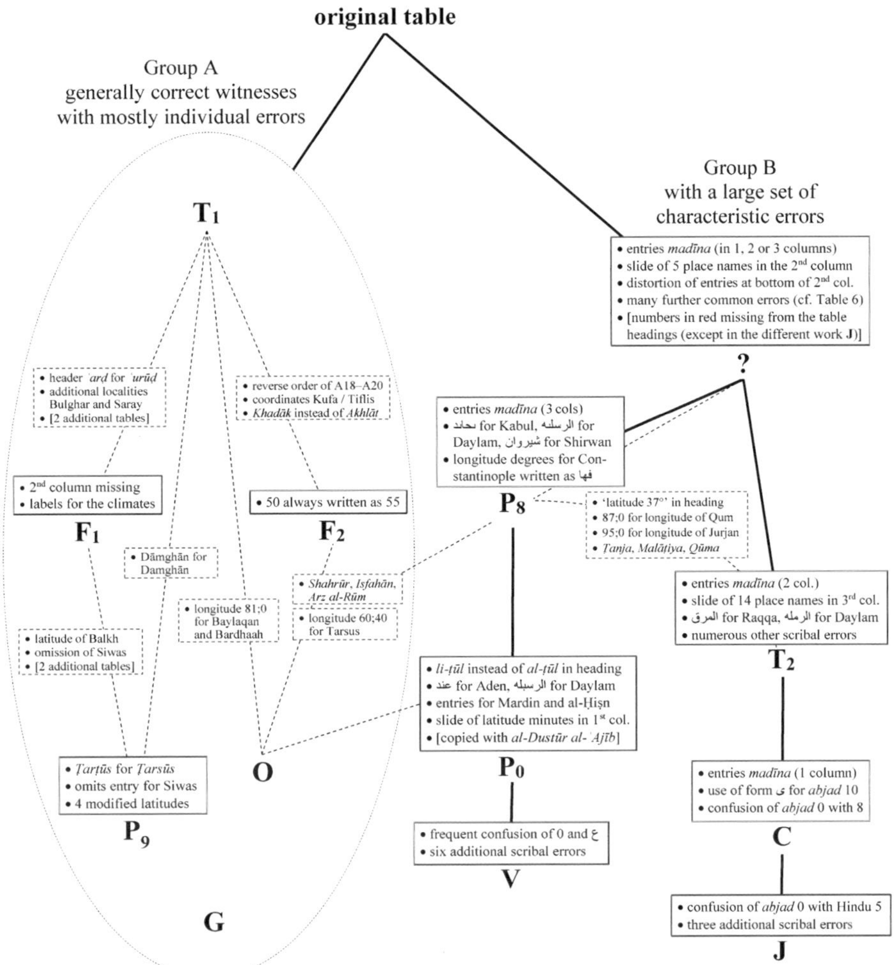

Figure 1: Stemma of the witnesses for the geographical table in the *Shāmil Zīj* on the basis of their errors and other characteristics and some additional information. Solid lines indicate direct dependences; the solid rectangles just above the sigla list the characteristics of the witness concerned that were passed on to its descendants. Dashed lines indicate the presence of shared characteristics, which are listed in dashed rectangles, in addition to individual errors. Characteristics given between square brackets relate to other tables in the manuscripts concerned. In addition to the sigla introduced in Section 2, **G** stands for the transcription of the table by Greaves with its unique errors.

ating longitudes and latitudes. Since **V** has six minor errors not found in $\mathbf{P_0}$ (namely, in the longitudes of Harran (B3), Bistam (B22) and Arzingan (C21) and the spelling of Sawa (B12), Baylaqan (C10) and Arzingan (C21)), but none the other way around, it is almost certain that **V** was copied from $\mathbf{P_0}$, which an investigation of the entire manuscripts should easily be able to confirm. Furthermore, it is likely that $\mathbf{P_0V}$ depend on $\mathbf{P_8}$ (which does not include the slide in the first column), since they follow this witness in the spelling of Sanaa (A2) as صنعه, the defective spellings of the names Kabul (A22) and Daylam (B18), the latitude for Egypt (A7), the incorrect latitude of Siwas (C22), and a peculiar writing of the degrees of longitude (59) of Constantinople (A27) (نط in $\mathbf{P}_8$, فيها in $\mathbf{P}_0$**V**). As already mentioned, **CJ** include a large number of scribal errors that are otherwise only found in $\mathbf{T}_2$ and add several further ones of their own. **J** was copied 18 years after **C** and also other parts of the *zīj* contained in it were derived from the *Shāmil Zīj*, so that it is more likely that of these two sources **C** was the original. This is confirmed by three small additional mistakes found in **J** but not in **C**.

Within Group A it is much more difficult to make out patterns, simply because the number of idiosyncratic and shared errors in each of the five witnesses is too small. $\mathbf{T}_1\mathbf{F}_2$ share the errors in the longitude of Kufa (A11) and the latitude of Tiflis (C8), the reversal of the order of entries A18 to A20, and an incorrect spelling of Akhlat (C19). Additionally, $\mathbf{T}_1$ has a scribal error in the longitude for Kirman (A21) and inexplicable errors in the longitudes of Baylaqan (C10) and Bardhaah (C11), which are further only found in **O**. $\mathbf{F}_2$ adds to the common errors the systematic confusion of 50 with 55′ and incorrect spellings of Abhar and several other localities. It thus seems clear that $\mathbf{T}_1$ and **F** cannot have been copied from each other directly, but must have derived from a common ancestor. $\mathbf{P}_9$ has seven small deviations in longitudes and latitudes that are not found in any of the other sources and are no common scribal errors. Near the end of the first column several coordinates appear to have been corrected, as was the latitude of Constantinople (C27), which was traditionally much too high in most Islamic sources. Besides the omission of the entire second column and seven deviations in coordinates that are all common scribal errors and only occasionally appear in other witnesses, $\mathbf{F}_1$ has several distorted place names and some mix-ups at the transition from the first to the third column and for Constantinople. Also **O** has mostly individual errors, most of which are common scribal errors. The inexplicable errors in the longitudes of Baylaqan (C10) and Bardhaah (C11) suggest that $\mathbf{T}_1$ and **O** had a common ancestor. The addition of interlinear entries for Mardin and al-Ḥiṣn with different longitudes from $\mathbf{P}_0$**V** and added latitudes, may suggest that the Priest Cyriacus at least consulted a copy of the *Shāmil Zīj* from the family of $\mathbf{P}_0$**V**.

The reference table also allows us to easily identify the main sources for the geographical table in the *Shāmil Zīj*. The latter has 50 of its 79 localities in common with the table in Kūshyār ibn Labbān's *Jāmiʿ Zīj*, for 49 (!) of which the coordinates are identical (the only exception being Baghdad (A12), for which Kūshyār has the anomalous longitude 75;0°). In 23 of these 49 cases the same coordinates are also found in the Maʾmūnic tradition, in four cases each in al-Bīrūnī and in the *ʿAlāʾī Zīj*. Because we already know that the author of the *Shāmil Zīj* adopted some of Kūshyār's tables for planetary equations, it is thus most probable that KUS was his main source. For seven localities the coordinates given in the *Shāmil Zīj* further only appear in the tradition of al-Bīrūnī's *al-Qānūn*. The geographical table from the *ʿAlāʾī Zīj* is unlikely to have been an important source since it has only 32 localities in common with the *Shāmil Zīj*, and among these the coordinates are identical, or differ merely by a common scribal error, for only 19. For 18 localities, the *Shāmil Zīj* presents coordinates that are not found in any of the seven main and smaller traditions considered in the reference table.

To conclude, we have seen in this article how geographical tables in Islamic sources can be riddled with scribal errors of different types, and how even the availability of multiple copies of the same table may not suffice to establish the original coordinates reliably. We had to make use of the collection of longitudes and latitudes in Kennedy and Kennedy, *Geographical Coordinates* (K&K) in order to be able to decide on the most likely original coordinates for more than a dozen of localities from the geographical table in the *Shāmil Zīj*. For this purpose, the reference table that I have set up of the coordinates in each of the main Islamic traditions turned out to be extremely helpful. The expansion of this table to a larger set of several hundreds of the most important localities, as well as its implementation in the form of a graphical computer database that can conveniently show all traditions and their variations, would be highly desirable. The implementation of a database of geographical coordinates based on K&K, taking into account corrections and additions that Mercè Comes and I have already made and including further ones, will not only facilitate the creation of a more extensive geographical reference table, but also analyses of the probability of scribal errors in numerical values in Arabic and Persian sources.

Acknowledgements

This article is dedicated to the memory of Edward S. Kennedy (1912–2009) and his wife Mary Helen Scanlon (1923–2018). Particularly by adding new sources to their database of geographical coordinates in Islamic sources and trying to identify hitherto unknown localities, I realised the enormous extent of their pioneering work in this direction. I am grateful to David A. King, Clemency Montelle and Matthieu Husson for their very useful comments on an earlier version of this article, and to Sonja Brentjes for last-minute corrections. I also very much appreciate the assistance of the Written Heritage Research Institute (Miras-e Maktoob, Tehran), Flora Vafea (Cairo) and José Bellver (at the time, Ptolemaeus Arabus et Latinus) in obtaining access to scans of manuscripts of the *Shāmil Zīj*.

Bibliography

Ali, Jamil, *The Determination of the Coordinates of Positions for the Correction of Distances between Cities. A Translation from the Arabic of al-Bīrūnī's Kitāb Taḥdīd nihāyāt al-amākin litaṣḥīḥ masāfāt al-masākin*, Beirut: American University of Beirut, 1967.

Assemanus, Josephus Simonius, *Bibliothecae Medicae Laurentianae et Palatinae codicum mss. orientalum catalogus*, Florentiae: ex Typographio Albiziano, 1742.

BEA = Hockey, Thomas (ed.), *The Biographical Encyclopedia of Astronomers*, 2 vols, Dordrecht: Springer, 2007. Second edition in 4 vols, New York: Springer, 2014.

Bellhouse, David R., 'An Analysis of Errors in Mathematical Tables', *Historia Mathematica* 42 (2015), pp. 280–95.

Berggren, J. Lennart, 'The Origins of al-Bīrūnī's 'Method of the Zījes' in the Theory of Sundials', *Centaurus* 28 (1985), pp. 1–16.

Bulgakov, Pavel G., and Imām Ibrāhīm Aḥmad, *Kitāb Nihāyāt al-amākin li-taṣḥīḥ masāfāt al-masākin li-Abī Rayḥān Muḥammad ibn Aḥmad al-Bīrūnī al-Khwārizmī*, Cairo: Maṭbaʿat Lajnat al-Taʾlīf wa-l-Tarjama wa-l-Nashr, 1964.

Carra de Vaux, Bernard, 'L'Almageste d'Abû'lwéfa Albûzdjâni', *Journal Asiatique. 8e série* 19 (1892), pp. 408–71.

Comes, Mercè, 'The "Meridian of Water" in the Tables of Geographical Coordinates of al-Andalus and North Africa', *Journal for the History of Arabic Science* 10 (1992–1994), pp. 41–51. Reprinted in Mònica Rius Piniés and Susana Gómez Muns (eds), *Homenaje a Mercè Comes. Coordenadas del Cielo y de la Tierra*, Barcelona: Publicacions i Edicions de la Universitat de Barcelona, 2014, pp. 377–88.

van Dalen, Benno, 'A Statistical Method for Recovering Unknown Parameters from Medieval Astronomical Tables', *Centaurus* 32 (1989), pp. 85–145. Reprinted in van Dalen, *Islamic Astronomical Tables*, chapter I.

van Dalen, Benno, *Ancient and Mediaeval Astronomical Tables: Mathematical Structure and Parameter Values*, PhD dissertation, University of Utrecht: Mathematical Institute, 1993.

van Dalen, Benno, 'The *Zīj-i Nāṣirī* by Maḥmūd ibn ʿUmar. The Earliest Indian-Islamic Astronomical Handbook with Tables and Its Relation to the *ʿAlāʾī Zīj*', in Charles Burnett, Jan P. Hogendijk, Kim Plofker and Michio Yano (eds), *Studies in the History of the Exact Sciences in Honour of David Pingree*, Leiden / Boston: Brill, 2004, pp. 825–62. Reprinted in van Dalen, *Islamic Astronomical Tables*, chapter VI (newly formatted with the original page numbers indicated).

van Dalen, Benno, *Islamic Astronomical Tables. Mathematical Analysis and Historical Investigation*, Farnham UK: Ashgate, 2013.

van Dalen, Benno, *Ptolemaic Tradition and Islamic Innovation: the Astronomical Tables of Kūshyār ibn Labbān*, Turnhout: Brepols, 2021.

Dallal, Ahmad S., 'Al-Bīrūnī on Climates', *Archives Internationales d'Histoire des Sciences* 34 (1984), pp. 3–18.

DSB = Gillispie, Charles C. (ed.), *Dictionary of Scientific Biography*, 14 vols plus 2 supplements, New York: Scribner's Sons, 1970–1980.

EI[2] = P. J. Bearman et al. (eds), *The Encyclopaedia of Islam. New Edition*, 11 vols plus supplement and index, Leiden: Brill, 1960–2004.

EI[3] = Kate Fleet et al. (eds), *Encyclopaedia of Islam Three*, Leiden: Brill, 2007–.

Flügel, Gustav, *Kashf al-ẓunūn ʿan asāmī al-kutub wa-l-funūn. Lexicon bibliographicum et encyclopaedicum a Mustafa ben Abdallah Katib Jelebi dicto et nomine Haji Khalifa celebrato compositum*, 7 vols, London: Bentley / Oriental Translation Fund of Great Britain and Ireland, 1835–1858.

GAS = Sezgin, Fuat, *Geschichte des arabischen Schrifttums*, 17 vols, Leiden: Brill / Frankfurt am Main: Institut für Geschichte der Arabisch-Islamischen Wissenschaften, 1967–2015.

Greaves, John, *Binae tabulae geographicae, una Nassir Eddini Persae, altera Ulug Beigi Tartari*, London: Jacob Flesher, 1652.

Haddad, Fuad I., and Edward S. Kennedy, 'Geographical Tables of Medieval Islam', *al-Abhath* 24 (1971), 87–102. Reprinted in *SIES*, pp. 636–51.

Hasse, Dag Nikolaus, 'Mosul and Frederick II Hohenstaufen: Notes on Aṯīraddīn al-Abharī and Sirāǧaddīn al-Urmawī', in Isabelle Draelants, Anne Tihon and Baudoin van den Abeele (eds), *Occident et Proche-Orient: contacts scientifiques au temps des Croisades. Actes du colloque de Louvain-la-Neuve, 24 et 25 mars 1997*, Turnhout: Brepols, 2000, pp. 145–63.

Honigmann, Ernst, 'Bemerkungen zu den geographischen Tabellen al-Battānī's', *Rivista degli studi orientali* 11 (1926–1928), pp. 169–75.

Honigmann, Ernst, *Die sieben Klimata und die πόλεις ἐπίσημοι. Eine Untersuchung zur Geschichte der Geographie und Astrologie im Altertum und Mittelalter*, Heidelberg: Winter, 1929.

Ḥusaynī Ashkawarī, Jaʿfar, *Fihrist-i nuskha'hā-i khaṭṭī-i Kitābkhānah-i Majlis-i Shūrā-i Islāmī*, vol. XX, Tehran: Kitābkhānah, Mūzih va Markaz-i Asnād, Majlis-i Shūrā-yi Islāmī, 1389 HS.

Irani, Rida A. K., 'Arabic Numeral Forms', *Centaurus* 4 (1955), pp. 1–12.

ʿIrfānian, Ghulāmʿalī , *Fihrist-i kutub-i khaṭṭī-i Kitābhānah-i Markazī-i Āstān-i Quds-i Raẓavī*, vol. X, Meshhed: Kitābhānah-i Markazī-i Āstān-i Quds-i Raẓavī, 1983.

Kennedy, Edward S., 'A Survey of Islamic Astronomical Tables', *Transactions of the American Philosophical Society. New Series* 46 (1956), pp. 123–77. Reprinted with pages renumbered 1–55, Philadelphia: American Philosophical Society, 1989.

Kennedy, Edward S., 'Comets in Islamic Astronomy and Astrology', *Journal of Near Eastern Studies* 16 (1957), pp. 44–51. Reprinted in *SIES*, pp. 311–18.

Kennedy, Edward S., *A Commentary upon Bīrūnī's Kitāb Taḥdīd al-Amākin. An 11th century treatise on mathematical geography*, Beirut: American University of Beirut, 1973.

Kennedy, Edward S., 'Mathematical Geography', in Roshdi Rashed and Régis Morelon (eds), *Encyclopedia of the History of Arabic Science*, vol. I: *Astronomy – Theoretical and Applied*, London / New York: Routledge, 1996, pp. 185–201.

Kennedy, Edward S., and Muhammad Agha, 'Planetary Visibility Tables in Islamic Astronomy', *Centaurus* 7 (1960), pp. 134–40. Reprinted in *SIES*, pp. 144–50.

Kennedy, Edward S., and Mary-Helen Kennedy, *Al-Kāshī's Geographical Table*, Philadelphia: American Philosophical Society, 1987 (*Transactions*, vol. 77).

Kennedy, Edward S., and Mary Helen Kennedy, *Geographical Coordinates of Localities from Islamic Sources* (also abbreviated as K&K), Frankfurt am Main: Institut für Geschichte der Arabisch-Islamischen Wissenschaften, 1987.

King, David A., *Fihris al-makhṭūṭāt al-ʿilmiyya al-maḥfūẓa bi-Dār al-Kutub al-Miṣriyya* (*A Catalogue of the Scientific Manuscripts in the Egyptian National Library*, in Arabic), 2 parts, Cairo: General Egyptian Book Organisation, 1981–1986.

King, David A., *A Survey of the Scientific Manuscripts in the Egyptian National Library*, Winona Lake: Eisenbrauns / Cairo: The American Research Center in Egypt, 1986.

King, David A., 'Some Early Islamic Tables for Determining Lunar Crescent Visibility', in David A. King and George Saliba (eds), *From Deferent to Equant: A Volume of Studies in the History of Science in the Ancient and Medieval Near East in Honor of E. S. Kennedy*, New York: The New York Academy of Sciences, 1987, pp. 185–225. Reprinted in *Id.*, *Astronomy in the Service of Islam*, Aldershot: Variorum, 1993, chapter II.

King, David A., 'Universal Solutions in Islamic Astronomy', in J. Lennart Berggren and Bernard R. Goldstein (eds), *From Ancient Omens to Statistical Mechanics. Essays on the Exact Sciences Presented to Asger Aaboe*, Copenhagen: University Library, 1987, pp. 121–32. Reprinted in *Id.*, *Astronomy in the Service of Islam*, Aldershot: Variorum, 1993, chapter VI.

King, David A., *World-Maps for Finding the Direction and Distance to Mecca. Innovation and Tradition in Islamic Science*, London: Al-Furqān Islamic Heritage Foundation / Leiden: Brill, 1999.

King, David A., 'Bringing Astronomical Instruments Back to Earth—The Geographical Data on Medieval Astrolabes (to ca. 1100)', in Lodi Nauta and Arjo Vanderjagt (eds), *Between Demonstration and Imagination. Essays in the History of Science and Philosophy Presented to John D. North*, Leiden: Brill, 1999, pp. 3–53.

King, David A., and Julio Samsó, 'Astronomical Handbooks and Tables from the Islamic World (750–1900): An Interim Report', *Suhayl* 2 (2001), pp. 9–105.

K&K: see Kennedy and Kennedy, *Geographical Coordinates*.

Krause, Max, 'Stambuler Handschriften islamischer Mathematiker', *Quellen und Studien zur Geschichte der Mathematik, Astronomie und Physik, Abteilung B: Studien* 3 (1936), pp. 437–532.

Lelewel, Joachim, *Géographie du moyen âge*, 4 tomes et une épilogue, Brussels: Pilliet, 1852–1857.

Le Strange, Guy, *The Lands of the Eastern Caliphate. Mesopotamia, Persia, and Central Asia from the Moslem conquest to the time of Timur*, Cambridge: Cambridge University Press, 1905.

Levi della Vida, Giorgio, *Secondo elenco dei manoscritti arabi islamici della Biblioteca Vaticana*, Città del Vaticano: Biblioteca Apostolica Vaticana, 1965.

Maddison, Francis, 'Greaves, John', in *Oxford Dictionary of National Biography*, https://doi.org/10.1093/ref:odnb/11371, published online 2004, accessed 4 March 2019.

MAOSIC = Boris A. Rosenfeld and Ekmeleddin İhsanoğlu, *Mathematicians, Astronomers, and other Scholars of Islamic Civilization and their Works (7th–19th c.)*, Istanbul: Research Centre for Islamic History, Art and Culture (IRCICA), 2003.

Mercier, Eric, 'Mathematical Geography in the Western Islamic World: Geographical Coordinates of Localities in the al-Maghreb and al-Andalus Localities (9th-18th Centuries)', *Suhayl* 18 (2020/21), pp. 25–49.

Mercier, Raymond, 'English Orientalists and Mathematical Astronomy', in Gül A. Russell (ed.), *The 'Arabick' Interest of the Natural Philosophers in Seventeenth-Century England*, Leiden: Brill, 1994, pp. 158–214.

Moussa, Ali, 'Mathematical Methods in Abū al-Wafā''s *Almagest* and the *qibla* Determinations', *Arabic Sciences and Philosophy* 21 (2011), pp. 1–56.

Nallino, Carlo Alfonso, 'Le tabelle geografiche d'al-Battânî', *Cosmos di Guido Cora, Serie II* 12 (1894–1896), pp. 161–83.

Nallino, Carlo Alfonso, *al-Battānī sive Albatenii opus astronomicum. Ad fidem codicis Escurialensis arabice editum. Latine versum, adnotationibus instructum*, 3 vols, Milan: Ulrico Hoepli, 1899–1907.

Nallino, Carlo Alfonso, 'Al-Khuwārizmī e il suo rifacimento della Geografia di Tolomeo', in Maria Nallino (ed.), *Raccolta di scritti editi e inediti, vol. 5 (Astrologia – astronomia – geografia)*, Rome: Istitut per l'Oriente, 1944, pp. 458–532.

Nicoll, Alexander, *Bibliothecae Bodleianae codicum manuscriptorum orientalium catalogi. Partis secundae volumen primum. Arabicos complectens*, Oxford: Clarendon, 1821.

OALT = Ekmeleddin İhsanoğlu (ed.), *Osmanlı Astronomi Literatürü Tarihi*, 2 vols, Istanbul: IRCICA, 1997.

Pedersen, Fritz S., *The Toledan Tables. A Review of the Manuscripts and the Textual Versions with an Edition*, 4 vols, Copenhagen: Reitzels, for Det Kongelige Danske Videnskabernes Selskab, 2002.

Pingree, David, *The Astronomical Works of Gregory Chioniades*, vol. I: *The Zīj al-'Alā'ī*, 2 parts, Amsterdam: Gieben, 1985–1986.

Regier, Mary H., 'Kennedy's Geographical Tables of Medieval Islam: An Exploratory Statistical Analysis', in David A. King and George Saliba (eds), *From Deferent to Equant: A Volume of Studies in the History of Science in the Ancient and Medieval Near East in Honor of E. S. Kennedy*, New York: The New York Academy of Sciences, 1987, pp. 357–72.

Reinaud, Joseph Toussaint, and Stanislas Guyard, *Géographie d'Aboulféda. Traduite de l'arabe en français*, 2 vols, Paris: Imprimerie Royale, 1848–1883.

Reinaud, Joseph Toussaint, and William MacGuckin de Slane, *Géographie d'Aboulféda: Texte Arabe*, Paris: Imprimerie Royale, 1840.

Robles Macías, Luis A., 'The Longitude of the Mediterranean Throughout History: Facts, Myths and Surprises', *e-Perimetron* 9 (2014), pp. 1–29.

Saliba, George, 'Easter Computation in Medieval Astronomical Handbooks', *al-Abhath* 23 (1970), pp. 179–212. Reprinted in *SIES*, pp. 677–709.

Saliba, George, 'The Double-Argument Lunar Tables of Cyriacus', *Journal for the History of Astronomy* 7 (1976), pp. 41–46.

Saliba, George, 'The Planetary Tables of Cyriacus', *Journal for the History of Arabic Science* 2 (1978), pp. 53–65.

al-Shanṭī, ʿIṣām Muḥammad, *Fihris al-makhṭūṭāt al-muṣawwara*, Part III: *al-ʿUlūm. al-Falak, al-tanjīm, al-mīqāt. al-Qism al-thānī* (*Catalogue of Microfilmed Manuscripts*, Part 3: *Science. Astronomy, Astrology, Timekeeping. Section 2*), Cairo: Institute of Arabic Manuscripts 1999.

SIES = David A. King and Mary-Helen Kennedy (eds), *Studies in the Islamic Exact Sciences by E. S. Kennedy, Colleagues and Former Students*, Beirut: American University, 1983.

de Slane, William [MacGuckin], *Catalogue des manuscrits arabes*, Paris: Imprimerie Nationale, 1883–1895.

Suter, Heinrich, 'Nachträge und Berichtigungen zu "Die Mathematiker und Astronomen der Araber und ihre Werke"', *Abhandlungen zur Geschichte der mathematischen Wissenschaften mit Einschluß ihrer Anwendungen* 14 (1902), pp. 157–85.

Thomann, Johannes, 'Scientific and Archaic Arabic Numerals: Origins, Usages and Scribal Traditions of the Two *Abjad* Systems', Sciamvs 19 (2018), pp. 167–200.

Togan, Ahmet Zeki Velidi, *Bīrūnī's Picture of the World*, Delhi: Archaeological Survey of India, 1941.

Weber, Claudia Michaela, 'Neue Analysen und Identifikationen zu al-Bīrūnīs (gest. 440/1048) Bestimmungen geographischer Koordinaten in Indien und Pakistan', *Zeitschrift für Geschichte der Arabisch-Islamischen Wissenschaften* 22 (2020), pp. 198–257.

Yaltkaya, Şerefettin, and Kilisli Rifat Bilge, *Kâtib Çelebi. Keşf-el-zunun*, 2 vols, İstanbul: Maarif Matbaasi, 1941.

Indexes

Historical Persons

The Arabic definite article *al-* is not considered in the alphabetisation; likewise, *ibn* 'son (of)' is not considered in the alphabetisation unless it is written with a capital.

Works

An initial *Kitāb* ('book') in Arabic titles is abbreviated as '*K.*' and is not considered in the alphabetisation. Likewise, the Arabic definite article *al-* and the Hebrew definite article *ha-* are not considered in the alphabetisation.

Manuscripts

Parameters

Trigonometry

Spherical astronomy

Sun

Geography

Others

Places

This index does not include localities that occur only in the tables of the article by van Dalen, but not in the main text. The Arabic definite article *al-* is not considered in the alphabetisation.

Modern Persons

Contributors

José Chabás is professor emeritus at the Universitat Pompeu Fabra, Barcelona, and has published extensively on the history of astronomy in the late Middle Ages and the early modern period.

Benno van Dalen is a researcher in the history of Islamic astronomy and is one of the two research leaders of the project Ptolemaeus Arabus et Latinus at the Bayerische Akademie der Wissenschaften in Munich.

Seb Falk is a Fellow of Girton College, University of Cambridge (UK). He researches the cultures and practices of astral and mathematical sciences, focusing on late-medieval England and its connections with the wider world.

Bernard R. Goldstein is University Professor Emeritus at the University of Pittsburgh (USA) and has recently completed a series of books and papers that began in the 1970s, some jointly authored by José Chabás, on medieval Hebrew astronomical tables.

Sho Hirose is an independent researcher based in Tokyo (Japan), who specializes in the history of astronomy in Sanskrit sources. He is the author of *A World History of Astronomy* (in Japanese).

Matthieu Husson is a researcher in the history of late medieval astronomy in Europe and is the principal investigator of the ERC project ALFA (Shaping a European scientific scene: Alfonsine astronomy) based at the Paris Observatory.

Richard L. Kremer studies medieval and early modern European astronomy and is professor emeritus of History at Dartmouth College in Hanover (USA). He is associate editor and reviews editor of the *Journal for the History of Astronomy*.

Li Liang is a researcher in the history of astronomy in early modern China and is a professor at the Institute for the History of Natural Sciences, Chinese Academy of Sciences, in Beijing.

Anuj Misra is a historian of mathematics who works on early modern sources in Islamicate and Sanskrit mathematical astronomy. He is currently a Marie Skłodowska-Curie Fellow at the Department of Cross-Cultural and Regional Studies at the University of Copenhagen.

Clemency Montelle is a researcher in the history of mathematics and astronomy in India and is a professor in the School of Mathematics and Statistics at the University of Canterbury, Christchurch (New Zealand).

Kailyn Pritchard is a researcher in the history of trigonometry in medieval and early modern Europe and a graduate student in the Mathematics Department at Simon Fraser University, in Burnaby, British Columbia (Canada).

Johannes Thomann is a researcher in the history of science in the Islamic world, and is a visiting scholar at the Institute of Asian and Oriental Studies at the University of Zurich.

Glen Van Brummelen is a researcher in the mathematical astronomy of medieval Europe and Islam, and is Dean of the Faculty of Natural and Applied Sciences at Trinity Western University, Vancouver (Canada).

Colour Plates

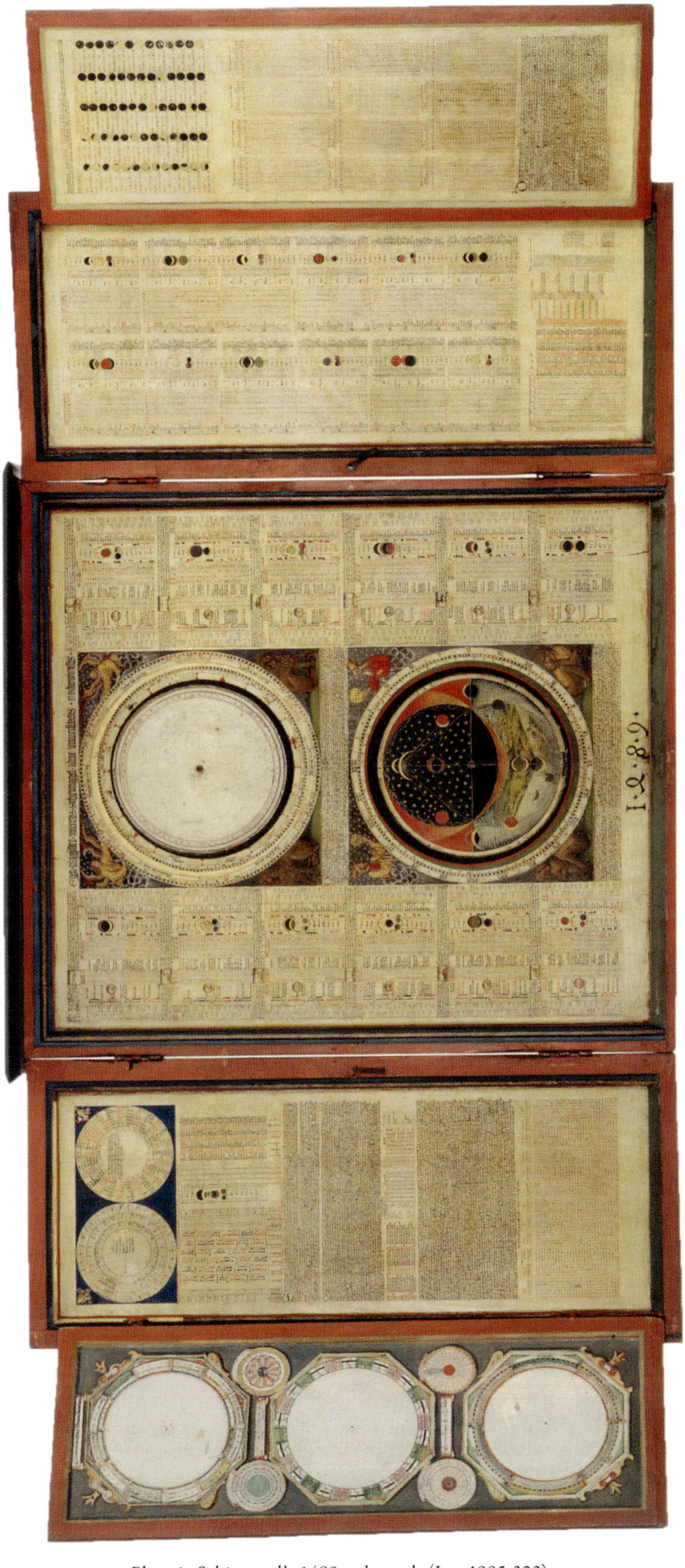

Plate 1: Schinnagel's 1489 polyptych (Inv. 1995-323).

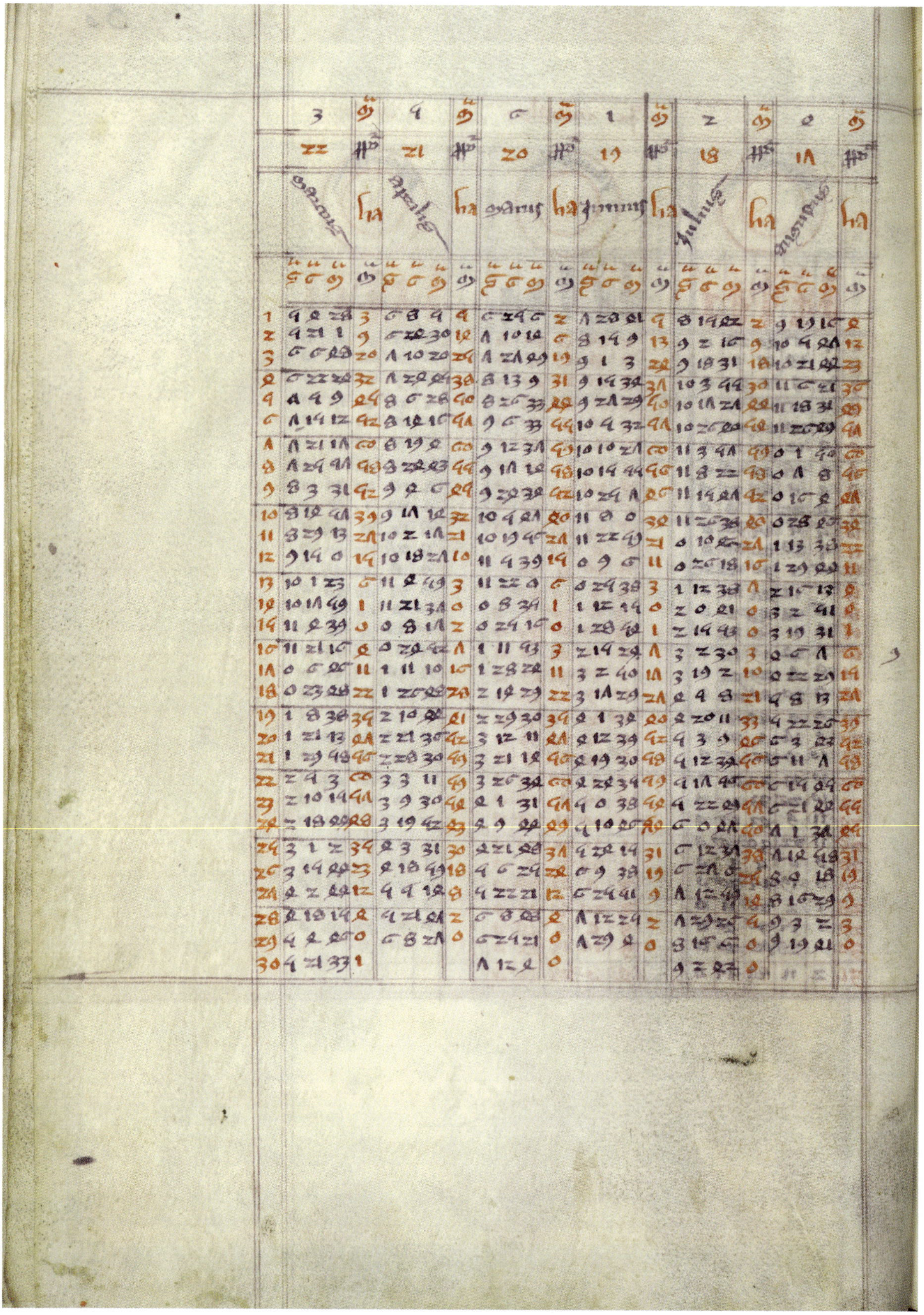

Plate 2: Jacob ben Makhir's *Almanac*, Table 9 (true lunar anomaly, excerpt). Madrid, Biblioteca Nacional de España, MS 9288, fol. 50v.

Plate 3: Jacob ben Makhir's *Almanac*, Table 10 (complete lunar equation with double arguments, excerpt). Madrid, Biblioteca Nacional de España, MS 9288, fol. 75r.

a)

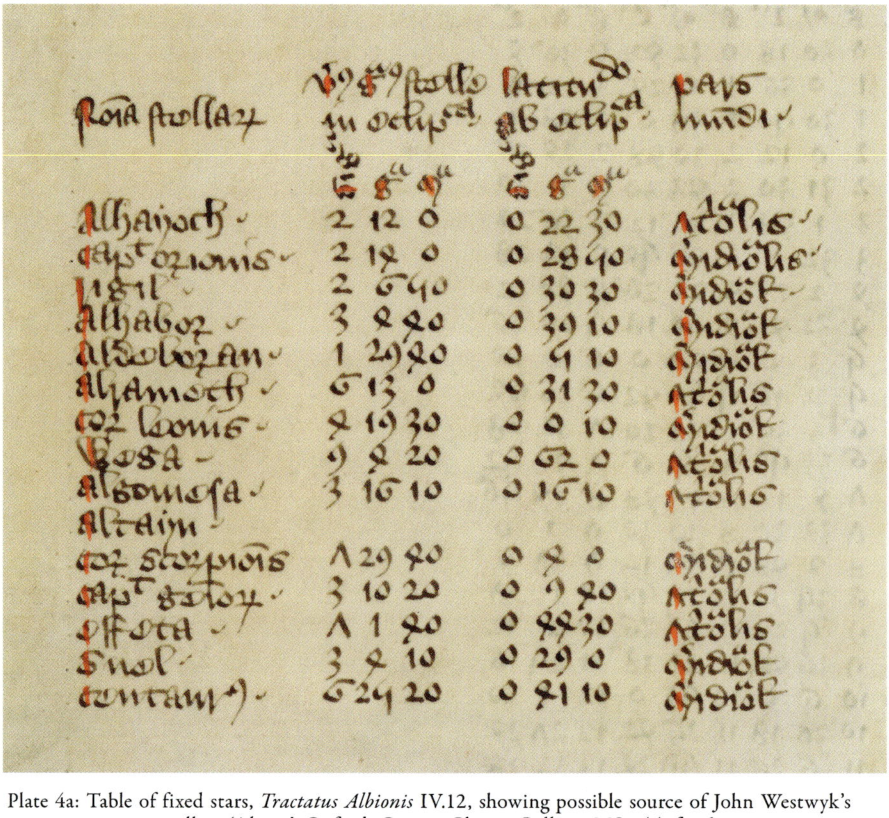

b)

Plate 4a: Table of fixed stars, *Tractatus Albionis* IV.12, showing possible source of John Westwyk's misspelling 'Altayn'. Oxford, Corpus Christi College, MS 144, f. 76v.
Plate 4b: As above. Oxford, Bodleian Library, MS Laud Misc. 657, f. 37v.

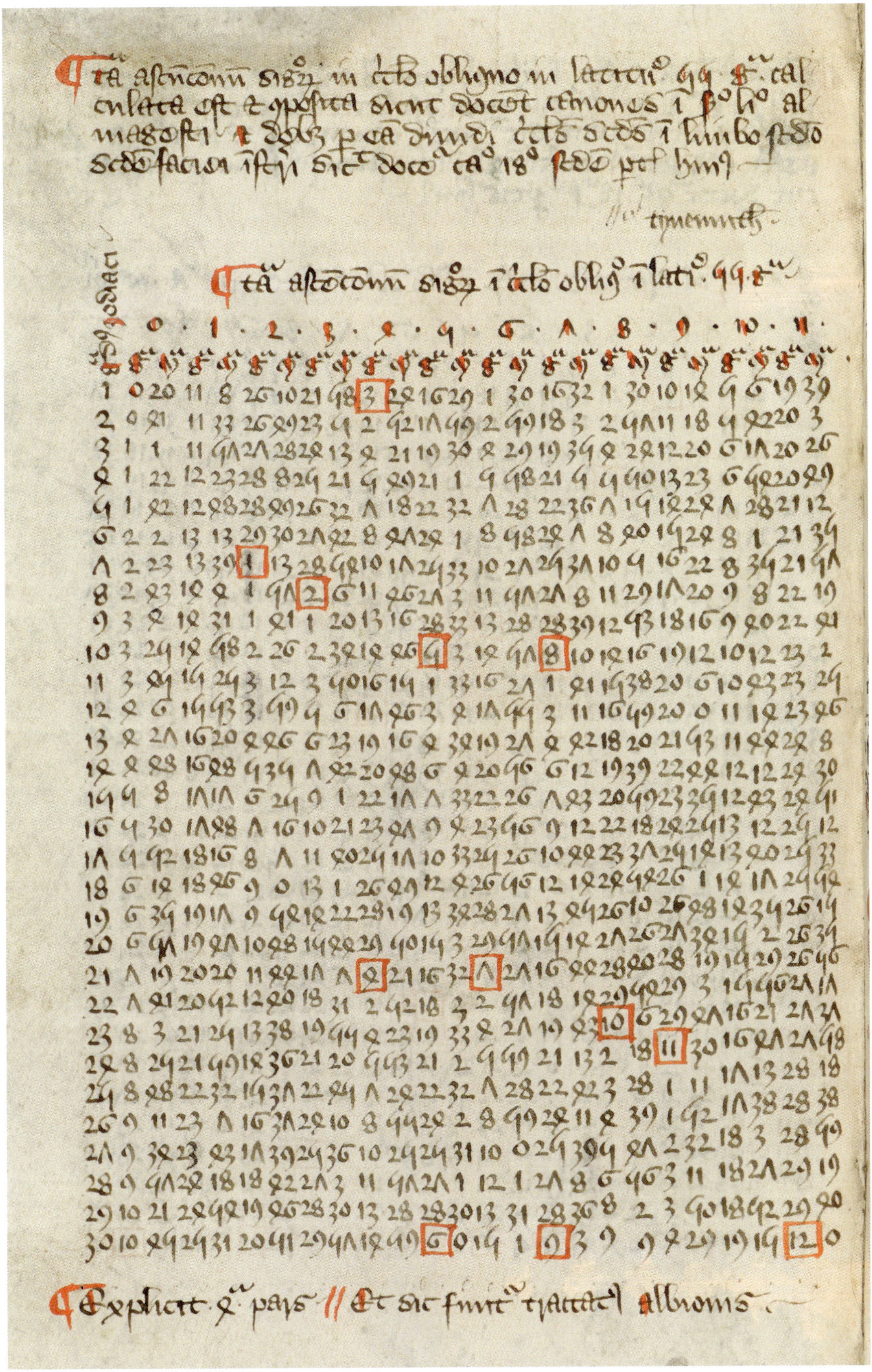

Plate 5: John Westwyk's table of oblique ascensions for 55°. Oxford, Bodleian Library, MS Laud Misc. 657, f. 42v.
By permission of the Bodleian Libraries, The University of Oxford.

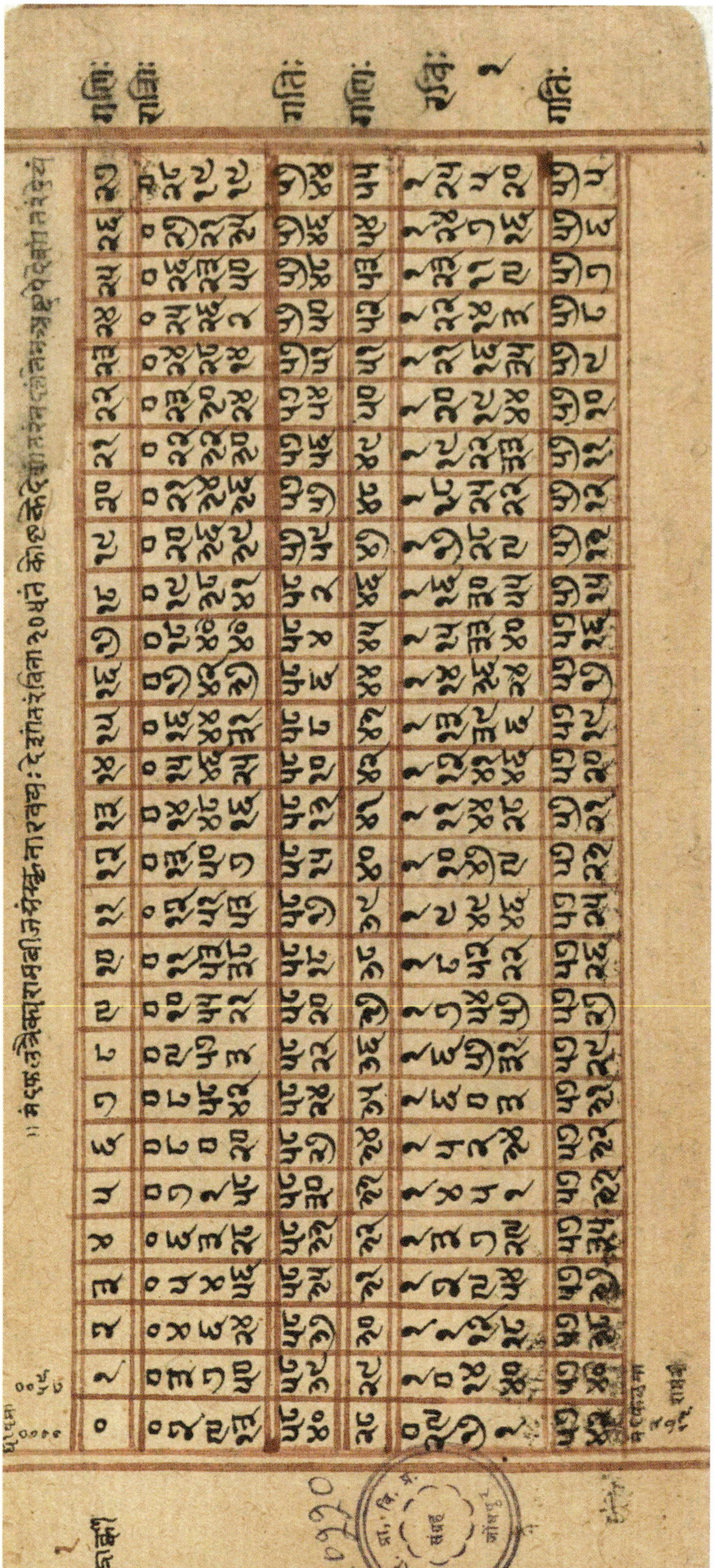

Plate 6: The first page of the solar equation table and associated true velocity (arguments 1–55) from Dinakara's *Candrārkī*. Jodhpur, Rajasthan Oriental Research Institute (RORI), MS 20220 (f. 1v).

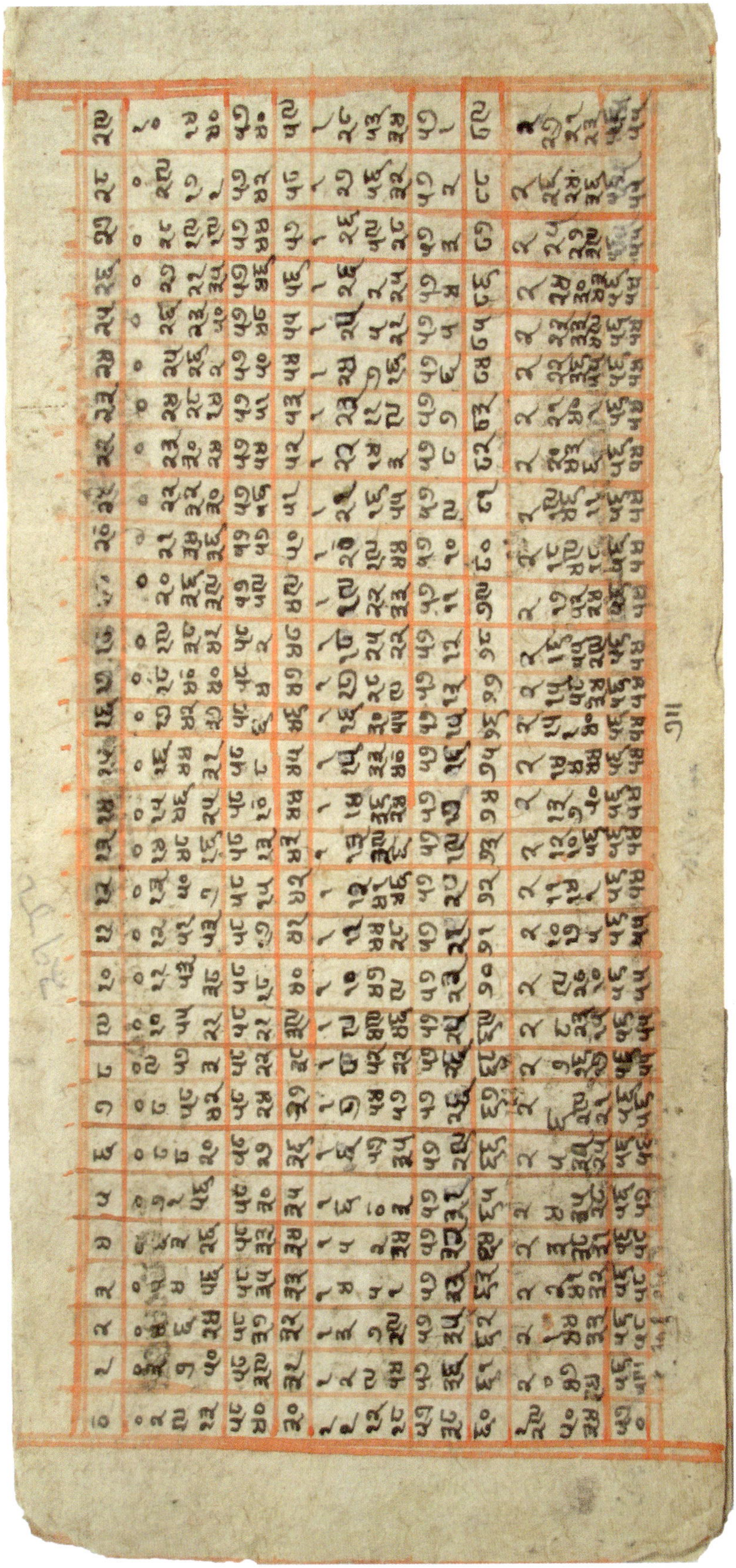

Plate 7: The first page of the solar equation table and associated true velocity (arguments 1–89) from Dinakara's *Candrārkī*. Jodhpur, Rajasthan Oriental Research Institute (RORI), MS 7752 (f. 1r).

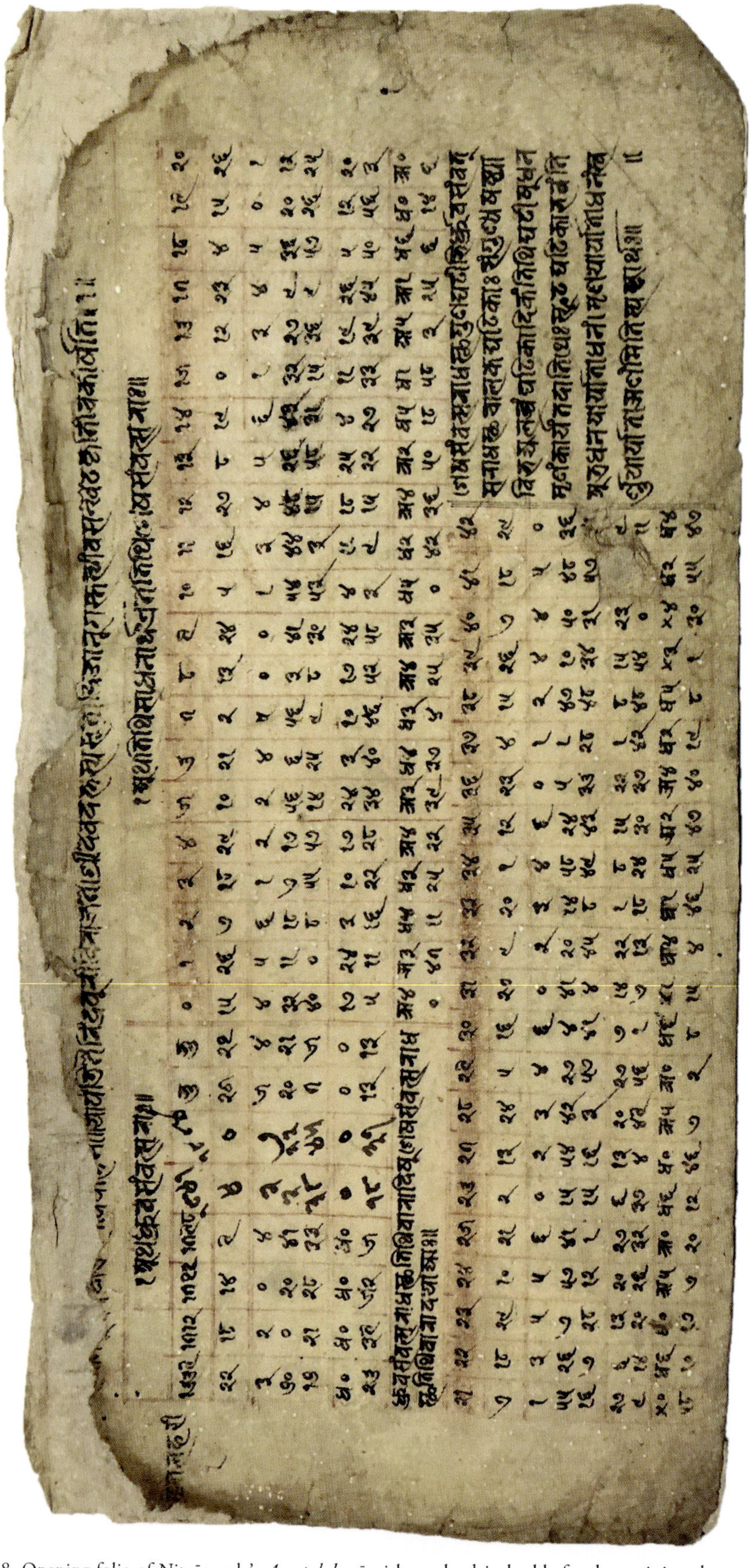

Plate 8: Opening folio of Nityānanda's *Amṛtalaharī* with a calendrical table for determining the current lunar day (tithi). Folio 1v of MS Sanskrit 19 from the collection of the University of Tokyo.

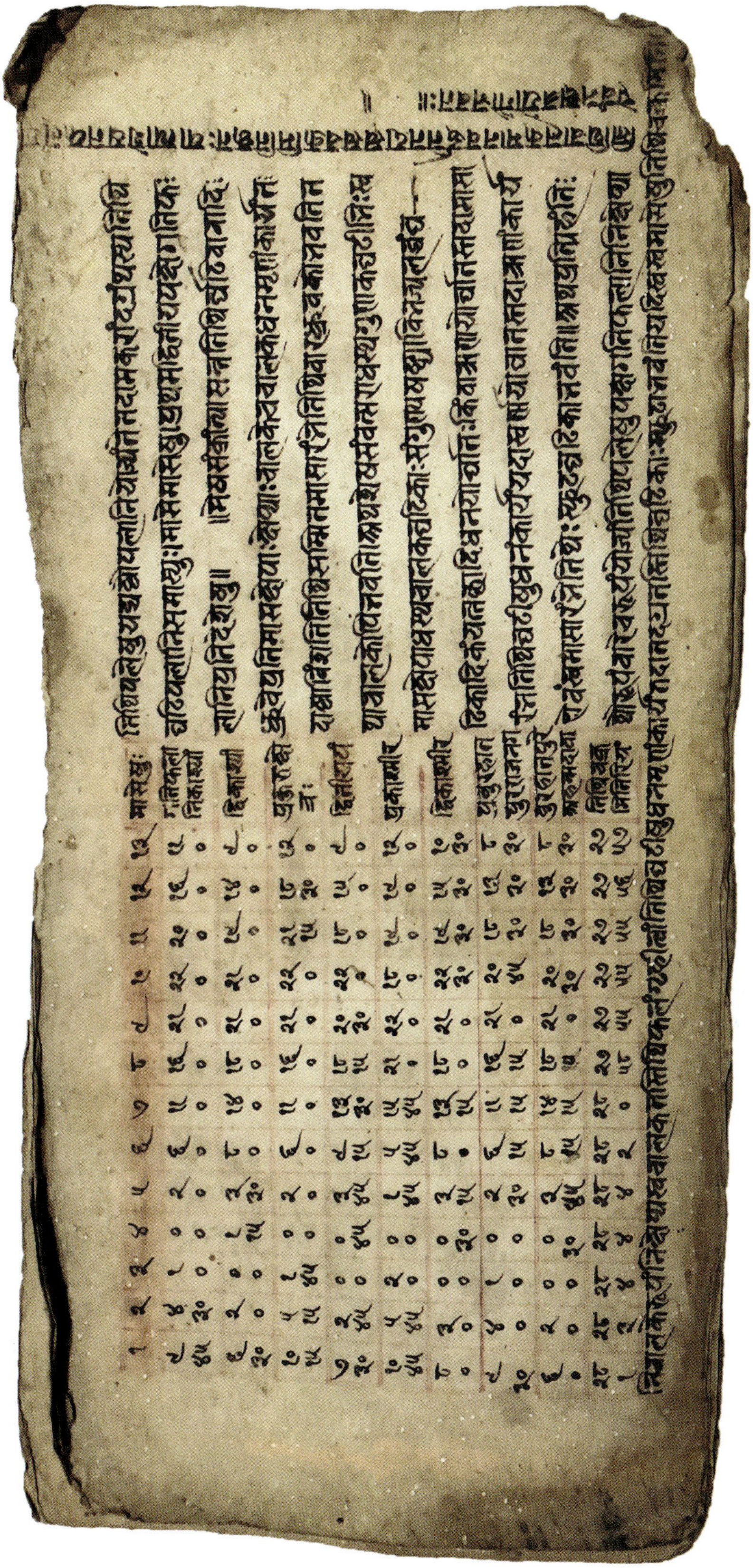

Plate 9: Second folio of Nityānanda's *Amṛtalaharī* with the continuation of a calendrical table. Folio 2r of MS Sanskrit 19 from the collection of the University of Tokyo.

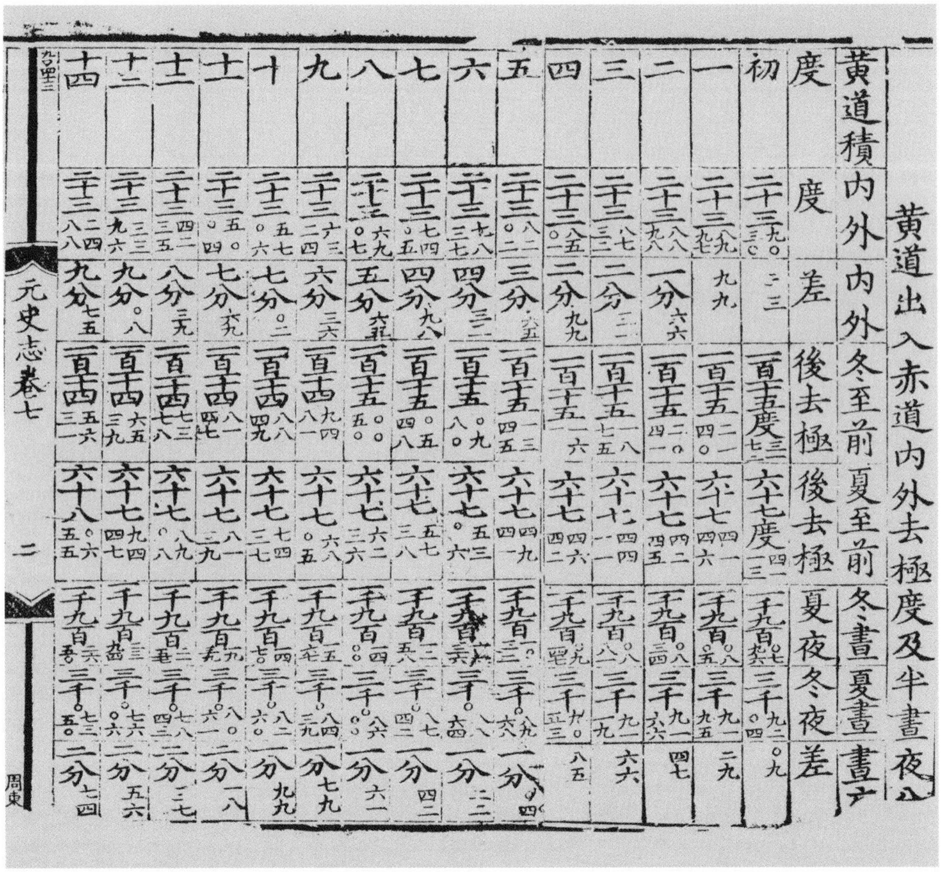
黃道出入赤道內外去極度及半晝夜分

a)

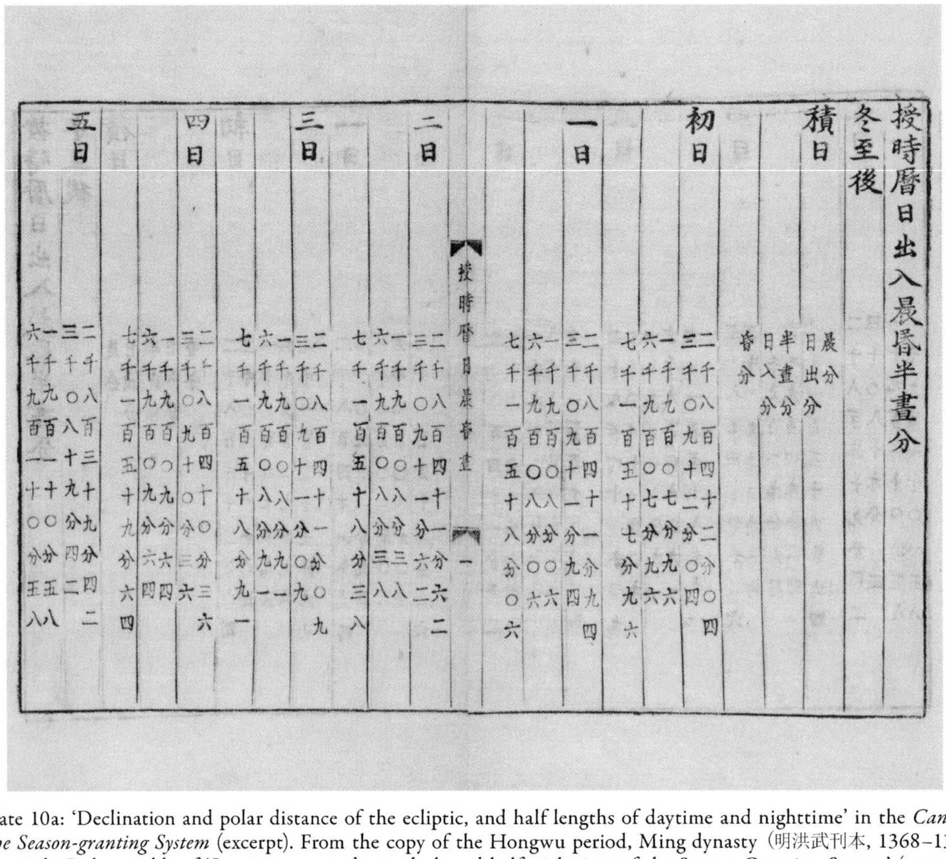
授時曆日出入晨昏半晝分

b)

Plate 10a: 'Declination and polar distance of the ecliptic, and half lengths of daytime and nighttime' in the *Canon of the Season-granting System* (excerpt). From the copy of the Hongwu period, Ming dynasty (明洪武刊本, 1368–1398).
Plate 10b: Pick-up table of 'Sunrise, sunset, dawn, dusk and half nighttime of the Season-Granting System' (excerpt). From the Gyujanggak Library of Seoul National University, collection no. 893.

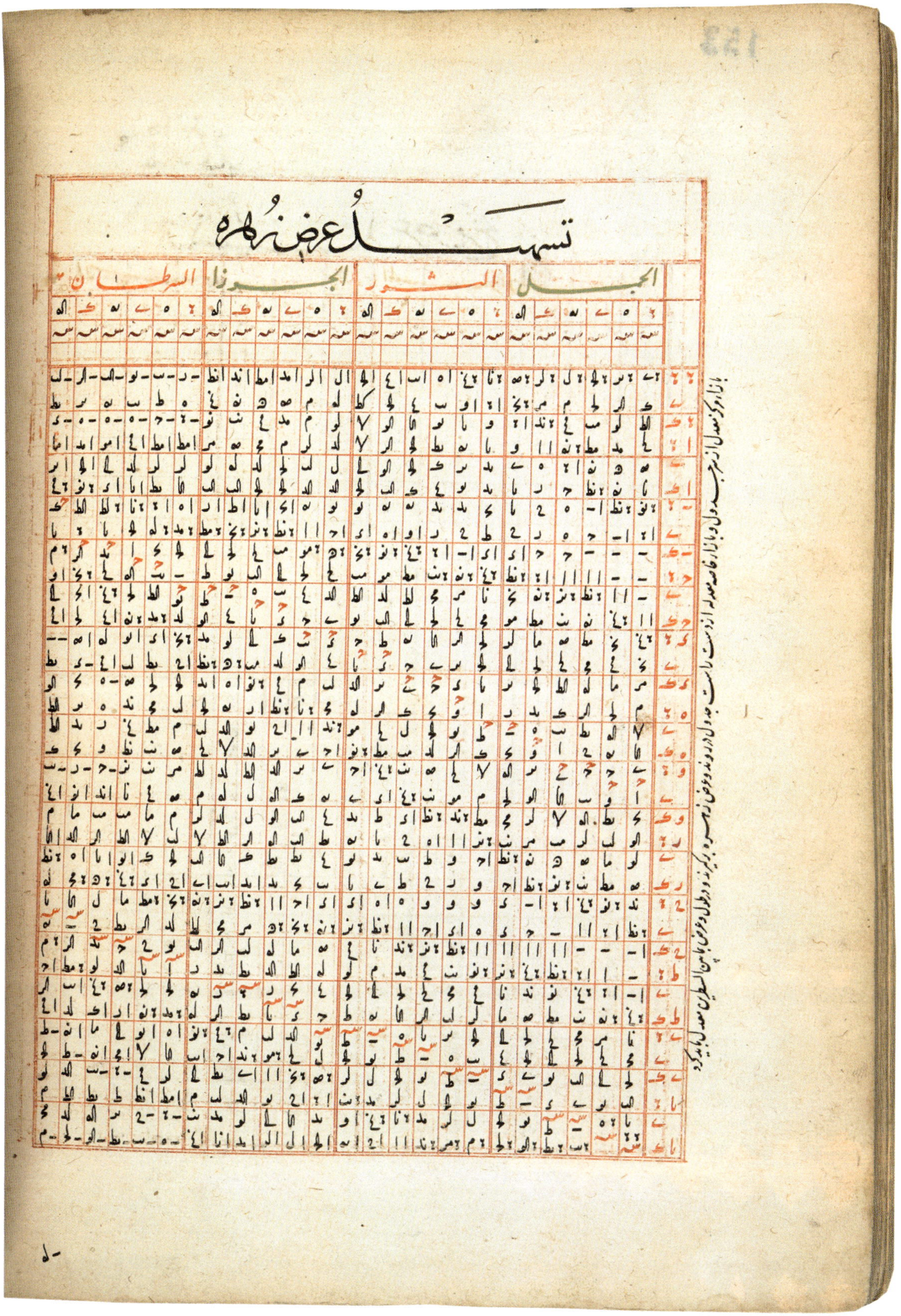

Plates 11: The first of the three pages of al-Kāshī's double-argument latitude table for Venus (see pp. 36 and 307 for the other two pages). © The British Library Board, MS India Office 430, fol. 153v.

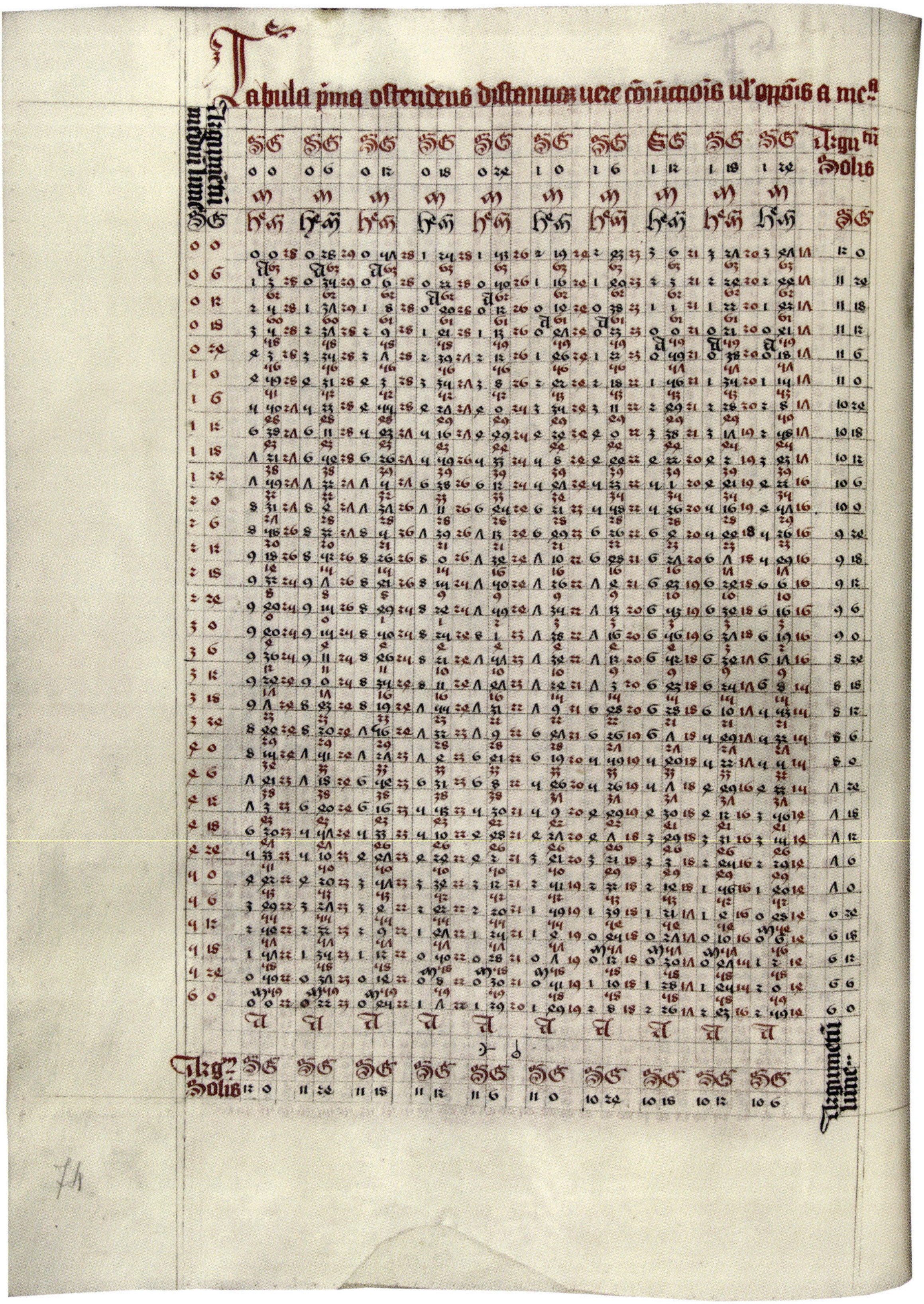

Plate 12: First page of the *Tabulae permanentes*, here entitled *Tabula … ostendens distanciae vere coniunctionis vel oppositionis a media*. Innsbruck, Universitäts- und Landesbibliothek Tirol, Servitenkloster I.b.62, p. 74.

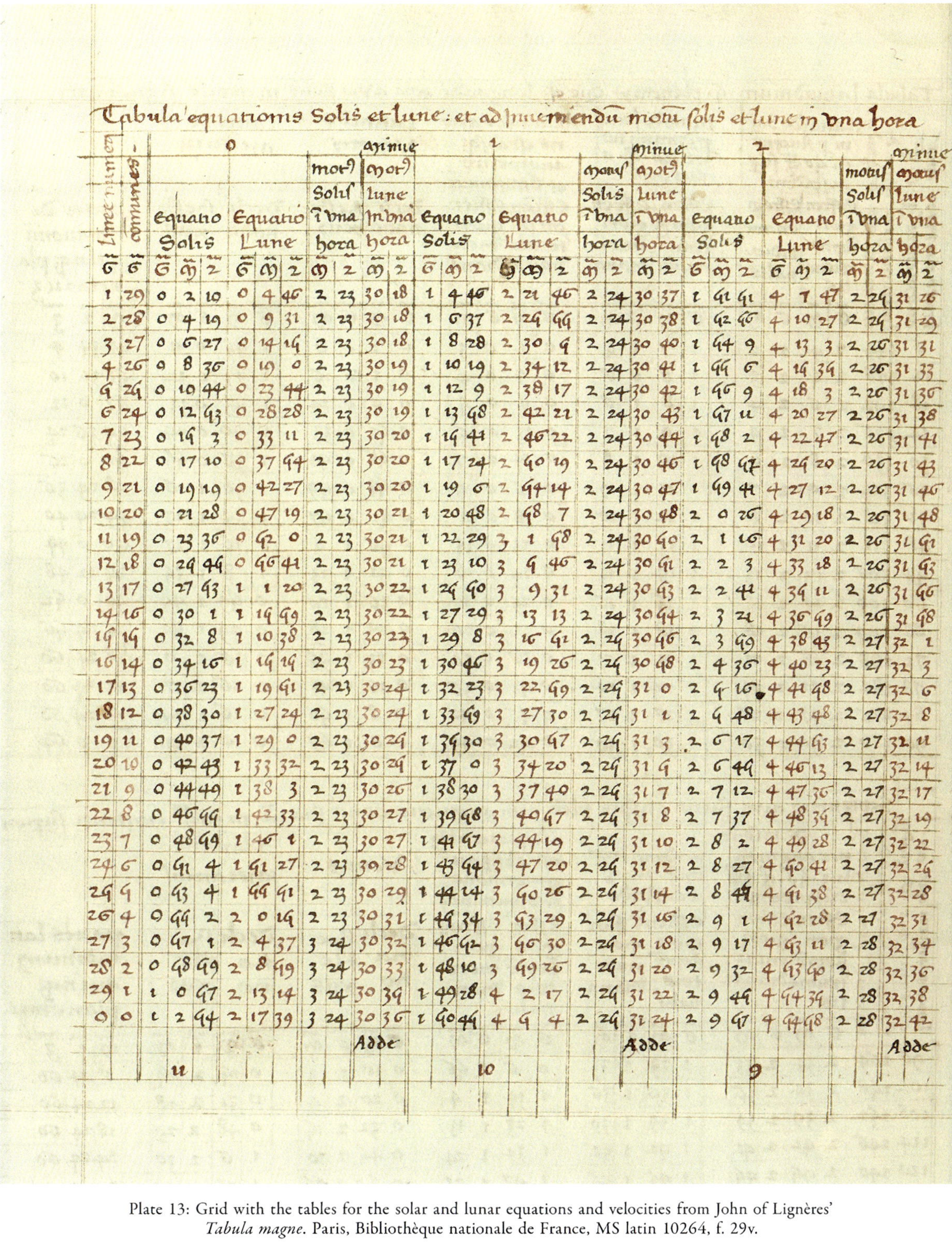

Tabula equationis solis et lune: et ad inveniendum motum solis et lune in una hora

Linee numeri	communes	0 Equatio solis			0 Equatio lune			Minute motus solis in una hora		Minute motus lune in una hora		1 Equatio solis			1 Equatio lune			Minute motus solis in una hora		Minute motus lune in una hora		2 Equatio solis			2 Equatio lune			Minute motus solis in una hora		Minute motus lune in una hora	
g	g	s	g	m	s	g	m	m	2	m	2	s	g	m	s	g	m	m	2	m	2	s	g	m	s	g	m	m	2	m	2
1	29	0	2	10	0	4	46	2	23	30	18	1	4	46	2	21	46	2	24	30	37	1	51	51	4	7	47	2	25	31	26
2	28	0	4	19	0	9	31	2	23	30	18	1	6	37	2	25	55	2	24	30	38	1	52	56	4	10	27	2	25	31	29
3	27	0	6	27	0	14	15	2	23	30	18	1	8	28	2	30	5	2	24	30	40	1	54	9	4	13	3	2	26	31	31
4	26	0	8	36	0	19	0	2	23	30	19	1	10	19	2	34	12	2	24	30	41	1	55	6	4	15	35	2	26	31	33
5	25	0	10	44	0	23	44	2	23	30	19	1	12	9	2	38	17	2	24	30	42	1	56	9	4	18	3	2	26	31	36
6	24	0	12	53	0	28	28	2	23	30	19	1	13	58	2	42	21	2	24	30	43	1	57	11	4	20	27	2	26	31	38
7	23	0	15	3	0	33	11	2	23	30	20	1	15	41	2	46	22	2	24	30	44	1	58	2	4	22	47	2	26	31	41
8	22	0	17	10	0	37	54	2	23	30	20	1	17	24	2	50	19	2	24	30	46	1	58	47	4	25	20	2	26	31	43
9	21	0	19	19	0	42	27	2	23	30	20	1	19	6	2	54	14	2	24	30	47	1	59	41	4	27	12	2	26	31	46
10	20	0	21	28	0	47	19	2	23	30	21	1	20	48	2	58	7	2	24	30	48	2	0	26	4	29	18	2	26	31	48
11	19	0	23	36	0	52	0	2	23	30	21	1	22	29	3	1	58	2	24	30	50	2	1	16	4	31	20	2	26	31	51
12	18	0	25	45	0	56	41	2	23	30	21	1	23	10	3	5	46	2	24	30	51	2	2	3	4	33	18	2	26	31	53
13	17	0	27	53	1	1	20	2	23	30	22	1	25	50	3	9	31	2	24	30	53	2	2	41	4	35	11	2	26	31	56
14	16	0	30	1	1	15	59	2	23	30	22	1	27	29	3	13	13	2	24	30	54	2	3	21	4	36	49	2	26	31	58
15	15	0	32	8	1	10	38	2	23	30	23	1	29	8	3	16	51	2	25	30	56	2	3	59	4	38	43	2	27	32	1
16	14	0	34	16	1	15	15	2	23	30	23	1	30	46	3	19	26	2	25	30	58	2	4	36	4	40	23	2	27	32	3
17	13	0	36	23	1	19	51	2	23	30	24	1	32	23	3	22	59	2	25	31	0	2	5	16	4	41	58	2	27	32	6
18	12	0	38	30	1	27	24	2	23	30	24	1	33	59	3	27	30	2	25	31	1	2	5	48	4	43	48	2	27	32	8
19	11	0	40	37	1	29	0	2	23	30	25	1	35	30	3	30	57	2	25	31	3	2	6	17	4	44	53	2	27	32	11
20	10	0	42	43	1	33	32	2	23	30	25	1	37	0	3	34	20	2	25	31	5	2	6	45	4	46	13	2	27	32	14
21	9	0	44	49	1	38	3	2	23	30	26	1	38	30	3	37	40	2	25	31	7	2	7	12	4	47	36	2	27	32	17
22	8	0	46	55	1	42	33	2	23	30	27	1	39	58	3	40	57	2	25	31	8	2	7	37	4	48	35	2	27	32	19
23	7	0	48	59	1	46	1	2	23	30	27	1	41	57	3	44	19	2	25	31	10	2	8	2	4	49	28	2	27	32	22
24	6	0	51	4	1	51	27	2	23	30	28	1	43	44	3	47	20	2	25	31	12	2	8	27	4	50	41	2	27	32	25
25	5	0	53	4	1	55	51	2	23	30	29	1	44	24	3	50	26	2	25	31	14	2	8	54	4	51	38	2	27	32	28
26	4	0	55	2	2	0	15	2	23	30	31	1	45	34	3	53	29	2	25	31	16	2	9	1	4	52	28	2	27	32	31
27	3	0	57	1	2	4	37	3	24	30	32	1	46	52	3	56	30	2	25	31	18	2	9	17	4	53	11	2	28	32	34
28	2	0	58	59	2	8	59	3	24	30	33	1	48	10	3	59	26	2	25	31	20	2	9	32	4	53	50	2	28	32	36
29	1	1	0	57	2	13	14	3	24	30	35	1	49	28	4	2	17	2	25	31	22	2	9	45	4	54	35	2	28	32	38
0	0	1	2	54	2	17	39	3	24	30	36	1	50	45	4	5	4	2	25	31	24	2	9	57	4	54	58	2	28	32	42
										Adde										Adde										Adde	
			11													10								9							

Plate 13: Grid with the tables for the solar and lunar equations and velocities from John of Lignères' *Tabula magne*. Paris, Bibliothèque nationale de France, MS latin 10264, f. 29v.

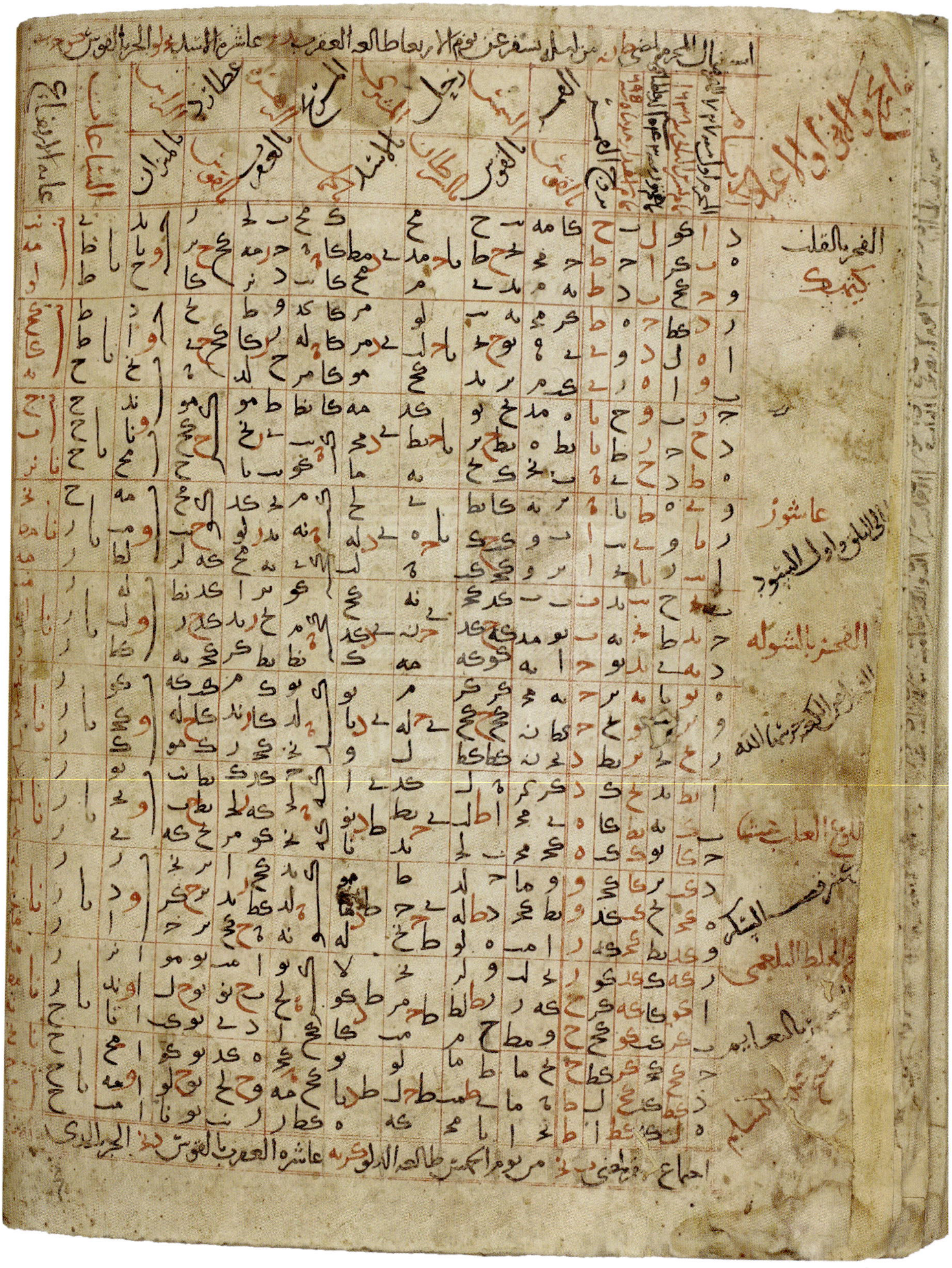

Plate 14: Ephemeris for the month Muḥarram 727 Hijra. Cairo, Dār al-kutub al-miṣriyya, MS *mīqāt* 817, fol. 69v.

Plate 15: Horoscopes preceding the ephemeris for 727 Hijra and mentioning the *Ḥākimī Zīj* of Ibn Yūnus. Cairo, Dār al-kutub al-miṣriyya, MS *mīqāt* 817, fol. 69r.

Plate 16: Geographical table from the *Shāmil Zīj*. Paris, Bibliothèque nationale de France, MS arabe 2529, fol. 20v.